SPECIAL FUNCTIONS

SPECIAL
FUNCTIONS

Z.X. Wang
D.R. Guo
Physics Department
Peking University

Translated by
D.R. Guo
X.J. Xia

World Scientific
Singapore • New Jersey • London • Hong Kong

Published by

World Scientific Publishing Co. Pte. Ltd.,
P O Box 128, Farrer Road, Singapore 9128
USA office: 687 Hartwell Street, Teaneck, NJ 07666
UK office: 73 Lynton Mead, Totteridge, London N20 8DH

SPECIAL FUNCTIONS

ISBN 9971-50-659-9
     9971-50-667-X (pbk)

Printed in Singapore by JBW Printers & Binders Pte. Ltd.

FOREWORD

In January 1980 when I visited Beijing, Professor Z. X. Wang, my M.Sc. thesis advisor of many years ago, gave me a copy of his book with Mr. D.R. Guo on *Special Functions*. On many occasions in later years, I had consulted it for various things such as the hypergeometric series and the elliptic functions. The book is systematic, clear and to the point, as one would expect from Professor Wang's style and personality.

It is great news that the book is now being published in an English translation. It will benefit many students and research workers who do not read Chinese.

C.N. Yang

PREFACE

This book is written for those who make constant use of special functions. As various special functions are frequently used in different branches of physics and engineering, there has always been the need for a book which can be referred to with convenience. Such a book could not be a mere summary nor a collection of materials of monographic nature. A summary may not be sufficient for the solution of actual problems and a collection of monographic topics will be too voluminous and inconvenient for reference purposes. Also, it should not be a compilation of formulae. This would not meet the needs of readers who want to master the mathematical manipulations and deductions of them. We hope that this book contains, in a not too large volume, the various principal special functions in common use and their basic properties and manipulations, in order that the readers may have a grasp of the basic methods for handling these functions so as to use them freely and with skill. To shorten the volume, some formulae which we take as somewhat less important are put in the Exercises at the end of the relevant chapters, and necessary hints for the more difficult ones are provided. These exercises may be profitable to the readers also as materials for practising.

There are many books on special functions, among which one widely used is *Modern Analysis* by Whittaker and Watson (mainly Part II). In many respects, the present book is based on that classic work. However, the scheme is not the same and the presentations of the subjects aare rather different too. Besides, some topics not covered in W & W's book are included here.

The organization of materials is as follows. Chapters 1 and 2 give a supplement of those topics necessary for the study of special functions, which are not provided in mathematical courses at the undergraduate level.[a] Chapter 3 is on the Gamma functions which is necessary for the discussion of almost all other special functions. Chapters 4 to 7 are concerned with hypergeometric and confluent hypergeometric functions including their important special cases: Legendre functions and Bessel functions. These are the functions most often encountered in practice. Their main properties are explored from the point of view of the singularities of the differential equations satisfied by them. Chapters 8 to 10 are about elliptic functions, being studied from the periodicity of these functions, with little concern in differential equations. Chapters 11 and 12 return again to linear differential equations and give discussions of the Lamé funcjtions and Mathieu functions, respectively; they are discussed after the elliptic functions as the former necessarily makes use of the elliptic functions and the latter is related to the elliptic cylinder coordinates. Various polynomials and the sundry generating functions are discussed in connection with the respective functions, and not taken out for a separate exposition. As examples, the Hermite polynomials, the Laguerre polynomials and their generating functions are all given as special cases of the confluent hypergeometric functions.

In the last part of the book, three Appendices are annexed. Appendices I and II are on the reductions of elliptic integrals. Appendix II gives expressions of the Laplace equation in different orthogonal curvilinear coordinate systems, and expressions of the gradient, the divergence and the curl of vectors.

The bibliography at the end of the book is short. It lists mainly some monographs for further reading.

Thanks are due to the late Professor S.H. Min who had given helpful suggestions for the section on Riemann's ς-function.

Z.X. Wang

D.R. Guo

---

[a]In China, these include courses on functions of a complex variable and differential equations in mathematical physics.

Translators' Note: The senior author, late Prof. J.S. (Z.X.) Wang, wished to have this book translated into English, in order for it to be useful to a wider circle of workers in physics and engineering. Unfortunately, his death frustrated the plan. The translators, D.R. Guo (co-author of the original text) and X.J. Xia, both being Prof. Wang's students, braving their linguistic handicap for the task, are glad to have carried out the unfulfilled wish of their highly esteemed professor.

We are also grateful to World Scientific Publishing for the opportunity to publish this book in English.

D.R. Guo
X.J. Xia

CONTENTS

Chapter 3. THE GAMMA FUNCTION

Chapter 6. CONFLUENT HYPERGEOMETRIC FUNCTIONS

SPECIAL FUNCTIONS

Chapter 1

THE EXPANSION OF FUNCTIONS IN INFINITE SERIES AND INFINITE PRODUCTS

In this chapter, we shall introduce to the readers some topics concerning the expansion of functions in infinite series and infinite products, and the asymptotic expansion of a function, which are usually not given in elementary courses in mathematics.

1.1. Bernoulli Polynomials and Bernoulli Numbers

Bernoulli polynomials $\varphi_n(x)$ $(n = 0, 1, 2, \ldots)$ are given by the following expansion:

$$\frac{te^{xt}}{e^t - 1} = \sum_{n=0}^{\infty} \frac{t^n}{n!} \varphi_n(x) \ . \tag{1}$$

The function on the l.h.s. is called the *generating function* of $\varphi_n(x)$. The series is convergent for $|t| < 2\pi$, since the singularities of the generating function nearest to $t = 0$ are $\pm 2\pi i$. In the literature, some authors employ the symbol $B_n(x)$ to denote the $\varphi_n(x)$ introduced here.

When $x = 0$, (1) reduces to

$$\frac{t}{e^t - 1} = \sum_{n=0}^{\infty} \frac{t^n}{n!} \varphi_n(0) \ . \tag{2}$$

1

This formula is often expressed in the following form:

$$t\left(\frac{1}{e^t - 1} + \frac{1}{2}\right) = \frac{t}{2}\frac{e^{t/2} + e^{-t/2}}{e^{t/2} - e^{-t/2}} = 1 + \sum_{n=1}^{\infty}(-)^{n-1}\frac{t^{2n}}{(2n)!}B_n . \qquad (3)$$

Only even powers of t occur in the series since the l.h.s. is an even function of t. Comparing (2) and (3) and writing simply φ_n for $\varphi_n(0)$, we obtain

$$\varphi_0 = 1, \qquad\qquad \varphi_1 = -\frac{1}{2} ,$$
$$\varphi_{2k} = (-)^{k-1}B_k , \quad \varphi_{2k+1} = 0 \qquad (k = 1, 2, \ldots) . \qquad (4)$$

B_k are called *Bernoulli numbers*. Sometimes, φ_n are also called Bernoulli numbers, and in the literature many authors use the symbol B_n to denote the φ_n here.

The following are some fundamental properties and formulae of the Bernoulli polynomials.

1. Explicit expressions of Bernoulli polynomials and the recurrence formula for Bernoulli numbers

From (2), we have

$$\frac{te^{xt}}{e^t - 1} = \sum_{k=0}^{\infty}\frac{t^k}{k!}\varphi_k \cdot \sum_{l=0}^{\infty}\frac{t^l}{l!}x^l = \sum_{n=0}^{\infty}\frac{t^n}{n!} \cdot \sum_{k=0}^{n}\binom{n}{k}\varphi_k x^{n-k} ,$$

where $\binom{n}{k} = n(n-1)(n-2)\ldots(n-k+1)/k!$. Comparing with (1), we obtain the explicit expression for $\varphi_n(x)$:

$$\varphi_n(x) = \sum_{k=0}^{n}\binom{n}{k}\varphi_k x^{n-k} \qquad (n = 0, 1, 2, \ldots) , \qquad (5)$$

where we have to calculate φ_k. For this purpose, we have from (2)

$$1 = \frac{e^t - 1}{t}\sum_{k=0}^{\infty}\frac{t^k}{k!}\varphi_k = \sum_{l=1}^{\infty}\frac{t^{l-1}}{l!}\sum_{k=0}^{\infty}\frac{t^k}{k!}\varphi_k = \sum_{n=1}^{\infty}t^{n-1}\sum_{k=0}^{n-1}\frac{\varphi_k}{k!(n-k)!} .$$

Comparing the two sides gives immediately

$$\varphi_0 = 1, \quad \sum_{k=0}^{n-1}\frac{1}{k!(n-k)!}\varphi_k = 0 \qquad (n \geq 2) . \qquad (6)$$

This is a recurrence formula for φ_k. Putting $n = 2, 3, \ldots$ successively, we can evaluate φ_n. (Note that, by (4), all φ_n with odd n, except φ_1, are equal to zero.)

(5) and (6) may be symbolically expressed in the form:

$$\varphi_n(x) = (\varphi + x)^n \qquad (n = 0, 1, 2, \ldots) \tag{7}$$

and

$$(\varphi + 1)^n - \varphi_n = 0 \qquad (n = 2, 3, \ldots) . \tag{8}$$

Here it is understood that after the binomial expressions have been developed, the symbols φ^k for powers are to be replaced by φ_k.

The first ten Bernoulli numbers and the first seven Bernoulli polynomials are given below:

$$
\begin{aligned}
&B_1 = \frac{1}{6}, && B_2 = \frac{1}{30}, && B_3 = \frac{1}{42}, && B_4 = \frac{1}{30}, \\
&B_5 = \frac{5}{66}, && B_6 = \frac{691}{2730}, && B_7 = \frac{7}{6}, && B_8 = \frac{3617}{510}, \\
&B_9 = \frac{43867}{798}, && B_{10} = \frac{1\,74611}{330} && .
\end{aligned}
\tag{9}
$$

$$\varphi_0(x) = 1, \quad \varphi_1(x) = x - \frac{1}{2}, \quad \varphi_2(x) = x^2 - x + \frac{1}{6},$$

$$\varphi_3(x) = x(x-1)(x - \frac{1}{2}) = x^3 - \frac{3}{2}x^2 + \frac{1}{2}x,$$

$$\varphi_4(x) = x^4 - 2x^3 + x^2 - \frac{1}{30}, \tag{10}$$

$$\varphi_5(x) = x(x-1)(x-\frac{1}{2})(x^2 - x - \frac{1}{3}) = x^5 - \frac{5}{2}x^4 + \frac{5}{3}x^3 - \frac{1}{6}x,$$

$$\varphi_6(x) = x^6 - 3x^5 + \frac{5}{2}x^4 - \frac{1}{2}x^2 + \frac{1}{42} .$$

2. Derivatives and integrals

Differentiating (5), we obtain

$$\frac{d}{dx}\varphi_n(x) = \sum_{k=0}^{n-1} \binom{n}{k}(n-k)\varphi_k x^{n-k-1} = n\varphi_{n-1}(x) \tag{11}$$

and

$$\frac{d^p}{dx^p}\varphi_n(x) = \frac{n!}{(n-p)!}\varphi_{n-p}(x) . \tag{12}$$

Changing n to $n+1$ in (11) and integrating give

$$\int_a^x \varphi_n(y)\,dy = \frac{1}{n+1}\left[\varphi_{n+1}(x) - \varphi_{n+1}(a)\right] . \tag{13}$$

3. Difference relations

$$\varphi_0(x+1) = \varphi_0(x), \qquad \varphi_1(x+1) = \varphi_1(x) + 1,$$
$$\varphi_n(x+1) = \varphi_n(x) + nx^{n-1} \qquad (n \geq 2) . \tag{14}$$

They can be proved as follows. From (1), we have

$$\frac{t\,e^{(x+1)t}}{e^t - 1} = \sum_{n=0}^{\infty} \frac{t^n}{n!} \varphi_n(x+1) = te^{xt} + \frac{t\,e^{xt}}{e^t-1} = \sum_{n=0}^{\infty} \frac{t^{n+1}}{n!} x^n + \sum_{n=0}^{\infty} \frac{t^n}{n!} \varphi_n(x) .$$

Comparing the coefficients of t^n in the series gives (14).

4. The complementary argument relation

$$\varphi_n(1-x) = (-)^n \varphi_n(x) . \tag{15}$$

Again, by (1) we have

$$\frac{t\,e^{(1-x)t}}{e^t - 1} = \sum_{n=0}^{\infty} \frac{t^n}{n!} \varphi_n(1-x) = \frac{-t\,e^{-xt}}{e^{-t}-1} = \sum_{n=0}^{\infty} \frac{(-t)^n}{n!} \varphi_n(x) .$$

Comparing the coefficients of t^n in the series, we obtain (15).

5. Addition formula

Replacing x in (1) by $x+y$, we obtain

$$\frac{t\,e^{(x+y)t}}{e^t - 1} = \sum_{n=0}^{\infty} \frac{t^n}{n!} \varphi_n(x+y) .$$

But the l.h.s. is equal to

$$\frac{t\,e^{yt}}{e^t - 1} e^{xt} = \sum_{k=0}^{\infty} \frac{t^k}{k!} \varphi_k(y) \cdot \sum_{l=0}^{\infty} \frac{t^l}{l!} x^l = \sum_{n=0}^{\infty} \frac{t^n}{n!} \sum_{k=0}^{n} \binom{n}{k} \varphi_k(y) x^{n-k} ,$$

therefore we have the addition formula:

$$\varphi_n(x+y) = \sum_{k=0}^{n} \binom{n}{k} \varphi_k(y) x^{n-k} . \tag{16}$$

6. Summation formula

$$\sum_{s=1}^{m} s^n = \frac{1}{n+1}[\varphi_{n+1}(m+1) - \varphi_{n+1}] \quad (n \geq 1) . \tag{17}$$

The proof is as follows. From (14), we have

$$s^n = \frac{1}{n+1}[\varphi_{n+1}(s+1) - \varphi_{n+1}(s)] ,$$

thus,

$$\sum_{s=1}^{m} s^n = \sum_{s=0}^{m} s^n = \frac{1}{n+1}[\varphi_{n+1}(m+1) - \varphi_{n+1}(0)] .$$

Besides, by (3), we can also obtain the *expansion formula for the cotangent function*:

$$\frac{t}{2} \cot \frac{t}{2} = \frac{it}{2} \frac{e^{it/2} + e^{-it/2}}{e^{it/2} - e^{-it/2}} = 1 - \sum_{n=1}^{\infty} \frac{B_n}{(2n)!} t^{2n}, \quad |t| < 2\pi \tag{18}$$

and that *for the tangent function*:

$$\frac{t}{2} \tan \frac{t}{2} = \frac{t}{2} \cot \frac{t}{2} - t \cot t = \sum_{n=1}^{\infty} \frac{(2^{2n} - 1)B_n}{(2n)!} t^{2n}, \quad |t| < \pi . \tag{19}$$

Combining (18) and (19) gives the *expansion formula for the cosecant function*:

$$t \csc t = \frac{t}{2} \cot \frac{t}{2} + \frac{t}{2} \tan \frac{t}{2}$$

$$= 1 + \sum_{n=1}^{\infty} \frac{2(2^{n-1} - 1)B_n}{(2n)!} t^{2n}, \quad |t| < \pi . \tag{20}$$

Some other formulae related to the Bernoulli polynomials are given in the exercises at the end of this chapter.

1.2. Euler Polynomials and Euler Numbers

Euler polynomials $E_n(x)$ $(n = 0, 1, 2, \ldots)$ are given by the following expansion:

$$\frac{2e^{xt}}{e^t + 1} = \sum_{n=0}^{\infty} \frac{t^n}{n!} E_n(x) . \tag{1}$$

The function on the l.h.s. is called the *generating function of the Euler polynomials*. The series converges in $|t| < \pi$ because the nearest singularities of the function from $t = 0$ are $t = \pm\pi i$.

Let $x = \frac{1}{2}$. The l.h.s. of (1) is an even function of t, and hence no odd powers of t could be present in the series on the r.h.s. Thus, we have

$$\frac{2e^{t/2}}{e^t + 1} = \mathrm{sech}\,\frac{t}{2} = \sum_{n=0}^{\infty} \frac{(-)^n E_n}{(2n)!}\left(\frac{t}{2}\right)^{2n}, \tag{2}$$

where $E_n = (-)^n 2^{2n} E_{2n}\left(\frac{1}{2}\right)$ are called the *Euler numbers*. In the literature, some authors define the Euler numbers to be $E_n = 2^n E_n\left(\frac{1}{2}\right)$; then their $E_{2n+1} = 0$ while E_{2n} is equal to $(-)^n E_n$ here.

The following are some of the fundamental properties of the Euler polynomials and Euler numbers. The proofs of them are similar to those in the preceding section on the properties of Bernoulli polynomials and numbers.

1. The explicit expressions of Euler polynomials and the recurrence formula for Euler numbers

The l.h.s. of (1) can be expanded into

$$\frac{2e^{t/2}e^{(x-\frac{1}{2})t}}{e^t + 1} = \sum_{k=0}^{\infty} \frac{(-)^k E_k}{(2k)!}\left(\frac{t}{2}\right)^{2k} \cdot \sum_{l=0}^{\infty} \frac{(x - \frac{1}{2})^l}{l!}t^l$$

$$= \sum_{n=0}^{\infty} \frac{t^n}{n!} \sum_{k=0}^{[n/2]} \frac{(-)^k E_k}{2^{2k}}\binom{n}{2k}\left(x - \frac{1}{2}\right)^{n-2k},$$

where $[n/2]$ denotes the largest integer not greater than $n/2$. Comparing with (1) gives the explicit expression of Euler polynomials:

$$E_n(x) = \sum_{k=0}^{[n/2]} (-)^k \frac{E_k}{2^{2k}}\binom{n}{2k}\left(x - \frac{1}{2}\right)^{n-2k}. \tag{3}$$

The Euler numbers E_k can be calculated successively with the following recurrence relation:

$$E_0 = 1, \qquad \sum_{l=0}^{k}(-)^l \binom{2k}{2l}E_l = 0 \qquad (k \geq 1). \tag{4}$$

(4) can be proved as follows. Changing t in (2) to $2t$, we have

$$1 = \frac{e^t + e^{-t}}{2} \sum_{l=0}^{\infty} (-)^l \frac{E_l}{(2l)!} t^{2l} = \sum_{r=0}^{\infty} \frac{t^{2r}}{(2r)!} \cdot \sum_{l=0}^{\infty} (-)^l \frac{E_l}{(2l)!} t^{2l}$$

$$= \sum_{k=0}^{\infty} \frac{t^{2k}}{(2k)!} \sum_{l=0}^{k} (-)^l \frac{E_l (2k)!}{(2l)!(2k-2l)!} \;.$$

Comparing the two sides, (4) results.

The first ten Euler numbers and the first seven Euler polynomials are given below:

$$E_0 = 1, \quad E_1 = 1, \quad E_2 = 5, \quad E_3 = 61, \quad E_4 = 1385,$$
$$E_5 = 50521, \quad E_6 = 2702765, \quad E_7 = 199360981, \tag{5}$$
$$E_8 = 19391512145, \quad E_9 = 2404879675441.$$

$$E_0(x) = 1, \qquad E_1(x) = x - \frac{1}{2}, \qquad E_2(x) = x(x-1),$$
$$E_3(x) = \left(x - \frac{1}{2}\right)\left(x^2 - x - \frac{1}{2}\right),$$
$$E_4(x) = x(x-1)(x^2 - x - 1), \tag{6}$$
$$E_5(x) = \left(x - \frac{1}{2}\right)(x^4 - 2x^3 - x^2 + 2x + 1),$$
$$E_6(x) = x(x-1)(x^4 - 2x^3 - 2x^2 + 3x + 3) \;.$$

2. Average

$$\frac{1}{2}[E_n(x+1) + E_n(x)] = x^n \;. \tag{7}$$

3. Derivative

$$\frac{d^p}{dx^p} E_n(x) = \frac{n!}{(n-p)!} E_{n-p}(x) \;. \tag{8}$$

4. The complementary argument relation

$$E_n(1-x) = (-)^n E_n(x) \;. \tag{9}$$

5. Summation formula

By (7), we have

$$\sum_{s=1}^{m} (-)^s s^n = \frac{1}{2} \sum_{s=1}^{m} (-)^s [E_n(s+1) + E_n(s)]$$

$$= \frac{1}{2}[(-)^m E_n(m+1) - E_n(1)] \;. \tag{10}$$

Again, replacing t by $2it$ in (2) we obtain the expansion

$$\sec t = \sum_{n=0}^{\infty} \frac{E_n}{(2n)!} t^{2n} , \quad |t| < \pi/2 \tag{11}$$

for the secant function.

Further formulae are given at the end of this chapter as exercises.

1.3. Euler-Maclaurin Formula

The Euler-Maclaurin formula (or, simply Euler formula) is important in numerical integration, asymptotic expansion and summation problems. We shall derive it from the

Darboux Formula: Let $f(z)$ be an analytic function along the straight line from a point a to the point z; $\varphi(t)$ is an arbitrary polynomial of degree n. Then

$$\varphi^{(n)}(0)\{f(z) - f(a)\}$$
$$= \sum_{m=1}^{n} (-)^{m-1}(z-a)^m \{\varphi^{(n-m)}(1)f^{(m)}(z) - \varphi^{(n-m)}(0)f^{(m)}(a)\}$$
$$+ (-)^n (z-a)^{n+1} \int_0^1 \varphi(t) f^{(n+1)}[a + (z-a)t] dt . \tag{1}$$

Taylor's formula is a special case of it, namely, $\varphi(t) = (t-1)^n$.

Proof: Let $0 \leq t \leq 1$. Multiply the following identity

$$\frac{d}{dt} \sum_{m=1}^{n} (-)^m (z-a)^m \varphi^{(n-m)}(t) f^{(m)}[a + (z-a)t]$$
$$= -(z-a)\varphi^{(n)}(t)f'[a + (z-a)t] + (-)^n (z-a)^{n+1} \varphi(t) f^{(n+1)}[a + (z-a)t] \tag{2}$$

with dt and integrate from 0 to 1. Notice that $\varphi(t)$ is a polynomial of degree n, so $\varphi^{(n)}(t) = \varphi^{(n)}(0)$ and we have (1).

Euler Formula: In the Darboux formula (1), putting in for $\varphi(t)$ the Bernoulli polynomial $\varphi_n(t)$ and replacing n by $2n$, we obtain

$$\varphi_{2n}^{(2n)}(0)\{f(z) - f(a)\}$$
$$= \sum_{m=1}^{2n} (-)^{m-1}(z-a)^m \{\varphi_{2n}^{(2n-m)}(1)f^{(m)}(z) - \varphi_{2n}^{(2n-m)}(0)f^{(m)}(a)\}$$
$$+ (z-a)^{2n+1} \int_0^1 \varphi_{2n}(t) f^{(2n+1)}[a + (z-a)t] dt . \tag{3}$$

Making the following substitutions

$$\varphi_{2n}^{(2n)}(0) = (2n)! \quad \text{[Eqs. (12) and (10) of Sec. 1.1]},$$

$$\varphi_{2n}^{(2n-m)}(x) = \frac{(2n)!}{m!}\varphi_m(x) \quad \text{[Eq. (12) of Sec. 1.1]},$$

$$\varphi_m(1) = (-)^m \varphi_m(0) = (-)^m \varphi_m \quad \text{[Eq. (15) of Sec. 1.1]},$$

$$\varphi_1 = -\frac{1}{2}, \quad \varphi_{2k+1} = 0 \quad (k \geq 1) \quad \text{[Eq. (4) of Sec. 1.1]},$$

in (3), we obtain

$$f(z) - f(a)$$

$$= \frac{z-a}{2}[f'(z) + f'(a)] + \sum_{k=1}^{n}(-)^k\frac{(z-a)^{2k}}{(2k)!}B_k[f^{(2k)}(z) - f^{(2k)}(a)]$$

$$+ \frac{(z-a)^{2n+1}}{(2n)!}\int_0^1 \varphi_{2n}(t)f^{(2n+1)}[a + (z-a)t]dt . \tag{4}$$

Let $F(z) = f'(z)$ and write h for $z - a$. (4) becomes

$$\int_a^{a+h} F(x)dx = \frac{h}{2}[F(a+h) + F(a)]$$

$$+ \sum_{k=1}^{n}\frac{(-)^k h^{2k}B_k}{(2k)!}[F^{(2k-1)}(a+h) - F^{(2k-1)}(a)]$$

$$+ \frac{h^{2n+1}}{(2n)!}\int_0^1 \varphi_{2n}(t)F^{(2n)}(a+ht)dt . \tag{5}$$

Replacing a in (5) successively by $a + h$, $a + 2h, \ldots, a + (m-1)h$, then adding, we obtain finally the *Euler formula*:

$$\int_a^{a+mh} F(x)dx$$

$$= h\left\{\frac{F(a)}{2} + F(a+h) + \ldots + F[a + (m-1)h] + \frac{F(a+mh)}{2}\right\}$$

$$+ \sum_{k=1}^{n}\frac{(-)^k B_k h^{2k}}{(2k)!}[F^{(2k-1)}(a+mh) - F^{(2k-1)}(a)] + R_n , \tag{6}$$

where

$$R_n = \frac{h^{2n+1}}{(2n)!}\int_0^1 \varphi_{2n}(t)\sum_{s=0}^{m-1}F^{(2n)}(a + hs + ht)dt . \tag{7}$$

It is seen immediately that the first term on the r.h.s. of (6) is nothing but the approximation of the value of the integral on the l.h.s. of (6) calculated by the trapezoidal method. Therefore, (6) is a formula for more accurate approximation. The last sum in (6) containing Bernoulli numbers is a correction term and the integral R_n given by (7) may be used to estimate the error (see Example 2 below).

The remainder R_n in the Euler formula (6) can be simplified. For this purpose, we define the periodic functions $P_\lambda(t)$ with period l as follows:

$$P_\lambda(t) = \varphi_\lambda(t)/\lambda! \qquad \text{for } 0 \le t < 1,$$
$$P_\lambda(t+1) = P_\lambda(t), \quad \lambda \text{ being non-negative integer}. \qquad (8)$$

Then

$$R_n = h^{2n+1} \int_0^1 P_{2n}(t) \sum_{s=0}^{m-1} F^{(2n)}[a + h(t+s)]dt .$$

Changing $t + s$ to t and making use of the periodicity of $P_\lambda(t)$, we obtain

$$R_n = h^{2n+1} \sum_{s=0}^{m-1} \int_s^{s+1} P_{2n}(t) F^{(2n)}(a + ht)dt$$
$$= h^{2n+1} \int_0^m P_{2n}(t) F^{(2n)}(a + ht)dt \qquad (9)$$

or, integrating by parts,

$$R_n = -h^{2n+2} \int_0^m P_{2n+1}(t) F^{(2n+1)}(a + ht)dt . \qquad (10)$$

In the above procedure, we have used the following formulae:

$$\frac{d}{dt} P_\lambda(t) = \frac{d}{dt} \varphi_\lambda(t)/\lambda! = \varphi_{\lambda-1}(t)/(\lambda-1)! = P_{\lambda-1}(t) , \qquad (11)$$

$$P_{2n+1}(1) = (-)^{2n+1} P_{2n+1}(0) = 0 \qquad (12)$$

which are derived from the properties of the Bernoulli polynomials [Eqs. (11), (15) and (4) of Sec. 1.1].

Formula for estimating the value of $P_\lambda(t)$

As $P_\lambda(t)$ is a periodic function with period 1, it can be expanded in a Fourier series. When $\lambda = 2n$, on account of $\varphi_{2n}(1 - t) = \varphi_{2n}(t)$ [Eq. (15) of

Sec. 1.1], we have $P_{2n}(1-t) = P_{2n}(t) = P_{2n}(-t)$ for $0 \le t < 1$, i.e., $P_{2n}(t)$ is an even function. Therefore

$$P_{2n}(t) = \sum_{k=0}^{\infty} a_k \cos 2k\pi t \ .$$

By (11) and (12) we find

$$a_0 = \int_0^1 P_{2n}(t)dt = P_{2n+1}(1) - P_{2n+1}(0) = 0 \qquad (n \ge 1) \ ,$$

$$a_k = 2 \int_0^1 P_{2n}(t) \cos 2k\pi t dt = \frac{(-)^{n+1}2}{(2k\pi)^{2n}} \ ,$$

so

$$P_{2n}(t) = (-)^{n+1} \sum_{k=1}^{\infty} \frac{2 \cos 2k\pi t}{(2k\pi)^{2n}} \quad (n \ge 1) \ . \tag{13}$$

Similarly,

$$P_{2n+1}(t) = (-)^{n+1} \sum_{k=1}^{\infty} \frac{2 \sin 2k\pi t}{(2k\pi)^{2n+1}} \quad (n \ge 0) \ . \tag{14}$$

From (13) and (14), we have

$$|P_\lambda(t)| \le \frac{2}{(2\pi)^\lambda} \sum_{k=1}^{\infty} \frac{1}{k^\lambda} \ .$$

For $\lambda \ge 2$,

$$1 + \frac{1}{2^\lambda} + \frac{1}{3^\lambda} + \dots < 1 + \int_1^\infty \frac{dx}{x^\lambda} = 1 + \frac{1}{\lambda - 1} \le 2 \ ,$$

hence

$$|P_\lambda(t)| \le \frac{4}{(2\pi)^\lambda} \ . \tag{15}$$

This also holds for $\lambda = 1$ since $|P_1(t)| = |\varphi_1(t)| = |t - \frac{1}{2}| \le \frac{1}{2} < \frac{4}{2\pi}$ [cf. Eq. (10) of Sec. 1.1].

When $\lambda = 2n$, we have further from (13)

$$|P_{2n}(t)| \le |P_{2n}(0)| = \frac{B_n}{(2n)!} \ . \tag{16}$$

In the last step, Eqs. (8) and (4) of Sec. 1.1 have been used.

If $F(x)$ possesses a definite sign (positive or negative) for $x > 0$, and $F(x)$, together with its derivatives to all orders, monotonically approaches zero when $x \to \infty$, it can be proved[a] that the remainder in the Euler formula has the form

$$R_n = \theta \frac{(-)^{n+1} B_{n+1}}{(2n+2)!}$$

$$\times h^{2n+2}[F^{(2n+1)}(a+mh) - F^{(2n+1)}(a)] \quad (0 \le \theta \le 1) .$$
(17)

Applications

Ex. 1. In the Euler formula (6), putting $F(x) = e^{tx}$, $a = 0$, $m = 1$, $h = 1$ and using the remainder (10), we have

$$\frac{e^t - 1}{t} = \frac{1}{2}(e^t + 1) + \sum_{k=1}^{n} \frac{(-)^k B_k}{(2k)!} t^{2k-1}(e^t - 1)$$

$$- t^{2n+1} \int_0^1 P_{2n+1}(s) e^{ts} ds .$$

Multiplying both sides with $t/(e^t - 1)$, removing terms, there results

$$\frac{t}{e^t - 1} = 1 - \frac{t}{2} - \sum_{k=1}^{n} \frac{(-)^k B_k}{(2k)!} t^{2k} + \frac{t^{2n+2}}{e^t - 1} \int_0^1 P_{2n+1}(s) e^{ts} ds .$$
(18)

This is the finite Taylor expansion of the function $t/(e^t - 1)$. Comparing with Eq. (3) of Sec. 1.1, we see that (18) is not restricted to $|t| < 2\pi$.

Ex. 2. Calculation of the *Euler constant γ*.

Euler constant γ is defined to be

$$\gamma = \lim_{m \to \infty} \left\{ 1 + \frac{1}{2} + \ldots + \frac{1}{m} - \ln m \right\} .$$
(19)

To show that the limit exists, let

$$u_n = \frac{1}{n} - \ln \frac{n+1}{n} = \int_0^1 \frac{t}{n(n+t)} dt \quad (n \ge 1) .$$

Since $u_n > 0$ and $u_n < \int_0^1 dt/n^2 = 1/n^2$, $\sum_{n=1}^{\infty} u_n$ converges. Now, the limit in (19) is equal to $\lim_{m \to \infty} \left\{ \sum_{n=1}^{m} u_n + \ln \frac{m+1}{m} \right\} = \sum_{n=1}^{\infty} u_n$, and hence exists.

---

[a] cf. for example, Kropp, *Unendliche Reihen*, pp. 550–552 (1931).

The convergence of the sequence in (19) is extremely slow. But if we calculate it with the Euler formula, we can obtain its approximate value within an error of order 10^{-7} when we use only three terms containing the Bernoulli numbers. The procedure is as follows.

In the Euler formula (6), let $F(x) = 1/(1+x)$, $a = 0$, $h = 1$, $n = 3$, and replace m by $m - 1$. With the remainder (10) we obtain

$$
\begin{aligned}
1 + \frac{1}{2} &+ \ldots + \frac{1}{m} - \ln m \\
&= \frac{1}{2}\left(1 + \frac{1}{m}\right) + \frac{B_1}{2}\left(1 - \frac{1}{m^2}\right) \\
&\quad - \frac{B_2}{4}\left(1 - \frac{1}{m^4}\right) + \frac{B_3}{6}\left(1 - \frac{1}{m^6}\right) \\
&\quad - 7! \int_1^m P_7(t)\,dt/t^8 \ .
\end{aligned}
\tag{20}
$$

Let $m \to \infty$, then we have

$$
\gamma = \frac{1}{2} + \frac{B_1}{2} - \frac{B_2}{4} - \frac{B_3}{6} - 7! \int_1^\infty P_7(t)\,dt/t^8 \ .
\tag{21}
$$

Combining with (20) and substituting the values of the Bernoulli numbers given in Eq. (10) of Sec. 1.1, we obtain

$$
\begin{aligned}
\gamma = 1 + \frac{1}{2} &+ \ldots + \frac{1}{m} - \ln m - \frac{1}{2m} + \frac{1}{12m^2} \\
&- \frac{1}{120m^4} + \frac{1}{252m^6} - 7! \int_m^\infty P_7(t)\,dt/t^8 \ .
\end{aligned}
\tag{22}
$$

To estimate the value of the integral, we have by (15)

$$
\left| 7! \int_m^\infty P_7(t)\,dt/t^8 \right| \le 7! \frac{4}{(2\pi)^7} \int_m^\infty dt/t^8 = \frac{45}{2}(m\pi)^{-7} \ .
$$

Take $m = 5$. We see that the absolute value of the integral is less than 10^{-7}. Assume that $\ln 5$ is known,[b] then (22) gives $(m = 5)$

$$
\gamma = 0.5772157 \ .
\tag{23}
$$

The value of γ upto 23 decimal places is

$$
\gamma = 0.57721\,56649\,01532\,86060\,651\ldots \ .
\tag{24}
$$

---

[b]$\ln 5 = 1.60943\,79124$.

1.4. Lagrange's Expansion Formula

Before we turn to the main subject of this section, let us first state and prove an important theorem in the theory of functions of a complex variable which deals with the zeros and poles of a function.

Theorem: Let $\psi(z)$ be a function analytic inside a simple contour C except at a number of poles b_j $(j = 1, 2, \ldots)$; a_k $(k = 1, 2, \ldots)$ be the zeros of $\psi(z)$ inside C; on C, $\psi(z) \neq 0$. Let $\varphi(z)$ be an analytic function in and on C. Then

$$\frac{1}{2\pi i} \oint_C \varphi(z) \frac{\psi'(z)}{\psi(z)} dz = \sum_k n_k \varphi(a_k) - \sum_j p_j \varphi(b_j) , \qquad (1)$$

where n_k and p_j are respectively the orders of the zero a_k and the pole b_j, the integral being taken along the positive direction of C (counter clockwise).

Proof: According to Cauchy's theorem,

$$\frac{1}{2\pi i} \oint_C \varphi(z) \frac{\psi'(z)}{\psi(z)} dz = \frac{1}{2\pi i} \Big\{ \sum_k \int_{(a_k)} + \sum_j \int_{(b_j)} \Big\} \varphi(z) \frac{\psi'(z)}{\psi(z)} dz , \qquad (2)$$

where (a_k) and (b_j) designate simple contours encircling (in positive direction) a_k and b_j respectively, each contour enclosing only one zero or one pole.

In the neighborhood of a_k,

$$\psi(z) = (z - a_k)^{n_k} \psi_k(z) , \quad \psi_k(a_k) \neq 0;$$

therefore,

$$\frac{\psi'(z)}{\psi(z)} = \frac{d}{dz} \ln \psi(z) = \frac{n_k}{z - a_k} + \frac{\psi_k'(z)}{\psi_k(z)} .$$

Since $\psi_k(z)$ is analytic in the neighborhood of a_k and $\psi_k(a_k) \neq 0$, the function $\psi_k'(z)/\psi_k(z)$ is also analytic within the contour (a_k), and its contribution to the integral in (2) is zero. Thus, by the residue theorem, we have

$$\frac{1}{2\pi i} \int_{(a_k)} \varphi(z) \frac{\psi'(z)}{\psi(z)} dz = \frac{n_k}{2\pi i} \int_{(a_k)} \varphi(z) \frac{dz}{z - a_k} = n_k \varphi(a_k) . \qquad (3)$$

Similarly,

$$\frac{1}{2\pi i} \int_{(b_j)} \varphi(z) \frac{\psi'(z)}{\psi(z)} dz = -p_j \varphi(b_j) . \qquad (4)$$

Substituting (3) and (4) in (2) gives (1).

Let $\varphi(z) \equiv 1$. We have an important particular case of (1), namely,

$$\frac{1}{2\pi i} \oint_C \frac{\psi'(z)}{\psi(z)} dz = N - P ,\tag{5}$$

where N is the number of zeros and P the number of poles of $\psi(z)$ within C; multiple zeros and poles are counted according to their orders (n_k and p_j).

If $\psi(z)$ has no poles in C, then $P = 0$, and

$$\frac{1}{2\pi i} \oint_C \frac{\psi'(z)}{\psi(z)} dz = N .\tag{6}$$

Now we come to *Lagrange's theorem*:

Let $f(z)$ and $\varphi(z)$ be analytic in and on the simple contour C, and a is a point within C. If for points ς on C, the parameter t satisfies

$$|t\varphi(\varsigma)| < |\varsigma - a|\tag{7}$$

then,

(i) As an equation for z,

$$z = a + t\varphi(z)\tag{8}$$

has one and only one root inside C; when $t = 0$, the root approaches a.

(ii) The function $f(z)$ can be expanded in a power series of t:

$$f(z) = f(a) + \sum_{n=1}^{\infty} \frac{t^n}{n!} \frac{d^{n-1}}{da^{n-1}} \{ f'(a)[\varphi(a)]^n \} .\tag{9}$$

This is *Lagrange's expansion formula* (in which, the variables z and t are connected by Eq. (8)).

Proof: Applying (6) to the function

$$\psi(z) \equiv z - a - t\varphi(z)$$

and taking note of condition (7), the number of zeros of $\psi(z)$ in C is

$$\begin{aligned}
N &= \frac{1}{2\pi i} \oint_C \frac{1 - t\varphi'(\varsigma)}{\varsigma - a - t\varphi(\varsigma)} d\varsigma \\
&= \frac{1}{2\pi i} \oint_C [1 - t\varphi'(\varsigma)] \sum_{n=0}^{\infty} \frac{[t\varphi(\varsigma)]^n}{(\varsigma - a)^{n+1}} d\varsigma \\
&= \sum_{n=0}^{\infty} \frac{t^n}{n!} \frac{d^n}{da^n} \{ [\varphi(a)]^n \} - \sum_{n=0}^{\infty} \frac{t^{n+1}}{(n+1)!} \frac{d^{n+1}}{da^{n+1}} \{ [\varphi(a)]^{n+1} \} \\
&= 1
\end{aligned}$$

which proves (i), i.e., $\psi(z) \equiv z - a - t\varphi(z) = 0$ has one and only one root inside C.

Next, let z be the only root of Eq. (8) in C. By (1), we have

$$\frac{1}{2\pi i} \oint_C f(\varsigma) \frac{\psi'(z)}{\psi(z)} dz = f(z) \ .$$

On the other hand,

$$\begin{aligned}
\frac{1}{2\pi i} \oint_C f(\varsigma) \frac{\psi'(\varsigma)}{\psi(\varsigma)} d\varsigma &= \frac{1}{2\pi i} \oint_C f(\varsigma) \frac{1 - t\varphi'(\varsigma)}{\varsigma - a - t\varphi(\varsigma)} d\varsigma \\
&= \frac{1}{2\pi i} \oint_C f(\varsigma)[1 - t\varphi'(\varsigma)] \sum_{n=0}^{\infty} t^n \frac{[\varphi(\varsigma)]^n}{(\varsigma - a)^{n+1}} d\varsigma \\
&= \sum_{n=0}^{\infty} \frac{t^n}{n!} \frac{d^n}{da^n} \{f(a)[\varphi(a)]^n\} \\
&\quad - \sum_{n=0}^{\infty} \frac{t^{n+1}}{(n+1)!} \frac{d^n}{da^n} \left\{f(a) \frac{d}{da} [\varphi(a)]^{n+1}\right\} \\
&= f(a) + \sum_{n=1}^{\infty} \frac{t^n}{n!} \frac{d^{n-1}}{da^{n-1}} \{f'(a)[\varphi(a)]^n\} \ ,
\end{aligned}$$

which is (9).

Example. Let $\varphi(z) = (z^2 - 1)/2$. Then the equation $z - x - t\varphi(z) = 0$ has the only root

$$z = \frac{1 - \sqrt{1 - 2xt + t^2}}{t}$$

which approaches x when $t \to 0$. According to Lagrange's formula (9), by putting $f(z) \equiv z$, we have

$$\frac{1 - \sqrt{1 - 2xt + t^2}}{t} = x + \sum_{n=1}^{\infty} \frac{t^n}{n!} \frac{d^{n-1}}{dx^{n-1}} \left\{ \left(\frac{x^2 - 1}{2} \right)^n \right\} \ .$$

Differentiating both sides with respect to x, we obtain

$$\frac{1}{\sqrt{1 - 2xt + t^2}} = \sum_{n=0}^{\infty} \frac{t^n}{2^n n!} \frac{d^n}{dx^n} (x^2 - 1)^n \ , \tag{10}$$

an important expansion formula in the theory of Legendre polynomial (see Chap. 5).

1.5. Expansion of Meromorphic Functions in Rational Fractions

Meromorphic function is a single-valued function which has, in a finite domain, no singularities other than poles.

Example 1. Any rational function $P_n(z)/Q_m(z)$, where $P_n(z)$ and $Q_m(z)$ are polynomials of degree n and m respectively, is a meromorphic function. If $P_n(z)$ and $Q_m(z)$ have no common factors, the only singularities in a finite domain are the zeros of $Q_m(z)$.

Example 2. csc z, cot z, etc.; their singularities are $z = \pm n\pi$, $n = 0, 1, 2, \ldots$, all being simple poles.

In a finite domain, the number of poles of a meromorphic function must be finite. Otherwise there would be an accumulation point (limit point) which cannot be a pole, since a pole is an isolated singularity.

If a meromorphic function has an infinite number of poles a_n, $n = 1, 2, \ldots$, as in Example 2 above, then we must have $\lim\limits_{n \to \infty} a_n = \infty$. For, if the limit is finite, there would be an infinite number of poles in the vicinity of it (thus in a finite domain).

We have enumerated above some basic properties of meromorphic functions. Now we shall discuss a special kind of expansion, namely, expansion in rational fractions, of this class of function. In such expansion, all the poles of the function, together with the nature of the singularity at each of them, are exhibited in a whole (compare with the Laurent expansion).

Mittag-Leffler theorem: Let $f(z)$ be a meromorphic function whose poles are $a_1, a_2, a_3, \ldots$, and $0 < |a_1| \le |a_2| \le |a_3| \le \ldots$. If there exists a sequence of contours $\{C_m\}$ such that

(i) when $m \to \infty$, the nearest distance from C_m to the origin $(z = 0)$, R_m, tends to infinity, but l_m/R_m remains finite, where l_m is the circumference of C_m;

(ii) on C_m

$$|z^{-p} f(z)| < M , \tag{1}$$

where p is certain smallest non-negative integer and M a constant independent of m.

Then $f(z)$ can be expanded in the form of the following series of rational fractions:

$$f(z) = \sum_{k=0}^{p} f^{(k)}(0) \frac{z^k}{k!} + \sum_{n=1}^{\infty} \left\{ G_n\left(\frac{1}{z - a_n}\right) - \varphi_{np}(z) \right\} , \tag{2}$$

where

$$G_n\left(\frac{1}{z-a_n}\right) = \frac{A_{n,s_n}}{(z-a_n)^{s_n}} + \frac{A_{n,s_n-1}}{(z-a_n)^{s_n-1}} + \ldots + \frac{A_{n,1}}{z-a_n} \qquad (3)$$

is the principal part of $f(z)$ at a_n; s_n is the order of a_n;

$$\varphi_{np}(z) = \sum_{k=0}^{p} \left[\frac{d^k}{d\varsigma^k} G_n\left(\frac{1}{\varsigma-a_n}\right)\right]_{\varsigma=0} \frac{z^k}{k!} . \qquad (4)$$

Proof: Let m be sufficiently large such that z lies within C_m, then, by Cauchy's theorem,

$$\frac{1}{2\pi i}\oint_{C_m} \frac{f(\varsigma)d\varsigma}{\varsigma-z} = \frac{1}{2\pi i}\int_{(z+)} \frac{f(\varsigma)d\varsigma}{\varsigma-z} + \sum_{r=1}^{r_m} \frac{1}{2\pi i}\int_{(a_r^+)} \frac{f(\varsigma)d\varsigma}{\varsigma-z}$$

$$= f(z) + \sum_{r=1}^{r_m} \frac{1}{2\pi i}\int_{(a_r^+)} \frac{d\varsigma}{\varsigma-z}\left\{G_r\left(\frac{1}{\varsigma-a_r}\right) + P_r(\varsigma-a_r)\right\}, \qquad (5)$$

where (a_r^+) denotes a small contour around a_r in the positive direction, within which $f(z)$ has no poles other than a_r; similar to (z^+). $P_r(\varsigma-a)$ is the regular part of $f(z)$ in the vicinity of a_r; r_m is the number of poles within C_m. Since $P_r(\varsigma-a_r)/(\varsigma-z)$ is analytic in the vicinity of a_r, its contour integral is equal to zero. Furthermore, according to Cauchy's theorem for a multiply connected domain,

$$\frac{1}{2\pi i}\int_{(a_r^+)} \frac{d\varsigma}{\varsigma-z}G_r\left(\frac{1}{\varsigma-a_r}\right)$$

$$= \frac{1}{2\pi i}\oint_{C_R} \frac{d\varsigma}{\varsigma-z}G_r\left(\frac{1}{\varsigma-a_r}\right) - \frac{1}{2\pi i}\int_{(z+)} \frac{d\varsigma}{\varsigma-z}G_r\left(\frac{1}{\varsigma-a_r}\right),$$

where C_R is a circle with center at $z=0$, radius R, enclosing the points a_r and z. When $|\varsigma| \to \infty$,

$$\frac{1}{\varsigma-z}G_r\left(\frac{1}{\varsigma-a_r}\right) = O(\varsigma^{-2})$$

therefore $\int_{C_R} \to 0$ when $R \to \infty$, and we have

$$\frac{1}{2\pi i}\int_{(a_r^+)} \frac{d\varsigma}{\varsigma-z}G_r\left(\frac{1}{\varsigma-a_r}\right) = -\frac{1}{2\pi i}\int_{(z+)} \frac{d\varsigma}{\varsigma-z}G_r\left(\frac{1}{\varsigma-a_r}\right)$$

$$= -G_r\left(\frac{1}{z-a_r}\right) . \qquad (6)$$

(5) then becomes

$$\frac{1}{2\pi i}\oint_{C_m}\frac{f(\varsigma)d\varsigma}{\varsigma-z}=f(z)-\sum_{r=1}^{r_m}G_r\left(\frac{1}{z-a_r}\right).\tag{7}$$

On the other hand, the l.h.s. integral in (7) can be expressed as

$$\frac{1}{2\pi i}\oint_{C_m}\frac{f(\varsigma)d\varsigma}{\varsigma-z}=\frac{1}{2\pi i}\oint_{C_m}f(\varsigma)d\varsigma\left[\sum_{k=0}^{p}\frac{z^k}{\varsigma^{k+1}}+\frac{z^{p+1}}{\varsigma^{p+1}(\varsigma-z)}\right]$$

$$=\sum_{k=0}^{p}f^{(k)}(0)\frac{z^k}{k!}+\sum_{k=0}^{p}z^k\sum_{r=1}^{r_m}\frac{1}{2\pi i}\int_{(a_r^+)}f(\varsigma)\varsigma^{-k-1}d\varsigma$$

$$+\frac{z^{p+1}}{2\pi i}\oint_{C_m}\frac{f(\varsigma)d\varsigma}{\varsigma^{p+1}(\varsigma-z)}.$$

Using the same method as in proving (6), we have

$$\frac{1}{2\pi i}\int_{(a_r^+)}f(\varsigma)\varsigma^{-k-1}d\varsigma=\frac{1}{2\pi i}\int_{(a_r^+)}G_r\left(\frac{1}{\varsigma-a_r}\right)\varsigma^{-k-1}d\varsigma$$

$$=-\frac{1}{2\pi i}\int_{(O+)}G_r\left(\frac{1}{\varsigma-a_r}\right)\varsigma^{-k-1}d\varsigma=-\frac{1}{k!}\left[\frac{d^k}{d\varsigma^k}G_r\left(\frac{1}{\varsigma-a_r}\right)\right]_{\varsigma=0}.$$

Substituting it in the preceding equation and using (7), we obtain

$$f(z)=\sum_{k=0}^{p}f^{(k)}(0)\frac{z^k}{k!}+\sum_{r=1}^{r_m}\left\{G_r\left(\frac{1}{z-a_r}\right)-\varphi_{rp}(z)\right\}+\frac{z^{p+1}}{2\pi i}\oint_{C_m}\frac{f(\varsigma)d\varsigma}{\varsigma^{p+1}(\varsigma-z)}.$$

When $m\to\infty$, by condition (i), we have, for $|z|<R$ (R being an arbitrary positive number),

$$\left|\oint_{C_m}\frac{f(\varsigma)d\varsigma}{\varsigma^{p+1}(\varsigma-z)}\right|<\frac{Ml_m}{R_m(R_m-|R|)}\to 0,$$

and hence (2). The series in (2)^c is uniformly convergent in $|z|<R$ (R arbitrary), provided $z\neq a_n$, $n=1,2,\ldots$.

When $z=0$ is also a pole of $f(z)$, the above theorem cannot be applied directly to $f(z)$. However, it can be applied to the function $F(z)\equiv f(z)-$

---

^{c} Whose terms are arranged according to the order of the sequence of contours $\{C_m\}$, and may be arranged arbitrarily if the series converges absolutely.

$G_0(1/z)$, $G_0(1/z)$ being the principal part of $f(z)$ at $z = 0$, provided that $F(z)$ satisfies the conditions for the validity of the theorem.

When all the a_r's are simple poles ($\neq 0$) of the meromorphic function $f(z)$, and $|f(z)| < M$ on C_m (i.e., the case $p = 0$), M being independent of m, we have the particularly simple expansion formula

$$f(z) = f(0) + \sum_{n=1}^{\infty} b_n \left\{ \frac{1}{z - a_n} + \frac{1}{a_n} \right\}, \tag{8}$$

where b_n is the residue of $f(z)$ at a_n.

The series on the r.h.s. of (2) and (8) represent, in the whole z plane, the function on the l.h.s.

Example 1. The rational-fraction expansion of cot z.

cot z is a meromorphic function, whose poles are 0 and $n\pi$, ($n = \pm 1$, $\pm 2, \dots$), all being simple and with residue equal to 1. Since $z = 0$ is also a pole, we have to consider the function $F(z) \equiv \cot z - 1/z$, whose poles are $n\pi$ ($n = \pm 1, \pm 2, \dots$), again with residue equal to 1. Let C_m be a square contour shown in Fig. 1.

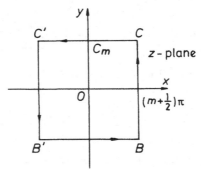

Fig. 1.

It can be shown that $|F(z)| < M$ on C_m, independent of m. For,

$$|\cot z|^2 = \frac{e^{2y} + e^{-2y} + 2\cos 2x}{e^{2y} + e^{-2y} - 2\cos 2x} \quad (z = x + iy) \,.$$

On the two sides BC and B'C' of C_m, $x = \pm(m + \frac{1}{2})\pi$, so $\cos 2x = -1$ and $|\cot z|^2 < 1$. On BB' and CC', $y = \pm(m + \frac{1}{2})\pi$ and

$$|\cot z|^2 < \frac{e^{2y} + e^{-2y} + 2}{e^{2y} + e^{-2y} - 2} = \left(\frac{1 + e^{-2y}}{1 - e^{-2y}} \right)^2 \longrightarrow 1 \quad \text{when } m \to \infty \,.$$

1.6. Infinite Product

In this section, we shall deal with the fundamental concepts and convergence problems of the infinite product, especially with the condition for its absolute convergence.

The infinite product

$$\prod_{n=1}^{\infty} u_n = u_1 u_2 u_3 \ldots \tag{1}$$

is said to be convergent when, and only when, there exists an m such that for all $n > m$, $u_n \neq 0$, and that the partial product

$$p_n = u_{m+1} u_{m+2} \ldots u_n \quad (n > m) \tag{2}$$

tends to a limit U_m not equal to zero when $n \to \infty$. Then,

$$U = \prod_{n=1}^{\infty} u_n = u_1 u_2 \ldots u_m U_m \tag{3}$$

is called the value of the infinite product, which is evidently independent of m.

Theorem 1: $\lim_{n \to \infty} u_n = 1$ is a necessary condition for the convergence of the infinite product (1).

Proof: For $n \to \infty$, we then have $u_n = p_n/p_{n-1} \to 1$, since p_n and p_{n-1} have the same limit U_m as $n \to \infty$.

Accordingly, we usually write $u_n = 1 + a_n$, and (1) then becomes

$$\prod_{n=1}^{\infty} (1 + a_n) . \tag{4}$$

It follows that $\lim_{n \to \infty} a_n = 0$ is a necessary condition for the convergence of (4).

Theorem 2: A necessary and sufficient condition for the convergence of (4) is that there exists an m for which the series

$$\sum_{n=m+1}^{\infty} \ln(1 + a_n) \tag{5}$$

converges; the logarithms are to take their principal values, i.e., $|\arg(1+a_n)| < \pi$. Denote this sum (5) by L, then

$$\prod_{n=1}^{\infty} (1 + a_n) = (a + a_1)(1 + a_2) \ldots (a + a_m) e^L . \tag{6}$$

Proof: The condition is sufficient. For, if (5) converges, the sequence

$$P_n = (1 + a_{m+1})(1 + a_{m+2}) \ldots (1 + a_n) = \exp\left\{ \sum_{r=m+1}^{n} \ln(1 + a_r) \right\}$$

will be convergent, and $\lim_{n \to \infty} p_n = e^L$. Moreover, $1/z$ is bounded on C_m. So we can apply (8) to $F(z)$, obtaining

$$\cot z - \frac{1}{z} = \left(\cot z - \frac{1}{z} \right)_{z \to 0} + \sum_{n=-\infty}^{\infty}{}' \left\{ \frac{1}{z - n\pi} + \frac{1}{n\pi} \right\},$$

where $\sum'$ designates that $n = 0$ is to be omitted in the summation. When $z \to 0$, $\cot z - 1/z \to 0$, and hence

$$\cot z = \frac{1}{z} + \sum_{n=-\infty}^{\infty}{}' \left\{ \frac{1}{z - n\pi} + \frac{1}{n\pi} \right\} = \frac{1}{z} + \sum_{n=1}^{\infty} \frac{2z}{z^2 - n^2\pi^2}. \tag{9}$$

Replacing z in (9) by $t/2$ gives

$$\frac{t}{2} \cot \frac{t}{2} = 1 + \sum_{n=1}^{\infty} \frac{2t^2}{t^2 - (2n\pi)^2}. \tag{10}$$

Comparing with Eq. (18) of Sec. 1.1, we obtain

$$\sum_{n=1}^{\infty} \frac{2t^2}{t^2 - (2n\pi)^2} = -\sum_{k=1}^{\infty} \frac{B_k}{(2k)!} t^{2k}. \tag{11}$$

When $|t| < 2\pi$, each term of the series on the l.h.s. can be expanded in a power series of t, namely,

$$\frac{2t^2}{t^2 - (2n\pi)^2} = -2 \sum_{k=1}^{\infty} \frac{t^{2k}}{(2n\pi)^{2k}}.$$

Substituting this result in (11) and changing the order of summation,[d] it results

$$\sum_{k=1}^{\infty} t^{2k} \left(2 \sum_{n=1}^{\infty} \frac{1}{(2n\pi)^{2k}} \right) = \sum_{k=1}^{\infty} \frac{B_k}{(2k)!} t^{2k}$$

---

[d] cf. Knopp, *Unendliche Reihen*, Sec. 16 [90], (1931).

from which we arrive at the summation formula

$$\sum_{n=1}^{\infty} \frac{1}{n^{2k}} = \frac{(2\pi)^{2k} B_k}{2(2k)!} , \quad k = 1, 2, \ldots \tag{12}$$

where B_k are the Bernoulli numbers [Sec. 1.1].

The condition is also necessary. For, if the product converges, then there exists an m for which $(1 + a_n) \neq 0$ $(n > m)$, and $p_n \to p \neq 0$ when $n \to \infty$. But

$$\sum_{r=m+1}^{\infty} \ln(1 + a_r) = \lim_{n \to \infty} \ln p_n = \ln p .$$

Therefore, the series (5) is convergent. However, its value depends upon the arguments of the factors in p. These arguments cannot be defined arbitrarily, since $\lim_{n \to \infty} \ln(1 + a_n) = 0$ is a necessary condition for the convergence of the series (5) and $\ln(1 + a_n) = \ln|1 + a_n| + i \arg(1 + a_n)$. Thus, we must have $a_n \to 0$ and $\arg(1 + a_n) \to 0$. Therefore, except for a finite number of terms, it is necessary to have $|\arg(1 + a_n)| < \pi$, i.e., the logarithms are to take their principal values.

Necessary and sufficient condition for the absolute convergence of an infinite product.

Definition of absolute convergence: The product $\prod_{n=1}^{\infty} (1 + a_n)$ is absolutely convergent if $\prod_{n=1}^{\infty} (1 + |a_n|)$ converges.

Theorem 3: If $\prod_{n=1}^{\infty} (1 + a_n)$ converges absolutely, it is also convergent.

Proof: Since $\prod_{n=1}^{\infty} (1 + |a_n|)$ converges, there exists an r for which

$$q_s = (1 + |a_{r+1}|)(1 + |a_{r+2}|) \ldots (1 + |a_s|) \longrightarrow q \neq 0 \tag{7}$$

when $s \to \infty$. For an arbitrary $k(> 0)$, when s is sufficiently large, we have by developing the product,

$$|(1 + a_{s+1})(1 + a_{s+2}) \ldots (1 + a_{s+k}) - 1|$$
$$\leq (1 + |a_{s+1}|)(1 + |a_{s+2}|) \ldots (1 + |a_{s+k}|) - 1$$
$$= \frac{q_{s+k}}{q_s} - 1 < \varepsilon . \tag{8}$$

Thus there exists an m such that for any $s > m$,

$$|(1 + a_{m+1})(1 + a_{m+2}) \ldots (1 + a_s) - 1| = |p_s - 1| < \frac{1}{2} ,$$

i.e.,

$$\frac{1}{2} < |p_s| < \frac{3}{2}$$

which shows that when $s > m$, $1 + a_s \neq 0$, and if p_s tends to a limit, the limit is different from zero. In fact, by (8), we have

$$\left| \frac{p_{s+k}}{p_s} - 1 \right| \leq \frac{q_{s+k}}{q_s} - 1 < \varepsilon \quad (s > m, k(>0) \text{ arbitrary})$$

or

$$|p_{s+k} - p_s| < \varepsilon |p_s| .$$

Then, according to Cauchy's criterion, p_s has a limit, i.e., $\prod\limits_{n=1}^{\infty} (1 + a_n)$ converges.

Theorem 4: A necessary and sufficient condition for the absolute convergence of the product $\prod\limits_{n=1}^{\infty} (1+a_n)$ is that the series $\sum\limits_{n=1}^{\infty} a_n$ converges absolutely.

Proof: Let $P_n = (1+|a_{m+1}|)(1+|a_{m+2}|) \ldots (1+|a_n|)$, $(n > m)$. Evidently, $P_n \neq 0$. Let $S_n = |a_{m+1}| + |a_{m+2}| + \ldots + |a_n|$. Since $1 + |a_\nu| \leq e^{|a_\nu|}$, it is seen that $S_n < P_n < e^{S_n}$ and the equivalence of the convergence of S_n and P_n follows.

Theorem 5: If the product $\prod\limits_{n=1}^{\infty} (1+a_n)$ converges absolutely, then the series $\sum\limits_{n=1}^{\infty} \ln(1 + a_n)$ is also absolutely convergent; and vice versa.

Proof: For the convergence of either the product or the series, we must have $a_n \to 0$ when $n \to \infty$. Therefore, for sufficiently large n such that $|a_n| < \frac{1}{2}$,

$$\left| \frac{\ln(1 + a_n)}{a_n} - 1 \right| = \left| -\frac{a_n}{2} + \frac{a_n^2}{3} - \frac{a_n^3}{4} + \ldots \right|$$
$$< \frac{1}{2^2} + \frac{1}{2^3} + \frac{1}{2^4} + \ldots = \frac{1}{2} ,$$

i.e.,

$$\frac{1}{2} < \left| \frac{\ln(1 + a_n)}{a_n} \right| < \frac{3}{2} .$$

Thus the convergence of $\sum\limits_{n=1}^{\infty} |\ln(1 + a_n)|$ and of $\sum\limits_{n=1}^{\infty} |a_n|$ are equivalent, and thus Theroem 5 follows from Theorem 4.

Uniform Convergence

The infinite product $\prod_{n=1}^{\infty}\{1 + u_n(z)\}$ is said to be uniformly convergent in a certain domain if for an arbitrary $\varepsilon > 0$ there exists an m, independent of z, such that for any $p > 0$

$$\left| \prod_{n=1}^{m+p}\{1 + u_n(z)\} - \prod_{n=1}^{m}\{1 + u_n(z)\} \right| < \varepsilon .$$

1.7. The Expansion of a Function in Infinite Product. Weierstrass Theorem

Consider first the case of an integral (entire) function.

A function analytic in the entire plane is called an integral (or entire) function; for example, polynomials, $\sin z$, $\cos z$, e^z, etc.

If an integral function is also analytic at infinity then, according to Liouville's theorem, it must be a constant. In general, the point at infinity may be a pole of an integral function, for example, as in the case of a polynomial, or an essential singularity, as in the case of e^z, $\sin z$, etc.

In algebra, we often express a polynomial in terms of a product of its prime factors. Similarly for an integral function, we can represent it by a product which, however, is infinite if the zeros of the function are infinite in number. For example, $\sin z$, $\cos z$, etc. The discussion of the convergence problem of such an expansion will appeal to the theorems proved in the last section.

Theorem 1: Suppose that $f(z)$ is an integral function with simple zeros $a_1, a_2, \ldots$, not equal to zero, and $\lim\limits_{n \to \infty} a_n = \infty$.[e] Assume that there exists a sequence of contours $\{C_m\}$ such that on them, $|f'(z)/f(z)| < M$, M being a positive number independent of m. Then $f(z)$ can be expanded in an infinite product:

$$f(z) = f(0)e^{f'(0)/f(0)\cdot z} \prod_{n=1}^{\infty} \left\{ \left(1 - \frac{z}{a_n}\right) e^{z/a_n} \right\} , \tag{1}$$

where each factor $(1 - z/a_n)e^{z/a_n}$ is equal to zero only at the point $z = a_n$ and is called a prime factor of the integral function $f(z)$.

---

[e] We shall not discuss the simple case of a finite number of zeros. In the case of an infinite number of zeros, the point at infinity must be a, and the only, limit (accumulation) point, since if there is a limit point of zeros other than $z = \infty$, $f(z)$ must be identically zero.

Proof: Let $F(z) = f'(z)/f(z)$. At the point $z = a_n$,

$$f(z) = (z - a_n)f'(a_n) + \frac{(z - a_n)^2}{2!}f''(a_n) + \dots ,$$
$$f'(z) = f'(a_n) + (z - a_n)f''(a_n) + \dots , \quad f'(a_n) \neq 0 .$$

Hence,

$$F(z) = \frac{f'(z)}{f(z)} = \frac{\varphi(z)}{z - a_n}, \quad \varphi(a_n) = 1 ,$$

i.e., $a_n (n = 1, 2, \dots)$ are simple poles of $F(z)$ with residues all equal to 1; elsewhere, $F(z)$ is analytic. Thus, $F(z)$ is a meromorphic function. Since $F(z)$ satisfies the condition of Eq. (8) of Sec. 1.5, it can be expanded in rational fractions:

$$F(z) = F(0) + \sum_{n=1}^{\infty} \left\{ \frac{1}{z - a_n} + \frac{1}{a_n} \right\} ,$$

i.e.,

$$\frac{f'(z)}{f(z)} = \frac{f'(0)}{f(0)} + \sum_{n=1}^{\infty} \left\{ \frac{1}{z - a_n} + \frac{1}{a_n} \right\} .$$

The series on the r.h.s. is uniformly convergent. We can multiply the equation with dz and integrate from 0 to z, obtaining (1).

Example. The infinite-product expansion of $\sin z$.

As $z = 0$ is a simple zero of $\sin z$, we cannot apply theorem 1 directly. Consider, however, the function $f(z) = \sin z/z$ whose zeros are $z = \pm n\pi, n = 1, 2, \dots$, all being simple. Now $f'(z)/f(z) = \cot z - 1/z$, and we have shown in Example 1 of Sec. 1.5 that it satisfies the condition stated for the expansion (1) to hold. Therefore we can expand $\sin z/z$ in an infinite product:

$$\frac{\sin z}{z} = \prod_{n=1}^{\infty} \left\{ \left(1 - \frac{z}{n\pi} \right) e^{z/n\pi} \right\} \left\{ \left(1 + \frac{z}{n\pi} \right) e^{-z/n\pi} \right\} \quad (2)$$

or,

$$\frac{\sin z}{z} = \prod_{n=1}^{\infty} \left(1 - \frac{z^2}{n^2\pi^2} \right) . \quad (3)$$

Since $\sum_{n=1}^{\infty} z^2/n^2\pi^2$ converges absolutely, this infinite product is absolutely convergent by theorem 4 of Sec. 1.6.

It is evident that the expansion of a function in infinite product is not confined to the case that the zeros of the function are all simple. Suppose that a_n is a zero of $f(z)$ of order m_n. It is not difficult to show that a_n is still a simple pole of $F(z) = f'(z)/f(z)$, but the corresponding residue is now m_n. Therefore, if other conditions are satisfied, we can obtain again, by means of Eq. (8) of Sec. 1.5, the infinite product expansion of $f(z)$, namely,

$$f(z) = f(0)e^{\frac{f'(0)}{f(0)}z} \prod_{n=1}^{\infty} \left\{ \left(1 - \frac{z}{a_n}\right) e^{z/a_n} \right\}^{m_n}, \tag{4}$$

where the prime factor involving a_n occurs m_n times.

Besides, the restriction that $|f'(z)/f(z)| < M$ (independent of m) can be released, if only the formula (1) or (4) is modified correspondingly. (See the general factor theorem of Weierstrass below.)

The expansion in infinite product is not restricted to integral functions. A meromorphic function $f(z)$ can always be expanded in an infinite product since it can be expressed as the quotient of two integral functions. To represent a meromorphic function as a quotient of two integral functions, we proceed as follows. Let $G(z)$ be an integral function whose zeros are the poles of $f(z)$. Then the product $f(z)G(z)$ has no singularities in a finite domain and is, therefore, an integral function. Let it be denoted by $G_1(z)$, then $f(z) = G_1(z)/G(z)$.

The general (prime) factor theorem of Weierstrass

Suppose that $f(z)$ has no essential singularities in a *finite* domain. Let its zeros or poles be $a_1, a_2, \ldots, 0 < |a_1| \le |a_2| \le \ldots$, then $f(z)$ can be expanded in an infinite product of the form:

$$f(z) = f(0)e^{G(z)} \prod_{n=1}^{\infty} \left\{ \left(1 - \frac{z}{a_n}\right) e^{g_n(z)} \right\}^{m_n}, \tag{5}$$

where $G(z)$ is an integral function with $G(0) = 0$; $g_n(z)$ is a properly chosen polynomial which plays the role of making the product converge absolutely and uniformly in any finite domain, except at the poles of $f(z)$. m_n is the order of the zero or pole a_n; m_n is negative if the corresponding a_n is a pole.

Proof: We first show that a polynomial $g_n(z)$ can be found such that the infinite product in (5) is to be convergent for any z not being a pole.

According to theorem 5 of the preceding section, the said convergence problem is equivalent to that of the series $\sum \ln\{(1 - z/a_n)e^{g_n(z)}\}^{m_n}$. The general term of this series is

$$u_n = \ln \left\{ \left(1 - \frac{z}{a_n} \right) e^{g_n(z)} \right\}^{m_n} = m_n \left\{ \ln \left(1 - \frac{z}{a_n} \right) + g_n(z) \right\}$$

$$= m_n \left\{ -\sum_{s=1}^{\infty} \frac{1}{s} \left(\frac{z}{a_n} \right)^s + g_n(z) \right\} \ .$$

If we take

$$g_n(z) = \sum_{s=1}^{k_n-1} \frac{1}{s} \left(\frac{z}{a_n} \right)^s \ , \tag{6}$$

where k_n is an integer (> 1) to be determined, we have

$$|u_n| = |m_n| \left| \sum_{s=k_n}^{\infty} \frac{1}{s} \left(\frac{z}{a_n} \right)^s \right| \le |m_n| \left| \frac{z}{a_n} \right|^{k_n} \sum_{s=0}^{\infty} \frac{1}{s+k_n} \left| \frac{z}{a_n} \right|^s$$

$$\le |m_n| \left| \frac{z}{a_n} \right|^{k_n} \sum_{s=0}^{\infty} \left| \frac{z}{a_n} \right|^s \ .$$

Let K be any positive number and $|z| < K$. Since $a_n \to \infty$, there exists an N such that $|a_n| < 2K (n > N)$ and $|z a_n^{-1}| < K |a_n|^{-1} < \frac{1}{2}$. Hence, when $n > N$,

$$|u_n| < 2 |m_n| \left| \frac{K}{a_n} \right|^{k_n} \ .$$

Now choose the least possible k_n to make $|u_n| < b_n$ and $\sum_{n=1}^{\infty} b_n$ is any convergent series with positive terms, for example, $b_n = 2^{-n}$. Then the series $\sum u_n$ not only converges, but also converges absolutely and uniformly (for $|z| < K$, except at the poles). Thus, for the $g_n(z)$ so chosen, the product in (5) is absolutely and uniformly convergent.

Let

$$F(z) = \prod_{n=1}^{\infty} \left\{ \left(1 - \frac{z}{a_n} \right) e^{g_n(z)} \right\}^{m_n} \ ,$$

then $f(z)/F(z) = G_1(z)$ is an integral function without zeros, and hence $G_1'(z)/G_1(z)$ is also an integral function. Denoting the latter by $G_2'(z)$, we have $G_1(z) = C e^{G_2(z)}$, C being an arbitrary constant. Thus

$$f(z) = C e^{G_2(z)} \prod_{n=1}^{\infty} \left\{ \left(1 - \frac{z}{a_n} \right) e^{g_n(z)} \right\}^{m_n} \ .$$

When $z = 0$, the infinite product on the r.h.s. is equal to 1. So $f(0) = Ce^{G_2(0)}$, and

$$f(z) = f(0)e^{G_2(z) - G_2(0)} \prod_{n=1}^{\infty} \left\{ \left(1 - \frac{z}{a_n} \right) e^{g_n(z)} \right\}^{m_n}$$

which is (5); the function $G(z) = G_2(z) - G_2(0)$ is to be determined by the behavior of the function $f(z)$ when $|z| \to \infty$. For example, if $F(z) = f'(z)/f(z)$ satisfies the condition (1) in Sec. 1.5, then, by applying Eq. (2) of that section to $F(z)$ and integrating from 0 to z, we find that

$$G(z) = \sum_{k=0}^{p} F^{(k)}(0) \frac{z^{k+1}}{(k+1)!} . \tag{7}$$

And, according to Eq. (4) of that section,

$$g_n(z) = \sum_{k=0}^{p} \left[\frac{d^k}{d\varsigma^k} G_n \left(\frac{1}{\varsigma - a_n} \right) \right]_{\varsigma = 0} \frac{z^{k+1}}{(k+1)!} , \tag{8}$$

where $G_n(1/(z - a_n))$ is the principal part of $F(z)$ at the point a_n.

1.8. Asymptotic Expansion

The asymptotic expansion of a function is commonly understood to be an approximate expression of it for large values of its argument or a parameter thereof.[f] For example, the approximation formula for the calculation of the error function [Eq. (1) of Sec. 6.15]: erf $x = \frac{2}{\sqrt{\pi}} \int_0^x e^{-t^2} dt$ when x is large; the expressions of the Bessel function $J_\nu(z)$ when $|z|$ or $|\nu|$, or both $|z|$ and $|\nu|$, are large [Secs. 7.10, 7.12]; the series solution of an ordinary differential equation at its irregular singular point [Sec. 2.10]; etc. In the present section we shall introduce only the fundamental concepts concerning asymptotic expansions. The important Watson Lemma for the asymptotic expansion of a function represented by the so called Laplace integral will be discussed in the following section. Other methods of obtaining asymptotic expansions are to be described later when we come to various special functions.

When $f(z)$ is analytic at infinity, it can be expressed as a Taylor series in descending powers of z:

$$f(z) = c_0 + c_1 z^{-1} + c_2 z^{-1} + \dots ,$$

---

[f]In general, an asymptotic expansion is not restricted to the case that the argument or the parameter is large; the concepts may be applied to any finite values of them as well.

where $c_n = f^{(n)}(\infty)/n!$. The degree of approximation can be improved indefinitely if only sufficient number of terms are taken on the r.h.s.

In general, if we can find a series, not necessary being convergent,

$$A_0 + A_1 z^{-1} + A_2 z^{-2} + \ldots \tag{1}$$

which has the property that for any *fixed n*, when arg z is confined in a certain range, $\Delta_1 < \arg z < \Delta_2$,

$$\lim_{|z| \to \infty} z^n \{ f(z) - S_n(z) \} = 0 , \tag{2}$$

i.e.,

$$f(z) = S_n(z) + o(z^{-n}) , \tag{3}$$

where

$$S_n = A_0 + A_1 z^{-1} + \ldots + A_n z^{-n} \tag{4}$$

is the partial sum, then the series (1) is said to be the asymptotic series representing $f(z)$ at $z = \infty;$[g] it is written,

$$f(z) \sim A_0 + A_1 z^{-1} + A_2 z^{-2} + \ldots$$
$$|z| \to \infty, \qquad \Delta_1 < \arg z < \Delta_2 . \tag{5}$$

It is important to point out that in the asymptotic expansion (5), we are using the partial sum $S_n(z)$, whose number of terms is finite, as an approximation for the function $f(z)$. By no means is $f(z) = \lim_{n \to \infty} S_n(z)$, since, the series may be a divergent one and the limit does not exist; even $A_k z^{-k}$ might not approach zero as $k \to \infty$.

In any asymptotic representation, it is important to estimate the error of the approximation introduced. According to (2) or (3), the order of magnitude of the error is seen to be $o(z^{-n})$. Hence, for a fixed n and a given range of arg z, the larger $|z|$ is, the smaller the error will be, i.e., the better the approximation. Besides, the degree of approximation also depends upon the number of terms taken for the partial sum. But if the series is divergent, then for a given z, *the approximation cannot be indefinitely improved by taking more and more terms, i.e., by increasing n.* This is quite different from representing a function approximately by the partial sum of a convergent series.

---

[g] We shall deal mainly with asymptotic expansion in the form of a power series. For other forms, cf. A. Erdélyi, *Asymptotic Expansions* (1956).

Example. Find the asymptotic expansion of the function

$$f(x) = \int_x^\infty \frac{e^{x-t}}{t} dt \tag{6}$$

for $x \to \infty, x > 0$.

Let $t - x = \lambda$, then

$$\int_x^\infty \frac{e^{x-t}}{t} dt = \int_0^\infty \frac{e^{-\lambda}}{x+\lambda} d\lambda = \frac{1}{x} \int_0^\infty e^{-\lambda} d\lambda \left[\sum_{k=0}^{n-1} \left(-\frac{\lambda}{x} \right)^k + \left(-\frac{\lambda}{x} \right)^n \frac{x}{x+\lambda} \right]$$

$$= \sum_{k=0}^{n-1} (-)^k \frac{k!}{x^{k+1}} + \frac{(-)^n}{x^n} \int_0^\infty \frac{e^{-\lambda} \lambda^n}{x+\lambda} d\lambda . \tag{7}$$

If we take the sum of the first n terms on the r.h.s. as an approximation for $f(x)$, the error made will be of the order

$$\left| \frac{(-)^n}{x^n} \int_0^\infty \frac{e^{-\lambda} \lambda^n}{x+\lambda} d\lambda \right| < \frac{1}{x^{n+1}} \int_0^\infty e^{-\lambda} \lambda^n d\lambda = \frac{n!}{x^{n+1}} . \tag{8}$$

Thus,

$$f(x) \sim \frac{1}{x} - \frac{1}{x^2} + \frac{2!}{x^3} - \ldots + (-)^{n-1} \frac{(n-1)!}{x^n} + \ldots . \tag{9}$$

From (8), it is seen that, for a fixed n, the error approaches zero as x^{-n-1} when $x \to \infty$; i.e., the larger x is, the better the approximation. But, on the contrary, if x is fixed, then when n exceeds a certain value ($\approx x$), the error grows larger as n increases further. We see thus, for asymptotic expansion in general, the degree of approximation cannot be indefinitely improved by increasing the number of terms in the partial sum.

It must be noticed here that in the case of a complex variable, the asymptotic expansion of a function depends on the range of the argument (phase) of the variable: in different ranges, the asymptotic expressions are generally different [cf. for Example Eq. (7) of Sec. 6.8].

Some fundamental properties of asymptotic expansions
1. Linear combination
   If

$$f(z) \sim \sum_0^\infty A_n z^{-n}, \qquad g(z) \sim \sum_0^\infty B_n z^{-n} , \tag{10}$$

then,

$$\alpha f(z) + \beta g(z) \sim \sum_0^\infty (\alpha A_n + \beta B_n) z^{-n} , \tag{11}$$

where α and β are arbitrary constants. This can be proved immediately by the definition of asymptotic expansion.

2. Multiplication[h]

Given (10), then

$$f(z)g(z) \sim \sum_{m=0}^{\infty} C_m z^{-m} , \qquad (12)$$

where

$$C_m = \sum_{k=0}^{m} A_k B_{m-k} . \qquad (13)$$

Proof: By definition,

$$f(z) = A_0 + A_1 z^{-1} + \ldots + A_n z^{-n} + \varepsilon z^{-n}$$
$$g(z) = B_0 + B_1 z^{-1} + \ldots + B_n z^{-n} + \eta z^{-n} ,$$

where ε and $\eta \to 0$ when $|z| \to \infty$. Therefore, when $|z| \to \infty$,

$$z^n \left\{ f(z)g(z) - \sum_{m=0}^{n} C_m z^{-m} \right\} = A_0 \eta + B_0 \varepsilon + O(z^{-1}) \to 0 .$$

3. Term by term integration

Let z be a real variable, or a complex variable with fixed argument. If

$$f(z) \sim A_2 z^{-2} + A_3 z^{-3} + \ldots = \sum_{k=2}^{\infty} A_k z^{-k} , \qquad (14)$$

then

$$\int_z^{\infty} f(z) dz \sim \sum_{k=2}^{\infty} \frac{A_k}{k-1} z^{-k+1} . \qquad (15)$$

Proof: Let $S_n(z)$ be the partial sum of the series on the r.h.s. of (14). Since $z^n \{f(z) - S_n(z)\} \to 0$ when $|z| \to \infty$, for an arbitrary $\varepsilon > 0$, there exists an R such that when $|z| > R$,

$$|f(z) - S_n(z)| < \varepsilon |z|^{-n} .$$

---

[h] Can also be used for division provided the denominator of the asymptotic expansion does not approach zero when $|z| \to \infty$.

Thus,

$$\left| \int_z^\infty f(z)dz - \int_z^\infty S_n(z)dz \right| \leq \int_z^\infty |f(z) - S_n(z)||dz| < \varepsilon \int_{|z|}^\infty |z|^{-n}d|z|$$

$$= \frac{\varepsilon}{n-1} \frac{1}{|z|^{n-1}}$$

and

$$\left| z^{n-1} \left\{ \int_z^\infty f(z)dz - \sum_{k=2}^n \frac{A_k}{k-1} z^{-k+1} \right\} \right| \rightarrow 0, \qquad \text{when} \qquad |z| \rightarrow \infty .$$

4. Differentiating term by term

In general, term by term differentiation is not permissible for an asymptotic expansion. For example, $e^{-x} \sin(e^x) \sim 0 + \frac{0}{x} + \frac{0}{x^2} + \ldots$, whereas its derivative, namely, $-e^{-x} \sin(e^x) + \cos(e^x)$, which oscillates when $x \rightarrow \infty$, has no asymptotic expansion at all. However, if the function does have a derivative and both of them have asymptotic expansions in the form of a power series, then, by applying the result of term by term integration given above to the asymptotic expansion of $f'(z)$, we see immediately that the asymptotic expansion of $f(z)$ can be differentiated term by term.

5. Uniqueness

For a given range of the arg z, a function of z can have at most one asymptotic expansion, i.e., the asymptotic expansion, if it exists, is unique. But, the same asymptotic expansion may represent different functions. We shall see these presently.

Let

$$f(z) \sim \sum_0^\infty A_k z^{-k}$$

and also

$$f(z) \sim \sum_0^\infty B_k z^{-k} .$$

Then, to a fixed n,

$$\lim_{|z| \to \infty} \{z^n (A_0 + A_1 z^{-1} + \ldots + A_n z^{-n} - B_0 - B_1 z^{-1} - \ldots - B_n z^{-n})\} = 0 ,$$

which requires $A_0 = B_0, A_1 = B_1, \ldots, A_n = B_n$. But as n is arbitrary, the two asymptotic expansions must be the same.

However, the same divergent series might be the asymptotic expansion of two different functions. For example,

$$f(z) \sim \sum_{k=0}^{\infty} A_k z^{-k} \qquad (|\arg z| < \pi/2)$$

$$f(z) + e^{-z} \sim \sum_{k=0}^{\infty} A_k z^{-k} .$$

For further theory concerning asymptotic expansion, cf. A. Erdélyi, *Asymptotic Expansions* (1956).

1.9. The Asymptotic Expansion of the Laplace Integral, Watson's Lemma

$$f(z) = \int_0^\infty e^{-zt} \varphi(t) dt \qquad (1)$$

is called the *Laplace integral* and $f(z)$ the *Laplace transform* of $\varphi(t)$. This form of integral often arises in solving differential equation by definite integral [cf. Sec. 2.13].

Watson's lemma:

Let $\varphi(t)$ be a single-valued analytic function in $|\arg t| < \theta$; when $t \to \infty, \varphi(t) = O(e^{bt}), b$ real, and when $t \to 0$,

$$t\varphi(t) \sim \sum_{n=1}^{\infty} a_n t^{n/r} \quad (r > 0) ,$$

i.e.,

$$\varphi(t) \sim \sum_{n=1}^{\infty} a_n t^{n/r-1} \quad (r > 0) . \qquad (2)$$

[Such $\varphi(t)$ often occurs in the problem of the inversion of a series; cf. for example, Sec. 7.11.] Then we have the important asymptotic expansion formula

$$f(z) = \int_0^\infty e^{-zt} \varphi(t) dt \sim \sum_{n=1}^{\infty} a_n \Gamma\left(\frac{n}{r}\right) z^{-n/r}$$

$$|z| \to \infty, \qquad |\arg z| \le \frac{\pi}{2} - \delta \quad (\delta > 0) , \qquad (3)$$

where $\Gamma(s)$ is the Gamma function [Chap. 3].

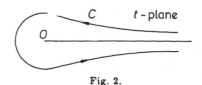

Fig. 2.

Proof: From (2) and the behavior of $\varphi(t)$ when $t \to \infty$, we know that to any fixed positive integer N, there exists a $K > 0$ such that

$$\left| \varphi(t) - \sum_{n=1}^{N-1} a_n t^{n/r-1} \right| < K t^{N/r-1} e^{bt} \ .$$

Therefore, by Eq. (1) of Sec. 3.1,

$$\int_0^\infty e^{-zt} \varphi(t) dt = \sum_{n=1}^{N-1} a_n \int_0^\infty e^{-zt} t^{n/r-1} dt + R_N$$

$$= \sum_{n=1}^{N-1} a_n \Gamma\left(\frac{n}{r}\right) z^{-n/r} + R_N \ ,$$

where

$$|R_N| < K \int_0^\infty e^{-xt+bt} t^{N/r-1} dt = K \Gamma\left(\frac{N}{r}\right) (x-b)^{-N/r}$$

$$= O(|z|^{-N/r}), \quad (x = \mathrm{Re}(z))$$

so far as $x - b > 0$, which is always satisfied when $|z| > b \csc \delta$. This proves (3).

The Watson lemma can be generalized to the following contour integral:

$$f(z) = \int_\infty^{(0+)} e^{-zt} \varphi(t) dt$$

$$0 < \arg t < 2\pi, \quad |\arg z| \le \pi/2 - \delta, \quad \delta > 0 \ . \tag{4}$$

The path of integration C is shown in Fig. 2; when $t \to \infty, \varphi(t) = O(e^{bt}), b$ real, and when $t \to 0$,

$$\left| t\varphi(t) - \sum_{n=1}^N a_n t^{\lambda_n} \right| = o(|t|^{\lambda_N}) \ ,$$

where $0 < \lambda_1 < \lambda_2 < \ldots$. Under these conditions, we have the asymptotic expansion formula:

$$f(z) = \int_\infty^{(0+)} e^{-zt} \varphi(t) dt \sim 2i \sum_{n=1}^\infty a_n \Gamma(\lambda_n) \sin \lambda_n \pi \cdot e^{i\lambda_n \pi} z^{-\lambda_n}$$

$$|z| \to \infty, \qquad |\arg z| < \frac{\pi}{2} - \delta, \qquad \delta > 0 . \tag{5}$$

The proof is similar to that given above; only make use of the contour expression of the Gamma function [Eq. (2) of Sec. 3.7].

Example. Find the asymptotic expansion of

$$f(x) = \int_\infty^{(0+)} e^{-xt-\beta t^2} (-t)^{-\mu} dt \qquad (|\arg(-t)| < \pi) \tag{6}$$

for $x \to \infty$; $|\arg x| \leq \frac{\pi}{2} - \delta, \delta > 0$; $f(x)$ is the solution of the Hermite equation in integral form [Eq. (17) of Sec. 2.13].

By (5), we have

$$f(x) = e^{i\mu\pi} \int_\infty^{(0+)} e^{-xt} \sum_{n=0}^\infty \frac{(-\beta)^n}{n!} t^{2n-\mu} dt$$

$$\sim -2ix^{\mu-1} \sin \mu\pi \sum_{n=0}^\infty \frac{(-\beta)^n}{n!} \Gamma(2n-\mu+1) x^{-2n} . \tag{7}$$

1.10. Expansion in Terms of Functions of an Orthonormal Set

Let $\{\varphi_n(x)\}(n = 1, 2, \ldots)$ be a set of continuous functions of a real variable $x, a \leq x \leq b$. If $\varphi_n(x)$ satisfy the following relations:

$$(\varphi_m, \varphi_n) = \int_a^b \overline{\varphi}_m(x)\varphi_n(x)\rho(x)dx = \delta_{mn} = \begin{cases} 0 & (m \neq n) \\ 1 & (m = n) , \end{cases}$$

$$(m, n = 1, 2, \ldots) \tag{1}$$

where $\overline{\varphi}_m$ is the complex conjugate of φ_m, the set of functions $\{\varphi_n(x)\}$ is said to form an *orthonormal system* with *weight* $\rho(x)(> 0)$. (φ_m, φ_n) is defined as the *inner product* of the two functions. Two functions with their inner product equal to zero is said to be orthogonal to each other.

An example of an orthonormal system is $e^{inx}/\sqrt{2\pi}, n = 0, \pm 1, \pm 2, \ldots,$ $0 \leq x \leq 2\pi$, because

$$\frac{1}{2\pi} \int_0^{2\pi} e^{-imx} e^{inx} dx = \delta_{mn} ; \tag{2}$$

$\rho(x) = 1$ here.

There are many orthonormal systems of this sort to be dealt with in this book; examples are: polynomials of Jacobi, Chebyshev, Legendre, Laguerre, Hermite, etc.

With a specific orthonormal system of functions as base, expressing a given function as a linear combination of them is an important kind of expansion. A famous example of such an expansion is the Fourier series expansion, whose base is the orthonormal system of trigonometric functions. We shall only treat the expansion of continuous functions or analytic functions. For the more general case of square integrable (in Lebeques sense) functions for instance, the reader is referred to any standard text on functional analysis.

Let $f(x)$ be a given function continuous in $a \leq x \leq b$. When we use the linear combination

$$f_k(x) = \sum_{n=1}^{k} C_n \varphi_n(x) \tag{3}$$

of orthonormal functions $\{\varphi_n(x)\}$ as an approximation for $f(x)$, we denote by

$$d_k(x) = f(x) - f_k(x) \tag{4}$$

the error made. The mean square error is defined to be

$$(d_k, d_k) = \int_a^b |d_k|^2 \rho(x) dx . \tag{5}$$

To make (d_k, d_k) minimal, we have to choose the coefficients in (3) such that

$$c_n = (\varphi_n, f) . \tag{6}$$

For,

$$(d_k, d_k) = (f - f_k, f - f_k) = (f, f) + (f_k, f_k) - (f, f_k) - (f_k, f)$$

$$= (f, f) + \sum_{n=1}^{k} |c_n|^2 - \sum_{n=1}^{k} (f, \varphi_n) c_n - \sum_{n=1}^{k} (\varphi_n, f) \bar{c}_n$$

$$= (f, f) - \sum_{n=1}^{k} |(\varphi_n, f)|^2 + \sum_{n=1}^{k} |c_n - (\varphi_n, f)|^2 . \tag{7}$$

When c_n change their values, only the last sum is affected, but it is never negative. Therefore, to make this sum a minimum (equal to zero), c_n must be given by (6).

The c_n given by (6) are called the *generalized Fourier coefficients* of $f(x)$ with respect to the orthonormal system $\{\varphi_n(x)\}$. In the following we shall deal only with those combinations (3) whose c_n are the generalized Fourier coefficients of $f(x)$.

Substituting (6) in (7) and noticing that $(d_k, d_k) \geq 0$, we have

$$\sum_{n=1}^{k} |c_n|^2 \leq (f, f) \qquad (k \quad \text{arbitrary}) \tag{8}$$

which is called the *Bessel inequality*. From this inequality we see that the series $\sum_{n=1}^{\infty} |c_n|^2$ made of generalized Fourier coefficients of $f(x)$ is always convergent, since (f, f) is finite.

If $(d_k, d_k) \to 0$ when $k \to \infty$, i.e.,

$$\lim_{k \to \infty} \int_a^b \left| f - \sum_{n=1}^{k} c_n \varphi_n \right|^2 \rho(x) dx = 0 , \tag{9}$$

where $c_n = (\varphi_n, f)$, then $\sum_{n=1}^{\infty} c_n \varphi_n$ is said to *converge in mean* to $f(x)$. In such a case, we have, by (7),

$$\sum_{n=1}^{\infty} |c_n|^2 = (f, f) \quad (c_n = (\varphi_n, f)) \tag{10}$$

which is called the *Parseval equality* or the *completeness relation*.

If (10) is satisfied by *any* continuous function $f(x)(a \leq x \leq b)$, then $\{\varphi_n(x)\}$ is called a complete system of functions in the space of continuous functions. For example, according to the theory of Fourier series, the set of functions $e^{in\pi}/\sqrt{2\pi}(n = 0, \pm1, \pm2, \dots)$ mentioned before is a complete system of functions.

Note that when the series $\sum_{n=1}^{\infty} c_n \varphi_n$ converges in mean to $f(x)$, it is not necessary for the sum of the series to exist; and even if it exists, it may not be equal to $f(x)$. But if the series is uniformly convergent in $a \leq x \leq b$, then the limit process in (9) can be carried out under the integration sign and we have

$$f(x) = \sum_{n=1}^{\infty} c_n \varphi_n(x), \quad c_n = (\varphi_n, f) . \tag{11}$$

Closure property of a complete system of functions

The set of functions $\{\varphi_n(x)\}$ is said to be *closed* if there exists no continuous function not identically equal to zero, which is orthogonal to all the $\varphi_n(x)$.

A complete system must also be closed. For, if there exists a continuous function $f(x)$ orthogonal to each member of the system, then its generalized Fourier coefficients will all be equal to zero and hence, according to (10), $(f, f) = 0$, i.e., $f(x) \equiv 0$.

Orthonormalization Process

Let $\psi_n(x)(n = 1, 2, \dots)$ be a set of linearly independent functions continuous in $a \leq x \leq b$. It is always possible to construct from them an orthonormal set as follows.

Let $\|\psi_1\|$, called the *norm* of the function ψ_1, represent $(\psi_1, \psi_1)^{1/2}$, then

$$\varphi_1(x) = \psi_1(x)/\|\psi_1\| \tag{12}$$

is a normalized function, since $(\varphi_1, \varphi_1) = 1$. Next, take

$$\varphi_2(x) = c_1\varphi_1(x) + c_2\psi_2(x) . \tag{13}$$

We can determine c_1 and c_2 such that

$$(\varphi_1, \varphi_2) = 0, \quad (\varphi_2, \varphi_2) = 1 . \tag{14}$$

These require that c_1 and c_2 satisfy the two equations

$$0 = (\varphi_1, \varphi_2) = c_1 + c_2(\varphi_1, \psi_2) \tag{15}$$

and

$$1 = (\varphi_2, \varphi_2) = c_1(\varphi_2, \varphi_1) + c_2(\varphi_2, \psi_2) = c_2(\varphi_2, \psi_2) .$$

To the last equation, applying (13), we have

$$1 = c_2\bar{c}_1(\varphi_1, \psi_2) + c_2\bar{c}_2(\psi_2, \psi_2) . \tag{16}$$

Solving (15) and (16) gives

$$|c_2|^2 = \frac{1}{\|\psi_2\| - |(\varphi_1, \psi_2)|^2} , \quad c_1 = -c_2(\varphi_1, \psi_2) . \tag{17}$$

Take again

$$\varphi_3(x) = c_1\varphi_1(x) + c_2\varphi_2(x) + c_3\psi_3(x) , \tag{18}$$

where c_1 and c_2 are not necessarily the same c_1 and c_2 in (13). We can proceed as above to determine c_1, c_2 and c_3 from the equations

$$(\varphi_1, \varphi_3) = 0, \quad (\varphi_2, \varphi_3) = 0, \quad (\varphi_3, \varphi_3) = 1 . \tag{19}$$

Continuing in this way, we shall arrive at a system of orthonormal functions $\{\varphi_n(x)\}, n = 1, 2, \ldots$.

Complete system of functions of several variables[i]

From complete systems of functions of a single variable, we can construct a complete system of functions of two or more variables according to the following theorem:

Theorem: Let $\{\varphi_n(s)\}(n = 1, 2, \ldots)$ be a complete orthonormal set of functions in the interval $a \le s \le b$, and for a fixed n let

$$\psi_{1n}(t), \quad \psi_{2n}(t), \quad \ldots$$

be a set of orthonormal functions complete in the interval $c \le t \le d$. Then the functions

$$\omega_{mn}(s, t) = \varphi_n(s)\psi_{mn}(t) \quad (m, n = 1, 2, \ldots)$$

form a complete orthonormal set in the rectangular region: $a \le s \le b, c \le t \le d$; for any function $f(s, t)$ continuous in this region, there is the completeness relation

$$\int\int |f(s,t)|^2 ds\, dt = \sum_{m,n=1}^{\infty} \left| \int\int \overline{\omega}_{mn}(s,t) f(s,t) ds\, dt \right|^2 , \tag{20}$$

where we have assumed that the weights of the two systems of orthonormal functions are both equal to 1.

To prove the theorem, we need only to note that by the completeness of $\{\varphi_n\}$, we have

$$\int_a^b |f(s,t)|^2 ds = \sum_{n=1}^{\infty} |g_n(t)|^2 , \tag{21}$$

where

$$g_n(t) = \int_a^b \overline{\varphi}_n(s) f(s,t) ds . \tag{22}$$

Since the series on the r.h.s. of (21) is a convergent series of non-negative continuous functions, it is also uniformly convergent by Dini's theorem[j] and can be integrated term by term, i.e.,

$$\int\int |f|^2 ds\, dt = \sum_{n=1}^{\infty} \int_c^d |g_n(t)|^2 dt . \tag{23}$$

---

[i] Cf. Courant & Hilbert, *Methods of Mathematical Physics*, Vol. I, p. 56 (1958).
[j] Cf. Bromwich, (1925), p. 141; Courant & Hilbert, *ibid.* p. 57, footnote.

As $\{\psi_{mn}(t)\}(m=1,2,\ldots)$ is also a complete set for each fixed n, we have

$$\int_c^d |g_n(t)|^2 dt = \sum_{m=1}^{\infty} \left| \int_c^d \overline{\psi}_{mn}(t)g_n(t)dt \right|^2$$

$$= \sum_{m=1}^{\infty} \left| \int\int \overline{\omega}_{mn}(s,t)f(s,t)dsdt \right|^2 .$$

Substituting in (23) gives (20).

Exercise 1

1. Prove the following relations of the Bernoulli polynomials $\varphi_n(x)$ [Sec. 1.1]:

  (i) $\displaystyle\sum_{r=0}^{n} \binom{n}{r}\varphi_r(x) = \varphi_n(x+1)$,

  (ii) $\displaystyle\sum_{r=0}^{n-1} \binom{n}{r}\varphi_r(x) = nx^{n-1}$, $(n \geq 2)$,

  (iii) $\int_x^{x+1} \varphi_n(t)dt = x^n$.

2. Prove the *multiplication formula* for the Bernoulli polynomials:

$$\varphi_n(mx) = m^{n-1} \sum_{r=0}^{m-1} \varphi_n\left(x + \frac{r}{m}\right) ,$$

where m is an arbitrary positive integer. Deduce from it the following formula:

$$\sum_{s=1}^{m-1} \varphi_n\left(\frac{s}{m}\right) = -\left(1 - \frac{1}{m^{n-1}}\right) \varphi_n ;$$

a special case of it is

$$\varphi_n\left(\frac{1}{2}\right) = -\left(1 - \frac{1}{2^{n-1}}\right) \varphi_n ,$$

where $\varphi_n = \varphi_n(0)$ [Sec. 1.1].

3. Prove the following integral representations for the Bernoulli polynomials:

  (i) $\varphi_{2n}(x) = (-)^{n+1}(2n) \int_0^{\infty} \frac{\cos(2\pi x)-e^{-2\pi t}}{\text{ch}(2\pi t)-\cos(2\pi x)} t^{2n-1} dt$
      $(0 < \text{Re}(x) < 1), \qquad n = 1, 2, \ldots$.

(ii) $\varphi_{2n+1}(x) = (-)^{n+1}(2n+1)\int_0^\infty \frac{\sin(2\pi x)}{\text{ch}(2\pi t) - \cos(2\pi x)}t^{2n}\,dt$
       $(0 < \text{Re}\,(x) < 1), \qquad n = 0,1,2,\ldots\,.$

[Hint: Use the residue theorem to calculate the integral along the rectangular contour with the vertices $-R, +R, R+i, -R+i, (R \to \infty)$, then compare the result with the Fourier expansions of the Bernoulli polynomials given in Eqs. (13) and (14) of Sec. 1.3.]

4. Bernoulli polynomial of order n and degree ν, $B_\nu^{(n)}(x)$, is defined by the expansion

$$\frac{t^n e^{xt}}{(e^t - 1)^n} = \sum_{\nu=0}^\infty \frac{t^\nu}{\nu!}B_\nu^{(n)}(x) \quad (n = 0,1,2,\ldots)\,.$$

The ordinary Bernoulli polynomial [Sec. 1.1] $\varphi_\nu(x) \equiv B_\nu^{(1)}(x)$, and $B_\nu^{(0)}(x) = x^\nu$. Prove the following formulae:

(i) $B_\nu^{(n)}(x) = \sum_{k=0}^\nu \binom{\nu}{k}B_k^{(n)}\cdot x^{\nu-k}$ where $B_k^{(n)} = B_k^{(n)}(0)$,

(ii) $\frac{d}{dx}B_\nu^{(n)}(x) = \nu B_{\nu-1}^{(n)}(x)$,

(iii) $\int_a^x B_\nu^{(n)}(t)\,dt = \frac{1}{\nu+1}[B_{\nu+1}^{(n)}(x) - B_{\nu+1}^{(n)}(a)]$,

(iv) $B_\nu^{(n)}(x+1) = B_\nu^{(n)}(x) + \nu B_{\nu-1}^{(n-1)}(x)$,

(v) $B_\nu^{(n)}(n-x) = (-)^\nu B_\nu^{(n)}(x)$.

5. Prove the following recurrence relations:

$$(\nu - n)B_\nu^{(n)}(x) - \nu x B_{\nu-1}^{(n)}(x) + n B_\nu^{(n+1)}(x+1) = 0\,,$$
$$B_\nu^{(n+1)}(x) = \left(1 - \frac{\nu}{n}\right)B_\nu^{(n)}(x) + \nu\left(\frac{x}{n} - 1\right)B_{\nu-1}^{(n)}(x)\,.$$

6. Show that

(i) $B_n^{(n+1)}(x) = (x-1)(x-2)\ldots(x-n) = \binom{x-1}{n}n!$,

(ii) $\int_0^1 (x-1)(x-2)\ldots(x-n)\,dx = B_n^{(n)}$,

(iii) $\int_0^1 x(x-1)\ldots(x-n+1)\,dx = B_n^{(n)}(1) = \frac{1}{1-n}B_n^{(n-1)}$.

7. Show that

$$x(x-1)\ldots(x-n+1) = \sum_{p=0}^n \binom{n-1}{p-1}B_{n-p}^{(n)}\cdot x^p \quad ..$$

8. Prove the following formulae for the Euler polynomials $E_n(x)$ [Sec. 1.2]:

(i) $\sum_{r=0}^{n} \binom{n}{r} E_r(x) = E_n(x+1)$,

(ii) $\sum_{r=0}^{n} \binom{n}{r} E_r(x) + E_n(x) = 2x^n$.

9. Prove the multiplication formulae

$$E_n(mx) = m^n \sum_{r=0}^{m-1} (-)^r E_n\left(x+\frac{r}{m}\right) \qquad (m \quad \text{odd}) ,$$

$$E_n(mx) = \frac{2m^n}{n+1} \sum_{r=0}^{m-1} (-)^{r+1} \varphi_{n+1}\left(x+\frac{r}{m}\right) \qquad (m \quad \text{even})$$

$(\varphi_n(x)$ is the Bernoulli polynomial) .

10. Show that

$$E_{n-1}(x) = \frac{2^n}{n}\left[\varphi_n\left(\frac{x+1}{2}\right) - \varphi_n\left(\frac{x}{2}\right)\right]$$

$$= \frac{2}{n}\left[\varphi_n(x) - 2^n\varphi_n\left(\frac{x}{2}\right)\right] .$$

11. Prove the following Fourier expansions for the Euler polynomials:

$$E_{2n}(x) = (-)^n 4(2n)! \sum_{r=0}^{\infty} \frac{\sin[(2r+1)\pi x]}{[(2r+1)\pi]^{2n+1}} ,$$

$$n = 1, 2, \ldots , \qquad 0 \le x \le 1 ;$$

$$E_{2n+1}(x) = (-)^{n+1} 4(2n+1)! \sum_{r=0}^{\infty} \frac{\cos[(2r+1)\pi x]}{[(2r+1)\pi]^{2n+2}} ,$$

$$n = 0, 1, 2, \ldots , \qquad 0 \le x \le 1 .$$

12. Prove the following integral representations of the Euler polynomials:

(i) $E_{2n}(x) = (-)^n 4 \int_0^{\infty} \frac{t^{2n} \sin \pi x \, \text{ch} \, \pi t}{\text{ch} \, 2\pi t - \cos 2\pi x} dt$,

$$n = 0, 1, 2, \ldots , \qquad (0 < \text{Re}(x) < 1) ;$$

(ii) $E_{2n+1}(x) = (-)^{n+1} 4 \int_0^{\infty} \frac{t^{2n+1} \cos \pi x \, \text{sh} \, \pi t}{\text{ch} \, 2\pi t - \cos 2\pi x} dt$,

$$n = 0, 1, 2 \ldots , \qquad (0 < \text{Re}(x) < 1).$$

[Hint: See Ex. 3 above.]

13. Euler polynomial of order n and degree ν, $E_{\nu}^{(n)}(x)$, is defined by the following expansion:

$$\frac{2^n e^{xt}}{(e^t+1)^n} = \sum_{\nu=0}^{\infty} \frac{t^\nu}{\nu!} E_{\nu}^{(n)}(x) .$$

Prove the addition formula

$$E_\nu^{(m+n)}(x+y) = \sum_{k=0}^{\nu} \binom{\nu}{k} E_k^{(m)}(x) E_{\nu-k}^{(n)}(y) \; .$$

14. By means of the Darboux formula [Sec. 1.3] or otherwise, prove the following expansion formula:

$$f(z) - f(a) = \sum_{n=1}^{\infty} \frac{(-)^{n-1}(z-a)^n}{n!(1-r)^n} \{f^{(n)}(z) - r^n f^{(n)}(a)\} \; .$$

Find the remainder after n terms and discuss the convergence of the series.

15. Show that

$$f(x+h) - f(x) = \sum_{m=1}^{n} (-)^{m-1} \frac{(2m)! h^m}{2^{2m}(m!)^3} \{f^{(m)}(x+h) - (-)^m f^{(m)}(x)\}$$

$$+ (-)^n h^{n+1} \int_0^1 \gamma_n(t) f^{(n+1)}(x+ht) dt \; ,$$

where

$$\gamma_n(x) = \frac{x^{n+\frac{1}{2}}(1-x)^{n+\frac{1}{2}}}{(n!)^2} \frac{d^n}{dx^n} [x^{-\frac{1}{2}}(1-x)^{-\frac{1}{2}}]$$

$$= \frac{1}{n!\pi} \int_0^1 (x-z)^n z^{-\frac{1}{2}}(1-z)^{-\frac{1}{2}} dz \; .$$

Also show that

$$\{(1-tx)[1+t(1-x)]\}^{-\frac{1}{2}} = \sum_{n=0}^{\infty} \gamma_n(x) n! t^n \; .$$

[Comparing the derivative expression of $\gamma_n(x)$ with Eq. (8) of Sec. 4.10, it is seen that $\gamma_n(x)$ differs from the Jacobi polynomial $F(-n, -n, -n+\frac{1}{2}, x)$ only by a constant factor.]

16. In the Darboux formula Eq. (1) of Sec. 1.3 putting

$$\varphi(x+1) = \frac{1}{n!} \left[\frac{d^n}{dt^n} \{ \frac{(1-r)e^{xt}}{1-re^{-t}} \} \right]_{t=0}$$

show that

$$f(x+h) - f(x) = -\sum_{m=1}^{n} a_m \frac{h^m}{m!} \left\{ f^{(m)}(x+h) - \frac{1}{r} f^{(m)}(x) \right\}$$

$$+ (-)^n h^{n+1} \int_0^1 \varphi(t) f^{(n+1)}(x+ht) dt ,$$

where a_m are the coefficients in the series

$$\frac{1-r}{1-re^{-t}} = 1 - a_1 \frac{t}{1!} + a_2 \frac{t^2}{2!} - a_3 \frac{t^3}{3!} + \dots .$$

17. Show that

$$f(z) - f(a) = \sum_{m=1}^{n} (-)^{m-1} \frac{2(2^{2n}-1)B_m}{(2m)!} (z-a)^{2m-1} \{ f^{(2m-1)}(z)$$

$$+ f^{(2m-1)}(a) \} + \frac{(z-a)^{2n+1}}{(2n)!}$$

$$\times \int_0^1 \psi_{2n}(t) f^{(2n+1)}[a + (z-a)t] dt ,$$

where

$$\psi_n(t) = \frac{2}{n+1} \left[\frac{d^{n+1}}{d\nu^{n+1}} \left(\frac{\nu e^{t\nu}}{e^\nu + 1} \right) \right]_{\nu=0} .$$

18. Show that

$$\lambda f(\lambda x) = \sum_{n=0}^{\infty} \frac{(1-\lambda^{-1})^n}{n!} \frac{d^n}{dx^n} [x^n f(x)] .$$

19. If $y - x - \varphi(y) = 0, \varphi(y)$ being a given function, show that

$$f(y) = f(x) + \sum_{m=1}^{\infty} \frac{1}{m!} [\varphi(x)]^m \left(\frac{1}{1 - \varphi'(x)} \frac{d}{dx} \right)^m f(x) .$$

20. Let

$$W(a, b, x) = x + \frac{a-b}{2!} x^2 + \frac{(a-b)(a-2b)}{3!} x^3 + \dots .$$

The series converges so long as $|x| < |b|^{-1}$. Show that

$$\frac{d}{dx} W(a, b, x) = 1 + (a-b)W(a-b, b, x)$$

and, if

$$y = W(a, b, x) ,$$

then

$$x = W(b, a, y) .$$

Examples of such function are

$$W(1, 0, x) = e^x - 1 ,$$
$$W(0, 1, x) = \ln(1 + x) ,$$
$$W(a, 1, x) = \frac{(1 + x)^a - 1}{a} .$$

21. Show that

$$\sec z = 4\pi \sum_{n=1}^{\infty} \frac{(-)^n (2n - 1)}{4z^2 - (2n - 1)^2 \pi^2} ,$$

$$\csc z = \frac{1}{z} + \sum_{n=1}^{\infty} (-)^n \frac{2z}{z^2 - n^2 \pi^2} ,$$

$$\tan z = \sum_{n=1}^{\infty} \frac{8z}{(2n - 1)^2 \pi^2 - 4z^2} .$$

22. Show that, when $0 < a < 1$,

$$\frac{e^{az}}{e^z - 1} = \frac{1}{z} + \sum_{n=1}^{\infty} \frac{2z \cos 2na\pi - 4n\pi \sin 2na\pi}{z^2 + 4n^2 \pi^2} .$$

23. Show that

$$\sum_{m=-\infty}^{\infty} \sum_{n=-\infty}^{\infty} \frac{1}{(m^2 + a^2)(n^2 + b^2)} = \frac{\pi^2}{ab} \coth \pi a \cdot \coth \pi b .$$

24. Prove the summation formula

$$\sum_{n=0}^{\infty} \frac{(-)^n}{(2n + 1)^{2k+1}} = \frac{\pi^{2k+1} E_k}{2^{2k+2} (2k)!} ,$$

where E_k are the Euler numbers [Sec. 1.2].

Chapter 2

LINEAR ORDINARY DIFFERENTIAL EQUATIONS OF THE SECOND ORDER

2.1. Singular Points of Linear Ordinary Differential Equations of the Second Order

Since most of the special functions discussed in this book are solutions of certain linear ordinary differential equations of the second order, we give, in this chapter, a brief introduction to the methods of solving such equations. Emphasis is put on two methods: solution in series and solution in integrals.

The standard form of a linear ordinary differential equation of the second order is taken to be

$$\frac{d^2w}{dz^2} + p(z)\frac{dw}{dz} + q(z)w(z) = 0 \,, \tag{1}$$

where $w(z)$ is the unknown function; $p(z)$ and $q(z)$, the coefficients of the equation, are known functions of the complex variable. It is required to find $w(z)$ which satisfies (1) in a definite domain under certain conditions, e.g. the initial conditions $w(z_0) = c_0$ and $w'(z_0) = c_1$.

The analytic property of the solution of (1) is completely determined by the analytic nature of the coefficients $p(z)$ and $q(z)$. Suppose each of $p(z)$ and $q(z)$ is a single-valued analytic function of z in a certain domain, except for a number of isolated singular points, then the points in the domain may be classified into the following two categories.

Ordinary point of the equation—If the coefficients $p(z)$ and $q(z)$ are both analytic at z_0 and its vicinity, the point z_0 is called an ordinary point of the equation.

Singular point of the equation—If *one* of the two coefficients $p(z)$ and $q(z)$ is not analytic at z_0, the point z is called a singular point of the equation.

2.2. Solution of the Equation in the Vicinity of an Ordinary Point

Theorem: If $p(z)$ and $q(z)$ are single-valued and analytic in the circular region $|z - z_0| < R$, the equation

$$w'' + pw' + qw = 0 \tag{1}$$

has a unique solution inside this circle satisfying the initial conditions

$$w(z_0) = c_0, \quad w'(z_0) = c_1 , \tag{2}$$

where c_0 and c_1 are arbitrary constants, and $w(z)$ is single-valued and analytic in this circle.

Proof: Introduce a new unknown function $u = w'$, then (1) is equivalent to the pair of equations[a]

$$w' = u, \quad u' = -pu - qw . \tag{3}$$

To facilitate discussion, we consider the set of equations in symmetric form:

$$u' = a(z)u + b(z)v, \quad v' = c(z)u + d(z)v , \tag{4}$$

in which the coefficients $a(z), b(z), c(z)$ and $d(z)$ are single-valued and analytic in the circular region $|z - z_0| < R$. We prove now by the method of successive approximations that (4) have a unique set of single-valued analytic solutions satisfying the initial conditions

$$u(z_0) = \alpha, \quad v(z_0) = \beta , \tag{5}$$

α and β being arbitrary constants.

---

[a]This method of proof is convenient for generalization to equations of higher order.

Equations (4) and the initial conditions (5) may be expressed in conjunction by the integral equations

$$\left.\begin{array}{l} u(z) = \alpha + \int_{z_0}^{z} [a(\varsigma)u + b(\varsigma)v]d\varsigma , \\ v(z) = \beta + \int_{z_0}^{z} [c(\varsigma)u + d(\varsigma)v]d\varsigma . \end{array}\right\} \tag{6}$$

Construct sequences of functions

$$\left.\begin{array}{l} u_0(z) = \alpha , \qquad v_0(z) = \beta , \\ u_{n+1}(z) = \alpha + \int_{z_0}^{z} [au_n + bv_n]d\varsigma , \\ v_{n+1}(z) = \beta + \int_{z_0}^{z} [cu_n + dv_n]d\varsigma , \\ (n = 0, 1, 2, \dots) . \end{array}\right\} \tag{7}$$

We shall show that these two sequences are convergent:

$$\lim_{n\to\infty} u_n(z) = u(z), \qquad \lim_{n\to\infty} v_n(z) = v(z) , \tag{8}$$

and $u(z)$ and $v(z)$ constitute the unique set of solutions satisfying (4) and initial conditions (5).

Since the integrands in (7) are analytic, the sequences (7) are those of analytic functions and the values of the integrals are independent of the path of integration. Take the path to be a straight line from z_0 to z and let $z_0 = 0$ to simplify writing. We have, with $\varsigma = \rho e^{i\theta}$, where $0 \le \rho \le R_1 < R$,

$$u_1(z) - \alpha = \int_0^{\rho} [a\alpha + b\beta]e^{i\theta}d\rho . \tag{9}$$

As a, b, c and d are all analytic functions in the circle $|z| \le R_1 < R$, there must exist an $M > 0$ such that

$$|a| < M, \quad |b| < M, \quad |c| < M, \quad |d| < M . \tag{10}$$

Hence we have from (9)

$$|u_1(z) - u_0(z)| < 2mM\rho , \tag{11}$$

where m denotes the larger of $|\alpha|$ and $|\beta|$. Similarly, we have

$$|v_1(z) - v_0(z)| < 2mM\rho . \tag{12}$$

Using (11) and (12), we obtain from (7)

$$|u_2(z) - u_1(z)| \le \int_0^{\rho} [|a||u_1 - u_0| + |b||v_1 - v_0|]d\rho$$

$$< 2^2 mM^2 \int_0^{\rho} \rho d\rho = m\frac{(2M\rho)^2}{2!}$$

and
$$|v_2(z) - v_1(z)| < m\frac{(2M\rho)^2}{2!} .$$

Continuing the above procedure and by induction, we get

$$|u_n - u_{n-1}| < m\frac{(2M\rho)^n}{n!}, \quad |v_n - v_{n-1}| < m\frac{(2M\rho)^n}{n!} . \tag{13}$$

The r.h.s. of either of these inequalities is the general term of the power series expansion of the exponential function $m\exp(2M\rho)$, hence the sequences

$$u_n = u_0 + (u_1 - u_0) + \ldots + (u_n - u_{n-1}) ,$$
$$v_n = v_0 + (v_1 - v_0) + \ldots + (v_n - v_{n-1})$$

converge uniformly in $|z| \leq R_1$. Then, by Weierstrass' theorem on sequences of analytic functions, the two limiting functions $u(z)$ and $v(z)$ of the sequences are analytic *in* $|z| < R$. Further, we have

$$u(z) = \lim_{n\to\infty} u_n(z) = \alpha + \lim_{n\to\infty} \int_0^z [au_{n-1} + bv_{n-1}]d\varsigma$$
$$= \alpha + \int_0^z [au + bv]d\varsigma ,$$

and similarly,

$$v(z) = \lim_{n\to\infty} v_n(z) = \beta + \int_0^z [cu + dv]d\varsigma .$$

From these we see that $u(z)$ and $v(z)$ satisfy (6), hence they are solutions of (4) and satisfy the initial conditions (5). This proves that the solutions exist and are analytic.

The above solutions are unique, for, if there exist two sets of solutions u, v and u^*, v^*, then, as they satisfy (6) respectively, we have

$$|u - u^*| \leq \int_0^\rho [|a||u - u^*| + |b||v - v^*|]d\rho . \tag{14}$$

And, as u, v, u^* and v^* are all analytic in $|z| < R$, there must exist an $A > 0$ such that

$$|u - u^*| < A, \quad |v - v^*| < A .$$

When these are substituted in (14) we have $|u - u^*| < 2MA\rho$ and similarly $|v - v^*| < 2MA\rho$. Substituting these results in (14) again, we have $|u - u^*| <$

$A(2M\rho)^2/2!$, and similarly $|v - v^*| < A(2M\rho)/2!$. Continuing this process, and by induction, we have

$$|u - u^*| < A\frac{(2M\rho)^n}{n!}, \quad |v - v^*| < A\frac{(2M\rho)^n}{n!}, \tag{15}$$

where n may be arbitrarily large. When $n \to \infty$, $(2M\rho)^n/n! \to 0$, and as u, v, u^* and v^* are all independent of n, we must have

$$|u - u^*| \equiv 0, \quad |v - v^*| \equiv 0,$$

i.e.,

$$u \equiv u^*, \quad v \equiv v^*.$$

The procedure of the proof also suggests a method of obtaining a solution of the equation, namely, we use sequence (7) and find the solution by successive approximations. In practice, however, we often make use of the analyticity of the solution, expand it in the form of a Taylor series

$$w(z) = \sum_{n=0}^{\infty} c_n (z - z_0)^n, \tag{16}$$

and substitute it in (1) to determine the coefficients of expansion c_n, with c_0 and c_1 given by the initial conditions.

2.3. Solutions of the Equation in the Vicinity of a Singular Point

According to the theorem in the preceding section, an ordinary point of the equation must also be an ordinary point of the solution, i.e., at such point and its vicinity, the solution is single-valued and analytic. But a singular point [Sec. 2.1] of the equation *may* yet be a singular point of the solution. Thus, if we still attempt to find the solution in the form of a power series, we have to employ the Laurent expansion. Before doing so, it is necessary to know whether this singular point is a branch point of the solution. We treat this problem by the method of analytic continuation.

We shall first prove that the analytic continuation of a solution of the equation $w'' + pw' + qw = 0$ is also a solution. Suppose $w(z)$ is a solution of the equation in the circle $K_b : |z - b| \leq R$, $w_1(z)$ is the analytic continuation of $w(z)$ in the circle $K_{b_1} : |z - b_1| \leq R_1$, and the common region of K_b and K_{b_1} is g. Then, since the function $F(z) \equiv w'' + pw' + qw$ is zero in K_b, its analytic continuation $F_1(z) \equiv w_1'' + pw_1' + qw_1$ has to be zero in g, and consequently is identically zero in K_{b_1}, i.e., $w_1(z)$ satisfies the equation.

Further, the analytic continuations $w_1^*(z), w_2^*(z)$ of two linearly independent solutions $w_1(z), w_2(z)$ will remain, respectively, to be linearly independent solutions. This can be shown as follows. Multiplying the equations

$$w_1'' + pw_1' + qw_1 = 0 , \quad w_2'' + pw_2' + qw_2 = 0$$

respectively by w_2, w_1 and subtracting one from the other, we have

$$w_2 w_1'' - w_1 w_2'' + p(w_2 w_1' - w_1 w_2') = 0 ,$$

or,

$$\frac{d}{dz}\Delta(z) + p\Delta(z) = 0 , \tag{1}$$

where $\Delta(z) \equiv w_1 w_2' - w_2 w_1'$ is the Wronski determinant of w_1 and w_2. Suppose $p(z)$ is analytic at a certain point b and its vicinity, we have from (1)

$$\Delta(z) \equiv w_1 w_2' - w_2 w_1' = \Delta(b) \exp\left\{-\int_b^z p(\varsigma)d\varsigma\right\} , \tag{2}$$

where $\Delta(b) \neq 0$ (otherwise we would have $\Delta(z) \equiv 0$, contradicting the assumption that w_1 and w_2 are linearly independent). Thus, when w_1 and w_2 are analytically continued to w_1^* and w_2^*, so far as the paths do not pass through any singular point of $p(z)$, the corresponding continuation $\Delta^*(z) \equiv w_1^* w_2^{*'} - w_2^* w_1^{*'}$ of $\Delta(z)$ is definitely non-vanishing, hence w_1^* and w_2^* are linearly independent.

Let z_0 be a singular point of the equation, b an ordinary point near z_0, and w_1, w_2 are two arbitrary linearly independent solutions in the neighborhood K_b of b. When w_1 and w_2 are continued along a path c encircling z_0 once (there being no other singularities of the equation within C, except z_0), their corresponding analytic continuations w_1^* and w_2^*, being also solutions of the equation respectively, must be related to w_1 and w_2 in the region K_b by

$$\left.\begin{array}{l} w_1^* = a_{11}w_1 + a_{12}w_2 , \\ w_2^* = a_{21}w_1 + a_{22}w_2 , \end{array}\right\} \tag{3}$$

where

$$a_{11}a_{22} - a_{12}a_{21} \neq 0 ; \tag{4}$$

otherwise, w_1^* and w_2^* would be linearly dependent.

Now let us find a solution w, such that its analytic continuation w^* around z_0 once satisfies the relation

$$w^* = \lambda w , \tag{5}$$

where λ is a constant. Such a solution is called a *multiplied solution*. Let

$$w = b_1 w_1 + b_2 w_2 \tag{6}$$

with constants b_1 and b_2 undetermined, we then have

$$w^* = b_1 w_1^* + b_2 w_2^* = (b_1 a_{11} + b_2 a_{21}) w_1 + (b_1 a_{12} + b_2 a_{22}) w_2 \ .$$

In order that (5) holds, we must have (since w_1 and w_2 are linearly independent)

$$b_1 a_{11} + b_2 a_{21} = \lambda b_1 \ , \quad b_1 a_{12} + b_2 a_{22} = \lambda b_2 \ ,$$

or,

$$b_1 (a_{11} - \lambda) + b_2 a_{21} = 0 \ , \quad b_1 a_{12} + b_2 (a_{22} - \lambda) = 0 \ . \tag{7}$$

This set of simultaneous homogeneous equations have non-zero solutions b_1 and b_2 only when λ satisfies the equation

$$\begin{vmatrix} a_{11} - \lambda & a_{21} \\ a_{12} & a_{22} - \lambda \end{vmatrix} = 0 \ . \tag{8}$$

Equation (8) is a quadratic equation, and has, on account of (4), two non-zero roots. For each root, (7) determine a set of solutions b_1 and b_2 (in fact, only the ratio of them is determined) and we obtain a multiplied solution. If (8) has two different roots λ_1 and λ_2, the two corresponding multiplied solutions w_1 and w_2 must be linearly independent, otherwise, we would have $w_2/w_1 = C$ and its analytic continuation is $w_2^*/w_1^* = \lambda_2 w_2/\lambda_1 w_1 = C\lambda_2/\lambda_1 = C$, i.e., $\lambda_1 = \lambda_2$. This is contrary to the postulate that λ_1 and λ_2 are different.

From relation (5), we see that a singular point z_0 of the equation is in general a branch point of the solution, except when $\lambda = 1$. Hence, we cannot yet express the solution at once in a Laurent series. We know, however, that the analytic continuation of the function $(z - z_0)^\rho$ when z varies around z_0 once is $e^{i2\pi\rho}(z - z_0)^\rho$. Thus, by taking ρ to satisfy $e^{i2\pi\rho} = \lambda$, we shall have $w/(z - z)^\rho$ single-valued and can then be expanded in a Laurent series:

$$\frac{w(z)}{(z - z_0)^\rho} = \sum_{n=-\infty}^{\infty} c_n (z - z_0)^n$$

in a certain neighborhood, $0 < |z - z_0| < R$, of z_0. Hence

$$w(z) = (z - z_0)^\rho \sum_{n=-\infty}^{\infty} c_n (z - z_0)^n \ . \tag{9}$$

If (8) has two distinct roots, the equation $w'' + pw' + qw = 0$ has two linearly independent solutions in the neighborhood $0 < |z - z_0| < R$ of its singular point z_0, which are of the forms

$$\left.\begin{array}{l}
w_1(z) = (z - z_0)^{\rho_1} \displaystyle\sum_{n=-\infty}^{\infty} c_n (z - z_0)^n \,, \\[2mm]
w_2(z) = (z - z_0)^{\rho_2} \displaystyle\sum_{n=-\infty}^{\infty} d_n (z - z_0)^n \,,
\end{array}\right\} \tag{10}$$

where $\rho_j = \ln \lambda_j / 2\pi i \, (j = 1, 2)$, and $\rho_1 - \rho_2 \neq integer$; otherwise $\lambda_1 = \lambda_2$. We must note that for a definite λ the corresponding ρ is not unique; an arbitrary integer may be added.

In the case $\lambda_1 = \lambda_2$, only one solution is obtained from the above. Let it be denoted by w_1 [Eq. (10)], we shall have

$$w_1^* = \lambda_1 w_1 \,. \tag{11}$$

Let w_2 be another solution, linearly independent of w_1, and its analytic continuation be

$$w_2^* = a_{21} w_1 + a_{22} w_2 \,. \tag{12}$$

Then (11) and (12) replace (3) $(a_{11} = \lambda_1, a_{12} = 0)$ and the equation corresponding to (8) is

$$\begin{vmatrix} \lambda_1 - \lambda & a_{21} \\ 0 & a_{22} - \lambda \end{vmatrix} = 0 \,. \tag{13}$$

As we have assumed that this equation has double roots, so $a_{22} = \lambda_1$ ånd

$$w_2^* = a_{21} w_1 + \lambda_1 w_2 \,. \tag{14}$$

Then from (11) and (14), we have

$$\left(\frac{w_2}{w_1}\right)^* = \frac{w_2}{w_1} + \frac{a_{21}}{\lambda_1} \,.$$

This shows that the function w_2/w_1, when continued along C for a complete round, is increased by a_{21}/λ_1. This is the same as the multivaluedness of the function $a_{21} \ln(z - z_0)/2\pi i \lambda_1$. Consequently $w_2/w_1 - a_{21} \ln(z - z_0)/2\pi i \lambda_1$ is single-valued, and can be expanded in a Laurent series:

$$\frac{w_2}{w_1} - \frac{a_{21}}{2\pi i \lambda_1} \ln(z - z_0) = \sum_{n=-\infty}^{\infty} d_n (z - z_0)^n \,,$$

whence we obtain the second solution[b] linearly independent of w_1,

$$w_2(z) = gw_1(z)\ln(z - z_0) + w_1(z)\sum_{n=-\infty}^{\infty} d_n(z - z_0)^n$$

$$= gw_1(z)\ln(z - z_0) + (z - z_0)^{\rho_1}\sum_{n=-\infty}^{\infty} e_n(z - z_0)^n .$$

The above results can be summarized in a theorem:

If z_0 is a singular point of equation $w'' + pw' + qw = 0$, then the two linearly independent solutions of the equation in the neighborhood $0 < |z - z_0| < R$ of z_0 (where R is sufficiently small so that there is no singular point in the annular region) are

$$w_1(z) = (z - z_0)^{\rho_1}\sum_{n=-\infty}^{\infty} c_n(z - z_0)^n , \tag{15}$$

$$w_2(z) = (z - z_0)^{\rho_2}\sum_{n=-\infty}^{\infty} d_n(z - z_0)^n , \tag{16}$$

$$(\rho_1 - \rho_2 \neq \text{integer})$$

or,

$$w_2(z) = gw_1(z)\ln(z - z_0) + (z - z_0)^{\rho_1}\sum_{n=-\infty}^{\infty} d_n(z - z_0)^n$$

$$(\rho_1 - \rho_2 = \text{integer}) . \tag{17}$$

In these solutions, ρ_1, ρ_2, c_n, d_n and g are coefficients to be determined. When these solutions are substituted in the equation to determine the coefficients, we obtain in general, however, an infinite number of simultaneous equations, each with an infinite number of unknowns. Thus, in such general cases, to find a solution in the series form is inconvenient.

2.4. Regular Solution. Regular Singularities

When the coefficients $p(z)$ and $q(z)$ of the equation

$$w'' + pw' + qw = 0 \tag{1}$$

---

[b]This second solution can also be deduced from the first by means of (2) [or (27) of the next section]. See, for example, Whittaker and Watson (1927) p. 200.

satisfy certain conditions, the Laurent series in (15), (16) or (17) of the last section happen to contain only a finite number of negative power terms. By extracting the lowest negative power factor and combining it in the factor $(z - z_0)^\rho$, we find

$$w_1(z) = (z - z_0)^{\rho_1} \sum_{n=0}^{\infty} c_n(z - z_0)^n \quad (c_0 \neq 0) , \tag{2}$$

$$w_2(z) = g w_1(z) \ln(z - z_0) + (z - z_0)^{\rho_2} \sum_{n=0}^{\infty} d_n(z - z_0)^n$$
$$(d_n \neq 0) , \tag{3}$$

where (3) includes the cases (16) and (17) of the last section. Solutions (2) and (3) are called *regular solutions*; they are characterized by the fact that the series in them are Taylor series. Substituting such solutions in (1), we obtain a sequence of recurrence relations [see (16) below] by which the undetermined coefficients in the series can be found one by one. The following is a discussion of the necessary and sufficient conditions for (1) to have regular solutions.

Theorem: The necessary and sufficient conditions for (1) to have two regular solutions in the vicinity $0 < |z - z_0| < R$ of its singular point z_0' are that

$$(z - z_0)p(z) \quad \text{and} \quad (z - z_0)^2 q(z) \quad \text{are analytic in} \quad |z - z_0| < R . \tag{4}$$

A singular point satisfying conditions (4) is called a *regular singular point*, otherwise, *irregular singular point*.

First, we show that conditions (4) are necessary. For convenience in writing, suppose $z_0 = 0$. From Eq. (1) of Sec. 2.3, we have

$$p(z) = -\frac{d}{dz} \ln(w_1 w_2' - w_2 w_1') = -\frac{d}{dz} \ln \left\{ w_1^2 \frac{d}{dz} \left(\frac{w_2}{w_1} \right) \right\} . \tag{5}$$

If (1) has in $0 < |z| < R$ two regular solutions

$$w_1(z) = z^{\rho_1} \sum_{n=0}^{\infty} c_n z^n \quad (c_0 \neq 0) , \tag{6}$$

$$w_2(z) = g w_1(z) \ln z + z^{\rho_2} \sum_{n=0}^{\infty} d_n z^n \quad (d_0 \neq 0) , \tag{7}$$

then, by substituting them in (5) and direct calculation, noting that $\rho_1 - \rho_2$ = integer when $g \neq 0$, we see that the point $z = 0$ is at most a pole of $p(z)$ of the first order. Thus $zp(z)$ is analytic at $z = 0$.

Next, we have from (1)

$$q(z) = -\frac{w_1''}{w_1} - p(z)\frac{w_1'}{w_1} \ . \tag{8}$$

Then by (6) and the result that $z = 0$ is at most a first ordered pole of $p(z)$, we see at once that $z = 0$ is at most a second ordered pole of $q(z)$, so $z^2 q(z)$ is analytic at $z = 0$.

Now, we prove that conditions (4) are sufficient. To do this, we only have to show that under these conditions, the coefficients c_n obtained by substituting (6) in (1) make the series convergent. Along with the proof, we shall also find the procedure of obtaining the solution.

Again, suppose $z_0 = 0$. Multiplying (1) by z^2, we have

$$z^2 w'' + z p_1(z) w' + q_1(z) w = 0 \ , \tag{9}$$

where

$$p_1(z) = zp(z) \ , \quad q_1(z) = z^2 q(z) \ . \tag{10}$$

According to conditions (4), $p_1(z)$ and $q_1(z)$ are analytic in $|z| < R$ and can be expanded in Taylor series

$$p_1(z) = \sum_0^\infty a_k z^k \ , \quad q_1(z) = \sum_0^\infty b_k z^k \ . \tag{11}$$

Assume the formal solution of (1) to be

$$w(z) = z^\rho \sum_0^\infty c_n z^n \quad (c_0 \neq 0) \ . \tag{12}$$

Substituting it in (9) gives

$$\sum_{n=0}^\infty c_n (\rho + n)(\rho + n - 1) z^n + \sum_{k=0}^\infty a_k z^k \sum_{n=0}^\infty c_n (\rho + n) z^n$$

$$+ \sum_{k=0}^\infty b_k z^k \sum_{n=0}^\infty c_n z^n = 0 \ . \tag{13}$$

Setting the coefficient of the lowest power of z in this equation to zero, we obtain

$$c_0[\rho(\rho - 1) + a_0\rho + b_0] = 0 .$$

But $c_0 \neq 0$, hence we must have

$$\rho(\rho - 1) + a_0\rho + b_0 = 0 , \qquad (14)$$

which is the equation determining the exponent ρ, and is called the *indicial equation*. We denote by ρ_1 and ρ_2, the two roots of this quadratic equation of ρ and assume

$$\mathrm{Re}(\rho_1) \geq \mathrm{Re}(\rho_2) . \qquad (15)$$

By putting the coefficients of $z^n (n \geq 1)$ in (13) equal to zero, we get the recurrence relations

$$[(\rho + n)(\rho + n - 1) + a_0(\rho + n) + b_0]c_n$$
$$+ \sum_{k=1}^{n} [a_k(\rho + n - k) + b_k]c_{n-k} = 0 ,$$
$$(n = 1, 2, \ldots) . \qquad (16)$$

By means of these relations we can express $c_n (n = 1, 2, \ldots)$ one by one in terms of c_0, so that the formal solution (12) is completely determined, except for the arbitrary constant c_0. We are then left with the sole problem of proving the convergence of the series.

Consider first the case of $\rho = \rho_1$, and let's estimate the coefficients c_n. Putting $\rho_1 - \rho_2 = s$, we have, according to (15), $\mathrm{Re}(s) \geq 0$. Since ρ_1 and ρ_2 are the roots of the indicial equation (14), $\rho_1 + \rho_2 = 1 - a_0, \rho_1\rho_2 = b_0$, and the coefficients of c_n in (16) $(\rho = \rho_1)$ are

$$(\rho_1 + n)(\rho_1 + n - 1) + a_0(\rho_1 + n) + b_0$$
$$= (\rho_1 + n)(\rho_1 + n - 1 + a_0) + b_0 = n(n + s) . \qquad (17)$$

As $p_1(z)$ and $q_1(z)$ are analytic, we have by Cauchy's inequality that there exist $M \geq 1$ and $r > 0$ such that

$$|a_k| < \frac{M}{r^k} , \quad |b_k| < \frac{M}{r^k} , \quad |a_k\rho_1 + b_k| < \frac{M}{r^k} . \qquad (18)$$

Hence, for the coefficient of c_{n-k} in (16),

$$|a_k(\rho_1 + n - k) + b_k| \leq |a_k\rho_1 + b_k| + (n - k)|a_k|$$
$$< Mr^{-k}(n - k + 1) . \qquad (19)$$

When $n = 1$, (16) becomes

$$(1 + s)c_1 + (a_1\rho_1 + b_1)c_0 = 0 ,$$

thus

$$|c_1| = |a_1\rho_1 + b_1||1 + s|^{-1}|c_0| < Mr^{-1}|c_0|$$
$$(\text{since } \mathrm{Re}(s) \geq 0) . \qquad (20)$$

Assume that

$$|c_\nu| < M^\nu r^{-\nu}|c_0|, \quad \nu = 2, 3, \ldots, n - 1 , \qquad (21)$$

then, using (17), (19), (20), we obtain from (16)

$$|c_n| < \frac{1}{n(n + s)} \sum_{k=1}^{n} Mr^{-k}(n - k + 1)|c_{n-k}|$$

$$\leq \frac{1}{n^2} \sum_{k=1}^{n} Mr^{-k}(n - k + 1)M^{n-k}r^{-n+k}|c_0|$$

$$= \frac{|c_0|r^{-n}}{n^2} \sum_{m=1}^{n} mM^m \leq \frac{|c_0|r^{-n}M^n}{n^2} \sum_{m=1}^{n} m$$

$$= |c_0|M^n r^{-n}\frac{n(n + 1)}{2n^2} < M^n r^{-n}|c_0|, \quad \text{when} \quad n \geq 2 ,$$

and by induction, relation (21) holds good for any $\nu(\geq 1)$. Consequently,

$$|c_n z^n| = |c_n||z|^n < |c_0|M^n r^{-n}|z|^n \quad (n \geq 1) .$$

Thus, provided $|z| \leq R_1 < M^{-1}r$, the series $\sum_0^\infty c_n z^n$ converges absolutely and uniformly. This proves that

$$w_1(z) = z^{\rho_1} \sum_{n=0}^{\infty} c_n z^n \qquad (22)$$

is indeed a solution of the equation.

For $\rho = \rho_2$, we have to distinguish between two cases:

1. $\rho_1 - \rho_2 \neq$ integer. In this case, putting $\rho = \rho_2$ and replacing c_n by $d_n(n = 0, 1, 2, \ldots)$ in (16), we obtain, in a similar way as before, the second solution

$$w_2(z) = z^{\rho_2} \sum_{n=0}^{\infty} d_n z^n , \qquad (23)$$

where the coefficients d_n are all expressed in terms of d_0, an arbitrary non-zero constant.

The two solutions expressed by (22) and (23) are obviously linearly independent of each other, since, when $z \to 0$, $w_1(z) \approx c_0 z^{\rho_1}$ and $w_2(z) \approx d_0 z^{\rho_2}$, whereas $\rho_1 \neq \rho_2$.

2. $\rho_1 - \rho_2 = m (m = 0, 1, 2, \ldots)$. This corresponds to the case $\lambda_1 = \lambda_2$ in the general theory in the last section.

If $m = 0$, i.e., $\rho_1 = \rho_2$, it is obvious that we can get only one solution $w_1(z)$ by the above given procedure. The second solution has to be found by other means [see (27) below, and Sec. 2.5].

When $m > 0$, the coefficients $d_1, d_2, \ldots, d_{m-1}$ can be determined in terms of d_0 from (16) as above (putting $\rho = \rho_2 = \rho_1 - m$ and changing c_n to d_n). But the equation for the determination of d_m is

$$[(\rho_2 + m)(\rho_2 + m - 1) + a_0(\rho_2 + m) + b_0]d_m$$
$$+ \sum_{k=1}^{m}[a_k(\rho_2 + m - k) + b_k]d_{m-k} = 0 , \qquad (24)$$

where the coefficient of d_m is $\rho_1(\rho_1 - 1) + a_0\rho_1 + b_0 = 0$, as ρ_1 is a root of the indicial equation (14). Thus, we shall have $d_m = \infty$ unless the sum

$$\sum_{k=1}^{m}[a_k(\rho_2 + m - k) + b_k]d_{m-k} = 0 \qquad (25)$$

in (24). Of course, $d_m = \infty$ is meaningless, and we have to find the second solution by other means [see (27) below, and Sec. 2.5]. However, if (25) holds, then d_m may be any constant, and the subsequent $d_n (n > m)$ can be sequentially expressed in terms of d_0 and d_m according to the recurrence relation (16), so that

$$w_2(z) = d_0 u(z) + d_m v(z) . \qquad (26)$$

It is easy to see that $v(z)$ on the r.h.s. of (26) differs from the first solution $w_1(z)$ by at most a constant factor, since the recurrence relations between the successive coefficients of both series are exactly the same (note that $\rho_2 + m = \rho_1$). Thus, we may take $d_0 u(z)$ as the second solution (i.e., putting $d_m = 0$), for which the proof of the convergence of the series is the same as before.

Whatever the case may be (whether, or not, $\rho_1 - \rho_2 =$ integer and, when $\rho_1 - \rho_2 =$ integer, whether, or not (25) holds), we can always use (5) to obtain the second solution:

$$w_2(z) = Aw_1(z) \int^z \frac{e^{-\int^z p(\varsigma)d\varsigma}}{[w_1(z)]^2} dz + Bw_1(z) \tag{27}$$

once the first solution $w_1(z)$ [Eq. (22)] has been found. However, this formula is inconvenient for direct calculation, as the integrand contains $[w_1(z)]^{-2}$ while $w_1(z)$ is, in general, an infinite series. In a concrete problem, when $\rho_1 - \rho_2 = m(m = 0, 1, 2, \ldots)$ we often put (7), or (3), into (1) to find the undetermined constant g and the coefficients $d_n(n = 0, 1, 2, \ldots)$. The series thus obtained can be shown to be convergent by analogous method as before.[c] However, (7) can also be deduced from (27)[d]; in the deduction only the convergence of the series in $w_1(z)$ is required. Hence, it is not necessary to give a separate proof of the convergence of the series in (7).

2.5. Frobenius Method

This is another method to solve for the regular solutions, especially suitable for the second solution.

Suppose $z = 0$ is a regular singular point of the equation $w'' + pw' + qw = 0$. Substituting the assumed solution

$$w(z) = z^\rho \sum_{n=0}^{\infty} c_n z^n = \sum_{n=0}^{\infty} c_n z^{\rho+n} \tag{1}$$

in the differential form $L[w] \equiv z^2 w'' + zp_1 w' + q_1 w$ with $p_1 = zp(z), q_1 = z^2 q(z)$, we have

$$L[w] \equiv z^\rho \left\{ \sum_{n=1}^{\infty} [c_n f_0(\rho+n) + \sum_{k=1}^{n} c_{n-k} f_k(\rho+n-k)]z^n + c_0 f_0(\rho) \right\}, \tag{2}$$

where

$$f_0(\lambda) \equiv \lambda(\lambda-1) + a_0\lambda + b_0, \quad f_k(\lambda) \equiv a_k\lambda + b_k \quad (k \geq 1),$$

and a_k, b_k are respectively the coefficients of the Taylor expansions of $p_1(z)$ and $q_2(z)$ at $z = 0$ [Eq. (11) of Sec. 2.4].

---

[c] See, for example, Whittaker and Watson (1927) p. 199.
[d] Whittaker and Watson, l.c., p. 200.

Choosing the c_n's to satisfy the recurrence relations

$$c_n f_0(\rho + n) + \sum_{k=1}^{n} c_{n-k} f_k(\rho + n - k) = 0 \quad (n = 1, 2, \ldots) \qquad (3)$$

[Eq. (16) of the last section], (2) becomes

$$L[w] \equiv c_0 z^\rho f_0(\rho) = c_0 z^\rho (\rho - \rho_1)(\rho - \rho_2) , \qquad (4)$$

where ρ_1 and ρ_2 are the two roots of the indicial equation $f_0(\rho) = 0$ [(14) of the last section].

If $\rho_1 - \rho_2 \neq$ integer, we can put $\rho = \rho_1$ and $\rho = \rho_2$ successively in (3) and obtain two sets of coefficients $c_n^{(1)}$ and $c_n^{(2)}$ which will give the two linearly independent solutions of $L[w] = 0$, respectively. This is just what we have done in the last section.

If $\rho_1 - \rho_2 = 0$, i.e., $\rho_1 = \rho_2$, (4) becomes

$$L[w] \equiv c_0 z^\rho (\rho - \rho_1)^2 . \qquad (5)$$

Differentiating it with respect to ρ, we have

$$L\left[\frac{\partial w}{\partial \rho}\right] \equiv c_0 \ln z \cdot z^\rho (\rho - \rho_1)^2 + c_0 z^\rho \cdot 2(\rho - \rho_1) . \qquad (6)$$

Then $\rho = \rho_1$, the r.h.s. of (5) and (6) are all zero. Consequently, when the coefficients $c_n (n = 1, 2, \ldots)$ in $w(z)$ are determined by the relation (3) as functions of ρ, we get instantly the two linearly independent solutions

$$w_1(z) = (w)_{\rho=\rho_1} = z^{\rho_1} \sum_{n=0}^{\infty} (c_n)_{\rho=\rho_1} z^n \quad (c_0 = \text{arbitrary constant}) \qquad (7)$$

$$w_2(z) = \left(\frac{\partial w}{\partial \rho}\right)_{\rho=\rho_1} = w_1(z) \ln z + z^{\rho_1} \sum_{n=1}^{\infty} \left(\frac{\partial c_n}{\partial \rho}\right)_{\rho=\rho_1} \cdot z^n . \qquad (8)$$

Note that, in $w_2(z)$, the last series begins with $n = 1$.

When $\rho_1 - \rho_2 = m (m = 1, 2, \ldots)$, the above procedure is not valid, and we have to put

$$c_0 = c_0'(\rho - \rho_2) , \qquad (9)$$

where c_0' is now an arbitrary constant. Use again (3) to determine the coefficients $c_n (n = 1, 2, \ldots)$, then (4) becomes

$$L[w] \equiv c_0' z^\rho (\rho - \rho_1)(\rho - \rho_2)^2 , \qquad (10)$$

whose derivative with respect to ρ is

$$L\left[\frac{\partial w}{\partial \rho}\right] \equiv c_0' \ln z \cdot z^\rho (\rho - \rho_1)(\rho - \rho_2)^2 + c_0' z^\rho (\rho - \rho_2)^2$$
$$+ c_0' z^\rho \cdot 2(\rho - \rho_1)(\rho - \rho_2) . \tag{11}$$

From (10) and (11) we see that $(w)_{\rho=\rho_1}$, $(w)_{\rho=\rho_2}$ and $(\partial w/\partial \rho)_{\rho=\rho_2}$ are all solutions of the equation $L[w] = 0$. The first solution is $w_1(z) = (w)_{\rho=\rho_1}$ as before. However, $w_2(z) = (w)_{\rho=\rho_2}$ differs from $(w)_{\rho=\rho_1}$ at most by a constant factor. This can be seen as follows. Since $c_0 = c_0'(\rho - \rho_2)$ now and all the $c_1, c_2, \ldots, c_{m-1}$ determined by (3) contain the factor $(\rho - \rho_2)$, we have

$$(c_1)_{\rho=\rho_2} = (c_2)_{\rho=\rho_2} = \ldots (c_{m-1})_{\rho=\rho_2} = 0 . \tag{12}$$

But, $(c_m)_{\rho=\rho_2}$ can be any constant, as the coefficient of c_m in (3) $(n = m)$ for the determination of c_m is $f_0(\rho + m) = (\rho + m - \rho_1)(\rho + m - \rho_2) = (\rho - \rho_2)(\rho + m - \rho_2)$, so it contains $(\rho - \rho_2)$ too. (This corresponds to the case when (25) of the last section holds.) The subsequent $(c_n)_{\rho=\rho_2} (n > m)$ can be expressed in terms of $(c_m)_{\rho=\rho_2}$; the relations obtained are exactly the same as that between $(c_n)_{\rho=\rho_1}$ and c_0 in $(w)_{\rho=\rho_1}$. Thus, $(w)_{\rho=\rho_2}$ and $(w)_{\rho=\rho_1}$ are not linearly independent of each other.

The second solution linearly independent of $(w)_{\rho=\rho_1}$ is

$$w_2(z) = \left(\frac{\partial w}{\partial \rho}\right)_{\rho=\rho_2}$$
$$= \ln z \cdot z^{\rho_2} \sum_{n=m}^{\infty} (c_n)_{\rho=\rho_2} z^n + z^{\rho_2} \sum_{n=0}^{\infty} \left(\frac{\partial c_n}{\partial \rho}\right)_{\rho=\rho_2} \cdot z^n , \tag{13}$$

where (12) has been used in the first sum. We can easily see that the first term in (13) differs from $w_1(z) \ln z$ by at most a constant factor, since the recurrence relations between the coefficients of both series are identical, as is shown above. It is to be noted that $c_0 = c'(\rho - \rho_2)$ is also a function of ρ now, whose derivative is not zero (but c_0'). Thus, the second series on the r.h.s. of (13) begins from $n = 0$, in contrast with (8).

For concrete examples of application of Frobenius method, see, e.g., Sec. 4.4.

2.6. Point at Infinity

In the previous discussions, the point in question, whether it is an ordinary point or a singular point of the equation, is assumed to be located at a finite region. For the point at infinity, we have to make the transformation of the variable: $z = 1/t$, and discuss the resulting equation

$$\frac{d^2 w}{dt^2} + \left\{ \frac{2}{t} - \frac{1}{t^2} p\left(\frac{1}{t}\right) \right\} \frac{dw}{dt} + \frac{1}{t^4} q\left(\frac{1}{t}\right) w(t) = 0 \tag{1}$$

for its properties and solutions at $t = 0$. We can easily find:

1. The conditions for $t = 0 (z = \infty)$ to be an ordinary point are that

$$p\left(\frac{1}{t}\right) = 2t + a_2 t^2 + a_3 t^3 + \dots ,$$

i.e.,

$$p(z) = \frac{2}{z} + \frac{a_2}{z^2} + \dots , \tag{2}$$

and

$$q\left(\frac{1}{t}\right) = b_4 t^4 + b_5 t^5 + \dots ,$$

i.e.,

$$q(z) = \frac{b_4}{z^4} + \frac{b_5}{z^5} + \dots . \tag{3}$$

2. The conditions for $t = 0 (z = \infty)$ to be a regular singular point are that

$$\left. \begin{array}{l} \frac{1}{t} p\left(\frac{1}{t}\right) = zp(z) \quad \text{and} \quad \frac{1}{t^2} q\left(\frac{1}{t}\right) = z^2 q(z) \\ \text{both to be analytic at } t = 0 \ (z = \infty). \end{array} \right\} \tag{4}$$

2.7. Equations of Fuchsian Type

An equation with all its singular points being regular is called an equation of Fuchsian type. The most important special case of this type of equations is that with three regular singularities, of which the prototype is the hypergeometric equation [see Sec. 2.9 and Chap. 4]; the commonly occurred Legendre equation belongs to this category.

First, let us look at the general characteristics of the coefficients of equations of Fuchsian type.

Let $a_r (r = 1, 2, \dots , n)$ and ∞ be the regular singular points of the equation

$$w'' + p(z)w' + q(z)w = 0 , \tag{1}$$

which has no other singularities. Since the a_r's are at the most poles of $p(z)$ of the first order, we have

$$p(z) = \sum_{r=1}^{n} \frac{A_r}{z - a_r} + \varphi(z) ,$$

where A_r is the residue of $p(z)$ at $z = a_r$, and $\varphi(z)$ is a function analytic on the whole z-plane. According to Eq. (4) of Sec. 2.6, when $z \to \infty$, $\varphi(z)$ should approach zero, because $z = \infty$ is a regular singular point. Consequently, by Liouville's theorem, $\varphi(z) \equiv$ constant $= 0$, and

$$p(z) = \sum_{r=1}^{n} \frac{A_r}{z - a_r} \ . \tag{2}$$

By analogous reasoning, we have

$$q(z) = \sum_{r=1}^{n} \left\{ \frac{B_r}{(z - a_r)^2} + \frac{C_r}{z - a_r} \right\} , \tag{3}$$

and, according to Eq. (4) of Sec. 2.6, we must have

$$\sum_{r=1}^{n} C_r = 0 \ . \tag{4}$$

From (2) and (3), we immediately obtain Eqs. (14) and (11) of Sec. 2.4, in which $a_0 = \lim_{z \to 0} zp(z)$, $b_0 = \lim_{z \to 0} z^2 q(z)$. At the regular singular point a_r, the indicial equation is

$$\rho^2 + (A_r - 1)\rho + B_r = 0 \quad (r = 1, 2, \ldots, n) \ . \tag{5}$$

At ∞, from Eq. (1) of Sec. 2.6, we have

$$\lim_{t \to 0} t \left\{ \frac{2}{t} - \frac{1}{t^2} p \left(\frac{1}{t} \right) \right\} = 2 - \lim_{z \to \infty} zp(z) = 2 - \sum_{r=1}^{n} A_r ,$$

$$\lim_{t \to 0} t^2 \cdot \frac{1}{t^4} q \left(\frac{1}{t} \right) = \lim_{z \to \infty} z^2 q(z) = \lim_{z \to \infty} z^2 \sum_{r=1}^{n} \left\{ \frac{B_r}{(z - a_r)^2} + \frac{C_r}{z - a_r} \right\}$$

$$= \sum_{r=1}^{n} B_r + \lim_{z \to \infty} z \sum_{r=1}^{n} C_r [1 + a_r z^{-1} + O(z^{-2})]$$

$$= \sum_{r=1}^{n} B_r + \sum_{r=1}^{n} a_r C_r \quad [\text{by (4)}] \ .$$

Hence, the indicial equation[e] is

$$\rho^2 + \left(1 - \sum_{r=1}^{n} A_r \right) \rho + \sum_{r=1}^{n} (B_r + a_r C_r) = 0 \ . \tag{6}$$

---

[e] It is to be noted that this equation concerns the exponents of t, and hence of z^{-1}.

From (5) and (6), we can draw the important conclusion:

$$\textit{The sum of the roots of the indicial equations} \; = \; n-1 \, . \qquad (7)$$

If $z = \infty$ is not a singular point but an ordinary point, we shall have, from Eqs. (2) and (3) of Sec. 2.6,

$$\left.\begin{aligned}
&\sum_{r=1}^{n} A_r = 2 \, , \\
&\sum_{r=1}^{n} C_r = 0 \, , \\
&\sum_{r=1}^{n} (B_r + a_r C_r) = 0 \, , \\
&\sum_{r=1}^{n} (2a_r B_r + a_r^2 C_r) = 0 \, .
\end{aligned}\right\} \qquad (8)$$

2.8. Equations of Fuchsian Type Having Five Regular Singular Points

Suppose the singular points are $a_r (r = 1, 2, 3, 4)$ and ∞. Denote the exponents at a_r by α_r, β_r and those at ∞ by μ_1 and μ_2. From (2), (3), (4) and (5) of the last section, the equation of Fuchsian type having these five singularities will be

$$\begin{aligned}
\frac{d^2 u}{dz^2} &+ \left\{ \sum_{r=1}^{4} \frac{1 - \alpha_r - \beta_r}{z - a_r} \right\} \frac{du}{dz} \\
&+ \left\{ \sum_{r=1}^{4} \frac{\alpha_r \beta_r}{(z - a_r)^2} + \frac{Az^2 + 2Bz + C}{\prod_{r=1}^{4}(z - a_r)} \right\} u = 0 \, ,
\end{aligned} \qquad (1)$$

where A, B and C are constants, $A = -\sum_{r=1}^{4} C_r \left(\sum_{k=1}^{4} a_k - a_r \right) = \sum_{r=1}^{4} a_r$ $\cdot C_r = \mu_1 \mu_2 - \sum_{r=1}^{4} \alpha_r \beta_r$; B and C are arbitrary, because C_r in the coefficient $q(z)$ of the equation cannot be uniquely determined by $a_r (r = 1, 2, 3, 4)$ and the exponents, *unless* $n = 2$ [see next section].

It has been shown by Klein and Bôcher that some linear ordinary differential equations [see Table 1] in certain branches of mathematical physics are various confluent forms of a special case of (1), for which the difference of the two exponents at every singular point is $\frac{1}{2}$. Let $\beta_r = \alpha_r + \frac{1}{2} (r = 1, 2, 3, 4), \mu_2 =$

$\mu_1 + \frac{1}{2}$, then, with z replaced by ς, (1) becomes

$$
\frac{d^2u}{d\varsigma^2} + \left\{ \sum_{r=1}^{4} \frac{\frac{1}{2} - 2\alpha_r}{\varsigma - a_r} \right\} \frac{du}{d\varsigma}
$$
$$
+ \left\{ \sum_{r=1}^{4} \frac{\alpha_r(\alpha_r + \frac{1}{2})}{(\varsigma - a_r)^2} + \frac{A\varsigma^2 + 2B\varsigma + C}{\prod_{r=1}^{4}(\varsigma - a_r)} \right\} u = 0 , \qquad (2)
$$

Table 1.

Type of Eq.	(a)	(b)	(c)	Name of Eq.
I	3	1	0	Lamé
II	2	0	1	Mathieu
III	1	2	0	Legendre
IV	0	1	1	Bessel
V	1	0	1	Weber
VI	0	0	1	Stokes

and is called the *generalized Lamé equation.* By means of the conclusion (7) of Sec. 2.7, we have

$$
A = \left(\sum_{r=1}^{4} \alpha_r \right)^2 - \sum_{r=1}^{4} \alpha_r^2 - \frac{3}{2} \sum_{r=1}^{4} \alpha_r + \frac{3}{16} . \qquad (3)
$$

Now, let us give an explanation of what is meant by the confluent form of a differential equation. A differential equation derivable from another differential equation by making two or more singular points of the latter equation to coincide is called the confluent form of the latter equation. After confluence, the nature of the singularities is in general altered.

If two of the singularities of (2) are made to coincide, for instance, $a_1 = a_2$, the new singular point is still regular; the corresponding exponents α and β, however, are given by the following relations:

$$
\left.
\begin{array}{l}
\alpha + \beta = 2(\alpha_1 + \alpha_2) , \\
\alpha\beta = \alpha_1\left(\alpha_1 + \frac{1}{2}\right) + \alpha_2\left(\alpha_2 + \frac{1}{2}\right) + D , \\
\text{where} \\
D = (Aa_1^2 + 2Ba_1 + C)/\{(a_1 - a_3)(a_1 - a_4)\} .
\end{array}
\right\} \qquad (4)
$$

Besides, the difference of the two exponents at the confluent singular point is no longer $\frac{1}{2}$, but may have any assigned value by suitably choosing the arbitrary constants B and C in D.

If we make three or more singularities of (2) confluent, i.e., coincident, the resultant singular point will be irregular, because terms like $(\varsigma - a_i)^{-3}$ or $(\varsigma - a_i)^{-4}$ will appear in the coefficient of u in the new equation, where a_i is the confluent singularity.

By different ways of confluence of the five singularities, we obtain six different types of equations (exhausting all possible forms of confluence), which are classified according to

(a) the number of singular points with exponents differed by $\frac{1}{2}$,[f]

(b) the number of other regular singularities,

(c) the number of irregular singularities.

These are tabulated as Ex. 2-7 at the end of this chapter.

All these equations will be given detailed investigations one by one in subsequent chapters, except for the last type which can be transformed easily into the Bessel equation [see Ex. 6 at the end of this chapter] and will not be discussed separately.

2.9. Equations of Fuchsian Type Having Three Regular Singularities

The general equation of Fuchsian type having three regular singularities $a, b, c (\neq \infty$, thus ∞ is an ordinary point) can be derived according to (2) to (8) of Sec. 2.7 to be

$$\frac{d^2w}{dz^2} + \left\{ \frac{1 - \alpha_1 - \alpha_2}{z - a} + \frac{1 - \beta_1 - \beta_2}{z - b} + \frac{1 - \gamma_1 - \gamma_2}{z - c} \right\} \frac{dw}{dz}$$

$$+ \left\{ \frac{\alpha_1\alpha_2(a - b)(a - c)}{z - a} + \frac{\beta_1\beta_2(b - c)(b - a)}{z - b} + \frac{\gamma_1\gamma_2(c - a)(c - b)}{z - c} \right\}$$

$$\times \frac{1}{(z - a)(z - b)(z - c)} w = 0 , \tag{1}$$

where $(\alpha_1, \alpha_2), (\beta_1, \beta_2), (\gamma_1, \gamma_2)$ denote respectively the exponent pairs at the three points $a, b, c \ (\neq \infty)$. They satisfy

$$\alpha_1 + \alpha_2 + \beta_1 + \beta_2 + \gamma_1 + \gamma_2 = 1 . \tag{2}$$

---

[f]Such singularities must be regular for if there exists a formal regular solution at an irregular singular point, the indicial equation is of the first degree, and hence there can only be one exponent [see Sec. 2.10].

If one of the points a, b, c, e.g. c is ∞, (1) reduces to

$$\frac{d^2 w}{dz^2} + \left\{ \frac{1 - \alpha_1 - \alpha_2}{z - a} + \frac{1 - \beta_1 - \beta_2}{z - b} \right\} \frac{dw}{dz}$$

$$+ \left\{ \frac{\alpha_1 \alpha_2 (a - b)}{z - a} + \frac{\beta_1 \beta_2 (b - a)}{z - b} + \gamma_1 \gamma_2 \right\} \frac{1}{(z - a)(z - b)} w = 0 . \tag{3}$$

If, further, we put $a = 0, b = 1$, then (3) is simplified to

$$z(1 - z)\frac{d^2 w}{dz^2} + [\gamma - (\alpha + \beta + 1)z]\frac{dw}{dz} - \alpha\beta w = 0 , \tag{4}$$

where $(0, 1 - \gamma), (0, \gamma - \alpha - \beta), (\alpha, \beta)$ are the exponent pairs at $z = 0, 1, \infty$ respectively. Note that the values of the exponents are changed. The reduction from (1) to (4) can be effected by the transformations

$$\varsigma = \frac{(b - c)(z - a)}{(b - a)(z - c)} , \tag{5}$$

$$w = \left(\frac{z - a}{z - c} \right)^{\alpha_1} \left(\frac{z - b}{z - c} \right)^{\beta_1} \omega ; \tag{6}$$

z and w in (4) are then replaced respectively by ς and ω. As the general equation of Fuchsian type having three regular singularities can always be transformed to the simpler equation (4) by the transformations (5) and (6), (4) is taken as the prototype of such equations and is called the hypergeometric equation, whose solutions will be investigated in detail in Chap. 4.

The solutions of (1) are characterized completely by the singularities and the corresponding exponents of the equation, and are thus often expressed with the help of the P-symbol:

$$w(z) = P \left\{ \begin{array}{ccc} a & b & c \\ \alpha_1 & \beta_1 & \gamma_1, \ z \\ \alpha_2 & \beta_2 & \gamma_2 \end{array} \right\} , \tag{7}$$

which is called the *Riemann P-equation*. The solutions of (4) are then represented by

$$w(z) = P \left\{ \begin{array}{ccc} 0 & 1 & \infty \\ 0 & 0 & \alpha, \ z \\ 1 - \gamma & \gamma - \alpha - \beta & \beta \end{array} \right\} . \tag{8}$$

It is not difficult to show that a fractional linear transformation from z to z_1, namely

$$z_1 = \lambda \frac{z - \mu}{z - \nu} , \tag{9}$$

leaves the form of (1) unchanged, only the singular points a, b, c now becoming a_1, b_1, c_1 respectively, corresponding to z to z_1 according to (9). The whole transformation scheme may be expressed as

$$P \left\{ \begin{array}{ccc} a & b & c \\ \alpha_1 & \beta_1 & \gamma_1, \\ \alpha_2 & \beta_2 & \gamma_2 \end{array} z \right\} = P \left\{ \begin{array}{ccc} a_1 & b_1 & c_1 \\ \alpha_1 & \beta_1 & \gamma_1, \\ \alpha_2 & \beta_2 & \gamma_2 \end{array} z_1 \right\} . \tag{10}$$

Note that the exponent pairs of the corresponding singularities remain unaltered under this sort of transformation.

However, if the function $w(z)$ is transformed to $w_1(z)$ by

$$w_1(z) = \left(\frac{z-a}{z-c} \right)^k \left(\frac{z-b}{z-c} \right)^l w(z) ,$$

then it can be shown that the exponents change according to the following equation

$$\left(\frac{z-a}{z-c} \right)^k \left(\frac{z-b}{z-c} \right)^l P \left\{ \begin{array}{ccc} a & b & c \\ \alpha_1 & \beta_1 & \gamma_1, \\ \alpha_2 & \beta_2 & \gamma_2 \end{array} z \right\}$$

$$= P \left\{ \begin{array}{ccc} a & b & c \\ \alpha_1 + k & \beta_1 + l & \gamma_1 - k - l, \\ \alpha_2 + k & \beta_2 + l & \gamma_2 - k - l \end{array} z \right\} . \tag{11}$$

If one of the singularities, say c, is ∞, then the transformation (11) has to be replaced by

$$(z - a)^k (z - b)^l P \left\{ \begin{array}{ccc} a & b & \infty \\ \alpha_1 & \beta_1 & \gamma_1, \\ \alpha_2 & \beta_2 & \gamma_2 \end{array} z \right\}$$

$$= P \left\{ \begin{array}{ccc} a & b & \infty \\ \alpha_1 + k & \beta_1 + l & \gamma_1 - k - l, \\ \alpha_2 + k & \beta_2 + l & \gamma_2 - k - l \end{array} z \right\} . \tag{12}$$

Applying (10) and (11) to the transformations (5) and (6), we find

$$\left(\frac{z-a}{z-c} \right)^{-\alpha_1} \left(\frac{z-b}{z-c} \right)^{-\beta_1} P \left\{ \begin{array}{ccc} a & b & c \\ \alpha_1 & \beta_1 & \gamma_1, \\ \alpha_2 & \beta_2 & \gamma_2 \end{array} z \right\}$$

$$= P \left\{ \begin{array}{ccc} 0 & 1 & \infty \\ 0 & 0 & \gamma_1 + \alpha_1 + \beta_1, \\ \alpha_2 - \alpha_1 & \beta_2 - \beta_1 & \gamma_2 + \alpha_1 + \beta_1 \end{array} \frac{(b-c)(z-a)}{(b-a)(z-c)} \right\} , \tag{13}$$

whence the parameters α, β, γ in (4) are related to the exponents of (1) by [compare (13) with (8)]

$$\left.\begin{array}{l} \alpha = \gamma_1 + \alpha_1 + \beta_1 \ , \\ \beta = \gamma_2 + \alpha_1 + \beta_1 \ , \\ \gamma = 1 + \alpha_1 - \alpha_2 \ . \end{array}\right\} \tag{14}$$

Example. The associate Legendre equation

$$(1 - z^2)\frac{d^2 w}{dz^2} - 2z\frac{dw}{dz} + \left[n(n+1) - \frac{m^2}{1 - z^2}\right] w = 0 \tag{15}$$

is of Fuchsian type with the three regular singularities $z = \pm 1, \infty$. Comparing to (3), we have

$$\alpha_1 = \beta_1 = \frac{m}{2}, \quad \alpha_2 = \beta_2 = -\frac{m}{2} \ ,$$
$$\gamma_1 = n + 1, \quad \gamma_2 = -n \ .$$

Hence the solutions of (15) can be represented by

$$w(z) = P\left\{\begin{array}{cccc} -1 & 1 & \infty & \\ \frac{m}{2} & \frac{m}{2} & n+1 \ , & z \\ -\frac{m}{2} & -\frac{m}{2} & -n & \end{array}\right\} \ , \tag{16}$$

which will be dealt with in detail in Chap. 5.

2.10. Irregular Singularities. Formal Regular Solution

According to the theorem in Sec. 2.4, a linear ordinary differential equation of the second order can have at most one solution in the regular form in the vicinity of its irregular singular point. The problem concerning the condition for the existence of such solution has been solved only for the case when the singularities are poles. Here we only give the necessary condition for the existence of such solution[g] and explain how to deal with the solution of the equation if this condition is not fulfilled.

Let $z = \infty$ be the irregular singular point to be discussed. This covers the majority of cases in practice, and such assumption does not invalidate the generality of the result, since any point $z = a$ can always be transformed to $t = \infty$ by the transformation $z - a = 1/t$.

---

[g] For the sufficient condition, see Bieberbach (1953), Sec. 6.8, p. 161.

Formal regular solution in the vicinity of an irregular singularity
Let $z = \infty$ be the irregular singularity of the equation

$$L[w] \equiv w'' + p(z)w' + q(z)w = 0 \tag{1}$$

and consider only the case when it is a pole of $p(z)$ or/and $q(z)$. Then we can write

$$p(z) = z^{n_1}\left(a_0 + \frac{a_1}{z} + \dots\right) = z^{n_1}\sum_{l=0}^{\infty} a_l z^{-l}, \tag{2}$$

$$q(z) = z^{n_2}\left(b_0 + \frac{b_1}{z} + \dots\right) = z^{n_2}\sum_{s=0}^{\infty} b_s z^{-s}, \tag{3}$$

where n_1 and n_2 are integers, called the *order* of $p(z)$ and $q(z)$ respectively. Suppose neither a_0 nor b_0 is zero (unless $p(z)$ or $q(z)$ is identically zero), then at least one of the relations

$$n_1 > -1, \quad n_2 > -2 \tag{4}$$

must hold; otherwise, according to the result of Sec. 2.6, $z = \infty$ would be a regular singularity or an ordinary point.

Now, let us find the necessary condition that there is a solution of the regular form at $z = \infty$. For this aim, putting

$$w(z) = z^{\rho}\sum_{k=0}^{\infty} c_k z^{-k}, \quad c_0 \neq 0, \tag{5}$$

and substituting it in the differential form on the l.h.s. of (1), we have

$$L[w] \equiv z^{\rho}\left\{\sum_{k=0}^{\infty} c_k(\rho - k)(\rho - k - 1)z^{-k-2}\right.$$
$$+ \sum_{l=0}^{\infty} a_l z^{n_1-l}\sum_{k=0}^{\infty} c_k(\rho - k)z^{-k-1}$$
$$\left. + \sum_{s=0}^{\infty} b_s z^{n_2-s}\sum_{k=0}^{\infty} c_k z^{-k}\right\}, \tag{6}$$

whence we see, if $n_2 \geq n_1$, the coefficient of the highest power of $z(z^{n_2})$ in the bracket on the right is $b_0 c_0$. As neither b_0 nor c_0 is zero by assumption,[h]

---

[h] By assumption, b_0 is zero only when $q(z) \equiv 0$, but then (1) can be reduced to an equation of the first order, which is easy to solve and need not be discussed.

so the equation has no solution of regular form in this case. If $n_1 > n_2$, the coefficient of the highest power term will be

$$a_0 c_0 \rho \quad \text{when} \quad n_1 > n_2 + 1 , \tag{7}$$

$$c_0(a_0 \rho + b_0) \quad \text{when} \quad n_1 = n_2 + 1 . \tag{8}$$

Hence the equations (all of the first order) for the exponent will be

$$\rho = 0 \quad (n_1 > n_2 + 1) , \tag{9}$$

$$\rho = -\frac{b_0}{a_0} \quad (n_1 = n_2 + 1) , \tag{10}$$

where we have assumed that $a_0 \neq 0$, unless $p(z) \equiv 0$. When $p(z) \equiv 0$, the highest power of z in (6) is again $b_0 c_0$, and there exists no solution of regular form. In consequence, we have the following theorem:

The necessary condition for the existence of one solution of the regular form in the vicinity of an irregular singularity of pole type is $n_1 > n_2$, i.e., the order of $p(z)$ must be larger than that of $q(z)$.

In contradistinction to the case of the regular singularity, the series in the formal regular solution at an irregular singularity is not always convergent (hence the name *formal* solution). However, even if the series diverges, the formal solution is still significant by that it is in fact the asymptotic expansion of the solution when $z \to \infty$. This point will be expounded in concrete cases in the following (see, e.g. the example in Sec. 2.11).

2.11. Irregular Singularities. Normal Solutions and Subnormal Solutions

If the equation

$$w'' + pw' + qw = 0 \tag{1}$$

has no regular solution (not even a formal one) at its irregular singularity $z = \infty$, then, according to the general theory of Sec. 2.3, the point $z = \infty$ must be an essential singularity of the solution (and may be a branch point of the solution at the same time), i.e., the corresponding Laurent expansion has an infinite number of positive powers of z. In this case, a more practical method is to look for the so called *normal solution* with the form

$$w(z) = e^{Q(z)} z^\rho \sum_{k=0}^{\infty} c_k z^{-k} , \tag{2}$$

where $Q(z)$ is a polynomial of z; the factor $e^{Q(z)}$ reflects the character that the point at infinity is an essential singularity of the solution, and the remaining part has the form of a regular solution.

The concept of normal solution comes from the investigation of the analogous problem for the first order equation

$$w' + R(z)w = 0 \ . \tag{3}$$

If there is a regular solution $w = z^\rho \sum_{k=0}^\infty c_k z^{-k}, c_0 \neq 0$, for the equation, then we must have

$$\begin{aligned} R(z) &= -\frac{w'}{w} = -\frac{d}{dz}\ln w = -\frac{\rho}{z} + \frac{c_1 z^{-2} + \dots}{c_0 + c_1 z^{-1} + \dots} \\ &= -\frac{\rho}{z} + \frac{a_2}{z^2} + \frac{a_3}{z^3} + \dots \ , \end{aligned} \tag{4}$$

i.e., $R(z) \to 0$ when $z \to \infty$. Hence, if $z = \infty$ is a pole singularity of $R(z)$, so that

$$R(z) = A_s z^s + A_{s-1} z^{s-1} + \dots + A_1 z + A_0 + \sum_{l=1}^\infty a_l z^{-l}$$
$$(s \geq 0, \quad A_s \neq 0) \ , \tag{5}$$

(3) cannot have a regular solution. Its solution is, however,

$$\begin{aligned} w(z) &= C \exp\left\{-\int^z R(\varsigma)d\varsigma\right\} \\ &= C \exp\left\{-\left(\frac{A_s}{s+1}z^{s+1} + \dots + A_0 z\right)\right\} \\ &\quad \times z^{-a_1} \exp\left\{a_2 z^{-1} + \frac{a_3}{2}z^{-2} + \dots\right\} \\ &= Ce^{Q(z)}z^\rho \psi(z) \ , \end{aligned} \tag{6}$$

where $Q(z)$ represents the polynomial $-(A_s z^{s+1}/(s+1) + \dots + A_0 z)$, $\rho = -a_1, \psi(z) = \exp\{a_2 z^{-1} + a_3 z^{-2}/2 + \dots\} = \sum_0^\infty c_k z^{-k}, c_0 = 1$. We see that (6) is just the normal solution referred above.

Now we inquire, under what condition will (1) of second order has normal solutions. Let

$$w(z) = e^{Q(z)}v(z) \tag{7}$$

and substitute it in (1), then we have the equation for $v(z)$:

$$v'' + p^*(z)v' + q^*(z)v = 0 \ , \tag{8}$$

where

$$p^*(z) = p(z) + 2Q'(z) , \tag{9}$$

$$q^*(z) = q(z) + p(z)Q'(z) + Q''(z) + [Q'(z)]^2 . \tag{10}$$

For (7) to be a normal solution, (8) should have at least one regular solution at $z = \infty$; the condition is that the order of $p^*(z)$ must be higher than that of $q^*(z)$, according to Sec. 2.10. This provides a condition for determining the polynomial $Q(z)$. We explicate this by the following example.

Example. Find the solutions of the Weber equation[i]

$$\psi'' + (\lambda - \alpha^2 x^2)\psi = 0 \tag{11}$$

near $x = \infty$, where λ and $\alpha(\alpha > 0)$ are constants.

The only singularity of this equation is at $x = \infty$, and it is irregular. As the coefficient $p(x) \equiv 0$, the equation has not even a formal regular solution [cf. Sec. 2.10, after Eq. (10)]. Let's see whether it has normal solutions. Putting

$$\psi(x) = e^{Q(x)}v(x) \tag{12}$$

and substituting it in (11), we obtain

$$v'' + p^*(x)v' + q^*(x)v = 0 , \tag{13}$$

where, by (9) and (10),

$$\left.\begin{array}{l} p^*(x) = 2Q'(x) , \\ q^*(x) = \lambda - \alpha^2 x^2 + Q''(x) + [Q'(x)]^2 . \end{array}\right\} \tag{14}$$

From the requirement that (13) is to have regular solutions — order of p^* higher than that of q^* — we see immediately that the degree of $Q'(x)$ cannot exceed 1, since q^* contains the term $Q'(x)$ squared and there is no other term to cancel it. Assuming

$$Q(x) = a_1 x + a_2 x^2 \tag{15}$$

we have

$$p^*(x) = 2(a_1 + 2a_2 x) ,$$
$$q^*(x) = \lambda - \alpha^2 x^2 + 2a_2 + (a_1 + 2a_2 x)^2$$
$$= \lambda + 2a_2 + a_1^2 + 4a_1 a_2 x - \alpha^2 x^2 + 4a_2^2 x^2 .$$

---

[i]We shall discuss the solutions of this equation in detail in Sec. 6.12.

Thus we see, only when $4a_2^2 = \alpha^2$, i.e., $a_2 = \pm\alpha/2$, and $a_1 = 0$, is the order of p^* higher than that of q^*. Consequently,

$$Q(x) = \pm\frac{1}{2}\alpha x^2, \quad \psi(x) = e^{\pm\frac{\alpha}{2}x^2}v(x),$$

and the equations for $v(x)$ are

$$v'' \pm 2\alpha xv' + (\lambda \pm \alpha)v = 0. \tag{16}$$

This is called the *Hermite equation*.[j] Consider the case of the negative sign; $Q(x) = -\alpha x^2/2$. Assume

$$v(x) = x^\rho \sum_{k=0}^\infty c_k x^{-k} \quad (c_0 \neq 0) \tag{17}$$

and put it in (16), then following the procedure of deducing a formal regular solution, we obtain

$$\rho = \frac{\lambda - \alpha}{2\alpha}, \tag{18}$$

$$c_{2k+1} = 0 \quad \text{for} \quad k = 0, 1, 2, \ldots, \tag{19}$$

and the recurrence relation

$$\begin{aligned}
c_{2k+2} &= \frac{(\rho - 2k)(\rho - 2k - 1)}{2\alpha(\rho - 2k - 2) - (\lambda - \alpha)}c_{2k} \\
&= -\frac{(\rho - 2k)(\rho - 2k - 1)}{4\alpha(k + 1)}c_{2k};
\end{aligned} \tag{20}$$

in the last step, (18) have been used for the denominator.

From the recurrence relation (20), it is seen that as $k \to \infty$, $|c_{2k+2}/c_{2k}| \to \infty$, hence the series $\sum_0^\infty c_{2k}x^{-2k}$ is divergent for any finite value of x. We shall see afterwards (cf. the example in Sec. 2.13), when we have obtained the integral solution of (16), that this series is the asymptotic expansion of the solution as $x \to \infty$, i.e.,

$$\psi(x) \sim e^{-\frac{\alpha}{2}x^2} x^{\frac{\lambda-\alpha}{2\alpha}} \sum_{k=0}^\infty c_{2k}x^{-2k} \quad (x \to \infty), \tag{21}$$

---

[j] The solutions of it will be further discussed in Sec. 6.13.

in which the coefficients are given by (20), namely

$$c_{2k} = (-)^k \frac{(-\rho)_{2k}}{(4\alpha)^k \cdot k!} c_0 .$$

In practical applications, an important special case is

$$\rho = \frac{\lambda - \alpha}{2\alpha} = n \quad (n = 0, 1, 2, \ldots) , \tag{22}$$

for which the series is terminated to become a polynomial of degree n. For example, when n is even $(n = 2m)$, we have from (20)

$$c_{2k+2} = -\frac{(2m - 2k)(2m - 2k - 1)}{4\alpha(k + 1)} c_{2k} .$$

This gives, for $k = m$, $c_{2m+2} = 0$, and hence the coefficients $c_{2k+2} = 0$ for all $k \geq m$, and the series becomes a polynomial of degree $2m(= n)$. The same is true when n is odd. The polynomial is usually denoted as $H_n(\xi)$, $\xi = \sqrt{\alpha}x$, with the coefficient of the highest power ξ^n assigned the value 2^n. Such polynomial is called the *Hermite polynomial*:

$$\begin{aligned} H_n(\xi) = (2\xi)^n &- \frac{n(n - 1)}{1!}(2\xi)^{n-2} \\ &+ \frac{n(n - 1)(n - 2)(n - 3)}{2!}(2\xi)^{n-4} \\ &- \ldots + (-)^{[n/2]} \frac{n!}{[n/2]!}(2\xi)^{n-2[n/2]} , \end{aligned} \tag{23}$$

where $[n/2] = n/2$ or $(n - 1)/2$ depending on n being even or odd.

Subnormal solution

When the attempt for the normal solution fails, we may sometimes transform the equation to one having normal solutions by means of transformation of the independent variable: $z = \varsigma^s$ (s = positive integer). For example, the equation

$$w'' + q(z)w = 0 \tag{24}$$

has no normal solution when the expansion of $q(z)$ at $z = \infty$ is

$$q(z) = \sum_{l=1}^{\infty} b_l z^{-l}, \quad b_1 \neq 0 . \tag{25}$$

By the transformation $z = \varsigma^2$, (24) becomes

$$\frac{d^2w}{d\varsigma^2} - \frac{1}{\varsigma}\frac{dw}{d\varsigma} + 4\varsigma^2 q(\varsigma^2)w = 0 \; , \tag{26}$$

which has the normal solutions

$$w(\varsigma) = e^{\pm 2\sqrt{-b_1}\varsigma}\varsigma^{\frac{1}{2}}\sum_{k=0}^{\infty} c_k \varsigma^{-k} \; .$$

Hence the solutions of (24) are

$$w(z) = e^{\pm 2\sqrt{-b_1}z^{\frac{1}{2}}}z^{\frac{1}{4}}\sum_{k=0}^{\infty} c_k z^{-k/2} \; , \tag{27}$$

which are called *subnormal solutions*.

2.12. Method of Solution by Integrals. Basic Principle

In order to explicate the basic principle of the method of solution by integrals, we introduce the concept of the adjoint of a differential form:

$$L[u] \equiv p_0(z)\frac{d^2u}{dz^2} + p_1(z)\frac{du}{dz} + p_2(z)u \; . \tag{1}$$

Multiplying both sides of (1) by $v(z)$, we have

$$vL[u] \equiv vp_0(z)u'' + vp_1(z)u' + vp_2(z)u \; . \tag{2}$$

By means of the relations

$$vp_0 u'' = u(vp_0)'' + \frac{d}{dz}[(vp_0)u' - (vp_0)'u] \; , \tag{3}$$

$$vp_1 u' = -u(vp_1)' + \frac{d}{dz}(vp_1 u) \; , \tag{4}$$

(2) can be written as

$$\begin{aligned}
vL[u] \equiv &u[(vp_0)'' - (vp_1)' + vp_2] \\
&+ \frac{d}{dz}[vp_0 u' - (vp_0)'u + vp_1 u] \; ,
\end{aligned} \tag{5}$$

or

$$vL[u] - u\overline{L}[v] \equiv \frac{d}{dz}Q[u,v] , \tag{6}$$

where

$$\overline{L}[v] \equiv \frac{d^2}{dz^2}(p_0 v) - \frac{d}{dz}(p_1 v) + p_2 v \tag{7}$$

is called the *adjoint* of $L[u]$; the equation $\overline{L}[v] = 0$ is called the *adjoint equation* of $L[u] = 0$, and $\overline{L}$ the *adjoint operator* of the differential operator L. The expression

$$Q[u,v] \equiv u' p_0 v - u(p_0 v)' + u p_1 v \tag{8}$$

is called the *bilinear concomitant*, which is a homogeneous differential form of second degree in u and v, but linear in u, u' or v, v' taken separately.

The basic principle of the method of solution by integrals consists of using the integral transform

$$u(z) = \int_C K(z,t)v(t)dt \tag{9}$$

to reduce the solution of the linear differential equation

$$L[u] \equiv p_0 \frac{d^2 u}{dz^2} + p_1 \frac{du}{dz} + p_2 u = 0 \tag{10}$$

to that for $v(t)$. The function $K(z,t)$ in (9) is called the *kernel of the integral transform* and C is the path of integration on the complex t-plane; both of them are to be suitably chosen in accordance with the properties of (10) (see below).

Suppose the integral (9) can be differentiated twice under the integral sign, then

$$L[u] = \int_C L_z[K(z,t)]v(t)dt , \tag{11}$$

where L_z is the operator L with the added subscript to mark its operation on z. If we take $K(z,t)$ to be a particular solution of the partial differential equation[k]

$$L_z[K(z,t)] = M_t[K(z,t)] , \tag{12}$$

where M_t is a partial differential operator on t, then by (6), (11) is transformed to

$$L[u] = \int_C v(t) M_t[K(z,t)]dt$$
$$= \int_C K(z,t)\overline{M}_t[v(t)]dt + \{Q[K,v]\}_C , \tag{13}$$

[k] The r.h.s. of (12) may be $M_t[G(z,t)]$, with $G(z,t)$ different from $K(z,t)$; the only change is to replace K by G in (13) and (15) below.

where $\overline{M}_t$ is the adjoint operator of M_t, $Q[K,v]$ is the corresponding bilinear concomitant, and $\{Q\}_C$ represents the change of Q, along the path of integration C, as a function of t.

By taking $v(t)$ to satisfy

$$\overline{M}_t[v(t)] = 0 \tag{14}$$

and choosing the path of integration such that

$$\{Q[K,v]\}_C = 0 , \tag{15}$$

i.e., the difference of $Q[K,v]$ at the starting point and the terminating point of C equals zero, we shall have the r.h.s. of (13) zero, and $u(z)$ of (9) is an integral solution of (10).

From the above deductions we see, the main difficulty in applying this method of solution resides in the choice of M_t, which must be able to facilitate the solution of both (12) and (14). Solving (14) is simple, as it is an ordinary differential equation. We usually require it to be easier to solve than the original equation, e.g., we require it to be an equation of the first order. As to (12), the problem of its solution is in general rather involved since it is a partial differential equation. Usually, we proceed by taking a suitable kernel $K(z,t)$ according to the nature of the singularities of the original equation, rather than choosing M_t first and then solving (12) for $K(z,t)$. The most commonly used kernels are

1. *Kernel of the Laplace transform*

$$K(z,t) = e^{zt} \tag{16}$$

usually applied to equations with an irregular singularity at $z = \infty$.

2. *Kernel of the Euler transform*

$$K(z,t) = (z-t)^\mu , \tag{17}$$

where μ is a parameter. This kernel is often used in solving equations of Fuchsian type.

3. *Kernel of the Mellin transform*

$$K(z,t) = z^t . \tag{18}$$

2.13. Equations of Laplacian Type and Laplace Transform

One of the most important applications of the method of solution by integrals is to solve equations of Laplacian type:

$$L[u] \equiv (a_0 z + b_0) u'' + (a_1 z + b_1) u' + (a_2 z + b_2) u = 0 , \tag{1}$$

all of whose coefficients being linear functions of z. In the finite region, there is at most one singularity $z = -b_0/a_0$ which is regular. If $a_0 = 0$, there is no singularity in the finite region. The point at $z = \infty$ is an irregular singular point of the equation.

In applying Laplace transform to solve (1), we put

$$u(z) = \int_C e^{zt} v(t) dt , \tag{2}$$

then

$$
\begin{aligned}
L[u] &= \int_C v(t) L_z[e^{zt}] dt \\
&= \int_C v(t) \{ (a_0 z + b_0) t^2 + (a_1 z + b_1) t + (a_2 z + b_2) \} e^{zt} dt \\
&= \int_C v(t) M_t[e^{zt}] dt ,
\end{aligned} \tag{3}
$$

where

$$
\begin{aligned}
M_t &\equiv t^2 \left(a_0 \frac{\partial}{\partial t} + b_0 \right) + t \left(a_1 \frac{\partial}{\partial t} + b_1 \right) + \left(a_2 \frac{\partial}{\partial t} + b_2 \right) \\
&= (a_0 t^2 + a_1 t + a_2) \frac{\partial}{\partial t} + (b_0 t^2 + b_1 t + b_2) .
\end{aligned} \tag{4}
$$

Integrating by parts, we obtain from (3)

$$
\begin{aligned}
L[u] = \int_C e^{zt} \left\{ -\frac{d}{dt} [(a_0 t^2 + a_1 t + a_2) v(t)] + (b_0 t^2 + b_1 t + b_2) v(t) \right\} dt \\
+ \{ (a_0 t^2 + a_1 t + a_2) v(t) e^{zt} \}_C .
\end{aligned} \tag{5}
$$

Taking $v(t)$ to satisfy [Eq. (14) of Sec. 2.12]

$$-\frac{d}{dt} [(a_0 t^2 + a_1 t + a_2) v(t)] + (b_0 t^2 + b_1 t + b_2) v(t) = 0 \tag{6}$$

and choosing C such that

$$\{P(z,t)\}_C \equiv \{(a_0 t^2 + a_1 t + a_2)v(t)e^{zt}\}_C = 0 , \tag{7}$$

(2) will then be a solution of (1), provided it is legitimate to differentiate under the integral sign.

As (6) is an equation of the first order, it is easy to find its solution $v(t)$ (assume that a_0, a_1, and a_2 are not all zero, otherwise (1) would be an equation of constant coefficients, and there is no need to apply the method):

$$\begin{aligned}
\frac{d}{dt}\ln v &= \frac{b_0 t^2 + (b_1 - 2a_0)t + b_2 - a_1}{a_0 t^2 + a_1 t + a_2} \\
&= \frac{b_0 t^2 + (b_1 - 2a_0)t + b_2 - a_1}{a_0(t - \alpha_1)(t - \alpha_2)} \\
&= \mu + \frac{\lambda_1}{t - \alpha_1} + \frac{\lambda_2}{t - \alpha_2} ,
\end{aligned} \tag{8}$$

$$v(t) = e^{\mu t}(t - \alpha_1)^{\lambda_1}(t - \alpha_2)^{\lambda_2} , \tag{9}$$

where α_1 and α_2 are the two roots of the equation $a_0 t^2 + a_1 t + a_2 = 0$. If $a_0 = 0$, or $\alpha_1 = \alpha_2$, (8) and (9) and the expressions followed have to be appropriately modified, but it causes not much difficulties (see the following examples).

Now let us see how to choose the path of integration C. Substituting (9) in the l.h.s. of (7), we have

$$P(z,t) \equiv a_0 e^{(\mu+z)t}(t - \alpha_1)^{\lambda_1+1}(t - \alpha_2)^{\lambda_2+1} . \tag{10}$$

(i) If $\mathrm{Re}(\lambda_1)$ and $\mathrm{Re}(\lambda_2)$ are both larger than -1, C can be any piece-wise smooth curve from $t = \alpha_1$ to $t = \alpha_2$ on the t-plane. At the two terminal points, $P(z,t) = 0$ and (7) is satisfied. Also, for such curves, the differentiation of (2) under the integral sign is legitimate. Hence, an integral solution of (1) is

$$u(z) = \int_{\alpha_1}^{\alpha_2} e^{(\mu+z)t}(t - \alpha_1)^{\lambda_1}(t - \alpha_2)^{\lambda_2} dt$$
$$(\mathrm{Re}(\lambda_1),\ \mathrm{Re}(\lambda_2) > -1) . \tag{11}$$

(ii) In case λ_1 and λ_2 are arbitrary non-integer constants, we can choose for C a double contour as that shown in Fig. 3, which starts from an arbitrary point $P(\neq \alpha_1, \alpha_2)$ on the t-plane, encircling each of α_1 and α_2 once in the positive sense, then encircling each of them once in the negative sense, and returns to P. This contour is closed on the Riemann surface of the function $(t - \alpha_1)^{\lambda_1} (t - \alpha_2)^{\lambda_2}$, hence $P(z, t)$ has the same value at the starting point and the terminating point, and (7) is satisfied. Also, for such a contour the integral (2) is not vanishing, and can be differentiated under the integral sign. Thus

$$u(z) = \int^{(\alpha_1+, \alpha_2+, \alpha_1-, \alpha_2-)} e^{(\mu+z)t} (t - \alpha_1)^{\lambda_1} (t - \alpha_2)^{\lambda_2} dt$$

$$(\lambda_1, \lambda_2 \neq \text{ integer}) \tag{12}$$

is an integral solution of (1). Here in (12) we have indicated the path of integration with explicit symbols.

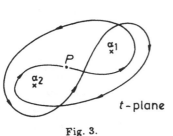

Fig. 3.

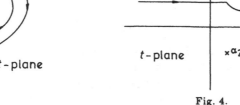

Fig. 4.

When λ_1 or λ_2 is an integer, the double contour integral vanishes and hence it is not applicable.

(iii) Another commonly used path of integration is the contour, starting from the point at infinity of the t-plane, encircling α_2 (if $\lambda_1 \neq$ integer) or encircling α_2 (if $\lambda_2 \neq$ integer) once, and then back to infinity. For example, suppose $\lambda_1 \neq$ integer, and $\text{Re}(\mu + z) > 0$, we can take C to be the contour shown in Fig. 4, which starts from $-\infty + i \, \text{Im}(\alpha_1)$, encircling α_1 once in the positive sense and then back to the starting point. Although this contour is not closed on the Riemann surface of $(t - \alpha_1)^{\lambda_1}$, but $P(z, t) \to 0$ at the starting and the terminal points, so (7) is satisfied. Also, for such contour, the integral (2) is uniformly convergent in $|\arg(\mu + z)| \leq \pi/2 - \delta(\delta > 0)$ and thus can be differentiated under the integral sign. Therefore,

$$u(z) = \int_{-\infty}^{(\alpha_1+)} e^{(\mu+z)t} (t - \alpha_1)^{\lambda_1} (t - \alpha_2)^{\lambda_2} dt$$

$$(\text{Re}(\mu + z) > 0, \quad \lambda_1 \neq \text{ integer}) \tag{13}$$

is an integral solution of (1).

As to the linear independency or dependency of the various integral solutions obtained by using different paths of integration, we shall give separate discussion in specific problems, and make no statement in general here.[1]

Example. Find the integral solutions of the Hermite equation $u'' - 2\alpha x u' + (\lambda - \alpha)u(x) = 0$ [Eq. (16) of Sec. 2.11], where $\alpha > 0$, and λ is real.

This is an equation of Laplacian type with $a_0 = 0, a_1 = -2\alpha, a_2 = 0, b_0 = 1, b_1 = 0, b_2 = \lambda - \alpha$. Putting

$$u(x) = \int_C e^{xt} v(t) dt , \tag{14}$$

we get from (6) easily

$$v(t) = A e^{-t^2/4\alpha} \, t^{-\lambda_1}, \quad \lambda_1 = \frac{\lambda + \alpha}{2\alpha} , \tag{15}$$

so

$$P(x,t) \equiv A' e^{xt} e^{-t^2/4\alpha} \, t^{-\lambda_1 + 1} . \tag{16}$$

As $P(x,t)$ contains the factor $\exp\{-t^2/4\alpha\}$ with $\alpha > 0$, we can take for C the contour starting from $-\infty$ on the real axis, encircling $t = 0$ once in the positive sense, and back to $-\infty$ on the real axis. The integral solution obtained is

$$u(x) = A \int_{-\infty}^{(0+)} e^{xt} e^{-t^2/4\alpha} \, t^{-\frac{\lambda+\alpha}{2\alpha}} dt , \tag{17}$$

where x is not subjected to restriction.

The integral solution of a Laplacian equation is especially convenient for finding the asymptotic expansion of the solution. Take (17) as an example. Suppose $|\arg x| \leq \pi/2 - \delta (\delta > 0)$, and, using Eq. (7) of Sec. 1.9, we get the asymptotic expansion immediately:

$$u(x) = -A \int_{\infty}^{(0+)} e^{-xt} e^{-t^2/4\alpha} (-t)^{-\frac{\lambda+\alpha}{2\alpha}} dt$$

$$\sim 2i A x^{\frac{\lambda-\alpha}{2\alpha}} \sin\frac{\lambda+\alpha}{2\alpha}\pi \sum_{n=0}^{\infty} \frac{(-)^n}{(4\alpha)^n n!} \Gamma\left(2n - \frac{\lambda-\alpha}{2\alpha}\right) x^{-2n} ,$$

$$|\arg x| \leq \pi/2 - \delta \quad (\delta > 0) . \tag{18}$$

---

[1]Cf. for example, Goursat and Hodrick, *A Course in Mathematical Analysis*, Vol. II, Part II, Differential Equations, Sec. 46. p. 124.

Comparing this with Eq. (21) of Sec. 2.11, it verifies that the series there is just the asymptotic expansion of the exact solution.

In principle, Laplace transform can be applied to equations whose coefficients are not linear functions of the independent variable, e.g., the equation

$$(a_0 z^2 + b_0 z + c_0)u'' + (a_1 z^2 + b_1 z + c_1)u'$$
$$+ (a_2 z^2 + b_2 z + c_2)u(z) = 0 . \tag{19}$$

However, the equation for the determination of $v(t)$ [Eq. (14) of Sec 2.12] is of the second order, and whether its solution is simpler than that of the original (19) depends on circumstances.

2.14. Euler Transform

The Euler transform

$$w(z) = \int_C (z - t)^\mu v(t)dt \quad (\mu \text{ to be determined}) \tag{1}$$

is suitable for solving equations of the following form:

$$L[w] \equiv p_0(z)w'' + p_1(z)w' + p_2(z)w(z) = 0 , \tag{2}$$

where

$$p_0(z) = a_0 z^2 + b_0 z + c_0 , \quad p_1(z) = b_1 z + c_1 , \quad p_2(z) = c_2 . \tag{3}$$

This is an equation of Fuchsian type having three regular singularities, including the point at infinity.

From (1) we have

$$L[w] \equiv \int_C \{p_0(z)\mu(\mu - 1)(z - t)^{\mu-2}$$
$$+ p_1(z)\mu(z - t)^{\mu-1} + p_2(z - t)^\mu\}v(t)dt . \tag{4}$$

Expressing all the coefficients (3) in powers of $(z - t)$,

$$\left.\begin{aligned}
p_0(z) &= p_0(t) + p_0'(t)(z - t) + \tfrac{1}{2}p_0''(t)(z - t)^2 \\
&= (a_0 t^2 + b_0 t + c_0) + (2a_0 t + b_0)(z - t) + a_0(z - t)^2 , \\
\\
p_1(z) &= p_1(t) + p_1'(t)(z - t) \\
&= (b_1 t + c_1) + b_1(z - t) \\
\\
p_2(z) &= p_2(t) = c_2 ,
\end{aligned}\right\} \tag{5}$$

and substituting them in (4), we get

$$
\begin{aligned}
L[w] &\equiv \int_C \bigg\{ p_0(t)\mu(\mu - 1)(z - t)^{\mu-2} \\
&\quad + [p_0'(t)\mu(\mu - 1) + p_1(t)\mu](z - t)^{\mu-1} \\
&\quad + \left[\frac{1}{2}p_0''(t)\mu(\mu - 1) + p_1'(t)\mu + p_2(t)\right](z - t)^{\mu} \bigg\} v(t)\,dt \\
&= \int_C v(t) \bigg\{ p_0(t)\frac{\partial^2}{\partial t} - [p_0'(t)(\mu - 1) + p_1(t)]\frac{\partial}{\partial t} \\
&\quad + \left[\frac{1}{2}p_0''(t)\mu(\mu - 1) + p_1'(t)\mu + p_2(t)\right] \bigg\}(z - t)^{\mu}\,dt \ .
\end{aligned}
$$

Now, take μ satisfying the equation

$$
\begin{aligned}
&\frac{1}{2}p_0''(t)\mu(\mu - 1) + p_1'(t)\mu + p_2(t) \\
&\equiv a_0\mu(\mu - 1) + b_1\mu + c_2 = 0 \ ,
\end{aligned}
\tag{6}
$$

then

$$
L[w] \equiv \int_C v(t) M_t[(z - t)^{\mu}]\,dt \ ,
\tag{7}
$$

where

$$
M_t \equiv \alpha\frac{\partial^2}{\partial t^2} - \beta\frac{\partial}{\partial t} \ ,
\tag{8}
$$

$$
\alpha = p_0(t), \quad \beta = p_0'(t)(\mu - 1) + p_1(t) \ .
\tag{9}
$$

Hence, by Eqs. (6) − (8) of Sec. 2.12, we obtain

$$
L[w] \equiv \int_C (z - t)^{\mu}\overline{M}_t[v(t)]\,dt + \{Q[(z - t)^{\mu}, v(t)]\}_C \ ,
\tag{10}
$$

where

$$
\overline{M}_t[v(t)] \equiv \frac{d^2}{dt^2}(\alpha v) + \frac{d}{dt}(\beta v) = \frac{d}{dt}[(\alpha v)' + \beta v] \ ,
\tag{11}
$$

$$
Q[(z - t)^{\mu}, v(t)] \equiv -\mu\alpha v(z - t)^{\mu-1} - [(\alpha v)' + \beta v](z - t)^{\mu} \ .
\tag{12}
$$

Take $v(t)$ satisfying the equation

$$
(\alpha v)' + \beta v = 0 \ ,
\tag{13}
$$

i.e.,

$$v(t) = \frac{A}{\alpha}e^{-\int^t \beta/\alpha \cdot dt} = \frac{A}{p_0(t)}\exp\int^t\left[-\frac{p_0'(\varsigma)}{p_0(\varsigma)}(\mu - 1) - \frac{p_1(\varsigma)}{p_0(\varsigma)}\right]d\varsigma$$

$$= A[p_0(t)]^{-\mu}\exp\left\{-\int^t\frac{p_1(\varsigma)}{p_0(\varsigma)}d\varsigma\right\} , \tag{14}$$

where A is an arbitrary constant, then $v(t)$ also satisfies $\overline{M}_t[v(t)] = 0$ and

$$Q[(z-t)^\mu, v(t)] \equiv -\mu p_0(t)v(t)(z-t)^{\mu-1} . \tag{15}$$

By choosing a path of integration C such that $\{Q\}_C = 0$, then (1) will be an integral solution of (2), provided differentiation under the integral sign is legitimate.

Noticing that Q, as a function of z and t, differs from the integrand in (1) only by a single-valued factor $p_0(t)(z-t)^{-1}$, we can always choose C to be such a contour that, as t varies along C and back to the starting point, the integrand in (1) resumes its value.

Example.   Find the integral solutions of the hypergeometric equation [Eq. (1) of Sec. 2.9].

The hypergeometric equation

$$z(1-z)\frac{d^2w}{dz^2} + [\gamma - (\alpha + \beta + 1)z]\frac{dw}{dz} - \alpha\beta w(z) = 0 \tag{16}$$

has just the form of (2) with $p_0(z) = -z^2 + z$, $p_1(z) = -(\alpha + \beta + 1)z + \gamma$, $p_2(z) = -\alpha\beta$. Assume the integral solution of (16) to be

$$w(z) = \int_C (z - t)^\mu v(t)dt . \tag{17}$$

From (6), we have

$$-\mu(\mu - 1) - (\alpha + \beta + 1)\mu - \alpha\beta = 0 , \tag{18}$$

its two roots being $\mu = -\alpha$ and $-\beta$. Taking $\mu = -\alpha$, we obtain from (14)

$$v(t) = At^{-\mu}(1-t)^{-\mu}\exp\left\{-\int^t\frac{\gamma - (\alpha + \beta + 1)\varsigma}{\varsigma(1-\varsigma)}d\varsigma\right\}$$

$$= At^{\alpha-\gamma}(1-t)^{\gamma-\beta-1} , \tag{19}$$

so

$$w(z) = A \int_C t^{\alpha-\gamma}(1-t)^{\gamma-\beta-1}(z-t)^{-\alpha}dt , \tag{20}$$

where the path of integration must be such that

$$Q[(z-t)^\mu, v] \equiv -A\alpha t^{\alpha-\gamma+1}(1-t)^{\gamma-\beta}(z-t)^{-\alpha-1} \tag{21}$$

has the same value at the starting point and the terminal point. For example, when $\text{Re}(\gamma) > \text{Re}(\beta) > 0$, we can take C to be the straight line from 1 to ∞ along the real axis on the t-plane, then at the starting point $t = 1, (1-t)^{\gamma-\beta} = 0$, and at the terminal point $t = \infty, Q \sim t^{-\beta} \to 0$. Thus

$$w(z) = A \int_1^\infty t^{\alpha-\gamma}(1-t)^{\gamma-\beta-1}(z-t)^{-\alpha}dt$$
$$(\text{Re}(\gamma) > \text{Re}(\beta) > 0) \tag{22}$$

is a solution of the hypergeometric equation (16), provided z is not on the path of integration.

Replacing t by $1/t$ in (22), we find another expression of the solution:

$$w(z) = A' \int_0^1 t^{\beta-1}(1-t)^{\gamma-\beta-1}(1-zt)^{-\alpha}dt , \tag{23}$$

where $\text{Re}(\gamma) > \text{Re}(\beta) > 0$, and z is not a real number between 1 to ∞.

Other choices of C and the other integral solutions obtained therefrom will be discussed in more detail in Sec. 4.5.

Exercise 2

1. Show that the value of the multiplier λ of the multiplied solution [Eq. (5) of Sec. 2.3] is independent of the choice of the fundamental set of solutions w_1 and w_2.

2. Derive from the generalized Lamé equation [Eq. (2) of Sec. 2.8] one of its confluent equations, the Lamé equation

$$\frac{d^2u}{d\varsigma^2} + \left\{ \sum_{r=1}^3 \frac{\frac{1}{2}}{\varsigma - a_r} \right\} \frac{du}{d\varsigma} - \frac{n(n+1)\varsigma + h}{4\prod_{r=1}^3(\varsigma - a_r)}u = 0 .$$

3. Derive from the generalized Lamé equation its another confluent equation

$$\frac{d^2u}{d\varsigma^2} + \left\{ \frac{\frac{1}{2}}{\varsigma} + \frac{\frac{1}{2}}{\varsigma - 1} \right\} \frac{du}{d\varsigma} - \frac{\lambda + 2q - 4q\varsigma}{4\varsigma(\varsigma - 1)} u = 0 \ ,$$

where λ and q are constants. Putting $\varsigma = \cos^2 z$ in this equation, we obtain the Mathieu equation

$$\frac{d^2u}{dz^2} + (\lambda - 2q\cos 2z)u = 0 \ .$$

4. Same as above, derive the equation

$$\frac{d^2u}{d\varsigma^2} + \left\{ \frac{\frac{1}{2}}{\varsigma} + \frac{1}{\varsigma - 1} \right\} \frac{du}{d\varsigma} + \frac{1}{4} \left\{ \frac{n(n+1)}{\varsigma} - \frac{m^2}{\varsigma - 1} \right\} \frac{u}{\varsigma(\varsigma - 1)} = 0 \ .$$

Putting in it $\varsigma = z^{-2}$, we obtain the associate Legendre equation

$$(1 - z^2)\frac{d^2u}{dz^2} - 2z\frac{du}{dz} + \left[n(n+1) - \frac{m^2}{1 - z^2} \right] u = 0 \ .$$

5. Also derive as above the equation

$$\varsigma^2\frac{d^2u}{d\varsigma^2} + \varsigma\frac{du}{d\varsigma} + \frac{1}{4}(\varsigma - n^2)u = 0 \ ;$$

putting $\varsigma = z^2$, we have the Bessel equation

$$z^2\frac{d^2u}{dz^2} + z\frac{du}{dz} + (z^2 - n^2)u = 0 \ .$$

6. Same as above, derive the equation

$$\varsigma\frac{d^2u}{d\varsigma^2} + \frac{1}{2}\frac{du}{d\varsigma} + \frac{1}{4}\left(n + \frac{1}{2} - \frac{1}{4}\varsigma \right) u = 0 \ ;$$

putting $\varsigma = z^2$, we obtain the Weber equation

$$\frac{d^2u}{dz^2} + \left(n + \frac{1}{2} - \frac{1}{4}z^2\right)u = 0 .$$

7. Same as above, derive the Stokes equation

$$\frac{d^2u}{d\varsigma^2} + (\alpha\varsigma + \beta)u = 0 .$$

By putting $u = (\alpha\varsigma + \beta)^{1/2}v, \alpha\varsigma + \beta = (3\alpha z/2)^{2/3}$, the equation is reduced to the Bessel equation (see Ex. 5 with $n = 1/3$).

8. Show that the generalized Lamé equation preserves its form under the following transformations:
   (i) the fractional linear transformations of the independent variable, with ∞ remaining a singularity;
   (ii) transformation of the dependent variable, $u = (z - a_r)^\lambda v$.

9. Verify: the general form of equations of Fuchsian type, having two regular singularities a and b, is

$$\frac{d^2u}{dz^2} + \left\{\frac{1 - \alpha - \alpha'}{z - a} + \frac{1 + \alpha + \alpha'}{z - b}\right\}\frac{du}{dz}$$
$$+ \frac{\alpha\alpha'(a - b)^2}{(z - a)^2(z - b)^2}u = 0 ,$$

where α, α' are the two exponents at the singularity $z = a$. The general solution of this equation is the elementary function

$$u = A\left(\frac{z - a}{z - b}\right)^\alpha + B\left(\frac{z - a}{z - b}\right)^{\alpha'} \quad (\alpha \neq \alpha')$$

or,

$$u = A\left(\frac{z - a}{z - b}\right)^\alpha + B\left(\frac{z - a}{z - b}\right)^\alpha \ln\frac{z - a}{z - b} \quad (\alpha = \alpha') ,$$

where A and B are two arbitrary constants.

10. Show that, when $\beta + \gamma + \beta' + \gamma' = \frac{1}{2}$,

$$P \left\{ \begin{array}{ccc} 0 & \infty & 1 \\ 0 & \beta & \gamma, \ z^2 \\ 1/2 & \beta' & \gamma' \end{array} \right\} = P \left\{ \begin{array}{ccc} -1 & \infty & 1 \\ \gamma & 2\beta & \gamma, \ z \\ \gamma' & 2\beta' & \gamma' \end{array} \right\} ;$$

both of them represent as a whole the solutions of the equation

$$\frac{d^2u}{dz^2} + \frac{2(1 - \gamma - \gamma')z}{z^2 - 1} \frac{du}{dz} + \left\{ \beta\beta' + \frac{\gamma\gamma'}{z^2 - 1} \right\} \frac{4u}{z^2 - 1} = 0 .$$

11. Show that, if $\gamma + \gamma' = 1/3$, and ω, ω^2 are the two complex roots of the equation $x^3 - 1 = 0$, we have

$$P \left\{ \begin{array}{ccc} 0 & \infty & 1 \\ 0 & 0 & \gamma, \ z^3 \\ 1/3 & 1/3 & \gamma' \end{array} \right\} = P \left\{ \begin{array}{ccc} 1 & \omega & \omega^2 \\ \gamma & \gamma & \gamma, \ z \\ \gamma' & \gamma' & \gamma' \end{array} \right\} ;$$

both of them represent as a whole the solutions of the equation

$$\frac{d^2u}{dz^2} + \frac{2z^2}{z^3 - 1} \frac{du}{dz} + \frac{9\gamma\gamma'z}{(z^3 - 1)^2} u = 0 .$$

12. Show that the general form of Fuchsian-type equations of the third order, having singularities $0, 1, \infty$ and exponents at each singularity all being $1, 1, -1$, is

$$\frac{d^3u}{dz^3} + \left\{ \frac{2}{z} + \frac{2}{z - 1} \right\} \frac{d^2u}{dz^2}$$

$$+ \left\{ -\frac{1}{z^2} + \frac{3}{z(z - 1)} - \frac{1}{(z - 1)^2} \right\} \frac{du}{dz}$$

$$+ \left\{ \frac{1}{z^3} - \frac{\alpha}{z^2(z - 1)} - \frac{3 - \alpha}{z(z - 1)^2} + \frac{1}{(z - 1)^3} \right\} u = 0 ,$$

where α is an arbitrary constant.

13. With the help of the Euler transform [Sec. 2.14] show that the general Fuchsian-type equation having three singularities $a, b, c \, (\neq \infty)$ [see Eq. (6) of Sec. 2.9] has the integral solution

$$w(z) = (z - a)^\alpha (z - b)^\beta (z - c)^\gamma$$
$$\times \int_C (t - a)^{\alpha' + \beta + \gamma - 1} (t - b)^{\alpha + \beta' + \gamma - 1}$$
$$\times (t - c)^{\alpha + \beta + \gamma' - 1} (z - t)^{-\alpha - \beta - \gamma} dt \,,$$

where $(\alpha, \alpha'), (\beta, \beta'), (\gamma, \gamma')$ are the exponent pairs at the singular points a, b, c respectively, and C is such a contour that the integrand assumes the same value at its starting point and the terminal point, z being outside the contour.

Chapter 3

THE GAMMA FUNCTION

3.1. Definition of the Gamma Function

The common definition of the Gamma function $\Gamma(z)$ is

$$\Gamma(z) = \int_0^\infty e^{-t} t^{z-1} dt \,, \tag{1}$$

which holds only when $\text{Re}(z) > 0$, as this is the condition for the convergence of the integral at its lower limit. In the following Secs. 3.3, 3.4 and 3.7, we shall present other definitions; the present definition, however, occurs most frequently in practice, and is called *Euler's integral of the second kind*.

The path of integration in (1) may be altered to a straight line starting from $t = 0$ to $\infty e^{i\alpha}$, so long as $|\alpha| < \pi/2$:

$$\Gamma(z) = \int_0^{\infty e^{i\alpha}} e^{-t} t^{z-1} dt \quad (\text{Re}(z) > 0) \,, \tag{2}$$

since the difference between the integrals (1) and (2) is equal to the integral on a circular arc in the right half-plane, which subtends an angle $|\alpha|$ and with its radius R tending to infinity; according to Jordan's lemma (or by direct calculation), this last integral vanishes when $R \to \infty$.

The symbol $\Gamma(z)$ is most commonly used for the Gamma function. Other two symbols are $\prod(z)$ and $z!$, both equal $\Gamma(z+1)$:

$$z! = \prod(z) = \Gamma(z+1) \ . \tag{3}$$

The notation $z!$ is used, in general, only for z being a positive integer [see Eq. (8) of the next section]. However, in this book, we shall not subject to this restriction and identify it with $\Gamma(z+1)$. The other notation, $\prod(z)$, will not occur in this book.

3.2. Recurrence Relation

$\Gamma(z)$ satisfies the following recurrence relation:

$$\Gamma(z+1) = z\Gamma(z) \tag{1}$$

or,

$$z! = z(z-1)! \tag{2}$$

which may be verified with integrating by parts the integral in Eq. (1) of Sec. 3.1:

$$\Gamma(z+1) = \int_0^\infty e^{-t} t^z \, dt$$

$$= [-e^{-t} t^z]_{t=0}^{t=\infty} + z \int_0^\infty e^{-t} t^{z-1} dt = z\Gamma(z) \ .$$

Let n be a positive integer. Equation (1) can be generalized to

$$\Gamma(z+n) = (z+n-1)(z+n-2)\ldots(z+1)z\Gamma(z) \tag{3}$$

or,

$$\Gamma(z) = \frac{\Gamma(z+n)}{z(z+1)\ldots(z+n-1)} = \frac{1}{(z)_n} \int_0^\infty e^{-t} t^{z+n-1} dt \ , \tag{4}$$

where

$$(z)_n = z(z+1)\ldots(z+n-1) \ . \tag{5}$$

Formula (4) extends the definition of $\Gamma(z)$ to $\text{Re}(z) > -n$ with n an arbitrary positive integer.

Equation (3) can also be written as

$$(z)_n = \frac{\Gamma(z+n)}{\Gamma(z)} = \frac{(z+n-1)!}{(z-1)!} \ . \tag{6}$$

In Eq. (1) of the last section, putting $z = 1$, we have

$$\Gamma(1) = 0! = \int_0^\infty e^{-t} dt = 1 . \tag{7}$$

In (3), putting $z = 1$, we have

$$\Gamma(n+1) = n! = n(n-1)\ldots 2.1 , \tag{8}$$

which shows that, when z is a positive integer n, $\Gamma(z+1)$ is just the factorial $n!$.

From (4) it is seen that $\Gamma(z)$ is a meromorphic function, of which all the singularities in a finite domain are simple poles; they are

$$z = 0, -1, -2, \ldots, -n, \ldots, \tag{9}$$

and the corresponding residue at $z = -n$ is

$$\lim_{z \to -n} (z+n)\Gamma(z) = \left.\frac{\Gamma(z+n+1)}{z(z+1)\ldots(z+n-1)}\right|_{z=-n} = \frac{(-)^n}{n!} . \tag{10}$$

3.3. The Infinite-Product Expression of Euler

With the limit expression for e^{-t}:

$$e^{-t} = \lim_{n\to\infty} \left(1 - \frac{t}{n}\right)^n ,$$

$\Gamma(z)$ may be regarded as the limit of the following integral

$$P_n(z) = \int_0^n \left(1 - \frac{t}{n}\right)^n t^{z-1} dt , \tag{1}$$

i.e., $P_n(z) \to \Gamma(z)$ when $n \to \infty$. This is proved as follows:

$$\Gamma(z) - P_n(z) = \int_0^n \left\{ e^{-t} - \left(1 - \frac{t}{n}\right)^n \right\} t^{z-1} dt + \int_n^\infty e^{-t} t^{z-1} dt .$$

It is readily seen that the limit of the second term is zero when $n \to \infty$. As for the first term, we may employ the following inequality

$$0 \le e^{-t} - \left(1 - \frac{t}{n}\right)^n \le \frac{t^2}{n} e^{-t} \tag{2}$$

(see below for justification) and obtain

$$\left| \int_0^n \left\{ e^{-t} - \left(1 - \frac{t}{n}\right)^n \right\} t^{z-1} dt \right| \leq \int_0^n \frac{1}{n} e^{-t} t^{x+1} dt$$

$$< \frac{1}{n} \int_0^\infty e^{-t} t^{x+1} dt \to 0 \, ,$$

where $x = \mathrm{Re}(z)$. Thus, $P_n(z) \to \Gamma(z)$.

Now, let us verify the inequality (2). From the series representations of e^y and $(1 - y)^{-1}$, it is seen that

$$1 + y \leq e^y \leq (1 - y)^{-1}$$

for $0 \leq y < 1$. Let $y = t/n$, then we have

$$\left(1 + \frac{t}{n}\right)^{-n} \geq e^{-t} \geq \left(1 - \frac{t}{n}\right)^n \, ,$$

so

$$0 \leq e^{-t} - \left(1 - \frac{t}{n}\right)^n = e^{-t} \left\{ 1 - e^t \left(1 - \frac{t}{n}\right)^n \right\}$$

$$\leq e^{-t} \left\{ 1 - \left(1 - \frac{t^2}{n^2}\right)^n \right\} \, ;$$

in the last step, the inequality $e^t \geq (1 + t/n)^n$ has been invoked. By induction, it can be shown that, if $0 \leq \alpha \leq 1$, we have $(1 - \alpha)^n \geq 1 - n\alpha$, whence

$$1 - \left(1 - \frac{t^2}{n^2}\right)^n \leq \frac{t^2}{n} \, ,$$

which proves (2).

In $P_n(z)$, let $t = n\tau$. Integrating by parts n times and noting that $\mathrm{Re}(z) > 0$, we have

$$P_n(z) = n^z \int_0^1 (1 - \tau)^n \tau^{z-1} d\tau$$

$$= n^z \left[\frac{\tau^z}{z} (1 - \tau)^n \right]_0^1 + \frac{n^z n}{z} \int_0^1 (1 - \tau)^{n-1} \tau^z d\tau$$

$$= \ldots$$

$$= \frac{n^z n(n - 1) \ldots 2.1}{z(z + 1) \ldots (z + n - 1)} \int_0^1 \tau^{z+n-1} d\tau$$

$$= \frac{1.2 \ldots n}{z(z + 1) \ldots (z + n)} n^z \, ,$$

i.e.,

$$\Gamma(z) = \lim_{n \to \infty} \frac{1.2 \ldots n}{z(z+1) \ldots (z+n)} n^z . \tag{3}$$

Since $\lim_{n \to \infty} n/(z+n) = 1$, (3) can be written as

$$\Gamma(z) = \lim_{n \to \infty} \frac{1.2 \ldots (n-1)}{z(z+1) \ldots (z+n-1)} n^z . \tag{4}$$

The last factor n^z in (4) can be written as

$$n^z = \prod_{m=1}^{n-1} \left(1 + \frac{1}{m}\right)^z$$

while the factor before it can be written as

$$\frac{1}{z} \prod_{m=1}^{n-1} \left(1 + \frac{z}{m}\right)^{-1} .$$

Hence we have

$$\Gamma(z) = \frac{1}{z} \prod_{n=1}^{\infty} \left\{ \left(1 + \frac{z}{n}\right)^{-1} \left(1 + \frac{1}{n}\right)^z \right\} , \tag{5}$$

which is the *Euler's infinite product expression*. This form of representation holds good for any z, $z = -n$ being the poles, and may be taken as the general definition of $\Gamma(z)$.

3.4. Weierstrass' Infinite Product

The last factor of Eq. (3) of the previous section can be written as

$$n^z = e^{z \ln n} = \exp \left\{ z \left[\ln n - \sum_{m=1}^{n} \frac{1}{m} \right] \right\} \prod_{m=1}^{n} e^{z/m} .$$

Hence,

$$\frac{1}{\Gamma(z)} = z e^{\gamma z} \prod_{n=1}^{\infty} \left\{ \left(1 + \frac{z}{n}\right) e^{-z/n} \right\} , \tag{1}$$

where

$$\gamma = \lim_{n \to \infty} \left\{ \sum_{m=1}^{n} \frac{1}{m} - \ln n \right\}$$
$$= 0.57721\ 56649\ 01532\ 86060\ 651\ldots , \tag{2}$$

and is named the *Euler constant* [cf. Sec. 1.3, example 2]. The infinite product of (1) gives the value of $\Gamma(z)$ for any z and at the same time manifests that the singularities of $\Gamma(z)$ are simple poles: $z = 0, -1, -2, \ldots$; there are no zeros. This is the definition after Weierstrass and is thus called *Weierstrass' infinite product*.

From

$$\sum_{m=1}^{n} \frac{1}{m} = \sum_{m=1}^{n} \int_0^1 x^{m-1} dx = \int_0^1 \frac{1-x^n}{1-x} dx$$

$$= \int_0^1 \frac{1-(1-y)^n}{y} dy = \int_0^n \left\{ 1 - \left(1 - \frac{t}{n}\right)^n \right\} \frac{dt}{t} , \qquad (3)$$

we find

$$\sum_{m=1}^{n} \frac{1}{m} - \ln n = \int_0^n \left\{ 1 - \left(1 - \frac{t}{n}\right)^n \right\} \frac{dt}{t} - \int_1^n \frac{dt}{t}$$

$$= \int_0^1 \left\{ 1 - \left(1 - \frac{t}{n}\right)^n \right\} \frac{dt}{t} - \int_1^n \left(1 - \frac{t}{n}\right)^n \frac{dt}{t} .$$

Letting $n \to \infty$, we have

$$\gamma = \int_0^1 \frac{1-e^{-t}}{t} dt - \int_1^\infty \frac{e^{-t}}{t} dt . \qquad (4)$$

3.5. Relation between the Γ-Function and the Trigonometric Function

From the Weierstrass' definition of $\Gamma(z)$, we have

$$\Gamma(z)\Gamma(-z) = -\frac{1}{z^2} \prod_{n=1}^{\infty} \left\{ \left(1 + \frac{z}{n}\right) e^{-z/n} \right\}^{-1} \prod_{n=1}^{\infty} \left\{ \left(1 - \frac{z}{n}\right) e^{z/n} \right\}^{-1}$$

$$= -\frac{1}{z^2} \prod_{n=1}^{\infty} \left(1 - \frac{z^2}{n^2}\right)^{-1} .$$

But, by Eq. (3) of Sec. 1.7

$$\frac{\sin \pi z}{\pi z} = \prod_{n=1}^{\infty} \left(1 - \frac{z^2}{n^2}\right) . \qquad (1)$$

Hence,

$$\Gamma(z)\Gamma(-z) = -\frac{\pi}{z \sin \pi z} .$$

Further, by using Eq. (1) of Sec. 3.2, we obtain

$$\Gamma(z)\Gamma(1-z) = \frac{\pi}{\sin \pi z} , \tag{2}$$

or, written in a more symmetrical form,

$$\Gamma(1+z)\Gamma(1-z) = z!(-z)! = \frac{\pi z}{\sin \pi z} . \tag{3}$$

Putting $z = \frac{1}{2}$ in (2), we have $\{\Gamma(\frac{1}{2})\}^2 = \pi$, of which the square root is

$$\Gamma\left(\frac{1}{2}\right) = \sqrt{\pi} , \tag{4}$$

since $\Gamma(\frac{1}{2}) > 0$ by definition.

Again, putting $z = \frac{1}{2}$ in (1), we have

$$\frac{\pi}{2} = \prod_{n=1}^{\infty} \frac{(2n)^2}{(2n-1)(2n+1)} = \lim_{m \to \infty} \left(\prod_{n=1}^{m} \frac{2n}{2n-1}\right)^2 \frac{1}{2m+1} \tag{5}$$

which is *Walli's product.*

3.6. Multiplication Formula

We now prove the following multiplication formula:

$$\Gamma(z)\Gamma\left(z+\frac{1}{n}\right)\Gamma\left(z+\frac{2}{n}\right)\ldots\Gamma\left(z+\frac{n-1}{n}\right)$$
$$= (2\pi)^{\frac{1}{2}(n-1)} n^{\frac{1}{2}-nz}\Gamma(nz) . \tag{1}$$

Let

$$\phi = \frac{n^{nz}}{n\Gamma(nz)} \prod_{r=0}^{n-1} \Gamma\left(z+\frac{r}{n}\right) . \tag{2}$$

Applying the limiting formula, Eq. (4) of Sec. 3.3 we obtain

$$\phi = n^{nz-1} \frac{\prod\limits_{r=0}^{n-1} \lim\limits_{m \to \infty} \frac{1.2\ldots(m-1)}{(z+\frac{r}{n})(z+\frac{r}{n}+1)\ldots(z+\frac{r}{n}+m-1)} m^{z+\frac{r}{n}}}{\lim\limits_{m \to \infty} \frac{1.2\ldots(nm-1)}{nz(nz+1)\ldots(nz+nm-1)} (nm)^{nz}}$$

$$= n^{nz-1} \lim_{m \to \infty} \frac{[(m-1)!]^n m^{nz+\frac{1}{2}(n-1)} n^{nm}}{(nm-1)!(nm)^{nz}}$$

$$= \lim_{m \to \infty} \frac{[(m-1)!]^n m^{\frac{1}{2}(n-1)} n^{nm-1}}{(nm-1)!} , \tag{3}$$

which shows that ϕ is independent of z. Putting in (2) $z = 1/n$, we have

$$\phi = \prod_{r=0}^{n-1} \Gamma\left(\frac{r+1}{n}\right) = \prod_{r=1}^{n-1} \Gamma\left(\frac{r}{n}\right) = \prod_{r=1}^{n-1} \Gamma\left(1 - \frac{r}{n}\right) ;$$

in the last step, r has been replaced by $n - r$. Thus, with Eq. (2) of Sec. 3.5,

$$\phi^2 = \prod_{r=1}^{n-1} \Gamma\left(\frac{r}{n}\right) \Gamma\left(1 - \frac{r}{n}\right) = \pi^{n-1} \prod_{r=1}^{n-1} \left(\sin \frac{\pi r}{n}\right)^{-1} . \tag{4}$$

Let the roots of $z^n - 1 = 0$ be $z = e^{2\pi r i/n}$, $r = 0, 1, 2, \ldots, n - 1$. We have

$$\frac{z^n - 1}{z - 1} = \sum_{r=0}^{n-1} z^r = \prod_{r=1}^{n-1} (z - e^{2\pi r i/n}) . \tag{5}$$

Putting $z = 1$ gives

$$n = \prod_{r=1}^{n-1} (1 - e^{2\pi i r/n}) = \prod_{r=1}^{n-1} e^{\pi r i/n} \left(-2i \sin \frac{\pi r}{n}\right)$$

$$= e^{\frac{\pi}{2}(n-1)i} 2^{n-1} (-i)^{n-1} \prod_{r=1}^{n-1} \sin \frac{\pi r}{n} = 2^{n-1} \prod_{r=1}^{n-1} \sin \frac{\pi r}{n} . \tag{6}$$

Substituting in (4), we have

$$\phi^2 = \frac{(2\pi)^{n-1}}{n} . \tag{7}$$

Taking the square root and substituting in (2), we arrive at (1).

In (1), let $n = 2$, then we have

$$2^{2z-1} \Gamma(z) \Gamma\left(z + \frac{1}{2}\right) = \pi^{1/2} \Gamma(2z) , \tag{8}$$

which can also be written as

$$2^{2z} z! \left(z - \frac{1}{2}\right)! = \pi^{1/2} (2z)! . \tag{9}$$

3.7. Contour Integral

Consider the contour integral

$$I = \int_\infty^{(0+)} e^{-t} t^{z-1} dt \; ; \tag{1}$$

the contour starts from ∞ near the positive real axis, running leftwards in the upper half plane, encircling the origin once in the positive sense, down to the lower half plane, then running rightwards to ∞ near the positive real axis [Fig. 5].

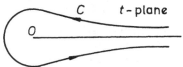

Fig. 5.

This contour integral exists for any value of z, and thus can be taken as a base for defining $\Gamma(z)$ in general. Let us find the relation between I and $\Gamma(z)$. Suppose first that z is confined in the domain $\text{Re}(z) > 0$ and is not an integer.

Deform the contour so that it consists of three parts: the first part starts from $t = +\infty$, running in the upper half plane along the real axis to the point $t = \delta (> 0)$, where δ could be arbitrarily small; the second part is a circle of radius δ around the origin in the positive sense; and the third part is a straight line running from $t = \delta$ in the lower half plane along the real axis to $t = +\infty$. We can write the integrals on the three parts of the contour as I_1, I_2, I_3. To specify the multi-valued function t^{z-1}, we take $\arg t = 0$ in I_1, then $\arg t$ is entirely determined in I_2 and I_3; in I_3, $\arg t = 2\pi$. These three integrals are then respectively

$$I_1 = \int_\infty^\delta e^{-t} t^{z-1} dt = -\int_\delta^\infty e^{-t} t^{z-1} dt \; ,$$

$$I_2 = \int_0^{2\pi} e^{-\delta e^{i\theta}} \left(\delta e^{i\theta}\right)^{z-1} \delta e^{i\theta} i d\theta$$

$$= \delta^z \int_0^{2\pi} e^{-\delta \cos\theta - i\delta \sin\theta + iz\theta} i d\theta \; ,$$

$$I_3 = e^{2\pi z i} \int_\delta^\infty e^{-t} t^{z-1} dt \; .$$

As we have assumed $\mathrm{Re}(z) > 0$, it is seen that $I_2 \to 0$, so $I_1 + I_3 \to I$ when $\delta \to 0$. Thus,

$$I = \left(e^{2\pi zi} - 1\right) \int_0^\infty e^{-t} t^{z-1} dt = \left(e^{2\pi zi} - 1\right) \Gamma(z)$$

$$= 2i e^{\pi zi} \sin \pi z \; \Gamma(z)$$

and we find

$$\Gamma(z) = -\frac{1}{2i \sin \pi z} \int_\infty^{(0+)} e^{-t} (-t)^{z-1} dt$$
$$(|\arg(-t)| < \pi) \; . \tag{2}$$

Although this relation is derived under the assumption $\mathrm{Re}(z) > 0$, but since the contour integral is unrestricted to the values of z, the condition $\mathrm{Re}(z) > 0$ can be negated according to the principle of analytic continuation.

However, (2) is invalid when z is an integer; for then the r.h.s. of it is an indeterminate form when z is a positive integer and becomes ∞ when z is a negative integer. But, by making use of Eq. (2) of Sec. 3.5, we have

$$\frac{1}{\Gamma(1-z)} = -\frac{1}{2\pi i} \int_\infty^{(0+)} e^{-t} (-t)^{z-1} dt \quad (|\arg(-t)| < \pi) \; , \tag{3}$$

which holds for any z, including z being an integer.

In (3), replacing $1 - z$ by z, we obtain

$$\frac{1}{\Gamma(z)} = -\frac{1}{2\pi i} \int_\infty^{(0+)} e^{-t} (-t)^{-z} dt \quad (|\arg(-t)| < \pi) \; . \tag{4}$$

Changing t to $-t$ gives

$$\frac{1}{\Gamma(z)} = \frac{1}{2\pi i} \int_{-\infty}^{(0+)} e^t t^{-z} dt \quad (|\arg t| < \pi) \; ; \tag{5}$$

the contour starts from $t = -\infty$ on the negative real axis, encircling the origin once in the positive sense and then goes back to the starting point.

The contour integrals (4) and (5) hold good for any values of z and thus can be taken as the general expression for $\Gamma(z)$.

Also, as in Eq. (2) of Sec. 3.1, the contours in the above formulae (2) – (5) can be rotated as a whole around the origin by an angle α without affecting the values of the integrals as long as $|\alpha| < \pi/2$. For example, from (5) we have

$$\frac{1}{\Gamma(z)} = \frac{1}{2\pi i}\int_{-\infty e^{\iota\alpha}}^{(0+)} e^t t^{-z}\,dt$$

$$(|\arg t - \alpha| < \pi)\ . \tag{6}$$

3.8. Euler Integral of the First Kind. B-Function

The Euler integral of the first kind, $B(p, q)$, is

$$B(p, q) = \int_0^1 x^{p-1}(1 - x)^{q-1}\,dx\ ; \tag{1}$$

the integral exists for $\mathrm{Re}(p) > 0$, $\mathrm{Re}(q) > 0$. By the transformation $x = 1 - t$, it can be shown that

$$B(p, q) = B(q, p)\ . \tag{2}$$

$B(p, q)$ can be expressed by the Γ-functions:

$$B(p, q) = \frac{\Gamma(p)\Gamma(q)}{\Gamma(p + q)}\ . \tag{3}$$

To prove it, consider the product

$$\Gamma(p)\Gamma(q) = \int_0^\infty e^{-u}u^{p-1}\,du \int_0^\infty e^{-v}v^{q-1}\,dv\ .$$

Putting $u = x^2, v = y^2$ leads to

$$\Gamma(p)\Gamma(q) = 4\int_0^\infty e^{-x^2}x^{2p-1}\,dx \int_0^\infty e^{-y^2}y^{2q-1}\,dy$$

$$= 4\int_0^\infty \int_0^\infty e^{-(x^2+y^2)}x^{2p-1}y^{2q-1}\,dx\,dy\ .$$

Using the plane polar coordinates r, θ given by $x = r\cos\theta, y = r\sin\theta$, it becomes

$$\Gamma(p)\Gamma(q) = 4\int_0^\infty e^{-r^2}r^{2(p+q)-1}\,dr \int_0^{\pi/2} (\cos\theta)^{2p-1}(\sin\theta)^{2q-1}\,d\theta\ . \tag{4}$$

Putting $r^2 = t$ in the first integral, we have

$$\int_0^\infty e^{-r^2}r^{2(p+q)-1}\,dr = \frac{1}{2}\int_0^\infty e^{-t}t^{p-q-1}\,dt = \frac{1}{2}\Gamma(p + q)\ . \tag{5}$$

In the second integral, let $\cos\theta = x$, then

$$\int_0^{\pi/2} (\cos\theta)^{2p-1}(\sin\theta)^{2q-1}d\theta = \frac{1}{2}\int_0^1 x^{p-1}(1-x)^{q-1}dx$$

$$= \frac{1}{2}B(p,q) .\qquad(6)$$

Substituting (5) and (6) into (4), we obtain $\Gamma(p)\Gamma(q) = \Gamma(p+q)B(p,q)$, which proves (3).

From (3), by the principle of analytic continuation, we obtain the function $B(p,q)$, called the *B-function*, which is not subjected to the restrictions $\mathrm{Re}(p) > 0$ and $\mathrm{Re}(q) > 0$.

In (6), replacing $2p$ and $2q$ respectively by $p+1$ and $q+1$, then applying (3), we obtain

$$\int_0^{\pi/2} (\cos\theta)^p(\sin\theta)^q d\theta = \frac{\Gamma\left(\frac{p+1}{2}\right)\Gamma\left(\frac{q+1}{2}\right)}{2\Gamma\left(\frac{p+q}{2}+1\right)} .\qquad(7)$$

In (5), changing $2(p+q)-1$ into p, it results

$$\int_0^\infty e^{-r^2}r^p dr = \frac{1}{2}\Gamma\left(\frac{p+1}{2}\right) .\qquad(8)$$

A special case of it is $(p=0)$

$$\int_0^\infty e^{-r^2} dr = \frac{1}{2}\Gamma\left(\frac{1}{2}\right) = \frac{1}{2}\sqrt{\pi} .\qquad(9)$$

In (1), putting $x = t/(1+t)$ and making use of (3), we have

$$B(p,q) = \int_0^\infty \frac{t^{p-1}dt}{(1+t)^{p+q}} = \frac{\Gamma(p)\Gamma(q)}{\Gamma(p+q)} .\qquad(10)$$

Putting $q = 1-p$ and assuming $0 < \mathrm{Re}(p) < 1$, we obtain, with Eq. (2) of Sec. 3.5,

$$\int_0^\infty \frac{t^{p-1}}{1+t}dt = \Gamma(p)\Gamma(1-p) = \frac{\pi}{\sin\pi p} ,\qquad(11)$$

which, of course, can also be obtained by invoking the residue theorem.

Now, let us prove the formula

$$\frac{(p+q)!}{p!q!} = \frac{\Gamma(p+q+1)}{\Gamma(p+1)\Gamma(q+1)} = \frac{1}{2\pi i}\int_{-\infty}^{(0+)} t^{-p-1}(1-t)^{-q-1}dt ,\qquad(12)$$

where $\text{Re}(p+q+1) > 0$ to ensure the convergence of the integral at the lower limit $t = -\infty; t = 1$ lies outside the contour.

It is readily shown [cf. Sec. 3.7] that, when $\text{Re}(p) < 0$, the r.h.s. of (12) (along the upper and lower banks of the negative real axis, setting respectively $t = xe^{\pi i}$ and $t = xe^{-\pi i}$) is equal to

$$-\frac{e^{p\pi i} - e^{-p\pi i}}{2\pi i} \int_0^\infty x^{-p-1}(1 + x)^{-q-1}dx = -\frac{\sin p\pi}{\pi}\frac{\Gamma(-p)\Gamma(p + q + 1)}{\Gamma(q + 1)} \,,$$

(10) being used in the last step. Applying (11), replacing p in it by $-p$, we have (12); the restriction $\text{Re}(p) < 0$ can be removed according to the principle of analytic continuation.

3.9. Double-Contour Integral

We shall prove the following formula:

$$\frac{1}{(2\pi i)^2} \int_P^{(1+,0+,1-,0-)} t^{\alpha-1}(1 - t)^{\beta-1}dt = \frac{e^{(\alpha+\beta)\pi i}}{\Gamma(1 - \alpha)\Gamma(1 - \beta)\Gamma(\alpha + \beta)} \; ; \quad (1)$$

the contour is shown in Fig. 6: P is a point lying between 0 and 1, at which $\arg t = \arg(1 - t) = 0$; starting from P the contour encircles $t = 1$ once in the positive sense first, then encircles $t = 0$ once in the positive sense, then $t = 1$

$$t\text{-plane}$$

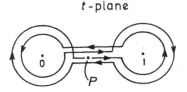

Fig. 6.

in the negative sense, and lastly, $t = 0$ in the negative sense, back to P. Such contour is called a *Pochhammer contour*.

If $\text{Re}(\alpha)$ and $\text{Re}(\beta) > 0$, then the values of the integrals around 0 and 1 approach zero as the radii of the circles tend to zero, and hence the integral on the r.h.s. of (1) reduces to several integrals over the straight line from 0 to 1:

$$\frac{1}{(2\pi i)^2}\left\{\int_0^1 t^{\alpha-1}(1-t)^{\beta-1}dt - e^{2\beta\pi i}\int_0^1 t^{\alpha-1}(1-t)^{\beta-1}dt\right.$$

$$\left. + e^{2(\alpha+\beta)\pi i}\int_0^1 t^{\alpha-1}(1-t)^{\beta-1}dt - e^{2\alpha\pi i}\int_0^1 t^{\alpha-1}(1-t)^{\beta-1}dt\right\}$$

$$= \frac{(1-e^{2\alpha\pi i})(1-e^{2\beta\pi i})}{(2\pi i)^2}\int_0^1 t^{\alpha-1}(1-t)^{\beta-1}dt$$

$$= \frac{e^{(\alpha+\beta)\pi i}\sin\alpha\pi\sin\beta\pi}{\pi^2}\frac{\Gamma(\alpha)\Gamma(\beta)}{\Gamma(\alpha+\beta)}$$

$$= \frac{e^{(\alpha+\beta)\pi i}}{\Gamma(1-\alpha)\Gamma(1-\beta)\Gamma(\alpha+\beta)} ,$$

which proves (1). But, as the two sides of (1) are both analytic functions of α and β, the result is not subjected to the conditions imposed in the proof, namely $\mathrm{Re}(\alpha) > 0$, $\mathrm{Re}(\beta) > 0$.

3.10. Dirichlet Integral

The Dirichlet integral is

$$I = \int\int\cdots\int f(t_1 + t_2 + \ldots + t_n)t_1^{\alpha_1-1}t_2^{\alpha_2-1}\ldots t_n^{\alpha_n-1}dt_1\,dt_2\ldots dt_n , \quad (1)$$

where the integration limits are $t_r \geq 0$ $(r = 1, 2, \ldots, n)$, $\sum_{r=1}^n t_r \leq 1$; f is a continuous function. To ensure the convergence of the integrals at $t_r = 0$, $\mathrm{Re}(\alpha_r)$ has to be positive.

We calculate first the iterated integral over t_1 and t_2. Let $\lambda = t_3 + t_4 + \ldots + t_n$, $\tau_2 = t_1 + t_2$, then

$$\int_0^{1-\lambda} dt_2 \int_0^{1-\lambda-t_2} dt_1 f(t_1 + t_2 + \lambda)t_1^{\alpha_1-1}t_2^{\alpha_2-1}$$

$$= \int_0^{1-\lambda} dt_2 \int_{t_2}^{1-\lambda} d\tau_2 f(\tau_2 + \lambda)(\tau_2 - t_2)^{\alpha_1-1}t_2^{\alpha_2-1} .$$

Interchanging the order of integrations and then putting $t_2 = \tau_2 t$, we obtain

$$\int_0^{1-\lambda} d\tau_2 \int_0^{\tau_2} dt_2 f(\tau_2 + \lambda)(\tau_2 - t_2)^{\alpha_1-1}t_2^{\alpha_2-1}$$

$$= \int_0^{1-\lambda} d\tau_2 f(\tau_2 + \lambda)\tau_2^{\alpha_1+\alpha_2-1}\int_0^1 dt(1-t)^{\alpha_1-1}t^{\alpha_2-1}$$

$$= \frac{\Gamma(\alpha_1)\Gamma(\alpha_2)}{\Gamma(\alpha_1+\alpha_2)}\int_0^{1-\lambda} d\tau_2 f(\tau_2 + \lambda)\tau_2^{\alpha_1+\alpha_2-1} .$$

Thus, the number of integrations is reduced by one while the form of the multiple integral is not altered. Applying again this process of reduction to τ_2 and t_3, the multiplicity of the integral is reduced by one once more; only a factor equal to

$$\frac{\Gamma(\alpha_1)\Gamma(\alpha_2)}{\Gamma(\alpha_1+\alpha_2)}\frac{\Gamma(\alpha_1+\alpha_2)\Gamma(\alpha_3)}{\Gamma(\alpha_1+\alpha_2+\alpha_3)} = \frac{\Gamma(\alpha_1)\Gamma(\alpha_2)\Gamma(\alpha_3)}{\Gamma(\alpha_1+\alpha_2+\alpha_3)}$$

occurs before the integral sign. Following this process of reduction, we shall arrive finally at the formula

$$\int\int\cdots\int f\left(\sum_{r=1}^{n}t_r\right)\prod_{r=1}^{n}t_r^{\alpha_r-1}dt_1dt_2\ldots dt_n$$
$$= \frac{\Gamma(\alpha_1)\Gamma(\alpha_2)\ldots\Gamma(\alpha_n)}{\Gamma(\alpha_1+\alpha_2+\ldots+\alpha_n)}\int_0^1 f(\tau)\tau^{\alpha_1+\alpha_2+\ldots+\alpha_n-1}d\tau ;$$

(2)

the limits of integration on the l.h.s. are $t_r \geq 0$ $(r = 1, 2, \ldots, n)$, $\sum_{r=1}^{n} t_r \leq 1$; and $\mathrm{Re}(\alpha_r) > 0$.

3.11. Logarithmic Derivative of the Γ-Function

Let

$$\psi(z) = \frac{d}{dz}\ln\Gamma(z) = \frac{\Gamma'(z)}{\Gamma(z)} .$$

(1)

From Eq. (2) of Sec. 3.2, taking the logarithmic derivative, we have

$$\psi(z+1) = \psi(z) + \frac{1}{z} .$$

(2)

From Eq. (2) of Sec. 3.5, taking the logarithmic derivative, we have

$$\psi(1-z) = \psi(z) + \pi\cot\pi z .$$

(3)

Taking the logarithmic derivative of Eq. (3) of Sec. 3.2, we have

$$\psi(z+n) = \psi(z) + \sum_{r=0}^{n-1}\frac{1}{z+r} ;$$

(4)

(2) is its special case.

By Eq. (1) of Sec. 3.4, we find

$$\psi(z) = -\gamma - \frac{1}{z} + \sum_{n=1}^{\infty} \left(\frac{1}{n} - \frac{1}{z+n} \right) . \tag{5}$$

Applying Eq. (2) of Sec. 3.4, and writing the series on the r.h.s. as a limit, we obtain

$$\psi(z) = -\frac{1}{z} + \lim_{m \to \infty} \left\{ \ln m - \sum_{n=1}^{m} \frac{1}{z+n} \right\} . \tag{6}$$

If $\mathrm{Re}(p) > 0$, we have

$$\frac{1}{p} = \int_0^{\infty} e^{-pt} dt . \tag{7}$$

Integrating with respect to p from $p = 1$ to $p = m$, the result is

$$\ln m = \int_0^{\infty} (e^{-t} - e^{-mt}) \frac{dt}{t} . \tag{8}$$

Substituting (7) and (8) in (6), we have

$$\psi(z) = \lim_{m \to \infty} \left\{ \int_0^{\infty} (e^{-t} - e^{-mt}) \frac{dt}{t} - \sum_{n=0}^{m} \int_0^{\infty} e^{-(z+n)t} dt \right\}$$

$$= \lim_{m \to \infty} \left\{ \int_0^{\infty} (e^{-t} - e^{-mt}) \frac{dt}{t} - \int_0^{\infty} \frac{e^{-zt} \left(1 - e^{-(m+1)t} \right)}{1 - e^{-t}} dt \right\} .$$

When $m \to \infty$, it can be shown that the values of the integrals containing the factor e^{-mt} approach zero. Thus,

$$\psi(z) = \int_0^{\infty} \left\{ \frac{e^{-t}}{t} - \frac{e^{-zt}}{1 - e^{-t}} \right\} dt . \tag{9}$$

Substituting (5) into the r.h.s. of (2), we get

$$\psi(z+1) = -\gamma + \sum_{n=1}^{\infty} \left(\frac{1}{n} - \frac{1}{z+n} \right) . \tag{10}$$

Putting $z = 0$, it results

$$\psi(1) = -\gamma . \tag{11}$$

Substituting in (1) and noting that $\Gamma(1) = 1$, we have

$$\Gamma'(1) = \psi(1) = -\gamma . \tag{12}$$

Putting $z = 1$ in (9), using (11), we have

$$\gamma = \int_0^\infty \left\{ \frac{1}{1 - e^{-t}} - \frac{1}{t} \right\} e^{-t} dt \ . \tag{13}$$

Replacing, in this integral, the lower limit 0 by δ and changing the variable in the first integral from t to u by the transformation $e^{-t} = u$, then setting $1 - u = v$, combining the second integral and letting $\delta \to 0$, we see that (13) is the same as Eq. (4) of Sec. 3.4.

Adding (13) to (9), we have

$$\psi(z) = -\gamma + \int_0^\infty \frac{e^{-t} - e^{-zt}}{1 - e^{-t}} dt \ . \tag{14}$$

Putting $m = z$ in (8) and subtracting (9) from it, we have

$$\psi(z) = \ln z + \int_0^\infty \left\{ \frac{1}{t} - \frac{1}{1 - e^{-t}} \right\} e^{-zt} dt \ . \tag{15}$$

Putting $p = z$ in (7), dividing by 2 and adding the result to (15), there results

$$\psi(z) = \ln z - \frac{1}{2z} + \int_0^\infty \left\{ \frac{1}{2} + \frac{1}{t} - \frac{1}{1 - e^{-t}} \right\} e^{-zt} dt \ . \tag{16}$$

Integrating with respect to z from $z = 1$ to z, we obtain

$$\ln \Gamma(z) = \left(z - \frac{1}{2} \right) \ln z - z + 1$$
$$+ \int_0^\infty \left\{ \frac{1}{2} + \frac{1}{t} - \frac{1}{1 - e^{-t}} \right\} \frac{e^{-t} - e^{-zt}}{t} dt \ . \tag{17}$$

The integral in (17), which is independent of z, can be evaluated as follows: Let

$$I = \int_0^\infty \left\{ \frac{1}{2} + \frac{1}{t} - \frac{1}{1 - e^{-t}} \right\} \frac{e^{-t}}{t} dt \ . \tag{18}$$

Putting $z = \frac{1}{2}$ in (17) and applying Eq. (4) of Sec. 3.5, we have

$$I - J = \frac{1}{2} \ln \pi - \frac{1}{2} \ , \tag{19}$$

where

$$J = \int_0^\infty \left\{ \frac{1}{2} + \frac{1}{t} - \frac{1}{1 - e^{-t}} \right\} \frac{e^{-t/2}}{t} dt \ . \tag{20}$$

Replacing t by $t/2$ in (18), we see that

$$I = \int_0^\infty \left\{ \frac{1}{2} + \frac{2}{t} - \frac{1}{1 - e^{-t/2}} \right\} \frac{e^{-t/2}}{t} dt \ .$$

Subtracting (20) from it gives

$$I - J = \int_0^\infty \left\{ \frac{1}{t} - \frac{e^{-t/2}}{1 - e^{-t}} \right\} \frac{e^{-t/2}}{t} dt \ .$$

Using (18) again, we have

$$J = \int_0^\infty \left\{ \left(\frac{1}{2} + \frac{1}{t} \right) e^{-t} - \frac{e^{-t/2}}{t} \right\} \frac{dt}{t} \ .$$

Integrating by parts the integrals in the last two terms, we obtain

$$J = -\frac{e^{-t} - e^{-t/2}}{t} \Bigg|_0^\infty - \frac{1}{2} \int_0^\infty \frac{e^{-t} - e^{-t/2}}{t} dt$$

$$= -\frac{1}{2} - \frac{1}{2} \ln \frac{1}{2} \ ,$$

(8) being used in the last step. Substituting the result in (19), we get

$$I = \frac{1}{2} \ln(2\pi) - 1 \ . \tag{21}$$

. Substituting in (17) leads to

$$\ln \Gamma(z) = \left(z - \frac{1}{2} \right) \ln z - z + \frac{1}{2} \ln(2\pi)$$

$$- \int_0^\infty \left\{ \frac{1}{2} + \frac{1}{t} - \frac{1}{1 - e^{-t}} \right\} \frac{e^{-zt}}{t} dt \ ,$$

$$(\mathrm{Re}(z) > 0) \tag{22}$$

which is called the *first formula of Binet.*

There is also the *second formula of Binet:*

$$\ln \Gamma(z) = \left(z - \frac{1}{2} \right) \ln z - z + \frac{1}{2} \ln(2\pi)$$

$$+ 2 \int_0^\infty \frac{\arctan(t/z)}{e^{2\pi t} - 1} dt \ . \tag{23}$$

[See Eq. (4) of Sec. 3.18].

3.12. Asymptotic Expansions

From Eq. (18) of Sec. 1.3, changing t to $-t$, we have

$$\frac{-t}{e^{-t}-1} = 1 + \frac{t}{2} + \sum_{r=1}^{n} \frac{(-)^{r-1}B_r}{(2r)!}t^{2r} + \frac{t^{2n+2}}{e^{-t}-1}\int_0^1 P_{2n+1}(x)e^{-tx}dx . \quad (1)$$

Substituting into the integral in Eq. (22) of the previous section, the contribution of the series term $\sum_r$ is

$$\sum_{r=1}^{n} \frac{(-)^{r-1}B_r}{(2r)!}\int_0^\infty t^{2r-2}e^{-zt}dt = \sum_{r=1}^{n} \frac{(-)^{r-1}B_r}{(2r)!}\frac{(2r-2)!}{z^{2r-1}} .$$

The contribution of the integral term may be calculated by Eq. (15) of Sec. 1.3

$$\left|\int_0^1 P_{2n+1}(x)e^{-tx}dx\right| \le \frac{4}{(2\pi)^{2n+1}}\int_0^1 e^{-tx}dx = \frac{4}{(2\pi)^{2n+1}}\frac{e^{-t}-1}{-t} ;$$

it is of the order $O(|z|^{-2n})$. This holds for any n, so we obtain for $\Gamma(z)$ the asymptotic expansion:

$$\ln\Gamma(z) = \left(z - \frac{1}{2}\right)\ln z - z + \frac{1}{2}\ln(2\pi)$$
$$+ \sum_{r=1}^{n} \frac{(-)^{r-1}B_r}{2r(2r-1)}z^{-2r+1} + O(z^{-2n-1}) . \quad (2)$$

By the relation $\Gamma(z+1) = z\Gamma(z) = z!$, (2) may be written as

$$\ln z! = \left(z + \frac{1}{2}\right)\ln z - z + \frac{1}{2}\ln(2\pi)$$
$$+ \sum_{r=1}^{n} \frac{(-)^{r-1}B_r}{2r(2r-1)}z^{-2r+1} + O(z^{-2n-1}) , \quad (3)$$

or,

$$\ln z! \sim z(\ln z - 1) + \frac{1}{2}\ln(2\pi z)$$
$$+ \frac{1}{12z} - \frac{1}{360z^3} + \frac{1}{1260z^5} - \frac{1}{1680z^7} + \cdots . \quad (4)$$

Taking its exponential function, we have

$$z! \sim z^z e^{-z} (2\pi z)^{1/2} \left\{ 1 + \frac{1}{12z} + \frac{1}{288z^2} - \frac{139}{51840z^3} - \frac{571}{24\,88320z^4} \right. $$
$$\left. + \ldots \right\} . \tag{5}$$

These asymptotic expansions are derived under the condition $\text{Re}(z) > 0$, as this is the condition for the convergence of the integral in (22) of the last section. This condition is equivalent to $|\arg z| < \pi/2$. We shall release it in Sec. 3.13 and Sec. 3.21. (3), (4) or (5) is called *Stirling's formula*.

Differentiating (2) gives

$$\psi(z) = \ln z - \frac{1}{2z} - \sum_{r=1}^{n} \frac{(-)^{r-1} B_r}{2r} z^{-2r} + O(z^{-2n-2}) , \tag{6}$$

which may also be derived directly by substituting (1) into Eq. (16) of the last section.

3.13. Another Derivation of the Asymptotic Expansion

In the Euler expansion formula, Eq. (6) of Sec. 1.3, let $F(x) = \ln(x+1)$ and take $a = 0, h = 1, m = n - 1$. Using the remainder in Eq. (10) of that section, we have

$$\ln 1 + \ln 2 + \ldots + \ln n = \int_1^n \ln x \, dx + \frac{1}{2} \ln n + \int_1^n \frac{P_1(x)}{x} dx ,$$

or,

$$\ln n! = \left(n + \frac{1}{2}\right) \ln n - (n-1) + \int_1^n \frac{P_1(x)}{x} dx . \tag{1}$$

Let

$$I_n = \int_1^n \frac{P_1(x)}{x} dx , \tag{2}$$

then, from (1), we have

$$\ln \prod_{r=1}^{n} (2r)^2 = 2n \ln 2 + 2 \ln n!$$
$$= (2n+1) \ln(2n) - 2n - \ln 2 + 2 + 2I_n ,$$

also,

$$\ln(2n+1)! = \left(2n + \frac{3}{2}\right) \ln(2n+1) - 2n + I_{2n+1} .$$

By subtraction, we obtain

$$\ln \left\{ \prod_{r=1}^{n} \left(\frac{2r}{2r-1} \right) \frac{1}{2n+1} \right\}$$

$$= (2n+1) \ln \left(1 - \frac{1}{2n+1} \right) - \frac{1}{2} \ln(2n+1) - \ln 2 + 2 + 2I_n - I_{2n+1} .$$

Multiplying with 2, it gives

$$\ln \left\{ \prod_{r=1}^{n} \left(\frac{2r}{2r-1} \right)^2 \frac{1}{2n+1} \right\}$$

$$= 2(2n+1) \ln \left(1 - \frac{1}{2n+1} \right) - 2\ln 2 + 4 + 4I_n - 2I_{2n+1} .$$

Let $n \to \infty$ and use Wallis' formula Eq. (5) of Sec. 3.5. We obtain

$$\ln \frac{\pi}{2} = 2 - 2\ln 2 + 2I_\infty ,$$

so,

$$I_\infty = \frac{1}{2} \ln(2\pi) - 1 . \tag{3}$$

Substituting in (1) and noting that $I_n = I_\infty - \int_n^\infty P_1(x)x^{-1}dx$, we find

$$\ln n! = \left(n + \frac{1}{2} \right) \ln n - n + \frac{1}{2} \ln(2\pi) - \int_n^\infty \frac{P_1(x)}{x} dx . \tag{4}$$

Integrating by parts, using Eqs. (11) and (12) of Sec. 1.3 and noting that $P_1(x)$ is a periodic function of period 1 and is equal to the Bernoulli polynomial in the interval $(0,1)$, we arrive at

$$\ln n! = \left(n + \frac{1}{2} \right) \ln n - n + \frac{1}{2} \ln(2\pi)$$

$$+ \sum_{r=1}^{k} \frac{(-)^{r-1} B_r}{2r(2r-1)n^{2r-1}} - (2k)! \int_n^\infty \frac{P_{2k+1}(x)}{x^{2k+1}} dx , \tag{5}$$

which is the same as Eq. (3) of Sec. 3.12 but n is now a positive integer.

In order to find a general asymptotic expansion formula, we put, in Eq. (6) of Sec. 1.3, $F(x) = \ln(x+z)$, $z \neq$ negative real number, and take as before, $a = 0, h = 1$, but $m = n$ now; the result is

$$\ln z + \ln(z+1) + \ldots + \ln(z+n)$$

$$= (z+n)\ln(z+n) - n - z\ln z + \frac{1}{2}[\ln(z+n) + \ln z] + \int_0^n \frac{P_1(x)}{x+z} dx . \tag{6}$$

Deducting (6) from (4), we obtain

$$\ln \frac{n!n^z}{z(z+1)\ldots(z+n)} = \left(z - \frac{1}{2}\right)\ln z - \left(z + n + \frac{1}{2}\right)\ln \frac{z+n}{n} + \frac{1}{2}\ln(2\pi)$$
$$- \int_0^n \frac{P_1(x)}{x+z}dx - \int_n^\infty \frac{P_1(x)}{x}dx \ .$$

Let $n \to \infty$. With Eq. (3) of Sec. 3.3, we have

$$\ln \Gamma(z) = \left(z - \frac{1}{2}\right)\ln z - z + \frac{1}{2}\ln(2\pi) - \int_0^\infty \frac{P_1(x)}{x+z}dx \ . \tag{7}$$

Integrating by parts gives

$$\ln \Gamma(z) = \left(z - \frac{1}{2}\right)\ln z - z + \frac{1}{2}\ln(2\pi)$$
$$+ \sum_{r=1}^n \frac{(-)^{r-1}B_r}{2r(2r-1)}z^{-2r+1} - (2n)! \int_0^\infty \frac{P_{2n+1}(x)}{(x+z)^{2n+1}}dx \ , \tag{8}$$

where z is not a negative real number, i.e., $|\arg z| < \pi$. This formula is the same as Eq. (2) of Sec. 3.12 (the integral in (8) is of order $O(z^{-2n})$) but more general since its range of validity is larger, $|\arg z| < \pi$.

3.14. Riemann ς-Function

Riemann ς-function is defined by

$$\varsigma(s) = \sum_{n=1}^\infty \frac{1}{n^s} \ , \tag{1}$$

which is valid when $\mathrm{Re}(s) = \sigma > 1$, since this is the condition for the convergence of the series. It is common to express s as $s = \sigma + it$, σ and t being real numbers.

The generalized ς-function is defined by

$$\varsigma(s, a) = \sum_{n=0}^\infty \frac{1}{(n+a)^s} \ , \tag{2}$$

where a is a constant, not equal to a negative integer, and $\mathrm{Re}(s) > 1$. For simplicity, we assume that a is a real number satisfying $0 < a \leq 1$, and that $\arg(n+a) = 0$. Evidently, $\varsigma(s) = \varsigma(s, 1)$. From

$$(n+a)^{-s}\Gamma(s) = \int_0^\infty x^{s-1}e^{-(n+a)x}dx \ ,$$

we obtain

$$\Gamma(s)\varsigma(s,a) = \int_0^\infty \frac{x^{s-1}e^{-ax}}{1-e^{-x}}dx \; ; \tag{3}$$

the exchange of the summation and the integration can be shown to be legitimate. This result also holds good for complex number a, so long as $\mathrm{Re}(a) > 0$ and $\mathrm{Re}(s) > 1$ at the same time, which are respectively the conditions for the convergence of the integral at its lower and upper limits.

Using the method in Sec. 3.7, it can be shown that

$$\varsigma(s,a) = -\frac{\Gamma(1-s)}{2\pi i}\int_\infty^{(0+)} \frac{(-z)^{s-1}e^{-az}}{1-e^{-z}}dz \; , \tag{4}$$

where $|\arg(-z)| < \pi$; no singularities of the integrand, namely, $z = 2n\pi i (n = \pm 1, \pm 2, \ldots)$, lie inside the contour. The integral in (4) is a single-valued analytic function for all values of s. Hence, the limitation $\mathrm{Re}(s) = \sigma > 1$ imposed on (4) can be negated. The result holds also for a being a complex number, so long as $\mathrm{Re}(a) > 0$.

We see from (4) that the singularities of $\varsigma(s,a)$ can only be those of $\Gamma(1-s)$, i.e., $s = 1, 2, \ldots,$[a] all of which are simple poles. But we already know that $\varsigma(s,a)$ has no singularities in $\mathrm{Re}(s) > 1$. Therefore, the only singular point of $\varsigma(s,a)$ is a simple pole at $s = 1$. When $s = 1$, the integral in (4) is

$$\frac{1}{2\pi i}\int_\infty^{(0+)} \frac{e^{-az}}{1-e^{-z}}dz \; ,$$

which is equal to the residue of the integrand at $z = 0$, i.e., equal to 1, and hence

$$\lim_{s\to\infty} \frac{\varsigma(s,a)}{\Gamma(1-s)} = -1 \; . \tag{5}$$

Now, the residue of $\Gamma(1-s)$ at $s = 1$ is -1; therefore, that of $\varsigma(s,a)$ at $s = 1$ is 1.

3.15. The Functional Equation of the ς-Function

Hurwitz has given a proof of the following formula:

$$\varsigma(s,a) = \frac{2\Gamma(1-s)}{(2\pi)^{1-s}}\left\{ \sin\frac{s\pi}{2}\sum_{n=1}^\infty n^{s-1}\cos(2n\pi a) \right.$$

$$\left. + \cos\frac{s\pi}{2}\sum_{n=1}^\infty n^{s-1}\sin(2n\pi a) \right\} \; ; \tag{1}$$

---

[a] When $s = 2, 3, \ldots$, the r.h.s. of (4) is an indeterminate form $(\infty, 0)$.

the series on the r.h.s. is convergent under the condition $\mathrm{Re}(s) < 0$.

To prove (1), consider a contour C, which is a circle with the origin as centre and of radius equal to $(2N+1)\pi$, N being a positive integer. In between C and the contour in (4) of the last section, the integrand $(-z)^{s-1}e^{-az}(1-e^{-z})^{-1}$ is single-valued and analytic, except for the simple poles $\pm 2n\pi i$ $(n = 1, 2, \ldots, N)$. Hence

$$\frac{1}{2\pi i} \int_C \frac{(-z)^{s-1}e^{-az}}{1-e^{-z}}\,dz - \frac{1}{2\pi i} \int_{(2N+1)\pi}^{(0+)} \frac{(-z)^{s-1}e^{-az}}{1-e^{-z}}\,dz$$

$$= \sum_{n=1}^{N}(R_n + R'_n) \, ,$$

where R_n and R'_n are respectively the residues at $z = 2n\pi i$ and $z = -2n\pi i$. The residues at $-z = 2n\pi e^{\pm \pi i/2}$ are $(2n\pi e^{\pm \pi i/2})^{s-1}e^{\pm 2n\pi ai}$ Therefore,

$$R_n + R'_n = (2n\pi)^{s-1}2\sin\left(\frac{s\pi}{2} + 2n\pi a\right) \, .$$

As it has been assumed that $0 < a \le 1$ and $\mathrm{Re}(s) < 0$, there must exist a positive number K, independent of N, such that $|e^{-az}(1-e^{-z})^{-1}| < K$ on C. Then, we have

$$\left|\frac{1}{2\pi i} \int_C \frac{(-z)^{s-1}e^{-az}}{1-e^{-z}}\,dz\right|$$

$$< \frac{K}{2\pi}\int_{-\pi}^{\pi}|[(2N+1)\pi]^s e^{is\theta}|\,d\theta < K[(2N+1)\pi]^\sigma e^{\pi|s|} \, ,$$

where $\sigma = \mathrm{Re}(s)$. For $\sigma < 0$, the r.h.s. $\to 0$ when $N \to \infty$, which proves (1).

Let $a = 1$ in (1), and replace s by $1 - s$. The result is

$$\varsigma(1 - s) = 2(2\pi)^{-s}\Gamma(s)\cos\frac{s\pi}{2}\cdot\varsigma(s) \, , \tag{2}$$

which is the functional equation of the ς-function. Since both sides of the equation are analytic functions except for isolated poles, it is not subjected to the restriction $\sigma < 0$ or $\sigma > 1$. The formula indicates that $s = -2m(m = 1, 2, \ldots)$ are zeros of $\varsigma(s)$.

From Eq. (2) of Sec. 3.5, we have

$$\cos\frac{s\pi}{2} = \sin\frac{(1-s)\pi}{2} = \frac{\pi}{\Gamma\left(\frac{1-s}{2}\right)\Gamma\left(\frac{1+s}{2}\right)} \, .$$

And, by Eq. (8) of Sec. 3.6, we have

$$\Gamma(s) = \pi^{-1/2} 2^{s-1} \Gamma\left(\frac{s}{2}\right) \Gamma\left(\frac{s+1}{2}\right) .$$

Substituting in (2) gives

$$\pi^{-s/2} \Gamma\left(\frac{s}{2}\right) \varsigma(s) = \pi^{-\frac{1-s}{2}} \Gamma\left(\frac{1-s}{2}\right) \varsigma(1-s) , \tag{3}$$

from which it is much easier to see that $s = -2m$ $(m = 1, 2, \ldots)$ are zeros of $\varsigma(s)$.

3.16. The Value of $\varsigma(s, a)$ when s is an Integer

If s is a negative integer, $-n$, the integrand in Eq. (4) of Sec. 3.14, namely, $(-z)^{s-1} e^{-az} (1 - e^{-z})^{-1}$ is single-valued, so the integral is equal to the residue at the origin. The value of this residue may be evaluated by employing the following series expression [Eq. (1) of Sec. 1.1]:

$$-\frac{z e^{-az}}{e^{-z} - 1} = \sum_{\nu=0}^{\infty} \frac{(-z)^{\nu}}{\nu!} \varphi_{\nu}(z) ,$$

where $\varphi_{\nu}(z)$ is the Bernoulli polynomial. From it we obtain

$$\varsigma(-n, a) = -\frac{\varphi_{n+1}(a)}{n+1} . \tag{1}$$

Putting $a = 1$, we have

$$\varsigma(-n) = -\frac{\varphi_{n+1}(1)}{n+1} .$$

Hence, by Eqs. (14) and (4) of Sec. 1.1,

$$\left. \begin{array}{l} \varsigma(0) = -\frac{1}{2}, \quad \varsigma(-2m) = 0 , \\ \varsigma(1 - 2m) = \frac{(-)^m B_m}{2m} \quad (m \geq 1) . \end{array} \right\} \tag{2}$$

Applying Eq. (2) of Sec. 3.15, we obtain

$$\varsigma(2m) = \frac{2^{2m-1} \pi^{2m} B_m}{(2m)!} . \tag{3}$$

Formula (1) holds also when a is a complex number with a real part greater than zero [see Sec. 3.14].

3.17. Hermite Formula

We prove first Plana's summation formula (m and n are both positive integers):

$$\sum_{k=m}^{n} \varphi(k) = \frac{1}{2}\{\varphi(m) + \varphi(n)\} + \int_{m}^{n} \varphi(x)dx$$

$$- i \int_{0}^{\infty} \frac{\varphi(n+iy) - \varphi(m+iy) - \varphi(n-iy) + \varphi(m-iy)}{e^{2\pi y} - 1} dy ,$$

$$(1)$$

where $\varphi(z)$ is a bounded analytic function in $m \leq \mathrm{Re}(z) \leq n$.

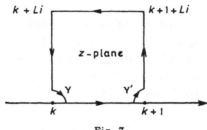

Fig. 7.

Proof: Consider the integration contour C as shown in Fig. 7; γ and γ' are circular arcs with centres respectively at $z = k$ and $z = k + 1$, and radii equal to ρ. Then,

$$\int_{C} \frac{\varphi(z)dz}{e^{-2\pi z i} - 1} = \int_{\gamma} + \int_{k+\rho}^{k+1-\rho} \frac{\varphi(x)dx}{e^{-2\pi x i} - 1} + \int_{\gamma'}$$

$$+ i \int_{\rho}^{L} \frac{\varphi(k+1+iy)dy}{e^{2\pi y} - 1} - \int_{k}^{k+1} \frac{\varphi(x+Li)dx}{e^{-2\pi x i + 2\pi L} - 1}$$

$$- i \int_{\rho}^{L} \frac{\varphi(k+iy)dy}{e^{2\pi y} - 1} = 0 .$$

When $\rho \to 0$,

$$\int_{\gamma} = - \int_{0}^{\pi/2} \frac{\varphi(k+\rho e^{i\theta})\rho e^{i\theta} id\theta}{e^{-2\pi i \rho e^{i\theta}} - 1} \to \frac{\varphi(k)}{2\pi} \int_{0}^{\pi/2} d\theta = \frac{1}{4}\varphi(k) ;$$

and similarly,

$$\int_{\gamma'} = - \int_{\pi/2}^{\pi} \frac{\varphi(k+1+\rho e^{i\theta})\rho e^{i\theta} id\theta}{e^{-2\pi i \rho e^{i\theta}} - 1} \to \frac{1}{4}\varphi(k+1) .$$

Let $L \to \infty$. We have

$$\frac{1}{4}\{\varphi(k) + \varphi(k+1)\} + \int_{k+\rho}^{k+1-\rho} \frac{\varphi(x)dx}{e^{-2\pi xi} - 1}$$

$$+ i\int_{\rho}^{\infty} \frac{\varphi(k+1+iy) - \varphi(k+iy)}{e^{2\pi y} - 1}dy + \varepsilon(\rho) = 0 , \tag{2}$$

where $\varepsilon(\rho) \to 0$ when $\rho \to 0$. Replacing i by $-i$ for the contour C, repeating the above procedure, we have

$$\frac{1}{4}\{\varphi(k) + \varphi(k+1)\} + \int_{k+\rho}^{k+1-\rho} \frac{\varphi(x)dx}{e^{2\pi xi} - 1}$$

$$- i\int_{\rho}^{\infty} \frac{\varphi(k+1-iy) - \varphi(k-iy)}{e^{2\pi y} - 1}dy + \eta(\rho) = 0 , \tag{3}$$

where $\eta(\rho) \to 0$ when $\rho \to 0$. Adding up (2) and (3) and putting $\rho \to 0$ in the result, we obtain

$$\frac{1}{2}\{\varphi(k) + \varphi(k+1)\} - \int_{k}^{k+1} \varphi(x)dx$$

$$+ i\int_{0}^{\infty} \frac{\varphi(k+1+iy) - \varphi(k+iy) - \varphi(k+1-iy) + \varphi(k-iy)}{e^{2\pi y} - 1}dy = 0 . \tag{4}$$

Summing this expression over k from m to $n-1$, we arrive at Plana's formula (1).

In (4), let $\varphi(z) = e^{-az}$, then set $k = 0$. The result is

$$\frac{1}{2}(1 + e^{-a}) - \int_{0}^{1} e^{-ax}dx$$

$$+ i\int_{0}^{\infty} \frac{e^{-a-iay} - e^{-iay} - e^{-a+iay} + e^{iay}}{e^{2\pi y} - 1}dy = 0 ,$$

from which we obtain

$$2\int_{0}^{\infty} \frac{\sin ay}{e^{2\pi y} - 1}dy = \frac{1}{2}\frac{e^a + 1}{e^a - 1} - \frac{1}{a} = \frac{1}{e^a - 1} - \frac{1}{a} + \frac{1}{2} . \tag{5}$$

Expressing $\sin ay$ in a power series, the l.h.s. of (5) can be expanded into

$$2\sum_{p=1}^{\infty} \frac{(-)^{p-1}a^{2p-1}}{(2p-1)!} \int_{0}^{\infty} \frac{y^{2p-1}dy}{e^{2\pi y} - 1} ,$$

while the r.h.s. can be expanded by Eq. (3) of Sec. 1.1 as

$$\sum_{p=1}^{\infty} (-)^{p-1} \frac{a^{2p-1}}{(2p)!} B_p .$$

So,

$$B_p = 4p \int_0^\infty \frac{y^{2p-1}}{e^{2\pi y} - 1} dy , \tag{6}$$

B_p being the Bernoulli number.

In (1), let $m = 0$ and $n \to \infty$. Assume that $\varphi(n) \to 0, \varphi(n \pm iy) \to 0$, then we have

$$\sum_{k=0}^{\infty} \varphi(k) = \frac{1}{2}\varphi(0) + \int_0^\infty \varphi(x) dx + i \int_0^\infty \frac{\varphi(iy) - \varphi(-iy)}{e^{2\pi y} - 1} dy . \tag{7}$$

Putting $\varphi(z) = (z + a)^{-s}$, and assuming $|\arg(z + a)| < \pi$, we have

$$\varphi(iy) - \varphi(-iy) = (a + iy)^{-s} - (a - iy)^{-s} = (re^{i\theta})^{-s} - (re^{-i\theta})^{-s}$$
$$= r^{-s}(e^{-is\theta} - e^{is\theta}) = -2ir^{-s} \sin(s\theta) .$$

Substituting into (7) gives

$$\varsigma(s, a) = \frac{a^{-s}}{2} + \frac{a^{1-s}}{s - 1} + 2 \int_0^\infty \frac{(a^2 + y^2)^{-s/2} \sin s\theta}{e^{2\pi y} - 1} dy , \tag{8}$$

where $\theta = \arctan(y/a)$. This is the *Hermite formula*. The integral in the formula is an analytic function of s for any s, since it can be proved that the integral is uniformly convergent in $|s| \le R$ for any $R > 0$.

3.18. Relation to the Γ-Function

We shall now find, by the Hermite formula of the last section, the values of $\varsigma(s, a)$ and $\partial \varsigma(s, a)/\partial s$ when $s \to 0$ and the relation between the ς-function and the Γ-function.

In (8) of the last section, putting $s = 0$ gives

$$\varsigma(0, a) = \frac{1}{2} - a , \tag{1}$$

which has been obtained before [cf. Eq. (1) of Sec. 3.16 where $\varphi_1(a) = a - \frac{1}{2}$ according to Eq. (10) of Sec. 1.1].

In Eq. (8) of the last section, putting $s = 2$, we obtain (note that $\theta = \arctan(y/a)$)

$$\varsigma(2, a) = \frac{1}{2a^2} + \frac{1}{a} + \int_0^\infty \frac{2\sin 2\theta}{a^2 + y^2} \frac{dy}{e^{2\pi y} - 1}$$

$$= \frac{1}{2a^2} + \frac{1}{a} + \int_0^\infty \frac{4ay}{(a^2 + y^2)^2} \frac{dy}{e^{2\pi y} - 1} ;$$

the r.h.s. is an analytic function in the domain $\text{Re}(a) > 0$, since the integral is uniformly convergent in any closed region within that domain.

From Eq. (5) of Sec. 3.11, we obtain by differentiation,

$$\psi'(z) = \sum_{n=0}^\infty \frac{1}{(z + n)^2} = \varsigma(2, z)$$

$$= \frac{1}{2z^2} + \frac{1}{z} + \int_0^\infty \frac{4zy\,dy}{(y^2 + z^2)^2(e^{2\pi y} - 1)} , \qquad (2)$$

where $\text{Re}(z) > 0$. Integrating with respect to z once, we find

$$\psi(z) = C - \frac{1}{2z} + \ln z - \int_0^\infty \frac{2y\,dy}{(y^2 + z^2)(e^{2\pi y} - 1)} ,$$

where C is an integration constant. To determine C, we note that the integral is of order $O(z^{-2})$ when z is a real number; for,

$$\left| \int_0^\infty \frac{2y\,dy}{(y^2 + z^2)(e^{2\pi y} - 1)} \right| \leq \int_0^\infty \frac{2y\,dy}{|z|^2(e^{2\pi y} - 1)} ,$$

and by Eq. (16) of Sec. 3.11 we can show that, when $|z| \to \infty$,

$$\left| \psi(z) + \frac{1}{2z} - \ln z \right| = \left| \int_0^\infty \left\{ \frac{1}{2} + \frac{1}{t} - \frac{1}{1 - e^{-t}} \right\} e^{-zt} dt \right| = O(z^{-2})$$

[cf. Eq. (2) of Sec. 3.12 for a similar estimation of the integral there]. Hence, by letting $z \to \infty$, we see $C = 0$, so

$$\psi(z) = \ln z - \frac{1}{2z} - \int_0^\infty \frac{2y\,dy}{(y^2 + z^2)(e^{2\pi y} - 1)} . \qquad (3)$$

Integrating gives

$$\ln \Gamma(z) = C + z \ln z - z - \frac{1}{2} \ln z + 2 \int_0^\infty \frac{\arctan(y/z)}{e^{2\pi y} - 1} dy ,$$

where C is an integration constant. According to the result just obtained, the integral is of order $O(z^{-1})$ when z is real and tends to ∞. Further, from Eq. (22) of Sec. 3.11 we have

$$\left| \ln \Gamma(z) - \left(z - \frac{1}{2}\right) \ln z + z - \frac{1}{2}\ln(2\pi) \right|$$

$$= \left| \int_0^\infty \left\{ \frac{1}{2} + \frac{1}{t} - \frac{1}{1 - e^{-t}} \right\} \frac{e^{-zt}}{t} dt \right| = O(z^{-1}) \, .$$

Hence, $C = \frac{1}{2}\ln(2\pi)$ and

$$\ln \Gamma(z) = \left(z - \frac{1}{2}\right) \ln z - z + \frac{1}{2}\ln(2\pi) + 2 \int_0^\infty \frac{\arctan(y/z)}{e^{2\pi y} - 1} dy \, , \qquad (4)$$

where $\mathrm{Re}(z) > 0, \arctan u = \int_0^u dt/(1 + t^2)$, the path of integration being a straight line from 0 to u. (4) is named the *second formula of Binet* [Eq. (23) of Sec. 3.11].

In Eq. (8) of the last section, putting $s \to 1$, we have

$$\lim_{s \to 1} \left\{ \varsigma(s, a) - \frac{1}{s - 1} \right\}$$

$$= \lim_{s \to 1} \frac{a^{1-s} - 1}{s - 1} + \frac{1}{2a} + \int_0^\infty \frac{2y \, dy}{(a^2 + y^2)(e^{2\pi y} - 1)} \, .$$

Applying (3), it becomes

$$\lim_{s \to 1} \left\{ \varsigma(s, a) - \frac{1}{s - 1} \right\} = -\psi(a) = -\frac{\Gamma'(a)}{\Gamma(a)} \, . \qquad (5)$$

Differentiating Eq. (8) of the last section, we have

$$\frac{\partial}{\partial s}\varsigma(s, a) = -\frac{a^{-s}}{2}\ln a - \frac{a^{1-s}}{s - 1}\ln a - \frac{a^{1-s}}{(s - 1)^2}$$

$$+ 2 \int_0^\infty \left\{ \theta \cos s\theta - \frac{1}{2}\ln(a^2 + y^2)\sin s\theta \right\} \frac{(a^2 + y^2)^{-s/2}}{e^{2\pi y} - 1} dy \, .$$

$$(6)$$

Setting $s = 0$, we get

$$\left\{ \frac{\partial}{\partial s}\varsigma(s, a) \right\}_{s=0} = \left(a - \frac{1}{2}\right)\ln a - a + 2 \int_0^\infty \frac{\arctan(y/a)}{e^{2\pi y} - 1} dy \, .$$

With (4), it gives

$$\left\{ \frac{\partial}{\partial s} \varsigma(s, a) \right\}_{s=0} = \ln \Gamma(a) - \frac{1}{2} \ln(2\pi) . \qquad (7)$$

In (5), putting $a = 1$, then by Eq. (11) of Sec. 3.11, we have

$$\lim_{s \to 1} \left\{ \varsigma(s) - \frac{1}{s-1} \right\} = -\psi(1) = \gamma . \qquad (8)$$

In (7), putting $a = 1$ gives

$$\varsigma'(0) = -\frac{1}{2} \ln(2\pi) . \qquad (9)$$

3.19. Euler Product of the ς-Function

Assume that $\sigma = \mathrm{Re}(s) > 1$ and that p are prime numbers, 2, 3, 5, ... From the series expression of $\varsigma(s)$, we have

$$\varsigma(s)(1 - 2^{-s}) = 1 + \frac{1}{3^s} + \frac{1}{5^s} + \frac{1}{7^s} + \cdots ; \qquad (1)$$

in the series on the r.h.s., terms with n equal to a multiple of 2 do not occur. Similarly, we obtain

$$\varsigma(s)(1 - 2^{-s})(1 - 3^{-s}) = 1 + \frac{1}{5^s} + \frac{1}{7^s} + \cdots ;$$

in the series on the r.h.s., terms with n equal to multiples of 2 and 3 do not occur. Continuing this way, we have

$$\varsigma(s)(1 - 2^{-s})(1 - 3^{-s}) \ldots (1 - p^{-s}) = 1 + \sum{}' n^{-s} , \qquad (2)$$

where the series on the r.h.s. starts from the term with prime number n greater than p, and all terms with n equal to multiple of 2, 3,... ,p do not occur. Let $p \to \infty$. As $\sigma > 1$, we obtain the *Euler product*:

$$\prod_p \left(1 - \frac{1}{p^s} \right) = \frac{1}{\varsigma(s)} . \qquad (3)$$

The l.h.s. of it converges when $\sigma = \mathrm{Re}(s) > 1$ [cf. Theorem 4 of Sec. 1.6]. Hence $\varsigma(s)$ has no zeros in $\sigma > 1$. We have shown in Sec. 3.15 that $s =$

$-2m(m = 1, 2, \ldots)$ are zeros of $\varsigma(s)$. Applying the result just obtained to Eq. (2) of Sec. 3.15, we see that when $\sigma < 0$, there are no zeros other than $s = -2m$; all other possible zeros must lie in the domain $0 \le \sigma \le 1$. Riemann had made the conjecture that all the zeros in this domain lie on the straight line $\sigma = \frac{1}{2}$, but it has not yet been proved.

3.20. Riemann Integral of the ς-Function

From the definition of the Γ-function, it is readily seen that

$$n^{-s}\pi^{-s/2}\Gamma\left(\frac{s}{2}\right) = \int_0^\infty e^{-n^2\pi x}x^{s/2-1}dx \ . \tag{1}$$

Let

$$\varpi(x) = \sum_{n=1}^\infty e^{-n^2\pi x} \ . \tag{2}$$

By (1), we have

$$\varsigma(s)\pi^{-s/2}\Gamma\left(\frac{s}{2}\right) = \int_0^\infty \varpi(x)x^{s/2-1}dx \ . \tag{3}$$

In the chapter on Theta functions [Chap. 9], we shall prove that

$$1 + 2\varpi(x) = x^{-1/2}\left\{1 + 2\varpi\left(\frac{1}{x}\right)\right\} \tag{4}$$

[Eq. (6) of Sec. 9.9]. Dividing the integral in (3) into two parts, one from 0 to 1 and the second, from 1 to ∞. Applying (4) to the first part, we obtain

$$\varsigma(s)\pi^{-s/2}\Gamma\left(\frac{s}{2}\right) = \int_0^1\left\{-\frac{1}{2} + \frac{1}{2}x^{-1/2} + x^{-1/2}\varpi\left(\frac{1}{x}\right)\right\}x^{s/2-1}dx$$

$$+ \int_1^\infty \varpi(x)x^{s/2-1}dx$$

$$= -\frac{1}{s} + \frac{1}{s-1} + \int_1^\infty \varpi(t)t^{\frac{1-s}{2}-1}dt + \int_1^\infty \varpi(x)x^{s/2-1}dx$$

$$= \frac{1}{s(s-1)} + \int_1^\infty (x^{s/2} + x^{(1-s)/2})\varpi(x)\frac{dx}{x} \ ,$$

or,

$$s(s-1)\varsigma(s)\pi^{-s/2}\Gamma\left(\frac{s}{2}\right)$$

$$= 1 + s(s-1)\int_1^\infty (x^{s/2} + x^{(1-s)/2})\varpi(x)\frac{dx}{x} \ . \tag{5}$$

Let $s = \frac{1}{2} + it$, $\xi(t) = \frac{1}{2}s(s-1)\zeta(s)\pi^{-s/2}\Gamma(s/2)$ and apply (5). It results in

$$\xi(t) = -\frac{1}{2}\left(\frac{1}{4} + t^2\right)\zeta\left(\frac{1}{2} + it\right)\pi^{-\frac{1}{4} - \frac{it}{2}}\Gamma\left(\frac{1}{4} + \frac{it}{2}\right)$$

$$= \frac{1}{2} - \left(t^2 + \frac{1}{4}\right)\int_1^\infty x^{-3/4}\varpi(x)\cos\left(\frac{t}{2}\ln x\right)dx \ . \tag{6}$$

Substituting the power series expression $\cos v = \sum_{n=0}^\infty (-)^n v^{2n}/(2n)!$ into the integral in (6), we obtain by integration,

$$\xi(t) = \sum_{n=0}^\infty a_n t^{2n} \ , \tag{7}$$

where

$$a_0 = \frac{1}{2} - \frac{b_0}{4} \ ,$$

$$a_n = \frac{(-)^n}{(2n)!}\left\{2n(2n-1)b_{n-1} - \frac{b_n}{4}\right\} \quad (n \geq 1) \ ,$$

$$b_n = \int_1^\infty \left(\frac{1}{2}\ln x\right)^{2n} x^{-3/4}\varpi(x)dx \ .$$

3.21. Another Derivation of the Asymptotic Expansion of the Γ-Function

From Eq. (1) of Sec. 3.4, we have

$$\frac{e^{-\gamma z}\Gamma(a)}{\Gamma(z+a)} = \left(1 + \frac{z}{a}\right)\prod_{n=1}^\infty \left\{\left(1 + \frac{z}{a+n}\right)e^{-z/n}\right\} \ .$$

Taking the principal value of the logarithms and expanding it in a series, we obtain

$$\ln\frac{e^{-\gamma z}\Gamma(a)}{\Gamma(z+a)} = \sum_{m=1}^\infty \frac{(-)^{m-1}}{m}\frac{z^m}{a^m} + \sum_{n=1}^\infty\left[\sum_{m=1}^\infty \frac{(-)^{m-1}}{m}\frac{z^m}{(a+n)^m} - \frac{z}{n}\right]$$

$$= \frac{z}{a} + \sum_{m=2}^\infty \frac{(-)^{m-1}}{m}\frac{z^m}{a^m}$$

$$+ \sum_{n=1}^\infty\left\{\sum_{m=2}^\infty \frac{(-)^{m-1}}{m}\frac{z^m}{(a+n)^m} - \frac{az}{n(n+a)}\right\}$$

$$= \frac{z}{a} - \sum_{n=1}^\infty \frac{az}{n(n+a)} + \sum_{m=2}^\infty \frac{(-)^{m-1}}{m}z^m\zeta(m,a) \ .$$

Applying Eq. (5) of Sec. 3.11, it results in

$$\ln \frac{\Gamma(z+a)}{\Gamma(a)} = z\psi(a) - \sum_{m=2}^{\infty} \frac{(-)^{m-1}}{m} z^m \varsigma(m, a) .$$ (1)

Consider the following integral

$$\frac{1}{2\pi i} \int_C z^s \Gamma(s)\Gamma(-s)\varsigma(s, a)ds = -\frac{1}{2\pi i} \int_C \frac{\pi z^s}{s \sin \pi s} \varsigma(s, a)ds .$$

The integration contour C is the straight line $\sigma = 3/2$ plus a semi-circle on the right of the straight line with center at $s = 3/2$, radius equal to N. The integral is equal to the sum of the residues at the poles lying inside the contour C; the poles being $s = m = 2, 3, \dots$. Thus, the sum of the residues is exactly equal to the series on the r.h.s. of (1). On the semi-circle, $\varsigma(s, a) = O(1)$, since the series representing $\varsigma(s, a)$ is convergent thereon. As $|z^s| = |z|^{\sigma} e^{-t \arg z}, (s = \sigma + it)$, the integrand is of order $O\{|z|^{\sigma} e^{-\pi|t| - t \arg z}\}$. Hence, when $|z| < 1$ and $|\arg z| < \pi$, that part of the integral over the semi-circle approaches zero as $N \to \infty$, so that there remains only the integral along the straight line $\sigma = 3/2$:

$$\ln \frac{\Gamma(z+a)}{\Gamma(a)} = z\psi(a) + \frac{1}{2\pi i} \int_{\frac{3}{2}-i\infty}^{\frac{3}{2}+i\infty} z^s \Gamma(s)\Gamma(-s)\varsigma(s, a)ds .$$ (2)

The integral on the r.h.s. is analytic for all values of z so long as $|\arg z| < \pi$; therefore, the result is not restricted to the condition $|z| < 1$.

It can be shown [cf. Whittaker and Watson (1927), p. 277] that when $R \to \infty$,

$$\int_{-n-\frac{1}{2}\pm Ri}^{\frac{3}{2}\pm Ri} \frac{\pi z^s}{s \sin \pi s} \varsigma(s, a)ds \to 0 \quad (n = 1, 2, \dots) .$$

Hence (2) can be reduced to

$$\ln \frac{\Gamma(z+a)}{\Gamma(a)} = z\psi(a) - \frac{1}{2\pi i} \int_{-n-\frac{1}{2}-i\infty}^{-n-\frac{1}{2}+i\infty} \frac{\pi z^s}{s \sin \pi s} \varsigma(s, a)ds$$

$$+ \sum_{m=-1}^{n} R_m ,$$ (3)

where R_m is the residue of the function $-\pi z^s \varsigma(s, a)/(s \sin \pi s)$ at $s = -m$. When $|z|$ is large, it can be shown that the integral in (3) is of order $O(|z|^{-n-\frac{1}{2}})$ [cf. Whittaker and Watson (1927), p. 277]. Therefore,

$$\ln \frac{\Gamma(z+a)}{\Gamma(a)} = z\psi(a) + \sum_{m=-1}^{n} R_m + O(|z|^{-n-\frac{1}{2}}) .$$ (4)

When $m > 0$, the residue is

$$R_m = \frac{(-)^m}{m} z^{-m} \varsigma(-m, a) = \frac{(-)^{m-1} z^{-m} \varphi_{m+1}(a)}{m(m+1)} ;$$

Eq. (1) of Sec. 3.16 has been applied in the last step, and $\varphi_{m+1}(a)$ is the Bernoulli polynomial [Sec. 1.1].

To find R_0, we expand the integrand in (3) at $s = 0$ as

$$-\frac{1}{s^2} \left(1 + \frac{\pi^2 s^2}{6} + \ldots \right)(1 + s \ln z + \ldots)\{\varsigma(0, a) + s\varsigma'(0, a) + \ldots \} .$$

Applying Eqs. (1) and (7) of Sec. 3.18, we find

$$R_0 = -\varsigma(0, a) \ln z - \varsigma'(0, a)$$
$$= \left(a - \frac{1}{2}\right) \ln z - \ln \Gamma(a) + \frac{1}{2} \ln(2\pi) .$$

Similarly, to find R_{-1}, we may expand the integrand at $s = 1$ by Eq. (5) of Sec. 3.18 as $(y = s - 1)$

$$(1 - y + y^2 + \ldots)\frac{1}{y}\left(1 + \frac{\pi^2 y^2}{6} + \ldots\right)$$
$$\times z(1 + y \ln z + \ldots)\left(\frac{1}{y} - \psi(a) + \ldots\right)$$

and obtain

$$R_{-1} = z\{\ln z - \psi(a) - 1\} .$$

Substituting these results in (4), we finally arrive at

$$\ln \Gamma(z + a) = \left(z + a - \frac{1}{2}\right) \ln z - z + \frac{1}{2} \ln(2\pi)$$
$$+ \sum_{m=1}^{n} \frac{(-)^{m-1} \varphi_{m+1}(a)}{m(m+1)z^m} + O(z^{-n-1}) , \qquad (5)$$

where $|\arg z| < \pi$. The last term of (5), namely, $O(z^{-n-1})$, is written according to the rule of decrement of the preceding terms and is seen to be more precise than the estimation given by the last term of (4).

3.22. Evaluation of the ς-Function

In Eq. (6) of Sec. 1.3, let $m \to \infty, h = 1, F(x) = x^{-s}$ and use for R_n the expression given by (9) of that section. The result is

$$\varsigma(s, a+1) = \frac{1}{s-1}\frac{1}{a^{s-1}} - \frac{1}{2a^s}$$
$$+ \sum_{k=1}^{n} \frac{(-)^{k-1}B_k(s)_{2k-1}}{(2k)!a^{s+2k-1}} - (s)_{2n}\int_0^\infty \frac{P_{2n}(t)dt}{(t+a)^{s+2n}} . \qquad (1)$$

Take $a = m$ (positive integer), then, since

$$\varsigma(s, m+1) = \sum_{r=0}^{\infty} \frac{1}{(r+m+1)^s} = \sum_{r=m+1}^{\infty} \frac{1}{r^s} = \varsigma(s) - \sum_{r=1}^{m} \frac{1}{r^s} ,$$

we have

$$\varsigma(s) = \sum_{r=1}^{m} \frac{1}{r^s} + \frac{1}{(s-1)m^{s-1}} - \frac{1}{2m^s} + \sum_{k=1}^{n} \frac{(-)^{k-1}B_k(s)_{2k-1}}{(2k)!m^{s+2k-1}}$$
$$-(s)_{2n}\int_0^\infty \frac{P_{2n}(t)dt}{(t+m)^{s+2n}} , \qquad (2)$$

which may be employed to compute the values of the ς-function.

Applying to $s = 3$, take $n = 4$ and $m = 5$. We find $\varsigma(3) = 1.20205\ 690$, which is precise to the eighth decimal place.

Exercise 3

1. Show that, under the condition $\sum_{m=1}^{k}(a_m - b_m) = 0$,

$$\prod_{n=1}^{\infty} \left\{ \frac{(n-a_1)\ldots(n-a_k)}{(n-b_1)\ldots(n-b_k)} \right\} = \prod_{m=1}^{k} \frac{\Gamma(1-b_m)}{\Gamma(1-a_m)} .$$

2. Show that

$$\frac{(a+b)!}{a!b!} = \prod_{s=1}^{\infty} \frac{(s+a)(s+b)}{s(s+a+b)} .$$

3. Show that, if $\omega = e^{2\pi i/n} = \cos(2\pi/n) + i\sin(2\pi/n), n$ being a positive integer greater than 1, then we have

$$x \prod_{k=1}^{\infty} \left(1 - \frac{x}{k^n}\right) = - \prod_{m=0}^{n-1} \{\Gamma(-\omega^m x^{1/n})\}^{-1} .$$

4. Show that

$$\Gamma(z) = \int_0^{\infty} t^{z-1} \left\{e^{-t} - 1 + t - \frac{t^2}{2!} + \ldots + (-)^{k-1}\frac{t^k}{k!}\right\} dt ,$$

where k is a positive integer, $-k > \mathrm{Re}(z) > -k - 1$.

5. Assume $\mathrm{Re}(z) > 0, a > 0$ and the path of integration is a quarter circle with centre at $-a$ and end points at ρ and $-a+i(\rho+a)$ or ρ and $-a-i(\rho+a)$. Show that, when $\rho \to \infty, \int (-t)^{-z}e^{-t}dt \to 0$, and thus derive that

$$\lim_{\rho \to \infty} \int_{-a+i\rho}^{-a-i\rho} (-t)^{-z}e^{-t} dt = \lim_{\rho \to \infty} \int_C (-t)^{-z}e^{-t} dt .$$

Then, putting $t = -a - iu$, obtain

$$\frac{1}{\Gamma(z)} = \frac{1}{2\pi} \int_{-\infty}^{\infty} e^{a+iu}(a + iu)^{-z} du .$$

6. Show that the Binet formula [Eq. (22) of Sec. 3.11] may assume the following form $(\mathrm{Re}(z) > 0)$:

$$\ln \Gamma(z) = \left(z - \frac{1}{2}\right) \ln z - z + \frac{1}{2} \ln(2\pi)$$
$$+ \int_0^{\infty} \left(\frac{1}{2} - \frac{1}{t} + \frac{1}{e^t - 1}\right) \frac{e^{-zt}}{t} dt .$$

7. Show that

$$\int_0^{\infty} \left(\frac{1}{2} - \frac{1}{t} + \frac{1}{e^t - 1}\right) \frac{e^{-t}}{t} dt = 1 - \frac{1}{2} \ln(2\pi) .$$

8. Show that

$$\gamma = \int_0^\infty \left\{ \frac{1}{1+t} - e^{-t} \right\} \frac{dt}{t} .$$

9. From

$$\ln \Gamma(z) = \int_0^\infty \left\{ z - 1 - \frac{1 - e^{-(z-1)t}}{1 - e^{-t}} \right\} \frac{e^{-t}}{t} dt \quad (\text{Re}(z) > 0)$$

and $\Gamma(z)\Gamma(1-z) = \pi/\sin \pi z$ [Eq. (2) of Sec. 3.5], show that

$$2 \ln \Gamma(z) = \ln \pi - \ln \sin(\pi z) + \int_0^\infty \left\{ \frac{\text{sh} \left(\frac{1}{2} - z \right) t}{\text{sh} \frac{t}{2}} - (1 - 2z) e^{-t} \right\} \frac{dt}{t}$$

for $0 < \text{Re}(z) < 1$.

10. Expanding $\text{sh}(\frac{1}{2} - x)t$ and $1 - 2x$ in Fourier sine series in the interval $0 < x < 1$, derive the *Kummer formula*

$$\ln \Gamma(x) = \frac{1}{2} \ln \pi - \frac{1}{2} \ln \sin(\pi x) + 2 \sum_{n=1}^\infty a_n \sin 2n\pi x$$

from the result of the last Ex., where

$$a_n = \int_0^\infty \left\{ \frac{2n\pi}{t^2 + 4n^2\pi^2} - \frac{e^{-t}}{2n\pi} \right\} \frac{dt}{t} = \frac{1}{2n\pi} [\gamma + \ln(2n\pi)] ,$$

and the result of Ex. 8 has been used in the last step.

11. From the Kummer formula of the last Ex., derive that

$$\ln \Gamma(x) = \frac{1}{2} \ln(2\pi) + \sum_{n=1}^\infty \left\{ \frac{1}{2n} \cos(2n\pi x) + \frac{\gamma + \ln(2n\pi)}{n\pi} \sin(2n\pi x) \right\} .$$

[In proving $\ln \sin \pi x = -\ln 2 - \sum_{n=1}^{\infty} \frac{1}{n} \cos(2n\pi x)$, it is necessary to show that

$$\int_0^1 \ln \sin \pi x \, dx = 2 \int_0^{\frac{1}{2}} \ln \sin \pi x \, dx = 2 \int_0^{\frac{1}{2}} \ln \cos \pi x \, dx = -\ln 2$$

$$\int_0^1 \sin 2n\pi x \, \cot \pi x \, dx = \int_0^1 \frac{\sin(2n-1)\pi x}{\sin \pi x} dx$$

$$= \int_0^1 \frac{\sin(2n+1)\pi x}{\sin \pi x} dx = 1.]$$

12. Under the conditions $\lambda > 0, x > 0, -\pi/2 < \alpha < \pi/2$, show that

$$\int_0^{\infty} t^{x-1} e^{-\lambda t \cos \alpha} \cdot \frac{\cos}{\sin} (\lambda t \sin \alpha) dt = \lambda^{-x} \Gamma(x) \frac{\cos}{\sin} (\alpha x) .$$

13. Show that, for $b > 0$,

$$\int_0^{\infty} x^{-z} \sin bx \, dx = \frac{\pi}{2} \frac{b^{z-1}}{\Gamma(z) \sin \frac{\pi z}{2}} \quad (0 < z < 2) ,$$

$$\int_0^{\infty} x^{-z} \cos bx \, dx = \frac{\pi}{2} \frac{b^{z-1}}{\Gamma(z) \cos \frac{\pi z}{2}} \quad (0 < z < 1) .$$

14. Show that

$$B(np, nq) = n^{-nq} \frac{B(p, q) B\left(p + \frac{1}{n}, q\right) \ldots B\left(p + \frac{n-1}{n}, q\right)}{B(q, q) B(2q, q) \ldots B\{(n-1)q, q\}} .$$

15. Show that

$$\int_0^1 \int_0^1 f(xy)(1-x)^{\mu-1} y^{\mu}(1-y)^{\nu-1} dx dy$$

$$= \frac{\Gamma(\mu)\Gamma(\nu)}{\Gamma(\mu+\nu)} \int_0^1 f(z)(1-z)^{\mu+\nu+1} dz .$$

16. Let C be the contour consisting of a straight line from $-i$ to $+i$, detouring the three points $-i, 0, +i$, and a half unit circle on the right of the straight line. Consider the integral

$$\int_C z^{p-q-1}(z+z^{-1})^{p+q-2}\,dz\,,$$

where $p+q > 1, q < 1$. Show that — the Cauchy formula —

$$\int_0^{\pi/2} \cos^{p+q-2}\theta\, \cos(p-q)\theta\; d\theta = \frac{\pi}{(p+q-1)2^{p+q-1}B(p,q)}\,,$$

which holds for any p, q, satisfying the condition $p+q > 1$.

17. Show that:

(i) When $\mathrm{Re}(s) > 0$,

$$(1-2^{1-s})\varsigma(s) = \frac{1}{1^s} - \frac{1}{2^s} + \frac{1}{3^s} - \frac{1}{4^s} + \dots = \frac{1}{\Gamma(s)}\int_0^\infty \frac{x^{s-1}dx}{e^x+1}\;;$$

(ii) When $\mathrm{Re}(s) > 1$,

$$(2^s-1)\varsigma(s) = \varsigma\left(s,\frac{1}{2}\right) = \frac{2^{s-1}}{\Gamma(s)}\int_0^\infty \frac{x^{s-1}}{\mathrm{sh}\,x}\,dx\,.$$

18. Show that

$$\varsigma(s) = \frac{\Gamma(1-s)}{(2^{s-1}-1)}\frac{1}{2\pi i}\int_\infty^{(0+)} \frac{(-z)^{s-1}}{e^z+1}\,dz\,,$$

$z = n\pi i\,(n = \pm 1, \pm 3, \pm 5, \dots)$ lying outside the contour.

19. Show that

$$\ln\Gamma(z) = \left(z-\frac{1}{2}\right)\ln z - z + \frac{1}{2}\ln(2\pi)$$

$$+\frac{1}{2}\sum_{s=1}^\infty \frac{s}{(s+1)(s+2)}\varsigma(s+1, z+1)\,.$$

20. Let $\Phi(z, s, a) = \sum_{n=0}^{\infty}(n + a)^{-s}z^n$. Show that

$$\Phi(z, s, a) = \frac{1}{\Gamma(s)} \int_0^{\infty} \frac{t^{s-1}e^{-at}}{1 - ze^{-t}} dt .$$

21. Show that the function Φ in the last Ex. can be represented by

$$\Phi(z, s, a) = -\frac{\Gamma(1 - s)}{2\pi i} \int_{\infty}^{(0+)} \frac{(-t)^{s-1}e^{-at}}{1 - ze^{-t}} dt ;$$
$$(|\arg(-t)| < \pi) .$$

22. Using the method described in Sec. 3.15, we can obtain the series expansion of $\Phi(z, s, a)$ in a:

$$\Phi(z, s, a) = z^{-a}\Gamma(1 - s) \sum_{n=-\infty}^{\infty} (-\ln z + 2n\pi i)^{s-1}e^{2n\pi ai} ,$$
$$0 < a \le 1, \quad \mathrm{Re}(s) < 0, \quad |\arg(-\ln z + 2n\pi i)| < \pi .$$

Then, by Eq. (1) of Sec. 3.15, obtain

$$\Phi(z, s, a) = \Gamma(1 - s)z^{-a} \left[\ln\left(\frac{1}{z}\right)\right]^{s-1}$$
$$+z^{-a} \sum_{r=0}^{\infty} \varsigma(s - r, a)\frac{(\ln z)^r}{r!} ,$$
$$|\ln z| < 2\pi, \quad s \ne 1, 2, 3, \ldots ; \quad a \ne 0, -1, -2, \ldots .$$

23. From the first series expansion in the last Ex., derive that

$$\Phi(z, s, a) = iz^{-a}(2\pi)^{s-1}\Gamma(1 - s)$$
$$\times \left\{ e^{-i\frac{\pi s}{2}} \Phi\left(e^{-2\pi ni}, 1 - s, \frac{\ln z}{2\pi i}\right) \right.$$
$$\left. - e^{i\pi(\frac{s}{2}+2a)} \Phi\left(e^{2\pi ai}, 1 - s, 1 - \frac{\ln z}{2\pi i}\right) \right\} ,$$

which is called the *Lerch transformation formula*.

24. *Möbius function.* From Eq. (3) of Sec. 3.19, show that

$$\frac{1}{\varsigma(z)} = \sum_{n=1}^{\infty} \frac{\mu(n)}{n^z} \, ,$$

where $\mu(n)$ is called the Möbius function; by definition, $\mu(n) = 1$ if $n = 1$, $\mu(n) = (-1)^k$ if n is the product of k distinct primes, and $\mu(n) = 0$ if n is divisible by a square > 1. For example, $\mu(1) = 1, \mu(2) = -1, \mu(3) = -1, \mu(4) = 0, \mu(5) = -1, \mu(6) = 1$.

25. If $\mathrm{Re}(z) > 1$, the series $\varsigma(z) = \sum_{1}^{\infty} 1/n^z$ is absolutely convergent, so that it may be multiplied by itself and the terms combined arbitrarily. Show that

$$\varsigma^2(z) = \sum_{n=1}^{\infty} \frac{\tau_n}{n^z} \, ,$$

where τ_n is the number of factors of n.

26. Show that, if $\mathrm{Re}(s) > 1$, then we have

$$\ln \varsigma(s) = \sum_{p} \sum_{m=1}^{\infty} \frac{1}{mp^{ms}} \, ,$$

p being the prime number 2, 3, 5,... Further, prove that, under the same condition $\mathrm{Re}(s) > 1$,

$$-\frac{\varsigma'(s)}{\varsigma(s)} = \sum_{n=1}^{\infty} \frac{\Lambda(n)}{n^s} \, ,$$

where
$$\Lambda(n) = \begin{cases} 0, & \text{if } n \neq \text{power of a prime}, \\ \ln p, & \text{if } n = \text{power of the prime } p. \end{cases}$$

Chapter 4

HYPERGEOMETRIC FUNCTION

Hypergeometric functions form an important class of special functions. Any solution of the differential equation of Fuchsian type [Sec. 2.9] with 3 regular singular points can be expressed in terms of hypergeometric functions; for example, the Legendre functions, special spherical polynomials, Jacobi polynomial, Chebyshev polynomial, etc.

4.1. Hypergeometric Series and Hypergeometric Function

In Sec. 2.9, we have dealt with the hypergeometric equation

$$z(1-z)w'' + [\gamma - (\alpha + \beta + 1)z]w' - \alpha\beta w(z) = 0 \tag{1}$$

which is the proto-type of the Fuchsian equation with three regular singularities, namely, 0, 1, ∞. By the method of series solution [Sec. 2.4], assuming $w(z) = \sum_{0}^{\infty} c_k z^{k+\rho}$, $c_0 \neq 0$, we find, at the singular point $z = 0$, the two exponents (indices) $\rho = 0$ and $1 - \gamma$, and the recurrence relation for the coefficients:

$$c_k = \frac{(\rho + k - 1 + \alpha)(\rho + k - 1 + \beta)}{(\rho + k)(\rho + k - 1 + \gamma)} c_{k-1} . \tag{2}$$

When $\gamma \neq 0$ or negative integers,[a] the solution with exponent 0 is obtained by means of (2) to be

$$1 + \frac{\alpha\beta}{1 \times \gamma}z + \frac{\alpha(\alpha+1)\beta(\beta+1)}{1 \times 2 \times \gamma(\gamma+1)}z^2 + \ldots$$
$$+ \frac{\alpha(\alpha+1)\ldots(\alpha+n-1)\beta(\beta+1)\ldots(\beta+n-1)}{1 \times 2 \times \ldots \times n\gamma(\gamma+1) \times \ldots \times (\gamma+n-1)}z^n + \ldots . \qquad (3)$$

The ratio of the consecutive coefficients in the series is

$$\frac{c_{n-1}}{c_n} = \frac{n(n-1+\gamma)}{(n-1+\alpha)(n-1+\beta)} = 1 + \frac{\gamma-\alpha-\beta+1}{n} + O(n^{-2}) . \qquad (4)$$

Hence the radius of convergence of the series is 1, and the series represents an analytic function in the circle $|z| < 1$.

The series (3) is called the *hypergeometric series* and is usually denoted by the symbol

$$F(\alpha, \beta, \gamma, z) = \sum_{n=0}^{\infty} \frac{(\alpha)_n (\beta)_n}{n!(\gamma)_n}z^n, \quad |z| < 1 , \qquad (5)$$

where

$$\left. \begin{aligned} (\lambda)_0 &= 1 \\ (\lambda)_n &= \lambda(\lambda+1)\ldots(\lambda+n-1) = \frac{\Gamma(\lambda+n)}{\Gamma(\lambda)} \quad (n \geq 1) \end{aligned} \right\} . \qquad (6)$$

Note that

$$F(\alpha, \beta, \gamma, 0) = 1 \qquad (7)$$

and

$$F(\alpha, \beta, \gamma, z) = F(\beta, \alpha, \gamma, z) , \qquad (8)$$

i.e., $F(\alpha, \beta, \gamma, z)$ is symmetric with respect to α and β.

When γ is not an integer, the second solution, with exponent $1 - \gamma$, at $z = 0$ can also be expressed as a hypergeometric series [cf. Sec. 4.3], namely,

$$z^{1-\gamma}F(\alpha - \gamma + 1, \beta - \gamma + 1, 2 - \gamma, z) . \qquad (9)$$

According to the theory of differential equation [Sec. 2.1], only the singularities of the differential equation could be singularities of its solutions.

---

[a]The cases for which $\gamma = 0$ or a negative integer will be treated in Sec. 4.4.

Therefore, the analytic function represented by the series (3) in the unit circle $|z| < 1$ can be analytically continued to the whole z-plane, except possibly the two points, $z = 1$ and $z = \infty$. The function so defined is the *hypergeometric function* and will be denoted by the same symbol $F(\alpha, \beta, \gamma, z)$. In Sec. 4.5, we shall see that unless α or β is a negative integer, the points $z = 1$ and $z = \infty$ are two branch points of the function $F(\alpha, \beta, \gamma, z)$; and when α or β is a negative integer, the series degenerates into a polynomial. Hence, $F(\alpha, \beta, \gamma, z)$ is a single-valued analytic function in the z-plane with a cut from 1 to ∞ along the real axis; the power series (3) is a single-valued branch, the branch which is equal to 1 when $z = 0$, representing it in $|z| < 1$.

Many elementary functions can be expressed in terms of the hypergeometric functions, for example,

$$(1 + z)^{\alpha} = F(-\alpha, \beta, \beta, -z) , \tag{10}$$

$$\arcsin z = z F\left(\frac{1}{2}, \frac{1}{2}, \frac{3}{2}, z^2\right) , \tag{11}$$

$$\arctan z = z F\left(\frac{1}{2}, 1, \frac{3}{2}, -z^2\right) , \tag{12}$$

$$\ln(1 + z) = z F(1, 1, 2, -z) . \tag{13}$$

For other examples cf. Erdélyi (1953), Vol. I, p. 101; and the Exs. 4, 5, 22, etc. at the end of this chapter.

4.2. Recurrence Relations

Let l, m and n be arbitrary integers. The function

$$F(\alpha + l, \beta + m, \gamma + n, z)$$

is named the *consecutive*, or *associate*, function of the function $F(\alpha, \beta, \gamma, z)$. It has been shown by Gauss that among any three consecutive functions, F_1, F_2, F_3, there exists a relation of the form

$$A_1 F_1 + A_2 F_2 + A_3 F_3 = 0 , \tag{1}$$

where A_1, A_2 and A_3 are some rational functions of z. This theorem can be proved[b] by using the integral representation of the hypergeometric function

---

[b]cf. for example, Whittaker and Watson (1927) Sec. 14.7.

[Sec. 4.5]. Here we give two most simple relations of this sort:

$$(\gamma - 1)F(\gamma - 1) - \alpha F(\alpha + 1) - (\gamma - \alpha - 1)F = 0 , \tag{2}$$

$$\gamma F - \beta z F(\beta + 1, \gamma + 1) - \gamma F(\alpha - 1) = 0 , \tag{3}$$

where F is a short hand for $F(\alpha, \beta, \gamma, z); F(\alpha \pm 1), \ldots$ represent the close neighbors of F:

$$\left. \begin{array}{l} F(\alpha \pm 1) = F(\alpha \pm 1, \beta, \gamma, z) \\ F(\beta \pm 1) = F(\alpha, \beta \pm 1, \gamma, z) \\ F(\gamma \pm 1) = F(\alpha, \beta, \gamma \pm 1, z) \end{array} \right\} \tag{4}$$

$F(\beta + 1, \gamma + 1) = F(\alpha, \beta + 1, \gamma + 1, z)$, etc.

(2) and (3) can be justified by using the series expression of the hypergeometric function (3) or (5) of the preceding section. For example, (2) is

$$(\gamma - 1) \sum_{k=0}^{\infty} \frac{(\alpha)_k (\beta)_k}{k!(\gamma)_k} z^k - \alpha \sum_{k=0}^{\infty} \frac{(\alpha + 1)_k (\beta)_k}{k!(\gamma)_k} z^k$$

$$- (\gamma - \alpha - 1) \sum_{k=0}^{\infty} \frac{(\alpha)_k (\beta)_k}{k!(\gamma)_k} z^k$$

$$= \sum_{k=0}^{\infty} \frac{(\alpha)_k (\beta)_k}{k!(\gamma)_k} z^k \{(\gamma + k - 1) - (\alpha + k) - (\gamma - \alpha - 1)\} = 0 .$$

All other recurrence formulae can be deduced from these two simple ones together with (8) of the last section. For example, exchanging α and β in (2) gives

$$(\gamma - 1)F(\gamma - 1) - \beta F(\beta + 1) - (\gamma - \beta - 1)F = 0 . \tag{5}$$

Subtracting (2) from (5), we obtain

$$\alpha F(\alpha + 1) - \beta F(\beta + 1) - (\alpha - \beta)F = 0 . \tag{6}$$

If we change α in (2) to $\alpha - 1$ and γ in (3) to $\gamma - 1$, then adding, there results

$$(\gamma - 1)F(\gamma - 1) - \beta z F(\beta + 1) - (\alpha - 1)F - (\gamma - \alpha)F(\alpha - 1) = 0 .$$

Using (2) and (6) to eliminate $F(\gamma - 1)$ and $F(\beta + 1)$, we have

$$[\gamma - 2\alpha + (\alpha - \beta)z]F + \alpha(1 - z)F(\alpha + 1) - (\gamma - \alpha)F(\alpha - 1) = 0 . \tag{7}$$

(2), (5), (6) and (7) are relations between F and its close neighbors. There are a total of $\binom{6}{2} = 15$ such relations; all of them can be deduced in a similar way. [Cf. Ex. 2 at the end of this chapter.] Having obtained these relations between the close neighbors, relations among any three consecutive functions can be derived. For example, by changing the γ in (5) to $\gamma + 1$, we obtain

$$\gamma F - \beta F(\beta + 1, \gamma + 1) - (\gamma - \beta) F(\gamma + 1) = 0 . \tag{8}$$

Eliminating $F(\beta + 1, \gamma + 1)$ from it by (3), we have

$$\gamma(1 - z) F - \gamma F(\alpha - 1) + (\gamma - \beta) z F(\gamma + 1) = 0 . \tag{9}$$

Another kind of important recurrence relation is

$$\left. \begin{array}{l} \frac{d}{dz} F(\alpha, \beta, \gamma, z) = \frac{\alpha\beta}{\gamma} F(\alpha + 1, \beta + 1, \gamma + 1, z) \\[2mm] \frac{d^m}{dz^m} F(\alpha, \beta, \gamma, z) = \frac{(\alpha)_m (\beta)_m}{(\gamma)_m} F(\alpha + m, \beta + m, \gamma + m, z) \end{array} \right\} \tag{10}$$

which can be proved directly by differentiating the series expression of $F(\alpha, \beta, \gamma, z)$ term by term.

4.3. Other Solutions of the Hypergeometric Equation Expressed in Terms of Hypergeometric Functions

In Sec. 4.1, we have stated that when γ is not an integer, the second solution of the hypergeometric equation at the point $z = 0$ can also be expressed in terms of the hypergeometric function [Eq. (9) of Sec. 4.1]. Other solutions in such form can be obtained by the transformations of the P-equation [Sec. 2.9] as follows.

The whole of the solutions of the hypergeometric equation can be represented by P-equation depicted in Eq. (8) of Sec. 2.9, namely,

$$P \left\{ \begin{array}{ccc} 0 & 1 & \infty \\ 0 & 0 & \alpha; \quad z \\ 1-\gamma & \gamma-\alpha-\beta & \beta \end{array} \right\} ; \tag{1}$$

$F(\alpha, \beta, \gamma, z)$ is the solution at $z = 0$ with exponent 0. To obtain the solution at the same point with exponent $1 - \gamma(\gamma \neq \text{integer})$, we employ the following transformation given by Eq. (12) of Sec. 2.9:

$$P \left\{ \begin{array}{ccc} 0 & 1 & \infty \\ 0 & 0 & \alpha; \quad z \\ \boxed{1-\gamma} & \gamma-\alpha-\beta & \beta \end{array} \right\}$$

$$= z^{1-\gamma} P \left\{ \begin{array}{ccc} 0 & 1 & \infty \\ \gamma-1 & 0 & \alpha-\gamma+1; \quad z \\ \boxed{0} & \gamma-\alpha-\beta & \beta-\gamma+1 \end{array} \right\}$$

$$= z^{1-\gamma} P \left\{ \begin{array}{ccc} 0 & 1 & \infty \\ 1-\gamma' & 0 & \alpha'; \quad z \\ \boxed{0} & \gamma'-\alpha'-\beta' & \beta' \end{array} \right\} ,$$

where $\alpha' = \alpha - \gamma + 1, \beta' = \beta - \gamma + 1, \gamma' = 2 - \gamma$. This transformation carries the solution of the original equation at $z = 0$ with exponent $1 - \gamma$ into corresponding solution of the new equation at $z = 0$ with exponent 0. We have drawn in the above P-equation a rectangle around the exponent in order to distinguish it from the others. According to the result obtained in Sec. 4.1, the solution sought is $F(\alpha', \beta', \gamma', z)$. Hence, the corresponding solution of the original equation is $z^{1-\gamma} F(\alpha - \gamma + 1, \beta - \gamma + 1, 2 - \gamma, z)$.

If γ is an integer, then, in general, one of the two solutions at $z = 0$ would contain a logarithmic term [Sec. 2.5]. In such a case, the second solution cannot be derived from the first one in the simple way described above. We shall deal with it in Sec. 4.4.

In summary, when γ is not an integer, the two fundamental solutions at $z = 0$ of the hypergeometric equation, expressed in terms of hypergeometric functions, are

$$w_1(z) = F(\alpha, \beta, \gamma, z) , \tag{2}$$

$$w_2(z) = z^{1-\gamma} F(\alpha - \gamma + 1, \beta - \gamma + 1, 2 - \gamma, z) , \tag{3}$$

where w_1 is the only solution which is analytic at $z = 0$; for $w_2, z = 0$ is a branch point.

The solutions at the other singularities can be found similarly by transformations of the P-equation. For example, to derive the solution at $z = 1$ with exponent 0 [see (1)], we use the transformation Eq. (10) of Sec. 2.9, namely, $z_1 = 1 - z$, so

$$P \left\{ \begin{matrix} 0 & 1 & \infty & \\ 0 & \boxed{0} & \alpha; & z \\ 1-\gamma & \gamma-\alpha-\beta & \beta & \end{matrix} \right\}$$

$$= P \left\{ \begin{matrix} 1 & 0 & \infty & \\ 0 & \boxed{0} & \alpha; & z_1 \\ 1-\gamma & \gamma-\alpha-\beta & \beta & \end{matrix} \right\}$$

$$= P \left\{ \begin{matrix} 1 & 0 & \infty & \\ 0 & \boxed{0} & \alpha'; & z_1 \\ \gamma'-\alpha'-\beta' & 1-\gamma' & \beta' & \end{matrix} \right\},$$

where $\alpha' = \alpha, \beta' = \beta, \gamma' = 1 + \alpha + \beta - \gamma$. Hence, the solution (again distinguished by a rectangle around the corresponding exponent) can be written at once:

$$w_3(z) = F(\alpha, \beta, 1 + \alpha + \beta - \gamma, \ 1 - z) . \tag{4}$$

The other solution at $z = 1$, independent of w_3, can be deduced from (3) by replacing α, β, γ therein with α', β', γ' and z with $1 - z$, i.e.,

$$w_4(z) = (1 - z)^{\gamma-\alpha-\beta} F(\gamma - \alpha, \gamma - \beta, 1 - \alpha - \beta + \gamma, 1 - z)$$
$$(\gamma - \alpha - \beta \neq \text{ integer}) . \tag{5}$$

In the same way, we obtain the two fundamental solutions of the hypergeometric equation at $z = \infty$ as

$$w_5(z) = (-z)^{-\alpha} F(\alpha, \ \alpha - \gamma + 1, \ \alpha - \beta + 1, \ z^{-1}) , \tag{6}$$

$$w_6(z) = (-z)^{-\beta} F(\beta, \ \beta - \gamma + 1, \ \beta - \alpha + 1, \ z^{-1}), \tag{7}$$
$$(\alpha - \beta \neq \text{ integer}) ;$$

the negative sign of z in the factors preceding the F functions is introduced for later convenience in discussing the connections between the various solutions at different singular points [Sec. 4.8].

From the six solutions $w_1(z), \ldots, w_6(z)$ given above, we can derive further solutions of the hypergeometric equation in terms of hypergeometric functions in other forms by means of P-equation transformations. In the present section, we shall only show those obtained by rational linear transformations. The theory concerning quadratic transformations will be dealt with in Sec. 4.12.

Take $w_1(z) = F(\alpha, \beta, \gamma, z)$ as an example. Use again a rectangle to mark the solution we are interested in the P-equation, we have, by Eq. (12) of Sec. 2.9,

$$
w_1(z) = F(\alpha, \beta, \gamma, z)
$$

$$
= P \left\{ \begin{array}{ccc} 0 & 1 & \infty \\ \boxed{0} & 0 & \alpha; \quad z \\ 1-\gamma & \gamma-\alpha-\beta & \beta \end{array} \right\}
$$

$$
= (1-z)^{\gamma-\alpha-\beta} P \left\{ \begin{array}{ccc} 0 & 1 & \infty \\ \boxed{0} & \alpha+\beta-\gamma & \gamma-\beta; \quad z \\ 1-\gamma & 0 & \gamma-\alpha \end{array} \right\} .
$$

Thus

$$
F(\alpha, \beta, \gamma, z) = C(1-z)^{\gamma-\alpha-\beta} F(\gamma-\beta, \gamma-\alpha, \gamma, z) .
$$

If we stipulate that the multi-valued function $(1-z)^{\gamma-\alpha-\beta}$ takes the value 1 when $z = 0$, then, as the values of the hypergeometric functions on both sides of the equation are equal to 1 for $z = 0$, $C = 1$ and

$$
\begin{aligned}
w_1(z) &= F(\alpha, \beta, \gamma, z) \\
&= (1-z)^{\gamma-\alpha-\beta} F(\gamma-\alpha, \gamma-\beta, \gamma, z) \\
&\quad (|\arg(1-z)| < \pi) .
\end{aligned}
\tag{8}
$$

Also, with Eqs. (10) and (12) of Sec. 2.9, we have

$$
P \left\{ \begin{array}{ccc} 0 & 1 & \infty \\ \boxed{0} & 0 & \alpha; \quad z \\ 1-\gamma & \gamma-\alpha-\beta & \beta \end{array} \right\}
$$

$$
= P \left\{ \begin{array}{ccc} 0 & \infty & 1 \\ \boxed{0} & 0 & \alpha; \quad \frac{z}{z-1} \\ 1-\gamma & \gamma-\alpha-\beta & \beta \end{array} \right\}
$$

$$
= \left(1 - \frac{z}{z-1} \right)^{\alpha} P \left\{ \begin{array}{ccc} 0 & \infty & 1 \\ \boxed{0} & \alpha & 0; \quad \frac{z}{z-1} \\ 1-\gamma & \gamma-\beta & \beta-\alpha \end{array} \right\} .
$$

Hence,

$$
F(\alpha,\beta,\gamma,z) = (1-z)^{-\alpha} F\left(\alpha, \gamma-\beta, \gamma, \frac{z}{z-1} \right)
$$
$$
(|\arg(1-z)| < \pi) . \tag{9}
$$

Exchanging α and β gives

$$
F(\alpha,\beta,\gamma,z) = (1-z)^{-\beta} F\left(\beta, \gamma-\alpha, \gamma, \frac{z}{z-1} \right)
$$
$$
(|\arg(1-z)| < \pi) . \tag{10}
$$

To summarize, we have the various expressions for $w_1(z)$:

$$
\left. \begin{array}{l}
w_1(z) = F(\alpha,\beta,\gamma,z) \\
\quad = (1-z)^{\gamma-\alpha-\beta} F(\gamma-\alpha, \gamma-\beta, \gamma, z) \\
\quad = (1-z)^{-\alpha} F\left(\alpha, \gamma-\beta, \gamma, \frac{z}{z-1} \right) \\
\quad = (1-z)^{-\beta} F\left(\beta, \gamma-\alpha, \gamma, \frac{z}{z-1} \right)
\end{array} \right\} \quad (|\arg(1-z)| < \pi) . \tag{11}
$$

By (11), we can derive for each of the six fundamental solutions $w_1(z), \ldots, w_6(z)$ four equivalent expressions. Thus, there are a total of 24 such expressions[c] for the solutions of the hypergeometric equation. Different expressions of the solutions are to be applied in different regions (relevant to z, or to α, β, γ). For example, we can use the series expression on the r.h.s. of (9) or (10), to calculate the value of $F(\alpha, \beta, \gamma, z)$ in the region $|z/(z-1)| < 1$, i.e., in the half plane $\mathrm{Re}(z) < \frac{1}{2}$.

There are six kinds of rational linear transformations in total which transform a hypergeometric equation again into a hypergeometric equation. These transformations correspond to permutations of the singular points 0, 1, ∞ among the exponents. They are tabulated in the following:

Table 2.

Transformation	$\varsigma = z$	$\varsigma = 1 - z$	$\varsigma = z^{-1}$	$\varsigma = \frac{1}{1-z}$	$\varsigma = 1 - z^{-1}$	$\varsigma = \frac{z}{z-1}$
Corresponding	0	1	∞	1	∞	0
Singularity	1	0	1	∞	0	∞
	∞	∞	0	0	1	1

The last three transformations can be derived from the second and the third.

4.4. The Second Solution of the Hypergeometric Equation when the Difference of the Exponents is an Integer

It is only necessary to discuss the second solution at $z = 0$ when γ is a positive integer; other cases may be treated by means of P-equation transformations.

Consider first *the case when* $\gamma = 1$. The two exponents at $z = 0$ are then equal to each other, $\rho_1 = \rho_2 = 0$. The first solution is still $w_1(z) = F(\alpha, \beta, \gamma, z)$. The second solution can be obtained from Eq. (8) of Sec. 2.5, which now reads (with $\rho_1 = 0$)

$$w_2(z) = w_1(z) \ln z + \sum_{k=1}^{\infty} \left(\frac{\partial c_k}{\partial \rho} \right)_{\rho=0} z^k , \tag{1}$$

where c_k satisfy the recurrence relation [Eq. (2) of Sec. 4.1]:

$$\begin{aligned}
c_k &= \frac{(\rho + k - 1 + \alpha)(\rho + k - 1 + \beta)}{(\rho + k)(\rho + k - 1 + \gamma)} c_{k-1} \\
&= \frac{(\rho + \alpha)_k (\rho + \beta)_k}{(\rho + 1)_k (\rho + \gamma)_k} c_0 \\
&= \frac{\Gamma(\rho + 1)\Gamma(\rho + \gamma)}{\Gamma(\rho + \alpha)\Gamma(\rho + \beta)} \frac{\Gamma(\rho + \alpha + k)\Gamma(\rho + \beta + k)}{\Gamma(\rho + 1 + k)\Gamma(\rho + \gamma + k)} c_0 .
\end{aligned} \tag{2}$$

---

[c] Cf. for example, Erdélyi (1953), Vol. I, p. 105, Eqs. (1) – (24).

Let $c_0 = 1$. Substituting c_k in (1) and note that $\partial c_k/\partial\rho = c_k\partial\ln c_k/\partial\rho$, we obtain

$$w_2(z) = F(\alpha,\beta,\gamma,z)\ln z + \sum_{k=0}^{\infty} \frac{(\alpha)_k(\beta)_k}{k!(\gamma)_k}z^k$$
$$\times \{\psi(\alpha+k) + \psi(\beta+k) - \psi(\gamma+k) - \psi(1+k)$$
$$- \psi(\alpha) - \psi(\beta) + \psi(\gamma) + \psi(1)\}, \qquad (3)$$

where $\psi(t) = \Gamma'(t)/\Gamma(t) = d\ln\Gamma(t)/dt, \gamma = 1$.

If α and β are not negative integers, the second solution can be taken as

$$w_2(z) = F(\alpha,\beta,1,z)\ln z$$
$$+ \sum_{k=0}^{\infty} \frac{(\alpha)_k(\beta)_k}{k!(\gamma)_k}z^k\{\psi(\alpha+k) + \psi(\beta+k) - 2\psi(1+k)\}, \qquad (4)$$

which differs from (3) by a constant multiple, $\psi(\alpha)+\psi(\beta)-2\psi(1)$, of the first solution.

If α or β is a negative integer, for example, $\alpha = -n$, then, since $\psi(-n+k)$ becomes infinite when $k \leq n$ [see Eq. (5) of Sec. 3.11], (4) loses its meaning and we have to turn back to (3) for the second solution. By Eq. (3) of Sec. 3.11, $\psi(z) = \psi(1-z) - \pi\cot\pi z$, we have

$$\lim_{\alpha\to -n}\{\psi(\alpha+k) - \psi(\alpha)\} = \lim_{\alpha\to -n}\{\psi(1-\alpha-k) - \psi(1-\alpha)\}$$
$$= \psi(1+n-k) - \psi(1+n) \quad (k \leq n) .$$

Substituting in (3) and neglecting the term which is a constant multiple of the first solution, we obtain

$$w_2(z) = F(-n,\beta,1,z)\ln z$$
$$+ \sum_{k=0}^{n} \frac{(-n)_k(\beta)_k}{(k!)^2}z^k\{\psi(1+n-k) + \psi(\beta+k) - 2\psi(1+k)\}$$
$$+ (-)^n n! \sum_{k=n+1}^{\infty} \frac{(k-n-1)!(\beta)_k}{(k!)^2}z^k , \qquad (5)$$

in which β is not a negative integer.

The case *when both α and β are negative integers* can be treated similarly.

Consider next the case *when* $\gamma = m$ (≥ 2). Here, we need formula (13) of Sec. 2.5 to find the second solution. The said formula is ($\rho_2 = 1 - m$)

$$w_2(z) = \ln z \, z^{\rho_2} \sum_{k=m-1}^{\infty} (c_k)_{\rho=\rho_2} z^k + z^{\rho_2} \sum_{k=0}^{\infty} \left(\frac{\partial c_k}{\partial \rho} \right)_{\rho=\rho_2} z^k , \qquad (6)$$

where c_k again satisfy (2), but

$$c_0 = c_0'(\rho - \rho_2) , \qquad (7)$$

c_0' being an arbitrary constant independent of ρ.

From (2) we have

$$\begin{aligned}
\left(\frac{\partial c_k}{\partial \rho} \right)_{\rho=1-m} &= \left(c_k \frac{\partial}{\partial \rho} \ln c_k \right)_{\rho=1-m} \\
&= \frac{(\alpha - m + 1)_k (\beta - m + 1)_k}{k!} \lim_{\rho \to 1-m} \Gamma(\rho + 1) \frac{c_0}{\Gamma(\rho + 1 + k)} \\
&\quad \times \{ \psi(\alpha + \rho + k) + \psi(\beta + \rho + k) - \psi(m + \rho + k) \\
&\quad - \psi(\alpha + \rho) - \psi(\beta + \rho) + \psi(m + \rho) + \psi(\rho + 1) \\
&\quad - \psi(\rho + 1 + k) + c_0'/c_0 \} .
\end{aligned} \qquad (8)$$

When $0 \leq k \leq m - 2$, by Eq. (2) of Sec. 3.5, $\Gamma(z)\Gamma(1 - z) = \pi/\sin \pi z$, thus,

$$\lim_{\rho \to 1-m} \frac{\Gamma(\rho + 1)}{\Gamma(\rho + 1 + k)} = (-)^k \frac{\Gamma(m - 1 - k)}{\Gamma(m - 1)} ,$$

and

$$\lim_{\rho \to 1-m} \{ \psi(\rho + 1) - \psi(\rho + 1 + k) \} = \psi(m - 1) - \psi(m - 1 - k) .$$

Substituting in (8) and noting that when $\rho \to \rho_2 (= 1-m)$, $c_0 = c_0'(\rho - \rho_2) \to 0$, we obtain

$$\left(\frac{\partial c_k}{\partial \rho} \right)_{\rho=1-m} = c_0'(-)^k \frac{\Gamma(m - 1 - k)}{\Gamma(m - 1)} \frac{(\alpha - m + 1)_k (\beta - m + 1)_k}{k!}$$
$$(0 \leq k \leq m - 2) . \qquad (9)$$

For $k \geq m - 1$, $\Gamma(\rho + 1 + k)$ and $\psi(\rho + 1 + k)$ are bounded when $\rho \to 1 - m$, so by Eq. (10) of Sec. 3.2, we have

$$\begin{aligned}
\lim_{\rho \to 1-m} [\Gamma(\rho + 1) c_0] &= c_0' \lim_{\rho \to 1-m} [\Gamma(\rho + 1)(\rho - 1 + m)] \\
&= c_0' \frac{(-)^m}{(m - 2)!} .
\end{aligned} \qquad (10)$$

Also,

$$\lim_{\rho \to 1-m} [\psi(\rho + 1) + c_0'/c_0] = \lim_{\rho \to 1-m} \left[\psi(-\rho) + \pi \cot \pi(-\rho) + \frac{1}{\rho - 1 + m} \right]$$
$$= \psi(m - 1) . \tag{11}$$

Thus, from (8) we find

$$\left(\frac{\partial c_k}{\partial \rho} \right)_{\rho=1-m} = c_0' \frac{(-)^m}{(m - 2)!} \frac{(\alpha - m + 1)_k (\beta - m + 1)_k}{k!(k - m + 1)!}$$
$$\times \{ \psi(\alpha - m + 1 + k) + \psi(\beta - m + 1 + k)$$
$$- \psi(1 + k) - \psi(2 - m + k) + \psi(1) + \psi(m - 1)$$
$$- \psi(\alpha - m + 1) - \psi(\beta - m + 1) \}$$
$$(k \geq m - 1) . \tag{12}$$

In calculating the first term on the r.h.s. of (6), we only need to evaluate the first coefficient; other coefficients can be deduced by the recurrence formula (2). From (2) and (10), we obtain

$$(c_{m-1})_{\rho=1-m} = c_0' \frac{(-)^m}{(m - 2)!} \frac{\Gamma(\alpha)\Gamma(\beta)}{\Gamma(m)\Gamma(\alpha - m + 1)\Gamma(\beta - m + 1)} . \tag{13}$$

Choosing c_0' so as to make this coefficient equal to 1, the first term on the r.h.s. of (6) is just $\ln z \times F(\alpha, \beta, \gamma, z)$. Substituting (9) and (12) with c_0' so determined into (6), we have

$$w_2(z) = F(\alpha, \beta, m, z) \ln z$$
$$+ z^{1-m} \frac{(-)^m \Gamma(m)}{\Gamma(\alpha)\Gamma(\beta)} \sum_{k=0}^{m-2} \frac{(-)^k}{k!} \Gamma(m - 1 - k)$$
$$\times \Gamma(\alpha - m + 1 + k)\Gamma(\beta - m + 1 + k)z^k + z^{1-m} \frac{\Gamma(m)}{\Gamma(\alpha)\Gamma(\beta)}$$

$$\times \sum_{k=m-1}^{\infty} \frac{\Gamma(\alpha - m + 1 + k)\Gamma(\beta - m + 1 + k)}{k!(k - m + 1)!} z^k$$

$$\times \{\psi(\alpha - m + 1 + k) + \psi(\beta - m + 1 + k) - \psi(1 + k)$$
$$- \psi(2 - m + k) - \psi(\alpha - m + 1) - \psi(\beta - m + 1)$$
$$+ \psi(1) + \psi(m - 1)\}$$

$$= F(\alpha, \beta, m, z) \ln z + \frac{(m - 1)!}{\Gamma(\alpha)\Gamma(\beta)} \sum_{s=1}^{m-1} (-)^{s-1}(s - 1)! \frac{\Gamma(\alpha - s)\Gamma(\beta - s)}{(m - s - 1)!} z^{-s}$$

$$+ \sum_{s=0}^{\infty} \frac{(\alpha)_s (\beta)_s}{s!(m)_s} z^s \{\psi(\alpha + s) + \psi(\beta + s) - \psi(m + s) - \psi(1 + s)$$
$$- \psi(\alpha - m + 1) - \psi(\beta - m + 1) + \psi(1) + \psi(m - 1)\} . \qquad (14)$$

If *both* α *and* β *are not negative integers*, the second solution can be taken
as

$$w_2(z) = F(\alpha, \beta, m, z) \ln z + \frac{(m - 1)!}{\Gamma(\alpha)\Gamma(\beta)}$$

$$\times \sum_{s=1}^{m=1} (-)^{s-1}(s - 1)! \frac{\Gamma(\alpha - s)\Gamma(\beta - s)}{(m - s - 1)!} z^{-s}$$

$$+ \sum_{s=0}^{\infty} \frac{(\alpha)_s (\beta)_s}{s!(m)_s} z^s \{\psi(\alpha + s) + \psi(\beta + s) - \psi(m + s) - \psi(1 + s)\} . \qquad (15)$$

If α *or* β *is a negative integer*, for example, $\alpha = -n$, (15) loses its meaning.
However, we can calculate, as in the case $\gamma = 1$ before, the indeterminate
forms, and, leaving aside the term which is a constant multiple of the first
solution, obtain the second solution in the form

$$w_2(zj = F(-n, \beta, m, z) \ln z - \frac{(m - 1)!}{\Gamma(\beta)} \sum_{s=1}^{m-1} \frac{(s - 1)!\Gamma(\beta - s)}{(m - s - 1)!(n + 1)_s} z^{-s}$$

$$+ \sum_{s=0}^{n} \frac{(-n)_s (\beta)_s}{s!(m)_s} z^s \{\psi(1 + n - s) + \psi(\beta + s) - \psi(m + s) - \psi(1 + s)\}$$

$$(\beta \neq \text{negative integer}) . \qquad (16)$$

Note that (15) and (16) include the case $\gamma = 1$, if only the finite sums
involving negative powers of z in them are negated.

When $\gamma = 0, -1, -2, \ldots$, since $1 - \gamma = 0$, the first exponent (with larger real part) of the hypergeometric equation at $z = 0$ is not 0 but $1 - \gamma$, and hence the first solution is

$$w_1(z) = z^{1-\gamma} F(\alpha - \gamma + 1, \beta - \gamma + 1, 2 - \gamma, z) \ . \tag{17}$$

The second solution can be obtained from (14) by replacing α and β in it by $\alpha - \gamma + 1$ and $\beta - \gamma + 1$ respectively, changing m to $2 - \gamma$ (positive integer) and finally multiplying with $z^{1-\gamma}$.

When $\gamma - \alpha - \beta$ *is an integer*, the second solution at $z = 1$ can also be derived from (15) (or from (16) if $\alpha = -n$). For example, $\gamma - \alpha - \beta = -m, m = 0, 1, 2, \ldots$. According to Eq. (4) of Sec. 4.3, replacing $m(= \gamma)$ in (15) by $1 + \alpha + \beta - \gamma = 1 + m$ and z by $1 - z$, we have

$$\begin{aligned}
w_2(z) = {} & F(\alpha, \beta, 1 + m, 1 - z) \ln(1 - z) \\
& + \frac{m!}{\Gamma(\alpha)\Gamma(\beta)} \sum_{s=1}^{m} (-)^{s-1}(s - 1)! \frac{\Gamma(\alpha - s)\Gamma(\beta - s)}{(m - s)!}(1 - z)^{-s} \\
& + \sum_{s=0}^{\infty} \frac{(\alpha)_s (\beta)_s}{s!(1 + m)_s} \{\psi(\alpha + s) + \psi(\beta + s) \\
& \qquad - \psi(1 + m + s) - \psi(1 + s)\}(1 - z)^s \ .
\end{aligned} \tag{18}$$

When $m = 0$, again, the finite sum in the second term is negated.

When $\alpha - \beta = m(m = 0, 1, 2, \ldots)$, by Eq. (6) of Sec. 4.3, changing β in (15) (or (16) if $\alpha = -n$) to $\alpha - \gamma - 1$, m to $m+1$, z to z^{-1}, and then multiplying with $(-z)^{-\alpha}$, we obtain the second solution at $z = \infty$ in the form:

$$\begin{aligned}
w_6(z) = {} & - (-z)^{-\alpha} F(\alpha, \alpha - \gamma + 1, m + 1, z^{-1}) \ln(-z) \\
& + (-z)^{-\alpha} \frac{m!}{\Gamma(\alpha)\Gamma(\alpha - \gamma + 1)} \sum_{s=1}^{m} (-)^{s-1}(s - 1)! \\
& \times \frac{\Gamma(\alpha - s)\Gamma(\alpha - \gamma + 1 - s)}{(m - s)!} z^s \\
& + (-z)^{-\alpha} \sum_{s=0}^{\infty} \frac{(\alpha)_s (\alpha - \gamma + 1)_s}{s!(m + 1)_s} z^{-s} \\
& \times \{\psi(\alpha + s) + \psi(\alpha - \gamma + 1 + s) - \psi(m + 1 + s) - \psi(1 + s)\} \ .
\end{aligned} \tag{19}$$

When $m = 0$, again omit the second term involving the finite sum. The fact that we use $-z$ instead of z in the logarithmic function is for the sake of convenience in defining the argument; it concerns only an additive term which is a constant multiple of the first solution.

4.5. Integral Representations of the Hypergeometric Function

The series expression of the hypergeometric function $F(\alpha, \beta, \gamma, z)$ holds only in $|z| < 1$. In many problems, such as analytic continuation, to determine the connections between the various solutions of the hypergeometric equation, etc., the series representation is not apt to work with; we need the various integral representations of the function.

There are two kinds of integral representation of the hypergeometric function. One is given by the integral solution of the hypergeometric equation, and the other, called Barnes integral representation [Sec. 4.6], is derived from the series expression.

In the example shown in Sec. 2.14, we have obtained from the Euler transformation an integral solution of the hypergeometric equation, namely,

$$w(z) = A \int_C t^{\alpha-\gamma}(1-t)^{\gamma-\beta-1}(z-t)^{-\alpha} dt \tag{1}$$

or, changing t to t^{-1},

$$w(z) = A \int_C t^{\beta-1}(1-t)^{\gamma-\beta-1}(1-zt)^{-\alpha} dt , \tag{2}$$

where A is an arbitrary constant and the path of integration C is such that the integrand or the corresponding concomitant [Eq. (21) of Sec. 2.14] has equal values, or vanishes, at the end points.

If $\mathrm{Re}(\gamma) > \mathrm{Re}(\beta) > 0$, one integral solution is

$$w(z) = A \int_0^1 t^{\beta-1}(1-t)^{\gamma-\beta-1}(1-zt)^{-\alpha} dt \tag{3}$$

[Eq. (23) of Sec. 2.14]. We shall show that this integral is uniformly convergent in the following domain:

$$|z| \le R, \ |z-1| \ge \rho , \quad |\arg(1-z)| \le \pi - \delta , \ \delta > 0 , \tag{4}$$

where R and ρ are arbitrary positive numbers, $R > \rho$. By drawing a diagram, it can be shown that $|1 - zt| \ge \rho \sin \delta/(1 + \rho)$, and, on the other hand, $|1 - zt| \le 1 + |zt| \le 1 + |z| \le 1 + R(0 \le t \le 1)$; therefore,

$$\left| t^{\beta-1}(1-t)^{\gamma-\beta-1}(1-zt)^{-\alpha} \right| \le M t^{\mathrm{Re}(\beta)-1}(1-t)^{\mathrm{Re}(\gamma-\beta)-1} ,$$

where

$$M = \left(\frac{\rho \sin \delta}{1 + \rho}\right)^{-\mathrm{Re}(\alpha)} e^{\pi|\mathrm{Im}(\alpha)|}, \quad \text{if } \mathrm{Re}(\alpha) > 0 \, ;$$

$$= (1 + R)^{-\mathrm{Re}(\alpha)} e^{\pi|\mathrm{Im}(\alpha)|}, \quad \text{if } \mathrm{Re}(\alpha) < 0 \, .$$

But the integral

$$\int_0^1 t^{\mathrm{Re}(\beta)-1}(1 - t)^{\mathrm{Re}(\gamma-\beta)-1} dt$$

converges when $\mathrm{Re}(\gamma) > \mathrm{Re}(\beta) > 0$. Hence, the integral (3) converges uniformly in the domain specified by (4), and represents a single-valued analytic function in the domain $|\arg(1 - z)| < \pi$.

By properly choosing the constant A and stipulating the value of the multi-valued function $(1 - zt)^{-\alpha}$, (3) can be made an integral representation of $F(\alpha, \beta, \gamma, z)$. We shall take $(1 - zt)^{-\alpha} = 1$ when $z = 0$ and expand $(1 - zt)^{-\alpha}$ in a series uniformly convergent in $|z| \leq R < 1$:

$$(1 - zt)^{-\alpha} = \sum_{n=0}^{\infty} \binom{-\alpha}{n}(-zt)^n = \sum_{n=0}^{\infty} \frac{(\alpha)_n}{n!} z^n t^n \, . \tag{5}$$

Substituting in (3) and using Eqs. (1) and (3) of Sec. 3.8, we have

$$w(z) = A \int_0^1 \sum_{n=0}^{\infty} \frac{(\alpha)_n}{n!} z^n t^{\beta+n-1}(1 - t)^{\gamma-\beta-1} dt$$

$$= A \sum_{n=0}^{\infty} \frac{(\alpha)_n}{n!} \frac{\Gamma(\beta + n)\Gamma(\gamma - \beta)}{\Gamma(\gamma + n)} z^n$$

$$= A \frac{\Gamma(\beta)\Gamma(\gamma - \beta)}{\Gamma(\gamma)} F(\alpha, \beta, \gamma, z) \, .$$

Hence,

$$F(\alpha, \beta, \gamma, z) = \frac{\Gamma(\gamma)}{\Gamma(\beta)\Gamma(\gamma - \beta)} \int_0^1 t^{\beta-1}(1 - t)^{\gamma-\beta-1}(1 - zt)^{-\alpha} dt \, , \tag{6}$$

where $\mathrm{Re}(\gamma) > \mathrm{Re}(\beta) > 0, |\arg(1 - z)| < \pi$, and $(1 - zt)^{-\alpha} = 1$ when $z = 0$. Here, according to the principle of analytic continuation, the restriction $|z| < 1$ can be released.

From (6) it can be seen that, in general, $z = 1$ and ∞ are branch points of $F(\alpha, \beta, \gamma, z)$. For, $t = 1/z$ is a singularity of the factor $(1 - zt)^{-\alpha}$ of the integrand (unless α is a negative integer). When z varies continuously along a path around the point $z = 1$ and returns to its initial value, the path of integration must be correspondingly deformed so as to ensure that $1/z$ does not cross it. This is necessary for the integral to represent a continuous function of z throughout. However, when z comes back to its initial value in this way, the integral does not necessarily take its original value. For example, in the manner shown in Figs. 8(a), (b), when $1/z$ varies along the dash path around $z = 1$ once and returns to the starting point, the integral

$$\int_0^1 \text{ changes into } \int_0^1 + \int_1^{(\frac{1}{z}+)} \; ,$$

whereas the last integral is, in general, not equal to zero. Thus, $z = 1$ is a branch point of $F(\alpha, \beta, \gamma, z)$ in general.

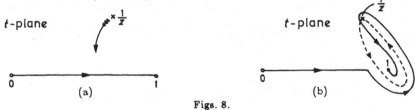

t-plane　　　　　　　　　　　　　　　　　　t-plane

(a)　　　　　　　　　　　　　　　　　　　(b)

Figs. 8.

In a similar way, it is seen that $z = \infty$ is, in general, also a branch point of $F(\alpha, \beta, \gamma, z)$, because the running of z around the point at infinity is equivalent to that $1/z$ running around the point $z = 0$. Thus the expression on the r.h.s. of (6) is a single-valued branch of $F(\alpha, \beta, \gamma, z)$, which takes the value 1 when $z = 0$.

The integral representation (6), though simple, is rather limited by the stringent conditions imposed on the parameters γ and β. When the condition $\text{Re}(\gamma) > \text{Re}(\beta) > 0$ is not satisfied, we have to resort to other representations. One of them is given by the contour integral [cf. (2)]

$$w(z) = A \int_P^{(1+,0+,1-,0-)} t^{\beta-1}(1-t)^{\gamma-\beta-1}(1-zt)^{-\alpha} dt \; ;$$

$1/z$ is to lie outside the contour. The integral is known as the Pochhammer double-contour integral, which has been depicted in Sec. 3.9 with the contour shown in Fig. 6. First let $|z|$ be sufficiently small such that $|zt| < 1$ on the contour. Then, as before, we can expand $(1 - zt)^{-\alpha}$ as (5) and substitute it in the last equation, obtaining

$$w(z) = A \sum_{n=0}^{\infty} \frac{(\alpha)_n}{n!} z^n \int_P^{(1+,0+,1-,0-)} t^{\beta+n-1}(1-t)^{\gamma-\beta-1} dt \; .$$

By Eq. (1) of Sec. 3.9, this becomes

$$
\begin{aligned}
w(z) &= A \sum_{n=0}^{\infty} \frac{(2\pi i)^2 e^{i\pi(\gamma+n)}}{\Gamma(1-\beta-n)\Gamma(1-\gamma+\beta)\Gamma(\gamma+n)} \frac{(\alpha)_n}{n!} z^n \\
&= A \frac{-4e^{i\pi\gamma}\Gamma(\beta)\Gamma(\gamma-\beta)\sin\pi\beta\sin\pi(\gamma-\beta)}{\Gamma(\gamma)} F(\alpha,\beta,\gamma,z) \ .
\end{aligned}
$$

Therefore,

$$
\begin{aligned}
F(\alpha,\beta,\gamma,z) &= -\frac{e^{-i\pi\gamma}\Gamma(\gamma)}{4\Gamma(\beta)\Gamma(\gamma-\beta)\sin\pi\beta\sin\pi(\gamma-\beta)} \\
&\times \int_P^{(1+,0+,1-,0-)} t^{\beta-1}(1-t)^{\gamma-\beta-1}(1-zt)^{-\alpha}dt \ , \\
&\qquad\qquad (\gamma \neq 0,-1,-2,\dots)
\end{aligned}
\tag{7}
$$

where $|\arg(1-z)| < \pi$; at the starting point P of the contour, $\arg t = \arg(1-t) = 0$, and $(1-zt)^{-\alpha} = 1$ when $z = 0$.

When β or $\gamma-\beta$ is a positive integer, the r.h.s. of (7) is indefinite. In such cases, the integral representation given in Ex. 7 at the end of this chapter is used.

When γ is zero or a negative integer. $\Gamma(\gamma) = \infty$, (7) loses its meaning. However, it furnishes an integral representation for the function $F(\alpha,\beta,\gamma,z)/\Gamma(\gamma)$, and still gives an integral solution of the hypergeometric equation.

From (6), (7) and the result given in Ex. 6 at the end of this chapter, it is seen that, when z and $\gamma(\neq 0,-1,-2,\dots)$ are fixed, $F(\alpha,\beta,\gamma,z)$ is an integral function of α and β; and, when z and α,β are fixed, it is a meromorphic function of γ having the same singularities of $\Gamma(\gamma)$.

4.6. Barnes' Integral Representation of the Hypergeometric Function

In $|z| < 1$, the hypergeometric function can be expressed as

$$
F(\alpha,\beta,\gamma,z) = \frac{\Gamma(\gamma)}{\Gamma(\alpha)\Gamma(\beta)} \sum_{n=0}^{\infty} \frac{G(n)}{n!} z^n \ ,
\tag{1}
$$

where

$$
G(n) = \frac{\Gamma(\alpha+n)\Gamma(\beta+n)}{\Gamma(\gamma+n)}
\tag{2}
$$

[cf. Eqs. (5) and (6) of Sec. 4.1]. We know that $\Gamma(-s)$ is a meromorphic function with $s = n = 0, 1, 2, \ldots$ as simple poles, the corresponding residues being $(-)^{n+1}/n!$ [see Eq. (10) of Sec. 3.2]. Hence, if α and β are not negative integers, we can write

$$F(\alpha, \beta, \gamma, z) = \frac{\Gamma(\gamma)}{\Gamma(\alpha)\Gamma(\beta)} \sum_{s=n=0}^{\infty} \text{Res} \left\{ -G(s)\Gamma(-s)(-z)^s \right\}$$

$$= \frac{\Gamma(\gamma)}{\Gamma(\alpha)\Gamma(\beta)} \frac{1}{2\pi i} \oint_C G(s)\Gamma(-s)(-z)^s ds , \tag{3}$$

where

$$G(s) = \frac{\Gamma(\alpha + s)\Gamma(\beta + s)}{\Gamma(\gamma + s)} . \tag{4}$$

The contour C encloses in it all the poles $s = n = 0, 1, 2, \ldots$ of $\Gamma(-s)$ but not a single pole of $G(s)$. Contour of this sort always exists when α and β are not zero or negative integers. For, in such cases, the poles of $G(s)$, namely, $s = -\alpha - n, -\beta - n(n = 0, 1, 2, \ldots)$, never coincide with the poles $s = n$ of $\Gamma(-s)$. So, we can choose for contour C the one shown in Fig. 9, which starts from $-iR$ on the imaginary axis, running along the imaginary axis, evading, when necessary, the singularities of the integrand, to iR, then goes back to the starting point along a semicircle C_R on the right, where $R = N + \frac{1}{2}, N$ being a positive integer $\rightarrow \infty$. The path along the imaginary axis is shown to be so detoured that it keeps all the poles of $\Gamma(-s)$ on its right and those of $G(s)$ on its left.

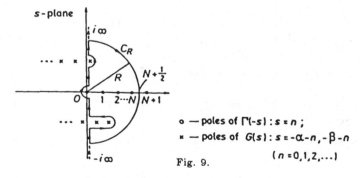

Fig. 9.

o — poles of $\Gamma(-s) : s = n$;

x — poles of $G(s) : s = -\alpha - n, -\beta - n$

$(n = 0, 1, 2, \ldots)$

We show first that the contour integral on the r.h.s. of (3) is meaningful. For this purpose we make use of Eq. (2) of Sec. 3.5 to write the integrand as

$$G(s)\Gamma(-s)(-z)^s = \frac{\Gamma(\alpha + s)\Gamma(\beta + s)}{\Gamma(\gamma + s)\Gamma(1 + s)} \frac{-\pi}{\sin \pi s} (-z)^s . \tag{5}$$

By the asymptotic expression for the Γ-function [Eq. (5) of Sec. 3.2], namely,

$$\ln \Gamma(\lambda + s) = \left(s + \lambda - \frac{1}{2}\right) \ln s - s + \frac{1}{2} \ln(2\pi) + O(s^{-1}) , \qquad (6)$$

we estimate that (5) is of the order

$$O(s^{\alpha+\beta-\gamma-1}) \frac{(-z)^s}{\sin \pi s} , \qquad |s| \to \infty, \quad |\arg s| < \pi . \qquad (7)$$

When $s \to \infty$ along the imaginary axis, (7) becomes

$$O(|s|^{\alpha+\beta-\gamma} \exp\{-\pi|\text{Im}(s)| - \arg(-z)\text{Im}(s)\}) . \qquad (8)$$

Hence, in the domain $|\arg(-z)| \leq \pi - \delta < \pi$, the integral $\int_{-i\infty}^{i\infty}$ converges uniformly.

On $C_R, s = Re^{i\theta}, R = N + \frac{1}{2}$. For large N, if $|z| < 1$, then $\ln |z| < 0$,

$$|(-z)^s \csc s\pi| = O\left(\exp\left\{s \ln |z| + is \arg(-z) - \pi \left(N + \frac{1}{2}\right) |\sin \theta|\right\}\right)$$

$$= O\left(\exp\left\{\left(N + \frac{1}{2}\right) \cos \theta \ln |z| - \left(N + \frac{1}{2}\right)\right.\right.$$

$$\left.\left. \times \sin \theta \arg(-z) - \pi \left(N + \frac{1}{2}\right) |\sin \theta|\right\}\right)$$

$$= O\left(\exp\left\{\left(N + \frac{1}{2}\right) \cos \theta \ln |z| - \left(N + \frac{1}{2}\right) \delta |\sin \theta|\right\}\right)$$

$$= \begin{cases} O\left(\exp\left\{(N + \frac{1}{2}) 2^{-1/2} \ln |z|\right\}\right) , & 0 \leq \theta \leq \frac{\pi}{4} , \\ O\left(\exp\left\{-(N + \frac{1}{2}) \delta 2^{-1/2}\right\}\right) , & \frac{\pi}{4} \leq \theta \leq \frac{\pi}{2} , \end{cases}$$

and

$$\lim_{N \to \infty} \int_{C_R} \longrightarrow 0 .$$

Therefore,

$$F(\alpha, \beta, \gamma, z)$$

$$= \frac{\Gamma(\gamma)}{\Gamma(\alpha)\Gamma(\beta)} \frac{1}{2\pi i} \int_{-i\infty}^{i\infty} \frac{\Gamma(\alpha + s)\Gamma(\beta + s)}{\Gamma(\gamma + s)} \Gamma(-s)(-z)^s ds , \qquad (9)$$

where $|\arg(-z)| < \pi$ and the contour is such that the poles of $\Gamma(-s)$ lie on its right and those of $\Gamma(\alpha + s)\Gamma(\beta + s)$ on its left. Of course, this requires that α

and β are not zero or negative integers. (9) is *Barnes' integral representation of the hypergeometric function, and the path of integration, Barnes' contour.*

Although (9) was proved under the condition $|z| < 1$, the result holds without this restriction, since both sides of the equation are not subjected to the restriction.

4.7. The Value of $F(\alpha, \beta, \gamma, 1)$

The value of $F(\alpha, \beta, \gamma, 1)$ can be interpreted in two ways: one is the value of the hypergeometric function at $z = 1$; the other is the sum of the hypergeometric series, which is only known to represent the hypergeometric function in $|z| < 1$, as $z \to 1$, namely,

$$\sum_{n=0}^{\infty} \frac{(\alpha)_n (\beta)_n}{n!(\gamma)_n} . \tag{1}$$

The value of the hypergeometric function at $z = 1$ can be evaluated from the integral representation of the function [Eq. (6) of Sec. 4.5]:

$$F(\alpha, \beta, \gamma, z) = \frac{\Gamma(\gamma)}{\Gamma(\beta)\Gamma(\gamma - \beta)} \int_0^1 t^{\beta-1}(1 - t)^{\gamma-\beta-1}(1 - zt)^{-\alpha} dt ,$$
$$(\text{Re}(\gamma) > \text{Re}(\beta) > 0) . \tag{2}$$

The result is simply

$$\begin{aligned} F(\alpha, \beta, \gamma, 1) &= \frac{\Gamma(\gamma)}{\Gamma(\beta)\Gamma(\gamma - \beta)} \int_0^1 t^{\beta-1}(1 - t)^{\gamma-\alpha-\beta-1} dt \\ &= \frac{\Gamma(\gamma)\Gamma(\gamma - \alpha - \beta)}{\Gamma(\gamma - \alpha)\Gamma(\gamma - \beta)} , \end{aligned} \tag{3}$$

so long as $\text{Re}(\gamma - \alpha - \beta) > 0, \text{Re}(\gamma) > \text{Re}(\beta) > 0$. (We shall see later that the last condition can be removed.)

As regards the sum of the series (1), it is necessary to investigate its condition of convergence, since it is the value of the hypergeometric series on its circle of convergence.

According to the criterion for the convergence of a series of complex numbers,[d] we know from Eq. (4) of Sec. 4.1 that, when $\text{Re}(\gamma - \alpha - \beta) > 0$, series

---

[d]Cf. for example, Bromwich (1925), p. 241, Sec. 79.

(1) is absolutely convergent (if $\text{Re}(\gamma - \alpha - \beta) \leq 0$, the series is divergent, unless it is a finite sum). Therefore, by Abel's second theorem,[e] and (3)

$$\sum_{n=0}^{\infty} \frac{(\alpha)_n (\beta)_n}{n!(\gamma)_n} = \lim_{z \to 1-0} F(\alpha, \beta, \gamma, z) = F(\alpha, \beta, \gamma, 1)$$

$$= \frac{\Gamma(\gamma)\Gamma(\gamma - \alpha - \beta)}{\Gamma(\gamma - \alpha)\Gamma(\gamma - \beta)}$$

$$(\text{Re}(\gamma - \alpha - \beta) > 0) \tag{4}$$

since the integral in (2) is uniformly convergent in the domain $|z| \leq 1$ and $|z - 1| \leq 1$.

Equations (3) and (4) were obtained under the conditions: $\text{Re}(\gamma - \alpha - \beta) > 0$ and $\text{Re}(\gamma) > \text{Re}(\beta) > 0$. We will now show that the last condition is not necessary.

Suppose $\text{Re}(\gamma) > \text{Re}(\beta) > -1$. From the recurrence relation

$$F(\alpha, \beta, \gamma, z) = F(\alpha, \beta + 1, \gamma + 1, z)$$

$$- \frac{\alpha(\gamma - \beta)}{\gamma(\gamma + 1)} z F(\alpha + 1, \beta + 1, \gamma + 2, z) , \tag{5}$$

applying (3) on the r.h.s. of it, we obtain

$$F(\alpha, \beta, \gamma, 1) = \frac{\Gamma(\gamma + 1)\Gamma(\gamma - \alpha - \beta)}{\Gamma(\gamma - \alpha + 1)\Gamma(\gamma - \beta)} - \frac{\alpha(\gamma - \beta)}{\gamma(\gamma + 1)} \frac{\Gamma(\gamma + 2)\Gamma(\gamma - \alpha - \beta)}{\Gamma(\gamma - \alpha + 1)\Gamma(\gamma - \beta + 1)}$$

$$= \frac{\Gamma(\gamma)\Gamma(\gamma - \alpha - \beta)}{\Gamma(\gamma - \alpha)\Gamma(\gamma - \beta)} ,$$

so that (3) holds for $\text{Re}(\gamma) > \text{Re}(\beta) > -1$. Repeating this process, we see by induction that (3) is valid so long as $\text{Re}(\gamma) > \text{Re}(\beta)$.

Now suppose $\text{Re}(\gamma) > \text{Re}(\beta) - 1$. From the recurrence relation

$$F(\alpha, \beta, \gamma, z) = F(\alpha + 1, \beta - 1, \gamma, z)$$

$$+ \frac{\alpha - \beta + 1}{\gamma} z F(\alpha + 1, \beta, \gamma + 1, z) , \tag{6}$$

applying again (3) on its r.h.s., we obtain

$$F(\alpha, \beta, \gamma, 1) = \frac{\Gamma(\gamma)\Gamma(\gamma - \alpha - \beta)}{\Gamma(\gamma - \alpha - 1)\Gamma(\gamma - \beta + 1)} + \frac{\alpha - \beta + 1}{\gamma} \frac{\Gamma(\gamma + 1)\Gamma(\gamma - \alpha - \beta)}{\Gamma(\gamma - \alpha)\Gamma(\gamma - \beta + 1)}$$

$$= \frac{\Gamma(\gamma)\Gamma(\gamma - \alpha - \beta)}{\Gamma(\gamma - \alpha)\Gamma(\gamma - \beta)} .$$

---

[e] Bromwich (1925), p. 252, Sec. 86.

Thus, (3) holds for $\text{Re}(\gamma) > \text{Re}(\beta) - 1$. By repeating this process and by induction, it is seen that the restriction $\text{Re}(\gamma) > \text{Re}(\beta)$ can be removed.

The recurrence relations (5) and (6) used above can be verified from the series expressions on the two sides of the equations, or, by deriving (5) from Eqs. (2) and (3) of Sec. 4.2, and (6) from Eqs. (3) and (6) of that section.

Another proof for the summation formula (4)

In the recurrence relation

$$
\begin{aligned}
\gamma[\gamma - 1 &- (2\gamma - \alpha - \beta - 1)z]F(\alpha, \beta, \gamma, z) \\
&+ (\gamma - \alpha)(\gamma - \beta)zF(\alpha, \beta, \gamma + 1, z) \\
&= \gamma(\gamma - 1)(1 - z)F(\alpha, \beta, \gamma - 1, z) \\
&= \gamma(\gamma - 1)\left\{ 1 + \sum_{n=1}^{\infty}(v_n - v_{n-1})z^n \right\},
\end{aligned}
\tag{7}
$$

[see Ex. 2 at the end of this chapter], where v_n is the coefficient of z^n in the series expression of $F(\alpha, \beta, \gamma - 1, z)$, let $z \to 1 - 0$ (along the real axis). If $1 + \sum_{n=1}^{\infty}(v_n - v_{n-1})$ converges to zero, then, by Abel's second theorem, the r.h.s. of (7) also tends to zero. Now, $1 + \sum_{n=1}^{\infty}(v_n - v_{n-1}) = \lim_{n \to \infty} v_n$. Using the asymptotic expression of the Γ-function [(6) of the preceding section], we have

$$
v_n = \frac{\Gamma(\gamma - 1)}{\Gamma(\alpha)\Gamma(\beta)}\frac{\Gamma(\alpha + n)\Gamma(\beta + n)}{\Gamma(\gamma - 1 + n)\Gamma(1 + n)} = O(n^{\alpha + \beta - \gamma}) .
$$

Thus, if $\text{Re}(\gamma - \alpha - \beta) > 0$, then $\lim_{n \to \infty} v_n = 0$, and

$$
\gamma(\alpha + \beta - \gamma)F(\alpha, \beta, \gamma, 1) + (\gamma - \alpha)(\gamma - \beta)F(\alpha, \beta, \gamma + 1, 1) = 0 ,
$$

or,

$$
F(\alpha, \beta, \gamma, 1) = \frac{(\gamma - \alpha)(\gamma - \beta)}{\gamma(\gamma - \alpha - \beta)}F(\alpha, \beta, \gamma + 1, 1) .
$$

Repeating this relation, we obtain

$$
\begin{aligned}
F(\alpha, \beta, \gamma, 1) &= \left\{ \prod_{n=0}^{m-1} \frac{(\gamma - \alpha + n)(\gamma - \beta + n)}{(\gamma + n)(\gamma - \alpha - \beta + n)} \right\} F(\alpha, \beta, \gamma + m, 1) \\
&= \left\{ \lim_{m \to 0} \prod_{n=0}^{m-1} \frac{(\gamma - \alpha + n)(\gamma - \beta + n)}{(\gamma + n)(\gamma - \alpha - \beta + n)} \right\} \lim_{m \to \infty} F(\alpha, \beta, \gamma + m, 1) ,
\end{aligned}
$$

provided that the two limits in it exist. By Eq. (4) of Sec. 3.3, so long as γ is not a negative integer, the first limit is found immediately to be $\Gamma(\gamma)\Gamma(\gamma - \alpha - \beta)/\Gamma(\gamma - \alpha)\Gamma(\gamma - \beta)$. The second limit can be evaluated as follows: Let $c_n(\alpha, \beta, \gamma)$ be the coefficient of z^n in the series expression of $F(\alpha, \beta, \gamma, z)$, i.e., $c_n(\alpha, \beta, \gamma) = (\alpha)_n(\beta)_n/n!(\gamma)_n$, and assume $m > |\gamma|$, then,

$$|F(\alpha, \beta, \gamma + m, 1) - 1| \leq \sum_{n=1}^{\infty} |c_n(\alpha, \beta, \gamma + m)|$$

$$\leq \sum_{n=1}^{\infty} c_n(|\alpha|, |\beta|, \ m - |\gamma|)$$

$$< \frac{|\alpha\beta|}{m - |\gamma|} \sum_{n=0}^{\infty} c_n(|\alpha| + 1, \ |\beta| + 1, \ m + 1 - |\gamma|) \ .$$

But the last series is convergent when $m > |\gamma| + |\alpha| + |\beta| + 1$ (which corresponds to the condition of convergence for series (1): $\mathrm{Re}(\gamma - \alpha - \beta) > 0$), and its value decreases as m increases. Therefore the r.h.s. of the last equation approaches zero when $m \to \infty$, so

$$\lim_{m \to \infty} F(\alpha, \beta, \gamma + m, 1) = 1 \qquad (8)$$

which proves (4).

4.8. Connections between the Fundamental Solutions at the Singular Points 0, 1, ∞. Analytic Continuation

An ordinary linear differential equation of the second order can only have two linearly independent solutions. Therefore, among any three of the six solutions, $w_1, w_2, \ldots, w_6$, obtained in Sec. 4.3 for the hypergeometric equation, there must exist a linear relation. There are in total $\binom{6}{3} = 20$ such relations.[f] Here we shall discuss the most fundamental ones of them, namely, that between w_1 and w_3, w_4, and also that between w_1 and w_5, w_6.

According to the theory of analytic continuation of the solution of a differential equation [Sec. 2.3], in the intersecting region of $|z| < 1$ and $|1 - z| < 1$, there should be a relation of the form $w_1 = Aw_3 + Bw_4$. If $\gamma - \alpha - \beta$ is not an integer, then, by Eqs. (4) and (5) of Sec. 4.3, we have

$$F(\alpha, \beta, \gamma, z) = AF(\alpha, \beta, \alpha + \beta - \gamma + 1, 1 - z)$$
$$+ B(1 - z)^{\gamma - \alpha - \beta} F(\gamma - \alpha, \gamma - \beta, \gamma - \alpha - \beta + 1, 1 - z) \ .$$
$$(1)$$

---

[f]Cf. for example, Erdélyi (1953), Vol. I, p. 106, (25) – (44).

Let $|\arg(1-z)| < \pi$ and $\mathrm{Re}(\gamma - \alpha - \beta) > 0$. Putting $z = 1$, by (3) of the preceding section, we have

$$A = \Gamma(\gamma)\Gamma(\gamma - \alpha - \beta)/\Gamma(\gamma - \alpha)\Gamma(\gamma - \beta) . \tag{2}$$

With A so determined and setting $z = 0$ in (1), assuming $\mathrm{Re}(1 - \gamma) > 0$, we may apply (3) of the last section and find also

$$B = \Gamma(\gamma)\Gamma(\alpha + \beta - \gamma)/\Gamma(\alpha)\Gamma(\beta) . \tag{3}$$

Hence,

$$\begin{aligned}
F(\alpha, \beta, \gamma, z) = {} & \frac{\Gamma(\gamma)\Gamma(\gamma - \alpha - \beta)}{\Gamma(\gamma - \alpha)\Gamma(\gamma - \beta)} F(\alpha, \beta, \alpha + \beta - \gamma + 1, 1 - z) \\
& + \frac{\Gamma(\gamma)\Gamma(\alpha + \beta - \gamma)}{\Gamma(\alpha)\Gamma(\beta)} (1 - z)^{\gamma - \alpha - \beta} \\
& \times F(\gamma - \alpha, \gamma - \beta, \gamma - \alpha - \beta + 1, 1 - z) . \\
& (|\arg(1 - z)| < \pi)
\end{aligned} \tag{4}$$

This connection formula is obtained under the assumptions $\mathrm{Re}(\gamma - \alpha - \beta) > 0$ and $\mathrm{Re}(1 - \gamma) > 0$. But both of these restrictions can be removed.[g]

If $\mathrm{Re}(\gamma - \alpha - \beta) < 0$, we can use Eq. (8) of Sec. 4.3 to transform (1) into

$$\begin{aligned}
F(\gamma - \alpha, \gamma - \beta, \gamma, z) = {} & A(1 - z)^{\alpha + \beta - \gamma} F(\alpha, \beta, \alpha + \beta - \gamma + 1, 1 - z) \\
& + B F(\gamma - \alpha, \gamma - \beta, \gamma - \alpha - \beta + 1, 1 - z) .
\end{aligned} \tag{5}$$

Assume $\mathrm{Re}(1 - \gamma) < 0$, by putting $z = 1$ we have

$$B = F(\gamma - \alpha, \gamma - \beta, \gamma, 1) = \frac{\Gamma(\gamma)\Gamma(\alpha + \beta - \gamma)}{\Gamma(\alpha)\Gamma(\beta)} ,$$

the same as (3). Putting $z = 0$, we find the same A as (2).

If $\mathrm{Re}(1 - \gamma) < 0$, i.e., $\mathrm{Re}(\gamma - 1) > 0$, we may apply Eq. (8) of Sec. 4.3 on the r.h.s. of (1) and obtain

$$\begin{aligned}
z^{\gamma - 1} F(\alpha, \beta, \gamma, z) = {} & A F(\alpha - \gamma + 1, \beta - \gamma + 1, \alpha + \beta - \gamma + 1, 1 - z) \\
& + B(1 - z)^{\gamma - \alpha - \beta} F(1 - \alpha, 1 - \beta, \gamma - \alpha - \beta + 1, 1 - z) .
\end{aligned} \tag{6}$$

---

[g] Of course, we still assume that γ is not zero or a negative integer, and that $\gamma - \alpha - \beta$ is not an integer (see the following section).

Assume $\text{Re}(\gamma - \alpha - \beta) > 0$, and setting z equal to 1 and 0 respectively in it, we again arrive at the same A and B.

In the case when $\text{Re}(\gamma - \alpha - \beta) < 0$, and $\text{Re}(1 - \gamma) < 0$, we only need to apply Eq. (8) of Sec. 4.3 to transform l.h.s. of (6). The same A and B will be found. Therefore, the validity of (4) is not restricted by the original assumption: $\text{Re}(\gamma - \alpha - \beta) > 0$ and $\text{Re}(1 - \gamma) > 0$.

(4) may be employed to evaluate the value of the hypergeometric function for $z \simeq 1$, since the series on the r.h.s. of it will converge rapidly.

Let us now turn to the connection formula between w_1 and w_5, w_6; it should be of the form $w_1 = Cw_5 + Dw_6$. If $\alpha - \beta$ is not an integer, then by Eqs. (6) and (7) of Sec. 4.3, we have

$$F(\alpha, \beta, \gamma, z) = C(-z)^{-\alpha} F(\alpha, \alpha - \gamma + 1, \alpha - \beta + 1, z^{-1})$$
$$+ D(-z)^{-\beta} F(\beta, \beta - \gamma + 1, \beta - \alpha + 1, z^{-1}) . \tag{7}$$

Let $|\arg(1 - z)| < \pi$, $\text{Re}(\beta) > \text{Re}(\alpha)$, then, when $|z| \to \infty$, the r.h.s. of (7) $\sim C(-z)^{-\alpha}$, and the l.h.s., by Eq. (9) of Sec. 4.3, is

$$F(\alpha, \beta, \gamma, z) = (1 - z)^{-\alpha} F(\alpha, \gamma - \beta, \gamma, z/(z - 1))$$
$$\sim (-z)^{-\alpha} F(\alpha, \gamma - \beta, \gamma, 1) = (-z)^{-\alpha} \frac{\Gamma(\gamma)\Gamma(\beta - \alpha)}{\Gamma(\gamma - \alpha)\Gamma(\beta)} .$$

Therefore,

$$C = \frac{\Gamma(\gamma)\Gamma(\beta - \alpha)}{\Gamma(\gamma - \alpha)\Gamma(\beta)} .$$

In (7), exchanging α and β, and noticing that both C and D depend on α and β, we see that

$$D = \frac{\Gamma(\gamma)\Gamma(\alpha - \beta)}{\Gamma(\gamma - \beta)\Gamma(\alpha)} .$$

Hence

$$F(\alpha, \beta, \gamma, z) = \frac{\Gamma(\gamma)\Gamma(\beta - \alpha)}{\Gamma(\gamma - \alpha)\Gamma(\beta)}(-z)^{-\alpha} F(\alpha, \alpha - \gamma + 1, \alpha - \beta + 1, z^{-1})$$
$$+ \frac{\Gamma(\gamma)\Gamma(\alpha - \beta)}{\Gamma(\gamma - \beta)\Gamma(\alpha)}(-z)^{-\beta} F(\beta, \beta - \gamma + 1, \beta - \alpha + 1, z^{-1})$$
$$(|\arg(-z)| < \pi) . \tag{8}$$

Obviously, (8) is not subjected to the restriction $\text{Re}(\alpha) > \text{Re}(\beta)$, since it is symmetric in α and β. This formula can be employed to calculate the value of the hypergeometric function when $|z| > 1$.

Equations (4) and (8), and also Eqs. (9) and (10) of Sec. 4.3, can be looked upon as the analytic continuation formulae for the hypergeometric function.

4.9. When $\gamma - \alpha - \beta, \alpha - \beta$ are Integers

Equation (4) of the preceding section holds only when $\gamma - \alpha - \beta$ is not an integer and (8) holds only when $\alpha - \beta$ is not an integer. This is because of the fact that when $\gamma - \alpha - \beta$ (or $\alpha - \beta$) is an integer, the second solution at $z = 1(z = \infty)$ usually contains a logarithmic term [Sec. 4.4]. In such cases, the simple method employed in the last section to derive the connection between the solutions at different points is not valid, since there is no formula corresponding to Eq. (3) or (4) of Sec. 4.7 for the solution with a logarithmic term. However, this difficulty may be overcome with Barnes' integral representation [Sec. 4.6].[h]

First, suppose $\alpha - \beta = m(m = 0, 1, 2, \ldots)$. We shall see how (8) of the preceding section is modified.

According to Eq. (9) of Sec. 4.6, we have

$$\frac{\Gamma(\alpha)\Gamma(\beta)}{\Gamma(\gamma)} F(\alpha, \beta, \gamma, z)$$
$$= \frac{1}{2\pi i} \int_{-i\infty}^{i\infty} \frac{\Gamma(\alpha + s)\Gamma(\beta + s)}{\Gamma(\gamma + s)} \Gamma(-s)(-z)^s ds$$
$$(|\arg(-z)| < \pi), \tag{1}$$

where the path of integration has to keep the singularities (poles) of $\Gamma(-s)$ on its right and those of $\Gamma(\alpha + s)\Gamma(\beta + s)$ on its left. When α and β are not negative integers, such a path always exists [Sec. 4.6]. Now, we supplement the path of integration in (1) with a semi-circle C_R on the left of the imaginary axis, centre at $s = 0$ and radius equal to R. When $R \to \infty$, and C_R is not to pass through any poles of $\Gamma(\alpha + s)\Gamma(\beta + s)$, it can be shown, in a similar way as in Sec. 4.6, that, for $|z| > 1$, the value of the integral along C_R approaches zero. Hence, the original integral in (1) is equal to the sum of the residues of the integrand at the poles of $\Gamma(\alpha + s)\Gamma(\beta + s)$. When $\alpha - \beta = m$, assuming that $\gamma - \beta$ is not an integer.

$$\beta + s = -k \ (k = 0, 1, 2, \ldots, m - 1) \quad \text{are simple poles;}$$
$$\beta + s = -k \ (k = m, m + 1, \ldots) \quad \text{are poles of order 2 .}$$

---

[h]Of course, this method is also applicable when $\gamma - \alpha - \beta$ (or $\alpha - \beta$) is not an integer; thus, it is a rather general method.

(if $m = 0$, all the poles are of order 2). Therefore,

$$\frac{\Gamma(\alpha)\Gamma(\beta)}{\Gamma(\gamma)} F(\alpha, \beta, \gamma, z) = \sum_{k=0}^{m-1} R_k + \sum_{k=m}^{\infty} R_k' , \tag{1'}$$

$$R_k = \lim_{s \to -\beta-k} (s + \beta + k) \frac{\Gamma(\beta + m + s)\Gamma(\beta + s)}{\Gamma(\gamma + s)} \Gamma(-s)(-z)^s$$

$$= \frac{\Gamma(m - k)(-)^k}{\Gamma(\gamma - \beta - k)k!} \Gamma(\beta + k)(-z)^{-\beta-k} ,$$

$$R_k' = \lim_{s \to -\beta-k} \frac{d}{ds} \left\{ (s + \beta + k)^2 \frac{\Gamma(\beta + m + s)\Gamma(\beta + s)}{\Gamma(\gamma + s)} \Gamma(-s)(-z)^s \right\}$$

$$= \frac{(-)^m}{(k - m)!k!} \frac{\Gamma(\beta + k)}{\Gamma(\gamma - \beta - k)} (-z)^{-\beta-k}$$
$$\times \{ \psi(k - m + 1) + \psi(k + 1) - \psi(\beta + k) - \psi(\gamma - \beta - k) + \ln(-z) \} .$$

In the last sum on the r.h.s. of (1'), replacing k by $k + m$ and noting that $\alpha = \beta + m$, we obtain

$$F(\alpha, \beta, \gamma, z) = \frac{\Gamma(\gamma)}{\Gamma(\alpha)} (-z)^{-\beta} \sum_{k=0}^{m-1} \frac{(\beta)_k \Gamma(m - k)}{k! \Gamma(\gamma - \beta - k)} z^{-k}$$
$$+ \frac{\Gamma(\gamma)}{\Gamma(\alpha)\Gamma(\gamma - \beta)} (-z)^{-\alpha} \sum_{k=0}^{\infty} \frac{(\beta)_{k+m} (1 - \gamma + \beta)_{k+m}}{k!(k + m)!} z^{-k}$$
$$\times \{ \psi(k + 1) + \psi(k + m + 1) - \psi(\alpha + k)$$
$$- \psi(\gamma - \alpha - k) + \ln(-z) \} . \tag{2}$$

$(\alpha - \beta = m = 0, 1, 2, \ldots ; \gamma - \beta \neq$ integer, $\alpha \neq$ negative integer; $|\arg(-z)| < \pi, |z| > 1$; when $m = 1$, drop the finite sum.)

A little manipulation will show that the r.h.s. of (2) is a linear combination of the two fundamental solutions of the hypergeometric equation at $z = \infty$ [cf. Eq. (19) of Sec. 4.4].

When $\gamma - \alpha - \beta$ *is an integer*, to obtain a connection formula for $F(\alpha, \beta, \gamma, z)$ and the two fundamental solutions at $z = 1$, we have to appeal to the *Barnes' lemma:*

$$\frac{1}{2\pi i} \int_{-i\infty}^{i\infty} \Gamma(\alpha + s)\Gamma(\beta + s)\Gamma(\gamma - s)\Gamma(\delta - s)ds$$

$$= \frac{\Gamma(\alpha + \gamma)\Gamma(\alpha + \delta)\Gamma(\beta + \gamma)\Gamma(\beta + \delta)}{\Gamma(\alpha + \beta + \gamma + \delta)} , \tag{3}$$

where the path of integration is such that the singularities of $\Gamma(\alpha + s)\Gamma(\beta + s)$ are kept on its left and those of $\Gamma(\gamma - s)\Gamma(\delta - s)$ on its right, assuming that the poles of the Γ-functions do not overlap.

The proof of (3) is as follows. Let C be a semi-circle on the right of the imaginary axis with center at $s = 0$ and radius ρ. If $\rho \to \infty$, and C is not to pass through any poles of $\Gamma(\gamma - s)\Gamma(\delta - s)$, then, when $s \to \infty$ along the imaginary axis or on C, by the asymptotic expansion formula for the Γ-function [Eq. (5) of Sec. 3.21, or Eq. (6) of Sec. 4.6], we have

$$\Gamma(\alpha + s)\Gamma(\beta + s)\Gamma(\gamma - s)\Gamma(\delta - s)$$

$$= \frac{\Gamma(\alpha + s)\Gamma(\beta + s)}{\Gamma(1 - \gamma + s)\Gamma(1 - \delta + s)}\pi^2 \csc \pi(\gamma - s) \csc \pi(\delta - s)$$

$$= O(s^{\alpha + \beta + \gamma + \delta - 2} \exp\{-2\pi|\mathrm{Im}(s)|\}) . \tag{4}$$

The integral in (3) is thus convergent and the integral along the semi-circle C approaches zero when $\rho \to \infty$, provided that $\mathrm{Re}(\alpha + \beta + \gamma + \delta - 1) < 0$. Hence, the value of the integral in (3), denoted by I, is equal to the sum of the residues of the integrand at the poles of $\Gamma(\delta - s)\Gamma(\gamma - s)$:

$$I = \sum_{n=0}^{\infty} \frac{\Gamma(\alpha + \gamma + n)\Gamma(\beta + \gamma + n)}{\Gamma(1 + n)\Gamma(1 - \delta + \gamma + n)} \frac{\pi}{\sin \pi(\delta - \gamma)}$$

$$+ \sum_{n=0}^{\infty} \frac{\Gamma(\alpha + \delta + n)\Gamma(\beta + \delta + n)}{\Gamma(1 + n)\Gamma(1 - \gamma + \delta + n)} \frac{\pi}{\sin \pi(\gamma - \delta)}$$

$$= \frac{\pi}{\sin \pi(\delta - \gamma)} \left\{ \frac{\Gamma(\alpha + \gamma)\Gamma(\beta + \gamma)}{\Gamma(1 - \delta + \gamma)} F(\alpha + \gamma, \beta + \gamma, 1 - \delta + \gamma, 1) \right.$$

$$\left. - \frac{\Gamma(\alpha + \delta)\Gamma(\beta + \delta)}{\Gamma(1 - \gamma + \delta)} F(\alpha + \delta, \beta + \delta, 1 - \gamma + \delta, 1) \right\}$$

$$= \frac{\pi}{\sin \pi (\delta - \gamma)} \left\{ \frac{\Gamma(\alpha + \gamma)\Gamma(\beta + \gamma)}{\Gamma(1 - \delta + \gamma)} \frac{\Gamma(1 - \delta + \gamma)\Gamma(1 - \alpha - \beta - \gamma - \delta)}{\Gamma(1 - \delta - \alpha)\Gamma(1 - \delta - \beta)} \right.$$
$$\left. - \frac{\Gamma(\alpha + \delta)\Gamma(\beta + \delta)}{\Gamma(1 - \gamma + \delta)} \frac{\Gamma(1 - \gamma + \delta)\Gamma(1 - \alpha - \beta - \gamma - \delta)}{\Gamma(1 - \gamma - \alpha)\Gamma(1 - \gamma - \beta)} \right\}$$

$$= \frac{\pi \Gamma(1 - \alpha - \beta - \gamma - \delta)}{\sin \pi (\delta - \gamma)} \left\{ \frac{\Gamma(\alpha + \gamma)\Gamma(\beta + \gamma)}{\Gamma(1 - \delta - \alpha)\Gamma(1 - \delta - \beta)} \right.$$
$$\left. - \frac{\Gamma(\alpha + \delta)\Gamma(\beta + \delta)}{\Gamma(1 - \gamma - \alpha)\Gamma(1 - \gamma - \beta)} \right\}$$

$$= \frac{\Gamma(\alpha + \gamma)\Gamma(\beta + \gamma)\Gamma(\alpha + \delta)\Gamma(\beta + \delta)}{\sin \pi (\delta - \gamma) \sin \pi (\alpha + \beta + \gamma + \delta) \Gamma(\alpha + \beta + \gamma + \delta)}$$
$$\times \left\{ \sin \pi (\alpha + \delta) \sin \pi (\beta + \delta) - \sin \pi (\alpha + \gamma) \sin \pi (\beta + \gamma) \right\}$$

$$= \frac{\Gamma(\alpha + \gamma)\Gamma(\beta + \gamma)\Gamma(\alpha + \delta)\Gamma(\beta + \delta)}{\Gamma(\alpha + \beta + \gamma + \delta)}$$
$$\times \frac{\cos \pi (\alpha - \beta) - \cos \pi (\alpha + \beta + 2\delta) - \cos \pi (\alpha - \beta) + \cos \pi (\alpha + \beta + 2\gamma)}{\cos \pi (\alpha + \beta + 2\gamma) - \cos \pi (\alpha + \beta + 2\delta)}$$

$$= \frac{\Gamma(\alpha + \gamma)\Gamma(\beta + \gamma)\Gamma(\alpha + \delta)\Gamma(\beta + \delta)}{\Gamma(\alpha + \beta + \gamma + \delta)} .$$

Although Barnes' lemma (3) is proved under the condition $\mathrm{Re}(\alpha + \beta + \gamma + \delta - 1) < 0$, yet the two sides of it are not subjected to this restriction, so (3) holds in general, as far as the poles of the Γ-functions do not overlap.

Also, in (3), let $s = t + k$, k being any real number, and replace $\alpha, \beta, \gamma, \delta$ respectively by $\alpha - k, \beta - k, \gamma + k, \delta + k$. We have

$$\frac{1}{2\pi i} \int_{-k-i\infty}^{-k+i\infty} \Gamma(\alpha + t)\Gamma(\beta + t)\Gamma(\gamma - t)\Gamma(\delta - t)\,dt$$
$$= \frac{\Gamma(\alpha + \gamma)\Gamma(\alpha + \delta)\Gamma(\beta + \gamma)\Gamma(\beta + \delta)}{\Gamma(\alpha + \beta + \gamma + \delta)} ; \tag{5}$$

the path of integration is such that the poles of $\Gamma(\alpha + t)\Gamma(\beta + t)$ are kept on its left and those of $\Gamma(\gamma - t)\Gamma(\delta - t)$ on its right.

Next, let us find the connection between $F(\alpha, \beta, \gamma, z)$ and the two fundamental solutions at $z = 1$ when $\gamma - \alpha - \beta = -m(m = 0, 1, 2, \ldots)$. By (5) we can write (1) in the form

$$\frac{\Gamma(\alpha)\Gamma(\beta)}{\Gamma(\gamma)}F(\alpha,\beta,\gamma,z) = \frac{1}{2\pi i}\int_{-i\infty}^{i\infty}\frac{\Gamma(\alpha+s)\Gamma(\beta+s)\Gamma(-s)}{\Gamma(\gamma+s)}(-z)^s ds$$

$$= \frac{1}{2\pi i}\int_{-i\infty}^{i\infty}\frac{1}{2\pi i}\int_{-k-i\infty}^{-k+i\infty}\Gamma(\alpha+t)\Gamma(\beta+t)\Gamma(\gamma-\alpha-\beta-t)\Gamma(s-t)dt$$

$$\times \frac{\Gamma(-s)}{\Gamma(\gamma-\alpha)\Gamma(\gamma-\beta)}(-z)^s ds .$$

Interchanging the order of integrations (justified by properly choosing k) gives

$$\frac{\Gamma(\alpha)\Gamma(\beta)}{\Gamma(\gamma)}F(\alpha,\beta,\gamma,z)$$

$$= \frac{1}{2\pi i}\int_{-k-i\infty}^{-k+i\infty}\frac{\Gamma(\alpha+t)\Gamma(\beta+t)\Gamma(\gamma-\alpha-\beta-t)}{\Gamma(\gamma-\alpha)\Gamma(\gamma-\beta)}dt$$

$$\times \frac{1}{2\pi i}\int_{-i\infty}^{i\infty}\Gamma(s-t)\Gamma(-s)(-z)^s ds .$$

According to (1), with $\beta = \gamma$, we have

$$\frac{1}{2\pi i}\int_{-i\infty}^{i\infty}\Gamma(-t+s)\Gamma(-s)(-z)^s ds = \Gamma(-t)F(-t,\beta,\beta,z)$$

$$= \Gamma(-t)\sum_{n=0}^{\infty}\frac{(-t)_n}{n!}z^n = \Gamma(-t)\sum_{n=0}^{\infty}\binom{t}{n}(-z)^n = \Gamma(-t)(1-z)^t , \qquad (6)$$

where $(1-z)^t$ is to take the value 1 when $z = 0$. Hence,

$$\frac{\Gamma(\alpha)\Gamma(\beta)}{\Gamma(\gamma)}F(\alpha,\beta,\gamma,z)$$

$$= \frac{1}{2\pi i}\int_{-k-i\infty}^{-k+i\infty}\frac{\Gamma(\alpha+t)\Gamma(\beta+t)\Gamma(\gamma-\alpha-\beta-t)\Gamma(-t)}{\Gamma(\gamma-\alpha)\Gamma(\gamma-\beta)}(1-z)^t dt$$

$$(|\arg(1-z)| < \pi) . \qquad (7)$$

The integral on the r.h.s. can be calculated as a sum of the residues, similar to that in proving Barnes' lemma: When $|\arg(1-z)| < \pi$ and $|1-z| < 1$, the integral on the r.h.s. is equal to the negative value of the sum of the residues at the poles of the integrand $\Gamma(\gamma-\alpha-\beta-t)\Gamma(-t)$. If $\gamma-\alpha-\beta \neq$ integer, all these poles are simple $(t = 0, 1, 2, \ldots)$, and the result will be the same as that obtained in Eq. (4) of Sec. 4.8. When $\gamma - \alpha - \beta = -m(m = 0, 1, 2, \ldots)$, $t = n - m \; (n = 0, 1, 2, \ldots , m-1)$ are simple poles, while $t = n \; (n = 0, 1, 2, \ldots)$ are poles of order 2. The residues at the simple poles are

$$\frac{\Gamma(\alpha + n - m)\Gamma(\beta + n - m)\Gamma(m - n)(1 - z)^{n-m}}{\Gamma(\gamma - \alpha)\Gamma(\gamma - \beta)}$$

$$\times \lim_{t \to n-m} (t - n + m)\Gamma(-m - t)$$

$$= -\frac{\Gamma(\alpha - m)\Gamma(\beta - m)}{\Gamma(\gamma - \alpha)\Gamma(\gamma - \beta)}(\alpha - m)_n(\beta - m)_n\Gamma(m - n)(1 - z)^{n-m}\frac{(-)^n}{n!}$$

$$= -\frac{(\alpha - m)_n(\beta - m)_n}{n!(1 - m)_n}\Gamma(m)(1 - z)^{n-m} \quad (\gamma - \alpha - \beta = -m) \; ;$$

and those at the second-order poles are

$$\frac{d}{dt}\left\{ (t - n)^2 \frac{\Gamma(\alpha + t)\Gamma(\beta + t)\Gamma(m - t)\Gamma(-t)}{\Gamma(\gamma - \alpha)\Gamma(\gamma - \beta)} \right\}\bigg|_{t \to n}$$

$$= (-)^m \frac{\Gamma(\alpha + n)\Gamma(\beta + n)(1 - z)^n}{(m + n)!n!\Gamma(\gamma - \alpha)\Gamma(\gamma - \beta)}$$
$$\times \{\psi(\alpha + n) + \psi(\beta + n) - \psi(1 + m + n) - \psi(1 + n) + \ln(1 - z)\} \; .$$

Therefore,

$$F(\alpha, \beta, \gamma, z)$$
$$= \frac{\Gamma(m)\Gamma(\gamma)(1 - z)^{-m}}{\Gamma(\alpha)\Gamma(\beta)} \sum_{n=0}^{m-1} \frac{(\alpha - m)_n(\beta - m)_n}{n!(1 - m)_n}(1 - z)^n$$

$$+ \frac{(-)^{m+1}\Gamma(\gamma)}{\Gamma(\alpha - m)\Gamma(\beta - m)} \sum_{n=0}^{\infty} \frac{(\alpha)_n(\beta)_n}{n!(m + n)!}(1 - z)^n$$
$$\times \{\psi(\alpha + n) + \psi(\beta + n) - \psi(1 + m + n) - \psi(1 + n) + \ln(1 - z)\} \; .$$
$$\tag{8}$$

$(\gamma - \alpha - \beta = -m(m = 0, 1, 2, \ldots); \alpha$ and $\beta \neq 0, -1, -2, \ldots ; |\arg(1 - z)| < \pi, |1 - z| < 1;$ when $m = 0$, drop the finite sum.)

Simple manipulation will show that the r.h.s. of (8) is a linear combination of the two fundamental solutions at $z = 1$ [cf. Eq. (18) of Sec. 4.4].

When $\gamma - \alpha - \beta = m(m = 1, 2, \dots)$, we can use Eq. (8) of Sec. 4.3 to obtain from (8) given above

$$F(\alpha, \beta, \gamma, z) = (1 - z)^m F(\gamma - \alpha, \gamma - \beta, \gamma, z)$$

$$= \frac{\Gamma(m)\Gamma(\gamma)}{\Gamma(\alpha + m)\Gamma(\beta + m)} \sum_{n=0}^{m-1} \frac{(\alpha)_n (\beta)_n}{n!(1 - m)_n}(1 - z)^n$$

$$+ \frac{(-)^{m+1}\Gamma(\gamma)(1 - z)^m}{\Gamma(\alpha)\Gamma(\beta)} \sum_{n=0}^{\infty} \frac{(\alpha + m)_n (\beta + m)_n}{n!(m + n)!}(1 - z)^n$$

$$\times \{\psi(\alpha + m + n) + \psi(\beta + m + n)$$

$$-\psi(1 + m + n) - \psi(1 + n) + \ln(1 - z)\} , \qquad (9)$$

where $|\arg(1 - z)| < \pi, |1 - z| < 1; \alpha$ and β not zero or negative integers.

When α (or β) is also a negative integer, (8) and (9) derived above are simplified, since $F(\alpha, \beta, \gamma, z)$ is now a polynomial — the *Jacobi polynomial* [see the next section]. Take (9) as an example. Suppose $\alpha = -s$, where s is a positive integer, but β is not a negative integer. If $s < m$, then the second term on the r.h.s. of (9) is zero, since $1/\Gamma(\alpha) = 1/\Gamma(-s) = 0$. Thus,

$$F(-s, \beta, \gamma, z) = \frac{\Gamma(m)\Gamma(\gamma)}{\Gamma(m - s)\Gamma(m + \beta)} \sum_{n=0}^{s} \frac{(-s)_n (\beta)_n}{n!(1 - m)_n}(1 - z)^n$$

$$= \frac{\Gamma(m)\Gamma(\gamma)}{\Gamma(m - s)\Gamma(m + \beta)} F(-s, \beta, 1 - m, 1 - z)$$

$$(s = 0, 1, 2, \dots, m - 1; \ \gamma = \beta + m - s) . \qquad (10)$$

If $s \geq m$, then, since $1/\Gamma(\alpha + m) = 1/\Gamma(m - s) = 0$, the first term on the r.h.s. of (9) is zero and there remains in the second term only

$$\frac{(-)^{m+1}\Gamma(\gamma)(1 - z)^m}{\Gamma(\alpha)\Gamma(\beta)} \sum_{n=0}^{s-m} \frac{(\alpha + m)_n (\beta + m)_n}{n!(m + n)!}(1 - z)^n \psi(\alpha + m + n)\Big|_{\alpha \to -s} .$$

Under this condition,

$$\lim_{\alpha \to -s} \frac{\psi(\alpha + m + n)}{\Gamma(\alpha)} = \lim_{\alpha \to s} \frac{\psi(1 - \alpha - m - n) - \pi \cot \pi\alpha}{\pi/\Gamma(1 - \alpha) \sin \pi\alpha}$$
$$= (-)^{s+1}\Gamma(s + 1);$$

therefore,

$$F(-s, \beta, \gamma, z) = \frac{(-)^{s+m}\Gamma(\gamma)\Gamma(s + 1)}{\Gamma(\beta)}(1 - z)^m$$
$$\times \sum_{n=0}^{s-m} \frac{(-s + m)_n (\beta + m)_n}{n!(m + n)!}(1 - z)^n$$
$$= \frac{(-)^{s+m}\Gamma(\gamma)\Gamma(s + 1)}{m!\Gamma(\beta)}(1 - z)^m$$
$$\times F(-s + m, \beta + m, 1 + m, 1 - z)$$
$$(s = m, m + 1, \dots ; \ \gamma = \beta + m - s) \ . \tag{11}$$

4.10. Jacobi Polynomials

When α or β is a negative integer, $-n$, the hypergeometric function $F(\alpha, \beta, \gamma, z)$ becomes a polynomial, called the *Jacobi polynomial of degree n* (or, hypergeometric polynomial of degree n):

$$F(-n, \beta, \gamma, z) = \sum_{k=0}^{n} \frac{(-n)_k (\beta)_k}{k!(\gamma)_k} z^k = \sum_{k=0}^{n} (-)^k \binom{n}{k} \frac{(\beta)_k}{(\gamma)_k} z^k \ . \tag{1}$$

Many important polynomials, such as the Legendre polynomial, spherical polynomial [see Chap. 5], Chebyshev polynomial [see the next section], etc., are special cases of Jacobi polynomials.

Summation formula

Putting $z = 1$ in (1), with Eq. (4) of Sec. 4.7, we obtain the important summation formula:

$$\sum_{k=0}^{n} (-)^k \binom{n}{k} \frac{(\beta)_k}{(\gamma)_k} = F(-n, \beta, \gamma, 1)$$
$$= \frac{\Gamma(\gamma)\Gamma(\gamma - \beta + n)}{\Gamma(\gamma + n)\Gamma(\gamma - \beta)} = \frac{(\gamma - \beta)_n}{(\gamma)_n} \ . \tag{2}$$

Integral representation and generating function

Since the hypergeometric equation is symmetric in α and β, by exchanging α and β in Eq. (2) of Sec. 4.5, we obtain another solution in the integral form:

$$w(z) = \int_C t^{\alpha-1}(1-t)^{\gamma-\alpha-1}(1-zt)^{-\beta}dt . \tag{3}$$

Assume $\alpha = -n$, a negative integer, and take C to be a contour encircling the point $t = 0$ in the positive (counterclockwise) sense with the points $t = 1$ and $t = 1/z$ kept outside of it. Define $(1-zt)^{-\beta} = 1$ when $z = 0$, then, for sufficiently small z such that $|zt| < 1$ on C, we have

$$(1-zt)^{-\beta} = \sum_{k=0}^{\infty} \binom{-\beta}{k}(-zt)^k .$$

Substituting in (3) leads to

$$\int^{(0+)} t^{-n-1}(1-t)^{\gamma+n-1} \sum_{k=0}^{\infty} \binom{-\beta}{k}(-zt)^k dt$$

$$= \sum_{k=0}^{\infty} \binom{-\beta}{k}(-z)^k \int^{(0+)} t^{-n+k-1}(1-t)^{\gamma+n-1} dt$$

$$= \sum_{k=0}^{n} \binom{-\beta}{k}(-z)^k \frac{2\pi i}{(n-k)!} \frac{d^{n-k}}{dt^{n-k}}(1-t)^{\gamma+n-1}\bigg|_{t=0}$$

$$= 2\pi i \sum_{k=0}^{n} \binom{-\beta}{k}(-z)^k \frac{\Gamma(\gamma+n)}{(n-k)\Gamma(\gamma+k)}(-)^{n-k}$$

$$= \frac{(-)^n \Gamma(\gamma+n)2\pi i}{n!\Gamma(\gamma)} \sum_{k=0}^{n}(-)^k \binom{n}{k}\frac{(\beta)_k}{(\gamma)_k}z^k$$

$$= \frac{(-)^n \Gamma(\gamma+n)2\pi i}{n!\Gamma(\gamma)} F(-n,\beta,\gamma,z) .$$

Hence,

$$F(-n,\beta,\gamma,z) = \frac{(-)^n \Gamma(\gamma)n!}{\Gamma(\gamma+n)2\pi i} \int^{(0+)} t^{-n-1}(1-t)^{\gamma+n-1}(1-zt)^{-\beta}dt , \tag{4}$$

$t = 1$ and $t = 1/z$ lie outside the contour; $|\arg(1 - t)| < \pi; (1 - zt)^{-\beta} = 1$ when $z = 0; \gamma \neq 0, -1, -2, \dots, -n + 1$.

By the substitution $t = v/(v - 1)$, (4) becomes

$$(\gamma)_n F(-n, \beta, \gamma, z)$$
$$= \frac{n!}{2\pi i} \int^{(0+)} \frac{(1 - v)^{\beta - \gamma}[1 - (1 - z)v]^{-\beta}}{v^{n+1}} dv . \tag{5}$$

Thus,

$$(1 - v)^{\beta - \gamma}[1 - (1 - z)v]^{-\beta} = \sum_{n=0}^{\infty} \frac{v^n}{n!}(\gamma)_n F(-n, \beta, \gamma, z) . \tag{6}$$

The function on the l.h.s. is called the *generating function of the Jacobi polynomials*.

Differential expression

Replacing v in (5) by $(v - z)/v(1 - z)$, we have

$$(\gamma)_n F(-n, \beta, \gamma, z) = z^{1-z}(1 - \gamma)^{\gamma+n-\beta} \frac{n!}{2\pi i} \int^{(z+)} \frac{(1 - v)^{\beta - \gamma} v^{\gamma+n-1}}{(v - z)^{n+1}} dv \tag{7}$$

from which we arrive at the differential expression

$$F(-n, \beta, \gamma, z) = \frac{\Gamma(\gamma)}{\Gamma(\gamma + n)} z^{1-\gamma}(1 - z)^{\gamma+n-\beta} \frac{d^n}{dz^n}[z^{\gamma+n-1}(1 - z)^{\beta-\gamma}] . \tag{8}$$

Orthogonality

Writing $p + n$ for β, it is seen that the Jacobi polynomial $w_n = F(-n, p + n, \gamma, z)$ satisfies the differential equation [Eq. (1) of Sec. 4.1]

$$z(1 - z)\frac{d^2 w_n}{dz^2} + [\gamma - (p + 1)z]\frac{dw_n}{dz} + n(n + p)w_n = 0 ;$$

or, written in self-adjoint form,

$$\frac{d}{dz}[z^{\gamma}(1 - z)^{p-\gamma+1}w'_n] + n(n + p)z^{\gamma-1}(1 - z)^{p-\gamma}w_n = 0 . \tag{9}$$

Let $w_m = F(-m, p + m, \gamma, z)$ be another Jacobi polynomial *with the same p* as w_n, which satisfies the equation

$$\frac{d}{dz}[z^{\gamma}(1 - z)^{p-\gamma+1}w'_m] + m(m + p)z^{\gamma-1}(1 - z)^{p-\gamma}w_m = 0 . \tag{10}$$

Multiplying (9) with w_m and (10) with w_n, subtracting, and then integrating from 0 to 1, we obtain

$$\int_0^1 w_n w_m z^{\gamma-1}(1-z)^{p-\gamma}dz = \frac{z^\gamma(1-z)^{p-\gamma+1}[w_m w_n' - w_n w_m']}{m(m+p)-n(n+p)}\bigg|_{z=0}. \quad (11)$$

Assume $\text{Re}(\gamma) > 0$, $\text{Re}(p-\gamma+1) = \text{Re}(\beta-n-\gamma+1) > 0$, then, for $m \neq n$, the r.h.s. of (11) equals to zero and hence

$$\int_0^1 w_n w_m z^{\gamma-1}(1-z)^{p-\gamma}dz = 0 \quad (m \neq n). \quad (12)$$

This *orthogonality relation* can also be obtained by direct calculation, using (8) (see the following).

When $m = n$, by (8), noting that $\beta = p + n$, we have

$$N_n = \int_0^1 w_n^2 z^{\gamma-1}(1-z)^{p-\gamma}dz$$

$$= \frac{\Gamma(\gamma)}{\Gamma(\gamma+n)}\int_0^1 w_n \frac{d^n}{dz^n}[z^{\gamma+n-1}(1-z)^{p+n-\gamma}]dz.$$

Integrating by parts n times, we obtain, under the conditions $\text{Re}(\gamma) > 0$ and $\text{Re}(p-\gamma+1) > 0$,

$$N_n = \frac{\Gamma(\gamma)}{\Gamma(\gamma+n)}(-)^n \int_0^1 z^{\gamma+n-1}(1-z)^{p+n-\gamma}\frac{d^n w_n}{dz^n}dz$$

$$= \frac{\Gamma(\gamma)}{\Gamma(\gamma+n)}\frac{(p+n)_n}{(\gamma)_n}n! \int_0^1 z^{\gamma+n-1}(1-z)^{p+n-\gamma}dz \quad \text{(by (1))}$$

$$= \frac{(p+n)_n}{(\gamma)_n}\frac{\Gamma(\gamma)\Gamma(p+n-\gamma+1)}{\Gamma(p+2n+1)}n!. \quad (13)$$

According to the general theory of orthogonal polynomials,[i] Jacobi polynomials $F(-n, p+n, \gamma, z)(n = 0, 1, 2, \ldots)$ form a complete system of orthogonal functions in the interval $[0, 1]$ with weight $z^{\gamma-1}(1-z)^{p-\gamma}$, where $\text{Re}(\gamma) > 0$

---

[i]Cf., for example, Courant and Hilbert, *Methods of Mathematical Physics*, Vol. I, Chap. 2 (1955).

and $\text{Re}(p - \gamma + 1) > 0$. Any quadratically integrable function $f(z)$ in $[0, 1]$ can be represented by a series of $F(-n, p + n, \gamma, z)$ in mean:

$$f(z) = \sum_{n=0}^{\infty} a_n F(-n, p + n, \gamma, z) , \qquad (14)$$

where

$$a_n = \frac{1}{N_n} \int_0^1 f(z) F(-n, p + n, \gamma, z) z^{\gamma-1} (1 - z)^{p-\gamma} dz . \qquad (15)$$

Another definition of the Jacobi polynomial, often used in the literature, is

$$P_n^{(\alpha,\beta)}(x) = \binom{n + \alpha}{n} F\left(-n, n + \alpha + \beta + 1, \alpha + 1, \frac{1-x}{2}\right) . \qquad (16)$$

Putting in it $x = 1 - 2z, \alpha = \gamma - 1, \beta = p - \gamma$, we have

$$P_n^{(\alpha,\beta)}(x) = \binom{n + \gamma - 1}{n} F(-n, p + n, \gamma, z) \qquad (17)$$

$P_n^{(\alpha,\beta)}(x) (n = 0, 1, 2, \dots)$ again form a complete set of orthogonal functions in the interval $[-1, 1]$ with weight $(1 - x)^\alpha (1 + x)^\beta$.

4.11. Chebyshev Polynomials

Chebyshev polynomial (of the first kind) is defined as

$$T_n(x) = \cos(n \arccos x) . \qquad (1)$$

Hence, $T_0(x) = 1$. To obtain an explicit expression for $T_n(x)$, let $x = \cos \theta$, then

$$T_n(x) = \cos n\theta = \text{Re}(e^{in\theta}) = \text{Re}[(\cos \theta + i \sin \theta)^n]$$

$$= \text{Re} \sum_{r=0}^{n} \binom{n}{r} \cos^{n-r} \theta (i \sin \theta)^r$$

$$= \sum_{k=0}^{[n/2]} \binom{n}{2k} \cos^{n-2k} \theta \times (-)^k \sin^{2k} \theta = \sum_{k=0}^{[n/2]} (-)^k \binom{n}{2k} x^{n-2k} (1 - x^2)^k$$

$$= \sum_{k=0}^{[n/2]}(-)^k \binom{n}{2k} x^{n-2k} \sum_{l=0}^{k} \binom{k}{l}(-)^{k-l} x^{2(k-l)}$$

$$= \sum_{k=0}^{[n/2]}\sum_{l=0}^{k}(-)^l \binom{n}{2k}\binom{k}{l} x^{n-2l} = \sum_{l=0}^{[n/2]}(-)^l x^{n-2l} \sum_{k=l}^{[n/2]} \binom{n}{2k}\binom{k}{l}.$$

Using Eq. (8) of Sec. 3.6, the multiplication formula of the Γ-function, and the relation $(s = $ positive integer$)$

$$\frac{\Gamma(\lambda)}{\Gamma(\lambda - s)} = (-)^s (1 - \lambda)_s,$$

together with Eq. (4) of Sec. 4.7, we have

$$\sum_{k=l}^{[n/2]} \binom{n}{2k}\binom{k}{l} = \sum_{k=0}^{[n/2]-l} \binom{n}{2k+2l}\binom{k+l}{l}$$

$$= \frac{n!}{l!} \sum_{k=0}^{[n/2]-l} \frac{\Gamma(k+l+1)}{k!\,\Gamma(n-2k-2l+1)\Gamma(2k+2l+1)}$$

$$= \frac{n!\pi}{l!2^n} \sum_{k=0}^{[n/2]-l} \frac{1}{k!\,\Gamma\left(l+\frac{1}{2}+k\right)\Gamma\left(\frac{n+1}{2}-l-k\right)\Gamma\left(\frac{n}{2}+1-l-k\right)}$$

$$= \frac{n!\pi}{l!2^n\,\Gamma\left(l+\frac{1}{2}\right)\Gamma\left(\frac{n+1}{2}-l\right)\Gamma\left(\frac{n}{2}+1-l\right)} \sum_{k=0}^{[n/2]-l} \frac{\left(\frac{1-n}{2}+l\right)_k \left(-\frac{n}{2}+l\right)_k}{k!\left(l+\frac{1}{2}\right)_k}$$

$$= \frac{n!\pi}{l!2^n\,\Gamma\left(l+\frac{1}{2}\right)\Gamma(n-2l+1)} \frac{\Gamma\left(l+\frac{1}{2}\right)\Gamma(n-l)}{\Gamma\left(\frac{n}{2}\right)\Gamma\left(\frac{n+1}{2}\right)}$$

$$= \frac{n!(n-l-1)!}{l!2^{2l-n+1}(n-2l)!\Gamma(n)} = \frac{n}{2}\frac{(n-l-1)!2^{n-2l}}{l!(n-2l)!}.$$

Hence,

$$T_n(x) = \frac{n}{2}\sum_{l=0}^{[n/2]} \frac{(-)^l(n-l-1)!}{l!(n-2l)!}(2x)^{n-2l} \quad (n \geq 1). \tag{2}$$

Note that the coefficient of the highest term, x^n, is 2^{n-1}. The first six $T_n(x)$ are:

$$\left.\begin{array}{l} T_0(x) = 1 , \quad T_1(x) = x , \quad T_2(x) = 2x^2 - 1 , \\ T_3(x) = 4x^3 - 3x , \quad T_4(x) = 8x^4 - 8x^2 + 1 , \\ T_5(x) = 16x^5 - 20x^3 + 5x . \end{array}\right\} \tag{3}$$

The differential equation satisfied by Chebyshev polynomial can be derived simply as follows: Let $y = T_n(x) = T_n(\cos\theta) = \cos n\theta$, then y satisfies

$$\frac{d^2y}{d\theta^2} + n^2y = 0 .$$

Returning to the variable x leads to the equation

$$(1 - x^2)\frac{d^2y}{dx^2} - x\frac{dy}{dx}n^2y = 0 . \tag{4}$$

Putting $z = (1 - x)/2$, (4) reduces to the hypergeometric equation

$$z(1 - z)\frac{d^2y}{dz^2} + \left(\frac{1}{2} - z\right)\frac{dy}{dx} + n^2y = 0 \tag{5}$$

with $\alpha = -n, \beta = n, \gamma = \frac{1}{2}$. Hence, from Eqs. (2) and (3) of Sec. 4.3, we obtain for (5) two linearly independent solutions: $F(-n, n, \frac{1}{2}, z)$ and $z^{\frac{1}{2}}F(-n+\frac{1}{2}, n+\frac{1}{2}, \frac{3}{2}, z)$. The last one is not a polynomial, and hence it must be that

$$T_n(x) = CF\left(-n, n, \frac{1}{2}, \frac{1-x}{2}\right) .$$

Putting $x = 1$, we see that $C = T_n(1) = 1$. Thus

$$T_n(x) = F\left(-n, n, \frac{1}{2}, \frac{1-x}{2}\right) , \tag{6}$$

which shows that the Chebyshev polynomial is a special case of the Jacobi polynomial: $\beta = n(p = 0), \gamma = \frac{1}{2}, z = (1 - x)/2$.

The following are some important properties of the Chebyshev polynomial.

1. Differential expression

From (6) and Eq. (8) of Sec. 4.10, we have

$$T_n(x) = (-)^n\frac{2^n n!}{(2n)!}(1 - x^2)^{\frac{1}{2}}\frac{d^n}{dx^n}(1 - x^2)^{n-\frac{1}{2}} . \tag{7}$$

2. Orthogonality relation and normalization factor

According to Eqs. (12) and (13) of Sec. 4.6, we have

$$\int_{-1}^{1} T_n(x)T_m(x)(1-x^2)^{-\frac{1}{2}}\,dx = \begin{cases} 0 & n \neq m, \\ \frac{\pi}{2} & n = m > 0, \\ \pi & n = m = 0. \end{cases} \tag{8}$$

Same result may be obtained by the substitution: $x = \cos\theta$ and $T_n(x) = \cos n\theta$.

3. The generating function of $T_n(x)$

Equation (6) in Sec. 4.10 does not hold since $\beta = n$ now. Again let $x = \cos\theta$, so that $T_n(x) = \cos n\theta$, it is easy to show that

$$\frac{1-xt}{1-2xt+t^2} = \sum_{n=0}^{\infty} T_n(x)t^n . \tag{9}$$

For, when $|t| < 1$,

$$\sum_{n=0}^{\infty} T_n(x)t^n = \sum_{n=0}^{\infty} \cos n\theta \times t^n = \frac{1}{2}\sum_{n=0}^{\infty}[(te^{i\theta})^n + (te^{-i\theta})^n]$$

$$= \frac{1}{2}\left[\frac{1}{1-te^{i\theta}} + \frac{1}{1-te^{-i\theta}}\right]$$

$$= \frac{1-t\cos\theta}{1-2t\cos\theta+t^2} = \frac{1-xt}{1-2xt+t^2} .$$

The l.h.s. of (9) is called the *generating function* of $T_n(x)$.[j]

4. Recurrence relation

When $x \leq 1$, from $T_n(x) = \cos n\theta$, it is easy to deduce that

$$T_{n+1}(x) - 2xT_n(x) + T_{n-1}(x) = 0 , \tag{10}$$

since $\cos(n+1)\theta - \cos(n-1)\theta = 2\cos\theta\cos n\theta$. But, as (10) is an algebraic identity, it holds for any x.

By similar method, we can derive many other relations; for example,

$$(1-x^2)T_n'(x) = \frac{n}{2}[T_{n-1}(x) - T_{n+1}(x)]$$

$$= n[T_{n-1}(x) - xT_n(x)] . \tag{11}$$

[j] For other generating functions of $T_n(x)$, cf. Erdélyi (1953), Vol. II, p. 186; also, Ex. 11 at the end of this chapter.

5. *Theorem:* In the interval $-1 \leq x \leq 1$, the largest deviation of $T_n(x)$ from zero is the smallest in comparison with all other polynomials of degree n with real coefficients and with 2^{n-1} as the coefficient of x^n (same as $T_n(x)$).

The proof is as follows. In the interval $-1 \leq x \leq 1$, $T_n(x) = \cos n\theta (x = \cos \theta)$. Therefore, the largest deviation of $T_n(x)$ from zero is 1, which occurs at the points $\theta_k = k\pi/n$, $k = 0, 1, 2, \ldots, n$. Let $x_k = \cos \theta_k$, then, when k is even, $T_n(x_k) = 1$, and when k is odd, $T_n(x_k) = -1$. Let $R_n(x)$ be another polynomial of degree n with real coefficients and the coefficient of x^n in it be 2^{n-1}. Then $T_n(x) - R_n(x)$ will be a polynomial of degree $n-1$. If the largest deviation of $R_n(x)$ from zero is smaller than that of $T_n(x)$, then, at the points x_k, we would have

$$T_n(x_0) - R_n(x_0) > 0$$
$$T_n(x_1) - R_n(x_1) < 0, \ldots$$

which means that the polynomial $T_n(x) - R_n(x)$ changes sign at least n times in the interval $-1 \leq x \leq 1$, and hence has at least n zeros in the interval. But this is contradictory to the fact that $T_n(x) - R_n(x)$ is a polynomial of degree $n-1$. Therefore, the largest deviation of $R_n(x)$ from zero cannot be smaller than that of $T_n(x)$.

4.12. Quadratic Transformations

In Secs. 4.3 and 4.8, we have discussed the linear transformations of the hypergeometric function $F(\alpha, \beta, \gamma, z)$, where the parameters α, β and γ are arbitrary as far as the hypergeometric functions present in the transformations are significant. For higher-order transformations of the variable z, however, the parameters are restricted in general (see Exs. 10 and 11 at the end of Chap. 2).

Gauss and Kummer had given some transformations of this sort, which are of the form[k]

$$x^{-p}(1-x)^{-q} F(\alpha, \beta, \gamma, x) = t^{p'}(1-t)^{q'} F(\alpha', \beta', \gamma', t) , \tag{1}$$

where $x = \varphi(t)$ is an algebraic function. However, they did not state the condition for the existence of such transformations. Goursat[1] had made thorough study of the situation and show that the variables x and t must satisfy

---

[k] The exponents of the l.h.s., $-p$ and $-q$, with negative signs, are introduced for later convenience when we transfer the corresponding factors to the r.h.s.

[1] Goursat, *Ann. Sci. École Norm. Sup.* (2) **10** (1881) 3-142.

a functional relation of sixth degree, so the highest transformations possible are of sixth degree. We shall give in the following a brief introduction of his results in quadratic transformations.

Let

$$P = x^{-p}(1 - x)^{-q}F(\alpha, \beta, \gamma, x) \ . \tag{2}$$

From the equation satisfied by $y = F(\alpha, \beta, \gamma, x)$, namely,

$$x(1 - x)\frac{d^2y}{dx^2} + [\gamma - (\alpha + \beta + 1)x]\frac{dy}{dx} - \alpha\beta y = 0 \ , \tag{3}$$

we find the equation for P:

$$x^2(1 - x^2)\frac{d^2P}{dx^2} + [l - (l + m)x]x(1 - x)\frac{dP}{dx}$$
$$+ (Ax^2 + Bx + C)P = 0 \ , \tag{4}$$

where

$$\left.\begin{aligned}
l &= 2p + \gamma \ , \\
m &= 2q + \alpha + \beta - \gamma + 1 \ , \\
A &= (p + q + \alpha)(p + q + \beta) \ , \\
C &= p(p + \gamma - 1) \ , \\
A + B + C &= q(q + \alpha + \beta - \gamma) \ .
\end{aligned}\right\} \tag{5}$$

If (1) is to hold for the transformation $x = \varphi(t)$, then P, as a function of t, should obviously satisfy an equation similar to (4):

$$t^2(1 - t^2)\frac{d^2P}{dt^2} + [l' + (l' + m')t]t(1 - t)\frac{dP}{dt}$$
$$+ (A't^2 + B't + C')P = 0 \ , \tag{6}$$

where l', m', A', B', C' have the same relations to $p', q', \alpha', \beta', \gamma'$ as given by (5); only the p and q have to be replaced respectively by $-p'$ and $-q'$ due to the different signs of them in (1).

Goursat proved that the only quadratic transformations which realize (1) are

$$x = (2t - 1)^2 \tag{7}$$

and its inversion, together with other transformations derived from the rational linear transformations of x and t given in Table 2 of Sec. 4.3, which transform 0, 1, ∞ again into 0, 1, ∞. Besides, A, B, C, l and m, hence α, β

and γ must satisfy certain conditions. The following is a table of quadratic transformations (and their inversions) given by Goursat, where

$$\left.\begin{array}{l} \lambda = 1 - \gamma \\ \mu = \gamma - \alpha - \beta \\ \nu = \beta - \alpha \end{array}\right\} \tag{8}$$

represent respectively the differences of the exponents at the singularities 0, 1, ∞.

The first column of the table gives the conditions which must be satisfied by the parameters λ, μ and ν; thus, by α, β and γ. The various transformations under the same Roman numerals are those derived from rational linear transformations of the variable t, which bring 0, 1, ∞ again into 0, 1, ∞. For example, the second transformation $x = (2 - t)^2/t^2$ belonging to I can be derived from the first, $x = (2t - 1)^2$, by changing t into $1/t$; the third one, $x = (1 + t)^2/(1 - t)^2$, can be derived from the first by changing t into $1/(1 - t)$. Transformations in the same column but belonging to different Roman numerals are related to each other by rational linear transformations of the variables, which transform 0, 1, ∞ again into 0, 1, ∞. For example, the first one belonging to II, $x = (2t - 1)^2/4t(t - 1)$, can be obtained from the first of I, $x = (2t - 1)^2$, by changing x to $x/(x - 1)$.

The following are some important examples of quadratic transformations.

Example 1. $y = F(\alpha, \beta, \alpha + \beta + \frac{1}{2}, x)$ belongs to the case $\mu = \frac{1}{2}$, i.e., to III in the table; $A + B + C = 0, m = \frac{1}{2}$. Hence, from the second equation of (5), $m = 2q + \frac{1}{2}$, we determine $q = 0; p$ is, however, arbitrary. Let $p = 0$, then the function P given by (2) satisfies the equation satisfied by y, namely,

$$x(1 - x)\frac{d^2 P}{dt^2} + \left[\alpha + \beta + \frac{1}{2} - (\alpha + \beta + 1)x\right]\frac{dP}{dx} - \alpha\beta P = 0 .$$

Making the transformation $x = 4t(1 - t)$ [III in the table], this equation becomes

$$t(1 - t)\frac{d^2 P}{dt^2} + \left[\alpha + \beta + \frac{1}{2} - (2\alpha + 2\beta + 1)t\right]\frac{dP}{dt} - 4\alpha\beta P = 0 ,$$

of which one solution is $F(2\alpha, 2\beta, \alpha + \beta + \frac{1}{2}, t)$, being equal to 1 when $t = 0$. The other solution is singular at $t = 0$. Therefore, we have the transformation formula

$$F\left(\alpha, \beta, \alpha + \beta + \frac{1}{2}, 4t(1 - t)\right) = F\left(2\alpha, 2\beta, \alpha + \beta + \frac{1}{2}, t\right) . \tag{9}$$

Table of Quadratic Transformations.

$C = 0$ $l = \tfrac{1}{2}$	I.	$x = (2t - 1)^2,$	$x = \left(\dfrac{2 - t}{t}\right)^2,$	$x = \left(\dfrac{1 + t}{1 - t}\right)^2,$
$\lambda = \pm\tfrac{1}{2}$	II.	$x = \dfrac{(2t - 1)^2}{4t(t - 1)},$	$x = \dfrac{(2 - t)^2}{4(1 - t)},$	$x = \dfrac{(1 + t)^2}{4t},$
$A + B + C = 0$ $m = \tfrac{1}{2}$	III.	$x = 4t(1 - t),$	$x = \dfrac{4(t - 1)}{t^2},$	$x = \dfrac{-4t}{(1 - t)^2},$
$\mu = \pm\tfrac{1}{2}$	IV.	$x = \dfrac{1}{4t(1 - t)},$	$x = \dfrac{t^2}{4(t - 1)},$	$x = \dfrac{(1 - t)^2}{-4t},$
$A = 0$ $l + m = \tfrac{3}{2}$	V.	$x = \dfrac{1}{(2t - 1)^2},$	$x = \left(\dfrac{t}{2 - t}\right)^2,$	$x = \left(\dfrac{1 - t}{1 + t}\right)^2,$
$\nu = \pm\tfrac{1}{2}$	VI.	$x = \dfrac{4t(t - 1)}{(2t - 1)^2},$	$x = \dfrac{4(1 - t)}{(2 - t)^2},$	$x = \dfrac{4t}{(1 + t)^2}.$

VII.	$x=\dfrac{1+\sqrt{t}}{2}$,	$x=\dfrac{1+\sqrt{1-t}}{2}$,	$x=\dfrac{1+\sqrt{t}}{2\sqrt{t}}$,
VIII.	$x=\dfrac{1+\sqrt{1-t}}{2\sqrt{1-t}}$,	$x=\dfrac{\sqrt{t-1}+\sqrt{t}}{2\sqrt{t-1}}$,	$x=\dfrac{\sqrt{t}+\sqrt{t-1}}{2\sqrt{t}}$,
	$x=\dfrac{(\sqrt{t}+\sqrt{t-1})^2}{4\sqrt{t(t-1)}}$,	$x=\dfrac{(1+\sqrt{1-t})^2}{4\sqrt{1-t}}$,	$x=\dfrac{(1+\sqrt{t})^2}{4\sqrt{t}}$,
IX.	$x=\dfrac{2}{1+\sqrt{t}}$,	$x=\dfrac{2}{1+\sqrt{1-t}}$,	$x=\dfrac{2\sqrt{t}}{1+\sqrt{t}}$,
X.	$x=\dfrac{2\sqrt{1-t}}{1+\sqrt{1-t}}$,	$x=\dfrac{2\sqrt{t-1}}{\sqrt{t-1}+\sqrt{t}}$,	$x=\dfrac{2\sqrt{t}}{\sqrt{t}+\sqrt{t-1}}$,
	$x=\dfrac{4\sqrt{t(t-1)}}{(\sqrt{t}+\sqrt{t-1})^2}$,	$x=\dfrac{4\sqrt{1-t}}{(1+\sqrt{1-t})^2}$,	$x=\dfrac{4\sqrt{t}}{(1+\sqrt{t})^2}$,
XI.	$x=\dfrac{\sqrt{t}-1}{\sqrt{t}+1}$,	$x=\dfrac{\sqrt{1-t}-1}{\sqrt{1-t}+1}$,	$x=\dfrac{1-\sqrt{t}}{1+\sqrt{t}}$,
XII.	$x=\dfrac{1-\sqrt{1-t}}{1+\sqrt{1-t}}$,	$x=\dfrac{\sqrt{t}-1}{\sqrt{t}+\sqrt{t-1}}$,	$x=\dfrac{\sqrt{t-1}-\sqrt{t}}{\sqrt{t-1}+\sqrt{t}}$,
	$x=\left(\dfrac{\sqrt{t-1}+\sqrt{t}}{\sqrt{t-1}-\sqrt{t}}\right)^2$,	$x=\left(\dfrac{1+\sqrt{1-t}}{1-\sqrt{1-t}}\right)^2$,	$x=\left(\dfrac{1+\sqrt{t}}{1-\sqrt{t}}\right)^2$.

Conditions for the groups:

- **VII, VIII:** $A+B=0$, $\quad l=m$, $\quad \overline{}\quad \lambda=\pm\mu$
- **IX, X:** $B+C=0$, $\quad l+2m=2$, $\quad \overline{}\quad \mu=\pm\nu$
- **XI, XII:** $A-C=0$, $\quad 2l+m=2$, $\quad \overline{}\quad \lambda=\pm\nu$

The Inversion Transformations.

[The sign of the radical in the formulae is to be taken according to the transformation required.]

Going back to the variable x, and stipulating that $t = 0$ when $x = 0$, (9) becomes

$$F\left(\alpha, \beta, \alpha + \beta + \frac{1}{2}, x\right) = F\left(2\alpha, 2\beta, \alpha + \beta + \frac{1}{2}, \frac{1 - \sqrt{1-x}}{2}\right). \qquad (10)$$

The same formula may be obtained from the inversion transformation [see VII in the table].

Example 2. $y = F(\alpha, \beta, 2\beta, x)$ belongs to the case $\mu = \nu$, i.e., to IX in the table; $B + C = 0, l + 2m = 2$. By (5), we find $q = -(p + \alpha)/2$. Take $p = 0$, then $q = -\alpha/2$ and $A = \frac{\alpha}{2}(\beta - \frac{\alpha}{2}), -B = C = 0; l = 2\beta, m = 1 - \beta$. Hence, if we put $y = (1 - x)^{-\alpha/2}P$, (4) satisfied by P will be

$$x^2(1 - x^2)\frac{d^2 P}{dx^2} + [2\beta - (\beta + 1)x]x(1 - x)\frac{dP}{dx} + \frac{\alpha}{2}\left(\beta - \frac{\alpha}{2}\right)x^2 P = 0 .$$

Making the transformation $x = 2\sqrt{t}/(\sqrt{t} + \sqrt{t - 1})$ [see IX in the table], i.e., $t = x^2/4(x - 1)$, we obtain

$$t(1 - t)\frac{d^2 P}{dt^2} + \left[\beta + \frac{1}{2} - (\beta + 1)t\right]\frac{dP}{dt} - \frac{\alpha}{2}\left(\beta - \frac{\alpha}{2}\right)P = 0 .$$

$F(\frac{\alpha}{2}, \beta - \frac{\alpha}{2}, \beta + \frac{1}{2}, x^2/4(x - 1))$ is the solution of this equation, which takes the value 1 when $t = 0$. The other solution is singular at $t = 0$. Hence, we have

$$F(\alpha, \beta, 2\beta, x) = (1 - x)^{-\alpha/2}F\left(\frac{\alpha}{2}, \beta - \frac{\alpha}{2}, \beta + \frac{1}{2}, \frac{x^2}{4(x - 1)}\right), \qquad (11)$$

where $(1 - x)^{-\alpha/2} = 1$ when $x = 0$. This formula can also be obtained in turn from the corresponding transformation in IV.

The right hand side of (11) belongs to the case $\mu = \frac{1}{2}$. Applying (10) gives

$$F(\alpha, \beta, 2\beta, x) = (1 - x)^{-\frac{\alpha}{2}}F\left(\alpha, 2\beta - \alpha, \beta + \frac{1}{2}, \frac{(1 - \sqrt{1-x})^2}{-4\sqrt{1-x}}\right)$$

$$((1 - x)^{-\alpha/2} = 1 \text{ when } x = 0) . \qquad (12)$$

From (12), using Eq. (9) of Sec. 4.3, we also obtain

$$F(\alpha, \beta, 2\beta, x)$$

$$= \left(\frac{1 + \sqrt{1-x}}{2}\right)^{-2\alpha} F\left(\alpha, \alpha - \beta + \frac{1}{2}, \beta + \frac{1}{2}, \left(\frac{1 - \sqrt{1-x}}{1 + \sqrt{1-x}}\right)^2\right)$$

$$\left(\text{when } x = 0, \sqrt{1-x} = 1, \left(\frac{1 + \sqrt{1-x}}{2}\right)^{-2\alpha} = 1\right) . \qquad (13)$$

Setting $z = (1 - \sqrt{1-x})/(1 + \sqrt{1-x})$ in (13), we have

$$F\left(\alpha, \beta, 2\beta, \frac{4z}{(1+z)^2}\right)$$

$$= (1+z)^{2\alpha} F\left(\alpha, \alpha - \beta + \frac{1}{2}, \beta + \frac{1}{2}, z^2\right), \tag{14}$$

where $(1+z)^{2\alpha} = 1$ when $z = 0$.

Further, by using the various rational linear transformations [Sec. 4.3] and the analytic continuation relations [Sec. 4.8], we can derive many other quadratic transformation formulae from those obtained above. For example, by Eq. (9) of Sec. 4.3, setting $\beta = \alpha + \frac{1}{2}$, we have

$$F\left(\alpha, \alpha + \frac{1}{2}, \gamma, x\right) = (1-x)^{-\alpha} F\left(\alpha, \gamma - \alpha - \frac{1}{2}, \gamma, \frac{x}{x-1}\right).$$

Its r.h.s. belongs to the case $\mu = \frac{1}{2}$. Applying (10) above leads to

$$F\left(\alpha, \alpha + \frac{1}{2}, \gamma, x\right)$$

$$= (1-x)^{-\alpha} F\left(2\alpha, 2\gamma - 2\alpha - 1, \gamma, \frac{\sqrt{1-x} - 1}{2\sqrt{1-x}}\right), \tag{15}$$

where $(1-x)^{-\alpha} = 1$ when $x = 0$.

Using again Eq. (9) of Sec. 4.3, we obtain

$$F\left(\alpha, \alpha + \frac{1}{2}, \gamma, x\right)$$

$$= \left(\frac{1 + \sqrt{1-x}}{2}\right)^{-2\alpha} F\left(2\alpha, 2\alpha + 1 - \gamma, \gamma, \frac{1 - \sqrt{1-x}}{1 + \sqrt{1-x}}\right) \tag{16}$$

(when $x = 0$, $\sqrt{1-x} = 1$, $\left(\frac{1+\sqrt{1-x}}{2}\right)^{-2\alpha} = 1$).

Another example is: From (9), using Eq. (4) of Sec. 4.8, we arrive at

$$F\left(2\alpha, 2\beta, \alpha + \beta + \frac{1}{2}, \frac{1+x}{2}\right)$$

$$= \frac{\Gamma\left(\alpha + \beta + \frac{1}{2}\right)\Gamma\left(\frac{1}{2}\right)}{\Gamma\left(\alpha + \frac{1}{2}\right)\Gamma\left(\beta + \frac{1}{2}\right)} F\left(\alpha, \beta, \frac{1}{2}, x^2\right)$$

$$- x\frac{\Gamma\left(\alpha + \beta + \frac{1}{2}\right)\Gamma\left(-\frac{1}{2}\right)}{\Gamma(\alpha)\Gamma(\beta)} F\left(\alpha + \frac{1}{2}, \beta + \frac{1}{2}, \frac{3}{2}, x^2\right); \tag{17}$$

the negative sign in the second term on the r.h.s. is determined from the values of the two sides at $x = 1$.

Concerning other quadratic transformations and some higher transformations, cf. Erdélyi (1953), Vol. I, pp. 110-114, formulae (1)–(47); Goursat, l.c. (at the beginning of this section).

4.13. Kummer's Formula and Summation Formula Derived from It

Kummer's formula is

$$F(\alpha, \beta, 1 + \alpha - \beta, -1) = \frac{\Gamma(1 + \alpha - \beta)\,\Gamma\!\left(1 + \frac{\alpha}{2}\right)}{\Gamma(1 + \alpha)\Gamma\!\left(1 + \frac{\alpha}{2} - \beta\right)}, \tag{1}$$

which can be deduced from the following quadratic transformation formula:

$$\begin{aligned} F(\alpha, \beta, 1 &+ \alpha - \beta, x) \\ &= (1 - x)^{-\alpha} F\left(\frac{\alpha}{2}, \frac{1 + \alpha - 2\beta}{2}, 1 + \alpha - \beta, \frac{-4x}{(1 - x)^2}\right), \end{aligned} \tag{2}$$

where $(1 - x)^{-\alpha} = 1$ when $x = 0$. Putting $x = -1$ in (2), its r.h.s. is equal to

$$2^{-\alpha} F\left(\frac{\alpha}{2}, \frac{1 + \alpha - 2\beta}{2}, 1 + \alpha - \beta, 1\right) = 2^{-\alpha} \frac{\Gamma(1 + \alpha - \beta)\Gamma\left(\frac{1}{2}\right)}{\Gamma\left(1 + \frac{\alpha}{2} - \beta\right)\Gamma\left(\frac{\alpha+1}{2}\right)}$$

by Eq. (3) of Sec. 4.7. But, by Eq. (8) of Sec. 3.6,

$$\Gamma\left(\frac{\alpha + 1}{2}\right) = \frac{\Gamma(\alpha + 1)2^{-\alpha}\pi^{\frac{1}{2}}}{\Gamma\left(\frac{\alpha}{2} + 1\right)},$$

and hence (1).

The proof of (2) is as follows. Its l.h.s. belongs to the case $\lambda = \nu$ described in the preceding section on quadratic transformations, i.e., the case XI therein; $A - C = 0, 2l + m = 2$. From the last condition, using (5) of the preceding section, we find $q = -(2p + \alpha)$. Take $p = 0$, then $q = -\alpha$. Let $F(\alpha, \beta, 1 + \alpha - \beta, x) = (1 - x)^{-\alpha} P$. The equation for P is

$$\begin{aligned} x^2(1 - x)^2 \frac{d^2 P}{dx^2} &+ [1 + \alpha - \beta - (1 - \alpha + \beta)x]x(1 - x)\frac{dP}{dx} \\ &- \alpha(2\beta - \alpha + 1)xP = 0. \end{aligned}$$

Let $t = -4x/(1-x)^2$, i.e., $x = (\sqrt{1-t}-1)/(\sqrt{1-t}+1)$. The last equation becomes

$$t(1-t)\frac{d^2P}{dt^2} + \left[1+\alpha-\beta - \left(\alpha-\beta+\frac{3}{2}\right)t\right]\frac{dP}{dt}$$
$$+ \frac{\alpha(2\beta-\alpha-1)}{4}P = 0 \ .$$

$F\left(\frac{\alpha}{2}, \frac{1+\alpha-2\beta}{2}, 1+\alpha-\beta, t\right)$ is the solution of it, which has the value 1 when $t = 0$, hence we have (2).

From Kummer's formula (1) we can derive the following two summation formulae:

$$F\left(\alpha, \beta, \frac{1+\alpha+\beta}{2}, \frac{1}{2}\right) = \frac{\Gamma\left(\frac{1}{2}\right)\Gamma\left(\frac{1+\alpha+\beta}{2}\right)}{\Gamma\left(\frac{1+\alpha}{2}\right)\Gamma\left(\frac{1+\beta}{2}\right)} \tag{3}$$

$$F\left(\alpha, 1-\alpha, \gamma, \frac{1}{2}\right) = \frac{\Gamma\left(\frac{\alpha}{2}\right)\Gamma\left(\frac{1+\gamma}{2}\right)}{\Gamma\left(\frac{\gamma+\alpha}{2}\right)\Gamma\left(\frac{1+\gamma-\alpha}{2}\right)}$$

$$= 2^{1-\gamma}\frac{\Gamma\left(\frac{1}{2}\right)\Gamma(\gamma)}{\Gamma\left(\frac{\gamma+\alpha}{2}\right)\Gamma\left(\frac{1+\gamma-\alpha}{2}\right)} \ . \tag{4}$$

For, from Eq. (9) of Sec. 4.3, we have

$$F(\alpha, \gamma-\beta, \gamma, z) = (1-z)^{-\alpha}F\left(\alpha, \beta, \gamma, \frac{z}{z-1}\right) \ .$$

Putting $z = -1$ gives

$$F\left(\alpha, \beta, \gamma, \frac{1}{2}\right) = 2^{\alpha}F(\alpha, \gamma-\beta, \gamma-1) \ . \tag{5}$$

Let $\gamma = \frac{1}{2}(1+\alpha+\beta)$. By Kummer's formula (1) and the multiplication formula of Γ-function [Eq. (8) of Sec. 3.6], (3) follows.

(3) can also be obtained from (17) of the last section by putting in it $x = 0$ and changing α to $\alpha/2$, β to $\beta/2$.

In (5), let $\beta = 1-\alpha$. We obtain

$$F\left(\alpha, 1-\alpha, \gamma, \frac{1}{2}\right) = 2^{\alpha}F(\alpha, \gamma+\alpha-1, \gamma, -1)$$
$$= 2^{\alpha}F(\gamma+\alpha-1, \alpha, \gamma, -1) \ .$$

Using (1) and the multiplication formula of the Γ-function we arrive at (4).

Other similar formulae can be derived from the transformations of the hypergeometric functions. Cf. Erdélyi (1953), Vol. I, p. 104, (46) – (56).

4.14. Asymptotic Expansions for Large Parameters

The behavior of the hypergeometric function $F(\alpha, \beta, \gamma, z)$ for $z \to \infty$ can be easily found by Eq. (8) of Sec. 4.8, or Eq. (2) of Sec. 4.9, since the series in them are convergent.

We shall discuss in this section the cases when the parameters α, β, γ tend to ∞.

Let us consider first the case when α, β and z are fixed while $|\gamma| \to \infty$. If $|z| < 1$, the hypergeometric series always converges for any value of γ which is not zero or a negative integer. Assume $|\arg \gamma| \leq \pi - \delta, \delta > 0$. By means of the asymptotic expansion of the Γ-function [Eq. (5) of Sec. 3.21], we see immediately that, when $|\gamma| \to \infty$,

$$F(\alpha, \beta, \gamma, z) = 1 + \frac{\alpha \times \beta}{1\gamma} z + \ldots + \frac{(\alpha)_n (\beta)_n}{n!(\gamma)_n} z^n + O(\gamma^{-n-1}) . \tag{1}$$

When the case is $\mathrm{Re}(\gamma) \to \infty$, it can be shown that (1) also holds for $|z| > 1$ and $|\arg(1 - z)| \leq \pi - \delta < \pi$. Let us assume first that $|z| < 1$, then

$$F(\alpha, \beta, \gamma, z) - \sum_{k=0}^{n} \frac{(\alpha)_k (\beta)_k}{k!(\gamma)_k} z^k = \sum_{k=n+1}^{\infty} \frac{(\alpha)_k (\beta)_k}{k!(\gamma)_k} z^k$$

$$= \sum_{k=0}^{\infty} \frac{(\alpha)_{k+n+1}(\beta)_{k+n+1}}{(k+n+1)!(\gamma)_{k+n+1}} z^{k+n+1} = \rho_{n+1}(\alpha, \beta, \gamma, z) .$$

By the relation $(\gamma)_{k+n+1} = \Gamma(\lambda + k + n + 1)/\Gamma(\lambda) = (\lambda + n + 1)_k \Gamma(\lambda + n + 1)/\Gamma(\lambda)$, we obtain

$$\rho_{n+1}(\alpha, \beta, \gamma, z) = \frac{\Gamma(\gamma)}{\Gamma(\alpha)\Gamma(\beta)} \frac{\Gamma(\alpha + n + 1)\Gamma(\beta + n + 1)}{\Gamma(\gamma + n + 1)} z^{n+1}$$

$$\times \sum_{k=0}^{\infty} \frac{(\alpha + n + 1)_k (\beta + n + 1)_k}{(k+n+1)!(\gamma + n + 1)_k} z^k .$$

From Eqs. (1) and (3) of Sec. 3.8, we have

$$\int_0^1 (1 - s)^n (sz)^k ds = \frac{\Gamma(n+1)\Gamma(k+1)}{\Gamma(n+k+2)} z^k .$$

Substituting in the last equation gives

$$\rho_{n+1} = \frac{\Gamma(\gamma)}{\Gamma(\alpha)\Gamma(\beta)} \frac{\Gamma(\alpha+n+1)\Gamma(\beta+n+1)}{n!\Gamma(\gamma+n+1)} z^{n+1}$$

$$\times \int_0^1 (1-s)^n F(\alpha+n+1,\beta+n+1,\gamma+n+1,sz)ds \ .$$

When $\text{Re}(\gamma) \to \infty, \text{Re}(\gamma) > \text{Re}(\beta)$, if n is taken sufficiently large such that $\text{Re}(\beta+n) > 0$, then, by Eq. (6) of Sec. 4.5, we obtain

$$\rho_{n+1} = \frac{\Gamma(\gamma)\Gamma(\alpha+n+1)z^{n+1}}{n!\Gamma(\alpha)\Gamma(\beta)\Gamma(\gamma-\beta)}$$

$$\times \int_0^1 \int_0^1 t^{\beta+n}(1-t)^{\gamma-\beta-1}(1-s)^n(1-stz)^{-\alpha-n-1}dsdt \ ,$$

where $|\arg(1-z)| \le \pi - \varepsilon, \varepsilon > 0, |z| \le R, |z-1| \ge r, R$ and r being arbitrary positive numbers, $R > r$. By a process similar to that given in Sec. 4.5, it can be shown that, in the above domain of z,

$$|t^{\beta+n}(1-t)^{\gamma-\beta-1}(1-stz)^{-\alpha-n-1}| \le Mt^{b+n}(1-t)^{c-b-1} \ ,$$

where b and c are the real parts of β and γ respectively, M being the upper bound $|(1-stz)^{-\alpha-n-1}|$, which depends only on α and n. Hence

$$|\rho_{n+1}| \le \left| \frac{\Gamma(\gamma)(\alpha)_{n+1}z^{n+1}}{(n+1)!\Gamma(\beta)\Gamma(\gamma-\beta)} \right| M \int_0^1 t^{b+n}(1-t)^{c-b-1}dt$$

$$= \left| \frac{(\alpha)_{n+1}\Gamma(b+n+1)Mz^{n+1}}{(n+1)!\Gamma(\beta)} \right| \left| \left| \frac{\Gamma(\gamma)}{\Gamma(\gamma-\beta)} \right| \frac{\Gamma(c-b)}{\Gamma(c+n+1)} \ .$$

Applying the asymptotic expansion of the Γ-function [Eq. (5) of Sec. 3.21] to the last two factors involving Γ-functions, we obtain

$$|\rho_{n+1}| \le \mu(\alpha,\beta,n)|z|^{n+1}|\gamma|^b c^{-b-n-1} \ .$$

Let $\varphi = \arg\gamma$. Assuming $|\arg\gamma| \le \frac{\pi}{2} - \delta, \delta > 0$, it is seen that $c/|\gamma| = \cos\varphi \ne 0$. Hence, when $\gamma \to \infty, |c/\gamma|^{-b}$ is bounded and $|\rho_{n+1}| = O(c^{-n-1}) = O(|\gamma|^{-n-1})$, which shows that (1) holds in the z-plane with a cut from $z = 1$ to ∞ along the real axis as $\gamma \to \infty$ in the region: $|\arg\gamma| \le \pi/2 - \delta < \pi/2$.

Although the above result is obtained under the condition for sufficiently large n, yet, since each term in the asymptotic expansion on the r.h.s. of (1) is $O(\gamma^{-k}), k = 1, 2, \ldots, n$, this restriction may be removed.

It can be shown[m] that (1) has a wider range of validity, namely, $-\frac{\pi}{2} - \varphi <$ arg $\gamma < \frac{\pi}{2} + \varphi$, where φ is an acute angle (> 0) dependent on z, $\varphi < |\arg z|$.

For fixed α, $\gamma (\neq 0, -1, -2, \ldots)$ and z in $0 < z < 1$, we have, when $\beta \to \infty$,

$$F(\alpha, \beta, \gamma, z) = F(\alpha, \beta, \gamma, \beta z/\beta)$$

$$= \sum_{n=0}^{\infty} \frac{(\alpha)_n (\beta z)^n}{n!(\gamma)_n}[1 + O(\beta^{-1})] .$$

Applying the asymptotic expansion formula for the confluent hypergeometric function $F(\alpha, \gamma, \beta z)$, which is equal to the sum of the series on the r.h.s., for $|\beta z| \to \infty$ [Eq. (7) of Sec. 6.8], we find the asymptotic expression of $F(\alpha, \beta, \gamma, z)$ for $\beta \to \infty$:

$$F(\alpha, \beta, \gamma, z) = e^{\pm i\pi\alpha} \frac{\Gamma(\gamma)}{\Gamma(\gamma - \alpha)}(\beta z)^{-\alpha}[1 + O(|\beta z|^{-1})]$$

$$+ \frac{\Gamma(\gamma)}{\Gamma(\alpha)}e^{\beta z}(\beta z)^{\alpha - \gamma}[1 + O(|\beta z|^{-1})] , \qquad (2)$$

where the factor $e^{\pm i\pi\alpha}$ takes the positive sign when $-\pi/2 < \arg(\beta z) < 3\pi/2$, and the negative sign when $-3\pi/2 < \arg(\beta z) < \pi/2$.

Watson[n] has obtained, by means of the method of steepest descent [cf. Sec. 7.11], the asymptotic expansions of $F(\alpha, \beta, \gamma, z)$ when two or three of the parameters are simultaneously large. His results are given in the following. Let

$$e^{\pm \xi} = z \pm \sqrt{z^2 - 1} ,$$

and stipulate that

$$(1 - e^\xi) = (e^\xi - 1)e^{\mp \pi i} ,$$

the factor $e^{\mp \pi i}$ taking the negative sign for $\text{Im}(z) > 0$, and the positive sign for $\text{Im}(z) < 0$. When $|\lambda| \to \infty$, there result

$$\left(\frac{z-1}{2}\right)^{-\alpha-\lambda} F\left(\alpha + \lambda, \alpha - \gamma + 1 + \lambda, \alpha - \beta + 1 + 2\lambda, \frac{2}{1-z}\right)$$

$$= \frac{2^{\alpha+\beta}\Gamma(\alpha - \beta + 1 + 2\lambda)\Gamma\left(\frac{1}{2}\right)\lambda^{-\frac{1}{2}}}{\Gamma(\alpha - \gamma + 1 + \lambda)\Gamma(\gamma - \beta + \lambda)}e^{-(\alpha+\lambda)\xi}$$

$$\times (1 - e^{-\xi})^{\frac{1}{2}-\gamma}(1 + e^{-\xi})^{\gamma-\alpha-\beta-\frac{1}{2}}[1 + O(\lambda^{-1})]$$

$$(|\arg \lambda| \leq \pi - \delta < \pi) \qquad (3)$$

---

[m]MacRobert, *Proc. Edinburgh Math. Soc.* **42** (1923) pp. 84-87.

[n]Watson, *Trans. Camb. Phil. Soc.* **22** (1918) 277.

and

$$F\left(\alpha + \lambda, \beta - \lambda, \gamma, \frac{1-z}{2}\right)$$

$$= \frac{\Gamma(1-\beta+\lambda)\Gamma(\gamma)}{\Gamma\left(\frac{1}{2}\right)\Gamma(\gamma-\beta+\lambda)} 2^{\alpha+\beta-1}(1-e^{-\xi})^{\frac{1}{2}-\gamma}(1+e^{-\xi})^{\gamma-\alpha-\beta-\frac{1}{2}}\lambda^{-\frac{1}{2}}$$

$$\times [e^{(\lambda-\beta)\xi} + e^{\mp i\pi(\frac{1}{2}-\gamma)}e^{-(\lambda+\alpha)\xi}][1+O(\lambda^{-1})] , \tag{4}$$

where the factor $e^{\mp i\pi(\frac{1}{2}-\gamma)}$ takes the negative sign for $\text{Im}(z) > 0$ and the positive sign for $\text{Im}(z) < 0$; and,

$$-\frac{\pi}{2} - w_2 + \delta < \arg \lambda < \frac{\pi}{2} + w_1 - \delta , \quad (\delta > 0)$$

where

$$w_2 = \arctan \frac{\eta}{\varsigma} , \qquad -w_1 = \arctan \frac{\eta - \pi}{\varsigma} \quad (\eta \geq 0) ,$$

$$w_2 = \arctan \frac{\eta + \pi}{\varsigma} \qquad -w_1 = \arctan \frac{\eta}{\varsigma} \quad (\eta \leq 0) ,$$

$$\xi = \varsigma + i\eta, |\arctan x| < \frac{\pi}{2} .$$

More about asymptotic expansion formulae in various other cases may be found in the literature given in Erdélyi (1953), Vol. I, p. 78.

4.15. Generalized Hypergeometric Series

By definition, the generalized hypergeometric series is

$$_pF_q(\alpha_1, \alpha_2, \ldots, \alpha_p; \gamma_1, \gamma_2, \ldots, \gamma_q; z)$$

$$= \sum_{n=0}^{\infty} \frac{(\alpha_1)_n(\alpha_2)_n \ldots (\alpha_p)_n}{n!(\gamma_1)_n(\gamma_2)_n \ldots (\gamma_q)_n} z^n . \tag{1}$$

When $p \leq q$, the series is convergent for any value of z. When $p > q + 1$, the series is divergent for any value of z other than zero, unless the series breaks off into a polynomial, which is, of course, meaningful.

When $p = q + 1$ (which is the case we are most interested in, for example, the hypergeometric series), the series converges in $|z| < 1$. If $\text{Re}(\sum \gamma - \sum \alpha) > 0$, then it converges also at the point $z = 1$.

The hypergeometric series $F(\alpha, \beta, \gamma, z)$ is a special case of (1), namely, $p = 2, q = 1(p = q + 1)$, and hence it is often written as $_2F_1(\alpha, \beta; \gamma; z)$.

The differential equation satisfied by $_{p+1}F_p$ can be expressed as

$$\{\vartheta(\vartheta + \gamma_1 - 1)\ldots(\vartheta + \gamma_p - 1) - z(\vartheta + \alpha_1)\ldots(\vartheta + \alpha_{p+1})\}y = 0 , \quad (2)$$

where $\vartheta = zd/dz$. It is not difficult to prove (2).

Another solution of (2) is

$$z^{1-\gamma_1}{}_{p+1}F_p(1 + \alpha_1 - \gamma_1, 1 + \alpha_2 - \gamma_1, \ldots, 1 + \alpha_{p+1} - \gamma_1 ;$$
$$2 - \gamma_1, 1 + \gamma_2 - \gamma_1, \ldots, 1 + \gamma_p - \gamma_1; z) \qquad (3)$$

provided that γ_1 is not an integer. There are $p - 1$ more solutions of this form with γ_1 replaced respectively by $\gamma_2, \gamma_3, \ldots, \gamma_p$ (all not integers). If the difference between any two of the values $1, \gamma_1, \ldots, \gamma_p$ is not an integer, then all these p solutions are linearly independent; otherwise, there will be solutions containing a logarithmic term. [Cf. Ince, (1927), Chap. XV.]

Saalschutz' summation formula

$$_3F_2(\alpha, \beta, -n; \gamma, 1 + \alpha + \beta - \gamma - n; 1) = \frac{(\gamma - \alpha)_n(\gamma - \beta)_n}{(\gamma)_n(\gamma - \alpha - \beta)_n} \qquad (4)$$

can be used to evaluate the sum of the series $_3F_2(\alpha_1, \alpha_2, \alpha_3; \gamma_1, \gamma_2; 1)$, where $\gamma_1 + \gamma_2 = \alpha_1 + \alpha_2 + \alpha_3 + 1$, and one of the α's is a negative integer $(-n)$; the series is then a polynomial.

The proof of (4) is as follows. From Eq. (8) of Sec. 4.3, we have

$$(1 - z)^{\alpha+\beta-\gamma}{}_2F_1(\alpha, \beta; \gamma; z) = {}_2F_1(\gamma - \alpha, \gamma - \beta; \gamma; z) . \qquad (5)$$

Expanding $(1-z)^{\alpha+\beta-\gamma}$ in a power series of z, then comparing the coefficients of z^n on both sides of (5), we obtain

$$\sum_{s=0}^{n} \frac{(\alpha)_s(\beta)_s(\gamma - \alpha - \beta)_{n-s}}{s!(\gamma)_s(n - s)!} = \frac{(\gamma - \alpha)_n(\gamma - \beta)_n}{n!(\gamma)_n} . \qquad (6)$$

With the formula

$$(\lambda)_{n-s} = (-)^s \frac{(\lambda)_n}{(1 - \lambda - n)_s} \qquad (7)$$

the l.h.s. of (6) reduces to

$$\frac{(\gamma - \alpha - \beta)_n}{n!} \sum_{s=0}^{n} \frac{(\alpha)_s(\beta)_s(-n)_s}{s!(\gamma)_s(1 - \gamma + \alpha + \beta - n)_s} ,$$

and we arrive at (4).

Formula (4) can be written as

$$_3F_2(\alpha, \beta, \gamma; \delta, \varepsilon; 1) = \frac{\Gamma(\delta)\Gamma(1+\alpha-\varepsilon)\Gamma(1+\beta-\varepsilon)\Gamma(1+\gamma-\varepsilon)}{\Gamma(1-\varepsilon)\Gamma(\delta-\alpha)\Gamma(\delta-\beta)\Gamma(\delta-\gamma)}, \qquad (8)$$

where one of the parameters α, β, γ is a negative integer, and $\delta + \varepsilon = \alpha + \beta + \gamma + 1$.

For further reference about the generalized hypergeometric series, cf., for example, Bailey (1953); Erdélyi (1953) Vol. I, Chap. 4.

4.16. Hypergeometric Series with Two Variables

By multiplying two hypergeometric series $F(\alpha, \beta, \gamma, x)$ and $F(\alpha', \beta', \gamma', y)$ together, we arrive at a double series of two variables x and y; its general term is

$$\frac{(\alpha)_m(\alpha')_n(\beta)_m(\beta')_n}{m!n!(\gamma)_m(\gamma')_n}x^m y^n \ .$$

In this expression, replacing one, two or three of the products $(\alpha)_m(\alpha')_n$, $(\beta)_m(\beta')_n$, $(\gamma)_m(\gamma')_n$ respectively by $(\alpha)_{m+n}$, $(\beta)_{m+n}$, $(\gamma)_{m+n}$, we obtain five different kinds of double series. The one with all three of the products replaced, namely,

$$\sum_{m,n=0}^{\infty} \frac{(\alpha)_{m+n}(\beta)_{m+n}}{m!n!(\gamma)_{m+n}}x^m y^n$$

is in fact the expansion of the hypergeometric series $F(\alpha, \beta, \gamma, x+y)$, since

$$F(\alpha, \beta, \gamma, x+y) = \sum_{s=0}^{\infty} \frac{(\alpha)_s(\beta)_s}{s!(\gamma)_s}(x+y)^s$$

$$= \sum_{s=0}^{\infty} \frac{(\alpha)_s(\beta)_s}{s!(\gamma)_s} \sum_{m=0}^{s} \binom{s}{m} x^m y^{s-m} = \sum_{m=0}^{\infty} C_m x^m \ ,$$

$$C_m = \sum_{s=m}^{\infty} \frac{(\alpha)_s(\beta)_s}{s!(\gamma)_s}\binom{s}{m} y^{s-m} = \sum_{n=0}^{\infty} \frac{(\alpha)_{m+n}(\beta)_{m+n}}{m!n!(\gamma)_{m+n}} y^n \ .$$

Leaving this trivial case apart, the remaining four kinds of double series are

$$F_1(\alpha; \beta, \beta'; \gamma; x, y) = \sum\sum \frac{(\alpha)_{m+n}(\beta)_m(\beta')_n}{m!n!(\gamma)_{m+n}} x^m y^n \ , \tag{1}$$

$$F_2(\alpha; \beta, \beta'; \gamma, \gamma'; x, y) = \sum\sum \frac{(\alpha)_{m+n}(\beta)_m(\beta')_n}{m!n!(\gamma)_m(\gamma')_n} x^m y^n \ , \tag{2}$$

$$F_3(\alpha, \alpha'; \beta, \beta'; \gamma; x, y) = \sum\sum \frac{(\alpha)_m(\alpha')_n(\beta)_m(\beta')_n}{m!n!(\gamma)_{m+n}} x^m y^n \ , \tag{3}$$

$$F_4(\alpha; \beta; \gamma, \gamma'; x, y) = \sum\sum \frac{(\alpha)_{m+n}(\beta)_{m+n}}{m!n!(\gamma)_m(\gamma')_n} x^m y^n \ . \tag{4}$$

All the m, n in them are to run from 0 to ∞. The functions represented by these four series are named *Appell's hypergeometric functions of two variables.*

These double series are absolutely convergent respectively in the following domains:[°]

$$|x| < 1, \quad |y| < 1 \ ; \tag{1a}$$

$$|x| + |y| < 1 \ ; \tag{2a}$$

$$|x| < 1, \quad |y| < 1 \ ; \tag{3a}$$

$$|x|^{\frac{1}{2}} + |y|^{\frac{1}{2}} < 1 \ . \tag{4a}$$

When $y = 0$, all these functions reduce to the ordinary hypergeometric function $F(\alpha, \beta, \gamma, x)$. The first three also become $F(\alpha, \beta, \gamma, x)$ when $\beta' = 0$.

Partial differential equations satisfied by Appell's functions

Writing $F_1(\alpha; \beta, \beta'; \gamma; x, y)$ as $\sum A_{m,n} x^m y^n$, it is easy to see that

$$A_{m+1,n} = \frac{(\alpha + m + n)(\beta + m)}{(m+1)(\gamma + m + n)} A_{m,n} \ .$$

Hence, we find that F_1 satisfies the equation

$$\{(\vartheta + \varphi + \alpha)(\vartheta + \beta) - \frac{1}{x}\vartheta(\vartheta + \varphi + \gamma - 1)\}F_1 = 0 \ , \tag{5}$$

[°]Cf. Bailey (1935), Chap. IX, pp. 74-75; Regards the general convergence problem of double series, cf. Bromwich (1925), Chap. V, pp. 78-97.

where $\vartheta = x\frac{\partial}{\partial x}, \varphi = y\frac{\partial}{\partial y}$. With reference to the relation between $A_{m,n+1}$ and $A_{m,n}$, we may arrive at another partial differential equation satisfied by F_1.

With the usual notations: $p = \frac{\partial z}{\partial x}, q = \frac{\partial z}{\partial y}$ and r, s, t representing respectively $\frac{\partial^2 z}{\partial x^2}, \frac{\partial^2 z}{\partial x \partial y}, \frac{\partial^2 z}{\partial y^2}$, the partial differential equations satisfied by the various F functions are respectively (with z representing F_1, F_2, F_3, F_4):

$$F_1: \left.\begin{aligned} x(1-x)r + y(1-x)s + \{\gamma - (\alpha + \beta + 1)x\}p \\ -\beta yq - \alpha\beta z = 0\ , \\ y(1-y)t + x(1-y)s + \{\gamma - (\alpha + \beta' + 1)y\}q \\ -\beta' xp - \alpha\beta' z = 0\ ; \end{aligned}\right\} \tag{6}$$

$$F_2: \left.\begin{aligned} x(1-x)r - xys + \{\gamma - (\alpha + \beta + 1)x\}p \\ -\beta yq - \alpha\beta z = 0\ , \\ y(1-y)t - xys + \{\gamma' - (\alpha + \beta' + 1)y\}q \\ -\beta' xp - \alpha\beta' z = 0\ ; \end{aligned}\right\} \tag{7}$$

$$F_3: \left.\begin{aligned} x(1-x)r + ys + \{\gamma - (\alpha + \beta + 1)x\}p - \alpha\beta z = 0\ , \\ y(1-y)t + xs + \{\gamma - (\alpha' + \beta' + 1)y\}q - \alpha'\beta' z = 0\ ; \end{aligned}\right\} \tag{8}$$

$$F_4: \left.\begin{aligned} x(1-x)r - y^2 t - 2xys + \{\gamma - (\alpha + \beta + 1)x\}p \\ -(\alpha + \beta + 1)yq - \alpha\beta z = 0\ , \\ y(1-y)t - x^2 r - 2xys + \{\gamma' - (\alpha + \beta + 1)y\}q \\ -(\alpha + \beta + 1)xp - \alpha\beta z = 0\ . \end{aligned}\right\} \tag{9}$$

Integral Representations

$$\frac{\Gamma(\beta)\Gamma(\beta')\Gamma(\gamma - \beta - \beta')}{\Gamma(\gamma)} F_1(\alpha; \beta, \beta'; \gamma; x, y)$$
$$= \int\int u^{\beta-1} v^{\beta'-1}(1 - u - v)^{\gamma-\beta-\beta'-1}(1 - ux - vy)^{-\alpha} du\, dv\ , \tag{10}$$

the domain of integration being the triangular area: $u \geq 0, v \geq 0, u + v \leq 1$;

$$\frac{\Gamma(\beta)\Gamma(\beta')\Gamma(\gamma - \beta)\Gamma(\gamma' - \beta')}{\Gamma(\gamma)\Gamma(\gamma')} F_2(\alpha; \beta, \beta'; \gamma, \gamma'; x, y)$$

$$= \int_0^1 \int_0^1 u^{\beta-1} v^{\beta'-1} (1 - u)^{\gamma-\beta-1} (1 - v)^{\gamma'-\beta'-1} (1 - ux - vy)^{-\alpha} du dv ; \tag{11}$$

$$\frac{\Gamma(\beta)\Gamma(\beta')\Gamma(\gamma - \beta - \beta')}{\Gamma(\gamma)} F_3(\alpha, \alpha'; \beta, \beta'; \gamma; x, y)$$

$$= \int \int u^{\beta-1} v^{\beta'-1} (1 - u - v)^{\gamma-\beta-\beta'-1} (1 - ux)^{-\alpha} (1 - vy)^{-\alpha'} du dv ; \tag{12}$$

the domain of integration being the triangular area: $u \geq 0, v \geq 0, u + v \leq 1$. The values of the parameters have to be such that the integrals are convergent.

These integral representations can be proved by expressing the integrands in powers of x and y, and then integrating term by term.

The integral representations of F_4 are somewhat more complicated. A simpler one is

$$\frac{\Gamma(\alpha)\Gamma(\beta)\Gamma(\gamma - \alpha)\Gamma(\gamma' - \beta)}{\Gamma(\gamma)\Gamma(\gamma')} F_4(\alpha; \beta; \gamma, \gamma'; x(1 - y), y(1 - x))$$

$$= \int_0^1 \int_0^1 u^{\alpha-1} v^{\beta-1} (1 - u)^{\gamma-\alpha-1} (1 - v)^{\gamma'-\beta-1}$$

$$\times (1 - ux)^{\alpha-\gamma-\gamma'+1} (1 - vy)^{\beta-\gamma-\gamma'+1} (1 - ux - vy)^{\gamma+\gamma'-\alpha-\beta-1} du dv , \tag{13}$$

where $\mathrm{Re}(\alpha) > 0, \mathrm{Re}(\beta) > 0, \mathrm{Re}(\gamma - \alpha) > 0, \mathrm{Re}(\gamma' - \beta) > 0$.

The function F_1 can also be represented by the more simple integral

$$\frac{\Gamma(\alpha)\Gamma(\gamma - \alpha)}{\Gamma(\gamma)} F_1(\alpha; \beta, \beta'; \gamma; x, y)$$

$$= \int_0^1 u^{\alpha-1} (1 - u)^{\gamma-\alpha-1} (1 - ux)^{-\beta} (1 - uy)^{-\beta'} du . \tag{14}$$

The proof is same as above.

Appell's functions F_1, F_2, F_3, F_4 can also be represented by Barnes' double integrals. For example, applying the relation $(\lambda)_{m+n} = (\lambda)_m (\lambda + m)_n$ to (1), we have

$$F_1(\alpha; \beta, \beta'; \gamma; x, y) = \sum_{m=0}^{\infty} \frac{(\alpha)_m (\beta)_m}{m! (\gamma)_m} x^m F(\alpha + m, \beta'; \gamma + m; y)$$

$$(|x| < 1, |y| < 1) . \tag{15}$$

Applying Eq. (9) of Sec. 4.6 to the hypergeometric function occurred in the sum, we find

$$\frac{\Gamma(\alpha)\Gamma(\beta')}{\Gamma(\gamma)} F_1(\alpha; \beta, \beta'; \gamma; x, y)$$

$$= \frac{1}{2\pi i} \sum_{m=0}^{\infty} \frac{(\beta)_m}{m!} x^m \int_{-i\infty}^{i\infty} \frac{\Gamma(\alpha+m+t)\Gamma(\beta'+t)\Gamma(-t)}{\Gamma(\gamma+m+t)} (-y)^t dt$$

$$= \frac{1}{2\pi i} \int_{-i\infty}^{i\infty} \sum_{m=0}^{\infty} \frac{(\alpha+t)_m(\beta)_m}{m!(\gamma+t)_m} x^m \frac{\Gamma(\alpha+t)\Gamma(\beta'+t)\Gamma(-t)}{\Gamma(\gamma+t)} (-y)^t dt ,$$

the series in which is equal to $F(\alpha+t, \beta; \gamma+t, x)$. Using again Eq. (9) of Sec. 4.6, we finally obtain

$$\frac{\Gamma(\alpha)\Gamma(\beta)\Gamma(\beta')}{\Gamma(\gamma)} F_1(\alpha; \beta, \beta'; \gamma; x, y)$$

$$= -\frac{1}{4\pi} \int_{-i\infty}^{i\infty} \int_{-i\infty}^{i\infty} \frac{\Gamma(\alpha+s+t)\Gamma(\beta+s)\Gamma(\beta'+t)\Gamma(-s)\Gamma(-t)}{\Gamma(\gamma+s+t)}$$
$$\times (-x)^s (-y)^t \, ds \, dt .$$

Similarly, we can derive the integral expressions of F_2, F_3, F_4 in this form. [Cf. Appell and Kampé de Fériet (1926)]. So long as $\alpha, \alpha', \beta, \beta'$ are not negative integers, all these formulae are valid. Therefore, it is convenient to use them for analytic continuation as in the case of one variable. [Cf. Sec. 4.9 and Ex. 21 at the end of this chapter.]

4.17. The Transformation Formulae of F_1 and F_2

By the integral representations of F_1 and F_2, we can derive some transformation formulae for them. For example, the integral in (14) of the last section, namely,

$$\int_0^1 u^{\alpha-1}(1-u)^{\gamma-\alpha-1}(1-ux)^{-\beta}(1-uy)^{-\beta'} du$$

keeps unchanged under the following five transformations:

$$\left.\begin{array}{ll} u = 1-v, & u = v/(1-x+vx) , \\ u = v/(1-y+vy), & u = (1-v)/(1-vx) , \\ u = (1-v)/(1-vy) . & \end{array}\right\} \tag{1}$$

Thus, we have the five transformations:

$$F_1(\alpha; \beta, \beta'; \gamma; x, y)$$

$$= (1-x)^{-\beta}(1-y)^{-\beta'} F_1\left(\gamma - \alpha; \beta, \beta'; \gamma; \frac{x}{x-1}, \frac{y}{y-1}\right) \tag{2}$$

$$= (1-x)^{-\alpha} F_1\left(\alpha; \gamma - \beta - \beta', \beta'; \gamma; \frac{x}{x-1}, \frac{x-y}{x-1}\right) \tag{3}$$

$$= (1-y)^{-\alpha} F_1\left(\alpha; \beta, \gamma - \beta - \beta'; \gamma; \frac{y-x}{y-1}, \frac{y}{y-1}\right) \tag{4}$$

$$= (1-x)^{\gamma-\alpha-\beta}(1-y)^{-\beta'} F_1\left(\gamma - \alpha; \gamma - \beta - \beta', \beta'; \gamma; x, \frac{y-x}{y-1}\right) \tag{5}$$

$$= (1-x)^{-\beta}(1-y)^{\gamma-\alpha-\beta'} F_1\left(\gamma - \alpha; \beta, \gamma - \beta - \beta'; \gamma; \frac{x-y}{x-1}, y\right) \cdot \tag{6}$$

Similarly, from the integral in (11) of the last section, namely

$$\int_0^1 \int_0^1 u^{\beta-1} v^{\beta'-1}(1-u)^{\gamma-\beta-1}(1-v)^{\gamma'-\beta'-1}(1-ux-vy)^{-\alpha} du\,dv,$$

by using the transformations

$$\left.\begin{array}{lll} (a) & u = 1 - u', & v = v'; \\ (b) & u = u', & v = 1 - v'; \\ (c) & u = 1 - u', & v = 1 - v'; \end{array}\right\} \tag{7}$$

we obtain respectively the transformation formulae for F_2,

$$F_2(\alpha; \beta, \beta'; \gamma, \gamma'; x, y)$$

$$= (1-x)^{-\alpha} F_2\left(\alpha; \gamma - \beta, \beta'; \gamma, \gamma'; \frac{x}{x-1}, \frac{-y}{x-1}\right) \tag{8}$$

$$= (1-y)^{-\alpha} F_2\left(\alpha; \beta, \gamma' - \beta'; \gamma, \gamma'; \frac{-x}{y-1}, \frac{y}{y-1}\right) \tag{9}$$

$$= (1 - x - y)^{-\alpha} F_2 \left(\alpha; \gamma - \beta, \gamma' - \beta'; \gamma, \gamma'; \frac{-x}{1 - x - y}, \frac{-y}{1 - x - y} \right) \tag{10}$$

4.18. Reducible Cases

Under certain conditions, Appell's functions can be reduced to the ordinary hypergeometric functions. For example, from (5) of the last section, putting $y = x$ and noticing that when one of the variables is zero, Appell's function is reduced to an ordinary hypergeometric function, we have

$$\begin{aligned}
F_1(\alpha; \beta, \beta'; \gamma; x, x) \\
= (1 - x)^{\gamma - \alpha - \beta - \beta'} F(\gamma - \alpha, \gamma - \beta - \beta'; \gamma; x) \\
= F(\alpha, \beta + \beta'; \gamma; x) .
\end{aligned} \tag{1}$$

Equation (8) of Sec. 4.3 has been employed in the last step.

Also, from (4) of the last section, by putting $\gamma = \beta + \beta'$, we obtain

$$\begin{aligned}
F_1(\alpha; \beta, \beta'; \beta + \beta'; x, y) \\
= (1 - y)^{-\alpha} F \left(\alpha, \beta; \beta + \beta'; \frac{y - x}{y - 1} \right) .
\end{aligned} \tag{2}$$

And, from (8) of the last section, putting $\gamma = \beta$, we obtain

$$F_2(\alpha; \beta, \beta'; \beta, \gamma'; x, y) = (1 - x)^{-\alpha} F \left(\alpha, \beta'; \gamma'; \frac{-y}{x - 1} \right) . \tag{3}$$

(2) shows that, when $\gamma = \beta + \beta'$, F_1 reduces to an ordinary hypergeometric function; (3) shows that, when $\gamma = \beta$ (or $\gamma' = \beta'$), F_2 also reduces to an ordinary hypergeometric function.

Besides, F_1, F_2 and F_3 can be reduced among themselves. For example,

$$F_1(\alpha, \beta, \beta'; \gamma; x, y) = (1 - y)^{-\beta'} F_3 \left(\alpha, \gamma - \alpha; \beta, \beta'; \gamma; x, \frac{y}{y - 1} \right) , \tag{4}$$

i.e., F_1 can always be expressed in terms of F_3. Conversely, when $\gamma = \alpha + \alpha'$, F_3

can also be expressed in terms of F_1. The proof of (4) is as follows:

$$F_1(\alpha; \beta, \beta'; \gamma; x, y)$$

$$= \sum_{m=0}^{\infty} \frac{(\alpha)_m (\beta)_m}{m!(\gamma)_m} F(\alpha + m, \beta'; \gamma + m; y) x^m$$

$$= \sum_{m=0}^{\infty} \frac{(\alpha)_m (\beta)_m}{m!(\gamma)_m} (1-y)^{-\beta'} F\left(\gamma - \alpha, \beta'; \gamma + m; \frac{y}{y-1}\right) x^m$$

$$= (1-y)^{-\beta'} \sum_{m,n=0}^{\infty} \frac{(\alpha)_m (\beta)_m (\gamma - \alpha)_n (\beta')_n}{m!n!(\gamma)_{m+n}} x^m \left(\frac{y}{y-1}\right)^n ,$$

which is exactly the r.h.s. of (4); Eq. (10) of Sec. 4.3 has been used in the proof.

According to (2) above, F_1 can be reduced to an ordinary hypergeometric function when $\gamma = \beta + \beta'$. Therefore, F_3 can also be reduced provided that $\gamma = \alpha + \alpha' = \beta + \beta'$; the reducing formula is:

$$F_3(\alpha, \gamma - \alpha; \beta, \gamma - \beta; \gamma; x, y) = (1-y)^{\alpha+\beta-\gamma} F(\alpha, \beta; \gamma; x + y - xy) . \qquad (5)$$

Also, F_1 can always be expressed in terms of F_2. For,

$$(1-y)^{-\beta'} F_2\left(\alpha; \beta, \beta'; \gamma, \alpha; x, \frac{y}{y-1}\right)$$

$$= (1-y)^{-\beta'} \sum_{m=0}^{\infty} \frac{(\alpha)_m (\beta)_m}{m!(\gamma)_m} x^m F\left(\alpha + m, \beta'; \alpha; \frac{y}{y-1}\right)$$

$$= \sum_{m=0}^{\infty} \frac{(\alpha)_m (\beta)_m}{m!(\gamma)_m} x^m F(\beta', -m; \alpha; y) \quad \text{[by Eq. (10) of Sec. 4.3]}$$

$$= \sum_{m=0}^{\infty} \frac{(\alpha)_m (\beta)_m}{m!(\gamma)_m} x^m \frac{(\alpha - \beta')_m}{(\alpha)_m} F(\beta', -m; 1 + \beta' - \alpha - m; 1 - y)$$
[by Eq. (4) of Sec. 4.8]

$$= \sum_{m=0}^{\infty} \sum_{n=0}^{m} \frac{(\beta)_m (\alpha - \beta')_m (\beta')_n (-m)_n}{m! n! (\gamma)_m (1 + \beta' - \alpha - m)_n} x^m (1 - y)^n$$

$$= \sum_{n=0}^{\infty} \sum_{m=n}^{\infty} \frac{(\beta)_m (\alpha - \beta')_m (\beta')_n (-m)_n}{m! n! (\gamma)_m (1 + \beta' - \alpha - m)_n} x^m (1 - y)^n$$

$$= \sum_{n=0}^{\infty} \sum_{s=0}^{\infty} \frac{(\beta)_{s+n} (\alpha - \beta')_{s+n} (\beta')_n (-s - n)_n}{(s + n)! n! (\gamma)_{s+n} (1 + \beta' - \alpha - s - n)_n} x^{s+n} (1 - y)^n$$

$$= \sum_{n=0}^{\infty} \sum_{s=0}^{\infty} \frac{(\beta)_{s+n} (\alpha - \beta')_s (\beta')_n}{s! n! (\gamma)_{s+n}} x^s [x(1 - y)]^n \ ,$$

i.e.,

$$(1 - y)^{-\beta'} F_2 \left(\alpha; \beta, \beta'; \gamma, \alpha; x, \frac{y}{y - 1} \right)$$

$$= F_1(\beta; \alpha - \beta', \beta'; \gamma; x, x(1 - y)) \ . \tag{6}$$

This formula shows at the same time that, if $\gamma' = \alpha$, F_2 can also be expressed in terms of F_1.

When $\gamma = \alpha$, F_1 on the r.h.s. of (6) can be reduced to an ordinary hypergeometric function by (2). Hence, when $\gamma' = \gamma = \alpha$, F_1 can be reduced; the result is:

$$F_2(\alpha; \beta, \beta'; \alpha, \alpha; x, y) = (1-x)^{-\beta}(1-y)^{-\beta'} F \left(\beta, \beta'; \alpha; \frac{xy}{(x - 1)(y - 1)} \right) \tag{7}$$

Under special conditions, F_4 can also be reduced to ordinary hypergeometric functions; the formula is:

$$F_4(\alpha; \beta; \gamma, \alpha + \beta - \gamma + 1; z(1 - Z), Z(1 - z))$$
$$= F(\alpha, \beta; \gamma; z) F(\alpha, \beta; \alpha + \beta - \gamma + 1; Z) \tag{8}$$

which holds in the neighborhoods of $z = 0$ and $Z = 0$ under the condition: $|z(1 - Z)|^{\frac{1}{2}} + |Z(1 - z)|^{\frac{1}{2}} < 1$; i.e., in the domain where the series on both sides are convergent [cf. Eq. (4a) of Sec. 4.16].

Replacing z and Z in (8) respectively by $1 - Z$ and $1 - z$, we find

$$F_4\left(\alpha;\beta;\gamma,\alpha+\beta-\gamma+1;z(1-Z),Z(1-z)\right)$$
$$= F(\alpha,\beta;\gamma;1-Z)F(\alpha,\beta;\alpha+\beta-\gamma+1;1-z) \qquad (9)$$

which holds in the neighborhoods of $z = 1$ and $Z = 1$ under the condition: $|z(1-Z)|^{1/2} + |Z(1-z)|^{\frac{1}{2}} < 1$.

Formulae (8) and (9) show that, when $\gamma + \gamma' = \alpha + \beta + 1$, F_4 can be represented by products of ordinary hypergeometric functions.

Now we turn to the proof of (8). Consider the function

$$(1-x)^{-\alpha}(1-y)^{-\beta}F_4\left(\alpha;\beta;\gamma,\gamma';\frac{-x}{(1-x)(1-y)},\frac{-y}{(1-x)(1-y)}\right).$$

When $|x|$ and $|y|$ are sufficiently small, this function can be expanded in a double series of x and y, in which the coefficient of $x^m y^n$ is

$$A_{mn} = \sum_{t=0}^{m}\sum_{s=0}^{n}\frac{(\alpha)_{t+s}(\beta)_{t+s}}{t!s!(\gamma)_t(\gamma')_s}\frac{(-)^{t+s}(\alpha+t+s)_{m-t}(\beta+t+s)_{n-s}}{(m-t)!(n-s)!}$$

$$= \frac{(\alpha)_m(\beta)_n}{m!n!}\sum_{t=0}^{m}\sum_{s=0}^{n}\frac{(\alpha+m)_s(\beta+n)_t(-m)_t(-n)_s}{t!s!(\gamma)_t(\gamma')_s}$$

$$= \frac{(\alpha)_m(\beta)_n}{m!n!}F(\alpha+m,-n;\gamma';1)F(\beta+n,-m;\gamma;1)$$

$$= \frac{(\alpha)_m(\beta)_n}{m!n!}\frac{(\gamma-\beta-n)_m(\gamma'-\alpha-m)_n}{(\gamma)_m(\gamma')_n} \qquad \text{[by Eq.(2) of Sec. 4.10]}.$$

By the formulae

$$(\lambda-m)_n = \frac{\Gamma(\lambda-m+n)}{\Gamma(\lambda-m)} = \frac{\Gamma(1-\lambda+m)\sin\pi(\lambda-m)}{\Gamma(1-\lambda+m-n)\sin\pi(\lambda-m+n)}$$

$$= (-)^n\frac{(1-\lambda)_m}{(1-\lambda)_{m-n}},$$

$$(\lambda-n)_m = \frac{\Gamma(\lambda-n+m)}{\Gamma(\lambda-n)} = \frac{\Gamma(\lambda)(\lambda)_{m-n}}{\Gamma(\lambda-n)}$$

$$= (-)^n(1-\lambda)_n(\lambda)_{m-n},$$

we obtain

$$A_{mn} = \frac{(\alpha)_m (\beta)_n}{m!n!} \frac{(1+\alpha-\gamma')_m (1+\beta-\gamma)_n (\gamma-\beta)_{m-n}}{(\gamma)_m (\gamma')_n (1+\alpha-\gamma')_{m-n}}.$$

If $\gamma + \gamma' = \alpha + \beta + 1$, then

$$A_{mn} = \frac{(\alpha)_m (\beta)_n}{m!n!} \frac{(\gamma-\beta)_m (\gamma'-\alpha)_n}{(\gamma)_m (\gamma')_n},$$

and it follows that

$$(1-x)^{-\alpha} (1-y)^{-\beta} F_4 \left(\alpha; \beta; \gamma, \gamma'; \frac{-x}{(1-x)(1-y)}, \frac{-y}{(1-x)(1-y)} \right)$$

$$= \sum_{m,n=0}^{\infty} \frac{(\alpha)_m (\beta)_n (\gamma-\beta)_m (\gamma'-\alpha)_n}{m!n!(\gamma)_m (\gamma')_n} x^m y^n$$

$$= F(\alpha, \gamma-\beta; \gamma; x) F(\beta, \gamma'-\alpha; \gamma'; y)$$

$$= (1-x)^{-\alpha} F\left(\alpha, \beta; \gamma; \frac{x}{x-1} \right) \times (1-y)^{-\beta} F\left(\beta, \alpha; \gamma'; \frac{y}{y-1} \right)$$

[by Eq. (9) of Sec. 4.3] .

Putting $x/(x-1) = z, y/(y-1) = Z$, we have (8).

When $\alpha = -n$, a negative integer, by changing β to $p+n$ and Z to $1-Z$ in (8), we obtain

$$F_4(-n; p+n, \gamma, p-\gamma+1; zZ, (1-z)(1-Z))$$

$$= F(-n, p+n; \gamma; z) F(-n, p+n; p-\gamma+1; 1-Z)$$

$$= \frac{(-)^n (\gamma)_n}{(p-\gamma+1)_n} F(-n, p+n; \gamma; z) F(-n, p+n; \gamma; Z) . \tag{10}$$

In the last step, Eq. (4) of Sec. 4.8 has been used. The ordinary hypergeometric functions on the r.h.s. of (10) are, in fact, two Jacobi polynomials [Sec. 4.10].

There are other cases when Appell's functions can be reduced to ordinary hypergeometric functions. The examples given above are those with least restrictions on the parameters.

For further discussions and results concerning Appell's functions, cf. Appell and Kampé de Fériet (1926); Erdélyi (1953), Vol. I, Chap. 5.

Exercise 4

1. With the integral representation of the hypergeometric function given in Sec. 4.5, derive the two fundamental recurrence relations [Eqs. (2) and (3) of Sec. 4.2]:

$$(\gamma - 1)F(\gamma - 1) - \alpha F(\alpha + 1) - (\gamma - \alpha - 1)F = 0 \,,$$
$$\gamma F - \beta z F(\beta + 1, \gamma + 1) - \gamma F(\alpha - 1) = 0 \,.$$

2. Derive the following 15 recurrence relations of $F(\alpha, \beta, \gamma, z)$ with its close neighbors:

$$(\gamma - \alpha - 1)F + \alpha F(\alpha + 1) - (\gamma - 1)F(\gamma - 1) = 0 \,,$$

$$(\gamma - \beta - 1)F + \beta F(\beta + 1) - (\gamma - 1)F(\gamma - 1) = 0 \,,$$

$$(\alpha - \beta)F - \alpha F(\alpha + 1) + \beta F(\beta + 1) = 0 \,,$$

$$[\gamma - 2\alpha + (\alpha - \beta)z]F + \alpha(1 - z)F(\alpha + 1) - (\gamma - \alpha)F(\alpha - 1) = 0 \,,$$

$$[\gamma - 2\beta + (\beta - \alpha)z]F + \beta(1 - z)F(\beta + 1) - (\gamma - \beta)F(\beta - 1) = 0 \,,$$

$$(\gamma - \alpha - \beta)F + \alpha(1 - z)F(\alpha + 1) - (\gamma - \beta)F(\beta - 1) = 0 \,,$$

$$(\gamma - \alpha - \beta)F + \beta(1 - z)F(\beta + 1) - (\gamma - \alpha)F(\alpha - 1) = 0$$

$$(\alpha - \beta)(1 - z)F + (\gamma - \alpha)F(\alpha - 1) - (\gamma - \beta)F(\beta - 1) = 0 \,,$$

$$\gamma(1-z)F - \gamma F(\alpha-1) + (\gamma-\beta)zF(\gamma+1) = 0 \, ,$$

$$\gamma(1-z)F - \gamma F(\beta-1) + (\gamma-\alpha)zF(\gamma+1) = 0 \, ,$$

$$\gamma[\alpha-(\gamma-\beta)z]F - \alpha\gamma(1-z)F(\alpha+1) + (\gamma-\alpha)(\gamma-\beta)zF(\gamma+1) = 0 \, ,$$

$$\gamma[\beta-(\gamma-\alpha)z]F - \beta\gamma(1-z)F(\beta+1) + (\gamma-\beta)(\gamma-\alpha)zF(\gamma+1) = 0 \, ,$$

$$[\alpha-1-(\gamma-\beta-1)z]F + (\gamma-\alpha)F(\alpha-1)$$
$$- (\gamma-1)(1-z)F(\gamma-1) = 0 \, ,$$

$$[\beta-1-(\gamma-\alpha-1)z]F + (\gamma-\beta)F(\beta-1)$$
$$- (\gamma-1)(1-z)F(\gamma-1) = 0 \, ,$$

$$\gamma[\gamma-1-(2\gamma-\alpha-\beta-1)z]F + (\gamma-\alpha)(\gamma-\beta)zF(\gamma+1)$$
$$- \gamma(\gamma-1)(1-z)F(\gamma-1) = 0 \, .$$

3. Prove the following recurrence formulae for the derivatives:

$$(\alpha)_n z^{\alpha-1} F(\alpha+n,\beta,\gamma,z) = \frac{d^n}{dz^n}[z^{\alpha+n-1} F(\alpha,\beta,\gamma,z)] \, ,$$

$$(-)^n (1-\gamma)_n z^{\gamma-n-1} F(\alpha,\beta,\gamma-n,z) = \frac{d^n}{dz^n}[z^{\gamma-1} F(\alpha,\beta,\gamma,z)] \, ,$$

$$(\gamma-\alpha)_n z^{\gamma-\alpha-1}(1-z)^{\alpha+\beta-\gamma-n} F(\alpha-n,\beta,\gamma,z)$$
$$= \frac{d^n}{dz^n}[z^{\gamma-\alpha+n-1}(1-z)^{\alpha+\beta-\gamma} F(\alpha,\beta,\gamma,z)] \, ,$$

$$\frac{(\gamma-\alpha)_n(\gamma-\beta)_n}{(\gamma)_n}(1-z)^{\alpha+\beta-\gamma-n} F(\alpha,\beta,\gamma+n,z)$$
$$= \frac{d^n}{dz^n}[(1-z)^{\alpha+\beta-\gamma} F(\alpha,\beta,\gamma,z)] \, ,$$

$$\frac{(-)^n(\alpha)_n(\gamma-\beta)_n}{(\gamma)_n}(1-z)^{\alpha-1} F(\alpha+n,\beta,\gamma+n,z)$$
$$= \frac{d^n}{dz^n}[(1-z)^{\alpha+n-1} F(\alpha,\beta,\gamma,z)] \, ,$$

$$(-)^n (1 - \gamma)_n z^{\gamma-1-n} (1 - z)^{\beta-\gamma} F(\alpha - n, \beta, \gamma - n, z)$$
$$= \frac{d^n}{dz^n} [z^{\gamma-1} (1 - z)^{\beta-\gamma+n} F(\alpha, \beta, \gamma, z)] ,$$

$$(-)^n (1 - \gamma)_n z^{\gamma-1-n} (1 - z)^{\alpha+\beta-\gamma-n} F(\alpha - n, \beta - n, \gamma - n, z)$$
$$= \frac{d^n}{dz^n} [z^{\gamma-1} (1 - z)^{\alpha+\beta-\gamma} F(\alpha, \beta, \gamma, z)] .$$

4. Show that

$$\cos \mu z = F\left(\frac{\mu}{2}, -\frac{\mu}{2}, \frac{1}{2}, \sin^2 z\right) ,$$

$$\sin \mu z = \mu \sin z\, F\left(\frac{1+\mu}{2}, \frac{1-\mu}{2}, \frac{3}{2}, \sin^2 z\right) .$$

5. Show that

$$\ln \frac{1+z}{1-z} = 2z F\left(\frac{1}{2}, 1, \frac{3}{2}, z^2\right) .$$

6. Show that

$$\lim_{\gamma \to -n} \{F(\alpha, \beta, \gamma, z)/\Gamma(\gamma)\}$$
$$= \frac{(\alpha)_{n+1}(\beta)_{n+1}}{(n+1)!} z^{n+1} F(\alpha + n + 1, \beta + n + 1, n + 2, z) .$$

7. Show that

(i) $F(\alpha, \beta, \gamma, z) = \dfrac{i\Gamma(\gamma)e^{-i\pi(\gamma-\beta)}}{2\Gamma(\beta)\Gamma(\gamma - \beta)\sin \pi(\gamma - \beta)}$
$$\times \int_0^{(1+)} t^{\beta-1}(1 - t)^{\gamma-\beta-1}(1 - zt)^{-\alpha} dt ,$$

where $\mathrm{Re}(\beta) > 0, 0 < \arg(1 - t) < 2\pi; 1/z$ lies outside the contour; $|\arg(1 - z)| < \pi; (1 - zt)^{-\alpha} = 1$ when $z = 0; \gamma - \beta \neq$ positive integer (otherwise the r.h.s. is an indeterminate form).

(ii) $F(\alpha, \beta, \gamma, z) = \dfrac{i\Gamma(\gamma)e^{-i\pi\beta}}{2\Gamma(\beta)\Gamma(\gamma - \beta)\sin \pi\beta}$
$$\times \int_1^{(0+)} t^{\beta-1}(1 - t)^{\gamma-\beta-1}(1 - zt)^{-\alpha} dt ,$$

where $\text{Re}(\gamma - \beta) > 0, 0 < \arg t < 2\pi; 1/z$ lies outside the contour; $|\arg(1-z)| < \pi; (1-zt)^{-\alpha} = 1$ when $z = 0; \beta \neq$ positive integer (otherwise the r.h.s. is an indeterminate form).

8. Show that, if $\text{Re}(\gamma - \alpha - \beta) < 0$, then, when $n \to \infty$,

$$S_n \div \frac{\Gamma(\gamma) n^{\alpha+\beta-\gamma}}{(\alpha + \beta - \gamma)\Gamma(\alpha)\Gamma(\beta)} \longrightarrow 1 \,,$$

where S_n denotes the sum of the first n terms of the series $F(\alpha, \beta, \gamma, 1)$.

9. Show that, if $\gamma - \alpha - \beta < 0$, then

$$\lim_{x \to 1-0} F(\alpha, \beta, \gamma, x) \div \left\{ \frac{\Gamma(\gamma)\Gamma(\alpha + \beta - \gamma)}{\Gamma(\alpha)\Gamma(\beta)} (1 - x)^{\gamma-\alpha-\beta} \right\} = 1 \,;$$

if $\gamma - \alpha - \beta = 0$, then

$$\lim_{x \to 1-0} F(\alpha, \beta, \gamma, x) \div \left\{ \frac{\Gamma(\alpha + \beta)}{\Gamma(\alpha)\Gamma(\beta)} \ln \frac{1}{1 - x} \right\} = 1 \,.$$

10. Show that

$$\lim_{x \to 1-0} \left\{ F(\alpha, \beta, \gamma, x) \right.$$
$$- \sum_{n=0}^{k} (-)^n \frac{\Gamma(\alpha + \beta - \gamma - n)\Gamma(\gamma - \alpha + n)\Gamma(\gamma - \beta + n)\Gamma(\gamma)}{n!\Gamma(\gamma - \alpha)\Gamma(\gamma - \beta)\Gamma(\alpha)\Gamma(\beta)}$$
$$\times \left. (1 - x)^{n+\gamma-\alpha-\beta} \right\}$$
$$= \frac{\Gamma(\gamma - \alpha - \beta)\Gamma(\gamma)}{\Gamma(\gamma - \alpha)\Gamma(\gamma - \beta)} \,,$$

where k is an integer which satisfies $k \leq \text{Re}(\alpha + \beta - \gamma) < k + 1$. (This formula shows the way in which the hypergeometric function approaches infinity when $x \to 1 - 0$ and $\alpha + \beta - \gamma$ is not an integer.)

11. Show that

$$\ln(1 - 2xt + t^2)^{-1} = 2 \sum_{n=1}^{\infty} n^{-1} T_n(x) t^n \,,$$

where $T_n(x)$ are the Chebyshev polynomials [Sec. 4.11].

12. **Show that**

$$\int_{-1}^{1} [T_n(x)]^2 dx = 1 - \frac{1}{4n^2 - 1}, \quad n = 0, 1, 2, \ldots .$$

13. *Chebyshev function of the second kind*, $U_n(x)$, is defined by

$$U_n(x) = \sin(n \arccos x) .$$

(The Chebyshev polynomial of the second kind is also defined as

$$\sin[(n + 1) \arccos x]/\sqrt{1 - x^2} = U_{n+1}(x)/\sqrt{1 - x^2} ,$$

but it is not a solution of Eq. (4) of Sec. 4.11.)
Show that

(i) $(1 - 2xt + t^2)^{-1} = (1 - x^2)^{-1/2} \sum_{n=0}^{\infty} U_{n+1}(x)t^n$.

(ii) $\int_{-1}^{1} U_m U_n (1 - x^2)^{-\frac{1}{2}} dx = \begin{cases} 0 & (m \neq n, \text{ or } m = n = 0) . \\ \frac{\pi}{2} & (m = n \neq 0) . \end{cases}$

(iii) $U_{n+1}(x) = \dfrac{(n + 1)!(-)^n 2^n}{(2n + 1)!} \dfrac{d^n}{dx^n}(1 - x^2)^{n+\frac{1}{2}}$.

(iv) $U_{n+1}(x) = (1 - x^2)^{\frac{1}{2}} \displaystyle\sum_{k=0}^{[n/2]} \dfrac{(-)^k (n - k)!}{k!(n - 2k)!}(2x)^{n-2k}$.

14. **Prove the following asymptotic expression for the Jacobi polynomial when $n \to \infty$:**

$$F(-n, p + n, \gamma, x) = \frac{\Gamma(\gamma)}{n^{\gamma - \frac{1}{2}}\sqrt{\pi}}(\sin \varphi)^{\frac{1}{2} - \gamma}(\cos \varphi)^{\gamma - p - \frac{1}{2}}$$
$$\times \cos\left[(2n + p)\varphi - \frac{\pi}{4}(2\gamma - 1)\right] + O(n^{-\gamma - \frac{1}{2}}) ,$$

where $0 < x < 1, \sin^2 \varphi = x$.

15. **Show that**

$$2\frac{\Gamma\left(\frac{1}{2}\right)\Gamma\left(\alpha+\beta+\frac{1}{2}\right)}{\Gamma\left(\alpha+\frac{1}{2}\right)\Gamma\left(\beta+\frac{1}{2}\right)} F\left(\alpha,\beta,\frac{1}{2},x\right)$$

$$= F\left(2\alpha,2\beta,\alpha+\beta+\frac{1}{2},\frac{1+\sqrt{x}}{2}\right) + F\left(2\alpha,2\beta,\alpha+\beta+\frac{1}{2},\frac{1-\sqrt{x}}{2}\right),$$

$$4\frac{\Gamma\left(\frac{1}{2}\right)\Gamma\left(\alpha+\beta+\frac{1}{2}\right)}{\Gamma(\alpha)\Gamma(\beta)} x^{\frac{1}{2}} F\left(\alpha+\frac{1}{2},\beta+\frac{1}{2},\frac{3}{2},x\right)$$

$$= F\left(2\alpha,2\beta,\alpha+\beta+\frac{1}{2},\frac{1+\sqrt{x}}{2}\right) - F\left(2\alpha,2\beta,\alpha+\beta+\frac{1}{2},\frac{1-\sqrt{x}}{2}\right).$$

[Hint: Note that the l.h.s. of the formulae are two linearly independent regular solutions of the same hypergeometric equation at the point $x = 0$; thus, each term on the r.h.s. is also a solution of the same equation.]

16. **Show that**

$$\left\{F\left(\alpha,\beta,\alpha+\beta+\frac{1}{2},z\right)\right\}^2 = \sum_{n=0}^{\infty} \frac{(2\alpha)_n(2\beta)_n(\alpha+\beta)_n}{n!\left(\alpha+\beta+\frac{1}{2}\right)_n(2\alpha+2\beta)_n} z^n.$$

17. **Show that**

$$F(1,\alpha,\alpha+1,-1) = \frac{\alpha}{2}\left\{\psi\left(\frac{\alpha+1}{2}\right) - \psi\left(\frac{\alpha}{2}\right)\right\},$$

$\psi(z)$ being the logarithmic derivative of the Γ-function [Sec. 3.11].

18. **Show that**

$$(\alpha+1)F(-\alpha,1,\beta+2,-1) + (\beta+1)F(-\beta,1,\alpha+2,-1)$$
$$= 2^{\alpha+\beta+1}\frac{\Gamma(\alpha+2)\Gamma(\beta+2)}{\Gamma(\alpha+\beta+2)}.$$

19. **Show that**

$$F\left(-\frac{n}{2},\frac{1-n}{2},n+\frac{3}{2},-\frac{1}{3}\right) = \left(\frac{8}{9}\right)^n \frac{\Gamma\left(\frac{4}{3}\right)\Gamma\left(n+\frac{3}{2}\right)}{\Gamma\left(\frac{3}{2}\right)\Gamma\left(n+\frac{4}{3}\right)}.$$

20. Show that, if $\operatorname{Re}(\delta - \varepsilon - \frac{3}{2}\alpha - 1) > 0$, then

$$_3F_2(\alpha, \alpha - \delta + 1, \alpha - \varepsilon + 1; \delta, \varepsilon; 1)$$

$$= 2^{-\alpha} \frac{\Gamma\left(\frac{1}{2}\right) \Gamma(\delta)\Gamma(\varepsilon)\Gamma\left(\delta + \varepsilon - \frac{3}{2}\alpha - 1\right)}{\Gamma\left(\delta - \frac{\alpha}{2}\right)\Gamma\left(\varepsilon - \frac{\alpha}{2}\right)\Gamma\left(\frac{1+\alpha}{2}\right)\Gamma(\delta + \varepsilon - \alpha - 1)} .$$

21. Show that, if $\alpha, \alpha', \beta, \beta'$ are not negative integers, then the Appell function F_3[Sec. 4.16] can be represented by the double Barnes' integral:

$$\frac{\Gamma(\alpha)\Gamma(\alpha')\Gamma(\beta)\Gamma(\beta')}{\Gamma(\gamma)} F_3(\alpha, \alpha'; \beta, \beta'; \gamma, x, y)$$

$$= -\frac{1}{4\pi^2} \int_{-i\infty}^{i\infty} \int_{-i\infty}^{i\infty} \frac{\Gamma(\alpha + s)\Gamma(\beta + s)\Gamma(\alpha' + t)\Gamma(\beta' + t)}{\Gamma(\gamma + s + t)}$$

$$\times \Gamma(-s)\Gamma(-t)(-x)^s(-y)^t \, ds \, dt ,$$

where the poles of $\Gamma(-s)$ and $\Gamma(-t)$ lie on the right of the corresponding contours; the poles of $\Gamma(\alpha+s)\Gamma(\beta+s)$ and $\Gamma(\alpha'+t)\Gamma(\beta'+t)$ lie separately on the left of the corresponding contours.

Assume that the poles of the integrand above do not overlap, then we have the continuation relation below:

$$F_3(\alpha, \alpha'; \beta, \beta'; \gamma; x, y)$$
$$= f(\alpha, \alpha', \beta, \beta')(-x)^{-\alpha}(-y)^{-\alpha'}$$
$$\times F_2\left(\alpha + \alpha' + 1 - \gamma; \alpha, \alpha'; \alpha + 1 - \beta, \alpha' + 1 - \beta'; \frac{1}{x}, \frac{1}{y}\right)$$
$$+ f(\alpha, \beta', \beta, \alpha')(-x)^{-\alpha}(-y)^{-\beta'}$$
$$\times F_2\left(\alpha + \beta' + 1 - \gamma; \alpha, \beta'; \alpha + 1 - \beta, \beta' + 1 - \alpha'; \frac{1}{x}, \frac{1}{y}\right)$$
$$+ f(\beta, \alpha', \alpha, \beta')(-x)^{-\beta}(-y)^{-\alpha'}$$
$$\times F_2\left(\beta + \alpha' + 1 - \gamma; \beta, \alpha'; \beta + 1 - \alpha, \alpha' + 1 - \beta'; \frac{1}{x}, \frac{1}{y}\right)$$
$$+ f(\beta, \beta', \alpha, \alpha')(-x)^{-\beta}(-y)^{-\beta'}$$
$$\times F_2\left(\beta + \beta' + 1 - \gamma; \beta, \beta'; \beta + 1 - \alpha, \beta' + 1 - \alpha'; \frac{1}{x}, \frac{1}{y}\right) ,$$

where

$$F(\lambda, \mu, \rho, \sigma) = \frac{\Gamma(\gamma)\Gamma(\rho - \lambda)\Gamma(\sigma - \mu)}{\Gamma(\rho)\Gamma(\sigma)\Gamma(\gamma - \lambda - \mu)} ,$$

$$|\arg(-x)| < \pi, \quad |\arg(-y)| < \pi .$$

22. Show that

$$\frac{(1 + \sqrt{z})^{-2\alpha} + (1 - \sqrt{z})^{-2\alpha}}{2} = F\left(\alpha, \alpha + \frac{1}{2}, \frac{1}{2}, z\right) ,$$

$$\left(\frac{1 + \sqrt{1 - z}}{2}\right)^{1 - 2\alpha} = F\left(\alpha - \frac{1}{2}, \alpha, 2\alpha, z\right)$$

$$= (1 - z)^{\frac{1}{2}} F\left(\alpha, \alpha + \frac{1}{2}, 2\alpha, z\right) ,$$

$$(1 - z)^{-2\alpha - 1}(1 + z) = F(2\alpha, \alpha + 1, \alpha, z) .$$

Chapter 5

LEGENDRE FUNCTIONS

5.1. Legendre Equation

Legendre functions are solutions of the following differential equation:

$$(1 - x^2)\frac{d^2 y}{dx^2} - 2x\frac{dy}{dx} + \nu(\nu + 1)y = 0 \; ; \tag{1}$$

ν and x may be any complex numbers.

Equation (1) is called the *Legendre equation of order ν*. This equation often gives rise from the solution of the Laplace equation or other similar equations by the method of separation of variables in spherical polar coordinates or rotational ellipsoidal coordinates. For example, in spherical polar coordinates, the Laplace equation is

$$\frac{1}{r^2}\frac{\partial}{\partial r}\left(r^2\frac{\partial V}{\partial r}\right) + \frac{1}{r^2\sin\theta}\frac{\partial}{\partial\theta}\left(\sin\theta\frac{\partial V}{\partial\theta}\right) + \frac{1}{r^2\sin^2\theta}\frac{\partial^2 V}{\partial\varphi^2} = 0 \; . \tag{2}$$

Let $V(r, \theta, \varphi) = R(r)\Theta(\theta)\Phi(\varphi)$. We obtain three ordinary differential equations

$$\frac{1}{r^2}\frac{d}{dr}\left(r^2\frac{dR}{dr}\right) - \frac{\lambda}{r^2}R = 0 \; , \tag{3}$$

$$\frac{d^2\Phi}{d\varphi^2} + \mu^2\Phi = 0 \; , \qquad (4)$$

$$\frac{1}{\sin\theta}\frac{d}{d\theta}\left(\sin\theta\frac{d\Theta}{d\theta}\right) + \left(\lambda - \frac{\mu^2}{\sin^2\theta}\right)\Theta = 0 \; , \qquad (5)$$

where λ and μ are parameters introduced during the process of separation of variables.

In (5), putting $x = \cos\theta, y(x) = \Theta(\theta)$, and writing $\nu(\nu+1)$ for λ, we obtain

$$\frac{d}{dx}\left[(1-x^2)\frac{dy}{dx}\right] + \left[\nu(\nu+1) - \frac{\mu^2}{1-x^2}\right]y = 0 \; . \qquad (6)$$

This is called the *associate Legendre equation*; Eq. (1) is its particular case, $\mu = 0$.

Equation (6) has three singularities: $-1, 1, \infty$, all of which are regular singularities with exponents $(\mu/2, -\mu/2), (\mu/2, -\mu/2), (\nu+1, -\nu)$ respectively [cf. Eq. (7) of Sec. 2.9, and the examples in that section]. Therefore, the equation belongs to the type of hypergeometric equation; its solutions are called *associate Legendre functions*, which can be expressed in terms of hypergeometric functions. According to Eqs. (12) and (16) of Sec. 2.9, we have

$$P\left\{\begin{array}{ccc} -1 & 1 & \infty \\ \frac{\mu}{2} & \frac{\mu}{2} & \nu+1; \;\; x \\ -\frac{\mu}{2} & -\frac{\mu}{2} & -\nu \end{array}\right\} = P\left\{\begin{array}{ccc} 1 & 0 & \infty \\ \frac{\mu}{2} & \frac{\mu}{2} & \nu+1; \;\; \frac{1-x}{2} \\ -\frac{\mu}{2} & -\frac{\mu}{2} & -\nu \end{array}\right\}$$

$$= \left(\frac{1-x}{2}\right)^{\mu/2}\left(1 - \frac{1-x}{2}\right)^{\mu/2} P\left\{\begin{array}{ccc} 1 & 0 & \infty \\ 0 & 0 & \nu+\mu+1; \;\; \frac{1-x}{2} \\ -\mu & -\mu & -\nu+\mu \end{array}\right\}$$

$$= 2^{-\mu}(1-x^2)^{\mu/2}P\left\{\begin{array}{ccc} 1 & 0 & \infty \\ 0 & 0 & \nu+\mu+1; \;\; \frac{1-x}{2} \\ -\mu & -\mu & -\nu+\mu \end{array}\right\} \; . \qquad (7)$$

In practice, the case mostly encountered is when ν and μ are both integers. The first fifteen sections of the present chapter are mainly concerned with such case. We shall employ here elementary methods, as possible as we can, without resorting to the theory of hypergeometric functions.

When μ and ν are not integers, the theory of hypergeometric functions is invoked as we can then use, to the full extent, the plentiful results of Chap. 4, such as the recurrence relations, transformation formulae and asymptotic expansions, etc.

5.2. Legendre Polynomials

Legendre polynomials are the polynomial solutions of the equations

$$(1 - x^2)\frac{d^2y}{dx^2} - 2x\frac{dy}{dx} + n(n + 1)y = 0$$
$$(n = 0, 1, 2, \dots) \, . \tag{1}$$

By the method of solution in series, we may assume, at the ordinary point $x = 0$ [Sec. 2.2] of Eq. (1), the solution to be

$$y = \sum_{k=0}^{\infty} a_k x^k \, . \tag{2}$$

Substituting in (1), we obtain the recurrence relation between the coefficients:

$$a_{k+2} = -\frac{(n - k)(n + k + 1)}{(k + 2)(k + 1)} a_k \, . \tag{3}$$

Thus, the two linearly independent solutions of (1), in the form of ascending powers of x, are

$$y_1 = a_0 \left\{ 1 - \frac{n(n + 1)}{1 \cdot 2}x^2 + \frac{n(n - 2)(n + 1)(n + 3)}{1 \cdot 2 \cdot 3 \cdot 4}x^4 - \dots \right\}$$
$$= a_0 F\left(-\frac{n}{2}, \frac{n + 1}{2}, \frac{1}{2}, x^2 \right) \tag{4}$$

and

$$y_2 = a_1 \left\{ x - \frac{(n - 1)(n + 2)}{2 \cdot 3}x^3 + \frac{(n - 1)(n - 3)(n + 2)(n + 4)}{2 \cdot 3 \cdot 4 \cdot 5}x^5 - \dots \right\}$$
$$= a_1 x F\left(\frac{1 - n}{2}, \frac{2 + n}{2}, \frac{3}{2}, x^2 \right) \, , \tag{5}$$

where $F(\alpha, \beta, \gamma, z)$ is the hypergeometric function.

From (3) we see that $a_{n+2} = a_{n+4} = \dots = 0$, and hence, for $n =$ even, the series representing y_1 breaks up into a polynomial of degree n, while y_2 remains as an infinite series; for $n =$ odd, y_2 becomes a polynomial of degree n

while y_1 remains as an infinite series. The radii of convergence of both power series are equal to 1.

In the said polynomials, we stipulate that the coefficient of the highest degree of x, x^n, is

$$a_n = \frac{(2n)!}{2^n (n!)^2} \ . \tag{6}$$

(The reason for this is explained after (13), or after (3) of the following section.) The other coefficients, $a_{n-2}, a_{n-4}, \ldots$, are determined by (3); the polynomials thus obtained are called *Legendre Polynomials* and denoted by $P_n(x)$:

$$P_n(x) = \frac{(2n)!}{2^n (n!)^2} \left\{ x^n - \frac{n(n-1)}{2(2n-1)} x^{n-2} \right.$$

$$\left. + \frac{n(n-1)(n-2)(n-3)}{2 \cdot 4 \cdot (2n-1)(2n-3)} x^{n-4} - \cdots \right\}$$

$$= \frac{1}{2^n} \sum_{r=0}^{[\frac{n}{2}]} (-)^r \frac{(2n-2r)!}{r!(n-r)!(n-2r)!} x^{n-2r}$$

$$= \frac{(2n)!}{2^n (n!)^2} x^n F\left(-\frac{n}{2}, \frac{1-n}{2}, \frac{1}{2} - n, x^{-2} \right) \ , \tag{7}$$

where $[n/2] = n/2$ when n is an ·even integer and $(n-1)/2$ when n is odd. The last expression with the hypergeometric function can be obtained by the continuation formula, Eq. (8) of Sec. 4.8, from (4) for even n and from (5) for odd n; it is only necessary to note that, by (6), we have from (3)

$$a_0 = (-)^{n/2} \frac{n!}{2^n \left(\frac{n}{2}! \right)^2} \quad (n \text{ even}) \tag{8}$$

$$a_1 = (-)^{(n-1)/2} \frac{n!}{2^{n-1} \left(\frac{n-1}{2}! \right)^2} \quad (n \text{ odd}) \ . \tag{9}$$

The first eight Legendre polynomials are

$$P_0(x) = 1, \quad P_1(x) = x ,$$

$$P_2(x) = \frac{1}{2}\left(3x^2 - 1\right) , \quad P_3(x) = \frac{1}{2}(5x^3 - 3x) ,$$

$$P_4(x) = \frac{1}{8}(35x^4 - 30x^2 + 3) ,$$

$$P_5(x) = \frac{1}{8}(63x^5 - 70x^3 + 15x) ,$$

$$P_6(x) = \frac{1}{16}(231x^6 - 315x^4 + 105x^2 - 5) ,$$

$$P_7(x) = \frac{1}{16}(429x^7 - 693x^5 + 315x^3 - 35x) .$$

$$(10)$$

Murphy's Expression for $P_n(x)$

Equation (1) is a special case of Eq. (6) in the last section with $\mu = 0, \nu = n$ (integer). Therefore, according to (7) in that section, all solutions of (1) can be represented by the P-symbol:

$$P\left\{ \begin{matrix} 1 & 0 & \infty \\ 0 & 0 & n+1; & \frac{1-x}{2} \\ 0 & 0 & -n \end{matrix} \right\} .$$

$$(11)$$

From there, we obtain the polynomial solution $F(n+1, -n, 1, (1-x)/2)$; the other solution contains a logarithmic term [Sec. 4.4]. Hence,

$$P_n(x) = AF\left(n+1, -n, 1, \frac{1-x}{2}\right) .$$

The coefficient of x^n on the r.h.s. is

$$A\frac{(n+1)_n(-n)_n}{n!(1)_n}\left(-\frac{1}{2}\right)^n = \frac{(2n)!}{2^n(n!)^2}A ,$$

while that in $P_n(x)$ is given by (6) to be $(2n)!/2^n(n!)^2$. Thus, $A = 1$, and we have

$$P_n(x) = F\left(n+1, -n, 1, \frac{1-x}{2}\right) .$$

$$(12)$$

This is *Murphy's expression* for $P_n(x)$. We see from this expression that $P_n(x)$ is a special case of the Jacobi polynomial given in Sec. 4.10.

From (12), we have

$$P_n(1) = 1 . \tag{13}$$

This is one of the reasons for assigning the coefficient a_n of the highest degree term in $P_n(x)$ the value $(2n)!/2^n(n!)^2$ [see (6)].

5.3. The Generating Function of $P_n(x)$. Rodrigues Formula

Legendre polynomials were introduced by Legendre in the theory of potential. They are related to the expansion of the reciprocal of the distance, $1/R$, in the Newtonian theory of potential or Coulomb potential; R is the distance between the two points **r** and **r'** (Fig. 10):

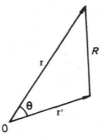

Fig. 10.

$R = |\mathbf{r} - \mathbf{r'}| = (r^2 + r'^2 - 2rr' \cos\theta)^{1/2}, \theta$ being the angle between **r** and **r'**.

Let $t = r'/r, x = \cos\theta$, then

$$\frac{1}{R} = \frac{1}{r}(1 - 2xt + t^2)^{-1/2} , \tag{1}$$

where the square root takes the value 1 when $t = 0$.

Writing $(1 - 2xt + t^2)^{-1/2}$ as $(t - x - \sqrt{x^2 - 1})^{-1/2}(t - x + \sqrt{x^2 - 1})^{-1/2}$, we see that, as a function of t, the square root has two singular points (branch points), $x \pm \sqrt{x^2 - 1}$, in the finite region. Therefore, so long as $|t| < \min|x \pm \sqrt{x^2 - 1}|$, we have the following Taylor expansion:

$$(1 - 2xt + t^2)^{-1/2} = \sum_{n=0}^{\infty} P_n(x)t^n . \tag{2}$$

From Eq. (10) of Sec. 1.4,[a] it is known that the expansion coefficients in (2) are

$$P_n(x) = \frac{1}{2^n n!} \frac{d^n}{dx^n} (x^2 - 1)^n$$

$$= \frac{1}{2^n n!} \frac{d^n}{dx^n} \sum_{r=0}^{n} \binom{n}{r} (-)^r x^{2n-2r}$$

$$= \frac{1}{2^n} \sum_{r=0}^{[\frac{n}{2}]} (-)^r \frac{(2n-2r)!}{r!(n-r)!(n-2r)!} x^{n-2r} . \tag{3}$$

According to (7) of the last section, these coefficients are exactly the Legendre polynomials. Therefore, the expression on the l.h.s. of (2) is called the *generating function of the Legendre polynomials*. The derivative representation (3) of $P_n(x)$ is named the *Rodrigues formula*. Here, we see another reason for assigning the value given by (6) of the last section to the coefficient a_n of x^n in $P_n(x)$; it makes $P_n(x)$ to be just the coefficients in the expansion (2).

Under the condition $-1 \leq x \leq 1$, the domain of convergence of the expansion (2) is $|t| < 1$, since, from $(x + \sqrt{x^2 - 1})(x - \sqrt{x^2 - 1}) = 1$, we know that when $x = \pm 1$, $\min |x \pm \sqrt{x^2 - 1}|$ takes its maximum value, 1.

In (2), putting separately $x = 1, -1, 0$, we have

$$\left. \begin{array}{ll} P_n(1) = 1, & P_n(-1) = (-1)^n , \\ P_n(0) = 0, & (n \text{ odd}) , \\ P_n(0) = (-)^{\frac{n}{2}} \dfrac{1 \cdot 3 \cdot 5 \dots (n-1)}{2 \cdot 4 \cdot 6 \dots n} & (n \text{ even}) , \end{array} \right\} \tag{4}$$

where the first equation is a repeat of (13) of the previous section.

Also, by (3), it is readily seen that

$$P_n(-x) = (-)^n P_n(x) . \tag{5}$$

5.4. Integral Representations of $P_n(x)$

Applying the Euler transformation [Sec. 2.14] to the Legendre equation

$$(1 - x^2)y'' - 2xy' + n(n+1)y = 0 \tag{1}$$

---

[a](3) can be proved directly; see Ex. 2 at the end of this chapter.

by assuming

$$y = \int_C (x - t)^\mu v(t) dt \ , \tag{2}$$

we obtain from Eq. (6) of Sec. 2.14, the equation determining μ:

$$\mu(\mu - 1) + 2\mu - n(n + 1) = 0 \ , \tag{3}$$

of which the two solutions are $\mu = n$ and $\mu = -n - 1$. Assume $\mu = -n - 1$. From Eq. (14) of Sec. 2.14, we have

$$v(t) = A(1 - t^2)^n \ . \tag{4}$$

Choose C to be a contour encircling the point $t = x$ in the t-plane. Since n is an integer, the integrand in (2) assumes the same value at the starting point and the end point of C [cf. text after Eq. (15) of Sec. 2.14], and hence

$$y = A \int_C (1 - t^2)^n (x - t)^{-n-1} dt \tag{5}$$

is an integral solution of Eq. (1). But the integral on the r.h.s. is equal to

$$-\frac{2\pi i}{n!} \frac{d^n}{dx^n} (x^2 - 1)^n \ , \tag{6}$$

so, comparing with (3) of the last section, we see that

$$P_n(x) = \frac{1}{2\pi i} \int_C \frac{(t^2 - 1)^n}{2^n (t - x)^{n+1}} dt \ . \tag{7}$$

This is the *Schläfli formula*.

If we take C to be a circle with centre at x and radius equal to $|x^2 - 1|^{1/2}$, then $t = x + \sqrt{x^2 - 1} \, e^{i\varphi}$ on C, and we have

$$P_n(x) = \frac{1}{2\pi i} \int_{-\pi}^{\pi} \frac{(x - 1 + \sqrt{x^2 - 1} \, e^{i\varphi})^n (x + 1 + \sqrt{x^2 - 1} \, e^{i\varphi})^n}{2^n (x^2 - 1)^{(n+1)/2} e^{i(n+1)\varphi}}$$

$$\times \sqrt{x^2 - 1} \, e^{i\varphi} i d\varphi$$

$$= \frac{1}{2\pi} \int_{-\pi}^{\pi} (x + \sqrt{x^2 - 1} \cos \varphi)^n d\varphi = \frac{1}{\pi} \int_0^{\pi} (x + \sqrt{x^2 - 1} \cos \varphi)^n d\varphi \ , \tag{8}$$

which is the *first Laplace integral representation of* $P_n(x)$. The multi-valued function $\sqrt{x^2 - 1}$ in it can be any single-valued branch of the function, because the integral on the r.h.s. of (8) becomes $\int_0^\pi (x - \sqrt{x^2 - 1}\cos\theta)^n d\theta$ when we change the variable of integration from φ to $\theta = \pi - \varphi$.

By Eq. (12) of Sec. 5.2, using the relation $F(n + 1, -n, 1, (1 - x)/2) = F(-n, n + 1, 1, (1 - x)/2)$, we arrive at

$$P_n(x) = \frac{1}{\pi} \int_0^\pi \frac{d\varphi}{(x + \sqrt{x^2 - 1}\cos\varphi)^{n+1}} , \tag{9}$$

which is the *second Laplace integral representation* of $P_n(x)$.

5.5. Recurrence Relations of $P_n(x)$

From the expansion formula (2) of the last section,

$$(1 - 2xt + t^2)^{-1/2} = \sum_{n=0}^{\infty} P_n(x)t^n , \tag{1}$$

differentiating both sides with respect to t, we have

$$(x - t)(1 - 2xt + t^2)^{-3/2} = \sum_{n=0}^{\infty} nP_n(x)t^{n-1} .$$

Multiplying the two sides with $(1 - 2xt + t^2)$ and applying (1) again on the l.h.s., there results

$$(x - t)\sum_{n=0}^{\infty} P_n(x)t^n = (1 - 2xt + t^2)\sum_{n=0}^{\infty} nP_n(x)t^{n-1} .$$

Comparing the coefficients of t^n on both sides, we obtain the recurrence relations

$$\left. \begin{array}{l} P_1(x) - xP_0(x) = 0 , \\ (n + 1)P_{n+1}(x) - (2n + 1)xP_n(x) + nP_{n-1}(x) = 0 \quad (n \geq 1) . \end{array} \right\} \tag{2}$$

Also, by differentiating both sides of (1) with respect to x, we have

$$t(1 - 2xt + t^2)^{-3/2} = \sum_{n=0}^{\infty} P_n'(x)t^n ,$$

or,

$$t\sum_{n=0}^{\infty} P_n(x)t^n = (1 - 2xt + t^2)\sum_{n=0}^{\infty} P_n'(x)t^n .$$

Comparing the coefficients of t^{n+1} on both sides, we get

$$P_n(x) = P'_{n+1}(x) - 2xP'_n(x) + P'_{n-1}(x) . \tag{3}$$

Differentiating (2) and eliminating $P'_{n-1}(x)$ with (3), we obtain the recurrence relation for the derivatives:

$$P'_{n+1}(x) = xP'_n(x) + (n+1)P_n(x) . \tag{4}$$

Eliminating $P'_{n+1}(x)$ from (3) and (4), it results

$$xP'_n(x) - P'_{n-1}(x) = nP_n(x) . \tag{5}$$

Also, eliminating $P'_n(x)$ from (4) and (5), we have

$$P'_{n+1}(x) - P'_{n-1}(x) = (2n+1)P_n(x) . \tag{6}$$

Replacing n by $n-1$ in (4), then eliminating $P'_{n-1}(x)$ with (5), we find

$$(x^2 - 1)P'_n(x) = nxP_n(x) - nP_{n-1}(x) . \tag{7}$$

(2) and (4) – (7) are the commonly used recurrence formulae. It can be shown that these formulae are also valid when n is not an integer [see Ex. 6 at the end of this chapter].

5.6. Legendre Polynomials as a Complete Set of Orthonormal Functions

The whole Legendre polynomials form a complete set of orthonormal functions in the interval $[-1, 1]$ with weight 1. To show this, we prove first the following important theorem:

Let $f_k(x)$ be a polynomial of degree k: $P_n(x)$ is the Legendre polynomial of degree n. If $k < n$, then

$$\int_{-1}^{1} f_k(x)P_n(x)dx = 0 . \tag{1}$$

Proof: From the Rodriques formula [Eq. (3) of Sec. 5.5], noting that, when $1 \le r \le n$,

$$\frac{d^{n-r}}{dx^{n-r}}(x^2 - 1)^n \bigg|_{x=\pm 1} = 0 , \tag{2}$$

we see, after integrating by parts k times,

$$\int_{-1}^{1} f_k(x) P_n(x) dx = \frac{1}{2^n n!} \int_{-1}^{1} f_k(x) \frac{d^n}{dx^n} (x^2 - 1)^n dx$$

$$= \frac{(-)^k}{2^n n!} f_k^{(k)}(x) \int_{-1}^{1} \frac{d^{n-k}}{dx^{n-k}} (x^2 - 1)^n dx$$

$$= \frac{(-)^k}{2^n n!} f_k^{(k)}(x) \frac{d^{n-k-1}}{dx^{n-k-1}} (x^2 - 1)^n \Big|_{-1}^{+1} = 0 . \qquad (3)$$

From (1) we obtain at once the orthogonality relation between the Legendre polynomials:

$$\int_{-1}^{1} P_m(x) P_n(x) dx = 0 \quad (m \neq n) . \qquad (4)$$

This relation can also be derived from the differential equations satisfied by $P_m(x)$ and $P_n(x)$ respectively, namely,

$$\frac{d}{dx}[(1 - x^2) P_m'] + m(m + 1) P_m = 0 ,$$

$$\frac{d}{dx}[(1 - x^2) P_n'] + n(n + 1) P_n = 0 .$$

Multiplying them respectively with P_n and P_m, subtracting, and then integrating from -1 to 1, we obtain

$$\int_{-1}^{1} \left\{ P_m \frac{d}{dx}[(1 - x^2) P_n'] - P_n \frac{d}{dx}[(1 - x^2) P_m'] \right\} dx$$

$$+ [n(n + 1) - m(m + 1)] \int_{-1}^{1} P_m P_n dx = 0 .$$

The first integral in it is equal to

$$\int_{-1}^{1} \frac{d}{dx} \{ (1 - x^2)(P_m P_n' - P_n P_m') \} dx$$

$$= (1 - x^2)(P_m P_n' - P_n P_m')|_{-1}^{1} = 0 ,$$

hence,

$$[n(n+1) - m(m+1)] \int_{-1}^{1} P_m P_n \, dx = 0 \; .$$

When $m \neq n, n(n+1) - m(m+1) \neq 0$; therefore, we have (4).

When $m = n$, we have, by (3),

$$\int_{-1}^{1} [P_n(x)]^2 dx = \frac{(-)^n}{2^n n!} P_n^{(n)}(x) \int_{-1}^{1} (x^2 - 1)^n dx \; .$$

Now,

$$\int_{-1}^{1} (x^2 - 1)^n \, dt = 2 \int_{0}^{1} (x^2 - 1)^n \, dx$$

$$= \int_{0}^{1} (t - 1)^n t^{-1/2} dt = (-)^n \int_{0}^{1} (1 - t)^n t^{-1/2} dt$$

$$= (-)^n \frac{\Gamma(n+1)\Gamma\left(\frac{1}{2}\right)}{\Gamma\left(n + \frac{3}{2}\right)} = (-)^n \frac{2^{2n+1}(n!)^2}{2n + 1}$$

by Eqs. (1) and (3) of Sec. 3.8, and Eq. (8) of Sec. 3.6. Furthermore,

$$P_n^{(n)}(x) = \frac{(2n)!}{2^n (n!)^2} n! = \frac{(2n)!}{2^n n!} \; .$$

Therefore,

$$\int_{-1}^{1} [P_n(x)]^2 dx = \frac{2}{2n + 1} \; . \tag{5}$$

(4) and (5) can be united into a single formula:

$$\int_{-1}^{1} P_m(x) P_n(x) dx = \frac{2}{2n + 1} \delta_{mn} \; , \tag{6}$$

where $\delta_{nn} = 1, \delta_{mn} = 0$ when $n \neq m$.

We have shown above that $\{P_n(x)\}(n = 0, 1, 2, \ldots)$ form an orthonormal set in the interval $[-1, 1]$ with weight 1. Now we come to the proof of the completeness of the set. For this purpose, we appeal to Weierstrass' theorem for approximation by polynomials.[b] It states that an arbitrary function, which is

---

[b] Cf. for example, Courant and Hilbert, *Methods of Mathematical Physics*, Vol. I, Chap. 2, Sec. 4.1, p. 65.

continuous in the interval $[-1,1]$ can be uniformly approximated from a polynomial sequence $\{f_n(x)\}$, i.e., to any $\varepsilon > 0$, there exists an $N(\varepsilon)$ independent of x, such that

$$|f(x) - f_n(x)| < \varepsilon, \quad n > N(\varepsilon) . \tag{7}$$

The polynomial $f_n(x)$ can always be expressed in terms of a linear combination of the Legendre polynomials:

$$f_n(x) = c_n P_n(x) + c_{n-1} P_{n-1}(x) + \ldots + c_0 P_0(x) . \tag{8}$$

$f_n(x)$ can evidently be represented by a linear combination of $P_n(x)$ and $x^{n-1}, x^{n-2}, \ldots$; and again, x^{n-1} can be represented by a linear combination of $P_{n-1}(x)$ and $x^{n-2}, x^{n-3}, \ldots$; etc. Therefore, the function $f(x)$ can be approximated uniformly by the special sequence of the form (8), and

$$\int_{-1}^{1} |f(x) - f_n(x)|^2 dx < 2\varepsilon^2 . \tag{9}$$

From the result in Sec. 1.10 [cf. (6) of that section], if we take the polynomial

$$q_n(x) = a_n P_n(x) + a_{n-1} P_{n-1}(x) + \ldots + a_0 P_0(x) \tag{10}$$

as an approximation of $f(x)$, then, when the values of the coefficients are given by

$$a_k = \frac{2k+1}{2} \int_{-1}^{1} f(x) P_k(x) dx , \tag{11}$$

the mean square error $\|f(x) - q_n(x)\|^2 = \|f\|^2 - \sum_{k=0}^{n} |a_k|^2 = \varepsilon_n$ will assume its minimum value. Hence, $\varepsilon_n < 2\varepsilon^2$. Moreover, according to Eq. (8) of Sec. 1.10, when n increases, $\varepsilon_n (> 0)$ is a monotonically decreasing sequence and there exists a limit ε_∞ such that $\varepsilon_\infty < 2\varepsilon^2$. But $\varepsilon (> 0)$ can be arbitrarily small, so we must have $\varepsilon_\infty = 0$, i.e.,

$$\|f\|^2 = \sum_{k=0}^{\infty} |a_k|^2 , \tag{12}$$

which proves the completeness of the set $\{P_n(x)\}$.

The series

$$a_0 P_0(x) + a_1 P_1(x) + \ldots + a_k P_k(x) + \ldots , \tag{13}$$

in which the coefficients are determined by (11), is called the *Legendre series of the function* $f(x)$. We have shown above that this series converges in mean to $f(x)$ under the condition that $f(x)$ is continuous in $[-1, 1]$. However, it is only necessary for $f(x)$ to be integrable quadratically in that interval. But then, whether the series (13) converges, and, even it does converge, whether its sum represents $f(x)$, are still problems to be analyzed with subtlety. The relevant theory is almost the same as the theory of Fourier series. We cite here only the conclusions:[c]

If $(1 - x^2)^{-1/4} f(x)$ is integrable in $[-1, 1]$, then the Legendre series

$$\sum_{k=0}^{\infty} \frac{2k+1}{2} P_k(x) \int_{-1}^{1} f(x') P_k(x') dx'$$

converges to $\frac{1}{2}\{f(x+0) + f(x-0)\}$ for any interior point x, of the interval $[-1, 1]$, which is such that $f(x)$ is of bounded variation in some neighborhood of x, or at which $f(x)$ has bounded derivatives, or at which the function $\sin^{\frac{1}{2}} \theta f(\cos \theta)$ satisfies any other condition for the convergence of the corresponding Fourier series.

Moreover, if $f(x)$ is continuous in any interval I (the continuity at the end points being on both sides) and the interval is interior to an interval in which the function is of bounded variation, then the Legendre series of $f(x)$ converges uniformly throughout I. Any other condition which is sufficient for the Fourier series of the function $\sin^{1/2} \theta f(\cos \theta)$ to be uniformly convergent will afford a corresponding sufficient condition for the Legendre series of $f(x)$ to converge uniformly throughout an interval interior to the interval $[-1, 1]$.

Another theorem for the expansion of a function in a Legendre series under more stringent condition will be given in Sec. 5.10.

5.7. Zeros of $P_n(x)$

All the n zeros of $P_n(x)$ are simple (i.e., of order 1) and lie entirely interior to the interval $[-1, 1]$, and hence, all being real. Also, the zeros of $P_n(x)$ and $P_{n-1}(x)$ are interlacing, i.e., between two consecutive zeros of $P_n(x)$, there must be one and only one zero of $P_{n-1}(x)$; and vice versa. These properties of the zeros are common to all orthogonal polynomials in general [cf. for example, Szegö, *Orthogonal Polynomials*, Sec. 33, p. 43]. For $P_n(x)$, however, we can show this by employing the Rodriques formula in Sec. 5.3 and the recurrence relations in Sec. 5.5.

Firstly, $P_n(x)$ cannot have multiple zeros (zeros of order higher than 1), because it is a solution of a second order ordinary differential equation; if α is a zero of it with order $m, m \geq 2$, then $P_n(\alpha) = P'_n(\alpha) = 0$, and $P_n(x) \equiv 0$.

---

[c] Cf. Hobson (1931), Chap. VII, Sec. 205, p. 329.

Next, we shall show that all the zeros of $P_n(x)$ lie in the interior of the interval $[-1, 1]$. According to Eq. (3) of Sec. 5.3,

$$P_n(x) = \frac{1}{2^n n!} \frac{d^n}{dx^n} (x^2 - 1)^n .$$

-1 and $+1$ are two nth-order zeros of $(x^2 - 1)^n$, then, by Rolle's theorem, the first derivative $d(x^2 - 1)^n/dx$ has at least one zero lying between -1 and $+1$. If $n = 1$, this is just the zero of $P_1(x)$. If $n > 1$, then, besides this zero, ± 1 are also zeros of $d(x^2 - 1)^n/dx$, and hence, in between -1 and $+1$, there are at least two zeros of $d^2(x^2 - 1)^n/dx^2$, which do not coincide. Following such arguments, we are led to the conclusion that $d^n(x^2 - 1)^n/dx^n$, and hence $P_n(x)$, has n and only n non-coincident zeros lying within $[-1, 1]$; ± 1 are no longer zeros [cf. Eq. (4) of Sec. 5.3: $P_n(1) = 1, P_n(-1) = (-1)^n$].

Finally, we shall prove that the zeros of $P_n(x)$ and $P_{n-1}(x)$ are interlacing. From the recurrence relation

$$(x^2 - 1)P_n'(x) = n[xP_n(x) - P_{n-1}(x)] \quad (n \geq 1) \tag{1}$$

[Eq. (7) of Sec. 5.5], we see that if $P_n(\alpha) = 0$, then $P_{n-1}(\alpha) \neq 0$; otherwise, as $\alpha \neq \pm 1$ (see above), we would have $P_n'(\alpha) = 0$, and hence α is a multiple zero, which is contradictory to the previous result. Besides, since $-1 < \alpha < +1$, we see from (1) that $P_{n-1}(\alpha)$ and $P_n'(\alpha)$ are of the same sign. Let α and β be two consecutive zeros of $P_n(x)$, then both $P_n'(\alpha)$ and $P_n'(\beta)$ cannot vanish and are of different signs; otherwise, α and β would not be contiguous. Hence, $P_{n-1}(\alpha)$ and $P_{n-1}(\beta)$ are of different signs according to the previous conclusion. However, this means that, between α and β, there is at least one zero of $P_{n-1}(x)$. But $P_{n-1}(x)$ can only have $n - 1$ zeros. Therefore, in between every two consecutive ones of the n zeros of $P_n(x)$, there must be one and only one zero of $P_{n-1}(x)$. This, of course, tells that the zeros of $P_n(x)$ and those of $P_{n-1}(x)$ are interlacing.

5.8. Legendre Functions of the Second Kind, $Q_n(x)$

By Eq. (27) of Sec. 2.4, we can derive from $P_n(x)$ the second solution,

$$Q_n(x) = P_n(x) \int_x^\infty \frac{dx}{(x^2 - 1)[P_n(x)]^2} \quad (|x| > 1) , \tag{1}$$

of the Legendre equation

$$(1 - x^2)y'' - 2xy' + n(n + 1)y = 0 . \tag{2}$$

$Q_n(x)$ is called the *Legendre function of the second kind*.

It is not easy to find the general term of the expansion of $Q_n(x)$ by direct calculation when we substitute the expansion of $P_n(x)$ into the integral on the r.h.s. of (1). Therefore, we use in the following the method of undetermined coefficients. Use the expression of $P_n(x)$ in descending powers [Eq. (7) of Sec. 5.2]

$$P_n(x) = \frac{(2n)!}{2^n (n!)^2} x^n \left\{ 1 - \frac{n(n-1)}{2 \cdot (2n-1)} x^{-2} + \dots \right\} .$$

Substituting into the integral on the r.h.s. of (1), we have

$$Q_n(x) = \frac{2^n (n!)^2}{(2n)!} x^n \left\{ 1 - \frac{n(n-1)}{2 \cdot (2n-1)} x^{-2} + \dots \right\}$$
$$\times \int_x^\infty \frac{dx}{x^{2n+2} (1 - x^{-2}) \left\{ 1 - \frac{n(n-1)}{2 \cdot (2n-1)} x^{-2} + \dots \right\}^2}$$
$$= \frac{2^n (n!)^2}{(2n+1)!} x^{-n-1} \left\{ 1 + \sum_{k=1}^\infty a_k x^{-2k} \right\} ,$$

where a_k are undetermined coefficients. Putting this expression in (2) and proceeding as in the procedure of seeking a series solution, we get

$$a_1 = \frac{(n+1)(n+2)}{2 \cdot (2n+3)} , \quad a_k = \frac{(n+2k-1)(n+2k)}{2k \cdot (2n+2k+1)} a_{k-1} \quad (k \geq 2) ,$$

from which it is found that

$$a_k = \frac{\left(\frac{n+1}{2} \right)_k \left(\frac{n+2}{2} \right)_k}{k! \left(n + \frac{3}{2} \right)_k} .$$

Thus,

$$Q_n(x) = \frac{2^n (n!)^2}{(2n+1)!} x^{-n-1} \left\{ 1 + \frac{(n+1)(n+2)}{2 \cdot (2n+3)} x^{-2} + \dots \right\}$$
$$= \frac{2^n (n!)^2}{(2n+1)!} x^{-n-1} F \left(\frac{n+1}{2}, \frac{n+2}{2}, n + \frac{3}{2}, x^{-2} \right) . \quad (3)$$

This is the formula corresponding to Eq. (7) of Sec. 5.2. The series in (3) converges only for $|x| > 1$. Nevertheless, its analytic continuation obtained through the hypergeometric function, represents a function $Q_n(x)$ which is a single-valued analytic function in the x-plane cut along the real axis from -1 to $+1$.

For other expressions of $Q_n(x)$ in terms of hypergeometric functions, cf. Secs. 5.17, 5.19 and the relevant Exercises at the end of this chapter.

Integral Representations

When $|x| > 1$, we have from Eq. (5) of Sec. 5.4,

$$y(x) = A \int_C (1 - t^2)^n (x - t)^{-n-1} dt \ .$$

Choosing C to be the straight line from $t = -1$ to $t = 1$, we get another integral solution of the Legendre equation (2):

$$y(x) = A \int_{-1}^{1} (1 - t^2)^n (x - t)^{-n-1} dt \ .$$

The function represented by this integral is single-valued and analytic in the x-plane cut along the real axis from -1 to $+1$ [cf. Sec. 4.5 about the discussion on the multi-valuedness of (6) of that section. Now, we show that, by properly choosing the constant A, the r.h.s. of the last equation is equal to $Q_n(x)$.

Since we have assumed that $|x| > 1$, so

$$y(x) = A x^{-n-1} \int_{-1}^{1} (1 - t^2)^n \left(1 - \frac{t}{x}\right)^{-n-1} dt$$

$$= A x^{-n-1} \int_{-1}^{1} (1 - t^2)^n \sum_{k=0}^{\infty} \binom{-n-1}{k} \left(-\frac{t}{x}\right)^k dt$$

$$= A x^{-n-1} \sum_{k=0}^{\infty} \binom{-n-1}{k} (-)^k x^{-k} \int_{-1}^{1} t^k (1 - t^2)^n dt \ .$$

As $|t/x| < 1$, the term by term integration here is legitimate. The last integral does not vanish only when k is an even number, while

$$\int_{-1}^{1} t^{2k} (1 - t^2)^n dt = 2 \int_{0}^{1} t^{2k} (1 - t^2)^n dt$$

$$= \int_{0}^{1} v^{k - \frac{1}{2}} (1 - v)^n dv = \frac{\Gamma\left(k + \frac{1}{2}\right) \Gamma(n + 1)}{\Gamma\left(n + \frac{3}{2} + k\right)} \ ;$$

therefore,

$$y(x) = A x^{-n-1} \sum_{k=0}^{\infty} \binom{-n-1}{2k} x^{-2k} \frac{\Gamma\left(k+\frac{1}{2}\right) \cdot n!}{\Gamma\left(n+\frac{3}{2}+k\right)}$$

$$= A x^{-n-1} \sum_{k=0}^{\infty} \frac{(n+2k)!}{(2k)!} \frac{\Gamma\left(k+\frac{1}{2}\right)}{\Gamma\left(n+\frac{3}{2}+k\right)} x^{-2k} \ .$$

Using Eqs. (8) and (9) of Sec. 3.6, we have

$$y(x) = A \frac{2^{2n+1}(n!)^2}{(2n+1)!} x^{-n-1} \sum_{k=0}^{\infty} \frac{\left(\frac{n+1}{2}\right)_k \left(\frac{n+2}{2}\right)_k}{k!\,\left(n+\frac{3}{2}\right)_k} x^{-2k}$$

$$= A \frac{2^{2n+1}(n!)^2}{(2n+1)!} x^{-n-1} F\left(\frac{n+1}{2},\frac{n+2}{2},n+\frac{3}{2},x^{-2}\right) \ .$$

Comparing with (3), we see that

$$Q_n(x) = \frac{1}{2^{n+1}} \int_{-1}^{1} (1-t^2)^n (x-t)^{-n-1} dt \ . \tag{4}$$

From (4), integrating by parts n times, noting that, when $k < n$, $(1-t^2)^{n-k} = 0$ for $t = \pm 1$, we find

$$Q_n(x) = \frac{1}{2} \int_{-1}^{1} \frac{1}{2^n n!} \left\{ \frac{d^n}{dt^n}(t^2-1)^n \right\} (x-t)^{-1} dt \ .$$

With Eq. (3) of Sec. 5.3, we obtain another integral expression for $Q_n(x)$:

$$Q_n(x) = \frac{1}{2} \int_{-1}^{1} \frac{P_n(t)}{x-t} dt \ , \tag{5}$$

which is called the *Neumann expression for $Q_n(x)$*.

Although the two integral expressions, (4) and (5), are derived under the condition $|x| > 1$, but, as we have mentioned above, they are both single-valued functions and analytic in the x-plane cut along the real axis from -1 to $+1$, and hence the restriction $|x| > 1$ can be removed.

Finite Representations of $Q_n(x)$

With (5), we can derive the following expressions for $Q_n(x)$:

$$\left.\begin{aligned}
Q_0(x) &= \frac{1}{2} \ln \frac{x+1}{x-1} \, , \\
Q_n(x) &= \frac{1}{2} P_n(x) \ln \frac{x+1}{x-1} - W_{n-1}(x) \quad (n \geq 1) \, ,
\end{aligned}\right\} \tag{6}$$

where the logarithmic function assumes real values when $x > 1$, and $W_{n-1}(x)$ is a polynomial of degree $n - 1$:

$$W_{n-1} = \frac{(2n)!}{2^n (n!)^2} \left\{ x^{n-1} + \left[\frac{1}{3} - \frac{n(n-1)}{2 \cdot (2n-1)} \right] x^{n-3} + \ldots \right\}$$

$$= \frac{(2n)!}{2^n (n!)^2} \sum_{k=0}^{[\frac{n-1}{2}]} x^{n-2k-1} \sum_{r=0}^{k} \frac{(-)^r}{2k-2r+1} \binom{n}{k} \frac{n!(2n-2r)!}{(n-2r)!(2n)!} \cdot \tag{7}$$

The proof is as follows:

$$Q_n(x) = \frac{1}{2} \int_{-1}^{1} \frac{P_n(t)}{x-t} dt$$

$$= \frac{1}{2} \int_{-1}^{1} \frac{P_n(x)}{x-t} dt - \frac{1}{2} \int_{-1}^{1} \frac{P_n(x) - P_n(t)}{x-t} dt$$

$$= \frac{1}{2} P_n(x) \ln \frac{x+1}{x-1} - W_{n-1}(x) \, ,$$

where

$$W_{n-1}(x) = \frac{1}{2} \int_{-1}^{1} \frac{P_n(x) - P_n(t)}{x-t} dt \, .$$

Write $P_n(x)$ as $\sum_{r=0}^{[n/2]} a_{n-2r} x^{n-2r}$ [Eq. (7) of Sec. 5.2]. Since the constant terms of $P_n(x)$ and $P_n(t)$ cancel with each other when n is even, the upper limit in the summation over r is $[(n-1)/2]$. Then

$$W_{n-1}(x) = \frac{1}{2} \int_{-1}^{1} \sum_{r=0}^{[\frac{n-1}{2}]} a_{n-2r} \frac{x^{n-2r} - t^{n-2r}}{x - t} dt$$

$$= \frac{1}{2} \sum_{r=0}^{[\frac{n-1}{2}]} a_{n-2r} \sum_{s=0}^{n-2r-1} x^{n-2r-s-1} \int_{-1}^{1} t^s dt$$

$$= \frac{1}{2} \sum_{r=0}^{[\frac{n-1}{2}]} a_{n-2r} \sum_{s=0}^{n-2r-1} x^{n-2r-s-1} \frac{1 - (-)^{s+1}}{s+1}$$

$$= \sum_{r=0}^{[\frac{n-1}{2}]} a_{n-2r} \sum_{k=r}^{[\frac{n-1}{2}]} \frac{x^{n-2k-1}}{2k - 2r + 1} \quad (\text{ by putting } s = 2k - 2r)$$

$$= \sum_{k=0}^{[\frac{n-1}{2}]} x^{n-2k-1} \sum_{r=0}^{k} \frac{a_{n-2r}}{2k - 2r + 1} .$$

Using the values a_{n-2r} given by Eq. (7) of Sec. 5.3, we obtain (7) above.

(6) manifests clearly the singularity and multi-valuedness of $Q_n(x)$: $x = \pm 1$ are its branch points. According to the stipulation on the logarithmic functions in it, namely, they assume real values when $x > 1$, we can prove that the $Q_n(x)$ represented by (6) is the same branch as that represented by (3) with hypergeometric series. For this sake, we only have to expand the logarithmic functions in (6) in descending powers of x, and to compare with (3).

When $-1 < x < 1$, i.e., when x lies on the two banks of the cut, the values of $Q_n(x)$ are not unique. On the upper bank, $\arg\{(x+1)/(x-1)\}$ is decreased by π, so

$$Q_n(x + i0) = -i\frac{\pi}{2} P_n(x) + \frac{1}{2} P_n(x) \ln \frac{1+x}{1-x} - W_{n-1}(x) ; \qquad (8)$$

while on the lower bank, $\arg\{(x+1)/(x-1)\}$ is increased by π, so

$$Q_n(x - i0) = i\frac{\pi}{2} P_n(x) + \frac{1}{2} P_n(x) \ln \frac{1+x}{1-x} - W_{n-1}(x) . \qquad (9)$$

It is common to take, when $-1 < x < 1$,

$$Q_n(x) = \frac{1}{2}\{Q_n(x+i0) + Q_n(x-i0)\}$$

$$= \frac{1}{2}P_n(x)\ln\frac{1+x}{1-x} - W_{n-1}(x) \tag{10}$$

as the definition of $Q_n(x)$. The $Q_n(x)$ so defined satisfies the Legendre equation (2) in $-1 < x < 1$, and is a second solution linearly independent of $P_n(x)$.

5.9. Recurrence Relations of $Q_n(x)$

By (5) of the last section, we can obtain at once the corresponding relations for $Q(x)$ from the recurrence relations of $P_n(x)$

$$\left.\begin{array}{l} Q_1(x) - xQ_0(x) + 1 = 0 \ , \\ (n+1)Q_{n+1}(x) - (2n+1)xQ_n(x) + nQ_{n-1}(x) = 0 \quad (n \geq 1) \ , \end{array}\right\} \tag{1}$$

$$Q'_{n+1}(x) - xQ'_n(x) = (n+1)Q_n(x) \ , \tag{2}$$

$$xQ'_n(x) - Q'_{n-1}(x) = nQ_n(x) \ , \tag{3}$$

$$Q'_{n+1}(x) - Q'_{n-1}(x) = (2n+1)Q_n(x) \ , \tag{4}$$

$$(x^2-1)Q'_n(x) = nxQ_n(x) - nQ_{n-1}(x) \ . \tag{5}$$

For example,

$$Q_1(x) - xQ_0(x) = \frac{1}{2}\int_{-1}^{1}\frac{P_1(t) - xP_0(t)}{x-t}dt$$

$$= -\frac{1}{2}\int_{-1}^{1}dt = -1 \ ,$$

which is the first formula in (1). Also

$$(n+1)Q_{n+1}(x) - (2n+1)xQ_n(x) + nQ_{n-1}(x)$$

$$= \frac{1}{2}\int_{-1}^{1}\frac{(n+1)P_{n+1}(t) - (2n+1)tP_n(t) + nP_{n-1}(t)}{x-t}dt$$

$$- \frac{2n+1}{2}\int_{-1}^{1}P_n(t)dt \ .$$

By Eq. (2) of Sec. 5.5, it is seen that the numerator in the first integral is equal to zero, while by Eq. (1) of Sec. 5.6, the second integral also vanishes for $n \geq 1$. Thus we have the second formula in (1).

(2)–(5) can be derived similarly from Eqs. (4)–(7) of Sec. 5.5.

5.10. Expansion of the Function $\frac{1}{x-t}$ in Terms of Legendre Functions. Neumann Expansion

When both x and t are real and $|x| > 1$, $1/(x - t)$ as a function of t is continuous in an interval slightly larger than $[-1, 1]$. Hence, according to the theorem stated at the end of Sec. 5.6, this function can be expanded in terms of $P_n(x)$:

$$\frac{1}{x-t} = \sum_{n=0}^{\infty} a_n P_n(t) ; \tag{1}$$

the series is uniformly convergent in $[-1, 1]$. Multiplying both sides of (1) with $P_m(t)$ and integrating from -1 and 1, we obtain, by using Eq. (6) of Sec. 5.6 and Eq. (5) of Sec. 5.8,

$$a_m = \frac{2m+1}{2} \int_{-1}^{1} \frac{P_m(t)}{x-t} dt = (2m+1)Q_m(x) ; \tag{2}$$

therefore,

$$\frac{1}{x-t} = \sum_{n=0}^{\infty} (2n+1)Q_n(x)P_n(t) . \tag{3}$$

Actually, the expansion (3) is valid under a less stringent condition. For, from Eq. (1) of Sec. 5.9 and Eq. (2) of Sec. 5.5.

$$xQ_0(x) - Q_1(x) - 1 = 0 ,$$
$$(2r+1)xQ_r(x) - (r+1)Q_{r+1}(x) - rQ_{r-1}(x) = 0 \quad (r \geq 1) ,$$
$$tP_0(t) - P_1(t) = 0 ,$$
$$(2r+1)tP_r(t) - (r+1)P_{r+1}(t) - rP_{r-1}(t) = 0 \quad (r \geq 1) ,$$

we have

$$(x - t)\sum_{r=0}^{n}(2r+1)P_r(t)Q_r(x) - \sum_{r=0}^{n}(r+1)[P_r(t)Q_{r+1}(x) - P_{r+1}(t)Q_r(x)]$$

$$- \sum_{r=1}^{n} r[P_r(t)Q_{r-1}(x) - P_{r-1}(t)Q_r(x)] - 1 = 0 .$$

After a combination of terms in the second and the third summations, only terms with $r = n$ in the second sum are left. Hence,

$$\frac{1}{x - t} = \sum_{r=0}^{n} (2r + 1) Q_r(x) P_r(t)$$

$$+ \frac{n + 1}{x - t} [P_{n+1}(t) Q_n(x) - P_n(t) Q_{n+1}(x)] . \tag{4}$$

It can be shown [cf. Hobson (1931), Sec. 3.8, pp. 59-62] that, if t is a point situated inside the ellipse with foci at ± 1 and passing through the point x, i.e., if[d]

$$|t + \sqrt{t^2 - 1}| < |x + \sqrt{x^2 - 1}| , \tag{5}$$

then the remainder on the r.h.s. of (4) approaches to zero uniformly when $n \to \infty$, and hence (3); the series is also uniformly convergent for all the points x on the ellipse.

By this result and the Cauchy formula, we have the following expansion theorem:

Let $f(x)$ be a function analytic in and on the ellipse C with foci at ± 1, then for points t in a confocal ellipse interior to C, we have

$$f(t) = \sum_{n=0}^{\infty} a_n P_n(t) , \tag{6}$$

where

$$a_n = \frac{2n + 1}{2\pi i} \oint_C f(x) Q_n(x) dx , \tag{7}$$

and the series converges uniformly. The proof is as follows. By the Cauchy formula, we have

$$f(t) = \frac{1}{2\pi i} \int_C \frac{f(x) dx}{x - t} = \sum_{n=0}^{\infty} (2n + 1) P_n(t) \cdot \frac{1}{2\pi i} \int_C f(x) Q_n(x) dx ,$$

and hence (6). This is called the *Neumann expansion*.

---

[d]Let $x + \sqrt{x^2 - 1} = r e^{i\theta}$, then $x - \sqrt{x^2 - 1} = r^{-1} e^{-i\theta}$, from which we have $x = \frac{1}{2}(r e^{i\theta} + r^{-1} e^{-i\theta})$. Let $x = \xi + i\eta$, then $\xi = \frac{1}{2}(r + r^{-1})\cos\theta, \eta = \frac{1}{2}(r - r^{-1})\sin\theta$. Therefore, $r = |x + \sqrt{x^2 - 1}| = $ constant represents an ellipse passing through the point x.

Multiplying the two sides of (6) with $P_m(t)$, integrating and using the orthogonality relation of $P_n(t)$, we find

$$a_n = \frac{2n+1}{2} \int_{-1}^{1} f(t) P_n(t) dt ,\tag{8}$$

same as Eq. (11) of Sec. 5.6. Now, as $f(t)$ is an analytic function, we can use the Rodriques formula [Eq. (3) of Sec. 5.3] and integrate by parts n times to obtain another formula for determining the coefficients:

$$a_n = \frac{2n+1}{2^{n+1} n!} \int_{-1}^{1} f^{(n)}(t) (1-t^2)^n dt .\tag{9}$$

5.11. Associate Legendre Functions $P_l^m(x)$

Associate Legendre functions are solutions of the differential equation [Eq. (6) of Sec. 5.1]

$$(1 - x^2) \frac{d^2 y}{dx^2} - 2x \frac{dy}{dx} + \left[l(l+1) - \frac{m^2}{1-x^2} \right] y = 0 .\tag{1}$$

In the present section, we shall deal only with the case when $-1 \le x \le 1$, $l = 0, 1, 2, \ldots$, and m is an arbitrary integer. The general case will be treated in Sec. 5.16.

We have noted in Sec. 5.1 that the solution of (1) can be represented by the P-symbol

$$(1 - x^2)^{\frac{m}{2}} P \left\{ \begin{array}{ccc} 1 & 0 & \infty \\ 0 & 0 & l+m+1; \ \frac{1-x}{2} \\ -m & -m & -l+m \end{array} \right\} ,\tag{2}$$

from which we find at once a solution expressed in terms of the hypergeometric function, namely,

$$y_1(x) = A(1 - x^2)^{\frac{m}{2}} F \left(l + m + 1, -l + m, 1 + m, \frac{1-x}{2} \right) ,\tag{3}$$

where A is an arbitrary constant. By Eq. (10) of Sec. 4.2, and Eq. (12) of Sec. 5.2, this solution can be written as

$$y_1(x) = A'(1 - x^2)^{m/2} \frac{d^m}{dx^m} F \left(l + 1, -l, 1, \frac{1-x}{2} \right)$$

$$= A'(1 - x^2)^{m/2} \frac{d^m}{dx^m} P_l(x) ,$$

where $P_l(x)$ is the Legendre polynomial of degree l.

Hobson defines the *associate Legendre function of the first kind* $P_l^m(x)$ of *order m and degree l* to be

$$P_l^m(x) = (-)^m (1 - x^2)^{m/2} \frac{d^m}{dx^m} P_l(x)$$

$$(l \geq m \geq 0, \ -1 \leq x \leq 1) ,$$ (4)

where the square root assumes positive values. Another definition is that of Ferrer, in which the factor $(-)^m$ is absent.

According to the Rodriques formula [Eq. (3) of Sec. 5.3], (4) can also be written as

$$P_l^m(x) = (-)^m \frac{(1 - x^2)^{m/2}}{2^l \, l!} \frac{d^{l+m}}{dx^{l+m}} (x^2 - 1)^l .$$ (5)

This form also holds when m is a negative integer, so long as $|m| \leq l$ (see below).

We can infer from the above results that if $v(x)$ is a solution of the Legendre equation of degree l, then $y = (1 - x^2)^{m/2} \times v^{(m)}(x)$ satisfies (1). In fact, this can be proved by direct calculation. Thereby, we obtain at once another solution of (1), namely,

$$Q_l^m(x) = (-)^m (1 - x^2)^{m/2} \frac{d^m}{dx^m} Q_l(x)$$

$$(-1 \leq x \leq 1) ,$$ (6)

$Q_l(x)$ being the Legendre function of the second kind of degree l defined by Eq. (10) of Sec. 5.8. $Q_l^m(x)$ is called the *associate Legendre function of the second kind of order m and degree l*. This is also Hobson's definition; that of Ferrer's but without the factor $(-)^m$.

It is easy to see that $P_l^m(x)$ is bounded in the interval $-1 \leq x \leq 1$, while $Q_l^m(x) \rightarrow \infty$ when $x \rightarrow \pm 1$. For, according to Eq. (6) of Sec. 5.8, $Q_l(x)$ contains a term with the logarithmic factor $\ln\{(x + 1)/(x - 1)\}$, so its mth derivative will involve terms with the factors $(x+1)^{-m}$ and $(x-1)^{-m} (m \geq 1)$.

In the present section, we shall treat only $P_l^m(x)$; for $Q_l^m(x)$, see Secs. 5.17 and 5.19, which deal with the general case.

Equation (1) is unaltered when m is replaced by $-m$; therefore, it is conceivable that the function

$$P_l^{-m}(x) = (-)^m \frac{(1 - x^2)^{-m/2}}{2^l l!} \frac{d^{l-m}}{dx^{l-m}} (x^2 - 1)^l \quad (m > 0)$$ (7)

is also a solution of (1). Indeed, $P_l^{-m}(x)$ and $P_l^m(x)$ are differed only by a constant multiple. Here is the proof:

$$\frac{d^{l+m}}{dx^{l+m}}(x^2-1)^l = \sum_{r=0}^{l+m} \binom{l+m}{r} \frac{d^r}{dx^r}(x-1)^l \frac{d^{l+m-r}}{dx^{l+m-r}}(x+1)^l$$

$$= \sum_{r=m}^{l} \binom{l+m}{r} \frac{l!}{(l-r)!}(x-1)^{l-r} \frac{l!}{(r-m)!}(x+1)^{r-m}$$

$$= \sum_{r=0}^{l-m} \binom{l+m}{r+m} \frac{l!}{(l-m-r)!}(x-1)^{l-m-r} \frac{l!}{r!}(x+1)^r$$

$$= \frac{(l+m)!}{(l-m)!}(x^2-1)^{-m} \sum_{r=0}^{l-m} \binom{l-m}{r}$$

$$\times \frac{l!}{(l-r)!}(x-1)^{l-r} \frac{l!}{(r+m)!}(x+1)^{r+m}$$

$$= (-)^m \frac{(l+m)!}{(l-m)!}(1-x^2)^{-m} \frac{d^{l-m}}{dx^{l-m}}(x^2-1)^l \; ;$$

therefore,

$$P_l^{-m}(x) = (-)^m \frac{(l-m)!}{(l+m)!} P_l^m(x) \; . \tag{8}$$

5.12. Orthogonality Relations of $P_l^m(x)$

The functions $P_l^m(x)$ satisfy the following orthogonality relations: $(m, m' \geq 0)$

$$\int_{-1}^{1} P_l^m P_{l'}^m \, dx = \frac{2}{2l+1} \frac{(l+m)!}{(l-m)!} \delta_{ll'} \; , \tag{1}$$

$$\int_{-1}^{1} P_l^m P_l^{m'} \frac{dx}{1-x^2} = \frac{1}{m} \frac{(l+m)!}{(l-m)!} \delta_{mm'} \; . \tag{2}$$

Proof: From the differential equations satisfied respectively by P_l^m and

$P_l^{m'}$ we have

$$[l(l+1) - l'(l'+1)]P_l^m P_{l'}^{m'} - \frac{m^2 - m'^2}{1-x^2} P_l^m P_{l'}^{m'}$$

$$= \frac{d}{dx}\left[(1-x^2)\left(P_l^m \frac{dP_{l'}^{m'}}{dx} - P_{l'}^{m'} \frac{dP_l^m}{dx}\right)\right] .$$

Integrating gives

$$[l(l+1) - l'(l'+1)]\int_{-1}^{1} P_l^m P_{l'}^{m'}\, dx = (m^2 - m'^2)\int_{-1}^{1} P_l^m P_{l'}^{m'} \frac{dx}{1-x^2} .$$

Hence,

$$\int_{-1}^{1} P_l^m P_{l'}^m\, dx = 0 \qquad (l \neq l') ,$$

$$\int_{-1}^{1} P_l^m P_l^{m'} \frac{dx}{1-x^2} = 0 \quad (m \neq m') .$$

Now

$$\int_{-1}^{1} [P_l^m]^2 dx = \frac{1}{2^{2l}(l!)^2} \int_{-1}^{1} (1-x^2)^m \left[\left(\frac{d}{dx}\right)^{l+m}(x^2-1)^l\right]^2 dx .$$

Let

$$G(x) = (1-x^2)^m \left(\frac{d}{dx}\right)^{l+m}(x^2-1)^l ,$$

which is a polynomial of degree $l + m$. Integrating the preceding integral by parts m times and noticing that $G^{(k)}(\pm 1) = 0$ when $k = 0, 1, 2, \ldots, m-1$, we have

$$\int_{-1}^{1} [P_l^m]^2 dx = \frac{(-)^m}{2^{2l}(l!)^2} \int_{-1}^{1} G^{(m)}(x)\left(\frac{d}{dx}\right)^l (x^2-1)^l dx .$$

Further integrating by parts l times, we have

$$\int_{-1}^{1} [P_l^m]^2 dx = \frac{(-)^{l+m}}{2^{2l}(l!)^2} G^{(l+m)}(x) \int_{-1}^{1} (x^2-1)^l dx ,$$

where

$$G^{(l+m)}(x) = \left(\frac{d}{dx}\right)^{l+m}\left[(-)^m\frac{(2l)!}{(l-m)!}x^{l+m} + \ldots\right]$$

$$= (-)^m\frac{(2l)!(l+m)!}{(l-m)!}$$

and

$$\int_{-1}^{1}(x^2-1)^l dx = \frac{(-)^l(l!)^2}{(2l)!}\frac{2^{2l+1}}{2l+1} \tag{3}$$

[cf. text before Eq. (5) of Sec. 5.6]. Therefore,

$$\int_{-1}^{1}[P_l^m]^2 dx = \frac{2}{2l+1}\frac{(l+m)!}{(l-m)!} , \tag{4}$$

which proves (1).

For (2), we have

$$\int_{-1}^{1}[P_l^m]^2\frac{dx}{1-x^2} = \frac{1}{2^{2l}(l!)^2}\int_{-1}^{1}(1-x^2)^{m-1}\left[\left(\frac{d}{dx}\right)^{l+m}(x^2-1)^l\right]^2 dx .$$

Let

$$G(x) = (1-x^2)^{m-1}\left(\frac{d}{dx}\right)^{l+m}(x^2-1)^l ,$$

which is a polynomial of degree $l+m-2$. Integrating the preceding integral by parts $m-1$ times and noticing that $G^{(k)}(\pm 1) = 0$ for $k = 0, 1, 2, \ldots, m-2$, we obtain

$$\int_{-1}^{1}[P_l^m]^2\frac{dx}{1-x^2} = \frac{(-)^{m-1}}{2^{2l}(l!)^2}\int_{-1}^{1}G^{(m-1)}(x)\frac{d^{l+1}}{dx^{l+1}}(x^2-1)^l dx .$$

Integrating by parts once more, it results

$$\int_{-1}^{1}[P_l^m]^2\frac{dx}{1-x^2} = \frac{(-)^{m-1}}{2^{2l}(l!)^2}\left[G^{(m-1)}(x)\frac{d^l}{dx^l}(x^2-1)^l\right]_{-1}^{1}$$

$$+ \frac{(-)^m}{2^l l!}\int_{-1}^{1}G^{(m)}(x)P_l(x)dx .$$

The last integral is equal to zero, because $G^{(m)}(x)$ is a polynomial of degree $l - 2$ [cf. Eq. (1) of Sec. 5.6]. Hence,

$$\int_{-1}^{1} [P_l^m]^2 \frac{dx}{1 - x^2} = \frac{(-)^{m-1}}{2^l l!} [G^{(m-1)}(1) P_l(1) - G^{(m-1)}(-1) P_l(-1)]$$

$$= \frac{(-)^{m-1}}{2^l l!} [G^{(m-1)}(1) - (-)^l G^{(m-1)}(-1)] \; ;$$

Eq. (4) of Sec. 5.3 has been used in the last step. Now,

$$G^{(m-1)}(1) = \frac{d^{m-1}}{dx^{m-1}} \left[(1 - x^2)^{m-1} \left(\frac{d}{dx} \right)^{l+m} (x^2 - 1)^l \right] \Bigg|_{x=1}$$

$$= \frac{d^{m-1}}{dx^{m-1}} (1 - x^2)^{m-1} \Bigg|_{x=1} \cdot \frac{d^{l+m}}{dx^{l+m}} (x^2 - 1)^l \Bigg|_{x=1}$$

$$= (-)^{m-1} (m - 1)! 2^{m-1} \binom{l + m}{l} \frac{d^l}{dx^l} (x - 1)^l \frac{d^m}{dx^m} (x + 1)^l \Bigg|_{x=1}$$

$$= \frac{(-)^{m-1}}{m} \frac{(l + m)!}{(l - m)!} 2^{l-1} l! \; , \tag{5}$$

while, by (5), $G^{(m-1)}(-1) = (-)^{l-1} G^{(m-1)}(1)$; therefore,

$$= \int_{-1}^{1} [P_l^m]^2 \frac{dx}{1 - x^2} = \frac{(-)^{m-1}}{2^l l!} \cdot 2 \frac{(-)^{m-1}}{m} \frac{(l + m)!}{(l - m)!} 2^{l-1} \cdot l!$$

$$= \frac{1}{m} \frac{(l + m)!}{(l - m)!} \; , \tag{6}$$

which proves (2).

With formula (8) of the last section, we obtain from (1) and (2)

$$\int_{-1}^{1} P_l^m P_{l'}^{-m} dx = (-)^m \frac{2}{2l+1} \delta_{ll'} , \tag{7}$$

$$\int_{-1}^{1} P_l^m P_l^{-m'} \frac{dx}{1-x^2} = \frac{(-)^m}{m} \delta_{mm'} . \tag{8}$$

The Completeness of $P_l^m(x)$

For a definite m, $\{P_l^m(x)\}(l \geq m)$ form a complete orthonormal set in the interval $[-1, 1]$ with weight 1. Any function $f(x)$, continuous in the interval $[-1, 1]$ and equal to zero at the end points, can be expanded, in the sense of convergence in mean, in terms of the associate Legendre functions $P_l^m(x)$ of an arbitrary order m:

$$f(x) = \sum_{l \geq m} a_l P_l^m(x) , \tag{9}$$

where

$$a_l = \frac{2l+1}{2} \frac{(l-m)!}{(l+m)!} \int_{-1}^{1} f(x) P_l^m(x) dx . \tag{10}$$

For a simple proof of this theorem, cf., for example, Chihonov and Samarski, *Equations of Mathematical Physics* (in Russian), App. II, Sec. 1.9.

5.13. Recurrence Relations for $P_l^m(x)$ and $Q_l^m(x)$

The following are four fundamental recurrence relations [*according to Hobson's definition of $P_l^m(x)$*]:

$$(2l+1)x P_l^m = (l+m)P_{l-1}^m + (l-m+1)P_{l+1}^m , \tag{1}$$

$$(2l+1)(1-x^2)^{\frac{1}{2}} P_l^m = P_{l-1}^{m+1} - P_{l+1}^{m+1} , \tag{2}$$

$$(2l+1)(1-x^2)^{\frac{1}{2}} P_l^m = (l-m+2)(l-m+1)P_{l+1}^{m-1} - (l+m)(l+m-1)P_{l-1}^{m-1} , \tag{3}$$

$$(2l+1)(1-x^2)\frac{dP_l^m}{dx} = (l+1)(l+m)P_{l-1}^m - l(l-m+1)P_{l+1}^m . \tag{4}$$

Many other recurrence relations may be derived from these four [cf. Ex. 24 at the end of this chapter].

To prove (1), we write Eq. (2) of Sec. 5.5 as

$$(2l+1)xP_l = (l+1)P_{l+1} + lP_{l-1} .$$

Differentiating both sides m times gives

$$(2l+1)xP_l^{(m)} + m(2l+1)P_l^{(m-1)} = (l+1)P_{l+1}^{(m)} + lP_{l-1}^{(m)} .$$

Applying Eq. (6) of Sec. 5.5 to the second term on the l.h.s., then multiplying the equation by $(-)^m(1-x^2)^{m/2}$, we have (1).

To prove (2), we differentiate both sides of Eq. (6) of Sec. 5.5 m times, and multiply with $(-)^m(1-x^2)^{m/2}$.

(3) is proved from Eq. (7) of Sec. 5.5, $(1-x^2)P_l' = lP_{l-1} - lxP_l$, by applying (1) (with $m = 0$) to the second term on the r.h.s. gives

$$(1-x^2)P_l' = lP_{l-1} - \frac{l}{2l+1}[lP_{l-1} + (l+1)P_{l+1}] ,$$

or,

$$(2l+1)(1-x^2)P_l' = l(l+1)P_{l-1} - l(l+1)P_{l+1} .$$

Differentiating both sides $m-1$ times and multiplying with $(-)^m(1-x^2)^{(m-1)/2}$, we have

$$(2l+1)(1-x^2)^{1/2}P_l^m + 2(m-1)(2l+1)xP_l^{m-1}$$
$$- (m-1)(m-2)(2l+1)(1-x^2)^{1/2}P_l^{m-2}$$
$$= l(l+1)P_{l+1}^{m-1} - l(l+1)P_{l-1}^{m-1} .$$

Applying (1) and (2) respectively to the second and third terms on the l.h.s. gives (3).

To prove (4), we have

$$(1-x^2)\frac{d}{dx}P_l^m = (1-x^2)\frac{d}{dx}[(-)^m(1-x^2)^{m/2}P_l^{(m)}]$$
$$= (-)^m(1-x^2)[(1-x^2)^{m/2}P_l^{(m+1)}$$
$$- m(1-x^2)^{m/2-1}xP_l^{(m)}] ,$$

i.e.,

$$(1-x^2)\frac{dP_l^m}{dx} = -(1-x^2)^{1/2}P_l^{m+1} - mxP_l^m . \tag{5}$$

Multiplying both sides with $2l+1$, then applying (3) and (1) respectively to the two terms on the r.h.s. gives the result.

The above formulae can also be derived from (9) and (10) of the last section.

Since $Q_l(x)$ satisfies the same recurrence relations as $P_l(x)$ [see Sec. 5.9; except for the relation between $Q_1(x)$ and $Q_0(x)$], the above recurrence relations also hold for $Q_l^m(x)$ according to Eq. (6) of Sec. 5.11.

Besides, these recurrence relations are also applicable to the case when l and m are not integers, *so long as x is confined in the interval $-1 < x < 1$* [cf. Ex. 37 at the end of this chapter].

5.14. Addition Formula

When the direction of the polar axis of the spherical polar coordinates is altered, there is the addition formula

$$P_l(\cos \gamma) = \sum_{m=-l}^{l} (-)^m P_l^m(\cos \theta) P_l^{-m}(\cos \theta') e^{im(\varphi-\varphi')} \tag{1}$$

$$= \sum_{m=-l}^{l} \frac{(l-m)!}{(l+m)!} P_l^m(\cos \theta) P_l^m(\cos \theta') e^{im(\varphi-\varphi')} \tag{2}$$

$$= P_l(\cos \theta) P_l(\cos \theta')$$

$$+ 2 \sum_{m=1}^{l} \frac{(l-m)!}{(l+m)!} P_l^m(\cos \theta) P_l^m(\cos \theta') \cos m(\varphi - \varphi') , \tag{3}$$

where

$$\cos \gamma = \cos \theta \cos \theta' + \sin \theta \sin \theta' \cos(\varphi - \varphi') , \tag{4}$$

i.e., γ is the angle between OP (in the direction θ, φ) and OP' (in the direction θ', φ'). [See Fig. 11.]

There are various methods for proving this formula. We shall give here the one embarking on the solutions of a differential equation.

When we solve, in spherical polar coordinates, the Laplace equation

$$\frac{1}{r^2} \left(\frac{\partial}{\partial r} \right) \left(r^2 \frac{\partial V}{\partial r} \right) + \frac{1}{r^2 \sin \theta} \frac{\partial}{\partial \theta} \left(\sin \theta \frac{\partial V}{\partial \theta} \right) + \frac{1}{r^2 \sin^2 \theta} \frac{\partial^2 V}{\partial \varphi^2} \tag{5}$$

by the method of separation of variables, we set $V(r, \theta, \varphi) = R(r) S(\theta, \varphi)$ and obtain

$$\frac{1}{r^2} \frac{d}{dr} \left(r^2 \frac{dR}{dr} \right) - \frac{\lambda}{r^2} R = 0 \tag{6}$$

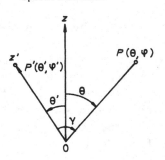

Fig. 11.

and

$$\frac{1}{\sin\theta}\frac{\partial}{\partial\theta}\left(\sin\theta\frac{\partial S}{\partial\theta}\right) + \frac{1}{\sin^2\theta}\frac{\partial^2 S}{\partial\varphi^2} + \lambda S = 0 , \tag{7}$$

where λ is a constant introduced during the separation of variables. It is well known in mathematical physics [cf. the discussion at the end of Sec. 5.19] that Eq. (7) has periodic and bounded solutions in $0 \le \varphi \le 2\pi$ and $0 \le \theta \le \pi$ only when $\lambda = l(l+1), l = 0, 1, 2, \ldots$. We denote such solutions by $S_l(\theta, \varphi)$ and call them *(spherical) surface harmonics of degree l*.

To (7), with $\lambda = l(l+1)$, applying again the variables separation by putting $S_l(\theta, \varphi) = \Theta(\theta)\Phi(\varphi)$, we obtain

$$\frac{d^2\Phi}{d\varphi^2} + m^2\Phi = 0 \tag{8}$$

and

$$\frac{1}{\sin\theta}\frac{d}{d\theta}\left(\sin\theta\frac{d\Theta}{d\theta}\right) + \left[l(l+1) - \frac{m^2}{\sin^2\theta}\right]\Theta = 0 . \tag{9}$$

The periodic solutions of (8) are $e^{im\varphi}, m = 0, \pm1, \pm2, \ldots$. The bounded solutions of (9) are $P_l^m(\cos\theta)$. For, putting $x = \cos\theta$, (9) becomes the associate Legendre equation [Eq. (1) of Sec. 5.11]. Thus, when $\lambda = l(l+1)$, the periodic bounded solutions of (7) are

$$S_l(\theta, \varphi) = P_l^m(\cos\theta)e^{im\varphi} \quad (m = 0, \pm1, \pm2, \ldots, \pm l) . \tag{10}$$

Now, let us alter the polar axis of the spherical coordinates from Oz to $Oz'(OP')$. This corresponds to a transformation from the spherical polar coordinates (r, θ, φ) to (r, γ, δ), γ being given by (4) while δ is the new azimuthal angle. Obviously, the form of Eq. (7) is unaltered under such a transformation. However, its periodic (and bounded) solutions are now

$$S_l(\gamma, \delta) = P_l^m(\cos\gamma)e^{im\delta} \quad (m = 0, \pm1, \pm2, \ldots, \pm l) ; \tag{11}$$

and $P_l(\cos\gamma)$ is one of them. Hence, expressing γ in terms of the original angle coordinates (θ, φ), i.e., (4), $P_l(\cos\gamma)$ must be a linear combination of $S_l(\theta, \varphi)$:

$$P_l(\cos\gamma) = \sum_{m=-l}^{l} A_m P_l^m(\cos\theta)e^{im\varphi} . \qquad (12)$$

Utilizing the orthogonality relations of $e^{im\varphi}$ [Eq. (2) of Sec. 1.10] and of $P_l^m(\cos\theta)$ [Eq. (7) of Sec. 5.12], we obtain at once

$$A_m = (-)^m \frac{2l+1}{4\pi} \int_0^\pi \int_0^{2\pi} P_l(\cos\gamma) P_l^{-m}(\cos\theta)e^{-im\varphi} d\omega , \qquad (13)$$

where $d\omega = \sin\theta\, d\varphi\, d\theta$ is the solid angle element.

But, since $P_l^{-m}(\cos\theta)e^{-im\varphi}$ is a solution of (7), it can be, on the inverse, expressed in terms of a linear combination of $S_l(\gamma, \delta)$:

$$P_l^{-m}(\cos\theta)e^{-im\varphi} = \sum_{m=-l}^{l} B_m' P_l^{m'}(\cos\gamma)e^{im'\delta} . \qquad (14)$$

Substituting in (13) and noticing that the solid angle element is invariant under the change of the polar axis, i.e., $d\omega = \sin\theta\, d\varphi\, d\theta = \sin\gamma\, d\delta\, d\gamma = d\Omega$, we have

$$A_m = (-)^m \frac{2l+1}{4\pi} \int_0^\pi \int_0^{2\pi} P_l(\cos\gamma) \sum_{m'=-l}^{l} B_{m'} P_l^{m'}(\cos\gamma)e^{im'\delta} d\Omega$$

$$= (-)^m \frac{2l+1}{4\pi} \cdot 2\pi \frac{2}{2l+1} B_0 = (-)^m B_0 . \qquad (15)$$

B_0 can be calculated from (14) as follows. When $\gamma = 0$, we have $\theta = \theta'$, $\varphi = \varphi'$ and $P_l^{m'}(\cos\gamma) = P_l^{m'}(1)$. But according to Eq. (5) of Sec. 5.11, when $m' \neq 0$, we have $P_l^{m'}(1) = 0$, while $P_l(1) = 1$. Therefore,

$$B_0 = P_l^{-m}(\cos\theta')e^{im\varphi'} .$$

Putting this result in (15), we obtain A_m; then by substituting them into (12), we arrive at (1). Formula (2) is derived from (1) by using Eq. (8) of Sec. 5.11.

In Sec. 5.21 we shall give another proof of (1).

5.15. Spherical Surface Harmonics $Y_{lm}(\theta, \varphi)$

In the preceding section, we have obtained for the equation

$$\frac{1}{\sin\theta}\frac{\partial}{\partial\theta}\left(\sin\theta\frac{\partial S}{\partial\theta}\right) + \frac{1}{\sin^2\theta}\frac{\partial^2 S}{\partial\varphi^2} + l(l+1)S = 0 \tag{1}$$

or,

$$\frac{\partial^2 S}{\partial\theta^2} + \cot\theta\frac{\partial S}{\partial\theta} + \frac{1}{\sin^2\theta}\frac{\partial^2 S}{\partial\varphi^2} + l(l+1)S = 0 , \tag{1'}$$

$2l + 1$ solutions [(10) of the last section]:

$$P_l^m(\cos\theta)e^{im\varphi} , \quad m = 0, \pm1, \pm2, \ldots, \pm l ,$$

which are bounded (with respect to θ) and periodic (with respect to φ) in $0 \le \theta \le \pi, 0 \le \varphi \le 2\pi$.

In many applications, it often adopts

$$Y_{lm}(\theta, \varphi) = \sqrt{\frac{(2l+1)(l-m)!}{4\pi(l+m)!}} P_l^m(\cos\theta)e^{im\varphi}$$

$$(m = 0, \pm1, \pm2, \ldots, \pm l) \tag{2}$$

as the periodic and bounded solutions of Eq. (1). Such solutions satisfy the following orthonormal relation:

$$\int_0^\pi \int_0^{2\pi} Y_{lm}^* Y_{l'm'} \sin\theta d\varphi d\theta = \delta_{mm'}\delta_{ll'} , \tag{3}$$

where $Y_{lm}^*(\theta, \varphi)$ is the complex conjugate of $Y_{lm}(\theta, \varphi)$, and

$$Y_{lm}^* = (-)^m Y_{l,-m} , \tag{4}$$

which may be proved by using Eq. (8) of Sec. 5.11. (3) can be shown by Eq. (1) of Sec. 5.12 and Eq. (2) of Sec. 1.10; but when m or $m' < 0$, we need Eq. (8) of Sec. 5.11 in addition.

Expressing in terms of $Y_{lm}(\theta, \varphi)$, the addition formula of the last section takes the form

$$Y_{l0}(\gamma) = \sqrt{\frac{4\pi}{2l+1}} \sum_{m=-l}^{l} (-)^m Y_{lm}(\theta, \varphi)Y_{l,-m}(\theta', \varphi')$$

$$= \sqrt{\frac{4\pi}{2l+1}} \sum_{m=-1}^{l} Y_{lm}(\theta, \varphi)Y_{lm}^*(\theta', \varphi') , \tag{5}$$

where $Y_{l0}(\gamma) = \sqrt{(2l+1)/4\pi}\; P_l(\cos\gamma)$.

From Eqs. (1)-(5) of Sec. 5.13, we find the following important recurrence relations:

$$\cos\theta\, Y_{lm} = \sqrt{\frac{(l+m)(l-m)}{(2l+1)(2l-1)}}\, Y_{l-1,m}$$
$$+ \sqrt{\frac{(l+m+1)(l-m+1)}{(2l+1)(2l+3)}}\, Y_{l+1,m}\; , \tag{6}$$

$$\sin\theta\, e^{i\varphi}\, Y_{lm} = \sqrt{\frac{(l-m)(l-m-1)}{(2l+1)(2l-1)}}\, Y_{l-1,m+1}$$
$$- \sqrt{\frac{(l+m+1)(l+m+2)}{(2l+1)(2l+3)}}\, Y_{l+1,m+1}\; , \tag{7}$$

$$\sin\theta\, e^{-i\varphi}\, Y_{lm} = -\sqrt{\frac{(l+m)(l+m-1)}{(2l+1)(2l-1)}}\, Y_{l-1,m-1}$$
$$+ \sqrt{\frac{(l-m+1)(l-m+2)}{(2l+1)(2l+3)}}\, Y_{l+1,m-1}\; , \tag{8}$$

$$-\sin\theta\, \frac{\partial}{\partial\theta} Y_{lm} = (l+1)\sqrt{\frac{(l+m)(l-m)}{(2l+1)(2l-1)}}\, Y_{l-1,m}$$
$$- \sqrt{\frac{(l+m+1)(l-m+1)}{(2l+1)(2l+3)}}\, Y_{l+1,m}\; , \tag{9}$$

$$\sin\theta\, \frac{\partial}{\partial\theta} Y_{lm} = \sin\theta\sqrt{(l-m)(l+m+1)}\, Y_{l,m+1} e^{-i\varphi}$$
$$+ m\cos\theta\, Y_{lm}\; . \tag{10}$$

With these relations, further recurrence formulae may be derived [cf. Exs. 26 and 27 at the end of this chapter].

Completeness of the set $\{Y_{lm}(\theta, \varphi)\}$

According to the theorem of constructing a complete set of functions of several variables from complete sets of functions of a single variable given at the end of Sec. 1.10, we see that Y_{lm}, $(l = 0, 1, 2, \ldots ; m = 0, \pm 1, \pm 2, \ldots, \pm l)$, form a complete set of functions. For $Y_{lm} = N_{lm} P_l^m(\cos\theta)e^{im\varphi}$ [N_{lm} being the normalization factor; cf. (2)], in which $\{e^{im\varphi}\}(m = 0, \pm 1, \pm 2, \ldots)$ is known to be a complete set while for a fixed m, $\{P_l^m(\cos\theta)\}(l \geq m)$ is also a complete set [Sec. 5.12].

Therefore, any function $f(\theta, \varphi)$, continuous on a spherical surface, can be expanded in a series of $Y_{lm}(\theta, \varphi)$:

$$f(\theta, \varphi) = \sum_{l=0}^{\infty} \sum_{m=-l}^{l} A_{lm} Y_{lm}(\theta, \varphi) , \tag{11}$$

where

$$A_{lm} = \int_0^\pi \int_0^{2\pi} Y_{lm}^*(\theta, \varphi) f(\theta, \varphi) \sin\theta d\varphi d\theta , \tag{12}$$

and the series is convergent *in mean* to $f(\theta, \varphi)$.

Changing the integration variables θ, φ in (12) into θ', φ', substituting in (11) and making use of the addition formula (5), we have

$$f(\theta, \varphi) = \sum_{l=0}^{\infty} \frac{2l+1}{4\pi} \int_0^\pi \int_0^{2\pi} f(\theta', \varphi') P_l(\cos\gamma) \sin\theta' d\varphi' d\theta' , \tag{13}$$

where

$$\cos\gamma = \cos\theta \cos\theta' + \sin\theta \sin\theta' \cos(\varphi - \varphi') .$$

The r.h.s. of (13) is called the *Laplace series* of $f(\theta, \varphi)$. It can be shown that if $f(\theta, \varphi)$ is absolutely integrable on a spherical surface, then at the point (θ, φ) where $f(\theta, \varphi)$ is continuous, the Laplace series (13) converges and the sum is equal to $f(\theta, \varphi)$. If the function is not continuous at a certain point (θ, φ), but there exists on the surface, a line passing through the point (θ, φ) such that when the variable point approaches to the point (θ, φ) from either side along the line, the limiting values of $f(\theta, \varphi)$ exist, and equal respectively to $f_1(\theta, \varphi)$ and $f_2(\theta, \varphi)$, then

$$\frac{1}{2}\{f_1(\theta, \varphi) + f_2(\theta, \varphi)\}$$

$$= \sum_{l=0}^{\infty} \frac{2l+1}{4\pi} \int_0^\pi \int_0^{2\pi} f(\theta', \varphi') P_l(\cos\gamma) \sin\theta' d\varphi' d\theta' , \tag{14}$$

so long as $f(\theta, \varphi)$ is of bounded variation.

Regards the proof of this theorem and the conditions under which the Laplace series converges uniformly to $f(\theta, \varphi)$, cf. Hobson (1931), Chap. VII, Sec. 211.

5.16. The General Associate Legendre Functions $P_\nu^\mu(z)$

Functions $P_\nu^\mu(z)$ are the solutions of the Legendre equation [Eq. (6) of Sec. 5.1]

$$(1 - z^2)\frac{d^2u}{dz^2} - 2z\frac{du}{dz} + [\nu(\nu + 1) - \frac{\mu^2}{1 - z^2}]u = 0 , \tag{1}$$

where μ and ν may be any complex numbers. The other solutions of Eq. (1) are $Q_\nu^\mu(z)$, which will be discussed in the following section. We shall define these functions according to Hobson by double contour integrals, and let them be such that they conform to the functions P_l^m and Q_l^m introduced earlier when μ and ν are integers.

In Sec. 5.1 we have already seen that if we put $u = (z^2 - 1)^{\mu/2} v(z)$ and change the independent variable into $\varsigma = (1 - z)/2$, then $v(\varsigma)$ satisfies the hypergeometric equation

$$\varsigma(1 - \varsigma)\frac{d^2v}{d\varsigma^2} + [\mu + 1 - (2\mu + 2)\varsigma]\frac{dv}{d\varsigma}$$
$$- (\mu - \nu)(\mu + \nu + 1)v = 0 \tag{2}$$

(it is represented in that section by the P-symbol). Returning to the variable z, the equation becomes

$$(1 - z^2)\frac{d^2v}{dz^2} - 2(\mu + 1)z\frac{dv}{dz} + (\nu - \mu)(\nu + \mu + 1)v = 0 . \tag{3}$$

Applying the Euler transformation [Sec. 2.14] to solve (3) by assuming

$$v(z) = \int_C (z - t)^\lambda w(t)dt , \tag{4}$$

we find, according to Eq. (6) of Sec. 2.14, the two allowable values for $\lambda : \nu - \mu$ and $-\nu - \mu - 1$. Take the latter, then by Eq. (14) of Sec. 2.14, we have $w(t) = (t^2 - 1)^\nu$. Choose C to be the double contour shown in Fig. 12 (note that $t = -1$ is lying outside of C), then we obtain one integral solution of (3):

$$v(z) = A(z^2 - 1)^{\mu/2} \int_M^{(z+,1+,z-,1-)} (t^2 - 1)^\nu (t - z)^{-\nu-\mu-1}dt , \tag{5}$$

A being an arbitrary constant.

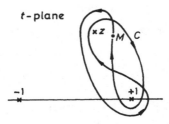

Fig. 12.

After Hobson, the *associate Legendre function of the first kind of order* μ *and degree* ν is defined to be[e]

$$P_\nu^\mu(z) = \frac{e^{-\nu\pi i}}{4\pi \sin\nu\pi} \frac{\Gamma(\nu+\mu+1)}{\Gamma(\nu+1)} (z^2-1)^{\mu/2}$$

$$\times \int_M^{(z+,1+,z-,1-)} \frac{1}{2^\nu} (t^2-1)^\nu (t-z)^{-\nu-\mu-1} dt ,$$

$$(\nu+\mu \neq \text{ negative integer})^e , \qquad (6)$$

where the phases of the various factors in the integrand are stipulated as follows: on the path of integration, when $t+1$ and $t-z$ are positive numbers, $\arg(t+1) = 0, \arg(t-z) = 0$; at the starting point M, $\arg(t-1) = \varphi, |\varphi| < \pi$. In addition, it is further stipulated that $|\arg(z-1)| < \pi, |\arg(z+1)| < \pi$, then, $P_\nu^\mu(z)$ is a single-valued analytic function in the z-plane cut from $-\infty$ to $+1$ along the real axis.

Now, let us investigate the relation between the function $P_\nu^\mu(z)$ defined by (6) and the hypergeometric function. For this purpose, let $t-1 = (z-1)s$, then $t-z = (z-1)(s-1)$, and the points $t = z$ and $t = 1$ are transformed respectively into the points $s = 1$ and $s = 0$; the integral in (6) is transformed into

$$I = (z-1)^{-\mu} \int_{M'}^{(1+,0+,1-,0-)} (s-1)^{-\nu-\mu-1} s^\nu \left(1 + \frac{z-1}{2} s\right)^\nu ds , \qquad (7)$$

with the contour of integration shown in Fig. 13, where M' is the point corresponding to M. The phases of the various factors or the values of the functions in the integrand can be deduced from the original stipulation for (6): when $s = 0$, the factor $[1 + (z-1)s/2]^\nu$ corresponding to $(t+1)^\nu$ takes the value

---

[e]When $\nu+\mu =$ negative integer, the r.h.s. of (6) is an indeterminate form [cf. Ex. 28 at the end of this chapter].

s-plane

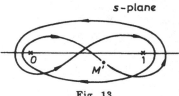

Fig. 13.

1; on the path of integration, when $s - 1$ is a positive number, $\arg(s - 1) = 0$; at the starting point M', $\arg s = \varphi - \arg(z - 1)$, for example, when M' is situated on the real axis from $s = 0$ to $s = 1$, $\arg s = 0$, since such points M' correspond to the points M lying on the straight line from $t = 1$ to $t = z$.

Now, deform the contour of integration in Fig. 13 into a Pochhammer contour shown in Fig. 6 of Sec. 3.9, and note that at the starting point of the path (situated on the real axis between $s = 0$ and $s = 1$), $\arg s = 0$, $\arg(s - 1) = -\pi$; we obtain at once from Eq. (7) of Sec. 4.5,

$$I = (z - 1)^{-\mu} e^{-i\pi(-\nu-\mu-1)}$$

$$\times \int^{(1+,0+,1-,0-)} s^{\nu}(1 - s)^{-\nu-\mu-1}\left(1 + \frac{z-1}{2}s\right)^{\nu} ds$$

$$= (z - 1)^{-\mu} e^{i\nu\pi} \frac{-4\Gamma(1+\nu)\Gamma(-\nu-\mu)\sin\pi(1+\nu)\sin\pi(-\nu-\mu)}{\Gamma(1-\mu)}$$

$$\times F\left(-\nu, \nu + 1, 1 - \mu, \frac{1-z}{2}\right).$$

Substituting in (6) and making use of Eq. (2) of Sec. 3.5 to simplify the result, we obtain

$$P_\nu^\mu(z) = \frac{1}{\Gamma(1-\mu)}\left(\frac{z+1}{z-1}\right)^{\mu/2} F\left(-\nu, \nu + 1, 1 - \mu, \frac{1-z}{2}\right), \qquad (8)$$

where $|\arg(z \pm 1)| < \pi$. This is a fundamental representation of $P_\nu^\mu(z)$, applicable for any values of μ and ν.

From (8), noting that $F(\alpha, \beta, \gamma, \varsigma) = F(\beta, \alpha, \gamma, \varsigma)$, we get at once the important relation:

$$P_\nu^\mu(z) = P_{-\nu-1}^\mu(z). \qquad (9)$$

When $\mu = 0$, again using $P_\nu(z)$ to represent $P_\nu^0(z)$, we have

$$P_\nu(z) = F\left(-\nu, \nu + 1, 1, \frac{1-z}{2}\right), \qquad (10)$$

which is a generalization of Murphy's expression of $P_n(z)$ [Eq. (12) of Sec. 5.2]. $P_\nu(z)$ is called the *Legendre function of the first kind* of degree ν. Note that

$$P_\nu(1) = 1 . \tag{11}$$

When $\mu = m$ $(m = 1, 2, \ldots)$, (8) becomes an indeterminate form, ∞/∞. Using the result of Ex. 6 at the end of Chap. 4, we have

$$P_\nu^m(z) = \frac{(z^2 - 1)^{m/2}}{2^m m!} \frac{\Gamma(\nu + m + 1)}{\Gamma(\nu - m + 1)} F\left(-\nu + m, \nu + 1 + m, 1 + m, \frac{1 - z}{2}\right) , \tag{12}$$

and, by the transformation (8) of Sec. 4.3,

$$P_\nu^m(z) = \frac{1}{m!}\left(\frac{z - 1}{z + 1}\right)^{m/2} \frac{\Gamma(\nu + m + 1)}{\Gamma(\nu - m + 1)} F\left(-\nu, \nu + 1, 1 + m, \frac{1 - z}{2}\right) . \tag{13}$$

Also, by Eq. (10) of Sec. 4.2, we obtain from (12),

$$P_\nu^m(z) = (z^2 - 1)^{m/2} \frac{d^m}{dz^m} P_\nu(z) , \tag{14}$$

which, of course, can be derived directly from (6).

When $\mu = -m(m = 1, 2, \ldots)$, (8) is still valid. By Eq. (8) of Sec. 4.3, we obtain

$$P_\nu^{-m}(z) = \frac{(z^2 - 1)^{m/2}}{2^m m!} F\left(-\nu + m, \nu + 1 + m, 1 + m, \frac{1 - z}{2}\right) . \tag{15}$$

Comparing with (12), we find at once

$$P_\nu^{-m}(z) = \frac{\Gamma(\nu - m + 1)}{\Gamma(\nu + m + 1)} P_\nu^m(z) . \tag{16}$$

Note that this relation holds only when $m =$ integers [cf. the second formula of Ex. 36 at the end of this chapter or Eq. (4) of Sec. 5.19].

When *both μ and ν are positive integers*, if $\mu > \nu$, then as $1/\Gamma(\nu - \mu + 1) = 0$, we have $P_\nu^\mu(z) = 0$. In such cases, we can take $\Gamma(\nu - \mu + 1)P_\nu^\mu(z)$ as a non-zero solution of (1), and make use of (12) or (13).

Schläfli's Integral Representation

The integral I in (6) can be expressed as

$$I = A + B - e^{2\nu\pi i} A - e^{2(\nu + \mu + 1)\pi i} B ,$$

where A and B are respectively those parts of the integral which encircle respectively the points $t = z$ and $t = 1$ once in the positive sense. When μ is an integer,

$$I = (1 - e^{2\nu\pi i})(A + B) .$$

It follows,

$$P_\nu^{\pm m}(z) = \frac{\Gamma(\nu \pm m + 1)}{\Gamma(\nu + 1)}(z^2 - 1)^{\pm m/2}\frac{1}{2\pi i}$$

$$\times \int_M^{(z+,1+)} \frac{1}{2^\nu}(t^2 - 1)^\nu(t - z)^{-\nu \mp m - 1}dt$$

$$(|\arg(z \pm 1)| < \pi) ; \tag{17}$$

at the starting point $M, |\arg(t - 1)| < \pi, |\arg(t - z)| < \pi$, and when $t > -1, \arg(t + 1) = 0$.

If $m = 0$, (17) reduces to

$$P_\nu(z) = \frac{1}{2\pi i}\int_M^{(z+,1+)}\frac{1}{2^\nu}(t^2 - 1)^\nu(t - z)^{-\nu - 1}dt , \tag{18}$$

which is the generalization of Eq. (7) of Sec. 5.4, and called the *Schläfli integral representation*. With this representation, it can be shown that the recurrence relations given in Sec. 5.5 are valid when n is not an integer [cf. Ex. 6 at the end of this chapter].

From (8), with the transformations for the hypergeometric functions, especially the linear and quadratic ones [Secs. 4.3, 4.8 and 4.12, respectively], we can deduce for $P_\nu^\mu(z)$ their recurrence relations [see Ex. 37 at the end of this chapter] and various expressions in terms of hypergeometric functions [Exs. 29–31 at the end of this chapter; also Erdélyi (1953), Vol. I, Sec. 3.2].

5.17. $Q_\nu^\mu(z)$

The *associate Legendre function of the second kind of order* μ *and degree* ν is defined, according to Hobson, by

$$Q_\nu^\mu(z) = \frac{e^{-(\nu+1)\pi i}}{4i\sin\nu\pi}\frac{\Gamma(\nu + \mu + 1)}{\Gamma(\nu + 1)}(z^2 - 1)^{\mu/2}$$

$$\times \int_M^{(-1+,1-)}\frac{1}{2^\nu}(t^2 - 1)^\nu(t - z)^{-\nu - \mu - 1}dt ; \tag{1}$$

the path of integration C is shown in Fig. 14a, $t = z$ lying outside the contour. The phases of the multi-valued factors in the integrand are stipulated as follows: At the point $B(t > 1)$ which the contour meets the real axis, $\arg(t - 1) = \arg(t + 1) = 0$, so, when C is deformed into the contour given in Fig. 14b, $\arg(t - 1) = \pi$ and $\arg(t + 1) = -2\pi$ for the starting point M lying on the real axis between -1 to $+1$; $\arg(t - z) = \arg(z - t) - \pi, |\arg(z - t)| < \pi$. Besides, it is again assumed that $|\arg(z - 1)| < \pi, |\arg(z + 1)| < \pi$. Then $Q_\nu^\mu(z)$ will be a single-valued analytic function in the z-plane cut along the real axis from $-\infty$ to $+1$.

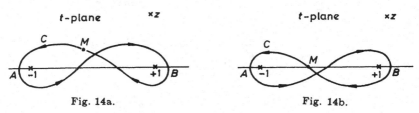

Fig. 14a.   Fig. 14b.

Let us first find the relation between $Q_\nu^\mu(z)$ and the hypergeometric function. Since $t - z = (z - t)e^{-\pi i} = z(1 - t/z)e^{-\pi i}$, the integral in (1) can be written as

$$I = 2^{-\nu}z^{-(\nu+\mu+1)}e^{(\nu+\mu+1)\pi i} \int_M^{(-1+,1-)} (t^2 - 1)^\nu \left(1 - \frac{t}{z}\right)^{-\nu-\mu-1} dt \, ,$$

where $|\arg z| < \pi$ and $(1 - t/z)^{-\nu-\mu-1} \to 1$ when $z \to \infty$.

Assume $|z| > 1$. We can change the path of integration so that $|t/z| < 1$ on it. Expanding the factor in the integrand which contains z in a binomial series, then integrating term by term, we find

$$I = 2^{-\nu}e^{(\nu+\mu+1)\pi i}z^{-(\nu+\mu+1)} \sum_{r=0}^{\infty} \binom{-\nu - \mu - 1}{r}(-z)^{-r}$$

$$\times \int_M^{(-1+,1-)} (t^2 - 1)^\nu t^r dt$$

$$= 2^{-\nu}e^{(\mu+1)\pi i}z^{-(\nu+\mu+1)} \sum_{r=0}^{\infty} \binom{-\nu - \mu - 1}{r}(-z)^{-r}$$

$$\times \int_M^{(-1+,1-)} (1 - t^2)^\nu t^r dt \, , \tag{2}$$

where the path of integration can be made to look like the one given in Fig. 15; at the starting point M, $\arg(1 - t) = \arg(1 + t) = 0$.

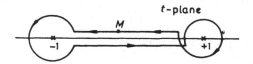

Fig. 15.

Assume that $\mathrm{Re}(\nu) > -1$. It is easy to show that the values of the circular path integrals near the ends of the contour approach to zero as the radii tend to zero. Hence,

$$\int_M^{(-1+,1-)} (1-t^2)^\nu t^r \, dt = -\int_{-1}^1 (1-t^2)^\nu t^r \, dt + e^{2\nu\pi i} \int_{-1}^1 (1-t^2)^\nu t^r \, dt$$

$$= (e^{2\nu\pi i} - 1) \int_{-1}^1 (1-t^2)^\nu t^r \, dt$$

$$= 2i e^{\nu\pi i} \sin \nu\pi \frac{1 + (-)^r}{2} \int_0^1 (1-s)^\nu s^{\frac{r-1}{2}} \, ds$$

$$= 2i e^{\nu\pi i} \sin \nu\pi \frac{1 + (-)^r}{2} \frac{\Gamma(\nu+1)\Gamma\left(\frac{r+1}{2}\right)}{\Gamma\left(\nu + \frac{r+3}{2}\right)},$$

Eqs. (1) and (3) of Sec. 3.8 have been used in the last step. Substituting this result in (2), we obtain

$$I = 2^{-\nu} e^{(\nu+\mu+1)\pi i} z^{-(\nu+\mu+1)} 2i \sin \nu\pi$$

$$\times \sum_{k=0}^\infty \binom{-\nu-\mu-1}{2k} \frac{\Gamma(\nu+1)\Gamma\left(k+\frac{1}{2}\right)}{\Gamma\left(\nu+k+\frac{3}{2}\right)} z^{-2k}. \qquad (3)$$

By Eq. (8) of Sec. 3.6,

$$\binom{-\nu-\mu-1}{2k} = \frac{\Gamma(\nu+\mu+1+2k)}{\Gamma(\nu+\mu+1)\Gamma(2k+1)}$$

$$= \frac{2^{\nu+\mu} \Gamma\left(\frac{\nu+\mu+1}{2}+k\right) \Gamma\left(\frac{\nu+\mu+2}{2}+k\right)}{\Gamma(\nu+\mu+1)\Gamma\left(k+\frac{1}{2}\right)\Gamma(k+1)}$$

$$= \frac{2^{\nu+\mu} \Gamma\left(\frac{\nu+\mu+1}{2}\right) \Gamma\left(\frac{\nu+\mu+2}{2}\right)}{\Gamma(\nu+\mu+1)\Gamma\left(k+\frac{1}{2}\right)} \frac{\left(\frac{\nu+\mu+1}{2}\right)_k \left(\frac{\nu+\mu+2}{2}\right)_k}{k!}$$

$$= \frac{\Gamma\left(\frac{1}{2}\right)}{\Gamma\left(k+\frac{1}{2}\right)} \frac{\left(\frac{\nu+\mu+1}{2}\right)_k \left(\frac{\nu+\mu+2}{2}\right)_k}{k!};$$

therefore,

$$I = 2^{-\nu}e^{(\nu+\mu+1)\pi i}z^{-(\nu+\mu+1)}2i \sin \nu\pi$$

$$\times \frac{\Gamma\left(\frac{1}{2}\right)\Gamma(\nu+1)}{\Gamma\left(\nu+\frac{3}{2}\right)}F\left(\frac{\nu+\mu+1}{2}, \frac{\nu+\mu+2}{2}, \nu+\frac{3}{2}, z^{-2}\right)$$

and it follows,

$$Q_\nu^\mu(z) = \frac{e^{\mu\pi i}}{2^{\nu+1}}\frac{\Gamma(\nu+\mu+1)\Gamma\left(\frac{1}{2}\right)}{\Gamma\left(\nu+\frac{3}{2}\right)}(z^2-1)^{\mu/2}z^{-\nu-\mu-1}$$

$$\times F\left(\frac{\nu+\mu+1}{2}, \frac{\nu+\mu+2}{2}, \nu+\frac{3}{2}, z^{-2}\right) , \tag{4}$$

where $|\arg(z \pm 1)| < \pi, |\arg z| < \pi$. Although this expression is derived under the conditions $|z| > 1$ and $\text{Re}(\nu) > -1$, but the integral in (1) and the functions on the r.h.s. of (4) are all free from these restrictions, and hence the conditions may be negated. (4) is a fundamental expression of $Q_\nu^\mu(z)$.

When $\mu = 0$, denoting again $Q_\nu^0(z)$ by $Q_\nu(z)$, we have

$$Q_\nu(z) = \frac{\Gamma(\nu+1)\Gamma\left(\frac{1}{2}\right)}{2^{\nu+1}\Gamma\left(\nu+\frac{3}{2}\right)}z^{-\nu-1}F\left(\frac{\nu+1}{2}, \frac{\nu+2}{2}, \nu+\frac{3}{2}, z^{-2}\right) , \tag{5}$$

which is a generalization of Eq. (3) of Sec. 5.8. For, when $\nu = n$,

$$\Gamma(\nu+1)\Gamma\left(\frac{1}{2}\right)/2^{\nu+1}\Gamma\left(\nu+\frac{3}{2}\right) = 2^n(n!)^2/(2n+1)! .$$

When $\mu = m(m = 1, 2, \ldots)$, we obtain at once from (1),

$$Q_\mu^m(z) = (z^2-1)^{m/2}\frac{d^m}{dz^m}Q_\nu(z) . \tag{6}$$

Also, from (4), by Eq. (8) of Sec. 4.3, we have

$$Q_\nu^\mu(z) = \frac{e^{\nu\pi i}}{2^{\nu+1}}\frac{\Gamma(\nu+\mu+1)\Gamma\left(\frac{1}{2}\right)}{\Gamma\left(\nu+\frac{3}{2}\right)}(z^2-1)^{-\mu/2}z^{-\nu+\mu-1}$$

$$\times F\left(\frac{\nu-\mu+1}{2}, \frac{\nu-\mu+2}{2}, \nu+\frac{3}{2}, z^{-2}\right) . \tag{7}$$

Changing μ to $-\mu$, then comparing with (4), we see

$$\frac{e^{\mu\pi i}Q_\nu^{-\mu}(z)}{\Gamma(\nu-\mu+1)} = \frac{e^{-\mu\pi i}Q_\nu^\mu(z)}{\Gamma(\nu+\mu+1)} . \qquad (8)$$

When $\mu = m$ (integer), it gives

$$Q_\nu^{-m}(z) = \frac{\Gamma(\nu-m+1)}{\Gamma(\nu+m+1)}Q_\nu^m(z) , \qquad (9)$$

which is the formula corresponding to Eq. (16) of Sec. 5.16. However, we note from (8) that, whether μ is an integer or not, $Q_\nu^{-\mu}(z)$ and $Q_\nu^\mu(z)$ are always linearly dependent of each other, whereas $P_\nu^{-\mu}(z)$ and $P_\nu^\mu(z)$ are linearly dependent of each other only when μ is an integer [cf. Ex. 36 (ii), or Eq. (4) of Sec. 5.19].

From (4), using the transformations of the hypergeometric functions, other expressions for $Q_\nu^\mu(z)$ can be found [cf. Sec. 5.19 and Ex. 35 at the end of this chapter; also Erdélyi (1953), Vol. I, Sec. 3.2].

5.18. Definition of $P_\nu^\mu(x)$ on the Cut: $-\infty < x < 1$

Let $-1 < x < 1$. From Eq. (8) of Sec. 5.16, namely

$$P_\nu^\mu(z) = \frac{1}{\Gamma(1-\mu)}\left(\frac{z+1}{z-1}\right)^{\mu/2} F\left(-\nu, \nu+1, 1-\mu, \frac{1-z}{2}\right) , \qquad (1)$$

where $\arg(z+1) = \arg(z-1) = 0$ for $z > 1$, we have

$$P_\nu^\mu(x+i0) = \frac{e^{-\mu\pi i/2}}{\Gamma(1-\mu)}\left(\frac{1+x}{1-x}\right)^{\mu/2} F\left(-\nu, \nu+1, 1-\mu, \frac{1-x}{2}\right) , \qquad (2)$$

$$P_\nu^\mu(x-i0) = \frac{e^{\mu\pi i/2}}{\Gamma(1-\mu)}\left(\frac{1+x}{1-x}\right)^{\mu/2} F\left(-\nu, \nu+1, 1-\mu, \frac{1-x}{2}\right) , \qquad (3)$$

where $\arg\{(1+x)/(1-x)\} = 0; x + i0$ represents the upper bank along the cut and $x - i0$ the lower bank. Thus, the values of $P_\nu^\mu(z)$ are discontinuous on the two sides of the cut.

We shall define, after Hobson, *in the interval* $-1 < x < 1$ *along the real axis*,

$$P_\nu^\mu(x) = e^{\mu\pi i/2}P_\nu^\mu(x+i0) = e^{-\mu\pi i/2}P_\nu^\mu(x-i0)$$

$$= \frac{1}{\Gamma(1-\mu)}\left(\frac{1+x}{1-x}\right)^{\mu/2} F\left(-\nu, \nu+1, 1-\mu, \frac{1-x}{2}\right) . \qquad (4)$$

This function clearly satisfies, in $-1 < x < 1$, the Legendre equation [Eq. (1) of Sec. 5.16].

When $\mu = m$, we find at once from this definition and Eqs. (12)–(14),(16) of Sec. 5.16,

$$P_\nu^m(x) = (-)^m \frac{1}{2^m m!} (1 - x^2)^{m/2} \frac{\Gamma(\nu + m + 1)}{\Gamma(\nu - m + 1)}$$

$$\times F\left(-\nu + m, \nu + 1 + m, 1 + m, \frac{1 - x}{2}\right) \tag{5}$$

$$= (-)^m \frac{1}{m!} \left(\frac{1 - x}{1 + x}\right)^{m/2} \frac{\Gamma(\nu + m + 1)}{\Gamma(\nu - m + 1)}$$

$$\times F\left(-\nu, \nu + 1, 1 + m, \frac{1 - x}{2}\right) , \tag{6}$$

$$P_\nu^m(x) = (-)^m (1 - x^2)^{m/2} \frac{d^m}{dx^m} P_\nu(x) , \tag{7}$$

and

$$P_\nu^{-m}(x) = (-)^m \frac{\Gamma(\nu - m + 1)}{\Gamma(\nu + m + 1)} P_\nu^m(x) , \tag{8}$$

where (7) and (8) are respectively the generalizations of Eqs. (4) and (8) of Sec. 5.11.

We see from (4) that if $\mathrm{Re}(\mu) > 0$, then $P_\nu^\mu(x) \to \infty$ in general when $x \to 1 - 0$, unless μ is an integer. By (5) and (8) it is seen that $P_\nu^{\pm m}(x)$ is bounded when $x \to 1 - 0$; its limiting value is zero for $m > 0$, and is equal to 1 when $m = 0$.

Next, consider the value of $P_\nu^\mu(x)$ when $x \to -1 + 0$. Assume that $\mu \neq$

integer. By Eq. (4) of Sec. 4.8, we obtain from (4)

$$P_\nu^\mu(x) = \frac{1}{\Gamma(1-\mu)} \left(\frac{1+x}{1-x}\right)^{\mu/2}$$

$$\times \left[\frac{\Gamma(1-\mu)\Gamma(-\mu)}{\Gamma(1+\nu-\mu)\Gamma(-\nu-\mu)} F\left(-\nu, \nu+1, 1+\mu, \frac{1+x}{2}\right) \right.$$

$$+ \frac{\Gamma(1-\mu)\Gamma(\mu)}{\Gamma(-\nu)\Gamma(\nu+1)} \left(\frac{1+x}{2}\right)^{-\mu}$$

$$\left. \times F\left(1+\nu-\mu, -\nu-\mu, 1-\mu, \frac{1+x}{2}\right) \right] .$$

Applying Eq. (8) of Sec. 4.3 on the second term, and making use of Eq. (2) of Sec. 3.5, we have

$$P_\nu^\mu(x) = \frac{\Gamma(-\mu)}{\Gamma(1+\nu-\mu)\Gamma(-\nu-\mu)} \left(\frac{1+x}{1-x}\right)^{\mu/2}$$

$$\times F\left(-\nu, \nu+1, 1+\mu, \frac{1+x}{2}\right)$$

$$- \frac{\sin\nu\pi}{\pi} \Gamma(\mu) \left(\frac{1-x}{1+x}\right)^{\mu/2} F\left(-\nu, \nu+1, 1-\mu, \frac{1+x}{2}\right) , \qquad (9)$$

from which we see that if $\nu \neq$ integer, $P_\nu^\mu(x) \to \infty$ in general when $x \to -1+0$, unless $1+\nu-\mu$ or $-\nu-\mu$ is equal to zero or a negative integer and also $\mathrm{Re}(\mu) < 0$; then $\lim_{x\to-1+0} P_\nu^\mu(x) \sim (1-x)^{-\mu/2}$. If ν is an integer, the second term in (9) vanishes and $\lim_{x\to-1+0} P_\nu^\mu(x) \sim (1+x)^{\mu/2}$.

When $\mu = m(= 0, 1, 2, \ldots)$, Eq. (4) of Sec. 4.8 is not applicable, and (9) also does not hold. However, by Eq. (8) of Sec. 4.9, we obtain from (5),

$$P_\nu^m(x) = -\frac{\sin\nu\pi}{\pi}(m-1)! \left(\frac{1-x}{1+x}\right)^{m/2} \sum_{k=0}^{m-1} \frac{(-\nu)_k(\nu+1)_k}{k!(1-m)_k} \left(\frac{1+x}{2}\right)^k$$

$$+ \frac{\sin\nu\pi}{\pi} \frac{\Gamma(\nu+m+1)}{2^m\Gamma(\nu-m+1)}(1-x^2)^{m/2}$$

$$\times \sum_{k=0}^{\infty} \frac{(-\nu+m)_k(\nu+1+m)_k}{k!(k+m)!} \left(\frac{1+x}{2}\right)^k$$

$$\times \{\psi(-\nu + m + k) + \psi(\nu + 1 + m + k)$$
$$- \psi(1 + m + k) - \psi(1 + k) + \ln \frac{1+x}{2}\}$$
$$(m = 0, 1, 2, \ldots) \; ; \tag{10}$$

when $m = 0$, the finite sum in it is to be negated. From this expression, we see that if $\nu \neq$ integer, then, when $x \to -1 + 0$, $P_\nu(x) \sim \ln\{(1 + x)/2\}$, $P_\nu^m(x) \sim (1 + x)^{-m/2}$, all becoming ∞. If $\nu = n(= 0, 1, 2, \ldots)$, then, in virtue of the factor $\sin \nu\pi$, the term containing the finite sum in (10) vanishes, and there remain in the series of the second term only those terms containing $\psi(-\nu + m + k)$ with $k \leq n - m$. By Eq. (3) of Sec. 3.11, we have

$$\lim_{\nu \to n} \sin \nu\pi \cdot \psi(-\nu + m + k)$$
$$= \lim_{\nu \to n} \sin \nu\pi \{\psi(1 + \nu - m - k) + \pi \cot \nu\pi\} = (-)^n \pi \quad (k \leq n - m) \; ;$$

therefore,

$$P_n^m(x) = (-)^n \frac{\Gamma(n + m + 1)}{\Gamma(n - m + 1)} \frac{(1 - x^2)^{m/2}}{2^m}$$
$$\times \sum_{k=0}^{n-m} \frac{(-n + m)_k (n + 1 + m)_k}{k!(k + m)!} \left(\frac{1+x}{2}\right)^k . \tag{11}$$

When $x \to -1 + 0$, $P_n^m(x) \sim (1 - x^2)^{m/2} \to 0(m > 0)$, $P_n(x) \to (-1)^n$; when $n < m$, $P_n^m(x) \equiv 0$.

If $\mu = -m(m = 1, 2, \ldots)$, we can use (8); if $\nu = -n$ $(n = 1, 2, \ldots)$, we can use Eq. (9) of Sec. 5.16: $P_\nu^\mu = P_{-\nu-1}^\mu$, so that it is unnecessary to discuss them separately.

When z is a real number less than -1, we may use the result of Ex. 31 at the end of this chapter to discuss the properties of $P_\nu^\mu(x)$ along the cut; we only need to note that on the upper bank, $\arg(z - 1) = \arg(z + 1) = \pi$, while on the lower bank, $\arg(z - 1) = \arg(z + 1) = -\pi$.

5.19. Definition of $Q_\nu^\mu(x)$ on the Cut: $-\infty < x < 1$

Same as in the previous section, we shall deal mainly with the segment $-1 < x < 1$ along the cut. Equation (4) of Sec. 5.17 may be used to deal with the segment $-\infty < x < -1$, but is not suitable for the segment $-1 < x < 1$, so some transformations are needed.

From Eq. (4) of Sec. 5.17, using Eq. (9) of Sec. 4.3, we have

$$Q_\nu^\mu(z) = \frac{e^{\mu\pi i}}{2^{\nu+1}} \frac{\Gamma(\nu+\mu+1)\Gamma\left(\frac{1}{2}\right)}{\Gamma\left(\nu+\frac{3}{2}\right)} (z^2-1)^{-\frac{\nu+1}{2}}$$

$$\times F\left(\frac{\nu+\mu+1}{2}, \frac{\nu-\mu+1}{2}, \nu+\frac{3}{2}, \frac{1}{1-z^2}\right). \tag{1}$$

Applying then Eq. (8) of Sec. 4.8 to this equation gives

$$Q_\nu^\mu(z) = \frac{e^{\mu\pi i}}{2} \left\{ \frac{\Gamma(\nu+\mu+1)\Gamma(-\mu)}{2^\mu\Gamma(\nu-\mu+1)} (z^2-1)^{\mu/2} \right.$$

$$\times F\left(\frac{\nu+\mu+1}{2}, \frac{-\nu+\mu}{2}, 1+\mu, 1-z^2\right)$$

$$\left. +\frac{\Gamma(\mu)}{2^{-\mu}}(z^2-1)^{-\mu/2}F\left(\frac{\nu-\mu+1}{2}, \frac{-\nu-\mu}{2}, 1-\mu, 1-z^2\right) \right\}. \tag{2}$$

Finally, with Eq. (9) of Sec. 4.12, substituting $t = (1-z)/2$, we obtain

$$Q_\nu^\mu(z) = \frac{e^{\mu\pi i}}{2} \left\{ \frac{\Gamma(\nu+\mu+1)\Gamma(-\mu)}{\Gamma(\nu-\mu+1)} \left(\frac{z-1}{z+1}\right)^{\mu/2} F\left(-\nu, \nu+1, 1+\mu, \frac{1-z}{2}\right) \right.$$

$$\left. +\Gamma(\mu)\left(\frac{z-1}{z+1}\right)^{-\mu/2} F\left(-\nu, \nu+1, 1-\mu, \frac{1-z}{2}\right) \right\}, \tag{3}$$

from which, by Eq. (8) of Sec. 5.16, we arrive at

$$Q_\nu^\mu(z) = \frac{\pi e^{\mu\pi i}}{2\sin\mu\pi} \left\{ P_\nu^\mu(z) - \frac{\Gamma(\nu+\mu+1)}{\Gamma(\nu-\mu+1)} P_\nu^{-\mu}(z) \right\}. \tag{4}$$

From (4), making use of (4) of the last section, we then obtain the values of $Q_\nu^\mu(z)$ on the cut from -1 to $+1$:

$$Q_\nu^\mu(x+i0) = \frac{\pi e^{\mu\pi i}}{2\sin\mu\pi} \left\{ e^{-\mu\pi i/2} P_\nu^\mu(x) - \frac{\Gamma(\nu+\mu+1)}{\Gamma(\nu-\mu+1)} e^{\mu\pi i/2} P_\nu^{-\mu}(x) \right\}, \tag{5}$$

$$Q_\nu^\mu(x-i0) = \frac{\pi e^{\mu\pi i}}{2\sin\mu\pi} \left\{ e^{\mu\pi i/2} P_\nu^\mu(x) - \frac{\Gamma(\nu+\mu+1)}{\Gamma(\nu-\mu+1)} e^{-\mu\pi i/2} P_\nu^{-\mu}(x) \right\}. \tag{6}$$

Eliminating $P_\nu^{-\mu}(x)$ from (5) and (6), we have

$$e^{-\mu\pi i/2}Q_\nu^\mu(x+i0) - e^{\mu\pi i/2}Q_\nu^\mu(x-i0) = -i\pi e^{\mu\pi i}P_\nu^\mu(x) . \qquad (7)$$

Hence, if we stipulate that, in $-1 < x < 1$,

$$e^{\mu\pi i}Q_\nu^\mu(x) = \frac{1}{2}\{e^{-\mu\pi i/2}Q_\nu^\mu(x+i0) + e^{\mu\pi i/2}Q_\nu^\mu(x-i0)\} , \qquad (8)$$

then $Q_\nu^\mu(x)$ and $P_\nu^\mu(x)$ are linearly independent of each other, and both satisfy the associate Legendre equation [Eq. (6) of Sec. 5.16] in $-1 < x < 1$.

Substituting (5) and (6) in (8) gives

$$Q_\nu^\mu(x) = \frac{\pi}{2\sin\mu\pi}\left\{\cos\mu\pi P_\nu^\mu(x) - \frac{\Gamma(\nu+\mu+1)}{\Gamma(\nu-\mu+1)}P_\nu^{-\mu}(x)\right\} . \qquad (9)$$

To discuss the value of $Q_\nu^\mu(x)$ for $x \to 1-0$, we may use (4) of the preceding section and obtain from (9)

$$Q_\nu^\mu(x) = \frac{1}{2}\left\{\Gamma(\mu)\cos\mu\pi\left(\frac{1+x}{1-x}\right)^{\mu/2} F\left(-\nu,\nu+1,1-\mu,\frac{1-x}{2}\right)\right.$$

$$\left. +\frac{\Gamma(\nu+\mu+1)\Gamma(-\mu)}{\Gamma(\nu-\mu+1)}\left(\frac{1+x}{1-x}\right)^{-\mu/2} F\left(-\nu,\nu+1,1+\mu,\frac{1-x}{2}\right)\right\} .$$

$$(10)$$

From (10) we see that, when $x \to 1-0$, $Q_\nu^\mu(x) \to \infty$ in general, unless μ is half a positive odd integer, then, the first term vanishes and $Q_\nu^\mu(x) \sim (1-x)^{\mu/2} \to 0$; or, $\nu - \mu$ is a negative integer, but μ is not an integer, also $\mathrm{Re}(\mu) < 0$, so the second term vanishes and $Q_\nu^\mu(x) \sim (1-x)^{-\mu/2} \to 0$.

To study the value of $Q_\nu^\mu(x)$ for $x \to -1+0$, we may apply Eq. (4) of Sec. 4.8 and Eq. (8) of Sec. 4.3 to (10), and obtain

$$Q_\nu^\mu(x) = -\frac{1}{2}\left\{\Gamma(\mu)\cos\nu\pi\left(\frac{1+x}{1-x}\right)^{-\mu/2} F\left(-\nu,\nu+1,1-\mu,\frac{1+x}{2}\right)\right.$$

$$+\frac{\Gamma(\nu+\mu+1)\Gamma(-\mu)}{\Gamma(\nu-\mu+1)}\cos(\nu+\mu)\pi\left(\frac{1+x}{1-x}\right)^{\mu/2}$$

$$\left. \times F\left(-\nu,\nu+1,1+\mu,\frac{1+x}{2}\right)\right\} . \qquad (11)$$

From (11), it is seen that, as $x \to -1+0$, $Q_\nu^\mu(x)$ also approaches ∞ in general, unless ν is half an odd integer and $\text{Re}(\mu) > 0$, or ν and μ are such that the second term is made equal to zero and $\text{Re}(\mu) < 0$.

When μ is an integer, (9) becomes an indeterminate form [cf. (8) of the last section]. Let $\mu = m(= 0, 1, 2, \dots)$. From (9), we have

$$Q_\nu^m(x) = \lim_{\mu \to m} \frac{\pi}{2 \sin \mu \pi} \left\{ \cos \mu\pi P_\nu^\mu(x) - \frac{\Gamma(\nu + \mu + 1)}{\Gamma(\nu - \mu + 1)} P_\nu^{-\mu}(x) \right\} .$$

Calculating the limiting value on the r.h.s., we find

$$Q_\nu^m(x) = \frac{1}{2} \Bigg\{ P_\nu^m(x) \left[\ln \frac{1+x}{1-x} - \psi(\nu + m + 1) - \psi(\nu - m - 1) \right]$$

$$+ (-)^m \left(\frac{1+x}{1-x} \right)^{m/2} \sum_{k=0}^{m-1} \frac{(-\nu)_k (\nu+1)_k (m-k-1)!}{k!}$$

$$\times (-)^k \left(\frac{1-x}{2} \right)^k + (-)^m 2^{-m} \frac{\Gamma(\nu + m + 1)}{\Gamma(\nu - m + 1)} (1 - x^2)^{m/2}$$

$$\times \sum_{k=0}^{\infty} \frac{(-\nu+m)_k (\nu+1+m)_k}{k!(k+m)!} \psi(k+1) \left(\frac{1-x}{2} \right)^k$$

$$+ (-)^m \frac{\Gamma(\nu + m + 1)}{\Gamma(\nu - m + 1)} \left(\frac{1-x}{1+x} \right)^{m/2}$$

$$\times \sum_{k=0}^{\infty} \frac{(-\nu)_k (\nu+1)_k}{k!(k+m)!} \psi(k+m+1) \left(\frac{1-x}{2} \right)^k \Bigg\} ;$$

$$(m = 0, 1, 2, \dots) \tag{12}$$

the finite sum disappears when $m = 0$. From this formula, it is seen that, when $x \to 1 - 0$, $Q_\nu^m(x)$ always tends to ∞, since, if $m = 0$, then as $P_\nu(1) = 1$[Eq. (11) of Sec. 5.16], $Q_\nu(x) \sim \ln(1-x)$; if $m > 0$, then $Q_\nu^m(x) \sim (1-x)^{-m/2}$. For $x \to -1 + 0$, similar conclusion is obtained.

When $\mu = -m(m = 1, 2, \dots)$, by Eq. (9) of Sec. 5.17 and the definition of $Q_\nu^\mu(x)$, we have

$$Q_\nu^{-m}(x) = (-)^m \frac{\Gamma(\nu + m + 1)}{\Gamma(\nu - m + 1)} Q_\nu^m(x) . \tag{13}$$

Hence, when $x \to 1 - 0$ or $-1 + 0$, $Q_\nu^{-m}(x)$ also tends to ∞.

Summarizing the above discussions on $P_\nu^\mu(x)$ and $Q_\nu^\mu(x)$, we see that, for μ an integer, the associate Legendre equation can have a solution which is bounded in $-1 \leq x \leq +1$, only when ν is also an integer; this solution is unique and equal to $P_n^m(x)$ (P_n^{-m} and P_{-n-1}^m are solutions linearly dependent on P_n^m).

5.20. Other Integral Expressions for $P_\nu(z)$ and $P_\nu^m(z)$

In Sec. 5.16, we have obtained the integral representation [Eq. (18) of Sec. 5.16]

$$P_\nu(z) = \frac{1}{2\pi i} \int_A^{(z+,1+)} \frac{(t^2 - 1)^\nu}{2^\nu (t-z)^{\nu+1}} dt .$$

(1)

The contour of integration is shown in Fig. 16, $t = -1$ lying outside the contour. At the starting point A, which lies on the real axis to the right of $t = 1$, $\arg(t-1) = \arg(t+1) = 0$, and $|\arg(t-z)| < \pi$.

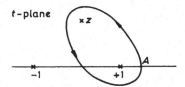

Fig. 16.

The $P_\nu(z)$ expressed by (1) is a single-valued analytic function in the z-plane cut along the real axis from $-\infty$ to -1. For, similar to the discussion on the integral representation of the hypergeometric function [Eq. (6) of Sec. 4.5], we find that $z = -1$ and ∞ are two branch points of $P_\nu(z)$. But if $\nu = n$ (positive integer), then the path of integration in (1) can be any contour which encircles the point $t = z$ once in the positive sense [see Eq. (7) of Sec. 5.4], because $t = \pm 1$ are no longer singularities of the integrand now so there is no need to make a cut in the z-plane.

Laplace's Integral Representations

Assume $\text{Re}(z) > 0$. The path of integration in (1) can be deformed into a circle with centre at $t = z$, radius equal to $|z^2 - 1|^{1/2}$, since under the condition $\text{Re}(z) > 0$, the distance from the point $t = z$ to the point $t = 1$ is shorter than that to the point $t = -1$, i.e., $|z - 1| < |z + 1|$, and hence $|z - 1| < |z^2 - 1|^{1/2} < |z + 1|$, which shows that $t = 1$ lies inside the circle while $t = -1$ lies outside. On the circle, we may write

$$t = z + \sqrt{z^2 - 1}\, e^{i\varphi} \quad (-\pi \leq \varphi \leq \pi) .$$

Substituting it into (1) gives

$$
\begin{aligned}
P_\nu(z) &= \frac{1}{2\pi} \int_{-\pi}^{\pi} \frac{(z-1+\sqrt{z^2-1}\,e^{i\varphi})^\nu (z+1+\sqrt{z^2-1}\,e^{i\varphi})^\nu}{2^\nu (z^2-1)^{\nu/2} e^{i\nu\varphi}} d\varphi \\
&= \frac{1}{2\pi} \int_{-\pi}^{\pi} (z+\sqrt{z^2-1}\cos\varphi)^\nu d\varphi \\
&= \frac{1}{\pi} \int_{0}^{\pi} (z+\sqrt{z^2-1}\cos\varphi)^\nu d\varphi \quad (|\arg z| < \pi/2) ,
\end{aligned} \tag{2}
$$

where $\sqrt{z^2-1}$ may take positive or negative value, as may be seen by changing the integration variable φ into $\pi-\varphi$. The phase of the integrand is so stipulated that $\arg(z+\sqrt{z^2-1}\cos\varphi) = \arg z$ when $\varphi = \pi/2$, because, according to the specification made in (1), $P_\nu(1) = 1$ [Eq. (11) of Sec. 5.16] when $z = 1$. (2) is called *Laplace's first integral representation*, which is the generalization of Eq. (8) of Sec. 5.4.

By Eq. (9) of Sec. 5.16, $P_\nu(z) = P_{-\nu-1}(z)$. Hence,

$$
P_\nu(z) = \frac{1}{\pi} \int_{0}^{\pi} \frac{d\varphi}{(z+\sqrt{z^2-1}\cos\varphi)^{\nu+1}} , \quad (|\arg z| < \pi/2) , \tag{3}
$$

where the phase of the integrand is specified as above. (3) is called *Laplace's second integral representation*.

When ν is an integer, the restriction $|\arg z| < \pi/2$ may be removed.

The Mehler-Dirichlet Integral Representation

In (2), putting $z+\sqrt{z^2-1}\cos\varphi = h$ (or, putting in (1), $t^2-1 = 2h(t-z)$), it results

$$
P_\nu(z) = \frac{1}{\pi i} \int_{z-\sqrt{z^2-1}}^{z+\sqrt{z^2-1}} h^\nu (h^2 - 2zh + 1)^{-1/2} dh, \quad |\arg z| < \pi/2 \tag{4}
$$

with the stipulation: $(h^2 - 2zh + 1)^{1/2} = -i\sqrt{z^2-1}\sin\varphi$; on the path of integration, $\arg(h^2 - 2zh + 1) = -\pi/2 + \arg\sqrt{z^2-1} = $ constant. The path of integration is a straight line leading from the lower limit of integration to the upper one [Fig. 17], because $\arg(dh) = \arg(-\sqrt{z^2-1}\sin\varphi d\varphi)$ is constant when φ varies from 0 to π. Further, since $h = z$ when $\varphi = \pi/2$ and $|\arg z| < \pi/2$, we have $\arg h = 0$ at the intersecting point A of the path of integration and the real axis.

If $z = \cos\theta, 0 < \theta < \pi/2$, and it is stipulated that $\sqrt{z^2-1} = i\sin\theta$, then (4) becomes

$$
P_\nu(\cos\theta) = \frac{1}{\pi i} \int_{e^{-i\theta}}^{e^{i\theta}} h^\nu (h^2 - 2\cos\theta \cdot h + 1)^{-1/2} dh ; \tag{5}
$$

the path of integration is the straight line leading from $h = e^{-i\theta}$ to $e^{i\theta}$ as shown in Fig. 18. The multi-valued integrand is specified as follows: At the point $A(h = \cos\theta)$ in the figure, $\arg h = 0$, whereas $(h^2 - 2\cos\theta \cdot h + 1)^{1/2} = \sin\theta$.

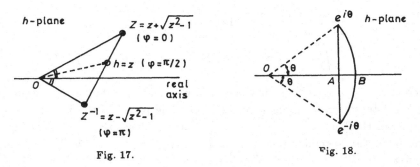

Fig. 17.   Fig. 18.

The path of integration in (5) can be deformed into a circular arc with centre at O, passing through the two points $e^{\pm i\theta}$ [see Fig. 18], since there is no singularity of the integrand[f] lying between the original path of integration and this circular arc. Then, by putting $h = e^{i\varphi}$, (5) reduces to

$$
\begin{aligned}
P_\nu(\cos\theta) &= \frac{1}{\pi} \int_{-\theta}^{\theta} \frac{e^{i(\nu+1)\varphi}}{(e^{2i\varphi} - 2\cos\theta e^{i\varphi} + 1)^{1/2}} d\varphi \\
&= \frac{1}{\pi} \int_{-\theta}^{\theta} \frac{e^{i(\nu+\frac{1}{2})\varphi}}{\sqrt{2(\cos\varphi - \cos\theta)}} d\varphi \\
&= \frac{2}{\pi} \int_{0}^{\theta} \frac{\cos\left(\nu+\frac{1}{2}\right)\varphi}{\sqrt{2(\cos\varphi - \cos\theta)}} d\varphi ,
\end{aligned}
\tag{6}
$$

where the square root takes the positive value, since it is to be positive when h is at the point A according to the above specification, i.e., $\arg(h^2 - 2h\cos\theta + 1) = 0$, whereas $\arg(h^2 - 2h\cos\theta + 1) = \arg(h - e^{i\theta})(h - e^{-i\theta})$ does not change when h varies from A to B. (6) is called the *Mehler-Dirichlet integral representation* of $P_\nu(\cos\theta)$.

(6) is derived under the condition $0 < \theta < \pi/2$. But the integral on its r.h.s., as a function of the complex variable $e^{i\theta}$, can be analytically continuated to $\pi/2 \le \theta < \pi$; the path of integration is the circular arc shown in Fig. 19. Thus, (6) is valid in $0 < \theta < \pi$.

---

[f]The integrand is not analytic at the end points of the path. But we may do as usual by detouring each of them along a small circular arc; the values of the integrals on the circles approach to zero when the radii tend to zero.

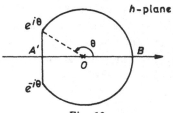

Fig. 19.

By the same reasoning, if we alter the path of integration in (5) into the circular arc shown in Fig. 19, or into other equivalent paths, (5) still holds in $0 < \theta < \pi$. However, such integral is in general not equal to the integral from $e^{-i\theta}$ to $e^{i\theta}$ along a straight line passing through the point A' (on the real axis; see Fig. 19) when $\pi/2 \leq \theta < \pi$. Because there is now a branch point $h = 0$ of the integrand in between the two paths, unless ν is an integer, then $h = 0$ is no longer a singularity and the two paths are equivalent.

We can also express $P_\nu(z)$ by the following contour integral:

$$P_\nu(z) = \frac{1}{2\pi i} \int_C h^\nu (h^2 - 2zh + 1)^{-1/2} dh ; \qquad (7)$$

C is a contour which encircles the two points $h = z \pm \sqrt{z^2 - 1}$ once in the positive sense (see Fig. 20), $h = 0$ being situated outside of C. The multivalued factors in the integrand are defined as follows: At the point A where the path of integration intersects the real axis, $\arg h = 0$, and when A tends to infinity along the positive real axis, $\arg(h^2 - 2zh + 1) = \arg\{(h - Z^{-1})(h - Z)\} = 0$.

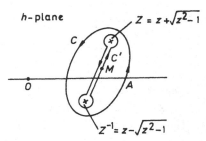

Fig. 20.

To prove (7), we deform C into C' as shown in Fig. 20, which starts from the point M on the right bank of the straight line leading from $Z^{-1} = z - \sqrt{z^2 - 1}$ to $Z = z + \sqrt{z^2 - 1}$, running at the vicinity of Z, encircling Z

once in the positive sense, then running at the vicinity of Z^{-1} along the left bank of the straight line, encircling Z^{-1} once in the positive sense and goes back to M along the right bank. It is readily shown that the values of the integrals on the circles around the point Z and Z^{-1} approach zero when the radii tend to zero. Hence, the integral on the r.h.s. of (7) is equal to

$$\frac{1}{2\pi i}(1 - e^{-\pi i}) \int_{Z^{-1}}^{Z} h^{\nu}(h^2 - 2zh + 1)^{-1/2}dh \ .$$

The path of integration is the straight line leading from Z^{-1} to Z; at the intersection of the path and the real axis, $\arg h = 0$, same as that stipulated in (4). Besides, at the starting point $M, \arg\{(h - Z^{-1})(h - Z)\} = -\pi + 2\arg\sqrt{z^2 - 1}$ (since the angle of inclination of the straight line from Z^{-1} to Z is $\arg(2\sqrt{z^2 - 1})$), and hence $\arg(h^2 - 2zh + 1)^{1/2} = -\pi/2 + \arg\sqrt{z^2 - 1}$, which is also in conformity with the specification in (4). Thus, (7) is proved.

Although (7) is proved under the condition $|\arg z| < \pi/2$, it can be analytically continuated to the left half plane. Then, the integration contour is that shown in Fig. 21; the phases are defined as before.

h-plane

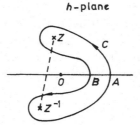

Fig. 21.

However, the path of integration can no longer be deformed into the dashed straight line from Z^{-1} to Z shown in the figure. Note that the intersecting points of the path and the real axis all lie on the right of $h = 0$.

From (1), differentiating with respect to z m times and multiplying with $(z^2 - 1)^{m/2}$, we obtain

$$P_\nu^m(z) = \frac{\Gamma(\nu + m + 1)}{\Gamma(\nu + 1)} \frac{(z^2 - 1)^{m/2}}{2\pi i} \int^{(z+,1+)} \frac{(t^2 - 1)^\nu}{2^\nu (t - z)^{\nu+m+1}} dt \ . \tag{8}$$

Putting $t = z + \sqrt{z^2 - 1}\, e^{i\varphi}$, we have [cf. the derivation of (2)]

$$P_\nu^m(z) = \frac{\Gamma(\nu + m + 1)}{\Gamma(\nu + 1)} \frac{1}{2\pi} \int_{-\pi}^{\pi} e^{-im\varphi} (z + \sqrt{z^2 - 1}\cos\varphi)^\nu \, d\varphi$$

$$= \frac{\Gamma(\nu + m + 1)}{\Gamma(\nu + 1)} \frac{1}{\pi} \int_0^{\pi} \cos m\varphi \cdot (z + \sqrt{z^2 - 1}\cos\varphi)^\nu \, d\varphi \, . \tag{9}$$

Furthermore, by Eq. (9) of Sec. 5.16, $P_\nu^m = P_{-\nu-1}^m$, we find

$$P_\nu^m(z) = \frac{\Gamma(-\nu + m)}{\Gamma(-\nu)} \frac{1}{\pi} \int_0^{\pi} \frac{\cos m\varphi \cdot d\varphi}{(z + \sqrt{z^2 - 1}\cos\varphi)^{\nu+1}}$$

$$= (-)^m \frac{\Gamma(\nu + 1)}{\Gamma(\nu - m + 1)} \frac{1}{\pi} \int_0^{\pi} \frac{\cos m\varphi \cdot d\varphi}{(z + \sqrt{z^2 - 1}\cos\varphi)^{\nu+1}}$$

$$(|\arg z| < \pi/2) \, . \tag{10}$$

Also, from (10), by Eq. (16) of Sec. 5.16, there results

$$P_\nu^{-m}(z) = (-)^m \frac{\Gamma(\nu + 1)}{\Gamma(\nu + m + 1)} \frac{1}{\pi} \int_0^{\pi} \frac{\cos m\varphi \cdot d\varphi}{(z + \sqrt{z^2 - 1}\cos\varphi)^{\nu+1}}$$

$$(|\arg z| < \pi/2) \, . \tag{11}$$

5.21. Addition Formulae

With (7) in the last section, we may obtain an addition formula which is more general than that given in Sec. 5.14, namely,

$$P_\nu(\varsigma) = P_\nu(z) P_\nu(z') + 2 \sum_{m=1}^{\infty} (-)^m P_\nu^m(z) P_\nu^{-m}(z') \cos\varphi \, , \tag{1}$$

where ν can be any complex number; z, z' are complex numbers not on the cut from $-\infty$ to 1; $\mathrm{Re}(z) > 0$, $\mathrm{Re}(z') > 0$; $\varsigma = zz' - (z^2 - 1)^{1/2}(z'^2 - 1)^{1/2}\cos\varphi$, φ being an arbitrary real number; the series in (1) is uniformly convergent. When $\nu = n$ (positive integer), the series breaks up into a finite sum over $m = 1$ to $m = n$; then, the limitations imposed on z and z' can be retracted.

Before giving a proof of (1), we derive first a formula for calculating an integral. Let A, B, C be any complex numbers, then

$$\int_{-\pi}^{\pi} \frac{d\omega}{A + B\cos\omega + C\sin\omega} = \frac{2\pi}{(A^2 - B^2 - C^2)^{1/2}} ; \qquad (2)$$

we take the square root so as to make

$$|A - (A^2 - B^2 - C^2)^{1/2}| < |B - iC| < |A + (A^2 - B^2 - C^2)^{1/2}| . \qquad (3)$$

(2) may be proved as follows. Let $z = e^{i\omega}$, then the integral I on the l.h.s. of (2) becomes a contour integral along the unit circle $|z| = 1$ in the z-plane:

$$I = \oint \frac{dz/iz}{A + B\frac{z+z^{-1}}{2} + C\frac{z-z^{-1}}{2i}} = \frac{2}{i(B - iC)} \oint \frac{dz}{z^2 + \alpha z + \beta} ,$$

$$\alpha = \frac{2A}{B - iC} , \qquad \beta = \frac{B + iC}{B - iC} .$$

The integrand of the last integral has two first-order poles:

$$z_1 = \frac{-A + (A^2 - B^2 - C^2)^{1/2}}{B - iC} , \qquad z_2 = \frac{-A - (A^2 - B^2 - C^2)^{1/2}}{B - iC} .$$

If both of these two poles lie within, or both lie outside, the unit circle I will be zero. If one of them is on the circle, then I does not exist. When (3) is fulfilled, z_1 must lie within the circle and z_2 outside of it. Then, by the theorem of residues,

$$I = \frac{2}{i(B - iC)} \frac{2\pi i}{z_1 - z_2} = \frac{2\pi}{(A^2 - B^2 - C^2)^{1/2}} .$$

Now, we come to prove (1). From Eq. (7) of Sec. 5.20, we have

$$P_\nu(\varsigma) = \frac{1}{2\pi i} \int_C h^\nu (h^2 - 2\varsigma h + 1)^{-1/2} dh ,$$

C being the contour in Fig. 20 and $h = 0$ lying outside of C. By (2) [concerning the fulfilment of the condition (3), see Hobson (1931), Sec. 220], putting $A = hz' - z$, $B = h(z'^2 - 1)^{1/2} - (z^2 - 1)^{1/2}\cos\varphi$, $C = -(z^2 - 1)^{1/2}\sin\varphi$, we obtain

$$P_\nu(\varsigma)$$

$$= \frac{1}{2\pi i} \int_C h^\nu dh \cdot \frac{1}{2\pi} \int_{-\pi}^{\pi} \frac{d\omega}{h[z' + \sqrt{z'^2 - 1}\cos\omega] - [z + \sqrt{z^2 - 1}\cos(\omega - \varphi)]}$$

$$= \frac{1}{2\pi} \int_{-\pi}^{\pi} \frac{d\omega}{z' + \sqrt{z'^2 - 1}\cos\omega} \frac{1}{2\pi i} \int_C \frac{h^\nu dh}{h - \frac{z + \sqrt{z^2 - 1}\cos(\omega - \varphi)}{z' + \sqrt{z'^2 - 1}\cos\omega}} .$$

$h_1 = \{z + \sqrt{z^2 - 1}\cos(\omega - \varphi)\}/\{z' + \sqrt{z'^2 - 1}\cos\omega\}$ is a bounded function of $\omega(-\pi \leq \omega \leq \pi)$, since the denominator in it is zero only when z' is purely imaginary, whereas it is assumed that $\mathrm{Re}(z') > 0$. It can be shown that when ω and φ assume the values in the interval $[-\pi, \pi]$, the corresponding points h_1 in the h-plane can all be embraced within the contour C [Hobson (1931), Sec. 220]. Hence, by Cauchy formula, we find from the above equation

$$P_\nu(\varsigma) = \frac{1}{2\pi} \int_{-\pi}^{\pi} \frac{[z + \sqrt{z^2 - 1}\cos(\omega - \varphi)]^\nu}{[z' + \sqrt{z'^2 - 1}\cos\omega]^{\nu+1}} d\omega \ . \tag{4}$$

Now, $P_\nu(\varsigma)$ is a function of $\cos\varphi$, and can thus be expanded in a Fourier cosine series:

$$P_\nu(\varsigma) = \frac{A_0}{2} + \sum_{m=1}^{\infty} A_m \cos m\varphi \ , \tag{5}$$

where

$$A_m = \frac{1}{\pi} \int_{-\pi}^{\pi} P_\nu(\varsigma) \cos m\varphi d\varphi \quad (m \geq 0)$$

$$= \frac{1}{2\pi^2} \int_{-\pi}^{\pi} d\varphi \int_{-\pi}^{\pi} \frac{[z + \sqrt{z^2 - 1}\cos(\omega - \varphi)]^\nu \cos m\varphi}{[z' + \sqrt{z'^2 - 1}\cos\omega]^{\nu+1}} d\omega$$

$$= \frac{1}{2\pi^2} \int_{-\pi}^{\pi} \frac{d\omega}{[z' + \sqrt{z'^2 - 1}\cos\omega]^{\nu+1}}$$

$$\times \int_{-\pi}^{\pi} [z + \sqrt{z^2 - 1}\cos\varphi]^\nu \cos m(\omega - \varphi) d\varphi$$

$$= \frac{1}{2\pi^2} \int_{-\pi}^{\pi} \frac{d\omega}{[z' + \sqrt{z'^2 - 1}\cos\omega]^{\nu+1}} \int_{-\pi}^{\pi} [z + \sqrt{z^2 - 1}\cos\varphi]^\nu \cos m\varphi d\varphi$$

$$= 2(-)^m P_\nu^{-m}(z') P_\nu^m(z) \ ;$$

in the last step, (9) and (11) of the last section have been applied. This proves (1).

When $\nu = n$ (positive integer), if $m > n$, then $P_n^m(z) = 0$, and the series on the r.h.s. of (1) becomes a finite sum, so (1) is an algebraic identity; the restrictions $\mathrm{Re}(z) > 0$ and $\mathrm{Re}(z') > 0$ may be retracted.

When z and z' are real numbers, we have the following addition formulae:

$$Q_\nu[xx' - (x^2 - 1)^{1/2}(x'^2 - 1)^{1/2} \cos \varphi]$$
$$= Q_\nu(x)P_\nu(x') + 2\sum_{m=1}^{\infty} (-)^m Q_\nu^m(x)P_\nu^{-m}(x') \cos m\varphi$$
$$(x > x' > 1, \quad \nu \neq -1, -2, \ldots), \tag{6}$$

$$P_\nu(\cos \theta \cos \theta' + \sin \theta \sin \theta' \cos \varphi)$$
$$= P_\nu(\cos \theta)P_\nu(\cos \theta')$$
$$+ 2\sum_{m=1}^{\infty} (-)^m P_\nu^{-m}(\cos \theta)P_\nu^m(\cos \theta') \cos m\varphi$$
$$= P_\nu(\cos \theta)P_\nu(\cos \theta')$$
$$+ 2\sum_{m=1}^{\infty} \frac{\Gamma(\nu - m + 1)}{\Gamma(\nu + m + 1)} P_\nu^m(\cos \theta)P_\nu^m(\cos \theta') \cos m\varphi$$
$$(0 \leq \theta' < \pi, 0 \leq \theta < \pi, \theta + \theta' < \pi, \varphi \text{ real}), \tag{7}$$

$$Q_\nu(\cos \theta \cos \theta' + \sin \theta \sin \theta' \cos \varphi)$$
$$= P_\nu(\cos \theta')Q_\nu(\cos \theta) + 2\sum_{m=1}^{\infty} (-)^m P_\nu^{-m}(\cos \theta')Q_\nu^m(\cos \theta) \cos m\varphi$$
$$= P_\nu(\cos \theta')Q_\nu(\cos \theta) + 2\sum_{m=1}^{\infty} \frac{\Gamma(\nu - m + 1)}{\Gamma(\nu + m + 1)}$$
$$\times P_\nu^m(\cos \theta')Q_\nu^m(\cos \theta) \cos m\varphi$$
$$(0 < \theta' < \pi/2, 0 < \theta < \pi, 0 < \theta + \theta' < \pi, \varphi \text{ real}). \tag{8}$$

Regarding their proofs, cf. Hobson (1931), Secs. 222–227.

5.22. Asymptotic Expansions of $P_\nu^\mu(\cos \theta)$ and $Q_\nu^\mu(\cos \theta)$ when $\nu \to \infty$

Since $P_\nu^\mu(z)$ and $Q_\nu^\mu(z)$ can be expressed in terms of hypergeometric functions, we may use the results of Sec. 4.14 to find the asymptotic expansions of these functions. In Sec. 4.14, we have discussed the asymptotic expansion of $F(\alpha, \beta, \gamma, z)$ when the parameters α, β, γ are large. The simplest is the case when α, β and z are fixed and $\gamma \to \infty$, because the hypergeometric series (may be divergent, e.g. when $|z| > 1$) is then the asymptotic expansion in itself for γ [Eq. (1) of Sec. 4.14]. When we make use of this result to find the asymptotic expansion of the Legendre functions for ν (or μ) $\to \infty$, the clue lies on that of making ν (or μ) to occur only in the third parameter γ of the hypergeometric function. Below, we shall only discuss the case when the argument $\cos\theta$ is real $(0 < \theta < \pi)$ and $\mathrm{Re}(\nu) \to \infty$.

From Eq. (4) of Sec. 5.17,

$$Q_\nu^\mu(z) = \frac{e^{\mu\pi i}}{2^{\nu+1}} \frac{\Gamma\left(\frac{1}{2}\right)\Gamma(\nu+\mu+1)}{\Gamma\left(\nu+\frac{3}{2}\right)} (z^2-1)^{\mu/2} z^{-\nu-\mu-1}$$

$$\times F\left(\frac{\nu+\mu+1}{2}, \frac{\nu+\mu+2}{2}, \nu+\frac{3}{2}, z^{-2}\right)$$

$$(|\arg z| < \pi, |\arg(z\pm1)| < \pi) .$$

Applying the transformations Eq. (15) of Sec. 4.12 and Eq. (8) of Sec. 4.3 to the hypergeometric function in it:

$$F\left(\frac{\nu+\mu+1}{2}, \frac{\nu+\mu+2}{2}, \nu+\frac{3}{2}, z^{-2}\right)$$

$$= (1-z^2)^{-\frac{\nu+\mu+1}{2}} F\left(\nu+\mu+1, \nu-\mu+1, \nu+\frac{3}{2}, \frac{\sqrt{z^2-1}-z}{2\sqrt{z^2-1}}\right)$$

$$= (1-z^{-2})^{-\frac{\nu+\mu+1}{2}} \left(1 - \frac{\sqrt{z^2-1}-z}{2\sqrt{z^2-1}}\right)^{-\nu-\frac{1}{2}}$$

$$\times F\left(\frac{1}{2}-\mu, \frac{1}{2}+\mu, \nu+\frac{3}{2}, \frac{\sqrt{z^2-1}-z}{2\sqrt{z^2-1}}\right) ,$$

we obtain

$$Q_\nu^\mu(z) = \frac{e^{\mu\pi i}}{\sqrt{2}} \frac{\Gamma\left(\frac{1}{2}\right)\Gamma(\nu+\mu+1)}{\Gamma\left(\nu+\frac{3}{2}\right)} (z^2-1)^{-1/4}(z+\sqrt{z^2+1})^{-\nu-\frac{1}{2}}$$

$$\times F\left(\frac{1}{2}-\mu, \frac{1}{2}+\mu, \nu+\frac{3}{2}, \frac{\sqrt{z^2-1}-z}{2\sqrt{z^2-1}}\right) , \tag{1}$$

where $|\arg(z\pm1)| < \pi, |\arg(z+\sqrt{z^2-1})| < \pi$. In this expression, ν occurs only in the third parameter of the hypergeometric function, hence, it can be used to find the asymptotic expansion of $Q_\nu^\mu(z)$ for $\mathrm{Re}(\nu) \to \infty$. In the following, we shall deal only with the case when the argument is $\cos\theta$.

From (1) we have

$$Q_\nu^\mu(\cos\theta \pm i0) = \left(\frac{\pi}{2\sin\theta}\right)^{1/2} \frac{e^{\mu\pi i}\Gamma(\nu+\mu+1)}{\Gamma\left(\nu+\frac{3}{2}\right)} e^{\mp i[(\nu+\frac{1}{2})\theta+\frac{\pi}{4}]}$$

$$\times F\left(\frac{1}{2}-\mu, \frac{1}{2}+\mu, \nu+\frac{3}{2}, \frac{\pm ie^{\mp i\theta}}{2\sin\theta}\right) . \tag{2}$$

Substituting in Eq. (7) of Sec. 5.19 and expressing the result in terms of the hypergeometric series, we have

$$P_\nu^\mu(\cos\theta) = \left(\frac{2}{\pi\sin\theta}\right)^{1/2} \frac{\Gamma(\nu+\mu+1)}{\Gamma\left(\nu+\frac{3}{2}\right)} \sum_{k=0}^\infty \frac{\left(\frac{1}{2}-\mu\right)_k \left(\frac{1}{2}+\mu\right)_k}{k!\left(\nu+\frac{3}{2}\right)_k (2\sin\theta)^k}$$

$$\times \sin\left[\left(\nu+k+\frac{1}{2}\right)\theta + \frac{2\mu-2k+1}{4}\pi\right] . \tag{3}$$

When $\mathrm{Re}(\nu) \to \infty$, the series on the r.h.s. is the asymptotic expansion of $P_\nu^\mu(\cos\theta)$ for fixed μ and $\varepsilon \le \theta \le \pi - \varepsilon(\varepsilon > 0)$.

In case $\mu = 0, \nu = n$ (positive integer), we obtain from (3)

$$P_n(\cos\theta) = \left(\frac{2}{\pi\sin\theta}\right)^{1/2} \frac{\Gamma(n+1)}{\Gamma\left(n+\frac{3}{2}\right)}$$

$$\times \sum_{k=0}^\infty \frac{\left[\left(\frac{1}{2}\right)_k\right]^2}{k!\left(n+\frac{3}{2}\right)_k (2\sin\theta)^k} \cos\left[\left(n+k+\frac{1}{2}\right)\theta - \frac{2k+1}{4}\pi\right] . \tag{4}$$

By Eq. (5) of Sec. 3.21, the asymptotic expansion of the Γ-function, we have

$$\ln\frac{\Gamma(n+1)}{\Gamma\left(n+\frac{3}{2}\right)} = -\frac{1}{2}\ln n + \frac{\varphi_2(1)-\varphi_2\left(\frac{3}{2}\right)}{2n} + O(n^{-2})$$

$$= -\frac{1}{2}\ln n - \frac{3}{8n} + O(n^{-2}) ,$$

where $\varphi_2(x)$ is calculated with Eq. (10) of Sec. 1.1. Thus,

$$\frac{\Gamma(n+1)}{\Gamma\left(n+\frac{3}{2}\right)} = n^{-1/2}\left[1 - \frac{3}{8n} + O(n^{-2})\right] . \tag{5}$$

Substituting in (4) gives

$$P_n(\cos\theta) = \left(\frac{2}{n\pi\sin\theta}\right)^{1/2}\left(1-\frac{3}{8n}\right)\left\{\cos\left[\left(n+\frac{1}{2}\right)\theta-\frac{\pi}{4}\right]\right.$$

$$+\frac{1}{2(2n+3)\cdot 2\sin\theta}\cos\left[\left(n+\frac{3}{2}\right)\theta-\frac{3\pi}{4}\right]\right\}+O(n^{-5/2})$$

$$= \left(\frac{2}{n\pi\sin\theta}\right)^{1/2}\left\{\left(1-\frac{1}{4n}\right)\cos\left[\left(n+\frac{1}{2}\right)\theta-\frac{\pi}{4}\right]\right.$$

$$+\frac{1}{8n}\cot\theta\cdot\sin\left[\left(n+\frac{1}{2}\right)\theta-\frac{\pi}{4}\right]\right\}+O(n^{-5/2})$$

$$(\varepsilon\le\theta\le\pi-\varepsilon)\,. \tag{6}$$

Also, by substituting (2) into Eq. (8) of Sec. 5.19, there results

$$Q_\nu^\mu(\cos\theta) = \left(\frac{\pi}{2\sin\theta}\right)^{1/2}\frac{\Gamma(\nu+\mu+1)}{\Gamma\left(\nu+\frac{3}{2}\right)}$$

$$\times\sum_{k=0}^{\infty}\frac{\left(\frac{1}{2}-\mu\right)_k\left(\frac{1}{2}+\mu\right)_k}{k!\left(\nu+\frac{3}{2}\right)_k(2\sin\theta)^k}\cos\left[\left(\nu+k+\frac{1}{2}\right)\theta+\frac{2\mu-2k+1}{4}\pi\right]$$

$$(\varepsilon\le\theta\le\pi-\varepsilon)\,, \tag{7}$$

of which a special case is

$$Q_n(\cos\theta) = \left(\frac{\pi}{2n\sin\theta}\right)^{1/2}\left\{\left(1-\frac{1}{4n}\right)\cos\left[\left(n+\frac{1}{2}\right)\theta+\frac{\pi}{4}\right]\right.$$

$$+\frac{1}{8n}\cot\theta\cdot\sin\left[\left(n+\frac{1}{2}\right)\theta+\frac{\pi}{4}\right]\right\}+O(n^{-5/2})$$

$$(\varepsilon\le\theta\le\pi-\varepsilon)\,. \tag{8}$$

Asymptotic expansions for other cases may be obtained by properly expressing the Legendre functions in terms of hypergeometric functions. Cf. for example, Erdélyi (1953), Vol. I, Sec. 3.9.1, p. 162.

5.23. Ultra-Spherical Polynomials $C_n^\lambda(x)$

These polynomials, also named *Gegenbauer polynomials*, can be obtained from the following expansion:

$$(1 - 2xt + t^2)^{-\lambda} = \sum_{n=0}^{\infty} C_n^\lambda(x)t^n \; ; \tag{1}$$

the function on the l.h.s. is the *generating function* of $C_n^\lambda(x)$, which is in general multi-valued, unless λ is an integer. We shall stipulate that when $t = 0$, the l.h.s. function of (1) takes the value 1, so

$$C_0^\lambda(x) \equiv 1 \; . \tag{2}$$

Legendre polynomial is a special case of the ultra-spherical polynomials: $P_n(x) = C_n^{1/2}(x)$ [cf. Eq. (2) of Sec. 5.3]. For the general relations between the Legendre functions and the ultra-spherical polynomials, cf. Exs. 45, 47 at the end of this chapter.

Recurrence Relations

Differentiating (1) with respect to t gives

$$2\lambda(x - t)(1 - 2xt + t^2)^{-\lambda-1} = \sum_{n=0}^{\infty} C_n^\lambda(x)nt^{n-1} \; .$$

Again applying (1), we have

$$2\lambda(x - t) \sum_{n=0}^{\infty} C_n^\lambda(x)t^n = (1 - 2xt + t^2) \sum_{n=0}^{\infty} nC_n^\lambda(x)t^{n-1} \; .$$

Comparing the coefficients of t^n on the two sides, we find

$$(n + 1)C_{n+1}^\lambda - 2(\lambda + n)xC_n^\lambda + (2\lambda + n - 1)C_{n-1}^\lambda = 0 \; . \tag{3}$$

By differentiating both sides of (1) with respect to x, we find

$$2\lambda t(1 - 2xt + t^2)^{-\lambda-1} = \sum_{n=0}^{\infty} \frac{d}{dx}C_n^\lambda(x) \cdot t^n \; .$$

Applying (1) again, there results

$$2\lambda t \sum_{n=0}^{\infty} C_n^\lambda(x)t^n = (1 - 2xt + t^2) \sum_{n=0}^{\infty} \frac{d}{dx}C_n^\lambda(x) \cdot t^n \; .$$

Comparing the coefficients of t^{n+1} on both sides leads to

$$\frac{dC_{n+1}^\lambda}{dx} - 2x\frac{dC_n^\lambda}{dx} - 2\lambda C_n^\lambda + \frac{dC_{n-1}^\lambda}{dx} = 0 . \tag{4}$$

From (3) and (4), eliminating C_{n+1}^λ, we find

$$x\frac{dC_n^\lambda}{dx} - \frac{dC_{n-1}^\lambda}{dx} = nC_n^\lambda ; \tag{5}$$

eliminating C_{n-1}^λ gives

$$\frac{dC_{n+1}^\lambda}{dx} - x\frac{dC_n^\lambda}{dx} = (2\lambda + n)C_n^\lambda ,$$

or

$$\frac{dC_n^\lambda}{dx} - x\frac{dC_{n-1}^\lambda}{dx} = (2\lambda + n - 1)C_{n-1}^\lambda . \tag{6}$$

Eliminating dC_{n-1}^λ/dx from (5) and (6) gives

$$(1 - x^2)\frac{dC_n^\lambda}{dx} + nxC_n^\lambda - (2\lambda + n - 1)C_{n-1}^\lambda = 0 . \tag{7}$$

(3) and (7) are two fundamental recurrence relations. Some other such relations are given in Ex. 46 at the end of this chapter.

Differential Equation. Expression in Terms of Hypergeometric Function

Differentiating (7), then applying (5) to eliminate dC_{n-1}^λ/dx, we find the differential equation satisfied by $C_n^\lambda(x)$:

$$(1 - x^2)\frac{d^2 C_n^\lambda}{dx^2} - (2\lambda + 1)x\frac{dC_n^\lambda}{dx} + n(2\lambda + n)C_n^\lambda = 0 , \tag{8}$$

of which the Legendre equation [Eq. (1) of Sec. 5.2] is a special case, $\lambda = \frac{1}{2}$.

Same as the Legendre equation, the solutions of Eq. (8), ± 1, and ∞ being its regular singularities, can be expressed in terms of hypergeometric functions. Comparing with Eq. (7) of Sec. 2.9, we obtain the solutions of (8) expressed in P-symbol to be

$$P\left\{\begin{array}{ccc} -1 & 1 & \infty \\ 0 & 0 & -n , \\ \frac{1}{2} - \lambda & \frac{1}{2} - \lambda & 2\lambda + n \end{array} x \right\}$$

$$= P\left\{\begin{array}{ccc} 0 & 1 & \infty \\ 0 & 0 & -n , \\ \frac{1}{2} - \lambda & \frac{1}{2} - \lambda & 2\lambda + n \end{array} \frac{1-x}{2} \right\} .$$

The polynomial solution is $F\left(-n, 2\lambda + n, \frac{1}{2} + \lambda, (1-x)/2\right)$. Hence,

$$C_n^\lambda(x) = AF\left(-n, 2\lambda + n, \frac{1}{2} + \lambda, \frac{1-x}{2}\right) .$$

To find the constant A, putting $x = 1$, we obtain $A = C_n^\lambda(1)$. Letting $x = 1$ in (1), we have

$$(1-t)^{-2\lambda} = \sum_{n=0}^\infty \binom{-2\lambda}{n}(-t)^n = \sum C_n^\lambda(1)t^n ;$$

therefore,

$$C_n^\lambda(1) = \binom{-2\lambda}{n}(-)^n = \frac{(2\lambda)_n}{n!} , \tag{9}$$

and it follows that

$$C_n^\lambda(x) = \frac{(2\lambda)_n}{n!}F\left(-n, 2\lambda + n, \frac{1}{2} + \lambda, \frac{1-x}{2}\right) . \tag{10}$$

Orthogonality

From (10) it is seen that $C_n^\lambda(x)$ is a special case of the Jacobi polynomial introduced in Sec. 4.10. Therefore, by (12) and (13) of that section, replacing z by $(1-x)/2$, we obtain immediately the orthogonality relation of $C_n^\lambda(x)$:

$$\int_{-1}^1 C_n^\lambda(x)C_m^\lambda(x)(1-x^2)^{\lambda-\frac{1}{2}}dx = \frac{\pi\Gamma(2\lambda+n)}{2^{2\lambda-1}n!(\lambda+n)[\Gamma(\lambda)]^2}\delta_{nm} . \tag{11}$$

Differentiation Formula. Explicit Expression

When $\lambda = m$ (positive integer), we have from (10), by Eq. (10) of Sec. 4.2, the formula

$$C_n^m(x) = \frac{1}{2^{m-1}(m-1)!(m+n)}\frac{d^m}{dx^m}T_{m+n}(x) , \tag{12}$$

where $T_{m+n}(x)$ is the Chebyshev polynomial [Sec. 4.11].

Using (12) and Eq. (2) of Sec. 4.11, we obtain at once the explicit expression for $C_n^m(x)$:

$$C_n^m(x) = \frac{1}{(m-1)!}\sum_{l=0}^{[n/2]}\frac{(-)^l(m+n-l-1)!}{l!(n-2l)!}(2x)^{n-2l} . \tag{13}$$

For general λ, we may start from (10), applying Eq. (17) of Sec. 4.12, to obtain the explicit expression for $C_n^\lambda(x)$:

$$C_n^\lambda(x) = \frac{1}{\Gamma(\lambda)} \sum_{l=0}^{[n/2]} \frac{(-)^l \Gamma(\lambda + n - l)}{l!(n - 2l)!} (2x)^{n-2l} . \tag{14}$$

The Fourier expansion of $C_n^\lambda(\cos\theta)$ and other formulae are given in the relevant Exs. at the end of this chapter.

Exercise 5

1. Let θ be the angle between the z-axis and the radius vector $\mathbf{r}$ leading from the origin $(0, 0, 0)$ to the point $P(x, y, z)$. Show that

$$P_n(\cos\theta) = \frac{(-)^n r^{n+1}}{n!} \frac{\partial^n}{\partial z^n} \left(\frac{1}{r}\right) , \quad r = |\mathbf{r}| .$$

2. From Eq. (2) of Sec. 5.3, we have

$$P_n(x) = \frac{1}{2\pi i} \int_C \frac{(1 - 2xt + t^2)^{-1/2}}{t^{n+1}} dt ,$$

where C is a contour encircling $t = 0$ once in the positive sense; there is no singularity of the square root function lying within the contour. Derive from it the Rodrigues formula [Eq. (3) of Sec. 5.3]. [Hint: Use the transformation: $(1 - 2xt + t^2)^{-1/2} = 1 - ut$.]

3. Write the generating function $(1 - 2\cos\theta \cdot t + t^2)^{-1/2}$ of $P_n(\cos\theta)$ as $(1 - t^{i\theta})^{-1/2}(1 - te^{-i\theta})^{-1/2}$. Show that $P_n(\cos\theta)$ can be expressed in a Fourier cosine series:

$$P_n(\cos\theta) = \sum_{k=0}^{n} (-)^n \binom{-\frac{1}{2}}{k} \binom{-\frac{1}{2}}{n-k} \cos(n - 2k)\theta$$

$$= \frac{1}{2^n} \sum_{k=0}^{n} \frac{1 \cdot 3 \cdot 5 \cdots (2k - 1) \cdot 1 \cdot 3 \cdot 5 \cdots (2n - 2k - 1)}{k!(n - k)!}$$

$$\times \cos(n - 2k)\theta$$

and hence $|P_n(\cos\theta)| \leq |P_n(1)| = 1$ when $0 \leq \theta \leq \pi$.

4. Show that

$$P_n(\cos\theta) = F\left(n+1, -n, 1, \sin^2\left(\frac{\theta}{2}\right)\right)$$

$$= (-)^n F\left(n+1, -n, 1, \cos^2\left(\frac{\theta}{2}\right)\right)$$

$$= \cos^{2n}\left(\frac{\theta}{2}\right) F\left(-n, -n, 1, -\tan^2\left(\frac{\theta}{2}\right)\right)$$

$$= \cos^n\theta F\left(-\frac{n}{2}, \frac{1-n}{2}, 1, -\tan^2\left(\frac{\theta}{2}\right)\right).$$

[Hint: The last formula can be proved by applying Eq. (11) of Sec. 4.9 to Eq. (7) of Sec. 5.2.]

5. Show that

$$P_n(z) = \frac{(2n)!}{2^{2n}(n!)^2} Z^n F\left(\frac{1}{2}, -n, \frac{1}{2} - n, Z^{-2}\right),$$

$$Q_n(z) = \frac{2^{2n+1}(n!)^2}{(2n+1)!} Z^{-n-1} F\left(\frac{1}{2}, 1+n, \frac{3}{2}+n, Z^{-2}\right),$$

$Z = z + \sqrt{z^2 - 1}$ and the square root assumes positive values when $z > 1$.

6. With the Schläfli integral expression [Eq. (18) of Sec. 5.16] show that the recurrence relations given in Sec. 5.5 also hold good when n is not an integer.

7. Show that

$$\int_0^1 P_m(x)P_n(x)\,dx = \begin{cases} \dfrac{1}{2n+1} & (m=n) \\[2mm] 0 & (m-n \text{ even}), \\[2mm] \dfrac{(-)^{\mu+\nu}}{2^{m+n-1}(n-m)(n+m+1)}\dfrac{n!m!}{(\nu!)^2(\mu!)^2} \\ \qquad (n=2\nu+1, m=2\mu). \end{cases}$$

8. Show that

$$\int_0^1 x^\nu P_n(x)dx = 2^{-n}\frac{\Gamma(\nu+1)\Gamma\left(\frac{\nu-n+3}{2}\right)}{\Gamma(\nu-n+2)\Gamma\left(\frac{\nu+n+3}{2}\right)}$$

$$= \frac{\nu(\nu-1)\ldots(\nu-n+2)}{(\nu+n+1)(\nu+n-1)\ldots(\nu-n+3)} \; ,$$

where $\mathrm{Re}(\nu) > -1$ when n is even and $\mathrm{Re}(\nu) > -2$ when n is odd so as to ensure the convergence of the integral at its lower limit. [Hint: Use Eq. (12) of Sec. 5.2 and Eq. (4) of Sec. 4.13.]

9. Show that

$$z^n = \sum_{m=0}^n a_m P_m(z) \; ,$$

where

$$a_m = 0 \quad (n-m \text{ odd}) \; ,$$

$$a_m = \frac{(2m+1)2^m n!\left(\frac{n+m}{2}\right)!}{\left(\frac{n-m}{2}\right)!(n+m+1)!} \quad (n-m \text{ even}) \; .$$

10. Show that

$$\frac{d^{m+1}}{dx^{m+1}}P_{m+n}(x)\bigg|_{x=1} = \frac{\Gamma(2m+n+2)}{2^{m+1}(m+1)!\Gamma(n)} \; ,$$

where n may be any complex number. [Hint: Use Eq. (8) of Sec. 5.16.]

11. Show that, when n and m are integers,

$$\int_0^\pi P_n(\cos\theta)\sin m\theta\, d\theta = 2\frac{(m-n+1)(m-n+3)\ldots(m+n-1)}{(m-n)(m-n+2)\ldots(m+n)} \; ,$$

provided that $m > n$ and $m+n$ is an odd number; in other cases, the integral vanishes [cf. Hobson (1931), Sec. 29].

With this result, show that

$$\frac{\pi}{4}P_n(\cos\theta) = \frac{2^{2n}(n!)^2}{(2n+1)!}\left[\sin(n+1)\theta + \frac{1\cdot(n+1)}{1\cdot(2n+3)}\sin(n+3)\theta\right.$$

$$\left. + \frac{1\cdot3\cdot(n+1)(n+2)}{1\cdot2\cdot(2n+3)(2n+5)}\sin(n+5)\theta + \ldots\right] \quad (0 < \theta < \pi)$$

and

$$\frac{4}{\pi} \frac{2 \cdot 4 \cdots (2n-2)}{1 \cdot 3 \cdots (2n-3)} \sin n\theta$$

$$= (2n-1)P_{n-1}(\cos\theta) + (2n+3)\frac{(n-1)^2 - n^2}{(n+2)^2 - n^2}P_{n+1}(\cos\theta)$$

$$+ (2n+7)\frac{[(n-1)^2 - n^2][(n+1)^2 - n^2]}{[(n+2)^2 - n^2][(n+4)^2 - n^2]}P_{n+3}(\cos\theta) + \cdots$$

$$(0 < \theta < \pi) .$$

12. Show that

$$2 \cdot \frac{3 \cdot 5 \cdots (2n+1)}{2 \cdot 4 \cdots (2n)} \cos n\theta$$

$$= (2n+1)P_n(\cos\theta) + (2n-3)\frac{n^2 - (n+1)^2}{n^2 - (n-2)^2}P_{n-2}(\cos\theta)$$

$$+ (2n-7)\frac{[n^2 - (n+1)^2][n^2 - (n-1)^2]}{[n^2 - (n-2)^2][n^2 - (n-4)^2]}P_{n-1}(\cos\theta)$$

$$+ \cdots$$

[Hint: Make use of the result of Ex. 3].

13. Show that, when $x > 1$,

$$Q_n(x) = 2^n n! \int_x^\infty \int_v^\infty \cdots \int_v^\infty \frac{(dv)^{n+1}}{(v^2 - 1)^{n+1}} ,$$

and from which, by using the formula

$$\int_x^\infty \int_v^\infty \cdots \int_v^\infty f(v)(dv)^{n+1} = \sum_{r=0}^n \frac{(-x)^{n-r}}{r!(n-r)!} \int_x^\infty v^r f(v)dv ,$$

obtain

$$Q_n(x) = \sum_{r=0}^n \frac{2^n n!}{r!(n-r)!}(-x)^{n-r} \int_x^\infty v^r (v^2 - 1)^{-n-1}dv .$$

14. Show that

$$Q_n(x) = (-)^n \frac{2^n n!}{(2n)!} \frac{d^n}{dx^n}\left\{ (x^2 - 1)^n \int_x^\infty (x^2 - 1)^{-n-1}dx \right\} .$$

[Hint: Note that the Legendre equation can be derived from the equation $(1 - x^2)w'' + 2(n - 1)w' + 2nw = 0$ by differentiating n times.]

15. Show that, when $z > 1$,

$$(1 - 2zt + t^2)^{-1/2} \ln \frac{z - t + (1 - 2zt + t^2)^{1/2}}{\sqrt{z^2 - 1}} = \sum_{n=0}^{\infty} Q_n(x)t^n$$

$$(\arg(z^2 - 1) = 0) \; ;$$

the function on the l.h.s. is called the generating function of $Q_n(z)$. [Cf. Hobson (1931), Sec. 4.3.]

16. When $-1 < t < 1$ and $Q_n(t)$ is specified according to Eq. (8) of Sec. 5.19, prove, with the result of the last problem, that

$$Q_n(t) = \frac{(-)^n r^{n+1}}{n!} \frac{\partial^n}{\partial z^n} \left\{ \frac{1}{2r} \ln \frac{r + z}{r - z} \right\} \, ,$$

where $r = (x^2 + y^2 + z^2)^{1/2}, t = z/r = \cos \theta$.

17. With the results of the last problem and Ex. 1, show that, from Eq. (6) of Sec. 5.8,

$$W_{n-1}(x) = \frac{1}{n} P_0(x) P_{n-1}(x) + \frac{1}{n - 1} P_1(x) P_{n-2}(x)$$

$$+ \ldots + P_{n-1}(x) P_0(x) \, .$$

18. Show that

$$W_{n-1}(x) = \frac{2n - 1}{1 \cdot n} P_{n-1}(x) + \frac{2n - 5}{3 \cdot (n - 1)} P_{n-3}(x)$$

$$+ \frac{2n - 9}{5 \cdot (n - 2)} P_{n-5}(x) + \ldots$$

$$= \sum_{r=0}^{[\frac{n-1}{2}]} \frac{2n - 4r - 1}{(2r + 1)(n - r)} P_{n-2r-1}(x) \, .$$

[Hint: W_{n-1} satisfies the differential equation $(1 - x^2)W''_{n-1} - 2xW'_{n-1} + n(n + 1)W_{n-1} = 2P'_n$. The r.h.s. of it can be expressed, with Eq. (6) of

Sec. 5.5, in a linear combination of the Legendre polynomials. Expanding $W_{n-1}(x)$ in terms of the Legendre polynomials and substituting in the equation, the coefficients of expansion are determined.]

19. With the result of Ex. 5, prove the following Fourier expansion

$$Q_n(\cos\theta) = \frac{2^{n+1}(n!)^2}{(2n+1)!}\left\{\cos(n+1)\theta + \frac{1\cdot(n+1)}{1\cdot(2n+3)}\cos(n+3)\theta\right.$$

$$\left. + \frac{1\cdot3\cdot(n+1)(n+2)}{1\cdot2\cdot(2n+3)(2n+5)}\cos(n+5)\theta + \ldots\right\},$$

where $Q_n(\cos\theta)$ is defined according to Eq. (8) of Sec. 5.19 $(0 < \theta < \pi)$.

20. Show that
$$Q_n(z) = \int_0^\infty \frac{d\psi}{(z + \sqrt{z^2 - 1}\ \mathrm{ch}\ \psi)^{n+1}},$$

z being an arbitrary complex number not lying in the interval $[-1, 1]$. [Hint: Use the result of Ex. 5 and Eq. (6) of Sec. 4.5 to obtain

$$Q_n(z) = Z^{-n-1}\int_0^1 v^{-1/2}(1-v)^n(1-vZ^{-2})^{-n-1}dv,$$

where $Z = z + \sqrt{z^2 - 1}$; then apply the transformation $v = (w-1)/(w+1)$.]

21. Show that

$$Q_n(z) = \int_0^{Z^{-1}} \frac{h^n\,dh}{(1 - 2zh + h^2)^{1/2}} \quad (Z = z + \sqrt{z^2 - 1}).$$

22. Show that

$$\text{(i)}\ P_n Q_{n-1} - Q_n P_{n-1} = \frac{1}{n},$$

$$\text{(ii)}\ P_{n+1}Q_{n-1} - Q_{n+1}P_{n-1} = \frac{2n+1}{n(n+1)}x.$$

23. Show that

$$P_n^m(\cos\theta) = (-)^m\frac{(2m)!}{2^m m!}\frac{\sin^m\theta}{(n-m)!}r^{n+m+1}\frac{\partial^{n-m}}{\partial z^{n-m}}\left(\frac{1}{r^{2m+1}}\right),$$

where $P_n^m(x)$ is the associate Legendre function [Eq. (4) of Sec. 5.11; Hobson's definition].

24. Prove the following recurrence relations $(-1 \leq x \leq 1)$:

(i) $xP_l^m = P_{l+1}^m + (l+m)(1-x^2)^{1/2}P_l^{m-1}$

$\quad = P_{l-1}^m - (l-m+1)(1-x^2)^{1/2}P_l^{m-1}$,

(ii) $(1-x^2)^{1/2}P_l^{m+1} = (l-m+1)P_{l+1}^m - (l+m+1)xP_l^m$

$\quad = (l-m)xP_l^m - (l+m)P_{l-1}^m$,

(iii) $(1-x^2)\dfrac{dP_l^m}{dx} = (l+1)xP_l^m - (l-m+1)P_{l+1}^m$

$\quad = (l+m)P_{l-1}^m - lxP_l^m$,

(iv) $(P_l^{m+1})^2 + (l-m)^2(P_l^m)^2 = (P_{l-1}^{m+1})^2 + (l+m)^2(P_{l-1}^m)^2$

or, more generally,

(v) $P_l^{m+1}P_{l'}^{m'+1} + (l-m)(l'-m')P_l^m P_{l'}^{m'}$

$\quad = P_{l-1}^{m+1}P_{l'-1}^{m'+1} + (l+m)(l'+m')P_{l-1}^m P_{l'-1}^{m'}$

[Hint: Use (i)].

25. Let $S_l(\theta, \varphi)$ be an arbitrary spherical surface harmonics of degree l [after Eq. (7) of Sec. 5.14]. Show that

$$\int_0^\pi \int_0^{2\pi} S_l(\theta, \varphi)P_l(\cos\theta\cos\theta' + \sin\theta\sin\theta'\cos(\varphi-\varphi'))\sin\theta\,d\varphi\,d\theta$$

$$= \frac{4\pi}{2l+1}S_l(\theta', \varphi') .$$

26. Show that

$$\cos\theta\frac{\partial}{\partial\theta}Y_{lm} - \frac{m}{\sin\theta}Y_{lm}$$
$$= \sqrt{(l-m)(l+m+1)}\cos\theta Y_{l,m+1}e^{-i\varphi} - m\sin\theta Y_{lm} ,$$
$$\cos\theta\frac{\partial}{\partial\theta}Y_{lm} + \frac{m}{\sin\theta}Y_{lm}$$
$$= -\sqrt{(l+m)(l-m+1)}\cos\theta Y_{l,m-1}e^{i\varphi} + m\sin\theta Y_{lm} ;$$

Y_{lm} is defined in Sec. 5.15.

27. By the transformation of coordinates

$$\frac{\partial}{\partial x} = \sin\theta\cos\varphi\frac{\partial}{\partial r} + \cos\theta\cos\varphi\frac{1}{r}\frac{\partial}{\partial\theta} - \frac{\sin\varphi}{r\sin\theta}\frac{\partial}{\partial\varphi} ,$$

$$\frac{\partial}{\partial y} = \sin\theta\sin\varphi\frac{\partial}{\partial r} + \cos\theta\sin\varphi\frac{1}{r}\frac{\partial}{\partial\theta} + \frac{\sin\varphi}{r\sin\theta}\frac{\partial}{\partial\varphi} ,$$

$$\frac{\partial}{\partial z} = \cos\theta\frac{\partial}{\partial r} - \sin\theta\frac{1}{r}\frac{\partial}{\partial\theta} ,$$

the first two of them being able to be combined into a single one:

$$\left(\frac{\partial}{\partial x} \pm i\frac{\partial}{\partial y}\right) = e^{\pm i\varphi}\left\{\sin\theta\frac{\partial}{\partial r} + \cos\theta\cdot\frac{1}{r}\frac{\partial}{\partial\theta} \pm \frac{i}{r\sin\theta}\frac{\partial}{\partial\varphi}\right\} ,$$

obtain, by using Eqs. (6)–(9) of Sec. 5.15, the following formulae:

$$\left(\frac{\partial}{\partial x} \pm i\frac{\partial}{\partial y}\right)\{f(r)Y_{lm}\}$$

$$= \pm\sqrt{\frac{(l\mp m)(l\mp m-1)}{(2l+1)(2l-1)}}Y_{l-1,m\pm 1}\left\{\frac{df}{dr} + (l+1)\frac{f}{r}\right\}$$

$$\mp\sqrt{\frac{(l\pm m+1)(l\pm m+2)}{(2l+1)(2l+3)}}Y_{l+1,m\pm 1}\left\{\frac{df}{dr} - l\frac{f}{r}\right\} ,$$

$$\frac{\partial}{\partial z}\{f(r)Y_{lm}\}$$

$$= \sqrt{\frac{(l+m)(l-m)}{(2l+1)(2l-1)}}Y_{l-1,m}\left\{\frac{df}{dr} + (l+1)\frac{f}{r}\right\}$$

$$+ \sqrt{\frac{(l+m+1)(l-m+1)}{(2l+1)(2l+3)}}Y_{l+1,m}\left\{\frac{df}{dr} - l\frac{f}{r}\right\} .$$

28. When $\nu + \mu$ is a negative integer, Eq. (6) of Sec. 5.16, is an indeterminate form. Show that

$$P_\nu^\mu(z) = \frac{e^{-\nu\pi i}}{4\pi \sin \nu\pi} \frac{1}{2^\nu \pi \cos(\nu + \mu)\pi} \frac{1}{\Gamma(\nu + 1)\Gamma(-\nu - \mu)}(z^2 - 1)^{\mu/2}$$

$$\times \int_M^{(z+,1+,z-,1-)} (t^2 - 1)^\nu (t - z)^{-\nu-\mu-1} \ln(t - z)dt ;$$

the contour is the one shown in Fig. 12 of Sec. 5.16; the multi-valued factors are specified as that for Eq. (6) of Sec. 5.16.

29. From Eq. (8) of Sec. 5.16, making use of Eq. (8) of Sec. 4.3 and Eq. (9) of Sec. 4.12, show that

$$P_\nu^\mu(z) = \frac{2^\mu}{\Gamma(1-\mu)}(z^2-1)^{-\mu/2}F\left(1+\nu-\mu, -\nu-\mu, 1-\mu, \frac{1-z}{2}\right)$$

$$= \frac{2^\mu}{\Gamma(1-\mu)}(z^2-1)^{-\mu/2}F\left(\frac{1+\nu-\mu}{2}, -\frac{\nu+\mu}{2}, 1-\mu, 1-z^2\right) .$$

30. Using the first formula of the last Ex. and Eq. (17) of Sec. 4.12, show that

$$P_\nu^\mu(z) = \frac{2^\mu \Gamma\left(\frac{1}{2}\right)(z^2-1)^{-\mu/2}}{\Gamma\left(\frac{2+\nu-\mu}{2}\right)\Gamma\left(\frac{1-\nu-\mu}{2}\right)}F\left(\frac{1+\nu-\mu}{2}, -\frac{\nu+\mu}{2}, \frac{1}{2}, z^2\right)$$

$$- \frac{2^{\mu+1}\Gamma\left(\frac{1}{2}\right)z(z^2-1)^{-\mu/2}}{\Gamma\left(\frac{1+\nu-\mu}{2}\right)\Gamma\left(-\frac{\nu+\mu}{2}\right)}F\left(\frac{2+\nu-\mu}{2}, \frac{1-\nu-\mu}{2}, \frac{3}{2}, z^2\right) .$$

31. From the second formula of Ex. 29, making use of Eq. (10) of Sec. 4.3 and

Eq. (4) of Sec. 4.8, show that

$$P_\nu^\mu(z) = \frac{2^\nu \Gamma\left(\nu + \frac{1}{2}\right) z^{\nu+\mu}}{\Gamma\left(\frac{1}{2}\right)\Gamma(1+\nu-\mu)}(z^2-1)^{-\mu/2}$$

$$\times \; F\left(\frac{1-\nu-\mu}{2}, -\frac{\nu+\mu}{2}, \frac{1}{2}-\nu, z^{-2}\right)$$

$$+ \frac{2^{-\nu-1}\Gamma\left(-\nu-\frac{1}{2}\right)}{\Gamma\left(\frac{1}{2}\right)\Gamma(-\nu-\mu)}z^{-\nu+\mu-1}(z^2-1)^{-\mu/2}$$

$$\times \; F\left(\frac{2+\nu-\mu}{2}, \frac{1+\nu-\mu}{2}, \frac{3}{2}+\nu, z^{-2}\right) .$$

32. Show that, if $\mathrm{Re}(\nu+1) > 0$,

$$Q_\nu^\mu(z) = \frac{e^{\mu\pi i}}{2^{\nu+1}}\frac{\Gamma(\nu+\mu+1)}{\Gamma(\nu+1)}(z^2-1)^{\mu/2}\int_{-1}^{1}(1-t^2)^\nu(z-t)^{-\nu-\mu-1}dt \; ;$$

on the path of integration, $\arg(l-t^2) = 0, |\arg(z-t)| < \pi$.

33. Show that

$$Q_\nu^\mu(z) = ie^{(\mu-\nu)\pi i}2^\mu\frac{\Gamma\left(\mu+\frac{1}{2}\right)\Gamma\left(\frac{1}{2}\right)}{4\pi\sin(\nu+\mu)\pi}\frac{(z^2-1)^{\mu/2}}{Z^{\nu+\mu+1}}$$

$$\times \; \int_{A}^{(1+,0+,1-,0-)} t^{\nu+\mu}(1-t)^{-\frac{1}{2}-\mu}(1-tZ^{-2})^{-\frac{1}{2}-\mu}dt \; ,$$

where $|\arg(z\pm 1)| < \pi, Z = z+\sqrt{z^2-1}, |\arg Z| < \pi, Z^2$ lies outside the contour; on the contour, $|\arg(1-tZ^{-2})| < \pi$; the starting point A of the contour is supposed to lie on the real axis between $t = 0$ and $t = 1$ $\arg t = \arg(1-t) = 0$ initially [cf. Hobson (1931), Sec. 152, p. 237].

34. Using the result of the last problem, show that

$$Q_\nu^\mu(z) = ie^{(\mu-\nu)\pi i}2^\mu\frac{\Gamma\left(\mu+\frac{1}{2}\right)\Gamma\left(\frac{1}{2}\right)}{4\pi\sin(\nu+\mu)\pi}(z^2-1)^{\mu/2}$$

$$\times \; \int_{A'}^{(Z^{-1}+,0+,Z^{-1}-,0-)} \frac{h^{\nu+\mu}}{(1-2zh+h^2)^{\mu+1/2}}dh \; ,$$

$Z = z+\sqrt{z^2-1}$ lying outside the contour. At the starting point A' on the real axis, $\arg h = 0$ initially; when A' moves along the real axis to infinity, $\arg(1-2zh+h^2) = \arg(h-Z)(h-Z^{-1}) = 0$.

35. **Show that**

$$Q_\nu^\mu(z) = \frac{\pi e^{\mu\pi i}}{2\sin(\nu+\mu)\pi} \frac{1}{\Gamma(1-\mu)}$$

$$\times \left\{ e^{\mp\nu\pi i} \left(\frac{z+1}{z-1}\right)^{\mu/2} F\left(-\nu, \nu+1, 1-\mu, \frac{1-z}{2}\right) \right.$$

$$\left. - \left(\frac{z-1}{z+1}\right)^{\mu/2} F\left(-\nu, \nu+1, 1-\mu, \frac{1+z}{2}\right) \right\}$$

$$= \frac{\Gamma(\nu+\mu+1)}{\Gamma(\nu-\mu+1)} \frac{\pi e^{\mu\pi i}}{2\sin(\nu-\mu)\pi} \frac{1}{\Gamma(1+\mu)}$$

$$\times \left\{ e^{\mp\nu\pi i} \left(\frac{z-1}{z+1}\right)^{\mu/2} F\left(-\nu, \nu+1, 1+\mu, \frac{1-z}{2}\right) \right.$$

$$\left. - \left(\frac{z+1}{z-1}\right)^{\mu/2} F\left(-\nu, \nu+1, 1+\mu, \frac{1+z}{2}\right) \right\},$$

where $e^{\mp\nu\pi i}$ takes the positive sign when $\text{Im}(z) < 0$ and the negative sign when $\text{Im}(z) > 0$. [Hint: From Eq. (3) Sec. 5.19, with Eq. (4) of Sec. 4.8, obtain the first formula, then apply Eq. (8) of Sec. 5.17 to derive the second formula.]

Also show that, when $\nu + \mu$ is a positive integer,

$$e^{\mp\nu\pi i} \left(\frac{z+1}{z-1}\right)^{\mu/2} F\left(-\nu, \nu+1, 1-\mu, \frac{1-z}{2}\right)$$

$$= \left(\frac{z-1}{z+1}\right)^{\mu/2} F\left(-\nu, \nu+1, 1-\mu, \frac{1+z}{2}\right).$$

[Cf. Eq. (4) of Sec. 5.17; note that when $\nu+\mu$ is equal to a positive integer, $Q_\nu^\mu(z)$ is in general finite.]

36. Show that

(i) $Q_\nu^\mu(z)\sin(\nu+\mu)\pi - Q_{-\nu-1}^\mu(z)\sin(\nu-\mu)\pi$
$$= \pi e^{\mu\pi i}\cos\nu\pi P_\nu^\mu(z) ,$$

(ii) $P_\nu^{-\mu}(z) = \dfrac{\Gamma(\nu-\mu+1)}{\Gamma(\nu+\mu+1)}\left\{P_\nu^\mu(z) - \dfrac{2}{\pi}e^{-\mu\pi i}\sin\mu\pi Q_\nu^\mu(z)\right\} ,$

(iii) $P_\nu^\mu(-z) = e^{\mp\nu\pi i}P_\nu^\mu(z) - \dfrac{2\sin(\nu+\mu)\pi}{\pi}e^{-\mu\pi i}Q_\nu^\mu(z)$

$(\nu+\mu \neq$ negative integer$)$,

where (ii) is just Eq. (4) of Sec. 5.19; the other two formulae can be proved with the result of the last exercise. When $\nu+\mu$ is a negative integer, (iii) has to be replaced, according to the second formula of Ex. 35, by

(iv) $P_\nu^\mu(-z) = e^{\mp\nu\pi i}P_\nu^\mu(z) + \dfrac{2}{\Gamma(\nu-\mu+1)\Gamma(-\nu-\mu)}e^{\mu\pi i}Q_\nu^{-\mu}(z)$.

From (iii), it can also be shown that

(v) $Q_\nu^\mu(-z) = -e^{\pm\nu\pi i}Q_\nu^\mu(z)$.

In (iii), (iv), (v), the exponential function is to take the upper sign when $\text{Im}(z) > 0$, and the lower sign when $\text{Im}(z) < 0$.

37. With the relation between $P_\nu^\mu(z)$ and the hypergeometric function, together with the recurrence relations of the latter [Sec. 4.2 and Exs. 2.3 at the end of Chap. 4], show that

(i) $(2\nu+1)zP_\nu^\mu(z) = (\nu+\mu)P_{\nu-1}^\mu(z) + (\nu+\mu+1)P_{\nu+1}^\mu(z)$,

(ii) $(2\nu+1)(z^2-1)^{1/2}P_\nu^\mu(z) = P_{\nu+1}^{\mu+1}(z) - P_{\nu-1}^{\mu+1}(z)$,

(iii) $(2\nu+1)(z^2-1)^{1/2}P_\nu^\mu(z)$
$$= (\nu-\mu+1)(\nu-\mu+2)P_{\nu+1}^{\mu-1}(z) - (\nu+\mu-1)(\nu+\mu)P_{\nu-1}^{\mu-1}(z) ,$$

(iv) $(2\nu+1)(z^2-1)\dfrac{dP_\nu^\mu(z)}{dz}$
$$= \nu(\nu-\mu+1)P_{\nu+1}^\mu(z) - (\nu+1)(\nu+\mu)P_{\nu-1}^\mu(z) .$$

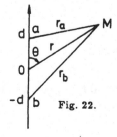

Fig. 22.

These relations also hold for $Q_\nu^\mu(z)$. Further, with the definitions of $P_\nu^\mu(x)$ [Eq. (4) of Sec. 5.18] and $Q_\nu^\mu(x)$ [Eq. (9) of Sec. 5.19] on the cut $-1 < x < 1$, we may show, by the above recurrence formulae, that those recurrence relations obtained in Sec. 5.13 are valid when l and m are not integers.

38. Prove the following expansion formula:

$$r^n P_n^m(\cos\theta) = d^n \sum_{l=m}^{\infty} a_l P_l^m(\xi) P_l^m(\eta) ,$$

$$a_l = \frac{\sqrt{\pi}}{2^{n+1}} \frac{\Gamma(n+m+1)\Gamma(l-m+1)(2l+1)}{\Gamma(l+m+1)\Gamma\left(\frac{n-l}{2}+1\right)\Gamma\left(\frac{n+l}{2}+\frac{3}{2}\right)} \delta_{n-l,2f} ,$$

$$(f = 0,1,2,\ldots)$$

where (ξ, η, φ) are the rotational prolate spheroidal coordinates [App. III, 8 and Fig. 22], $\cos\theta = z/r = \xi\eta/\lambda, \lambda = \sqrt{\xi^2 + \eta^2 - 1}$.

39. Using the differential equations satisfied by $P_\nu(x)$ and $Q_\nu(x)$, show that

(i) $\displaystyle\int_1^\infty P_\nu(x)P_\sigma(x)dx = [(\sigma-\nu)(\sigma+\nu+1)]^{-1}$

$$(\mathrm{Re}(\sigma) > \mathrm{Re}(\nu) > 0) ,$$

(ii) $\displaystyle\int_1^\infty Q_\nu(x)Q_\sigma(x)dx = \frac{\psi(\sigma+1)-\psi(\nu+1)}{(\sigma-\nu)(\sigma+\nu+1)}$

$(\mathrm{Re}(\sigma+\nu) > -1, \sigma+\nu+1 \neq 0, \nu,\sigma \neq$ negative integers$)$,

(iii) $\displaystyle\int_1^\infty \{Q_\nu(x)\}^2 dx = \frac{\psi'(\nu+1)}{2\nu+1}$ $\quad (\mathrm{Re}(\nu) > -1/2)$,

where $\psi(z)$ in (ii) and (iii) is the logarithmic derivative of the Γ-function [Sec. 3.11].

40. Show that

$$\left(n+\frac{1}{2}\right)(1+x)^{1/2}\int_{-1}^x (x-t)^{-1/2}P_n(t)dt = T_n(x) + T_{n+1}(x) ,$$

$$\left(n+\frac{1}{2}\right)(1-x)^{1/2}\int_x^1 (t-x)^{-1/2}P_n(t)dt = T_n(x) - T_{n+1}(x) ,$$

where $P_n(x)$ is the Legendre polynomial, $T_n(x)$ is the Chebyshev polynomial [Sec. 4.11]. These are the inversions of the Mehler-Schläfli integral formula $(\nu = n)$ given in Eq. (6) of Sec. 5.20.

41. By writing the generating function $(1-2t\cos\theta+t^2)^{-\lambda}$ of $C_n^\lambda(\cos\theta)$ [Eq. (1) of Sec. 5.23], in the form $(1-te^{i\theta})^{-\lambda}(1-te^{-i\theta})^{-\lambda}$, prove the following Fourier expansion formula:

$$C_n^\lambda(\cos\theta) = \sum_{l=0}^{n} \frac{(\lambda)_l(\lambda)_{n-l}}{l!(n-l)!}\cos(n-2l)\theta .$$

42. Show that

$$\int_0^\pi C_n^\lambda(\cos\theta)(\sin\theta)^{2\lambda}\,d\theta = \begin{cases} 0 , & n = 1, 2, 3, \ldots , \\ \frac{\pi\Gamma(2\lambda+1)}{2^{2\lambda}[\Gamma(\lambda+1)]^2} , & n = 0 . \end{cases}$$

43. Prove the following addition formula for $C_n^\lambda(x)$:

$$C_n^\lambda\{zz_1 - (z^2-1)^{1/2}(z_1^2-1)^{1/2}\cos\varphi\}$$

$$= \frac{\Gamma(2\lambda-1)}{[\Gamma(\lambda)]^2}\sum_{l=0}^{\infty}(-)^l\frac{4^l\Gamma(n-l+1)[\Gamma(\lambda+l)]^2(2\lambda+2l-1)}{\Gamma(2\lambda+n+l)}$$

$$\times (z^2-1)^{1/2}(z_1^2-1)^{1/2}C_{n-l}^{\lambda+l}(z)C_{n-l}^{\lambda+l}(z_1)C_l^{\lambda-\frac{1}{2}}(\cos\varphi) .$$

(Gegenbauer, Wiener Sitzungsberichte, CII (1893), p. 942.)

44. Using the result of the last two exercises show that

$$\int_0^\pi C_n^\lambda\{\cos\theta\cos\theta' + \sin\theta\sin\theta'\cos\varphi\}(\sin\varphi)^{2\lambda-1}d\varphi$$

$$= \frac{2^{2\lambda-1}n![\Gamma(\lambda)]^2}{\Gamma(2\lambda+n)}C_n^\lambda(\cos\theta)C_n^\lambda(\cos\theta') ,$$

$$(\mathrm{Re}(\lambda) > 0) .$$

45. Show that

$$C_{n-m}^{m+\frac{1}{2}}(z) = \frac{2^m m!}{(2m)!}\frac{d^m}{dz^m}P_n(z) = \frac{2^m m!}{(2m)!}(z^2-1)^{-m/2}P_n^m(z) ,$$

$P_n^m(z)$ being the associate Legendre function of order m and degree n [Eq. (14) of Sec. 5.16].

46. Prove the following recurrence relations of $C_n^\lambda(z)$:

$$zC_{n-1}^{\lambda+1} - C_{n-2}^{\lambda+1} - \frac{n}{2\lambda}C_n^\lambda = 0 ,$$

$$C_n^{\lambda+1} - zC_{n-1}^{\lambda+1} = \frac{2\lambda+n}{2\lambda}C_n^\lambda ,$$

$$nC_n^\lambda = (2\lambda + n - 1)zC_{n-1}^\lambda + 2\lambda(z^2 - 1)C_{n-2}^{\lambda-1} ,$$

$$\frac{dC_n^\lambda}{dz} = 2\lambda C_{n-1}^{\lambda+1} .$$

47. The *Gegenbauer function* $C_\alpha^\lambda(z)$ is a generalization of the polynomial $C_n^\lambda(z)$, which can be defined by Eq. (10) of Sec. 5.23 with the n in it replaced by an arbitrary parameter α:

$$C_\alpha^\lambda(z) = \frac{\Gamma(2\lambda + \alpha)}{\Gamma(\alpha + 1)\Gamma(2\lambda)} F\left(-\alpha, 2\lambda + \alpha, \frac{1}{2} + \lambda, \frac{1-z}{2}\right) .$$

Thus, by the first formula of Ex. 29 above, show that

$$C_\alpha^\lambda(z) = \frac{2^{\lambda-\frac{1}{2}}\Gamma(2\lambda + \alpha)\Gamma\left(\lambda + \frac{1}{2}\right)}{\Gamma(\alpha + 1)\Gamma(2\lambda)}(z^2 - 1)^{\frac{1}{4}-\frac{\lambda}{2}} P_{\alpha+\lambda-\frac{1}{2}}^{\frac{1}{2}-\lambda}(z) .$$

Then, it follows,

(i) $\quad C_\alpha^\lambda(z) = \dfrac{2^{1-2\lambda}\Gamma(2\lambda + \alpha)}{\Gamma(\alpha + 1)[\Gamma(\lambda)]^2} \displaystyle\int_0^\pi [z + \sqrt{z^2 - 1}\cos t]^\alpha (\sin t)^{2\lambda-1} dt$

$\quad (\mathrm{Re}(\lambda) > 0)$,

(ii) $\quad C_\alpha^\lambda(\cos\varphi) = \dfrac{2^{1-\lambda}\Gamma(2\lambda + \alpha)}{\Gamma(\alpha + 1)[\Gamma(\lambda)]^2}(\sin\varphi)^{1-2\lambda}$

$$\times \int_0^\infty \cos[(\lambda + \alpha)v](\cos v - \cos\varphi)^{\lambda-1} dv$$

$\quad (\mathrm{Re}(\lambda) > 0, 0 < \varphi < \pi)$,

(iii) $\quad C_\alpha^\lambda(z) = -\dfrac{\sin\alpha\pi}{\pi}\displaystyle\int_0^\infty \frac{(1 + 2zt + t^2)^{-\lambda}}{t^{\alpha+1}} dt$

$\quad (-2 < \mathrm{Re}(\lambda) < \mathrm{Re}(\alpha) < 0, \quad |\arg(z \pm 1)| < \pi)$.

Prove also the Mellin inversion of (iii):

$$(1 + 2zt + t^2)^{-\lambda} = \frac{i}{2}\int_{c-i\infty}^{c+i\infty} \frac{C_\alpha^\lambda(z)}{\sin\alpha\pi} t^\alpha d\alpha$$

$$(-2 < \mathrm{Re}(\lambda) < c < 0) .$$

48. Show that

$$\sin^n \theta P_n (\sin \theta) = \sum_{r=0}^{n} (-)^r \binom{n}{r} \cos^r \theta P_r (\cos \theta) .$$

49. Show that

$$P_n(z) = \left\{ \frac{(-)^n}{n!} \frac{d^n}{dz^n} (v^2 + z^2)^{-1/2} \right\}_{v^2 = 1 - z^2} .$$

50. Show that

$$P_n(z) = \frac{1}{n! \sqrt{\pi}} \int_{-\infty}^{\infty} e^{-(1-z^2)t^2} \left(-\frac{d}{dz} \right)^n e^{-z^2 t^2} dt$$

(Glaisher, Proc. London Math. Soc., VI).

51. Show that: when k is an integer and

$$(1 - 2zh + h^2)^{-k/2} = \sum_{n=0}^{\infty} a_n P_n(z)$$

then

$$a_n = \frac{h^n}{(1 - h^2)^{k-2}} \frac{2^{\frac{1}{2}(k-3)} (2n+1)}{1 \cdot 3 \cdot 5 \cdots (k-2)}$$

$$\times \left\{ \left(h^2 \frac{\partial}{\partial x} + \frac{\partial}{\partial y} \right)^{\frac{1}{2}(k-3)} x^{-n+\frac{k}{2}-2} y^{n+\frac{k}{2}-2} \right\}_{x=y=1}$$

52. Show that, when $|h|$ and $|z|$ are sufficiently small,

$$\frac{1 - h^2}{(1 - 2zh + h^2)^{3/2}} = \sum_{n=0}^{\infty} (2n+1) h^n P_n(z) .$$

53. Suppose that $f(x)$ can be expanded in the form:

$$f(x) = \sum_{n=0}^{\infty} a_n P_n(x)$$

and the series is uniformly convergent in a domain containing $x = 1$. Show that

$$\int_1^x f(x) dx = -a_0 - \frac{1}{3} a_1 + \sum_{n=1}^{\infty} \left(\frac{a_{n-1}}{2n-1} - \frac{a_{n+1}}{2n+3} \right) P_n(x) .$$

54. Show that

$$\frac{1}{2\pi} \int_{-\pi}^{\pi} \{1 - 2h[\cos\omega\cos\theta + \sin\omega\sin\theta\cos(\varphi' - \varphi)] + h^2\}^{-1/2} d\varphi$$

$$= \sum_{n=0}^{\infty} h^n P_n(\cos\omega) P_n(\cos\theta) \ .$$

55. Show that

$$\{1 - hx - (1 - 2hx + h^2)^{1/2}\}^m$$

$$= m(x^2 - 1)^m \sum_{n=m}^{\infty} \frac{k^{n+m}}{(n+m)!} \frac{1}{n} \frac{d^{m+n}}{dx^{n+m}} \left(\frac{x^2 - 1}{2}\right)^n \ .$$

56. Show that

$$\lim_{\lambda \to 0} [\Gamma(\lambda)(\lambda + n) C_n^\lambda(\cos\theta)] = 2\cos n\theta \ .$$

57. Prove the *Whipple transformation:*

$$e^{-\mu\pi i} Q_\nu^\mu(\mathrm{ch}\,\alpha) = \sqrt{\frac{\pi}{2}} \frac{\Gamma(\mu + \nu + 1)}{\sqrt{\mathrm{sh}\,\alpha}} P_{-\mu-\frac{1}{2}}^{-\nu-\frac{1}{2}}(\coth\alpha)$$

[cf. Hobson (1931), Sec. 159, p. 245].

58. For a certain type of boundary-value problem in a conical region, it is led to the special Legendre equation:

$$(1 - x^2)\frac{d^2 y}{dx^2} - 2x\frac{dy}{dx} - \left(p^2 + \frac{1}{4} + \frac{\mu^2}{1 - x^2}\right) y = 0 \ ,$$

where $|x| < 1$ and p a real number. The solutions of this equation, namely,

$$P_{-\frac{1}{2}+ip}^\mu(x) \ , \quad Q_{-\frac{1}{2}+ip}^\mu(x)$$

are called *conical functions*. Show that

$$P_{-\frac{1}{2}+ip}(\cos\theta) = 1 + \frac{4p^2 + 1^2}{2^2}\sin^2\frac{\theta}{2}$$

$$+ \frac{(4p^2 + 1^2)(4p^2 + 3^2)}{2^2\cdot 4^2}\sin^4\frac{\theta}{2} + \cdots$$

$$(0 \le \theta < \pi) \,,$$

$$P_{-\frac{1}{2}+ip}(\cos\theta) = 1 + \frac{p^2 + \left(\frac{1}{2}\right)^2}{2^2}\sin^2\theta$$

$$+ \frac{\left[p^2 + \left(\frac{1}{2}\right)^2\right]\left[p^2 + \left(\frac{3}{2}\right)^2\right]}{2^2\cdot 4^2}\sin^4\theta + \cdots$$

$$(0 \le \theta < \pi/2) \,,$$

from which we see that $P_{-\frac{1}{2}+ip}(1) = 1, P_{-\frac{1}{2}+ip}(-1) = \infty$.

59. Show that the reciprocal of the distance R between two points (r,θ,φ) and $(r',\theta',\varphi'), 1/R$, can be expressed in terms of the conical function as

$$\frac{1}{R} = e^{-\frac{1}{2}(\sigma+\sigma')}\int_0^\infty \frac{\cos p(\sigma - \sigma')}{\text{ch}\,\pi p}P_{-\frac{1}{2}+ip}(-\cos\gamma)dp \,,$$

where $e^\sigma = r, e^{\sigma'} = r', \cos\gamma = \cos\theta\cos\theta' + \sin\theta\sin\theta'\cos(\varphi - \varphi')$ [cf. Hobson (1931), Sec. 262, p. 446].

60. Let ξ, η, φ be the rotational prolate spheroidal coordinates, the focal distance being $2a$(see App. III, 8). Show that the reciprocal of r_{12}, the distance between two points, can be expressed as $(\xi_1 < \xi_2)$

$$a/r_{12} = \sum_{n=0}^\infty (2n+1)P_n(\xi_1)Q_n(\xi_2)P_n(\eta_1)P_n(\eta_2)$$

$$+ 2\sum_{n=1}^\infty (2n+1)\sum_{m=1}^n (-)^m\left[\frac{(n-m)!}{(n+m)!}\right]^2$$

$$\times\, P_n^m(\xi_1)Q_n^m(\xi_2)P_n^m(\eta_1)P_n^m(\eta_2)\cos m(\varphi - \varphi') \,.$$

[Hint: Express r_{12} as $r_{12}^2 = a^2(A^2 - B^2 - C^2)$, where

$$A = \xi_1 \xi_2 - \eta_1 \eta_2 \, ,$$

$$B = \sqrt{(\xi_1^2 - 1)(\xi_2^2 - 1)} + \sqrt{(1 - \eta_1^2)(1 - \eta_2^2)} \cos(\varphi_1 - \varphi_2) \, ,$$

$$C = \sqrt{(1 - \eta_1^2)(1 - \eta_2^2)} \sin(\varphi_1 - \varphi_2) \, .$$

From Eq. (2) of Sec. 5.21 and Eq. (3) of Sec. 5.10, obtain

$$\frac{a}{r_{12}} = \frac{1}{2\pi} \int_0^{2\pi} \frac{dv}{A - B \cos v - C \sin v}$$

$$= \frac{1}{2\pi} \int_0^{2\pi} \sum_{n=0}^{\infty} (2n + 1) P_n(\beta) Q_n(\alpha) dv \, ,$$

where

$$\alpha = \xi_1 \xi_2 - \sqrt{(\xi_1^2 - 1)(\xi_2^2 - 1)} \cos v \, ,$$

$$\beta = \eta_1 \eta_2 + \sqrt{(1 - \eta_1^2)(1 - \eta_2^2)} \cos(\varphi_1 - \varphi_2 - v) \, ,$$

then apply Eq. (7) of Sec 5.14 and Eq. (6) of Sec. 5.21. [Cf. Hobson (1931), p. 416.]

Chapter 6

CONFLUENT HYPERGEOMETRIC FUNCTIONS

Confluent hypergeometric functions are the solutions of the confluent hypergeometric equation. Such equation can be obtained from the hypergeometric equation by the confluence of two of its singularities. Confluent hypergeometric functions include as special cases the commonly used Bessel functions, Hermite functions, Laguerre functions, etc. We shall discuss them from the view of solutions of a differential equation.

6.1. Confluent Hypergeometric Functions

In Sec. 2.8, we have mentioned that the new equation obtained from a differential equation by the confluence of two or more of its singularities is called the confluent equation of the original equation. After confluence, the singularities of the new equation usually have properties more complicated than those of the original ones; it follows that the properties of the solutions are different. In the present chapter, we shall make the two singularities, 1 and ∞, of the hypergeometric equation confluent to bring about the confluent hypergeometric equation, and then discuss the solutions of the new equation — the confluent hypergeometric functions — and their properties.

In the hypergeometric equation

$$z(1-z)\frac{d^2y}{dz^2} + [\gamma - (\alpha+\beta+1)z]\frac{dy}{dz} - \alpha\beta y = 0,$$

replace z by z/b and divide it by b. We obtain

$$z\left(1-\frac{z}{b}\right)\frac{d^2y}{dz^2} + \left[\gamma - (\alpha+\beta+1)\frac{z}{b}\right]\frac{dy}{dz} - \alpha\frac{\beta}{b}y = 0.$$

The singularities of this new equation are $0, b, \infty$, all being regular. Now, let $b = \beta \to \infty$, we obtain

$$z\frac{d^2y}{dz^2} + (\gamma - z)\frac{dy}{dz} - \alpha y = 0. \tag{1}$$

This new equation has only two singularities now, 0 and ∞; the former is still a regular singularity, but the latter, being the confluence of the two original regular singularities, becomes an irregular singularity. Equation (1) is called the *confluent hypergeometric equation*, or *Kummer equation*.

The indicial equation of Eq. (1) at the regular singularity $z = 0$ [Eq. (14) of Sec. 2.4] is

$$\rho(\rho - 1) + \gamma\rho = 0.$$

Its roots are $\rho = 0$ and $1 - \gamma$, same as that of the hypergeometric equation. When $1 - \gamma$ is not an integer, by the method of solution in series [Sec. 2.4], we obtain two linearly independent solutions of Eq. (1):

$$y_1 = F(\alpha, \gamma, z) \tag{2}$$

$$y_2 = z^{1-\gamma}F(\alpha - \gamma + 1, 2 - \gamma, z), \tag{3}$$

where

$$F(\alpha, \gamma, z) = \sum_{n=0}^{\infty} \frac{(\alpha)_n}{n!(\gamma)_n}z^n \quad (\gamma \neq 0, 1, 2, \ldots) \tag{4}$$

is named the *confluent hypergeometric function (series)*; also called *Kummer function*.

When γ is an integer, the sign of γ will decide whether Eq. (2) or (3) is the first solution (the one with the larger exponent). After the first solution has been determined, the corresponding second solution can be obtained with the method depicted in Sec. 2.5 [cf. Sec. 6.9].

Obviously, the confluent hypergeometric function $F(\alpha, \gamma, z)$ can be formally derived from the hypergeometric function $F(\alpha, \beta, \gamma, z)$ by replacing z by z/β and then let $\beta \to \infty$. This kind of limiting process can be employed to derive from the numerous formulae on hypergeometric functions to the corresponding formulae on confluent hypergeometric functions. Of course, the results have to be justified afterwards.

$F(\alpha, \gamma, z)$ is also often written in the form $_1F_1(\alpha; \gamma; z)$, because it is a special case of the generalized hypergeometric functions [Sec. 4.15].

Either from the property of the differential equation (one of its singularities is 0, and the other, ∞), or from the series (4) by its own, we can infer that $F(\alpha, \gamma, z)$ is a single-valued analytic function in the whole z plane, which property is different from that of the hypergeometric function $F(\alpha, \beta, \gamma, z)$ [cf. the paragraph below Eq. (9) of Sec. 4.1].

In addition, for fixed z and $\gamma(\gamma \neq 0, -1, -2, \ldots)$, $F(\alpha, \gamma, z)$ is also an integral function of α, since when $|\alpha| < R$, R being an arbitrary positive number, the absolute value of the ratio between the consecutive terms of the series (4) approaches zero as $n \to \infty$:

$$\left| \frac{(\alpha + n)z}{(n + 1)(\gamma + n)} \right| < \frac{R + n}{n(n - |\gamma|)} |z| \to 0 \ .$$

Similarly, except for $\gamma = 0$ or a negative integer, $F(\alpha, \gamma, z)$ is also an analytic function of γ; $\gamma = -m(m = 0, 1, 2, \ldots)$ are its simple poles, because (4) can be written as

$$F(\alpha, \gamma, z) = 1 + \frac{\Gamma(\gamma)}{\Gamma(\alpha)} \sum_{n=1}^{\infty} \frac{\Gamma(\alpha + n)}{n!\Gamma(\gamma + n)} z^n \ , \tag{5}$$

and $\gamma = -m$ are simple poles of $\Gamma(\gamma)$; thus, except for the factor $\Gamma(\gamma)/\Gamma(\alpha)$, the series on the r.h.s. of (5), being equal to $\sum_{n=m+1}^{\infty} \Gamma(\alpha + n)z^n/n!\Gamma(n - m)$ for $\gamma = -m$, is convergent.

Summing up, $F(\alpha, \gamma, z)/\Gamma(\gamma)$ is a single-valued analytic function of α, γ, z; α, γ, z being arbitrary complex numbers.

Kummer's transformation—Replacing z in Eq. (10) of Sec. 4.3 by z/β, then taking the limit $\beta \to \infty$ formally, we obtain

$$F(\alpha, \gamma, z) = e^z F(\gamma - \alpha, \gamma, -z) \ . \tag{6}$$

This is an important transformation formula, named the *first Kummer for-*

mula, which can be proved directly as follows:

$$e^{-z}F(\alpha,\gamma,z) = \sum_{k=0}^{\infty} \frac{(-z)^k}{k!} \sum_{l=0}^{\infty} \frac{(\alpha)_l}{l!(\gamma)_l} z^l$$

$$= \sum_{n=0}^{\infty} (-z)^n \sum_{l=0}^{n} \frac{(-)^l(\alpha)_l}{l!(n-l)!(\gamma)_l}$$

$$= \sum_{n=0}^{\infty} \frac{(\gamma-\alpha)_n}{n!(\gamma)_n}(-z)^n = F(\gamma-\alpha,\gamma,-z) .$$

In the last step, Eq. (2) of Sec. 4.10 has been used. We shall give another proof of (6) in Sec. 6.4 by using the integral representation of the confluent hypergeometric function.

We also have the *second Kummer formula* [see Ex. 2 at the end of this Chapter]:

$$e^{-\frac{z}{2}}F(\alpha,2\alpha,z) = {}_0F_1\left(\frac{1}{2}+\alpha;\frac{z^2}{16}\right) . \tag{7}$$

6.2. Relations among the Consecutive Functions

Let l and m be arbitrary integers, then $F(\alpha+l,\gamma+m,z)$ is called the consecutive function of $F(\alpha,\gamma,z)$. The four close neighboring functions of $F(\alpha,\gamma,z)$ are usually denoted by the simple notations

$$\left.\begin{aligned} F(\alpha\pm1) &= F(\alpha\pm1,\gamma,z) , \\ F(\gamma\pm1) &= F(\alpha,\gamma\pm1,z) . \end{aligned}\right\} \tag{1}$$

Same as the hypergeometric functions, among any three consecutive functions F_1, F_2, F_3, there exists a relation

$$A_1 F_1 + A_2 F_2 + A_3 F_3 = 0 ,$$

where A_1, A_2, A_3 are rational functions of z. The following are the two most simple ones of such relations:

$$(\gamma-1)F(\gamma-1) - \alpha F(\alpha+1) - (\gamma-\alpha-1)F = 0 , \tag{2}$$

$$\gamma F - zF(\gamma+1) - \gamma F(\alpha-1) = 0 , \tag{3}$$

where F represents $F(\alpha, \gamma, z)$. These two relations can be formally derived, as mentioned in the last section, from Eqs. (2) and (3) of Sec. 4.2 and can also be derived from the integral representation of the confluent hypergeometric functions [cf. Ex. 3 at the end of this chapter]. Or, they may be proved directly from the series expressions.

The $\binom{4}{2} = 6$ relations among the four close neighboring functions and $F(\alpha, \gamma, z)$ itself can all be obtained from (2) and (3) [cf. Ex. 4 at the end of this chapter].

Besides, there is also a recurrence formula similar to that for the hypergeometric functions, namely,

$$\frac{d^m}{dz^m} F(\alpha, \gamma, z) = \frac{(\alpha)_m}{(\gamma)_m} F(\alpha + m, \gamma + m, z) .\tag{4}$$

Other differentiation formulae are given in Ex. 5 at the end of this chapter.

6.3. Whittaker Equation and Whittaker Functions $M_{k,m}(z)$

Another important form of the confluent hypergeometric equation is the Whittaker equation, which is obtained from Eq. (1) of Sec. 6.1 — the Kummer equation — by eliminating the first derivative from the equation. In the Kummer equation $zy'' + (\gamma - z)y' - \alpha y = 0$, put $y = e^{z/2} z^{-\gamma/2} w(z)$. We obtain

$$w'' + \left[-\frac{1}{4} + \left(\frac{\gamma}{2} - \alpha \right) \frac{1}{z} + \frac{\gamma}{2} \left(1 - \frac{\gamma}{2} \right) \frac{1}{z^2} \right] w = 0 .\tag{1}$$

The exponents of the two regular solutions of this equation at the point $z = 0$ are $1 - \frac{1}{2}\gamma$ and $\frac{1}{2}\gamma$. In order to make the two linearly independent solutions of the equation and some of the related formulae appearing in more symmetric forms, we put

$$\gamma = 1 + 2m, \quad \frac{\gamma}{2} - \alpha = k ,\tag{2}$$

i.e.,

$$m = \frac{\gamma - 1}{2} , \quad \alpha = \frac{1}{2} + m - k .\tag{3}$$

Equation (1) then becomes

$$w'' + \left[-\frac{1}{4} + \frac{k}{z} + \frac{\frac{1}{4} - m^2}{z^2} \right] w = 0 ,\tag{4}$$

which is the *Whittaker equation.*

From the relation between the Whittaker equation and the Kummer equation, we know immediately that, if *2m is not an integer*, the two linearly independent solutions of (4) at the point $z = 0$ are [cf. Eqs. (2), (3) of Sec. 6.1]

$$M_{k,m}(z) = e^{-\frac{z}{2}} z^{\frac{7}{2}} F(\alpha, \gamma, z)$$
$$= e^{-\frac{z}{2}} z^{\frac{1}{2}+m} F\left(\frac{1}{2} + m - k, 1 + 2m, z\right), \tag{5}$$

$$M_{k,-m}(z) = e^{-\frac{z}{2}} z^{1-\frac{7}{2}} F(\alpha - \gamma + 1, 2 - \gamma, z)$$
$$= e^{-\frac{z}{2}} z^{\frac{1}{2}-m} F\left(\frac{1}{2} - m - k, 1 - 2m, z\right). \tag{6}$$

$M_{k,\pm m}(z)$ are called *Whittaker functions*; their symmetric property with respect to *m* is manifest.

Owing to the factor $z^{\frac{1}{2}\pm m}$, $M_{k,\pm m}(z)$ are in general multi-valued. Ordinarily, they are defined by

$$-\pi < \arg z < \pi. \tag{7}$$

Then $M_{k,\pm m}(z)$ are single-valued analytic functions in the *z* plane cut along the negative real axis from $-\infty$ to 0.

From (5), together with (2) and (3), we have

$$F(\alpha, \gamma, z) = e^{\frac{z}{2}} z^{-\frac{7}{2}} M_{\frac{7}{2}-\alpha, \frac{\gamma-1}{2}}(z). \tag{8}$$

Expressing in terms of Whittaker functions, the first Kummer formula given in Eq. (6) of Sec. 6.1 becomes

$$z^{-\frac{1}{2}-m} M_{k,m}(z) = (-z)^{-\frac{1}{2}-m} M_{-k,m}(-z). \tag{9}$$

By (5) and (9), we have

$$M_{k,m}(z) = e^{\frac{z}{2}} z^{\frac{1}{2}+m} F\left(\frac{1}{2} + m + k, 1 + 2m, -z\right). \tag{10}$$

When $k = 0$, by the second Kummer formula [Eq. (7) of Sec. 6.1, or Ex. 2 at the end of this chapter], we have

$$M_{0,m}(z) = z_0^{\frac{1}{2}+m} F_1\left(1 + m; \frac{z^2}{16}\right). \tag{11}$$

6.4. Integral Representations

The Kummer equation

$$zy'' + (\gamma - z)y' - \alpha y = 0 \qquad (1)$$

is an equation of the Laplacian type [Sec. 2.13], so its solutions in integral form can be obtained with the method described in Sec. 2.13. Let

$$y(x) = \int_C e^{zt} v(t) dt .$$

From Eq. (6) of Sec. 2.13, we have

$$v(t) = At^{\alpha-1}(1-t)^{\gamma-\alpha-1} .$$

Hence,

$$y(z) = A \int_C e^{zt} t^{\alpha-1}(1-t)^{\gamma-\alpha-1} dt , \qquad (2)$$

where A is an arbitrary constant and the path of integration C is such that

$$\{Q(z,t)\}_C \equiv \{e^{zt} t^{\alpha}(1-t)^{\gamma-\alpha}\}_C = 0 \qquad (3)$$

[see Eqs. (7) and (9) of Sec. 2.13].

We shall now show that, by properly choosing A and the path of integration C, the r.h.s. of (2) can be identified with $F(\alpha, \gamma, z)$.

Assume $\mathrm{Re}(\gamma) > \mathrm{Re}(\alpha) > 0$. We may choose C to be a straight line from 0 to 1. Then (3) is satisfied and the integral in (2) is uniformly convergent in $|z| \leq R < \infty$. Thus, the integral solution is

$$\int_0^1 e^{zt} t^{\alpha-1}(1-t)^{\gamma-\alpha-1} dt = \int_0^1 \sum_{n=0}^{\infty} \frac{z^n}{n!} t^{\alpha+n-1}(1-t)^{\gamma-\alpha-1} dt$$

$$= \sum_{n=0}^{\infty} \frac{\Gamma(\alpha+n)\Gamma(\gamma-\alpha)}{n!\Gamma(\gamma+n)} z^n = \frac{\Gamma(\alpha)\Gamma(\gamma-\alpha)}{\Gamma(\gamma)} F(\alpha, \gamma, z) ,$$

and it follows that

$$F(\alpha, \gamma, z) = \frac{\Gamma(\gamma)}{\Gamma(\alpha)\Gamma(\gamma-\alpha)} \int_0^1 e^{zt} t^{\alpha-1}(1-t)^{\gamma-\alpha-1} dt , \qquad (4)$$

where $\mathrm{Re}(\gamma) > \mathrm{Re}(\alpha) > 0$, $\arg t = \arg(1-t) = 0$. Note that this result can be obtained from Eq. (6) of Sec. 4.5 by the formal limiting process mentioned in Sec. 6.1.

Replacing t in (4) by $1 - t$, we obtain another integral representation

$$F(\alpha, \gamma, z) = \frac{\Gamma(\gamma)e^z}{\Gamma(\alpha)\Gamma(\gamma - \alpha)} \int_0^1 e^{-zt}t^{\gamma-\alpha-1}(1 - t)^{\alpha-1}dt \tag{5}$$
$$\text{Re}(\gamma) > \text{Re}(\alpha) > 0, \quad \arg t = \arg(1 - t) = 0 .$$

Comparing (4) and (5), we come immediately to the Kummer formula $F(\alpha, \gamma, z) = e^z F(\gamma - \alpha, \gamma, -z)$ [Eq. (6) of Sec. 6.1].

The two integral representations given above, though comparatively simple, are rather restricted in its parameters α and γ. To evade the restrictions, we may take as the path of integration in (2) the Pochhammer contour shown in Fig. 6 in Sec. 3.9. Then

$$\int_P^{(1+,0+,1-,0-)} e^{zt}t^{\alpha-1}(1 - t)^{\gamma-\alpha-1}dt$$

$$= \sum_{n=0}^{\infty} \frac{z^n}{n!} \int_P^{(1+,0+,1-,0-)} t^{\alpha+n-1}(1 - t)^{\gamma-\alpha-1}dt$$

$$= \sum_{n=0}^{\infty} \frac{(2\pi i)^2 e^{(\gamma+n)\pi i}}{\Gamma(1 - \alpha - n)\Gamma(1 - \gamma + \alpha)\Gamma(\gamma + n)} \frac{z^n}{n!} \quad \text{[by Eq. (1) of Sec. 3.9]}$$

$$= \frac{(2\pi i)^2 e^{\gamma\pi i}}{\Gamma(1 - \alpha)\Gamma(\gamma)\Gamma(1 - \gamma + \alpha)} F(\alpha, \gamma, z)$$

(on account of $\Gamma(\lambda - n) = (-)^n\Gamma(\gamma)/(1 - \lambda)_n$), and it follows that

$$F(\alpha, \gamma, z) = \frac{\Gamma(1 - \alpha)\Gamma(\gamma)\Gamma(1 - \gamma + \alpha)e^{-\gamma\pi i}}{(2\pi i)^2}$$
$$\times \int_P^{(1+,0+,1-,0-)} e^{zt}t^{\alpha-1}(1 - t)^{\gamma-\alpha-1}dt , \tag{6}$$

at the starting point P on the real axis, $\arg t = \arg(1 - t) = 0$.

(6) can also be formally derived from Eq. (7) of Sec. 4.5 by interchanging α and β, replacing z by z/β, then let $\beta \to \infty$.

When α or $\gamma - \alpha$ is a positive integer, the r.h.s. of (6) is an indeterminate form. In such a case, the result of Ex. 6 at the end of this chapter may be used.

If $\alpha = -n(n = 0, 1, 2, \ldots)$, we may take the path C in (2) to be a simple contour encircling $t = 0$ once, keeping $t = 1$ outside of it. Then we have

$$\int^{(0+)} e^{zt} t^{-n-1} (1 - t)^{\gamma+n-1} dt$$

$$= \int^{(0+)} \sum_{k=0}^{\infty} \frac{z^k}{k!} t^{k-n-1} (1 - t)^{\gamma+n-1} dt$$

$$= \sum_{k=0}^{n} \frac{z^k}{k!} \frac{2\pi i}{(n-k)!} \frac{d^{n-k}}{dt^{n-k}} (1 - t)^{\gamma+n-1} \Big|_{t=0}$$

$$= \sum_{k=0}^{n} \frac{z^k}{k!} \frac{2\pi i}{(n-k)!} (-)^{n-k} (\gamma + n - 1)(\gamma + n - 2) \ldots (\gamma + k)$$

$$= \frac{2\pi i (-)^n \Gamma(\gamma + n)}{n! \Gamma(\gamma)} F(-n, \gamma, z) ,$$

so,

$$F(-n, \gamma, z) = \frac{(-)^n \Gamma(\gamma)}{\Gamma(\gamma + n)} \frac{n!}{2\pi i} \int^{(0+)} e^{zt} t^{-n-1} (1 - t)^{\gamma+n-1} dt \qquad (7)$$

$$|\arg(1 - t)| < \pi .$$

If $\gamma - \alpha$ is zero or a negative integer, the integral representation of $F(\alpha, \gamma, z)$ can be obtained from (7) by using Eq. (6) of Sec. 6.1.
(7) can also be derived directly from (6).

Barnes Integral Representation

Corresponding to Eq. (9) of Sec. 4.6, the confluent hypergeometric function can also be represented by Barnes' integral:

$$F(\alpha, \gamma, z) = \frac{\Gamma(\gamma)}{\Gamma(\alpha)} \frac{1}{2\pi i} \int_{-i\infty}^{i\infty} \frac{\Gamma(\alpha + s)\Gamma(-s)}{\Gamma(\gamma + s)} (-z)^s ds \qquad (8)$$

$$\left(\alpha \neq 0, -1, -2, \ldots ; |\arg(-z)| < \frac{\pi}{2} \right) ,$$

the poles of $\Gamma(\alpha + s)$ lying on the left of the path of integration and the poles of $\Gamma(-s)$ lying on its right; the condition subjected to $\arg(-z)$ is for the convergence of the integral. Equation (8) can be obtained formally from Eq. (9) of Sec. 4.6 by the limiting process described in Sec. 6.1 and the asymptotic formula of the Γ-function [Eq. (5) of Sec. 3.21]. The rigorous proof is similar to that for the hypergeometric function. We shall not repeat it here.

6.5. Whittaker Functions $W_{k,m}(z)$

$M_{k,\pm m}(z)$ are the solutions of Whittaker equation at the point $z = 0$. When $2m$ is an integer, one of them loses its meaning in general [cf. Eqs. (5) and (6) of Sec. 6.3 and Sec. 6.9]. Moreover, these solutions are not convenient for the discussion of the properties of the solutions for large $|z|$. To cope with these problems, Whittaker introduced another two functions, $W_{\pm k,m}(\pm z)$, which are linearly independent solutions of the Whittaker equation in any cases, and are apt to the investigation of the asymptotic behavior of the solutions.

From (2) of the preceding section, replacing zt by $-t$, multiplying with $e^{-z/2}z^{\gamma/2}$, and using Eqs. (2) and (3) of Sec. 6.3 expressing α and γ in terms of k and m, we obtain the solution of the Whittaker equation in integral form:

$$w(z) = Ae^{-\frac{z}{2}}z^k \int_{C'} e^{-t}(-t)^{-k-\frac{1}{2}+m} \left(1 + \frac{t}{z}\right)^{k-\frac{1}{2}+m} dt . \tag{1}$$

The path of integration C' is such as to make

$$\left\{ e^{-t}(-t)^{-k+\frac{1}{2}+m} \left(1 + \frac{t}{z}\right)^{k+\frac{1}{2}+m} \right\}_{C'} = 0 \tag{2}$$

[cf. Eq. (3) of Sec. 6.4]. C' can be taken as a contour which starts from the point at infinity on the positive real axis, encircling the point $t = 0$ once in the positive sense, then returns to the point at infinity, the point $t = -z$ being kept outside the contour.

The definition of the Whittaker function $W_{k,m}(z)$ is

$$W_{k,m}(z) = -e^{-\frac{z}{2}}z^k \frac{\Gamma\left(k + \frac{1}{2} - m\right)}{2\pi i} \int_{\infty}^{(0+)} e^{-t}(-t)^{-k-\frac{1}{2}+m} \left(1 + \frac{t}{z}\right)^{k-\frac{1}{2}+m} dt$$

$$\left(k + \frac{1}{2} - m \neq 0, -1, -2, \ldots\right) , \tag{3}$$

where $|\arg(z)| < \pi, |\arg(-t)| \leq \pi$, and when t approaches the point $t = 0$ along a path inside the contour, $\arg(1 + t/z) \to 0$. The asymptotic expression of this function for $|z| \to \infty$ is very simple (see next section).

Whittaker equation [Eq. (4) of Sec. 6.3] is unchanged when k and, in the mean time, z change their signs. Therefore, the following is also a solution of the equation:

$$W_{-k,m}(-z) = -e^{z/2}(-z)^{-k}\frac{\Gamma\left(-k+\frac{1}{2}-m\right)}{2\pi i}$$

$$\times \int_{\infty}^{(0+)} e^{-t}(-t)^{k-\frac{1}{2}+m}\left(1-\frac{t}{z}\right)^{-k-\frac{1}{2}+m} dt$$

$$\left(-k+\frac{1}{2}-m \neq 0, -1, -2, \ldots ; |\arg(-z)| < \pi, |\arg(-t)| \leq \pi\right). \quad (4)$$

Moreover, it is linearly independent of $W_{k,m}(z)$, a fact which will be clear when we come to their asymptotic expressions [(5) of the following section].

When $k+\frac{1}{2}-m = 0, -1, -2, \ldots$, the r.h.s. of (3) is an indeterminate form, since then $\Gamma\left(k+\frac{1}{2}-m\right) \to \infty$, and as $t = 0$ is no longer a branch point of the integrand, the value of the integral is zero. In such cases, we assume first that $k+\frac{1}{2}-m$ is not equal to zero or a negative integer. Then, we can deform the contour of integration into one which starts from infinity, runs along the upper "bank" of the positive axis to $\delta(>0)$, encircling $t = 0$ once in the positive sense, and then returns to infinity along the lower "bank" of the positive real axis. Assume $\text{Re}\left(k-\frac{1}{2}-m\right) < 0$, then the value of the integral along the circle around $t = 0$ approaches zero as δ tends to zero, and

$$\int_{\infty}^{(0+)} e^{-t}(-t)^{-k-\frac{1}{2}+m}\left(1+\frac{t}{z}\right)^{k-\frac{1}{2}+m} dt$$

$$= \left[e^{(-k-\frac{1}{2}+m)\pi i} - e^{-(-k-\frac{1}{2}+m)\pi i}\right]\int_0^{\infty} e^{-t}t^{-k-\frac{1}{2}+m}\left(1+\frac{t}{z}\right)^{k-\frac{1}{2}+m} dt$$

$$= -2i\sin\pi\left(k+\frac{1}{2}-m\right)\int_0^{\infty} e^{-t}t^{-k-\frac{1}{2}+m}\left(1+\frac{t}{z}\right)^{k-\frac{1}{2}+m} dt.$$

Substituting into (3) and using Eq. (2) of Sec. 3.5, we have

$$W_{k,m}(z) = \frac{e^{-z/2}z^k}{\Gamma\left(\frac{1}{2}-k+m\right)}\int_0^{\infty} e^{-t}t^{-k-\frac{1}{2}+m}\left(1+\frac{t}{z}\right)^{k-\frac{1}{2}+m} dt, \quad (5)$$

which can be taken as the representation of $W_{k,m}(z)$ under the condition $k+\frac{1}{2}-m = 0, -1, -2, \ldots$.

For $W_{-k,m}(-z)$, there is a similar expression. We need only to change z and k in (5) to $-z$ and $-k$, respectively.

In the literature, some use $\Psi(\alpha, \gamma, z)$ instead of $W_{k,m}(z)$, and call it *Tricomi function*; its definition is

$$\Psi(\alpha, \gamma, z) = e^{\frac{z}{2}} z^{-m-\frac{1}{2}} W_{k,m}(z)$$

$$\left(k = \frac{\gamma}{2} - \alpha, m = \frac{\gamma - 1}{2}, \alpha = \frac{1}{2} + m - k, \gamma = 2m + 1 \right) . \tag{6}$$

From (3), we have

$$\Psi(\alpha, \gamma, z) = \frac{e^{-\alpha \pi i} \Gamma(1 - \alpha)}{2\pi i} \int_{\infty e^{i\varphi}}^{(0+)} e^{-zt} t^{\alpha-1} (1 + t)^{\gamma - \alpha - 1} dt , \tag{7}$$

$t = -1$ lying outside the contour; at the starting point of the path of integration, $\arg t = \varphi, |\varphi + \arg z| < \pi/2$ to ensure the convergence of the integral; when t approaches 0 inside the contour, $\arg(1 + t) \to 0$.

Since, by definition, the function Ψ, in comparison with $W_{k,m}(z)$, does not contain the exponential factor and the power factor of z, some formulae, when referred to this function, are simpler [cf. Erdélyi (1953), Vol. I, Chap. 6].

6.6. Asymptotic Expansion of $W_{k,m}(z)$ when $z \to \infty$

In (3) of the last section, namely

$$W_{k,m}(z) = -\frac{e^{-z/2} z^k \Gamma\left(k + \frac{1}{2} - m\right)}{2\pi i}$$

$$\times \int_{\infty}^{(0+)} e^{-t} (-t)^{-k-\frac{1}{2}+m} \left(1 + \frac{t}{z} \right)^{k-\frac{1}{2}+m} dt$$

$$(| \arg z| < \pi, \ | \arg(-t)| \leq \pi) , \tag{1}$$

take the binomial expansion of the last factor of the integrand and assume $| \arg z| \leq \frac{\pi}{2} - \delta(\delta > 0)$, then, by Watson's lemma [Eq. (5) of Sec. 1.9], we arrive immediately at the asymptotic expansion

$$W_{k,m}(z) \sim e^{-z/2} z^k \left\{ 1 + \sum_{n=1}^{\infty} (-)^n \frac{\left(\frac{1}{2} - k + m\right)_n \left(\frac{1}{2} - k - m\right)_n}{n! z^n} \right\} \tag{2}$$

$$(|z| \to \infty) .$$

This formula is obtained under the condition: $| \arg z| \leq \frac{\pi}{2} - \delta(\delta > 0)$. We shall show in the following that it also holds in $| \arg z| \leq \pi - \delta$.

Let

$$\left(1 + \frac{t}{z}\right)^p = \sum_{n=0}^{N} \binom{p}{n}\left(\frac{t}{z}\right)^n + R_N(t,z) \tag{3}$$

$(p = k - \frac{1}{2} + m)$. Substituting into (1) and using Eq. (4) of Sec. 3.7, we have

$$W_{k,m}(z) = e^{-z/2} z^k \left\{ 1 + \sum_{n=1}^{N}(-)^n \frac{\left(\frac{1}{2} - k + m\right)_n \left(\frac{1}{2} - k - m\right)_n}{n! z^n} \right.$$

$$\left. - \frac{\Gamma\left(k + \frac{1}{2} - m\right)}{2\pi i} \int_{\infty}^{(0+)} e^{-t}(-t)^q R_N(t,z)\, dt \right\} \tag{4}$$

$(q = -k - \frac{1}{2} + m)$. It is only necessary to show that the value of the last integral is of the order $O(z^{-N-1})$ when $|z| \to \infty$. For this sake, we deform the path of integration and obtain

$$\int_{\infty}^{(0+)} e^{-t}(-t)^q R_N(t,z)\, dt$$

$$= \int_{C_r} e^{-t}(-t)^q R_N(t,z)\, dt + \left(e^{i\pi q} - e^{-i\pi q}\right) \int_{r}^{\infty} e^{-t} t^q R_N(t,z)\, dt \, ,$$

where C_r is a small circle with $t = 0$ as centre and radius equal to r. If $|z|$ is sufficiently large so that, on C_r, $|t/z| < 1$, we have the Taylor expansion

$$\left(1 + \frac{t}{z}\right)^p = \sum_{n=0}^{\infty} \binom{p}{n}\left(\frac{t}{z}\right)^n .$$

Then, according to (3),

$$R_N(t,z) = \sum_{n=N+1}^{\infty} \binom{p}{n}\left(\frac{t}{z}\right)^n = \sum_{N+1}^{\infty} a_n \left(\frac{t}{z}\right)^n .$$

But, by the Cauchy inequality if $|t/z| \le \lambda < \lambda_0 < 1$, then $|a_n| < M/\lambda_0^n$, M being the upper bound of $|(1 + t/z)^p|$, and

$$|R_N(t,z)| < \sum_{N+1}^{\infty} \frac{M}{\lambda_0^n}\left|\frac{t}{z}\right| = \frac{M}{\lambda_0^{N+1}(1 - \lambda/\lambda_0)}\left|\frac{t}{z}\right|^{N+1} .$$

From which we see

$$\left|\int_{C_r}\right| = O(|z|^{-N-1}) .$$

For the integral from r to ∞, the Taylor expansion is not available, so it requires some other means to estimate $R_N(t, z)$.

From the identity

$$(1 + \varsigma)^p = 1 + p\varsigma \int_0^1 (1 + \varsigma v)^{p-1} dv \ ,$$

integrating by parts N times gives

$$(1 + \varsigma)^p = \sum_{n=0}^{N} \binom{p}{n} \varsigma^n$$

$$+ \frac{p(p-1)\ldots(p-N)}{N!} \varsigma^{N+1} \int_0^1 (1-v)^N (1 + \varsigma v)^{-N-1} dv \ ,$$

and hence

$$R_N(t, z) = \frac{p(p-1)\ldots(p-N)}{N!} \varsigma^{N+1} \int_0^1 (1-v)^N (1 + \varsigma v)^{p-N-1} dv \ ,$$

where $\varsigma = t/z, r \le t < \infty, |\arg(1 + \varsigma v)| < \pi$. Let $\varsigma v = tv/z = \xi$.

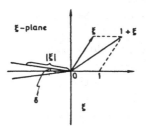

Fig. 23.

Since t and v are all real numbers and $|\arg z| \le \pi - \delta$, so $|\arg \xi| < \pi$. From Fig. 23, it is easily seen that

$$|1 + \xi| \ge \sqrt{1 + |\xi|^2 + 2|\xi| \cos(\pi - \delta)}$$
$$= \sqrt{1 + |\xi|^2 - 2|\xi| \cos \delta} \ .$$

When $|\xi| = \cos \delta$, the value of the radical is a minimum. Therefore, $|1 + \xi| \ge \sin \delta (\delta > 0)$, and

$$\left| \int_0^1 (1-v)^N (1 + \varsigma v)^{p-N-1} dv \right|$$

$$\le \int_0^1 (1-v)^N |1 + \xi|^{\mathrm{Re}(p)-N-1} e^{-\mathrm{Im}(p) \arg(1+\xi)} dv \ .$$

Assume $\mathrm{Re}(p) - N - 1 < 0$ (so long as N is large enough, this condition can always be satisfied), then the last integral is

$$\leq K \int_0^1 (1 - v)^N (\sin \delta)^{\mathrm{Re}(p) - N - 1} dv$$

$$= K (\sin \delta)^{\mathrm{Re}(p) - N - 1} / (N + 1) \, ,$$

K being the upper bound of $\exp\{-\mathrm{Im}(p) \arg(1 + \xi)\}$. Thus,

$$|R_N(t, z)| \leq K \frac{|p(p - 1) \ldots (p - N)|}{(N + 1)!} (\sin \delta)^{\mathrm{Re}(p) - N - 1} \left|\frac{t}{z}\right|^{N+1}$$

and

$$\left|\int_r^\infty e^{-t} t^q R_N(t, z) dt\right| = O(|z|^{N+1}) \, ,$$

which proves that (2) holds under the condition $|\arg z| \leq \pi - \delta$. Although we have assumed, in the proof, that $\mathrm{Re}(p) < N + 1$, so N has to be sufficiently large, this restriction, in fact, can be retracted. For, every term of the sum in (4) is of the order of $O(z^{-n})$.

In the following section, we shall show, by Barnes integral representation, that the *asymptotic expansion (2) even holds under the condition* $|\arg z| < 3\pi/2$. Nevertheless, we still give the above proof, because it is typical and widely used.

Moreover, from (2), we immediately have

$$W_{\pm k,m}(\pm z) = e^{\mp \frac{z}{2}} (\pm z)^{\pm k} \{1 + O(z^{-1})\} \, , \qquad (5)$$

from which we see that, for any k and m, $W_{k,m}(z)$ and $W_{-k,m}(-z)$ are always linearly independent of each other.

6.7. Barnes' Integral Representation of $W_{k,m}(z)$

From the asymptotic series representation, using the method similar to that described in Sec. 4.6, we can formally obtain the Barnes' integral representation for $W_{k,m}(z)$:

$$W_{k,m}(z) = \frac{e^{-\frac{z}{2}} z^k}{2\pi i} \int_{-i\infty}^{i\infty} \frac{\Gamma(s) \Gamma\left(-s - k - m + \frac{1}{2}\right) \Gamma\left(-s - k + m + \frac{1}{2}\right)}{\Gamma\left(-k - m + \frac{1}{2}\right) \Gamma\left(-k + m + \frac{1}{2}\right)} z^s ds \, .$$

$$(1)$$

The path of integration is such as to keep the poles of $\Gamma(s)$ on its left and those of $\Gamma(-s - k \pm m + \frac{1}{2})$ on its right. This requires that $k \pm m + \frac{1}{2}$ do not equal to positive integers.

Now, let us prove (1). Firstly, by the asymptotic expansion of the Γ-function [Eq. (5) of Sec. 3.21], we see that, when s approaches infinity along the imaginary axis,

$$\Gamma(s)\Gamma(-s + \lambda)\Gamma(-s + \mu)$$
$$= \frac{\pi^2 \Gamma(s)}{\Gamma(s + 1 - \lambda)\Gamma(s + 1 - \mu)} \csc \pi(s + 1 - \lambda) \csc \pi(s + 1 - \mu)$$
$$= O\left(|s|^{\lambda + \mu - \frac{3}{2}} e^{-3\pi |\mathrm{Im}(s)|/2}\right).$$

Hence, the integral on the r.h.s. of (1) is uniformly convergent in $|\arg z| \le 3\pi/2 - \delta < 3\pi/2$ and represents an analytic function on the Riemann sheets: $|\arg z| < 3\pi/2$.

Next, we are going to show that the asymptotic expansion of the r.h.s. of (1) is actually the same as that of $W_{k,m}(z)$ when $|z| \to \infty$. For this purpose, let N be a sufficiently large positive integer so that all the poles of $\Gamma(-s - k \pm m + \frac{1}{2})$ lie on the right of the straight line $\mathrm{Re}(s) = -N - \frac{1}{2}$. From the asymptotic expansion of the Γ-function, it can be shown that, in $|\arg z| \le 3\pi/2 - \delta$, the values of the integrals

$$\int_{-i\xi}^{-N-\frac{1}{2}-i\xi} \quad \text{and} \quad \int_{i\xi}^{-N-\frac{1}{2}+i\xi} \quad (\xi \text{ real})$$

approach zero as ξ tends to infinity. Then, by the theorem of residues, the r.h.s. of (1) is equal to

$$I = e^{-\frac{z}{2}} z^k \left\{ \sum_{n=0}^{N} R_n + \frac{1}{2\pi i} \right.$$
$$\left. \times \int_{-N-\frac{1}{2}-i\infty}^{-N-\frac{1}{2}+i\infty} \frac{\Gamma(s)\Gamma\left(-s - k - m + \frac{1}{2}\right)\Gamma\left(-s - k + m + \frac{1}{2}\right)}{\Gamma\left(-k - m + \frac{1}{2}\right)\Gamma\left(-k + m + \frac{1}{2}\right)} z^s \, ds \right\},$$

where R_n is the residue of the integrand at the point $s = -n$.

Let $s = -N - \frac{1}{2} + it$ and use the asymptotic expansion of the Γ-function. We obtain the modulus of the integrand to be

$$|z|^{-N-\frac{1}{2}} O\left\{ e^{-\delta|t|} |t|^{N-2k} \right\}$$

and hence

$$I = e^{-\frac{z}{2}} z^k \left\{ \sum_{n=0}^{N} R_n + O\left(|z|^{-N-\frac{1}{2}}\right) \right\} .$$

The residue R_n is found to be $(-)^n \left(\frac{1}{2} - k + m\right)_n \left(\frac{1}{2} - k - m\right)_n / n! \, z^n$, from which we see that I has the same asymptotic expansion as $W_{k,m}(z)$ [cf. (2) of the last section].

Finally, we come to show that the integral I satisfies the Whittaker equation [Eq. (4) of Sec. 6.3]. To this end, let $W(z) = e^{-z/2} z^k v(z)$. Substituting in the said equation gives

$$z^2 v'' + 2kzv' + \left(k - m - \frac{1}{2}\right)\left(k + m - \frac{1}{2}\right) v - z^2 v' = 0 .$$

Putting the integral

$$\int_{-i\infty}^{i\infty} \Gamma(s) \Gamma\left(-s - k - m + \frac{1}{2}\right) \Gamma\left(-s - k + m + \frac{1}{2}\right) z^s ds$$

into the l.h.s. of the equation for v, we obtain

$$\left(\int_{-i\infty}^{i\infty} - \int_{1-i\infty}^{1+i\infty}\right) \Gamma(s) \Gamma\left(-s - k - m + \frac{3}{2}\right) \Gamma\left(-s - k + m + \frac{3}{2}\right) z^s ds .$$

In between these two paths of integration, there are no singularities of the integrand and, as we have shown above, the value of the integrand approaches zero when s tends to infinity along the paths of integration. Therefore, by Cauchy theorem, the integral vanishes, i.e., I satisfies the Whittaker equation. Hence,

$$I = AW_{k,m}(z) + BW_{-k,m}(-z) .$$

Let $|z|$ approaches infinity, $\text{Re}(z) > 0$. From the asymptotic expressions of $W_{\pm k,m}(\pm z)$ [Eq. (5) of Sec. 6.6], we see that B must be zero. In addition, since I and $W_{k,m}(z)$ have the same asymptotic expression, $A = 1$ and (1) is justified.

From (1), the domain of $W_{k,m}(z)$ is extended to $|\arg z| < 3\pi/2$. The result also illustrates that Eq. (2) of Sec. 6.6 is valid in this domain.

6.8. Relations between $W_{\pm k, m}(\pm z)$ and $M_{\pm k, \pm m}(\pm z)$. Asymptotic Expansion of $F(\alpha, \gamma, z)$. Stokes Phenomenon

In (1) of Sec. 6.7, let

$$
\begin{aligned}
F(s) &\equiv \Gamma(s)\Gamma\left(-s-k-m+\frac{1}{2}\right)\Gamma\left(-s-k+m+\frac{1}{2}\right) \\
&= \frac{\pi^2 \Gamma(s)}{\Gamma\left(s+k+m+\frac{1}{2}\right)\Gamma\left(s+k-m+\frac{1}{2}\right)} \\
&\quad \times \frac{1}{\cos\pi(s+k+m)\cos\pi(s+k-m)} .
\end{aligned}
$$

When $|s| \to \infty$, $\mathrm{Re}(s) \geq 0$, using the asymptotic expansion of the Γ-function, we have

$$
F(s) = O\left[\exp\left\{\left(-s-\frac{1}{2}-2k\right)\ln s + s\right\}\right]\sec\pi(s+k+m)\sec\pi(s+k-m) .
$$

Let C_ρ be a semi-circle of radius ρ on the right of the imaginary axis; when $\rho \to \infty$, C_ρ is assumed not to pass through any singularities of $F(s)z^s$. Then, so long as $|\arg z| < 3\pi/2$, $\lim_{\rho\to\infty}\int_{C_\rho}F(s)z^s ds = 0$ [cf. Sec. 4.6 for similar discussion]. And, therefore,

$$
W_{k,m}(z) = \frac{-e^{-\frac{1}{2}z}z^k\left(\sum R'\right)}{\Gamma\left(-k-m+\frac{1}{2}\right)\Gamma\left(-k+m+\frac{1}{2}\right)} ,
$$

where $\sum R'$ is the sum of the residues of $F(s)z^s$ at its poles on the right of the Barnes' path of integration.

If $2m$ is not an integer, all the poles of $F(s)$ are simple. By calculating the corresponding residues, we find [cf. Eqs. (5) and (6) of Sec. 6.3]

$$
W_{k,m}(z) = \frac{\Gamma(-2m)}{\Gamma\left(\frac{1}{2}-m-k\right)}M_{k,m}(z) + \frac{\Gamma(2m)}{\Gamma\left(\frac{1}{2}+m-k\right)}M_{k,-m}(z) \tag{1}
$$
$$
(2m \neq \text{integer}, \quad |\arg z| < 3\pi/2) ,
$$

from which we obtain instantly the important relation

$$
W_{k,m}(z) = W_{k,-m}(z) . \tag{2}
$$

Changing the signs of k and z in (1) simultaneously, we obtain

$$
W_{-k,m}(-z) = \frac{\Gamma(-2m)}{\Gamma\left(\frac{1}{2}-m+k\right)}M_{-k,m}(-z) + \frac{\Gamma(2m)}{\Gamma\left(\frac{1}{2}+m+k\right)}M_{-k,-m}(-z)
$$
$$
(2m \neq \text{integer}, \quad |\arg(-z)| < 3\pi/2) .
$$
$$
\tag{3}
$$

Using Eq. (9) of Sec. 6.3, putting $-z = ze^{-\pi i}$, we obtain from (3)

$$
\begin{aligned}
W_{-k,m}(ze^{-\pi i}) = {} & \frac{\Gamma(-2m)}{\Gamma\left(\frac{1}{2}-m+k\right)} e^{-\left(\frac{1}{2}+m\right)\pi i} M_{k,m}(z) \\
& + \frac{\Gamma(2m)}{\Gamma\left(\frac{1}{2}+m+k\right)} e^{-\left(\frac{1}{2}-m\right)\pi i} M_{k,-m}(z) .
\end{aligned}
\tag{4}
$$

Eliminating $M_{k,-m}(z)$ from (1) and (4) gives

$$
\begin{aligned}
M_{k,m}(z) = {} & \frac{\Gamma(2m+1)}{\Gamma\left(\frac{1}{2}+m+k\right)} e^{\left(\frac{1}{2}+m-k\right)\pi i} W_{k,m}(z) \\
& + \frac{\Gamma(2m+1)}{\Gamma\left(\frac{1}{2}+m-k\right)} e^{-k\pi i} W_{-k,m}(ze^{-\pi i}) ,
\end{aligned}
\tag{5}
$$

where $2m \neq$ negative integer, $-\pi/2 < \arg z < 3\pi/2$, to ensure that the r.h.s. is significant.

If the relation between $M_{k,m}(z)$ and $W_{\pm k,m}(\pm z)$ in $-3\pi/2 < \arg z < \pi/2$ is desired, we only have to put $-z = ze^{\pi i}$ in (9) of Sec. 6.3. Then by the same manipulation as above, we obtain

$$
\begin{aligned}
M_{k,m}(z) = {} & \frac{\Gamma(2m+1)}{\Gamma\left(\frac{1}{2}+m+k\right)} e^{-\left(\frac{1}{2}+m-k\right)\pi i} W_{k,m}(z) \\
& + \frac{\Gamma(2m+1)}{\Gamma\left(\frac{1}{2}+m-k\right)} e^{k\pi i} W_{-k,m}(ze^{\pi i}) ,
\end{aligned}
\tag{6}
$$

where $2m \neq$ negative integer, $-3\pi/2 < \arg z < \pi/2$.

Substituting the asymptotic expansion of $W_{k,m}(z)$ [Eq. (2) of Sec. 6.6], in (5) or (6), we obtain respectively the asymptotic expansion of $M_{k,m}(z)$ in $-\pi/2 < \arg z < 3\pi/2$ and in $-3\pi/2 < \arg z < \pi/2$. Further, by Eq. (8) of Sec. 6.3, we obtain the asymptotic expansion of the Kummer function:

$$
\begin{aligned}
F(\alpha,\gamma,z) \sim {} & \frac{\Gamma(\gamma)}{\Gamma(\gamma-\alpha)} e^{\pm\alpha\pi i} z^{-\alpha} \left\{ 1 + \sum_{n=1}^{\infty} (-)^n \frac{(\alpha)_n(1-\gamma+\alpha)_n}{n!\, z^n} \right\} \\
& + \frac{\Gamma(\gamma)}{\Gamma(\alpha)} e^z z^{\alpha-\gamma} \left\{ 1 + \sum_{n=1}^{\infty} \frac{(\gamma-\alpha)_n(1-\alpha)_n}{n!\, z^n} \right\} ,
\end{aligned}
\tag{7}
$$

where the positive sign in the factor $e^{\pm\alpha\pi i}$ is taken when $-\pi/2 < \arg z < 3\pi/2$, and the negative sign is taken when $-3\pi/2 < \arg z < \pi/2$.

Stokes Phenomenon

In appearance, Eq. (7) is inconsistent. For, the function $F(\alpha, \gamma, z)$ on the l.h.s. is a single-valued analytic function in the whole plane, while that on the r.h.s. appears to be multi-valued in, for instance, $-\pi/2 < \arg z < \pi/2$ owing to the factor $e^{\pm \alpha \pi i}$. This occurs also in the second and the third quadrants, but owing to the second term. This paradox is called the *Stokes phenomenon*, which is a general characteristic of the asymptotic expansion: In different domains of validity, the asymptotic expressions are different, so that on the boundary of two domains or in their common region, there occurs discontinuity or multi-valuedness, even though original function is continuous and single-valued there.

However, such inconsistency is only apparent. For, the fact is that, in $-\pi/2 < \arg z < \pi/2$ for instance, since $\text{Re}(z) > 0$, the second term on the r.h.s. of (7) is the dominant term of the asymptotic expansion and the first term, which causes the apparent discontinuity or multi-valuedness, is the one to be neglected; i.e.,

$$F(\alpha, \gamma, z) \sim \frac{\Gamma(\gamma)}{\Gamma(\alpha)} e^z z^{\alpha - \gamma} \left\{ 1 + \sum_{n=1}^{\infty} \frac{(\gamma - \alpha)_n (1 - \alpha)_n}{n! \, z^n} \right\} \tag{8}$$

$$(|\arg z| < \pi/2) .$$

In the second and the third quadrants, since $\text{Re}(z) < 0$, the first term on the r.h.s. of (7) is the dominant one, and so

$$F(\alpha, \gamma, z) \sim \frac{\Gamma(\gamma)}{\Gamma(\gamma - \alpha)} e^{\pm \alpha \pi i} z^{-\alpha} \left\{ 1 + \sum_{n=1}^{\infty} (-)^n \frac{(\alpha)_n (1 - \gamma + \alpha)_n}{n! \, z^n} \right\} , \tag{9}$$

where the positive sign in the factor $e^{\pm \alpha \pi i}$ is taken when $\pi/2 < \arg z < 3\pi/2$ and the negative sign is taken when $-3\pi/2 < \arg z < -\pi/2$; in this way, then, the r.h.s. is actually single-valued.

Further, if desired, we can utilize the single-valuedness of $F(\alpha, \gamma, z)$ to deduce asymptotic expansions of it for other ranges of the phase of z. For example, for the asymptotic expansion in $\pi < \arg z < 2\pi$, we set $x = z e^{-2\pi i}$ so that $-\pi < \arg x < 0$. Then we have from (7)

$$F(\alpha, \gamma, z) = F\left(\alpha, \gamma, x e^{2\pi i}\right) = F(\alpha, \gamma, x)$$

$$\sim \frac{\Gamma(\gamma)}{\Gamma(\gamma - \alpha)} e^{-\alpha \pi i} x^{-\alpha} \{1 + \ldots\} + \frac{\Gamma(\gamma)}{\Gamma(\alpha)} e^x x^{\alpha - \gamma} \{1 + \ldots\}$$

$$= \frac{\Gamma(\gamma)}{\Gamma(\gamma-\alpha)}e^{\alpha\pi i}z^{-\alpha}\left\{1+\sum_{n=1}^{\infty}(-)^n\frac{(\alpha)_n(1-\gamma+\alpha)_n}{n!\,z^n}\right\}$$

$$+\frac{\Gamma(\gamma)}{\Gamma(\alpha)}e^{-2(\alpha-\gamma)\pi i}e^z z^{\alpha-\gamma}\left\{1+\sum_{n=1}^{\infty}\frac{(\gamma-\alpha)_n(1-\alpha)_n}{n!\,z^n}\right\} \tag{10}$$

$$(\pi < \arg z < 2\pi)\,.$$

6.9. The Case when γ (or $2m$) is an Integer

When γ is an integer, one of the two solutions of the confluent hypergeometric equation at $z = 0$, namely, $F(\alpha,\gamma,z)$ and $z^{1-\gamma}F(\alpha-\gamma+1,2-\gamma,z)$ is in general meaningless. When γ is a negative integer, $-m$, $F(\alpha,\gamma,z)$ is of no meaning, unless α is also a negative integer, $-n$, in which case $F(\alpha,\gamma,z)$ becomes a polynomial of degree n:

$$F(-n,-m,z) = \sum_{k=0}^{n}\frac{(-n)_k}{k!(-m)_k}z^k \quad (m \geq n) \tag{1}$$

provided $m \geq n$. The other solution $z^{1-\gamma}F(\alpha-\gamma+1,2-\gamma,z) = z^{1+m}$ $\times F(m-n+1,2+m,z)$ is still an infinite series. If $\gamma = 1$, then the two solutions are identical. If γ is a positive integer greater than, or equal to, 2, then $F(\alpha-\gamma+1,2-\gamma,z)$ loses its meaning, unless $\alpha-\gamma+1$ is a negative integer $\geq 2-\gamma$. Then, $F(\alpha-\gamma+1,2-\gamma,z)$ is a polynomial while $F(\alpha,\gamma,z)$ remains to be an infinite series (since $\alpha \geq 1$).

Therefore, when γ is a positive integer while α or $\alpha-\gamma+1$ is not an appropriate negative integer, we need to look for another solution. This solution may be found by the Frobenius method mentioned in Sec. 2.5. It is[a]

$$G(\alpha,\gamma,z) = F(\alpha,\gamma,z)\ln z$$
$$+\sum_{k=0}^{\infty}\frac{(\alpha)_k z^k}{k!(\gamma)_k}\{\psi(\alpha+k)-\psi(\gamma+k)-\psi(1+k)\}$$
$$+\frac{\Gamma(\gamma-1)\Gamma(\gamma)\Gamma(\alpha-\gamma+1)(-)^\gamma}{\Gamma(\alpha)}\sum_{k=0}^{\gamma-2}\frac{(\alpha-\gamma+1)_k}{k!(2-\gamma)_k}z^{k+1-\gamma}, \tag{2}$$

where $\gamma = 1,2,3,\ldots$; when $\gamma = 1$, the last finite sum is negated; also, it is assumed that both α and $\alpha-\gamma+1$ are not zero or negative integers.

---

[a]cf. Sec. 4.4 for detailed calculation.

If $\gamma = 0, -1, -2, \ldots$, whereas α and $\alpha - \gamma + 1$ are not zero or negative integers, the second solution can be obtained from (2). We only have to replace α in it by $\alpha - \gamma + 1$, γ by $2 - \gamma$, and to multiply the result with $z^{1-\gamma}$.

From (2) and Eq. (5) of Sec. 6.3, we find, as the second solution of the Whittaker equation for $2m = 0, 1, 2, \ldots$,

$$M_{k,m}(z) \ln z + e^{-\frac{z}{2}} z^{\frac{1}{2}+m} \sum_{n=0}^{\infty} \frac{\left(\frac{1}{2}-k+m\right)_n}{n!(1+2m)_n} z^n$$

$$\times \left\{ \psi\left(\frac{1}{2} - k + m + n\right) - \psi(1 + 2m + n) - \psi(1 + n) \right\}$$

$$+ (-)^{2m+1} e^{-\frac{z}{2}} z^{\frac{1}{2}+m} \frac{(2m-1)!(2m)! \Gamma\left(\frac{1}{2}-k-m\right)}{\Gamma\left(\frac{1}{2}-k+m\right)}$$

$$\times \sum_{n=0}^{2m-1} \frac{\left(\frac{1}{2}-k-m\right)_n}{n!(1-2m)_n} z^{n-2m} , \tag{3}$$

where $M_{k,m}(z)$ is the first solution; $\frac{1}{2} - k \pm m \neq$ zero or negative integer; when $m = 0$, the last finite sum is negated.

The relation between $W_{k,m}(z)$ and the two linearly independent solutions at the point $z = 0$ may be obtained from the Barnes' integral representation of $W_{k,m}(z)$ [Eq. (1) of Sec. 6.7] through the calculation of the residues. The result is

$$W_{k,m}(z) = \frac{(-)^{2m+1}}{(2m)! \Gamma\left(\frac{1}{2}-k-m\right)} \left[M_{k,m}(z) \ln z + e^{-\frac{z}{2}} z^{\frac{1}{2}+m} \right.$$

$$\times \sum_{n=0}^{\infty} \frac{\left(\frac{1}{2}-k+m\right)_n}{n!(1+2m)_n} z^n \left\{ \psi\left(\frac{1}{2} - k + m + n\right) - \psi(1 + 2m + n) \right.$$

$$\left. - \psi(1+n) \right\} + (-)^{2m+1} e^{-\frac{z}{2}} z^{\frac{1}{2}+m}$$

$$\times \frac{(2m-1)!(2m)! \Gamma\left(\frac{1}{2}-k-m\right)}{\Gamma\left(\frac{1}{2}-k+m\right)} \sum_{n=0}^{2m-1} \frac{\left(\frac{1}{2}-k-m\right)_n}{n!(1-2m)_n} z^{n-2m} \left. \right] \tag{4}$$

$$(2m = 0, 1, 2, \ldots)$$

which differs from (3) only by a constant multiple.

If $\frac{1}{2} - k + m$ is equal to zero or a negative integer, then $\frac{1}{2} - k - m$ is zero or a negative integer too (since $2m$ is supposed to be a positive integer), and (3) is meaningless. However, the r.h.s. of (4), being an indeterminate form, may have a limit. Calculation shows that $W_{k,m}(z)$ differs from $M_{k,m}(z)$ only by a constant factor. This, in fact, can be seen at once from (5) of the last section. Besides, in this case, the series in the first solution breaks up into a polynomial [cf. Eq. (5) of Sec. 6.3], while $M_{k,-m}(z)$ is meaningless [cf. the previous discussion on Kummer functions]. Then, we may take $W_{-k,m}(-z)$ as the second solution, since it is linearly independent of $W_{k,m}(z)$ in any cases [cf. Eq. (5) of Sec. 6.6].

If $\frac{1}{2} - k - m$ is zero or a negative integer, but $\frac{1}{2} - k + m$ is a positive integer, then, by evaluating (4) as an indeterminate form, it is found that $W_{k,m}(z)$ and $M_{k,-m}(z)$ differ from each other only by a constant factor.

6.10. The Asymptotic Expansions of $F(\alpha, \gamma, z)$ for Large $|\alpha|, |\gamma|$

Since the series expression of $F(\alpha, \gamma, z)$ is convergent in the whole z plane, the series is also the asymptotic expansion for $|\gamma| \to \infty$ when α and z take any finite values, i.e.,

$$F(\alpha, \gamma, z) = \sum_{n=1}^{N} \frac{(\alpha)_n}{n!(\gamma)_n} z^n + O(|\gamma|^{-N-1}) \tag{1}$$

$$(|\arg \gamma| \le \pi - \delta, \quad \delta > 0) \ .$$

This is readily shown by using the asymptotic expansion of the Γ-function [Eq. (5) of Secs. 3.21 and 4.14].

If $|\alpha|$ and $|\gamma|$ approach infinity simultaneously, but $|\gamma - \alpha|$ is finite, then, by using the first Kummer formula [Eq. (6) of Sec. 6.1], replacing α by $\gamma - \alpha$, z by $-z$, and multiplying with e^z, we obtain the asymptotic expansion of $F(\alpha, \gamma, z)$.

For asymptotic expansions in other cases, cf. Erdélyi (1953), Vol. I, p. 278, Sec. 6.13.2.

6.11. Differential Equations Reducible to the Confluent Hypergeometric Equation

Many differential equations which are important in applications, such as the Bessel equation, Hermite equation, etc., can be reduced to the confluent hypergeometric equation. In order to facilitate the discussion on the asymptotic expansions of the solutions of these equations and to keep the symmetry of the solutions with respect to the parameters in the equation, we shall start from the Whittaker equation and deduce the general form of the differential equation which can be reduced to the confluent hypergeometric equation.

Let $P_{k,m}(\varsigma)$ denote the solution of the Whittaker equation:

$$\frac{d^2 P_{k,m}}{d\varsigma^2} + \left[-\frac{1}{4} + \frac{k}{\varsigma} + \frac{\frac{1}{4} - m^2}{\varsigma^2} \right] P_{k,m}(\varsigma) = 0 \ . \tag{1}$$

Assume

$$y(z) = z^\beta e^{f(z)} P_{k,m}(h(z)) \ . \tag{2}$$

By direct calculation, we find the differential equation satisfied by $y(x)$ to be

$$\frac{d^2y}{dz^2} - \left[\frac{h''}{h'} + \frac{2\beta}{z} + 2f'(z)\right] \frac{dy}{dz}$$

$$+ \left\{ (f')^2 - f'' + 2\beta\frac{f'}{z} + \frac{\beta(\beta+1)}{z^2} + \frac{h''}{h'}\left(\frac{\beta}{z} + f'\right) \right.$$

$$+ \left(\frac{h'}{h}\right)^2 \left(\frac{1}{4} - m^2 + kh - \frac{h^2}{4}\right) \right\} y(z) = 0 \ . \tag{3}$$

The solutions of equations of the form (3) can readily be expressed in terms of the Whittaker functions.

An important special case of (3) is: $f(z) = \alpha z^\lambda, h(z) = Az^\lambda$ (usually, $\lambda = 1$ or 2). Then, (3) is simplified into

$$\frac{d^2y}{dz^2} + \left[\frac{1-\lambda-2\beta}{z} - 2\lambda\alpha z^{\lambda-1}\right] \frac{dy}{dz}$$

$$+ \left\{ \lambda^2\left(\alpha^2 - \frac{A^2}{4}\right) z^{2\lambda-2} + \lambda(2\alpha\beta + Ak\lambda)z^{\lambda-2} \right.$$

$$+ \frac{\beta(\beta+\lambda) - \lambda^2\left(\frac{1}{4} - m^2\right)}{z^2} \right\} y(z) = 0 \ , \tag{4}$$

óf which the solution is

$$y(z) = z^\beta e^{\alpha z^\lambda} P_{k,m}(Az^\lambda) \ . \tag{5}$$

Note that, when λ is not a negative integer, $z = 0$ is a regular singularity of Eq. (4), while $z = \infty$ is an irregular singularity and the equation has normal solutions at $z = \infty$ [cf. Sec. 2.11].

We enumerate below some important special cases of (4):

1. $\lambda = 1, \alpha = 0, \beta = -\frac{1}{2}, k = 0, A = 2i$; Eq. (4) reduces to the Bessel equation of order m:

$$\frac{d^2y}{dz^2} + \frac{1}{z}\frac{dy}{dz} + \left(1 - \frac{m^2}{z^2}\right) y = 0 \ , \tag{6}$$

whose solutions will be discussed in detail in the following chapter.

2. $\lambda = 2, \alpha = 0, \beta = -\frac{1}{2}, k = \frac{1}{2}n + \frac{1}{4}, m = \pm\frac{1}{2}, A = \frac{1}{2}$; (4) reduces to the Weber equation:

$$y'' + \left(n + \frac{1}{2} - \frac{z^2}{4}\right) y(z) = 0 \ . \tag{7}$$

3. $\lambda = 2, \alpha = \frac{1}{2}, \beta = -\frac{1}{2}, k = \frac{1}{2}n, m = \pm\frac{1}{4}, A = 1$; we obtain the Hermite equation:

$$y'' - 2zy' + 2ny(z) = 0 . \tag{8}$$

Besides, the Laguerre equation

$$zy'' + (\mu + 1 - z)y' + ny = 0 \tag{9}$$

is also a special case of Eq. (4); $\lambda = 1, \alpha = \frac{1}{2}, \beta = -\frac{1}{2}(1 + \mu), A = 1, m = \frac{1}{2}\mu, k = n + \frac{1}{2}(1 + \mu)$. Of course, more directly, Eq. (9) is a special case of the Kummer equation [Eq. (1) of Sec. 6.1] with $\gamma = 1 + \mu, \alpha = -n$.

In the following sections, we shall discuss respectively the solutions of Eqs. (7), (8) and (9). For other equations which can be reduced to confluent hypergeometric equation, cf. Erdélyi (1953), Vol. I, Sec. 6.2.

6.12. Weber Equation. Parabolic Cylinder Functions $D_n(z)$

When we solve, in parabolic cylinder coordinates [cf. App. III (5)], the wave equation or Laplace equation by the method of separation of variables, we are led to the Weber equation

$$y'' + \left(n + \frac{1}{2} - \frac{z^2}{4}\right) y(z) = 0 , \tag{1}$$

where n is a constant.

Using the result of the last section, we immediately obtain a solution of Eq. (1):

$$D_n(z) = 2^{\frac{n}{2}+\frac{1}{4}} z^{-\frac{1}{2}} W_{\frac{n}{2}+\frac{1}{4}, -\frac{1}{4}}\left(\frac{z^2}{2}\right) \quad (|\arg z| < 3\pi/4) ; \tag{2}$$

the constant factor $2^{\frac{n}{2}+\frac{1}{4}}$ is introduced for the sake of making the coefficient in the first term of the asymptotic expansion of $D_n(z)$ equal to 1 [see (5) below]. $D_n(z)$ is called the *parabolic cylinder function*.

From Eq. (1) of Sec. 6.8, we obtain

$$D_n(z) = \frac{\Gamma\left(\frac{1}{2}\right) 2^{\frac{n}{2}+\frac{1}{4}} z^{-\frac{1}{2}}}{\Gamma\left(\frac{1}{2} - \frac{n}{2}\right)} M_{\frac{n}{2}+\frac{1}{4}, -\frac{1}{4}}\left(\frac{z^2}{2}\right)$$

$$+ \frac{\Gamma\left(-\frac{1}{2}\right) 2^{\frac{n}{2}+\frac{1}{4}} z^{-\frac{1}{2}}}{\Gamma\left(-\frac{n}{2}\right)} M_{\frac{n}{2}+\frac{1}{4}, \frac{1}{4}}\left(\frac{z^2}{2}\right) \quad (|\arg z| < 3\pi/4) . \tag{3}$$

However, according to Eqs. (5) and (6) of Sec. 6.3,

$$z^{-\frac{1}{2}} M_{\frac{n}{2}+\frac{1}{4},-\frac{1}{4}} \left(\frac{z^2}{2} \right) = 2^{-\frac{1}{4}} e^{-\frac{z^2}{4}} F \left(-\frac{n}{2}, \frac{1}{2}, \frac{z^2}{2} \right) ,$$

$$z^{-\frac{1}{2}} M_{\frac{n}{2}+\frac{1}{4},\frac{1}{4}} \left(\frac{z^2}{2} \right) = 2^{-\frac{3}{4}} z e^{-\frac{z^2}{4}} F \left(\frac{1-n}{2}, \frac{3}{2}, \frac{z^2}{2} \right) , \tag{4}$$

and since $F(\alpha, \gamma, z)$ is a single-valued function of z, $D_n(z)$ is a single-valued analytic function of z in the whole plane.

By Eq. (2) of Sec. 6.6, we find at once the asymptotic expansion

$$D_n(z) \sim e^{-\frac{z^2}{4}} z^n \left\{ 1 - \frac{n(n-1)}{2z^2} + \frac{n(n-1)(n-2)(n-3)}{2 \times 4 z^4} - \ldots \right\} \tag{5}$$

$$(|\arg z| < 3\pi/4) .$$

The second solution — From the relation between $D_n(z)$ and the Whittaker functions above, we know that $D_{-n-1}(\pm iz)$ are solutions linearly independent of $D_n(z)$, as $W_{-k,m}(-z^2/2)$ is always linearly independent of $W_{k,m}(z^2/2)$ [see Eq. (5) of Sec. 6.6].

In certain boundary-value problems or eigenvalue problems, it is required to find a solution of Eq. (1) which is bounded on the entire real axis: $-\infty < z < \infty$. Since Eq. (1) has only one singularity, $z = \infty$, we need only to study the behavior of the solution of (1) for z approaching $\pm\infty$ along the real axis.

When $z \to +\infty$, we see at once from (5) that $D_n(z) \to 0$. To find the behavior of $D_n(z)$ for $z \to -\infty$, we need the asymptotic expansion of $D_n(z)$ in a range of the phase of z which includes the negative real values of z. For this, we find by using (3) and (4)

$$D_n(z) = \frac{\Gamma\left(\frac{1}{2}\right)}{\Gamma\left(\frac{1-n}{2}\right)} 2^{\frac{n}{2}} e^{\frac{-z^2}{2}} F \left(-\frac{n}{2}, \frac{1}{2}, \frac{z^2}{2} \right)$$

$$+ \frac{\Gamma\left(-\frac{1}{2}\right)}{\Gamma\left(-\frac{n}{2}\right)} 2^{\frac{n-1}{2}} z e^{-\frac{z^2}{4}} F \left(\frac{1-n}{2}, \frac{3}{2}, \frac{z^2}{2} \right) . \tag{6}$$

Assume $\pi/4 < \arg z < 5\pi/4$, then $\pi/2 < \arg z^2 < 5\pi/2$. Let $x = e^{-2\pi i} z^2/2$, then $-3\pi/2 < \arg x < \pi/2$. It follows, by Eq. (7) of Sec. 6.8, that

$$
\begin{aligned}
F\left(\alpha,\gamma,\frac{z^2}{2}\right) &= F(\alpha,\gamma,x) \\
&\sim \frac{\Gamma(\gamma)}{\Gamma(\gamma-\alpha)} e^{-\alpha\pi i} x^{-\alpha}\left\{1+\sum_{k=1}^{\infty}\frac{(\alpha)_k(1-\gamma+\alpha)_k}{k!(-x)^k}\right\} \\
&+ \frac{\Gamma(\gamma)}{\Gamma(\alpha)} e^x x^{\alpha-\gamma}\left\{1+\sum_{k=1}^{\infty}\frac{(\gamma-\alpha)_k(1-\alpha)_k}{k!x^k}\right\} \\
&(\pi/4 < \arg z < 5\pi/4)\,.
\end{aligned}
\tag{7}
$$

Applying this expansion on the r.h.s. of (6), we obtain

$$
\begin{aligned}
D_n(z) &\sim e^{-\frac{z^2}{4}} z^n\left\{1+\sum_{k=1}^{\infty}\frac{\left(-\frac{n}{2}\right)_k\left(\frac{1-n}{2}\right)_k}{k!}\left(-\frac{z^2}{2}\right)^{-k}\right\} \\
&- \frac{\sqrt{2\pi}}{\Gamma(-n)} e^{n\pi i} e^{\frac{z^2}{4}} z^{-n-1}\left\{1+\sum_{k=1}^{\infty}\frac{\left(\frac{1+n}{2}\right)_k\left(1+\frac{n}{2}\right)_k}{k!}\left(\frac{z^2}{2}\right)^{-k}\right\} \\
&(\pi/4 < \arg z < 5\pi/4)\,.
\end{aligned}
\tag{8}
$$

From it we see that, when $z \to -\infty$, $D_n(z) \to \infty$ (the second term) unless n is a positive integer or zero; then the second term on the r.h.s. of (8) vanishes owing to $1/\Gamma(-n) = 0$, while the series in the bracket of the first term breaks up into a polynomial of degree $n/2$ or $(n-1)/2$:

$$
\begin{aligned}
D_n(z) &= e^{-\frac{z^2}{4}} z^n \sum_{k=0}^{[n/2]}\frac{\left(-\frac{n}{2}\right)_k\left(\frac{1-n}{2}\right)_k}{k!}\left(-\frac{z^2}{2}\right)^{-k} \\
&= e^{-\frac{z^2}{4}} z^n\left\{1-\frac{n(n-1)}{2z^2}+\frac{n(n-1)(n-2)(n-3)}{2\times 4z^4}-\cdots\right\} \\
&(n = 0,1,2,\dots)\,.
\end{aligned}
\tag{9}
$$

In addition, $D_n(z)(n = 0,1,2,\dots)$ are the only solutions of the Weber Eq. (1) which are finite in $-\infty < z < \infty$. This is because of the fact that the other solution $D_{-n-1}(iz)$ approach infinity as $z \to +\infty$, which can be seen when we replace, in (5), n by $-n-1$ and z by iz.

Contour Integral Representation of $D_n(z)$

In order to represent $D_n(z)$ in the form of an integral which is suitable for deducing other formulae, we transform the Weber Eq. (1) to the Kummer equation. From the relation between the Whittaker equation and the Kummer equation [see Sec. 6.3], it is seen that the required transformation should be $y(z) = e^{-z^2/4}z$; the equation obtained is

$$v'' - zv' + nv = 0 . \tag{10}$$

This is a special case of the Kummer equation [cf. Eq. (1) of Sec. 6.1], and is also called the *Hermite equation* [see (8) of the preceding section and the example in Sec. 2.13].

In the example shown in Sec. 2.13, we have obtained an integral solution of Eq. (10):

$$v(z) = A \int_{\infty}^{(0+)} e^{-zt-\frac{t^2}{2}}(-t)^{-n-1} dt$$

together with its asymptotic expansion [cf. Eq. (18) of Sec. 2.13]:

$$v(z) \sim -2\pi i A \frac{z^n}{\Gamma(n+1)} \sum_{k=0}^{\infty} \frac{\left(-\frac{n}{2}\right)_k \left(\frac{1-n}{2}\right)_k}{k!} \left(-\frac{z^2}{2}\right)^{-k} .$$

Comparing with the asymptotic expansion of $D_n(z)$, we have

$$D_n(z) = -\frac{\Gamma(n+1)}{2\pi i} e^{-\frac{z^2}{4}} \int_{\infty}^{(0+)} e^{-zt-\frac{t^2}{2}}(-t)^{-n-1} dt \tag{11}$$
$$(|\arg z| < \pi/2, \quad |\arg(-t)| < \pi) .$$

When $n = 0, 1, 2, \ldots$, the integrand in (11) is single-valued, so that the path of integration can be deformed into any contour which encircles the point $t = 0$ once in the positive sense. Then

$$D_n(z) = (-)^n \frac{n!}{2\pi i} e^{-\frac{z^2}{4}} \int^{(0+)} e^{-zt-\frac{t^2}{2}} t^{-n-1} dt . \tag{12}$$

From (12), we obtain at once the *generating function* of $D_n(z)$:

$$e^{-\frac{t^2}{2}-zt-\frac{z^2}{4}} = \sum_{n=0}^{\infty} \frac{(-)^n D_n(z)}{n!} t^n \quad |t| < \infty , \tag{13}$$

and the *derivative expression*:

$$D_n(z) = (-)^n \left[\frac{d^n}{dt^n} e^{-\frac{1}{2}(t+z)^2 + \frac{z^2}{4}} \right]_{t=0} = (-)^n e^{\frac{z^2}{4}} \left[\frac{d^n}{dv^n} e^{-\frac{v^2}{2}} \right]_{v=z}$$

$$= (-)^n e^{\frac{z^2}{4}} \frac{d^n}{dz^n} \left(e^{-\frac{z^2}{2}} \right) . \tag{14}$$

Recurrence Relations

Differentiating the two sides of (13) with respect to t and expanding the l.h.s. again by (13), then comparing the coefficients of t^n on both sides, we find

$$D_{n+1}(z) - zD(z) + n D_{n-1}(z) = 0 . \tag{15}$$

Differentiating both sides of (13) with respect to z and expanding the l.h.s. again with (13), then comparing the coefficients on both sides, we obtain

$$D'_n(z) + \frac{1}{2} z D_n(z) - n D_{n-1}(z) = 0 . \tag{16}$$

Although the relations (15) and (16) are derived for integral values of n, it can be shown by the general representation of $D_n(z)$, (11), that they are valid for arbitrary n. For example, by (11), the l.h.s. of (15) is equal to

$$-\frac{e^{-\frac{z^2}{4}}}{2\pi i} \left\{ \Gamma(n+2) \int_\infty^{(0+)} e^{-zt-\frac{t^2}{2}} (-t)^{-n-2} dt \right.$$

$$- z\Gamma(n+1) \int_\infty^{(0+)} e^{-zt-\frac{t^2}{2}} (-t)^{-n-1} dt$$

$$\left. + n\Gamma(n) \int_\infty^{(0+)} e^{-zt-\frac{t^2}{2}} (-t)^{-n} dt \right\} .$$

Integrating by parts gives for the first term in the bracket

$$\Gamma(n+1) \int_\infty^{(0+)} e^{-zt-\frac{t^2}{2}} (-t)^{-n-1} (z+t) dt ,$$

which just cancels the other two terms. This proves that (15) holds for an arbitrary n.

Orthogonality

From (13), we have

$$e^{-\frac{s^2}{2} - zs - \frac{z^2}{4}} e^{-\frac{t^2}{2} - zt - \frac{z^2}{4}} = \sum_{m,n=0}^{\infty} \frac{(-)^{m+n} s^m t^n}{m! n!} D_m(z) D_n(z) .$$

Integrating both sides with respect to z, the result is

$$\int_{-\infty}^{\infty} e^{-\frac{z^2}{2}-(s+t)z-\frac{s^2+t^2}{2}} dz = e^{st} \int_{-\infty}^{\infty} e^{-\frac{1}{2}[z+(s+t)]^2} dz$$

$$= e^{st} \int_{-\infty}^{\infty} e^{-\frac{1}{2}u^2} du = e^{st}\sqrt{2\pi} = \sum_{n=0}^{\infty} \frac{(st)^n}{n!} \sqrt{2\pi}$$

$$= \sum_{m,n=0}^{\infty} \frac{(-)^{m+n}s^m t^n}{m!n!} \int_{-\infty}^{\infty} D_m D_n \, dz .$$

So, we have

$$\int_{-\infty}^{\infty} D_m(z) D_n(z) dz = n!\sqrt{2\pi}\delta_{mn} \quad (m,n = 0,1,2,\ldots) \tag{17}$$

which is the *orthogonality relation*.

The properties of $D_n(z)$ for non-negative integral values of n given above are of frequent use in the eigenvalue problem concerning the Weber equation. As eigenfunctions, $D_n(z)$ $(n = 0,1,2,\ldots)$ build up a complete set of orthonormal functions in the interval $[-\infty, +\infty]$; a function $f(z)$, which satisfies certain conditions, for example, having first and second continuous derivatives in $[-\infty, +\infty]$ and approaching zero when $|z| \to \infty$, can be expanded in terms of $D_n(z)$:

$$f(z) = \sum_{n=0}^{\infty} a_n D_n(z) , \tag{18}$$

$$a_n = \frac{1}{n!\sqrt{2\pi}} \int_{-\infty}^{\infty} f(z) D_n(z) dz . \tag{19}$$

6.13. Hermite Functions and Hermite Polynomials

In Eq. (8) of Sec. 6.11, the Hermite equation

$$y'' - 2zy' + 2ny = 0 , \tag{1}$$

putting $\xi = \sqrt{2}z$, we have

$$\frac{d^2y}{d\xi^2} - \xi\frac{dy}{d\xi} + ny = 0 , \tag{2}$$

which is exactly Eq. (10) in the last section. Therefore, apart from an arbitrary constant factor, the solution of the Hermite Eq. (2) differs from that of the Weber equation only by the factor $e^{\xi^2/4} (= e^{z^2/2})$.

Denote by $H_n(z)$ a solution of Eq. (1) and define its relation to $D_n(z)$ by

$$H_n(z) = 2^{\frac{n}{2}} e^{\frac{z^2}{2}} D_n(\sqrt{2}z) . \tag{3}$$

$H_n(z)$ is called the *Hermite function*.

When $n = 0, 1, 2, \ldots$, from (9) of the last section, we have

$$
\begin{aligned}
H_n(z) &= (2z)^n \sum_{k=0}^{[n/2]} \frac{\left(-\frac{n}{2}\right)_k \left(\frac{1-n}{2}\right)_k}{k!} (-z^2)^{-k} \\
&= \sum_{k=0}^{[n/2]} \frac{(-)^k n!}{k!(n-2k)!} (2z)^{n-2k} \\
&= (2z)^n - \frac{n(n-1)}{1!} (2z)^{n-2} + \frac{n(n-1)(n-2)(n-3)}{2!} (2z)^{n-4} - \cdots ,
\end{aligned}
\tag{4}
$$

which are called *Hermite polynomials* [cf. Eq. (23) of Sec. 2.11]. There is the other definition: $He_n(z) = e^{z^2/4} D_n(z)$, which is also named the Hermite polynomial. We have $H_n(z) = 2^{n/2} He_n(\sqrt{2}z)$, and the equation satisfied by $He_n(z)$ is $He_n''(z) - zHe_n'(z) + nHe_n(z) = 0$, i.e, Eq. (2).

By using the relation (3) between $H_n(z)$ and $D_n(\sqrt{2}z)$, and changing, in (13) and (14) of the last section, z to $\sqrt{2}z$, t to $-\sqrt{2}t$, we obtain the *generating function and the derivative expression* for $H_n(z)$:

$$e^{-t^2+2zt} = \sum_{n=0}^{\infty} \frac{H_n(z)}{n!} t^n \quad (|t| < \infty) \tag{5}$$

$$H_n(z) = (-)^n e^{z^2} \frac{d^n}{dz^n} (e^{-z^2}) . \tag{6}$$

From (5), we have

$$H_{2n}(0) = (-)^n \frac{(2n)!}{n!} , \quad H_{2n+1}(0) = 0 . \tag{7}$$

Also, from (15) and (16) of the last section (n not restricted to integers), we obtain the following recurrence relations:

$$H_{n+1}(z) - 2zH_n(z) + 2nH_{n-1}(z) = 0 , \tag{8}$$

$$H_n'(z) = 2nH_{n-1}(z) \ . \tag{9}$$

From (17) of the last section, we also obtain

$$\int_{-\infty}^{\infty} H_m(z)H_n(z)e^{-z^2}\,dz = \sqrt{\pi}2^n n! \delta_{mn} \ , \tag{10}$$

$$(n, m = 0, 1, 2, \dots) \ .$$

The Hermite polynomials thus also build up a complete set of orthonormal functions in the interval $[-\infty, \infty]$ with weight e^{-z^2}.

Other formulae relating to Hermite functions and polynomials may be obtained from their relations to $D_n(z)$. Cf. also Exs. 25 – Exs. 30 at the end of this chapter, and Erdélyi (1953), Vol. II, Sec. 10.13, p. 192.

6.14. Laguerre Polynomials

Laguerre equation [Eq. (9) of Sec. 6.11]

$$zy'' + (\mu + 1 - z)y' + ny = 0 \quad (n = 0, 1, 2, \dots) \tag{1}$$

is a special case of the Kummer equation $zy'' + (\gamma - z)y' - \alpha y = 0$ [Eq. (1) of Sec. 6.1] when $\alpha = -n (n = 0, 1, 2, \dots)$, and hence it has a polynomial solution $F(-n, \mu + 1, z)$ (μ is assumed to be a nonnegative integer). The definition of the *generalized Laguerre polynomial* $L_n^\mu(z)$ is

$$L_n^\mu(z) = \frac{\Gamma(\mu + 1 + n)}{n!\Gamma(\mu + 1)} F(-n, \mu + 1, z) \ . \tag{2}$$

It is a polynomial of degree n, μ being an arbitrary real or complex number but not a negative integer. This polynomial is sometimes called the *Sonine polynomial* and denoted by $S_\mu^n(z)$.

The special case of $L_n^\mu(z)$: $L_n^0(z) = L_n(z)$ is called the *Laguerre polynomial*.

Integral Representation and Derivative Expression

According to the definition (2) and Eq. (7) of Sec. 6.4, we obtain at once

$$L_n^\mu(z) = \frac{(-)^n}{2\pi i} \int^{(0+)} e^{zt}(1 - t)^{\mu+n} t^{-n-1}\,dt \ ; \tag{3}$$

$t = 1$ is kept outside the contour, and $|\arg(1 - t)| < \pi$.

In (3), let $t = 1 - v/z$, then we have

$$L_n^\mu(z) = e^z z^{-\mu} \frac{1}{2\pi i} \int^{(z+t)} \frac{e^{-v} v^{\mu+n}}{(v-z)^{n+1}} dv \tag{4}$$

$$= \frac{e^z z^{-\mu}}{n!} \frac{d^n}{dz^n} \left(z^{\mu+n} e^{-z} \right) , \tag{5}$$

which can also be derived from (8) below [see Ex. 31 at the end of this chapter].

When $\mu = m$ (a positive integer), (3) can be written as

$$L_n^m(z) = \frac{(-)^n}{2\pi i} \frac{d^m}{dz^m} \int^{(0+)} e^{zt} (1-t)^{m+n} t^{-m-n-1} dt .$$

Putting $t = 1 - v/z$, we get

$$L_n^m(z) = (-)^m \frac{d^m}{dz^m} \frac{e^z}{2\pi i} \int^{(z+)} \frac{e^{-v} v^{m+n}}{(v-z)^{m+n+1}} dv .$$

By (4), putting in it $\mu = 0$ and changing n to $n + m$, we obtain the derivative expression for $L_n^m(z)$:

$$L_n^m(z) = (-)^m \frac{d^m}{dz^m} L_{m+n}(z) . \tag{6}$$

Generating Function

Equation (3) can be written as

$$L_n^\mu(z) = -\frac{1}{2\pi i} \int^{(0+)} e^{zt} (1-t)^{\mu-1} \left(\frac{t}{t-1} \right)^{-n-1} dt .$$

Putting $t/(t-1) = v$ so that, in the integral, only the exponent of v contains n, and then writing again t for v, we have

$$L_n^\mu(z) = \frac{1}{2\pi i} \int^{(0+)} e^{\frac{-zt}{1-t}} (1-t)^{-\mu-1} t^{-n-1} dt , \tag{7}$$

from which we find

$$\frac{e^{-\frac{zt}{1-t}}}{(1-t)^{\mu+1}} = \sum_{n=0}^{\infty} L_n^\mu(z) t^n \quad (|t| < 1) . \tag{8}$$

The function on the l.h.s. is the *generating function* of $L_n^\mu(z)$.

Recurrence Relations

Differentiating both sides of (8) with respect to t, multiplying with $(1-t)^2$, and expanding the l.h.s. again by (8), we obtain, by comparing the coefficients of t^n on both sides,

$$(n+1)L_{n+1}^\mu + (z - \mu - 2n - 1)L_n^\mu + (\mu + n)L_{n-1}^\mu = 0 \quad (n \geq 1) . \quad (9)$$

Differentiating both sides of (8) with respect to z, multiplying by $1 - t$, and expanding the l.h.s. with (8), then comparing the coefficients, we obtain

$$\frac{d}{dz}L_n^\mu - \frac{d}{dz}L_{n-1}^\mu + L_{n-1}^\mu = 0 \quad (n \geq 1) . \quad (10)$$

Eliminating L_{n-1}^μ from (9) and (10) gives

$$(n+1)(L_{n+1}^\mu)' + (z - n - 1)(L_n^\mu)' - (n+1)L_{n+1}^\mu$$
$$+ (\mu + 2n + 2 - z)L_n^\mu = 0 \quad (n \geq 0) . \quad (11)$$

Changing n in (10) to $n+1$, then eliminating $(L_{n+1}^\mu)'$ and L_{n+1}^μ with (11) and (9), we have

$$z(L_n^\mu)' = nL_n^\mu - (\mu + n)L_{n-1}^\mu \quad (n \geq 1) . \quad (12)$$

Changing μ in (8) to $\mu + 1$, multiplying with $1 - t$, then expanding the l.h.s. by (8), we obtain by comparison of coefficients,

$$L_n^\mu(z) = L_n^{\mu+1}(z) - L_{n-1}^{\mu+1}(z) . \quad (13)$$

More general, replacing μ in (8) by $\mu - p, p$ being any real or complex number, we get

$$\frac{e^{-\frac{zt}{1-t}}}{(1-t)^{\mu+1}} = (1-t)^{-p} \sum_{l=0}^\infty t^l L_l^{\mu-p}(z), \quad |t| < 1 ,$$

or

$$\sum_{n=0}^\infty t^n L_n^\mu = \sum_{k=0}^\infty \binom{-p}{k}(-t)^k \sum_{l=0}^\infty t^l L_l^{\mu-p}$$
$$= \sum_{n=0}^\infty t^n \sum_{k=0}^n (-)^k \binom{-p}{k} L_{n-k}^{\mu-p} .$$

Thus, we have

$$L_n^\mu(z) = \sum_{k=0}^{n} (-)^k \binom{-p}{k} L_{n-k}^{\mu-p}(z) \quad (p \text{ arbitrary}) . \tag{14}$$

Further, from the relation between $L_n^\mu(z)$ and $F(-n, \mu + 1, z)$, (2), and Eq. (4) of Sec. 6.2, we obtain

$$\frac{d^r}{dz^r} L_n^\mu(z) = (-)^r L_{n-r}^{\mu+r}(z) \quad (r \le n) . \tag{15}$$

Integrals Containing a Product of Two Generalized Laguerre Polynomials

An important integral formula is

$$\int_0^\infty z^\lambda e^{-z} L_n^\mu(z) L_{n'}^{\mu'}(z) dz$$

$$= (-)^{n+n'} \Gamma(\lambda + 1) \sum_k \binom{\lambda - \mu}{n - k} \binom{\lambda - \mu'}{n' - k} \binom{\lambda + k}{k} , \tag{16}$$

where $\text{Re}(\lambda) > -1$ to ensure the convergence of the integral at the lower limit. The proof of this formula by means of (8) is as follows. From (8), we have

$$\sum_{n=0}^\infty t^n L_n^\mu(z) \sum_{n'=0}^\infty s^{n'} L_{n'}^{\mu'}(z) = \sum_{n,n'} t^n s^{n'} L_n^\mu(z) L_{n'}^{\mu'}(z)$$

$$= \frac{e^{-z\left(\frac{t}{1-t} + \frac{s}{1-s}\right)}}{(1-t)^{\mu+1}(1-s)^{\mu'+1}} \quad (|t| < 1, \ |s| < 1) .$$

Assume that s and t are positive real numbers smaller than 1, then

$$\sum_{n,n'} t^n s^{n'} \int_0^\infty z^\lambda e^{-z} L_n^\mu L_{n'}^{\mu'} dz = \int_0^\infty \frac{e^{-z\frac{1-ts}{(1-t)(1-s)}}}{(1-t)^{\mu+1}(1-s)^{\mu'+1}} z^\lambda dz$$

$$= (1-t)^{\lambda-\mu}(1-s)^{\lambda-\mu'}(1-ts)^{-\lambda-1} \int_0^\infty e^{-v} v^\lambda dv$$

$$= \Gamma(\lambda + 1) \sum_l \binom{\lambda - \mu}{l} (-t)^l \sum_{l'} \binom{\lambda - \mu'}{l'} (-s)^{l'}$$

$$\times \sum_k \binom{-\lambda - 1}{k} (-ts)^k$$

$$= \Gamma(\lambda + 1) \sum_{n,n'} t^n s^{n'} \sum_k \binom{\lambda - \mu}{n - k}$$

$$\times \binom{\lambda - \mu'}{n' - k} \binom{-\lambda - 1}{k} (-)^{n+n'+k} .$$

Comparing the coefficients of $t^n s^{n'}$ on both sides and noticing that $(-)^k \binom{-\lambda-1}{k}$ = $\binom{\lambda+k}{k}$, we have (16).

When $\lambda = \mu = \mu'$, we have the important special case of (16):

$$\int_0^\infty z^\mu e^{-z} L_n^\mu(z) L_{n'}^\mu(z) dz = \Gamma(\mu+1) \binom{\mu+n}{n} \delta_{nn'}$$

$$= \frac{\Gamma(\mu+n+1)}{n!} \delta_{nn'} , \qquad (17)$$

because only then the term with $k = n = n'$ in the sum on the r.h.s. of (16) does not vanish. (17) expresses the orthonormal relation among the generalized Laguerre polynomials.

Expansion Formula

By (16) and (17), we may obtain the following expansion formula:

$$z^s L_n^\mu(z) = \sum_{r=0}^{n+s} \alpha_r^s L_{n+s-r}^{\mu+p}(z) ; \qquad (18)$$

the coefficients in which are given by

$$\alpha_r^s = (-)^{s+r} \frac{(n+s-r)! \Gamma(s+\mu+p+1)}{\Gamma(n+s+\mu+p-r+1)}$$

$$\times \sum_k \binom{s+p}{n-k} \binom{s}{k+r-n} \binom{s+\mu+p+k}{k} , \qquad (19)$$

where s is an arbitrary non-negative integer and p is any real or complex number.

Equation (18) obviously holds as both sides of it are polynomials of degree $n + s$. Formula (19) for the expansion coefficients may be derived in the following way. Multiplying both sides of (18) with $z^{\mu+p} e^{-z} L_{n+s-r'}^{\mu+p}(z)$, and assuming $\text{Re}(\mu + p) > -1$, then integrating, we have by applying (16) and (17) respectively on both sides,

$$\int_0^\infty z^{s+\mu+p} e^{-z} L_{n+s-r'}^{\mu+p} L_n^\mu dz$$

$$= \sum_r \alpha_r^s \int_0^\infty z^{\mu+p} e^{-z} L_{n+s-r'}^{\mu+p} L_{n+s-r}^{\mu+p} dz$$

$$= \alpha_{r'}^{s} \frac{\Gamma(\mu + p + n + s - r' + 1)}{(n + s - r')!}$$

$$= (-)^{s - r'} \Gamma(s + \mu + p + 1)$$

$$\times \sum_{k} \binom{s}{n + s - r' - k} \binom{s + p}{n - k} \binom{s + \mu + p + k}{k} ,$$

from which we get (19), since $\binom{s}{n+s-r'-k} = \binom{s}{k+r'-n}$. Now, as (18) is an algebraic identity, the restriction, $\mathrm{Re}(\mu + p) > -1$, introduced to ensure the convergence of the integrals can be removed.

The following are two particular cases of the expansion coefficients:

$$\alpha_r^0 = (-)^r \binom{p}{r} \quad \text{(leads to (14))} \tag{20}$$

$$\alpha_r^1 = (-)^{r+1} \binom{p+2}{r} \left\{ (n+1) + \frac{(\mu-1)r}{p+2} \right\}$$

$$= (-)^{r+1} \frac{\Gamma(p+2)}{r!\,\Gamma(p-r+3)} \left\{ (n+1)(p+2) + (\mu-1)r \right\} . \tag{21}$$

From (18), take $p = 0$ and use (21). We obtain

$$zL_n^\mu = -(n+1)L_{n+1}^\mu + (\mu + 2n + 1)L_n^\mu - (\mu + n)L_{n-1}^\mu ,$$

which is (9). Taking $p = -1$ in (18) gives

$$zL_n^\mu = -(n+1)L_{n+1}^{\mu-1} + (\mu + n)L_n^{\mu-1} , \tag{22}$$

while $p = 1$ gives

$$zL_n^\mu = -(n+1)L_{n+1}^{\mu+1} + (\mu + 3n + 2)L_n^{\mu+1}$$
$$\quad - (2\mu + 3n + 1)L_{n-1}^{\mu+1} + (\mu + n)L_{n-2}^{\mu+1} , \tag{23}$$

etc.

6.15. Other Special Functions Expressible by Whittaker Functions

Here, we shall give as examples only few such functions represented by integrals. For more, cf. Ex. 39 at the end of this chapter and Erdélyi (1953), Vol I, Sec. 6.9.2, p. 266; H. Buchholz (1953), p. 208, Anhang I.

1. The *Error function* erf x

$$\operatorname{erf} \, x = \frac{2}{\sqrt{\pi}} \int_0^x e^{-t^2} \, dt = 1 - \operatorname{erfc} \, x \quad (x \text{ real}) \, , \tag{1}$$

where

$$\operatorname{erfc} \, x = \frac{2}{\sqrt{\pi}} \int_x^\infty e^{-t^2} \, dt \, .$$

The relation between the function (1) and the Whittaker function can be deduced from the following integral expression:

$$W_{k,m}(z) = \frac{e^{-\frac{1}{2}z} z^k}{\Gamma\left(\frac{1}{2} - k + m\right)} \int_0^\infty t^{-k-\frac{1}{2}+m} \left(1 + \frac{t}{z}\right)^{k-\frac{1}{2}+m} e^{-t} dt \, , \tag{2}$$

where $\frac{1}{2} - k + m$ is a positive integer [Eq. (5) of Sec. 6.5].

In the integral of (2), let $t = s^2 - x^2$, so that a factor e^{-s^2} appears in the integrand and in the meanwhile, the lower limit of the integral becomes x. Then,

$$W_{k,m}(z) = \frac{e^{-\frac{1}{2}z} z^k}{\Gamma\left(\frac{1}{2} - k + m\right)} 2e^{x^2}$$
$$\times \int_x^\infty (s^2 - x^2)^{-k-\frac{1}{2}+m} \left(\frac{z + s^2 - x^2}{z}\right)^{k-\frac{1}{2}+m} e^{-s^2} s \, ds \, .$$

Now, replacing z by x^2 and putting $-k - \frac{1}{2} + m = 0, k - \frac{1}{2} + m = -\frac{1}{2}$ in order to cancel all the factors involving s except e^{-s^2}, we see that $-k = m = \frac{1}{4}$ and

$$W_{-\frac{1}{4},\frac{1}{4}}(x^2) = 2e^{\frac{x^2}{2}} x^{\frac{1}{2}} \int_x^\infty e^{-s^2} ds \, .$$

Hence,

$$\operatorname{erf} \, x = 1 - \frac{1}{\sqrt{\pi}} e^{-\frac{x^2}{2}} x^{-\frac{1}{2}} W_{-\frac{1}{4},\frac{1}{4}}(x^2) \, . \tag{3}$$

2. *Incomplete Gamma Function* $\gamma(n, x)$

$$\gamma(n, x) = \int_0^x t^{n-1} e^{-t} dt \, . \tag{4}$$

Obviously,

$$\gamma(n, x) = \Gamma(n) - \int_x^\infty t^{n-1} e^{-t} dt \, . \tag{5}$$

Putting $t = s + x$, we have

$$\gamma(n, x) = \Gamma(n) - e^{-x}x^{n-1}\int_0^\infty \left(1 + \frac{s}{x}\right)^{n-1} e^{-s}ds .$$

Thus, if we put, in (2), $-k - \frac{1}{2} + m = 0, k - \frac{1}{2} + m = n - 1$, i.e., $k = (n-1)/2, m = n/2$, we obtain at once

$$\gamma(n, x) = \Gamma(n) - e^{-\frac{x}{2}} x^{\frac{n-1}{2}} W_{\frac{n-1}{2}, \frac{n}{2}}(x) . \qquad (6)$$

3. *Logarithmic Integral* li(z)

$$\mathrm{li}(z) = \int_0^z \frac{dt}{\ln t} \quad (|\arg(-\ln z)| < \pi) . \qquad (7)$$

Let $\ln t = s$, then

$$\mathrm{li}(z) = \int_{-\infty}^{\ln z} e^s s^{-1} ds = -\int_{-\ln z}^\infty e^{-s} s^{-1} ds .$$

Putting in it $t = s + \ln z$, we obtain

$$\mathrm{li}(z) = -e^{\ln z}\int_0^\infty e^{-t}(t - \ln z)^{-1} dt$$

$$= -z(-\ln z)^{-1}\int_0^\infty e^{-t}\left(1 + \frac{t}{-\ln z}\right)^{-1} dt .$$

Comparing with (2) as before gives

$$\mathrm{li}(z) = -(-\ln z)^{-\frac{1}{2}} z^{\frac{1}{2}} W_{-\frac{1}{2}, 0}(-\ln z)$$
$$(|\arg(-\ln z)| < \pi) . \qquad (8)$$

4. *Exponential Integral* $E_n(x)$

$$E_n(x) = \int_1^\infty e^{-x\xi} \xi^{-n} d\xi \quad (x > 0) . \qquad (9)$$

A special case of this function is

$$\mathrm{Ei}(-x) = -E_1(x) \quad (x > 0) . \qquad (10)$$

In (2), let $z = x, t = (\xi - 1)x$, and take $-k - \frac{1}{2} + m = 0, k - \frac{1}{2} + m = -n$, i.e., $k = -n/2, m = (1-n)/2$. We obtain

$$E_n(x) = e^{-\frac{x}{2}} x^{\frac{n}{2}-1} W_{-\frac{n}{2}, \frac{1-n}{2}}(x) . \qquad (11)$$

By their relations to the Whittaker function, we can extend the domains of definitions of these functions, indicate their multi-valuedness, clearly and obtain their asymptotic expansions.

Exercise 6

1. Show that, by proper transformation, the equation of Laplacian type,

$$(a_0 z + b_0)\frac{d^2 u}{dz^2} + (a_1 z + b_1)\frac{du}{dz} + (a_2 z + b_2)u = 0 ,$$

can be reduced to the Kummer equation

$$x\frac{d^2 y}{dx^2} + (\gamma - x)\frac{dy}{dx} - \alpha y = 0 .$$

Also show that, if we put $x = \lambda\xi$, $y = x^\rho e^{hx} w$, and choosing the parameters λ, ρ, h, properly the Kummer equation remains unchanged in form. Deduce from there the first Kummer formula [Eq. (6) of Sec. 6.1].

2. Prove the *Second formula of Kummer*:

$$e^{-\frac{z}{2}} F(\alpha, 2\alpha, z) = {}_0F_1\left(\frac{1}{2} + \alpha; \frac{z^2}{16}\right) .$$

[Hint: Use Eq. (4) of Sec. 6.4.]

3. From the integral representation of $F(\alpha, \gamma, z)$ [Eq. (4) or (6) of Sec. 6.4], deduce the following recurrence relations [Eqs. (2) and (3) of Sec. 6.2]:

$$(\gamma - 1)F(\gamma - 1) - \alpha F(\alpha + 1) - (\gamma - \alpha - 1)F = 0 ,$$
$$\gamma F - z F(\gamma + 1) - \gamma F(\alpha - 1) = 0 .$$

4. Using the results of the last exercise, deduce the following recurrence relations:

$$(\gamma - 2\alpha - z)F + \alpha F(\alpha + 1) - (\gamma - \alpha)F(\alpha - 1) = 0 ,$$
$$\gamma(\alpha + z)F - \alpha\gamma F(\alpha + 1) - (\gamma - \alpha)z F(\gamma + 1) = 0 ,$$
$$(\alpha - 1 + z)F - (\gamma - 1)F(\gamma - 1) + (\gamma - \alpha)F(\alpha - 1) = 0 ,$$
$$\gamma(\gamma - 1 + z)F - (\gamma - \alpha)z F(\gamma + 1) - \gamma(\gamma - 1)F(\gamma - 1) = 0 .$$

5. Prove the following derivative formulae

$$\frac{d^n}{dx^n}[x^{\alpha+n-1}F(\alpha,\gamma,x)] = (\alpha)_n x^{\alpha-1}F(\alpha+n,\gamma,x) \, ,$$

$$\frac{d^n}{dx^n}[x^{\gamma-1}F(\alpha,\gamma,x)] = (-)^n(1-\gamma)_n x^{\gamma-n-1}F(\alpha,\gamma-n,x) \, ,$$

$$\frac{d^n}{dx^n}[e^{-x}F(\alpha,\gamma,x)] = (-)^n\frac{(\gamma-\alpha)_n}{(\gamma)_n}e^{-x}F(\alpha,\gamma+n,x) \, ,$$

$$\frac{d^n}{dx^n}[e^{-x}x^{\gamma-\alpha+n-1}F(\alpha,\gamma,x)] = (\gamma-\alpha)_n e^{-x}x^{\gamma-\alpha-1}F(\alpha-n,\gamma,x) \, .$$

6. Show that

(i) $F(\alpha,\gamma,z) = \frac{\Gamma(\gamma)\Gamma(1+\alpha-\gamma)}{\Gamma(\alpha)}\frac{1}{2\pi i}\int_0^{(1+)}e^{zt}t^{\alpha-1}(t-1)^{\gamma-\alpha-1}dt \, ,$

where $\mathrm{Re}(\alpha) > 0, \gamma - \alpha \neq 1, 2, \ldots, |\arg t| < \pi, |\arg(t-1)| < \pi.$

(ii) $F(\alpha,\gamma,z) = \frac{\Gamma(\gamma)\Gamma(1-\alpha)}{\Gamma(\gamma-\alpha)}\frac{1}{2\pi i}\int_1^{(0+)}e^{zt}(-t)^{\alpha-1}(1-t)^{\gamma-\alpha-1}dt \, ,$

where $\mathrm{Re}(\gamma - \alpha) > 0, \alpha \neq 1, 2, \ldots, |\arg(-t)| < \pi, |\arg(1-t)| < \pi.$

7. Prove the following recurrence relations:

$$W_{k,m}(z) = z^{\frac{1}{2}}W_{k-\frac{1}{2},m-\frac{1}{2}}(z) + \left(\frac{1}{2}-k+m\right)W_{k-1,m}(z) \, ,$$

$$W_{k,m}(z) = z^{\frac{1}{2}}W_{k-\frac{1}{2},m+\frac{1}{2}}(z) + \left(\frac{1}{2}-k-m\right)W_{k-1,m}(z) \, ,$$

$$zW'_{k,m}(z) = \left(k-\frac{z}{2}\right)W_{k,m}(z) - \left\{m^2 - \left(k-\frac{1}{2}\right)^2\right\}W_{k-1,m}(z) \, .$$

8. $\Psi(\alpha,\gamma,x)$ is the *Tricomi function* defined by Eq. (6) or (7) of Sec. 6.5. Show that

$$\Psi(\alpha,\gamma,x) = x^{1-\gamma}\Psi(\alpha-\gamma+1,2-\gamma,x) \, ,$$

$$\Psi(\alpha,\gamma,x) = \frac{\Gamma(1-\gamma)}{\Gamma(\alpha-\gamma+1)}F(\alpha,\gamma,x)$$

$$+ \frac{\Gamma(\gamma-1)}{\Gamma(\alpha)}x^{1-\gamma}F(\alpha-\gamma+1,2-\gamma,x)$$

$$(\gamma \neq \text{integer}) \, .$$

9. Prove the following Laplace transforms:

$$\mathcal{L}\{t^{\beta-1}F(\alpha,\gamma,\lambda t)\} = \int_0^\infty e^{-st}t^{\beta-1}F(\alpha,\gamma,\lambda t)\,dt$$

$$= \Gamma(\beta)s^{-\beta}F\left(\alpha,\beta,\gamma,\frac{\lambda}{s}\right) \quad (|\lambda| < |s|)$$

$$= \Gamma(\beta)(s-\lambda)^{-\beta}F\left(\gamma-\alpha,\beta,\gamma,\frac{\lambda}{\lambda-s}\right)$$

$$(|\lambda| < |\lambda-s|) ,$$

$$\mathcal{L}\{e^{\lambda t}t^{m-\frac{1}{2}}M_{k,m}(t)\} = \Gamma(2m+1) \times \left(s-\lambda+\frac{1}{2}\right)^{-k-m-\frac{1}{2}}$$

$$\times \left(s-\lambda-\frac{1}{2}\right)^{k-m-\frac{1}{2}} ,$$

$$\left(\text{Re}(m) > -\frac{1}{2}, \quad \text{Re}(s) > \text{Re}(\lambda) - \frac{1}{2}\right) .$$

$$\mathcal{L}\{e^{\frac{t}{2}}t^{\lambda}W_{k,m}(t)\} = \frac{\Gamma\left(\lambda+m+\frac{3}{2}\right)\Gamma\left(\lambda-m+\frac{3}{2}\right)}{\Gamma(\lambda-k+2)}s^{-\lambda-m-\frac{3}{2}}$$

$$\times F\left(\lambda+m+\frac{3}{2}, m-k+\frac{1}{2}, \lambda-k+2, 1-s^{-1}\right) ,$$

$$\left(\text{Re}\left(\lambda\pm m+\frac{3}{2}\right) > 0, \quad \text{Re}(s) > 0\right) .$$

10. By the results of the last exercise, show that

$$\int_0^\infty t^{\beta-1}F(\alpha,\gamma,-t)\,dt = \frac{\Gamma(\beta)\Gamma(\gamma)\Gamma(\alpha-\beta)}{\Gamma(\alpha)\Gamma(\gamma-\beta)} ,$$

$$(\text{Re}(\alpha) > \text{Re}(\beta) > 0)$$

$$\int_0^\infty t^{\beta-1}\Psi(\alpha,\gamma,t)\,dt = \frac{\Gamma(\beta)\Gamma(\alpha-\beta)\Gamma(\beta-\gamma+1)}{\Gamma(\alpha)\Gamma(\alpha-\gamma+1)}$$

$$(\text{Re}(\alpha) > \text{Re}(\beta) > 0, \quad \text{Re}(\beta)+1 > \text{Re}(\gamma)) .$$

11. With the results of Ex. 9 and the inversion formula for Laplace transform:

$$f(t) = \mathcal{L}^{-1}\{F(s)\} = \frac{1}{2\pi i}\int_{b-i\infty}^{b+i\infty} e^{st}F(s)\,ds ,$$

where $F(s) = \mathcal{L}\{f(t)\}$, and b is a positive number such that $F(s)$ has no singularities for $\mathrm{Re}(s) > b$, show that

$$F(\alpha, \gamma, z) = \frac{n! z^{-n}}{2\pi i} \int_C e^{zs} s^{-n-1} F(\alpha, n+1, \gamma, s^{-1}) ds \ ,$$

where C is the circle $|s| = \rho > 1$, and $n = 0, 1, 2, \ldots$.

12. Use the convolution theorem of Laplace transform:

$$\mathcal{L}\{\int_0^t f_1(\tau) f_2(t-\tau) d\tau\} = F_1(s) F_2(s) \ ,$$

where $F_i(s) = \mathcal{L}\{f_i(t)\}(i = 1, 2)$, to prove the following addition formula of the integral

$$\int_0^t \frac{\tau^{\gamma-1}}{\Gamma(\gamma)} F(\alpha, \gamma, \tau) \frac{(t-\tau)^{\gamma'-1}}{\Gamma(\gamma')} F(\alpha', \gamma', t-\tau) d\tau$$

$$= \frac{t^{\gamma+\gamma'+1}}{\Gamma(\gamma+\gamma')} F(\alpha+\alpha', \gamma+\gamma', t)$$

$$(\mathrm{Re}(\gamma) > 0, \quad \mathrm{Re}(\gamma') > 0) \ .$$

13. Prove the following Laplace transform

$$\mathcal{L}\{e^{-t^2} t^{2\gamma-2} F(\alpha, \gamma, t^2)\}$$

$$= 2^{1-2\gamma} \Gamma(2\gamma - 1) \Psi\left(\gamma - \frac{1}{2}, \alpha + \frac{1}{2}, \frac{s^2}{4}\right)$$

$$\left(\mathrm{Re}(\gamma) > \frac{1}{2}, \quad \mathrm{Re}(s) > 0\right) \ ,$$

where $\mathcal{L}\{f(t)\} = \int_0^\infty e^{-st} f(t) dt$, $\Psi(\lambda, \mu, x)$ is the Tricomi function [Eq. (6) or (7) of Sec. 6.4].

14. Show that

$$\Gamma(\beta) \Psi(\alpha, \gamma, x) = x^{\alpha-\beta} \int_0^\infty e^{-xt} t^{\beta-1} F(\alpha, \alpha - \gamma + 1, \beta, -t) dt$$

$$(\mathrm{Re}(\beta) > 0, \quad \mathrm{Re}(x) > 0) \ .$$

15. Show that

$$\int_0^\infty \cos(2xy) F(\alpha, \gamma, -y^2) dy$$

$$= \frac{\sqrt{\pi}}{2} \frac{\Gamma(\gamma)}{\Gamma(\alpha)} x^{2\alpha-1} e^{-x^2} \Psi\left(\gamma - \frac{1}{2}, \alpha + \frac{1}{2}, x^2\right) \ .$$

16. Show that

$$\frac{1}{2\pi i} \int_{-i\infty}^{i\infty} \Gamma(-s)\Gamma(\gamma - s)\Psi(s, \gamma, x)\Psi(\gamma - s, \gamma, y)\,ds = \Gamma(\gamma)\Psi(\gamma, 2\gamma, x + y) \ ,$$

where the path of integration is the Barnes' contour [Sec. 4.6]. The formula proved is called the *Magnus Addition Formula* [cf. Erdélyi (1953), Vol. I, p. 285, (15)].

17. Show that

$$\int_0^\infty e^{-x} x^{\gamma + n - 1} (x + y)^{-1} F(\alpha, \gamma, x)\,dx$$
$$= (-)^n \Gamma(\gamma)\Gamma(1 - \alpha) y^{\gamma + n - 1}\Psi(\gamma - \alpha, \gamma, y) \ ,$$
$$-\mathrm{Re}(\gamma) < n < 1 - \mathrm{Re}(\alpha), \quad n = 0, 1, 2, \ldots \ ; \quad |\arg y| < \pi \ .$$

[cf. Erdélyi, p. 285, (16)].

18. With the Taylor expansion formula

$$f(\lambda x) = f[x + (\lambda - 1)x] = \sum \frac{(\lambda - 1)^n x^n}{n!} f^{(n)}(x)$$

together with the Lagrange expansion formula

$$\lambda f(\lambda x) = \sum_{n=0}^\infty \frac{(1 - \lambda^{-1})^n}{n!} \frac{d^n}{dx^n}[x^n f(x)]$$

[Ex. 18 in Chap. 1] and the results of Ex. 5 above, prove the following multiplication formulae:

(i) $F(\alpha, \gamma, \lambda x) = \sum_{n=0}^\infty \frac{(\alpha)_n}{n!(\gamma)_n}(\lambda - 1)^n x^n F(\alpha + n, \gamma + n, x)$,

(ii) $F(\alpha, \gamma, \lambda x) = \lambda^{1-\lambda} \sum_{n=0}^\infty \frac{(1-\gamma)_n}{n!}(1 - \lambda)^n F(\alpha, \gamma - n, x)$,

(iii) $F(\alpha, \gamma, \lambda x) = \lambda^{-\alpha} \sum_{n=0}^\infty \frac{(\alpha)_n}{n!}(1 - \lambda^{-1})^n F(\alpha + n, \gamma, x)$,
$$\left(\mathrm{Re}(\lambda) > \tfrac{1}{2}\right) \ .$$

If we put $\lambda = 1 + y/x$, i.e., $\lambda x = x + y$, then these formulae become addition formulae.

19. Show that

$$F(\alpha, \gamma, \lambda x)$$
$$= \sum_{n=0}^\infty \frac{(\alpha)_n}{n!(g + n)_n}(-x)^n \,{}_2F_1(-n, g + n; \gamma; \lambda) F(\alpha + n, g + 2n + 1, x) \ ,$$

where g is an arbitrary parameter not equal to $-2m - 1(m = 0, 1, 2, \ldots)$.
[Cf. Erdélyi (1953), Vol. I, p. 283, (7).]

20. Show that, when $|\arg \alpha| < \pi/2$,

$$\int_\infty^{(0+)} e^{(\frac{1}{4}-\alpha)z^2} z^m D_n(z) dz$$

$$= \frac{\pi^{\frac{3}{2}} 2^{\frac{n}{2}-m} e^{(m-\frac{1}{2})\pi i}}{\Gamma(-m)\Gamma\left(\frac{m}{2} - \frac{n}{2} + 1\right) \alpha^{\frac{m+1}{2}}} F\left(-\frac{n}{2}, \frac{m+1}{2}, \frac{m}{2} - \frac{n}{2} + 1, 1 - \frac{1}{2\alpha}\right).$$

21. With the result of the last Ex., prove that

$$\int_0^\infty e^{-\frac{3}{4}z^2} z^m D_{m+1}(z) dz = (\sqrt{2})^{-1-m} \Gamma(m+1) \sin(1-m)\frac{\pi}{4},$$

if the integral converges.

22. Let n be a positive integer. Show that

$$\int_{-\infty}^\infty e^{-\frac{z^2}{4}} (z-x)^{-1} D_n(z) dz = \pm i e^{\mp n\pi i/2} \sqrt{2\pi} \Gamma(n+1) e^{-\frac{x^2}{4}} D_{-n-1}(\mp ix);$$

the upper sign is taken when $\text{Im}(x) > 0$ and the lower sign is taken when $\text{Im}(x) < 0$.

23. Show that, when n is a positive integer,

$$D_n(x) = (-)^{[n/2]} 2^{n+2} (2\pi)^{-\frac{1}{2}} e^{\frac{x^2}{4}} \int_0^\infty e^{2t^2} t^n \frac{\cos}{\sin} (2xt) dt;$$

in the integrand, the cosine is taken when n is an even number, and the sine is taken when n is odd.

24. Show that

$$D_n(z) = \frac{\Gamma(n+1)}{\sqrt{2\pi}} \left[e^{\frac{n\pi i}{2}} D_{-n-1}(iz) + e^{-\frac{n\pi i}{2}} D_{-n-1}(-iz)\right],$$

$$D_n(z) = e^{-n\pi i} D_n(-z) + \frac{\sqrt{2\pi}}{\Gamma(-n)} e^{-\frac{n+1}{2}\pi i} D_{-n-1}(iz)$$

$$= e^{n\pi i} D_n(-z) + \frac{\sqrt{2\pi}}{\Gamma(-n)} e^{\frac{n+1}{2}\pi i} D_{-n-1}(-iz),$$

where $\pm i = e^{\pm \pi i/2}$.

25. Show that

$$\sum_{k=0}^n \frac{H_k(x) H_k(y)}{2^k k!} = \frac{H_{n+1}(x) H_n(y) - H_n(x) H_{n+1}(y)}{2^{n+1} n! (x-y)},$$

where $H_n(z)$ is the Hermite polynomial of degree n.

26. Show that

$$\sum_{n=0}^{\infty} \frac{\left(\frac{t}{2}\right)^n}{n!} H_n(x) H_n(y) = (1-t^2)^{-\frac{1}{2}} \exp\left\{\frac{2xyt - (x^2 + y^2)t^2}{1 - t^2}\right\} .$$

27. Prove the following integral formulae:

$$\int_0^x e^{-t^2} H_n(t) dt = H_{n-1}(0) - e^{-x^2} H_{n-1}(x) ,$$

$$\int_0^x H_n(t) dt = \frac{1}{2(n+1)} [H_{n+1}(x) - H_{n+1}(0)] ,$$

$$\int_{-\infty}^{\infty} e^{-t^2} H_{2n}(xt) dt = \sqrt{\pi} \frac{(2n)!}{n!} (x^2 - 1)^n ,$$

$$\int_{-\infty}^{\infty} e^{-t^2} t H_{2n+1}(xt) dt = \sqrt{\pi} \frac{(2n+1)!}{n!} x(x^2 - 1)^n ,$$

$$\int_{-\infty}^{\infty} e^{-t^2} t^n H_n(xt) dt = \sqrt{\pi} n! P_n(x) ,$$

$P(x)$ being the Legendre polynomial of degree n.

28. The Gauss transform of a function $F(t)$ is defined by

$$\mathcal{G}_x^\alpha \{F(t)\} = \frac{1}{\sqrt{2\pi\alpha}} \int_{-\infty}^{\infty} F(t) e^{-\frac{(x-t)^2}{2\alpha}} dt ,$$

α being a parameter. Show that

$$\mathcal{G}_x^\alpha \{H_n(t)\} = (1 - 2\alpha)^{\frac{n}{2}} H_n\left[(1 - 2\alpha)^{-\frac{1}{2}} x\right] \quad \left(0 \leq \alpha < \frac{1}{2}\right) ,$$

$$\mathcal{G}_x^{\frac{1}{2}} \{H_n(t)\} = (2x)^n ,$$

$$\mathcal{G}_x^{\frac{1}{2}} \{t^n\} = (2i)^{-n} H_n(ix) .$$

29. Show that

$$\sum_{k=0}^{n} (2^k k!)^{-1} [H_k(x)]^2 = (2^{n+1} n!)^{-1} \{[H_{n+1}(x)]^2 - H_n(x) H_{n+2}(x)\} ,$$

$$\sum_{k=0}^{\min(m,n)} (-2)^k k! \binom{m}{k} \binom{n}{k} H_{m-k}(x) H_{n-k}(x) = H_{m+n}(x) ,$$

$$\sum_{k=0}^{\min(m,n)} 2^k k! \binom{m}{k} \binom{n}{k} H_{m+n-2k}(x) = H_m(x) H_n(x) ,$$

$$\sum_{k=0}^{n} \binom{n}{k} H_k(\sqrt{2}x) H_{n-k}(\sqrt{2}y) = 2^{\frac{n}{2}} H_n(x+y) ,$$

$$\sum_{k=0}^{n} \binom{2n}{2k} H_{2k}(\sqrt{2}x) H_{2n-2k}(\sqrt{2}y) = 2^{n-1}\{H_{2n}(x+y) + H_{2n}(x-y)\} .$$

30. Show that

$$\sum_{k=0}^{n} \binom{n}{k} H_{2k}(x) H_{2n-2k}(y) = (-)^n n! L_n(x^2 + y^2) ,$$

$$\int_0^{\infty} e^{-t^2} [H_n(t)]^2 \cos(\sqrt{2}xt) dt = \sqrt{\pi} 2^{n-1} n! L_n(x^2) ,$$

$$\Gamma(n + \mu + 1) \int_{-1}^{1} (1 - t^2)^{\mu - \frac{1}{2}} H_{2n}(\sqrt{x}t) dt$$

$$= (-)^n \sqrt{\pi} (2n)! \Gamma\left(\mu + \frac{1}{2}\right) L_n^\mu(x) \quad \left(\operatorname{Re}(\mu) > -\frac{1}{2}\right) ,$$

where $L_n^\mu(z)$ is the generalized Laguerre polynomial.

31. Show that, if $\rho = 1/\xi$, then we have

$$\frac{d^n}{d\xi^n} f(\xi) = (-)^n \rho^{n+1} \frac{d^n}{d\rho^n} \left\{ \rho^{n-1} f\left(\frac{1}{\rho}\right) \right\} .$$

With this formula, derive the derivative expression of the Laguerre polynomial [Eq. (5) of Sec. 6.14] $L_n^\mu(z)$ from its generating function [Eq. (8) of Sec. 6.14].

32. Show that

$$F\left(\alpha, \gamma, \frac{xy}{x-1}\right) = (1-x)^\alpha \sum_{n=0}^{\infty} \frac{(\alpha)_n}{(\gamma)_n} L_n^{\gamma-1}(y) x^n$$

$$(|x| < 1, y > 0)$$

[Erdélyi (1953), Vol. I, p. 276, (5)]. Equation (8) of Sec. 6.14 is a special case of this expansion formula: $\alpha = \gamma = \mu + 1$.

33. Show that

$$L_n^\mu(x) = \sum_{k=0}^{n} \binom{n+\mu}{n-k} \frac{(-x)^k}{k!} .$$

34. Show that

$$\sum_{k=0}^{n} \frac{k!}{\Gamma(\mu + k + 1)} L_k^\mu(x) L_k^\mu(y)$$

$$= \frac{(n+1)!}{\Gamma(\mu + n + 1)} \frac{1}{x - y} \{L_n^\mu(x) L_{n+1}^\mu(y) - L_{n+1}^\mu(x) L_n^\mu(y)\} .$$

35. Show that

$$\sum_{n=0}^{\infty} L_n^{\mu-n}(x)y^n = e^{-xy}(1+y)^{\mu} \quad (|y| < 1) \ .$$

36. Show that

$$\frac{d^n}{dx^n}\left[x^{-\mu-1}e^{-\frac{1}{x}}\right] = (-)^n n! x^{-\mu-n-1} L_n^{\mu}(x^{-1})e^{-\frac{1}{x}} \ ,$$

$$\frac{d^m}{dx^m}[x^{\mu}L_n^{\mu}(x)] = (n-m+\mu+1)_m x^{\mu-m} L_n^{\mu-m}(x) \ ,$$

$$\frac{d^m}{dx^m}[e^{-x}x^{\mu}L_n^{\mu}(x)] = \frac{(m+n)!}{n!}e^{-x}x^{\mu-m}L_{m+n}^{\mu-m}(x) \ .$$

37. Show that

$$\int_x^{\infty} e^{-t}L_n^{\mu}(t)dt = e^{-x}\{L_n^{\mu}(x) - L_{n-1}^{\mu}(x)\} \ ,$$

$$\int_0^x (x-t)^{\beta-1}t^{\alpha}L_n^{\alpha}(t)dt = \frac{\Gamma(\alpha+n+1)\Gamma(\beta)}{\Gamma(\alpha+\beta+n+1)}x^{\alpha+\beta}L_n^{\alpha+\beta}(x) \ ,$$

$$(\mathrm{Re}(\alpha) > -1, \mathrm{Re}(\beta) > 0)$$

$$\int_0^x L_m(t)L_n(x-t)dt = \int_0^x L_{m+n}(t)dt = L_{m+n}(x) - L_{m+n+1}(x) \ .$$

38. Prove the following Laplace transforms:

$$\mathcal{L}\{t^{\mu}L_n^{\mu}(t)\} = \int_0^{\infty} e^{-st}t^{\mu}L_n^{\mu}(t)dt$$

$$= \frac{\Gamma(\mu+n+1)(s-1)^n}{n!s^{\mu+n+1}}$$

$$(\mathrm{Re}(\mu) > -1, \quad \mathrm{Re}(s) > 0) \ ,$$

$$\mathcal{L}\{t^{\beta}L_n^{\alpha}(t)\} = \frac{\Gamma(\beta+1)\Gamma(\alpha+n+1)}{n!\Gamma(\alpha+1)}s^{-\beta-1}F(-n,\beta+1,\alpha+1,s^{-1})$$

$$(\mathrm{Re}(\beta) > -1, \quad \mathrm{Re}(s) > 0) \ .$$

39. Show that

$$\sum_{k=0}^{n} L_k^{\alpha}(x)L_{n-k}^{\beta}(y) = L_n^{\alpha+\beta+1}(x+y) \ .$$

40. **Show that**

$$\mathrm{Ci}(z) = \int_{\infty}^{z} \frac{\cos t}{t} dt$$

$$= -\frac{z^{-\frac{1}{2}}}{2} \left\{ e^{i\left(\frac{z}{2}+\frac{\pi}{4}\right)} W_{-\frac{1}{2},0}\left(ze^{-\frac{\pi i}{2}}\right) + e^{-i\left(\frac{z}{2}+\frac{\pi}{4}\right)} W_{-\frac{1}{2},0}\left(ze^{\frac{\pi i}{2}}\right) \right\},$$

$$\mathrm{Si}(z) = \int_{0}^{z} \frac{\sin t}{t} dt$$

$$= \frac{\pi}{2} + \frac{iz^{-\frac{1}{2}}}{2} \left\{ e^{i\left(\frac{z}{2}+\frac{\pi}{4}\right)} W_{-\frac{1}{2},0}\left(ze^{-\frac{\pi i}{2}}\right) - e^{-i\left(\frac{z}{2}+\frac{\pi}{4}\right)} W_{-\frac{1}{2},0}\left(ze^{\frac{\pi i}{2}}\right) \right\}.$$

41. **Show that** $_0F_1(\gamma; x)$ **satisfies the following differential equation:**

$$x\frac{d^2y}{dx^2} + \gamma\frac{dy}{dx} - y = 0 .$$

42. **Show that**

$$E_1(x) = e^{-\frac{x}{2}} x^{-\frac{1}{2}} W_{-\frac{1}{2},0}(x) = -\gamma - \ln x - \sum_{1}^{\infty} \frac{(-x)^n}{n!n} .$$

Applying it to Ex. 40 to obtain

$$\mathrm{Ci}(z) = \gamma + \ln z + \sum_{1}^{\infty} \frac{(-)^n z^{2n}}{(2n)!2n} ,$$

$$\mathrm{Si}(z) = -\sum_{1}^{\infty} \frac{(-)^n z^{2n-1}}{(2n-1)!(2n-1)} .$$

Chapter 7

BESSEL FUNCTIONS

7.1. Bessel Equation. Its Relation to the Confluent Hypergeometric Equation

Bessel functions are the solutions of the following Bessel equation:

$$\frac{d^2y}{dz^2} + \frac{1}{z}\frac{dy}{dz} + \left(1 - \frac{\nu^2}{z^2}\right)y = 0 \tag{1}$$

where ν, which may be any real or complex number, is called the *order of the equation and of the solutions.*

Bessel functions, besides being solutions of Eq. (1), also occur, for example, in the expansion problem of certain functions [cf. Eq. (2) of Sec. 7.5, Eq. (3) of Sec. 7.18].

Bessel equation often gives rise in the solution of the boundary-value problem or eigenvalue problem of partial differential equations by the method of separation of variables. For example, for the wave equation in cylindrical coordinates,

$$\frac{1}{r}\frac{\partial}{\partial r}\left(r\frac{\partial u}{\partial r}\right) + \frac{1}{r^2}\frac{\partial^2 u}{\partial \theta^2} + \frac{\partial^2 u}{\partial z^2} - \frac{1}{c^2}\frac{\partial^2 u}{\partial t^2} = 0 ,$$

let $u(r, \theta, z, t) = R(r)\Theta(\theta)Z(z)e^{i\omega t}$, then we have for $R(r)$ the equation

$$\frac{1}{r}\frac{d}{dr}\left(r\frac{dR}{dr}\right) + \left(k^2 - \frac{m^2}{r^2}\right)R = 0 , \tag{2}$$

where k and m are constants introduced during the separation of the variables. Putting $\xi = kr$, $R(r) = y(\xi)$, (2) reduces to the Bessel Eq. (1).

Also, for the wave equation in spherical coordinates,

$$\frac{1}{r^2}\frac{\partial}{\partial r}\left(r^2\frac{\partial u}{\partial r}\right) + \frac{1}{r^2\sin\theta}\frac{\partial}{\partial\theta}\left(\sin\theta\frac{\partial u}{\partial\theta}\right) + \frac{1}{r^2\sin^2\theta}\frac{\partial^2 u}{\partial\varphi^2} - \frac{1}{c^2}\frac{\partial^2 u}{\partial t^2} = 0 ,$$

let $u(r, \theta, \varphi, t) = R(r)\Theta(\theta)\Phi(\varphi)e^{i\omega t}$, then we have for $R(r)$ the equation

$$\frac{1}{r^2}\frac{d}{dr}\left(r^2\frac{dR}{dr}\right) + \left[k^2 - \frac{l(l+1)}{r^2}\right]R = 0 , \tag{3}$$

where $l = 0, 1, 2, \ldots$ [cf. Sec. 5.14]. Putting $\xi = kr$ and employing the transformation $R(r) = \xi^{-\frac{1}{2}}y(\xi)$, we obtain

$$\frac{d^2y}{d\xi^2} + \frac{1}{\xi}\frac{dy}{d\xi} + \left[1 - \frac{(l+\frac{1}{2})^2}{\xi^2}\right]y = 0 \tag{4}$$

which is a Bessel equation of order half an odd integer, $(l + \frac{1}{2})$.

Relation between the Bessel equation and the confluent hypergeometric equation

The two equations have the same singularities, 0 and ∞; 0 is regular while ∞ is irregular. According to Eqs. (5) and (6) of Sec. 6.11, if we put, in (1),

$$y(z) = z^{-\frac{1}{2}}W(\xi) , \quad \xi = 2iz \tag{5}$$

we obtain

$$\frac{d^2W}{d\xi^2} + \left[-\frac{1}{4} + \frac{\frac{1}{4} - \nu^2}{\xi^2}\right]W(\xi) = 0 , \tag{6}$$

which is a special case of the Whittaker equation [Eq. (4) of Sec 6.3], $k = 0, m^2 = \nu^2$.

Then, by the relation between Whittaker equation and Kummer equation [Eq. (1) of Sec 6.1], let

$$y(z) = z^\nu e^{-\xi/2}u(\xi), \quad \xi = 2iz . \tag{7}$$

We find from (1)

$$\xi\frac{d^2u}{d\xi^2} + (2\nu + 1 - \xi)\frac{du}{d\xi} - \left(\nu + \frac{1}{2}\right)u(\xi) = 0 , \tag{8}$$

which is a special case of the Kummer equation, $\alpha = \nu + \frac{1}{2}$, $\gamma = 2\nu + 1 = 2\alpha$.

With these relations, we can utilize the results obtained in Chap. 6 to deduce the properties of the Bessel functions such as their integral representations, asymptotic expansions [Sec. 7.10], etc.

Besides, many equations connected with the Bessel equation may be derived from the following transformation:

$$u(z) = z^\alpha Z_\nu(\lambda z^\beta) , \tag{9}$$

where $Z_\nu(z)$ is the solution of the Bessel equation of order ν; then $u(z)$ satisfies the equation

$$z^2\frac{d^2u}{dz^2} + (1 - 2\alpha)z\frac{du}{dz} + (\lambda^2\beta^2 z^{2\beta} + \alpha^2 - \nu^2\beta^2)u = 0 . \tag{10}$$

7.2. Bessel Functions of the First Kind: $J_{\pm\nu}(z), 2\nu \neq$ integer

$J_{\pm\nu}(z)$ are two linearly independent solutions of the Bessel equation of order ν, namely,

$$\frac{d^2y}{dz^2} + \frac{1}{z}\frac{dy}{dz} + \left(1 - \frac{\nu^2}{z^2}\right)y = 0 \tag{1}$$

at the regular singularity $z = 0$ for $2\nu \neq$ integer. We shall find their series expressions.

Writing (1) as

$$z^2y'' + zy' + (z^2 - \nu^2)y = 0 \tag{1'}$$

and assuming

$$y = \sum_{k=0}^\infty c_k z^{k+\rho} \quad (c_0 \neq 0) ,$$

following the procedure of deducing the regular solution [Sec. 2.4], we find

$$c_0[\rho^2 - \nu^2] = 0 , \tag{2}$$

$$c_1[(\rho + 1)^2 - \nu^2] = 0 , \tag{3}$$

$$c_k = -\frac{c_{k-2}}{(\rho + k)^2 - \nu^2} \quad (k \geq 2) . \tag{4}$$

From (2), we obtain the two exponents at $z = 0$ to be $\rho = \pm \nu$. Suppose $\text{Re}(\nu) \geq 0$. Since $(\pm \nu + 1)^2 - \nu^2 = \pm 2\nu + 1 \neq 0$ (by assumption, $2\nu \neq$ integer), we have, from (3), $c_1 = 0$, and hence, by the recurrence relation (4),

$$c_{2k+1} = 0 \quad (k \geq 0) , \tag{5}$$

$$c_{2k} = -\frac{c_{2k-2}}{2k(\pm 2\nu + 2k)} = (-)^k \frac{c_0 \Gamma(\pm \nu + 1)}{2^{2k} k! \Gamma(\pm \nu + k + 1)} . \tag{6}$$

Taking $c_0 \Gamma(\pm \nu + 1) = 2^{\mp \nu}$, we obtain two series solutions for (1):

$$J_{\pm \nu}(z) = \sum_{k=0}^{\infty} \frac{(-)^k}{k!} \frac{1}{\Gamma(\pm \nu + k + 1)} \left(\frac{z}{2}\right)^{2k \pm \nu} \tag{7}$$

When 2ν is not an integer, these two solutions are evidently linearly independent of each other. For, letting $z \to 0$, the two solutions $J_{\pm \nu}(z) \sim z^{\pm \nu}$ respectively, and $J_{\nu}(z)/J_{-\nu}(z) \sim z^{2\nu}$ cannot be a constant.

When $\nu = n (n = 0, 1, 2, \ldots)$, owing to $\Gamma(-n + k + 1) = \infty$ for $k < n$, we have

$$J_{-n}(z) = \sum_{k=n}^{\infty} \frac{(-)^k}{k!} \frac{1}{\Gamma(-n + k + 1)} \left(\frac{z}{2}\right)^{2k-n}$$

$$= \sum_{k=0}^{\infty} \frac{(-)^{k+n}}{(k+n)!} \frac{1}{\Gamma(k+1)} \left(\frac{z}{2}\right)^{2k+n} = (-)^n J_n(z) , \tag{8}$$

which shows that $J_{-n}(z)$ and $J_n(z)$ are linearly dependent. In such cases, it is necessary to look for a second solution by some other means.

The cases of $2\nu = 2n + 1$, i.e., $\nu = n + \frac{1}{2} (n = 0, 1, 2, \ldots)$, will be dealt in the following section. We only point out here that, in such cases, the solutions given by (7) are still linearly independent of each other, because the Γ-functions will not become infinities in the series.

$J_{\pm \nu}(z)$ are called *Bessel functions of the first kind of order* ν.

Let u_k represent the general term of the series in (7) besides the common factor $(z/2)^{\pm \nu}$, then

$$\left| \frac{u_{k+1}}{u_k} \right| = \frac{1}{k+1} \left| \frac{\Gamma(\pm \nu + k + 1)}{\Gamma(\pm \nu + k + 2)} \right| \left| \frac{z}{2} \right|^2$$

$$= \frac{1}{(k+1)|\pm \nu + k + 1|} \left| \frac{z}{2} \right|^2 .$$

For $|z| \leq R_1, |\nu| \leq R_2$, where R_1 and R_2 are arbitrary positive numbers, we have, when k is sufficiently large,

$$\left| \frac{u_{k+1}}{u_k} \right| \leq \frac{1}{(k+1)(k+1-R_2)} \left(\frac{R_1}{2} \right)^2 \to 0 \ .$$

Therefore, the series converges uniformly in the domain mentioned, and hence, for non-integral ν, $J_{\pm\nu}(z)$, as functions of z, are single-valued and analytic in the whole z plane cut along the negative real axis; $|\arg z| < \pi$. While, as functions of ν, $J_{\pm\nu}(z)$ are integral functions.

Recurrence relations

From the series expression (7) of $J_\nu(z)$, multiplying both sides by z^ν, and differentiating with respect to z, we obtain

$$\frac{d}{dz}(z^\nu J_\nu) = 2^\nu \sum_{k=0}^\infty \frac{(-)^k}{k!} \frac{k+\nu}{\Gamma(k+\nu+1)} \left(\frac{z}{2} \right)^{2(k+\nu)-1}$$

$$= z^\nu \sum_{k=0}^\infty \frac{(-)^k}{k!} \frac{1}{\Gamma(\nu+k)} \left(\frac{z}{2} \right)^{2k+\nu-1} \ ,$$

i.e.,

$$\frac{d}{dz}(z^\nu J_\nu) = z^\nu J_{\nu-1} \ . \tag{9}$$

Similarly, we obtain

$$\frac{d}{dz}(z^{-\nu} J_\nu) = -z^{-\nu} J_{\nu+1} \ . \tag{10}$$

Carrying out the differentiations on the l.h.s. of (9) and (10), we obtain

$$\nu J_\nu + z J_\nu' = z J_{\nu-1} \ , \tag{11}$$

$$-\nu J_\nu + z J_\nu' = -z J_{\nu+1} \ . \tag{12}$$

Eliminating J_ν', we find

$$J_{\nu-1} + J_{\nu+1} = \frac{2\nu}{z} J_\nu \ ; \tag{13}$$

eliminating J_ν, we find

$$J_{\nu-1} - J_{\nu+1} = 2 J_\nu' \ . \tag{14}$$

When $\nu = 0$, (10) gives

$$J_0' = -J_1 . \tag{15}$$

(13) and (14) are two fundamental recurrence relations of the Bessel functions. Further, by repeatedly applying (9) and (10), we can show that

$$\left(\frac{d}{zdz}\right)^m \{z^\nu J_\nu\} = z^{\nu-m} J_{\nu-m} , \tag{16}$$

$$\left(\frac{d}{zdz}\right)^m \{z^{-\nu} J_\nu\} = (-)^m z^{-\nu-m} J_{\nu+m} . \tag{17}$$

All the above recurrence formulae for J_ν hold for any ν. Finally, it can be proved from (7) that

$$J_{\pm\nu}(ze^{\pi i}) = e^{\pm\nu\pi i} J_{\pm\nu}(z) . \tag{18}$$

7.3. Bessel Functions of Order Half an Odd Integer:
$J_{n+\frac{1}{2}}(z)(n = 0, \pm1, \pm2, \dots)$

One of the important properties of $J_{n+\frac{1}{2}}(z)$ is that they can be expressed in terms of elementary functions. For example,

$$J_{\frac{1}{2}}(z) = \sum_{k=0}^{\infty} \frac{(-)^k}{k!} \frac{1}{\Gamma\left(k+\frac{3}{2}\right)} \left(\frac{z}{2}\right)^{2k+\frac{1}{2}} .$$

By Eq. (8) of Sec. 3.6, $\Gamma\left(k+\frac{3}{2}\right) = \Gamma(2k+2)2^{-2k-1}\sqrt{\pi}/\Gamma(k+1)$, therefore,

$$J_{\frac{1}{2}}(z) = \sqrt{\frac{2}{\pi z}} \sum_{k=0}^{\infty} \frac{(-)^k}{(2k+1)!} z^{2k+1} = \sqrt{\frac{2}{\pi z}} \sin z . \tag{1}$$

Similarly, we can show that

$$J_{-\frac{1}{2}}(z) = \sqrt{\frac{2}{\pi z}} \cos z . \tag{2}$$

In general, we obtain from (1), by (17) of the last section,

$$J_{n+\frac{1}{2}}(z) = (-)^n z^{n+\frac{1}{2}} \left(\frac{d}{zdz}\right)^n \left\{z^{-\frac{1}{2}} J_{\frac{1}{2}}(z)\right\}$$

$$= (-)^n \sqrt{\frac{2}{\pi z}} z^{n+1} \left(\frac{d}{zdz}\right)^n \left\{\frac{\sin z}{z}\right\}$$

$$(n = 0, 1, 2, \dots) \tag{3}$$

and, from (2), by (16) of the last section,

$$J_{-n-\frac{1}{2}}(z) = \sqrt{\frac{2}{\pi z}} z^{n+1} \left(\frac{d}{zdz}\right)^n \left\{\frac{\cos z}{z}\right\} \quad (n = 0, 1, 2, \ldots). \tag{4}$$

By induction, we obtain respectively from (3) and (4) the following explicit expressions for $J_{\pm(n+\frac{1}{2})}(z)$:

$$J_{n+\frac{1}{2}}(z) = \sqrt{\frac{2}{\pi z}} \left\{ \sin\left(z - \frac{n\pi}{2}\right) \sum_{r=0}^{[n/2]} \frac{(-)^r (n+2r)!}{(2r)!(n-2r)!(2z)^{2r}} \right.$$
$$\left. + \cos\left(z - \frac{n\pi}{2}\right) \sum_{r=0}^{[(n-1)/2]} \frac{(-)^r (n+2r+1)!}{(2r+1)!(n-2r-1)!(2z)^{2r+1}} \right\} \tag{5}$$

$$J_{-n-\frac{1}{2}}(z) = \sqrt{\frac{2}{\pi z}} \left\{ \cos\left(z + \frac{n\pi}{2}\right) \sum_{r=0}^{[n/2]} \frac{(-)^r (n+2r)!}{(2r)!(n-2r)!(2z)^{2r}} \right.$$
$$\left. - \sin\left(z + \frac{n\pi}{2}\right) \sum_{r=0}^{[(n-1)/2]} \frac{(-)^r (n+2r+1)!}{(2r+1)!(n-2r-1)!(2z)^{2r+1}} \right\}. \tag{6}$$

In Sec. 7.10, we will prove (5) and (6) by the asymptotic expansion of $J_\nu(z)$.

7.4. Integral Representations of $J_\nu(z)$

There are many integral representations for $J_\nu(z)$. We shall give, in this section, several of them which are more important and fundamental [(5), (6), (9), (10), (13), (15), (17), (18) and (19) in this section].

The usual methods for obtaining the integral expressions are of two kinds: One is to derive directly from the integral solution of the corresponding differential equation; the other is to start from the series expression of the solution. We shall describe the former first.

According to the relation between the Bessel equation and the confluent hypergeometric equation given in Sec. 7.1, we can obtain an integral expression of $J_\nu(z)$. But we can also derive it directly from the integral solution of the Bessel equation. In the Bessel equation

$$y'' + \frac{1}{z}y' + \left(1 - \frac{\nu^2}{z^2}\right)y = 0 \tag{1}$$

let $y = z^\nu u(z)$. We get an equation of the Laplacian type [Eq. (1) of Sec. 2.13]:

$$z\frac{d^2u}{dz^2} + (2\nu + 1)\frac{du}{dz} + zu(z) = 0 . \qquad (2)$$

The general form of its integral solution is

$$u(z) = A \int_C e^{zt}(1 + t^2)^{\nu - \frac{1}{2}} dt$$

[cf. Sec. 2.13], or, by changing t to it,

$$u(z) = A \int_C e^{izt}(1 - t^2)^{\nu - \frac{1}{2}} dt ,$$

where A is an arbitrary constant and the path of integration C is such that

$$\left\{ e^{izt}(1 - t^2)^{\nu + \frac{1}{2}} \right\}_C = 0 . \qquad (3)$$

The integral solution of the Bessel equation (1) is thus

$$y(z) = Az^\nu \int_C e^{izt}(1 - t^2)^{\nu - \frac{1}{2}} dt . \qquad (4)$$

Assume $\text{Re}(\nu + \frac{1}{2}) > 0$, then we may take C to be the straight line segment from $t = -1$ to $t = 1$, since $(1 - t^2)^{\nu + \frac{1}{2}} = 0$ at the end points of this path, so (3) is satisfied. We now show that, by properly choosing the constant A and specifying the multi-valued factor $(1 - t^2)^{\nu - \frac{1}{2}}$, the r.h.s. of (4) is indeed equal to $J_\nu(z)$.

Assume $\arg(1 - t^2) = 0$ along the path of integration, then we have

$$y(z) = Az^\nu \int_{-1}^{1} e^{izt}(1 - t^2)^{\nu - \frac{1}{2}} dt$$

$$= Az^\nu \sum_{k=0}^{\infty} \frac{(iz)^k}{k!} \int_{-1}^{1} t^k(1 - t^2)^{\nu - \frac{1}{2}} dt$$

$$= Az^\nu \sum_{k=0}^{\infty} \frac{(-)^k}{(2k)!} z^{2k} \int_{-1}^{1} t^{2k}(1 - t^2)^{\nu - \frac{1}{2}} dt .$$

Now,

$$\int_{-1}^{1} t^{2k}(1 - t^2)^{\nu - \frac{1}{2}} dt = 2 \int_{0}^{1} t^{2k}(1 - t^2)^{\nu - \frac{1}{2}} dt$$

$$= \int_{0}^{1} s^{k - \frac{1}{2}}(1 - s)^{\nu - \frac{1}{2}} ds = \frac{\Gamma\left(k + \frac{1}{2}\right)\Gamma\left(\nu + \frac{1}{2}\right)}{\Gamma(\nu + k + 1)} ,$$

so, by Eq. (8) of Sec. 3.6, $\Gamma(k + \frac{1}{2}) = \Gamma(2k+1)2^{-2k}\sqrt{\pi}/\Gamma(k+1)$, we have

$$y(z) = A\sqrt{\pi}\Gamma\left(\nu + \frac{1}{2}\right) z^{\nu} \sum_{k=0}^{\infty} \frac{(-)^k}{k!} \frac{1}{\Gamma(\nu+k+1)} \left(\frac{z}{2}\right)^{2k} .$$

Comparing with the series expression of $J_\nu(z)$ given in Eq. (7) of Sec. 7.2, we see that $y(z) = A\sqrt{\pi}\Gamma(\nu + \frac{1}{2})2^\nu J_\nu(z)$. Therefore,

$$J_\nu(z) = \frac{1}{\sqrt{\pi}\Gamma\left(\nu+\frac{1}{2}\right)} \left(\frac{z}{2}\right)^\nu \int_{-1}^{1} e^{izt}(1-t^2)^{\nu-\frac{1}{2}} dt$$

$$(\mathrm{Re}(\nu) > -\frac{1}{2}, \quad \arg(1-t^2) = 0) . \tag{5}$$

In (5), putting $t = \cos\theta$, we obtain the *Poisson Integral Representation*

$$J_\nu(z) = \frac{1}{\sqrt{\pi}\Gamma\left(\nu+\frac{1}{2}\right)} \left(\frac{z}{2}\right)^\nu \int_0^\pi e^{iz\cos\theta} \sin^{2\nu}\theta\, d\theta , \tag{6}$$

from which we also have

$$J_\nu(z) = \frac{1}{\sqrt{\pi}\Gamma\left(\nu+\frac{1}{2}\right)} \left(\frac{z}{2}\right)^\nu \int_0^\pi \cos(z\cos\theta)\sin^{2\nu}\theta\, d\theta \tag{7}$$

since

$$\int_0^\pi \sin(z\cos\theta)\sin^{2\nu}\theta\, d\theta = -\int_0^\pi \sin(z\cos\theta)\sin^{2\nu}\theta\, d\theta = 0$$

(the last equation can be justified by changing θ to $\pi - \theta$).

The integral expression of $J_\nu(z)$ obtained above, though simple, is restricted by the condition $\mathrm{Re}(\nu + \frac{1}{2}) > 0$. In order to remove this restriction, we consider the path of integration C shown in Fig. 24 and write (4) as

$$y(z) = Az^\nu \int_P^{(1+,-1-)} e^{izt}(t^2-1)^{\nu-\frac{1}{2}} dt , \tag{8}$$

where the multi-valued factor $(t^2 - 1)^{\nu-\frac{1}{2}}$ is defined by that $\arg(t^2 - 1) = 0$ at the point P, the intersection of C with the real axis to the right of the point $t = 1$. For such contour C, (3) is evidently satisfied, since the values of $e^{izt}(t^2 - 1)^{\nu+\frac{1}{2}}$ are the same at the starting point and the end point.

The path of integration in (8) does not pass through any singularities of the integrand; the integral thus exists for any value of ν. Now, deform the path of integration into a contour shown in Fig. 25, which starts from a point

Fig. 24.   Fig. 25.

on the real axis at a distance δ_1 to the right of $t = -1$, running along the real axis to a point at a distance δ_2 to the left of $t = 1$, encircling $t = 1$ once in the positive sense, then goes back to the starting point, encircling the point $t = -1$ once in the negative sense, arriving again at the starting point. Assuming $\mathrm{Re}(\nu + \frac{1}{2}) > 0$, the values of the integrals around the points $t = \pm 1$ approach zero as the radii, δ_1 and $\delta_2 \to 0$. Since $\arg(t^2 - 1) = -\pi$ at the starting point (paying due regard to the fact that $\arg(t^2 - 1) = 0$ at the point P), we have

$$\int_P^{(1+,-1-)} e^{izt}(t^2-1)^{\nu-\frac{1}{2}}\,dt$$

$$= \int_{-1}^1 e^{izt}[(1-t^2)e^{-\pi i}]^{\nu-\frac{1}{2}}\,dt + \int_1^{-1} e^{izt}\left[(1-t^2)e^{\pi i}\right]^{\nu-\frac{1}{2}}\,dt$$

$$= [e^{-(\nu-\frac{1}{2})\pi i} - e^{(\nu-\frac{1}{2})\pi i}]\int_{-1}^1 e^{izt}(1-t^2)^{\nu-\frac{1}{2}}\,dt$$

$$= 2i\sin\pi\left(\frac{1}{2}-\nu\right)\frac{\Gamma\left(\nu+\frac{1}{2}\right)\sqrt{\pi}}{(z/2)^\nu}J_\nu(z)\quad[\text{by}\ (5)]$$

$$= \frac{2\pi i\sqrt{\pi}}{\Gamma\left(\frac{1}{2}-\nu\right)(z/2)^\nu}J_\nu(z)\ .$$

Thus,

$$J_\nu(z) = \frac{\Gamma\left(\frac{1}{2}-\nu\right)}{\sqrt{\pi}}\left(\frac{z}{2}\right)^\nu\frac{1}{2\pi i}\int_P^{(1+,-1-)} e^{izt}(t^2-1)^{\nu-\frac{1}{2}}\,dt\ ,\qquad(9)$$

where $\nu \neq \frac{1}{2}, \frac{3}{2}, \ldots$; $\arg(t^2 - 1) = 0$ at P. Though Eq. (9) is proved under the condition $\mathrm{Re}(\nu) > -\frac{1}{2}$, but as the functions on the two sides of it do not subject to this restriction, the said condition may be released according to the principle of analytic continuation.

When $\nu = n + \frac{1}{2}(n = 0, 1, 2, \ldots)$, the gamma functions $\Gamma(\frac{1}{2} - \nu)$ become infinity while the value of the integral is zero (since the integrand is then single-valued). The r.h.s. of (9) is indeterminate. However, from the above derivation, it can be shown that the limit of this indeterminate form when $\nu \to n + \frac{1}{2}$ is exactly the r.h.s. of (5).

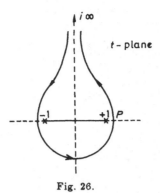

Fig. 26.

The path of integration in (4) can also be chosen as the one shown in Fig. 26. It starts from $t = i\infty$, running around the points $t = \pm 1$ once in the positive sense, then goes back to $i\infty$. So long as $\mathrm{Re}(z) > 0$, the condition (3) will hold and also the integral in (4) converges.

Again write (4) as

$$y(z) = Az^\nu \int_{i\infty}^{(-1+,1+)} e^{izt}(t^2 - 1)^{\nu - \frac{1}{2}} dt$$

and define $\arg(t^2 - 1) = 0$ at the point P which is the intersection of the contour with the real axis to the right of $t = 1$. We may assume that the contour lies entirely outside the circle $|t| = 1$. Then, the factor $(t^2 - 1)^{\nu - \frac{1}{2}}$ can be expanded in descending powers of t:

$$(t^2 - 1)^{\nu - \frac{1}{2}} = t^{2\nu - 1}(1 - t^{-2})^{\nu - \frac{1}{2}} = \sum_{k=0}^{\infty} \frac{\Gamma\left(\frac{1}{2} - \nu + k\right)}{k!\,\Gamma\left(\frac{1}{2} - \nu\right)} t^{2\nu - 1 - 2k} ,$$

where $-3\pi/2 < \arg t < \pi/2$, since the series expression is such that when $t \to \infty$, $(1 - t^{-2})^{\nu - \frac{1}{2}} \to 1$ and $\arg t^2 \sim \arg(t^2 - 1)$, and as $\arg(t^2 - 1) = 0$ at P, we should have $-3\pi < \arg(t^2 - 1) < \pi$.

Substituting the series expression into the integral, exchanging the order of summation and integration (can be justified; cf. the paragraph after (13)), we obtain

$$y(z) = A\frac{z^\nu}{\Gamma\left(\frac{1}{2} - \nu\right)} \sum_{k=0}^{\infty} \frac{\Gamma\left(\frac{1}{2} - \nu + k\right)}{k!} \int_{i\infty}^{(0+)} e^{izt} t^{2\nu - 1 - 2k} dt$$

$$(-3\pi/2 < \arg t < \pi/2) .$$

Let $u = e^{i\pi/2}zt$, $\alpha = \arg z$, and assume that $|\alpha| < \pi/2$, then

$$\int_{i\infty}^{(0+)} e^{izt} t^{2\nu-1-2k} \, dt = \left(e^{\frac{\pi i}{2}} z\right)^{-2\nu+2k} \int_{-\infty e^{i\alpha}}^{(0+)} e^u u^{2\nu-1-2k} \, du$$

$$= (-)^k e^{-\nu\pi i} z^{-2\nu+2k} \frac{2\pi i}{\Gamma(1 - 2\nu + 2k)} \quad \text{[by Eq. (6) of Sec. 3.7]}$$

and hence

$$y(z) = A \frac{2\pi i}{\Gamma\left(\frac{1}{2} - \nu\right)} e^{-\nu\pi i} z^{-\nu} \sum_{k=0}^{\infty} \frac{(-)^k}{k!} \frac{\Gamma\left(\frac{1}{2} - \nu + k\right)}{\Gamma(1 - 2\nu + 2k)} z^{2k} \ .$$

By Eq. (8) of Sec. 3.6 and comparing with Eq. (7) of Sec. 7.2, we see that

$$y(z) = A \frac{2\pi i}{\Gamma\left(\frac{1}{2} - \nu\right)} e^{-\nu\pi i} 2^\nu \sqrt{\pi} J_{-\nu}(z) \ .$$

Therefore,

$$J_{-\nu}(z) = \frac{e^{\nu\pi i} \Gamma\left(\frac{1}{2} - \nu\right) \left(\frac{z}{2}\right)^\nu}{\sqrt{\pi}} \frac{1}{2\pi i} \int_{i\infty}^{(-1+,1+)} e^{izt} (t^2 - 1)^{\nu-\frac{1}{2}} \, dt$$

$$\left(|\arg z| < \pi/2, \ -3\pi < \arg(t^2 - 1) < \pi, \ \nu + \frac{1}{2} \neq 1, 2, \dots\right) \ . \tag{10}$$

When $\nu + \frac{1}{2}$ is a positive integer, the r.h.s. of (10) is an indeterminate form [cf. the paragraph below Eq. (10) of Sec. 7.7].

If the condition $|\arg z| < \pi/2$ is not satisfied for z, say,

$$-\frac{\pi}{2} + \omega < \arg z < \frac{\pi}{2} + \omega \ , \quad |\omega| < \frac{\pi}{2} \ , \tag{11}$$

we can turn the integration contour as a whole around $t = 0$ by an angle $(-\omega)$. [Cf. the discussion after Eq. (2) of Sec. 3.1.] Then, by similar argument as above, we obtain

$$J_{-\nu}(z) = \frac{e^{\nu\pi i} \Gamma\left(\frac{1}{2} - \nu\right) \left(\frac{z}{2}\right)^\nu}{\sqrt{\pi}} \frac{1}{2\pi i} \int_{i\infty e^{-i\omega}}^{(-1+,1+)} e^{izt} (t^2 - 1)^{\nu-\frac{1}{2}} \, dt \ , \tag{12}$$

where $-\pi/2 + \omega < \arg z < \pi/2 + \omega$, and $\arg(t^2 - 1) = 0$ at the point P; $\nu + \frac{1}{2} \neq$ positive integers. By gradually turning the contour in this way, we may obtain such expressions of $J_{-\nu}(z)$ for any $\arg z$.

Another way of obtaining the integral representation of $J_\nu(z)$ is to start from its series representation.[a] From Eq. (7) of Sec. 7.2, namely,

$$J_\nu(z) = \left(\frac{z}{2}\right)^\nu \sum_{k=0}^{\infty} \frac{(-)^k}{k!} \frac{1}{\Gamma(\nu+k+1)} \left(\frac{z}{2}\right)^{2k},$$

by Eq. (5) of Sec. 3.7, we have

$$J_\nu(z) = \left(\frac{z}{2}\right)^\nu \sum_{k=0}^{\infty} \frac{(-)^k}{k!} \left(\frac{z}{2}\right)^{2k} \frac{1}{2\pi i} \int_{-\infty}^{(0+)} t^{-\nu-k-1} e^t \, dt$$

$$(|\arg t| < \pi) .$$

Interchanging the order of integration and summation, the r.h.s. becomes

$$\frac{\left(\frac{z}{2}\right)^\nu}{2\pi i} \int_{-\infty}^{(0+)} \sum_{k=0}^{\infty} \frac{1}{k!} \left(-\frac{z^2}{4t}\right)^k t^{-\nu-1} e^t \, dt = \frac{\left(\frac{z}{2}\right)^\nu}{2\pi i} \int_{-\infty}^{(0+)} e^{t-\frac{z^2}{4t}} t^{-\nu-1} \, dt .$$

Thus,

$$J_\nu(z) = \frac{\left(\frac{z}{2}\right)^\nu}{2\pi i} \int_{-\infty}^{(0+)} e^{t-\frac{z^2}{4t}} t^{-\nu-1} \, dt \quad (|\arg t| < \pi) . \qquad (13)$$

This is another important integral expression of $J_\nu(z)$.

To show that the interchange of the order of integration and summation is legitimate, we need only to note that the integral on the r.h.s. of (13) is uniformly convergent in $|z| \leq R$ (R being any positive number) and is hence an analytic function in the whole z plane. So, we can differentiate under the integral sign and obtain its Taylor expansion at $z = 0$ as

$$\sum_{n=0}^{\infty} \frac{z^n}{n!} \int_{-\infty}^{(0+)} e^t t^{-\nu-1} \frac{d^n}{dz^n} \left(e^{-z^2/4t}\right)\Big|_{z=0} dt .$$

But

$$\frac{d^n}{dz^n} e^{-z^2/4t}\Big|_{z=0} = \frac{d^n}{dz^n} \sum_{k=0}^{\infty} \frac{1}{k!} \left(-\frac{z^2}{4t}\right)^k \Big|_{z=0}$$

$$(k = 0, 1, 2, \ldots) \qquad = \begin{cases} \frac{(2k)!}{k!} \left(-\frac{1}{4t}\right)^k & (n = 2k) \\ 0 & (n \neq 2k) , \end{cases}$$

---

[a] This way of obtaining the integral expression of a function is typical in certain sense.

hence the r.h.s. of (13) is

$$\frac{\left(\frac{z}{2}\right)^{\nu}}{2\pi i} \sum_{k=0}^{\infty} \frac{(-)^k}{k!} \left(\frac{z}{2}\right)^{2k} \int_{-\infty}^{(0+)} e^t t^{-\nu-k-1} dt \ .$$

This is what we desire to prove.

Changing t into $zt/2$ in (13), we obtain a more simple expression

$$J_{\nu}(z) = \frac{1}{2\pi i} \int_{-\infty e^{-i\alpha}}^{(0+)} e^{\frac{z}{2}(t-t^{-1})} t^{-\nu-1} dt \ , \tag{14}$$

where $|\arg t + \alpha| < \pi, \alpha = \arg z$. If $|\alpha| < \pi/2$, we can turn the whole contour around $t = 0$ by an angle α without altering the value of the integral [cf. the similar operation in Sec. 3.1]. Therefore, we have

$$J_{\nu}(z) = \frac{1}{2\pi i} \int_{-\infty}^{(0+)} e^{\frac{z}{2}(t-t^{-1})} t^{-\nu-1} dt \tag{15}$$

$$(|\arg z| < \pi/2, \quad |\arg t| < \pi) \ .$$

The path of integration in (15) can also be taken to start from $t = -\infty$, running along the lower "bank" of the negative real axis to $t = -1$, encircling $t = 0$ once in the positive sense, then goes back to $t = -\infty$ along the upper "bank" of the negative real axis. Then

$$J_{\nu}(z) = \frac{1}{2\pi i} \Big\{ (e^{-\nu\pi i} - e^{\nu\pi i}) \int_{1}^{\infty} e^{-\frac{z}{2}(t-t^{-1})} t^{-\nu-1} dt$$

$$+ \int_{-\infty}^{(0+)} e^{\frac{z}{2}(t-t^{-1})} t^{-\nu-1} dt \Big\} \ .$$

In the first integral, putting $t = e^u$, and in the second integral, $t = i\theta$, we have, by rearranging the order,

$$J_{\nu}(z) = \frac{1}{2\pi} \int_{-\pi}^{\pi} e^{i(z \sin\theta - \nu\theta)} d\theta - \frac{\sin \nu\pi}{\pi} \int_{0}^{\infty} e^{-z \operatorname{sh} u - \nu u} du \tag{16}$$

$$(|\arg z| < \pi/2) \ ,$$

or

$$J_{\nu}(z) = \frac{1}{2\pi} \int_{-\pi}^{\pi} \cos(z \sin\theta - \nu\theta) d\theta - \frac{\sin \nu\pi}{\pi} \int_{0}^{\infty} e^{-z \operatorname{sh} u - \nu u} du \tag{17}$$

$$(|\arg z| < \pi/2) \ ,$$

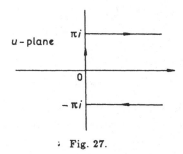

u - plane πi

0

$-\pi i$

; Fig. 27.

since $\sin(z \sin \theta - \nu\theta)$ is an odd function of θ, which contributes nothing to the value of the first integral.

From (15) we can derive another important integral expression for $J_\nu(z)$. Put $t = e^u$ in it, then we have

$$J_\nu(z) = \frac{1}{2\pi i} \int_{\infty-\pi i}^{\infty+\pi i} e^{z \, \mathrm{sh} u - \nu u} du \qquad (18)$$
$$(|\arg z| < \pi/2) \, ,$$

the path of integration being the one shown in Fig. 27 above.

Further, if $\mathrm{Re}(\nu) > -1$, we can use Jordan's lemma and deform the path of integration in (13) to a straight line parallel to the imaginary axis of the t-plane so that

$$J_\nu(z) = \frac{\left(\frac{z}{2}\right)^\nu}{2\pi i} \int_{c-i\infty}^{c+i\infty} e^{t - \frac{z^2}{4t}} t^{-\nu-1} dt \qquad (19)$$
$$(c > 0, \quad \mathrm{Re}(\nu) > -1) \, .$$

7.5. Bessel Functions of Integral Order $J_n(z)(n = 0, 1, 2, \dots)$

According to the analysis in Sec. 7.2, we know that, when n is an integer, $J_n(z)$ is an integral function of z. All the recurrence relations given in that section are, of course, applicable to $J_n(z)$.

The generating function of $J_n(z)$

In (15) of the last section, when $\nu = n$, an integer, the path of integration can be reduced to any contour which encircles $t = 0$ once in the positive sense, since the integrand is now a single-valued function. Thus

$$J_n(z) = \frac{1}{2\pi i} \int^{(0+)} e^{\frac{z}{2}(t-t^{-1})} t^{-n-1} dt \quad (n = 0, \pm 1, \pm 2, \dots) \qquad (1)$$

from which we see immediately that $J_n(z)$ are the coefficients of the Laurent expansion of the function $\exp\{z(t - t^{-1})/2\}$ in $0 < t < \infty$, i.e.,

$$e^{\frac{z}{2}(t-t^{-1})} = \sum_{n=-\infty}^{\infty} J_n(z) t^n . \tag{2}$$

The function on the l.h.s. is called the *generating function* of $J_n(z)$.

(2) is an important formula in the theory of Bessel functions. A series of expansion formulae can be deduced from it. Here are some examples.

In (2), putting $t = ie^{i\theta}$ and noting that, by Eq. (8) of Sec. 7.2, $J_{-n}(z) = (-)^n J_n(z)$, we find

$$e^{iz\cos\theta} = \sum_{n=-\infty}^{\infty} J_n(z) i^n e^{in\theta} \tag{3}$$

$$= J_0(z) + \sum_{n=1}^{\infty} \left[J_n(z) i^n e^{in\theta} + J_{-n}(z) i^{-n} e^{-in\theta} \right]$$

$$= J_0(z) + \sum_{n=1}^{\infty} J_n(z) i^n (e^{in\theta} + e^{-in\theta})$$

$$= J_0(z) + 2 \sum_{n=1}^{\infty} i^n J_n(z) \cos n\theta . \tag{4}$$

With the symbols

$$\varepsilon_0 = 1, \quad \varepsilon_n = 2 \quad (n = 1, 2, \ldots) , \tag{5}$$

(4) can be written neatly in the form

$$e^{iz\cos\theta} = \sum_{n=0}^{\infty} \varepsilon_n i^n J_n(z) \cos n\theta , \tag{6}$$

from which, by comparing the real and imaginary parts on the two sides, we obtain

$$\cos(z\cos\theta) = \sum_{n=0}^{\infty} \varepsilon_{2n} (-)^n J_{2n}(z) \cos 2n\theta , \tag{7}$$

$$\sin(z\cos\theta) = 2 \sum_{n=0}^{\infty} (-)^n J_{2n+1}(z) \cos(2n+1)\theta . \tag{8}$$

In (7), let $\theta = \pi/2$. We have

$$1 = \sum_{n=0}^{\infty} \varepsilon_n J_{2n}(z) = J_0(z) + 2 \sum_{n=1}^{\infty} J_{2n}(z) \ . \tag{9}$$

If we change t into $-t$ in (2), we have

$$e^{-\frac{z}{2}(t-t^{-1})} = \sum_{m=-\infty}^{\infty} J_m(z)(-)^m t^m \ .$$

Multiplying with (2), we obtain

$$1 = \sum_{k=-\infty}^{\infty} J_k(z) t^k \sum_{m=-\infty}^{\infty} J_m(z)(-)^m t^m$$

$$= \sum_{n=-\infty}^{\infty} t^n \sum_{m=-\infty}^{\infty} (-)^m J_m(z) J_{n-m}(z) \ .$$

Comparing both sides and using Eq. (8) of Sec. 7.2, we find

$$\sum_{m=-\infty}^{\infty} J_m^2(z) = J_0^2(z) + 2 \sum_{m=1}^{\infty} J_m^2(z) = 1 \ , \tag{10}$$

$$\sum_{m=-\infty}^{\infty} (-)^m J_m(z) J_{2n-m}(z) = 0 \tag{11}$$

or

$$\sum_{m=0}^{2n} (-)^m J_m(z) J_{2n-m}(z) + 2 \sum_{m=1}^{\infty} J_m(z) J_{2n+m}(z) = 0 \ . \tag{12}$$

From (10) we can draw the important conclusion that, if x is a real number, then

$$|J_0(x)| \leq 1, \ \ |J_m(x)| \leq \frac{1}{\sqrt{2}} \ \ (m = 1, 2, \ldots) \ . \tag{13}$$

Expansion of $(z/2)^m (m = positive\ integers)$ in terms of Bessel functions

From the Laurent expansion of the function $e^{z/u}$ in $0 < u < \infty$, it is readily seen that

$$z^m = \frac{m!}{2\pi i} \int^{(0+)} e^{\frac{z}{u}} u^{m-1} du \ . \tag{14}$$

Putting $u = -2t/(1 - t^2)$, we have

$$z^m = \frac{m!(-2)^m}{2\pi i} \int^{(0+)} e^{\frac{z}{2}(t - t^{-1})} \frac{1}{m} \frac{d}{dt} \left(\frac{t}{1 - t^2} \right)^m dt \ .$$

Assume $|t| < 1$. By the binomial expansion of $(1 - t^2)^{-m}$ and term by term differentiation, we obtain

$$z^m = \frac{(-2)^m}{2\pi i} \int^{(0+)} e^{\frac{z}{2}(t - t^{-1})} \sum_{n=0}^{\infty} \frac{(m + n - 1)!(m + 2n)}{n!} t^{m+2n-1} dt$$

$$= \frac{(-2)^m}{2\pi i} \sum_{n=0}^{\infty} \frac{(m + n - 1)!(m + 2n)}{n!} \int^{(0+)} e^{\frac{z}{2}(t - t^{-1})} t^{m+2n-1} dt$$

$$= (-2)^m \sum_{n=0}^{\infty} \frac{(m + n - 1)!(m + 2n)}{n!} J_{-m-2n}(z) \quad [\text{by} \quad (1)] \ .$$

Applying Eq. (8) of Sec. 7.2, we have

$$\left(\frac{z}{2} \right)^m = \sum_{n=0}^{\infty} \frac{(m + n - 1)!(m + 2n)}{n!} J_{m+2n}(z) \quad (m \geq 1) \ . \tag{15}$$

This expansion formula can be generalized to the case when m is not an integer: it is only necessary to replace the factorial $(m + n - 1)!$ in it by the corresponding Γ-function, $\Gamma(m + n)$ [see Ex. 20 at the end of this chapter].

Addition formulae

In (2), putting $z = x + y$, we have

$$\sum_{n=-\infty}^{\infty} J_n(x + y) t^n = e^{\frac{x}{2}(t - t^{-1})} e^{\frac{y}{2}(t - t^{-1})}$$

$$= \sum_{k=-\infty}^{\infty} J_k(x) t^k \sum_{l=-\infty}^{\infty} J_l(y) t^l = \sum_{n=-\infty}^{\infty} t^n \sum_{k=-\infty}^{\infty} J_k(x) J_{n-k}(y) \ ,$$

hence

$$J_n(x + y) = \sum_{k=-\infty}^{\infty} J_k(x) J_{n-k}(y) \ . \tag{16}$$

Another important addition formula is

$$J_0(R) = \sum_{m=-\infty}^{\infty} J_m(r_1) J_m(r_2) e^{im\theta}$$

$$= J_0(r_1) J_0(r_2) + 2 \sum_{m=1}^{\infty} J_m(r_1) J_m(r_2) \cos m\theta \ , \tag{17}$$

where $R = \sqrt{r_1^2 + r_2^2 - 2r_1 r_2 \cos \theta}$ represents the distance between any two points P_1 and P_2 in a plane; r_1 and r_2 are respectively the distances of P_1 and P_2 from the origin O, and θ is the angle between $\overline{OP}_1$ and $\overline{OP}_2$.

This formula can be proved as follows. By (1), we have

$$J_0(R) = \frac{1}{2\pi i} \int^{(0+)} e^{\frac{R}{2}(t - t^{-1})} t^{-1} dt \ . \tag{18}$$

Now,

$$r_1^2 + r_2^2 - 2r_1 r_2 \cos \theta = (r_1 e^{i\theta} - r_2)(r_1 e^{-i\theta} - r_2) \ ,$$

so

$$\frac{R}{2}(t - t^{-1}) = \frac{1}{2}\sqrt{(r_1 e^{i\theta} - r_2)(r_1 e^{-i\theta} - r_2)} \ (t - t^{-1})$$

$$= \frac{1}{2}(r_1 e^{i\theta} - r_2)\sqrt{\frac{r_1 e^{-i\theta} - r_2}{r_1 e^{i\theta} - r_2}} \ (t - t^{-1}) \ .$$

Let

$$u = \sqrt{\frac{r_1 e^{-i\theta} - r_2}{r_1 e^{i\theta} - r_2}} \ t \ ,$$

substituting it in the last equation, we find after some calculation

$$\frac{R}{2}(t - t^{-1}) = \frac{r_1}{2}\left(u e^{i\theta} - \frac{1}{u e^{i\theta}}\right) - \frac{r_2}{2}\left(u - \frac{1}{u}\right) \ .$$

Therefore, using (1) and (2), we have

$$J_0(R) = \frac{1}{2\pi i} \int^{(0+)} e^{\frac{r_1}{2}\left(u e^{i\theta} - \frac{1}{u e^{i\theta}}\right)} e^{-\frac{r_2}{2}(u - u^{-1})} u^{-1} du$$

$$= \frac{1}{2\pi i} \sum_{m=-\infty}^{\infty} J_m(r_1) e^{im\theta} \int^{(0+)} e^{-\frac{r_2}{2}(u - u^{-1})} u^{m-1} du$$

$$= \sum_{m=-\infty}^{\infty} J_m(r_1) J_{-m}(-r_2) e^{im\theta}$$

$$= \sum_{m=-\infty}^{\infty} J_m(r_1) J_m(r_2) e^{im\theta} \ .$$

In the last step we have used Eq. (8) of Sec. 7.2, $J_{-m}(z) = (-)^m J_m(z)$, and Eq. (18) of Sec. 7.2, $J_m(-z) = (-)^m J_m(z)$.

Formula (17) also holds when r_1, r_2 and θ are complex numbers [cf. Sec. 7.13; and, Watson (1944), Sec. 11.2, p. 358].

Integral representations of $J_n(z)$

Among the various integral representations of $J_n(z)$, the most important and most fundamental are, besides the contour integral expression given by (1) above, the *Poisson representation*

$$J_n(z) = \frac{\left(\frac{z}{2}\right)^n}{\sqrt{\pi}\,\Gamma\left(n + \frac{1}{2}\right)} \int_0^\pi \cos(z \cos\theta) \sin^{2n}\theta\, d\theta \qquad (19)$$

and the *Bessel representation*:

$$J_n(z) = \frac{1}{2\pi} \int_{-\pi}^\pi \cos(n\theta - z\sin\theta)\, d\theta \; . \qquad (20)$$

(19) and (20) are evidently special cases of Eqs. (7) and (17) of Sec. 7.4 respectively. Besides, (20) can also be derived from (19) [cf. Ex. 5 at the end of this chapter].

Inequalities about $J_n(z)$

For any real or complex values of z, we have, from the series expression [Eq. (7) of Sec. 7.2],

$$J_n(z) = \sum_{k=0}^\infty \frac{(-)^k}{k!} \frac{1}{(n+k)!} \left(\frac{z}{2}\right)^{2k+n} , \qquad (21)$$

of $J_n(z)$,

$$|J_n(z)| \le \left|\frac{z}{2}\right|^n \sum_{k=0}^\infty \frac{1}{k!(n+k)!} \left|\frac{z}{2}\right|^{2k}$$

$$\le \frac{1}{n!} \left|\frac{z}{2}\right|^n \sum_{k=0}^\infty \frac{1}{k!(n+1)^k} \left|\frac{z}{2}\right|^{2k} \; .$$

Hence, for $n \ge 0$,

$$|J_n(z)| \le \frac{1}{n!} \left|\frac{z}{2}\right|^n \exp\left\{\frac{1}{4}\frac{|z|^2}{n+1}\right\} \le \frac{1}{n!} \left|\frac{z}{2}\right|^n e^{|z|^2/4} \; . \qquad (22)$$

Also, by similar derivation, we obtain from (21)

$$J_n(z) = \frac{1}{n!} \left(\frac{z}{2}\right)^n (1 + \theta) , \qquad (23)$$

where

$$|\theta| \leq \exp\left\{\frac{1}{4}\frac{|z|^2}{n+1}\right\} - 1 \leq \frac{1}{n+1}\left[\exp\left\{\frac{|z|^2}{4}\right\} - 1\right]. \qquad (24)$$

These two inequalities are of some importance in the discussion of series expansion of functions in terms of Bessel functions.

7.6. Bessel Functions of the Second Kind $Y_\nu(z)$

In Sec. 7.2, we have seen that, when ν is not an integer, $J_\nu(z)$ and $J_{-\nu}(z)$ are two linearly independent solutions of the Bessel equation. But, for integral values of n, $J_n(z)$ and $J_{-n}(z)$ are differed only by a constant multiple $(-)^n$ [Eq. (8) of Sec. 7.2], and hence we have to find a second, independent solution. Such solution can be obtained by the method of Frobenius described in Sec. 2.5, just as we treated the similar problems in dealing with the hypergeometric and confluent hypergeometric equations. In the following, we shall give another approach to this sort of problems. It begins with the consideration of the linear independency of $J_\nu(z)$ and $J_{-\nu}(z)$.

From the differential equations satisfied by $J_{\pm\nu}(z)$, namely,

$$\frac{d}{dz}\left(z\frac{dJ_\nu}{dz}\right) + \left(z - \frac{\nu^2}{z}\right)J_\nu = 0,$$

$$\frac{d}{dz}\left(z\frac{dJ_{-\nu}}{dz}\right) + \left(z - \frac{\nu^2}{z}\right)J_{-\nu} = 0,$$

multiplying them by $J_{-\nu}(z)$ and $J_\nu(z)$ respectively and subtracting, we have

$$J_{-\nu}\frac{d}{dz}\left(z\frac{dJ_\nu}{dz}\right) - J_\nu\frac{d}{dz}\left(z\frac{dJ_{-\nu}}{dz}\right) = 0,$$

or

$$z[J_{-\nu}J'_\nu - J_\nu J'_{-\nu}] = \text{const.} \qquad (1)$$

Putting in it the series expressions of $J_{\pm\nu}(z)$ [Eq. (7) of Sec. 7.2] and letting $z \to 0$, we find the constant in (1) to be $2\sin\nu\pi/\pi$. Hence

$$J_\nu J'_{-\nu} - J_{-\nu}J'_\nu = -\frac{2\sin\nu\pi}{\pi z}. \qquad (2)$$

This makes it manifest that only when ν is an integer are $J_\nu(z)$ and $J_{-\nu}(z)$ linearly dependent on each other.

Now, let us try to find a linear combination of J_ν and $J_{-\nu}$,

$$Z_\nu(z) = aJ_\nu(z) + bJ_{-\nu}(z)$$

such that the limit of $Z_\nu(z)$ for $\nu \to n$, $Z_n(z)$, is a solution linearly independent of $J_n(z)$. By (2) we can calculate the Wronskians forms of Z_ν and J_ν and of Z_ν and $J_{-\nu}$

$$W[Z_\nu, J_\nu] = Z_\nu J_\nu' - J_\nu Z_\nu' = b\frac{2\sin\nu\pi}{\pi z} ,$$

$$W[Z_\nu, J_{-\nu}] = -a\frac{2\sin\nu\pi}{\pi z} .$$

Thus, we see that, in general, Z_n and J_n (or J_{-n}) are linearly dependent unless the constants a and b are so chosen that they contain the factor $(\sin\nu\pi)^{-1}$; for example,

$$a = \alpha(\nu)/\sin\nu\pi , \quad b = \beta(\nu)/\sin\nu\pi \quad (\alpha(n), \beta(n) \neq 0) .$$

Then,

$$Z_\nu(z) = \frac{1}{\sin\nu\pi}[\alpha(\nu)J_\nu(z) + \beta(\nu)J_{-\nu}(z)] .$$

To have such an expression of Z_ν significant for $\nu \to n$, it is necessary that

$$\lim_{\nu \to n}[\alpha(\nu)J_\nu(z) + \beta(\nu)J_{-\nu}(z)] = 0 ,$$

i.e.,

$$\alpha(n)J_n(z) + \beta(n)J_{-n}(z) \equiv 0 .$$

But, by Eq. (8) of Sec. 7.2, $J_{-n} = (-)^n J_n$; therefore, we must have $\alpha(n) + (-)^n\beta(n) = 0$. The most frequently used convention is to put $\beta(n) = -1$, and hence $\alpha(n) = (-)^n$, $\alpha(\nu) = \cos\nu\pi$, and to denote the Z_ν so defined by Y_ν.

$$Y_\nu(z) = \frac{\cos\nu\pi J_\nu(z) - J_{-\nu}(z)}{\sin\nu\pi} , \tag{3}$$

which are called the *Bessel functions of the second kind*. They are often otherwise denoted by the symbol $N_\nu(z)$ and named *Neumann functions*. The function $Y_\nu(z)$ was first introduced by Weber and Schläfli. [Cf. Watson (1944), pp. 63-64; also p. 70 and p. 71, Eq. (8) of Sec. 3.58.]

Now, we come to show that, when $\nu \to n$, the limit of the r.h.s. of (3) exists and satisfies the Bessel equation of order n. From (3), we have

$$\lim_{\nu \to n} Y_\nu(z) = \lim_{\nu \to n} \frac{\cos\nu\pi J_\nu - J_{-\nu}}{\sin\nu\pi} = \frac{1}{\pi}\left\{ \frac{\partial J_\nu}{\partial\nu} - (-)^n\frac{\partial J_{-\nu}}{\partial\nu} \right\}_{\nu \to n} .$$

It has been shown in Sec. 7.2 that $J_{\pm\nu}(z)$ are both integral functions of ν. Therefore, the limit exists and

$$Y_n(z) = \frac{1}{\pi} \left\{ \frac{\partial J_\nu}{\partial \nu} - (-)^n \frac{\partial J_{-\nu}}{\partial \nu} \right\}_{\nu \to n} . \tag{4}$$

To show that $Y_n(z)$ is indeed a solution of the Bessel equation of order n, it is only necessary to note that $J_{\pm\nu}(z)$ are also analytic functions of z (on the corresponding Riemann sheets) so that we are legitimate to interchange the order of differentiations with respect to ν and z. For details, cf. Watson (1944), pp. 58-59.

Series expression of $Y_n(z)$

Substituting the series expressions of $J_{\pm\nu}(z)$ [Eq. (7) of Sec. 7.2] into (4), we obtain, by direct calculation,

$$Y_n(z) = \frac{2}{\pi} J_n(z) \ln \frac{z}{2} - \frac{1}{\pi} \sum_{k=0}^{n-1} \frac{(n-k-1)!}{k!} \left(\frac{z}{2}\right)^{2k-n}$$
$$- \frac{1}{\pi} \sum_{k=0}^{\infty} \frac{(-)^k}{k!(n+k)!} [\psi(n+k+1) + \psi(k+1)] \left(\frac{z}{2}\right)^{2k+n} \tag{5}$$
$$(n = 0, 1, 2, \ldots ; \ |\arg z| < \pi) ,$$

where $\psi(z) = \Gamma'(z)/\Gamma(z)$; when $n = 0$, the finite sum in the second term is to be negated.

From (5) we see that, when $z \to 0$,

$$Y_0(z) \sim \frac{2}{\pi} \ln \frac{z}{2} , \quad Y_n(z) \sim -\frac{(n-1)!}{\pi} \left(\frac{z}{2}\right)^{-n} \quad (n \geq 1) . \tag{6}$$

According to definition (3) of $Y_\nu(z)$, it can be shown that $Y_\nu(z)$ also satisfies Eqs. (9) and (10) of Sec. 7.2. Thus, all the recurrence formulae for $J_\nu(z)$ [Eqs. (13) and (14) of Sec. 7.2 being the basic ones] are valid for $Y_\nu(z)$. Further, since all the terms occurring in the recurrence relations are continuous functions of ν, the relation also apply when ν is an integer n. For citing convenience, we list them as follows:

$$\frac{d}{dz}(z^\nu Z_\nu) = z^\nu Z_{\nu-1} \, ,$$

(7)

$$\frac{d}{dz}(z^{-\nu} Z_\nu) = -Z^{-\nu} Z_{\nu+1} \, ,$$

(8)

$$Z_{\nu-1} + Z_{\nu+1} = \frac{2\nu}{z} Z_\nu \, ,$$

(9)

$$Z_{\nu-1} - Z_{\nu+1} = 2Z'_\nu \, ,$$

(10)

where $Z_\nu(z)$ represents $J_\nu(z)$ or $Y_\nu(z)$.

Cylinder functions

Functions satisfying the recurrence relations (9) and (10), or their equivalents, (7) and (8), are called *cylinder functions* and denoted by $Z_\nu(z)$. Cylinder functions definitely satisfy the Bessel equation. The proof is as follows. By canceling $Z_{\nu+1}$ from (9) and (10), we obtain $Z'_\nu + (\nu/z)Z_\nu = Z_{\nu-1}$. Differentiating gives $Z''_\nu - (\nu/z^2)Z_\nu + (\nu/z)Z'_\nu = Z'_{\nu-1}$. Again, by canceling $Z_{\nu-1}$ from (9) and (10), and changing ν into $\nu-1$, we obtain $Z_\nu = [(\nu-1)/z]Z_{\nu-1} - Z'_{\nu-1}$. Eliminating $Z_{\nu-1}$ and $Z'_{\nu-1}$ from these three equations, we see that Z_ν satisfies the Bessel equation. Thus,

$$Z_\nu(z) = a_\nu J_\nu(z) + b_\nu Y_\nu(z) \, ,$$

(11)

where a_ν and b_ν are independent of z but may be functions of ν.

It can be shown also that the necessary and sufficient conditions for the linear combination $a_\nu J_\nu(z) + b_\nu Y_\nu(z)$ to be a cylinder function (i.e., to satisfy (9) and (10)) are

$$a_\nu = a_{\nu+1} \, , \quad b_\nu = b_{\nu+1} \, .$$

(12)

7.7. Bessel Functions of the Third Kind (Hankel Functions) $H_\nu^{(1)}(z), H_\nu^{(2)}(z)$

By definition, *Bessel functions of the third kind* are

$$H_\nu^{(1)}(z) = J_\nu(z) + iY_\nu(z) \, ,$$

(1)

$$H_\nu^{(2)}(z) = J_\nu(z) - iY_\nu(z) \, ;$$

(2)

$H_\nu^{(1)}(z)$ and $H_\nu^{(2)}(z)$ are also called *Hankel functions of the first and the second kind* respectively. As $J_\nu(z)$ and $Y_\nu(z)$ are linearly independent solutions of the Bessel equation, therefore, so are $H_\nu^{(1)}(z)$ and $H_\nu^{(2)}(z)$, including the case when $\nu = n$, an integer. In other words, whatever the value of ν may be, any two of the four functions $J_\nu(z), Y_\nu(z), H_\nu^{(1)}(z), H_\nu^{(2)}(z)$ are linearly independent solutions of the Bessel equation, and thus can be taken as the fundamental set.

Since both $J_\nu(z)$ and $Y_\nu(z)$ satisfy the recurrence relations (7)–(10) of the preceding section, so do $H_n^{(1)}(z)$ and $H_n^{(2)}(z)$.

Note that when $\nu = n(n = 0, 1, 2, \ldots), H_n^{(1)}(z)$ and $H_n^{(2)}(z)$ have, at the point $z = 0$, the same singular properties as $Y_n(z)$ [(6) of the preceding section].

Integral representations

The integral representation of $H_\nu^{(1)}(z)$ can be obtained from those of J_ν and $J_{-\nu}$ by definition (1) and Eq. (3) of Sec. 7.6 as follows. From Eq. (9) of Sec. 7.4, deforming the path of integration [Fig. 24 in Sec. 7.4] into the contour shown in Fig. 28. Then, since those parts of the path which are parallel to the real axis, when moved to infinity as shown, contribute zero values to the integral, we have

$$J_\nu(z) = \frac{\Gamma\left(\frac{1}{2} - \nu\right)}{\sqrt{\pi}\,2\pi i} \left(\frac{z}{2}\right)^\nu \left\{ \int_{1+i\infty}^{(1+)} e^{izt}(t^2 - 1)^{\nu - \frac{1}{2}} dt \right.$$
$$\left. + \int_{-i+i\infty}^{(-1-)} e^{izt}(t^2 - 1)^{\nu - \frac{1}{2}} dt \right\}, \tag{3}$$

where $\operatorname{Re}(z) > 0$ and $\arg(t^2 - 1) = 0$ at the point P. Similarly, from Eq. (10) of Sec. 7.4, by deforming the path of integration into the contour as shown in Fig. 29, we obtain

$$J_{-\nu}(z) = e^{\nu\pi i}\frac{\Gamma\left(\frac{1}{2} - \nu\right)}{\sqrt{\pi}\,2\pi i} \left(\frac{z}{2}\right)^\nu \left\{ \int_{1+i\infty}^{(1+)} e^{izt}(t^2 - 1)^{\nu - \frac{1}{2}} dt + e^{-2\pi(\nu - \frac{1}{2})i} \right.$$
$$\left. \times \int_{-1+i\infty}^{(-1+)} e^{izt}(t^2 - 1)^{\nu - \frac{1}{2}} dt \right\}, \tag{4}$$

where $\operatorname{Re}(z) > 0; \arg(t^2 - 1) = 0$ at P, while in the second integral $\arg(t^2 - 1)$ is increased by 2π so that it becomes 0 at the point Q.

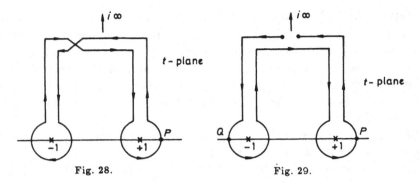

Fig. 28.　　　　　　　　　　　Fig. 29.

From (3) of the last section and (1), we have

$$H_\nu^{(1)}(z) = \frac{i}{\sin \nu \pi} \left[e^{-\nu \pi i} J_\nu(z) - J_{-\nu}(z) \right] . \tag{5}$$

Substituting (3) and (4) in (5) gives

$$H_\nu^{(1)}(z) = \frac{\Gamma\left(\frac{1}{2} - \nu\right)}{\sqrt{\pi}\pi i} \left(\frac{z}{2}\right)^\nu \int_{1+i\infty}^{(1+)} e^{izt}(t^2 - 1)^{\nu - \frac{1}{2}} dt , \tag{6}$$

where $\arg(t^2 - 1) = -\pi$ at the starting point $1 + i\infty$, and $\mathrm{Re}(z) > 0$, $\nu + \frac{1}{2} \neq$ positive integers.

Similarly, we obtain from (2)

$$H_\nu^{(2)}(z) = \frac{-i}{\sin \nu \pi} \left[e^{\nu \pi i} J_\nu(z) - J_{-\nu}(z) \right] . \tag{7}$$

Then, by substituting (3) and (4) into it, we find the integral representation for $H_\nu^{(2)}(z)$. But it is simpler to eliminate $Y_\nu(z)$ from (1) and (2), so that

$$J_\nu(z) = \frac{1}{2}[H_\nu^{(1)}(z) + H_\nu^{(2)}(z)] . \tag{8}$$

Then, from (3) and (6), we see immediately

$$H_\nu^{(2)}(z) = \frac{\Gamma\left(\frac{1}{2} - \nu\right)}{\sqrt{\pi}\pi i} \left(\frac{z}{2}\right)^\nu \int_{-1+i\infty}^{(-1-)} e^{izt}(t^2 - 1)^{\nu - \frac{1}{2}} dt , \tag{9}$$

where $\arg(t^2 - 1) = \pi$ at the starting point $-1 + i\infty$, and $\mathrm{Re}(z) > 0$, $\nu + \frac{1}{2} \neq$ positive integers.

From (1) and (2) eliminating $J_\nu(z)$, we have

$$Y_\nu(z) = \frac{1}{2i}[H_\nu^{(1)}(z) - H_\nu^{(2)}(z)] . \tag{10}$$

Then, with (6) and (9), we can derive the integral representation for $Y_\nu(z)$.

When $\nu + \frac{1}{2}$ is a positive integer, $\Gamma\left(\frac{1}{2} - \nu\right)$ becomes infinity and $(t^2 - 1)^{\nu - \frac{1}{2}}$ is a single-valued analytic function. Thus, the r.h.s. of (6) is an indeterminate form. To evaluate this indeterminate form, let us discuss first the general case of $\text{Re}\left(\nu + \frac{1}{2}\right) > 0$. In such case, the value of the circular path integral around $t = 1$ approaches zero as $t \to 1$. Thus,

$$\int_{1+i\infty}^{(1+)} e^{izt}(t^2 - 1)^{\nu - \frac{1}{2}} dt$$

$$= \left[e^{-i\pi(\nu - \frac{1}{2})} - e^{i\pi(\nu - \frac{1}{2})} \right] \int_{1+i\infty}^{1} e^{izt}(1 - t^2)^{\nu - \frac{1}{2}} dt$$

$$= 2i \sin\left(\frac{\pi}{2} - \nu\pi\right) \int_{1+i\infty}^{1} e^{izt}(1 - t^2)^{\nu - \frac{1}{2}} dt ,$$

where $-\pi/2 < \arg(1 - t^2) < 0$. Hence,[b]

$$H_\nu^{(1)}(z) = \frac{2}{\sqrt{\pi}\Gamma\left(\nu + \frac{1}{2}\right)} \left(\frac{z}{2}\right)^\nu \int_{1+i\infty}^{1} e^{izt}(1 - t^2)^{\nu - \frac{1}{2}} dt , \qquad (11)$$

where $-\pi/2 < \arg(1 - t^2) < 0, \text{Re}(z) > 0, \text{Re}\left(\nu + \frac{1}{2}\right) > 0$.

If $\nu = n + \frac{1}{2}, n = 0, 1, 2, \ldots$, (11) becomes

$$H_{n+\frac{1}{2}}^{(1)}(z) = \sqrt{\frac{2}{\pi z}} \frac{2}{n!} \left(\frac{z}{2}\right)^{n+1} \int_{1+i\infty}^{1} e^{izt}(1 - t^2)^n dt \qquad (12)$$

$$(\text{Re}(z) > 0) .$$

Similarly, we have

$$H_{n+\frac{1}{2}}^{(2)}(z) = -\sqrt{\frac{2}{\pi z}} \frac{2}{n!} \left(\frac{z}{2}\right)^{n+1} \int_{-1+i\infty}^{-1} e^{izt}(1 - t^2)^n dt . \qquad (13)$$

If $-\pi/2 + \omega < \arg z < \pi/2 + \omega$, where $|\omega| < \pi/2$, we need only to rotate the corresponding contour about $t = 0$ by an angle $(-\omega)$ [cf. the paragraph after Eq. (10) of Sec. 7.4].

Other integral expressions for Hankel functions [(14) and (15) below] can be obtained from Eq. (16) of Sec. 7.5. Applying this latter formula to (5), we obtain

$$H_\nu^{(1)}(z) = \frac{i}{\sin\nu\pi}\left\{ \frac{1}{2\pi}\int_{-\pi}^{\pi} e^{i[z\sin\theta - \nu(\theta + \pi)]} d\theta - \frac{1}{2\pi}\int_{-\pi}^{\pi} e^{i(z\sin\theta + \nu\theta)} d\theta \right.$$
$$\left. - \frac{\sin\nu\theta}{\pi}\int_0^\infty e^{-z\,\text{shu} - \nu(u + \pi i)} du - \frac{\sin\nu\pi}{\pi}\int_0^\infty e^{-z\,\text{shu} + \nu u} du \right\} .$$

---

[b]Compare Watson (1944), p. 170, Sec. 6.14, where the factor 2 is slipped.

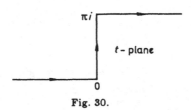

Fig. 30.

Now,

$$\int_{-\pi}^{\pi} e^{i[z \sin \theta - \nu(\theta+\pi)]} d\theta$$
$$= \int_{-\pi}^{0} e^{i[z \sin \theta - \nu(\theta+\pi)]} d\theta + \int_{0}^{\pi} e^{i[z \sin \theta - \nu(\theta+\pi)]} d\theta .$$

Changing $\theta + \pi$ into $-\theta$ in the first integral on the r.h.s. and θ to $\pi - \theta$ in the second integral, then combining to the second integral of the preceding equation, we obtain

$$\left(e^{-2\nu\pi i} - 1 \right) \int_{0}^{\pi} e^{i(z \sin \theta + \nu\theta)} d\theta .$$

Putting $\theta = \pi + it$ and substituting the result into the above expression of $H_{\nu}^{(1)}(z)$, letting $t = u + \pi i$ in the third integral and replacing u in the fourth integral by $-t$, we find

$$H_{\nu}^{(1)}(z) = \frac{1}{\pi i} \left\{ \int_{0}^{\pi i} + \int_{\pi i}^{\infty+\pi i} + \int_{-\infty}^{0} \right\} e^{z \, \text{sh} t - \nu t} dt$$
$$= \frac{1}{\pi i} \int_{-\infty}^{\infty+\pi i} e^{z \, \text{sh} t - \nu t} dt \quad (|\arg z| < \pi/2) . \qquad (14)$$

Figure 30 shows the path of integration.

   Similarly, we can obtain such integral representation for $H_{\nu}^{(2)}(z)$ which is merely a change of i, which occurs explicitly in (14), into $-i$ (but not that in the complex variable z):

$$H_{\nu}^{(2)}(z) = -\frac{1}{\pi i} \int_{-\infty}^{\infty-\pi i} e^{z \, \text{sh} t - \nu t} dt \quad (|\arg z| < \pi/2) . \qquad (15)$$

   From (14) and (15), many other integral representations can be derived [cf. Exs. 42, 43 at the end of this chapter].

Relations between Hankel functions and Whittaker functions

In the integral of (6), replace t by $s + 1$. It results

$$\int_{1+i\infty}^{(1+)} e^{izt}(t^2 - 1)^{\nu - \frac{1}{2}}\,dt = e^{iz}\int_{i\infty}^{(0+)} e^{izs}\,s^{\nu - \frac{1}{2}}(s + 2)^{\nu - \frac{1}{2}}\,ds\;,$$

where $-3\pi/2 < \arg s < \pi/2, \arg(s + 2)|_{s=0} = 0$. Further, let $u = e^{-\pi i/2}zs$, then the r.h.s. of it becomes

$$-e^{iz}(e^{\pi i/2}z)^{-\nu - \frac{1}{2}}2^{\nu - \frac{1}{2}}\int_{\infty e^{i\alpha}}^{(0+)} e^{-u}(-u)^{\nu - \frac{1}{2}}\left(1 + \frac{u}{2e^{-\pi i/2}z}\right)^{\nu - \frac{1}{2}}\,du\;,$$

where $|\alpha| = |\arg z| < \pi/2, -\pi + \alpha < \arg(-u) < \pi + \alpha, \arg(1 - u/2iz)|_{u=0} = 0$. Under the condition $|\alpha| < \pi/2$, we can turn around the whole contour about $u = 0$ by an angle $(-\alpha)$ without affecting the value of the integral [cf. the discussion after Eq. (2) of Sec. 3.1]. Thus, the last expression is equal to

$$-e^{-i\left(\frac{\nu\pi}{2} + \frac{\pi}{4}\right)}\left(\frac{z}{2}\right)^{-\nu - \frac{1}{2}}\frac{e^{iz}}{2}\int_{\infty}^{(0+)} e^{-u}(-u)^{\nu - \frac{1}{2}}\left(1 + \frac{u}{2e^{-\pi i/2}z}\right)^{\nu - \frac{1}{2}}\,du$$

$$= e^{-i\left(\frac{\nu\pi}{2} + \frac{\pi}{4}\right)}\left(\frac{z}{2}\right)^{-\nu - \frac{1}{2}}\frac{\pi i}{\Gamma\left(\frac{1}{2} - \nu\right)}W_{0,\nu}(2e^{-\pi i/2}z)$$

$$(|\arg(-u)| < \pi)\;,$$

[cf. Eq. (3) of Sec. 6.5], and we have

$$H_\nu^{(1)}(z) = \sqrt{\frac{2}{\pi z}}\,e^{-i\left(\frac{\nu\pi}{2} + \frac{\pi}{4}\right)}W_{0,\nu}(2e^{-\pi i/2}z)\;. \tag{16}$$

Similarly,

$$H_\nu^{(2)}(z) = \sqrt{\frac{2}{\pi z}}\,e^{i\left(\frac{\nu\pi}{2} + \frac{\pi}{4}\right)}W_{0,\nu}(2e^{\pi i/2}z)\;. \tag{17}$$

These two relations are obtained under the condition that $|\arg z| < \pi/2$. But as the Whittaker functions on the r.h.s. are defined respectively in the domain $|\arg(e^{\mp\pi i/2}z)| < 3\pi/2$, (16) can be taken as the analytic continuation of $H_\nu^{(1)}(z)$ to $-\pi < \arg z < 2\pi$, and (17) as that of $H_\nu^{(2)}(z)$ to $-2\pi < \arg z < \pi$.

Since we have already found in Sec. 6.6, the asymptotic expansion of the Whittaker function for $|z| \to \infty$, (16) and (17) can be used to obtain the asymptotic expansions of the Bessel functions for $|z| \to \infty$ [cf. Sec. 7.10].

7.8. Modified (or, Imaginary Argument) Bessel Functions $I_\nu(z)$ and $K_\nu(z)$. Thomson Functions $\text{ber}_\nu(z)$ and $\text{bei}_\nu(z)$; etc.

In many boundary-value problems, there occurs frequently the differential equation

$$\frac{d^2y}{dx^2} + \frac{1}{x}\frac{dy}{dx} - \left(1 + \frac{\nu^2}{x^2}\right)y = 0 \,, \tag{1}$$

where x is a real variable. Let $z = ix$, then (1) is reduced to the Bessel equation of order ν. Therefore, when ν is not an integer, $J_{\pm\nu}(ix)$ are two linearly independent solutions of (1).

In order that the solution of (1) to be real in case $\nu = n$, an integer, we introduce a new function $I_\nu(z)$, called the *Modified (or Imaginary Argument) Bessel function of the first kind:*

$$\begin{aligned}
I_\nu(z) &= e^{-\nu\pi i/2}J_\nu(ze^{\pi i/2}) && (-\pi < \arg z \leq \pi/2) \,, \\
&= e^{3\nu\pi i/2}J_\nu(ze^{-3\pi i/2}) && (\pi/2 < \arg z \leq \pi) \,, \tag{2}
\end{aligned}$$

$$= \left(\frac{z}{2}\right)^\nu \sum_{k=0}^\infty \frac{1}{k!}\frac{1}{\Gamma(\nu+k+1)}\left(\frac{z}{2}\right)^{2k} \,. \tag{3}$$

$I_\nu(z)$ satisfies the equation

$$\frac{d^2y}{dz^2} + \frac{1}{z}\frac{dy}{dz} - \left(1 + \frac{\nu^2}{z^2}\right)y = 0 \,, \tag{4}$$

called the *Bessel equation with imaginary argument.*

If ν is not an integer, $I_{\pm\nu}(z)$ are two linearly independent solutions of (4). But when $\nu = n$, an integer, we have, on account of $J_n(z) = (-)^n J_{-n}(z)$ [Eq. (8) of Sec. 7.2] and definition (2),

$$I_{-n}(z) = I_n(z) \,. \tag{5}$$

Then we have to look for a second solution for (4).

After the approach we took in Sec. 7.6, let

$$K_\nu(z) = \frac{\pi}{2\sin\nu\pi}[I_{-\nu}(z) - I_\nu(z)] \,. \tag{6}$$

When $\nu \neq n, K_\nu(z)$ is obviously a second solution linearly independent of $I_\nu(z)$. We call it the *Modified Bessel function of the second kind.*

From (2) and Eqs. (5) and (7) of Sec. 7.7, we find

$$K_\nu(z) = \frac{\pi i}{2}e^{\nu\pi i/2}H_\nu^{(1)}(ze^{\pi i/2}) = -\frac{\pi i}{2}e^{-\nu\pi i/2}H_\nu^{(2)}(ze^{-\pi i/2}) \,. \tag{7}$$

It is seen that, for any ν, including $\nu = n$, $K_\nu(z)$ is a second solution linearly independent of $I_\nu(z)$ [cf. the discussion after Eq. (2) of Sec. 7.7].

Further, from Eq. (16) of Sec. 7.7 we obtain a very simple relation:

$$K_\nu(z) = \sqrt{\frac{\pi}{2z}} W_{0,\nu}(2z) . \tag{8}$$

When $\nu = n$ $(n = 0, 1, 2, \ldots)$, by (7) and Eq. (1) of Sec. 7.7, we have

$$K_n(z) = \frac{\pi i}{2} e^{n\pi i/2} [J_n(ze^{\pi i/2}) + i Y_n(ze^{\pi i/2})] . \tag{9}$$

Using Eq. (5) of Sec. 7.6, it results

$$\begin{aligned}
K_n(z) &= \frac{\pi i}{2} e^{n\pi i/2} \left\{ i \frac{2}{\pi} J_n(ze^{\pi i/2}) \ln \frac{z}{2} - \frac{i}{\pi} e^{-n\pi i/2} \sum_{k=0}^{n-1} \frac{(-)^k (n-k-1)!}{k!} \right. \\
&\quad \times \left(\frac{z}{2} \right)^{2k-n} - \frac{i}{\pi} e^{n\pi i/2} \sum_{k=0}^{\infty} \frac{1}{k!(n+k)!} \\
&\quad \left. \times [\psi(n+k+1) + \psi(k+1)] \left(\frac{z}{2} \right)^{2k+n} \right\} \\
&= (-)^{n+1} \sum_{k=0}^{\infty} \frac{1}{k!(n+k)!} \left[\ln \frac{z}{2} - \frac{1}{2} \psi(n+k+1) - \frac{1}{2} \psi(k+1) \right] \\
&\quad \times \left(\frac{z}{2} \right)^{2k+n} + \frac{1}{2} \sum_{k=0}^{n-1} \frac{(-)^k (n-k-1)!}{k!} \left(\frac{z}{2} \right)^{2k-n} ,
\end{aligned} \tag{10}$$

where $|\arg z| < \pi$, and when $n = 0$, the finite sum in the second term is to be negated. From (10), we see that $z = 0$ is a singularity of $K_n(z)$ in the same way as it is of $Y_n(z)$ [cf. Eq. (6) of Sec. 7.6]:

$$K_0 \sim -\ln \frac{z}{2}, \quad K_n(z) \sim \frac{(n-1)!}{2} \left(\frac{z}{2} \right)^{-n} \quad (n \geq 1) . \tag{11}$$

On the other hand, $I_n(z)$ is, however, analytic at $z = 0$.

Thomson (Kelvin) functions

These functions are merely Bessel functions of arguments with phases equal to $\pm\pi/4$ or $\pm 3\pi/4$.

Thomson introduced in his works on certain electricity problems, the functions $\text{ber}(x)$ and $\text{bei}(x)$, which are respectively the real and imaginary parts of the modified Bessel function $I_0(x\sqrt{i})$ (x real), i.e.,

$$\text{ber}(x) + i\,\text{bei}(x) = I_0(x\sqrt{i}) = J_0(xi\sqrt{i}) , \tag{12}$$

where ber(x) and bei(x) are real functions of the real variable x. The series expressions for these two functions are respectively

$$\text{ber}(x) = 1 - \frac{1}{(2!)^2}\left(\frac{x}{2}\right)^4 + \frac{1}{(4!)^2}\left(\frac{x}{2}\right)^8 - \cdots , \tag{13}$$

$$\text{bei}(x) = \frac{1}{(1!)^2}\left(\frac{x}{2}\right)^2 - \frac{1}{(3!)^2}\left(\frac{x}{2}\right)^6 + \frac{1}{(5!)^2}\left(\frac{x}{2}\right)^{10} - \cdots . \tag{14}$$

The generalization of the above definitions are

$$\text{ber}_\nu(z) \pm i\,\text{bei}_\nu(z) = J_\nu(ze^{\pm 3\pi i/4}) . \tag{15}$$

Besides, there are also functions ker$_\nu(z)$, kei$_\nu(z)$, her$_\nu(z)$ and hei$_\nu(z)$, which are defined by

$$\text{ker}_\nu(z) \pm i\,\text{kei}_\nu(z) = e^{\mp\nu\pi i/2}K_\nu(ze^{\pm\pi i/4}) , \tag{16}$$

$$\text{her}_\nu(z) + i\,\text{hei}_\nu(z) = H_\nu^{(1)}(ze^{3\pi i/4}) , \tag{17}$$

$$\text{her}_\nu(z) - i\,\text{hei}_\nu(z) = H_\nu^{(2)}(ze^{-3\pi i/4}) . \tag{18}$$

From the relation (7) between $K_\nu(z)$ and $H_\nu^{(1)}(z)$, $H_\nu^{(2)}(z)$, we may deduce

$$\text{ker}_\nu(z) = -\frac{\pi}{2}\text{hei}_\nu(z) , \quad \text{kei}(z) = \frac{\pi}{2}\text{her}_\nu(z) . \tag{19}$$

All the above newly introduced functions are real when ν is real and z is positive (arg $z = 0$).

7.9. Spherical Bessel Functions $j_l(z), n_l(z), h_l^{(1)}(z), h_l^{(2)}(z)$

Spherical Bessel functions are solutions of the equation [Eq. (3) of Sec. 7.1].

$$\frac{d^2y}{dz^2} + \frac{2}{z}\frac{dy}{dz} + \left[1 - \frac{l(l+1)}{z^2}\right]y = 0 , \tag{1}$$

where $l = 0, 1, 2, \ldots$ in most cases; but we shall not be subjected to this restriction in the following.

It was seen in Sec. 7.1 that, by the transformation $y(z) = z^{-\frac{1}{2}}v(z)$, we find $v(z)$ satisfying the Bessel equation of order $l + \frac{1}{2}$:

$$\frac{d^2v}{dz^2} + \frac{1}{z}\frac{dv}{dz} + \left[1 - \frac{\left(l + \frac{1}{2}\right)^2}{z^2}\right]v = 0 . \tag{2}$$

Thus, the solutions of (1) can be expressed in terms of Bessel functions of order $l + \frac{1}{2}$ (using lower cases to represent the solutions of (1)). Now the commonly used definitions and symbols in physics are

$$j_l(z) = \sqrt{\frac{\pi}{2z}} J_{l+\frac{1}{2}}(z) , \tag{3}$$

$$n_l(z) = \sqrt{\frac{\pi}{2z}} Y_{l+\frac{1}{2}}(z) , \tag{4}$$

$$h_l^{(1)}(z) = \sqrt{\frac{\pi}{2z}} H_{l+\frac{1}{2}}^{(1)}(z) , \tag{5}$$

$$h_l^{(2)}(z) = \sqrt{\frac{\pi}{2z}} H_{l+\frac{1}{2}}^{(2)}(z) , \tag{6}$$

where l is arbitrary. Also, the following symbols are employed:
$\psi_l(z) = j_l(z), \varsigma_l^{(1,2)} = h_l^{(1,2)}(z)$.

If we denote any one of the $j_l, n_l, h_l^{(1)}, h_l^{(2)}$ by ψ_l, then from the recurrence relations of the Bessel functions [Eqs. (9) and (10) of Sec. 7.7] we can deduce the following basic recurrence formulae:

$$\psi_{l-1} + \psi_{l+1} = \frac{2l+1}{z} \psi_l , \tag{7}$$

$$l\psi_{l-1} - (l+1)\psi_{l+1} = (2l+1)\frac{d\psi_l}{dz} . \tag{8}$$

When l is an integer, the spherical Bessel functions can be expressed in terms of elementary functions [Sec. 7.3]. For example, by Eqs. (1), (2) and (3) of Sec. 7.3, we have

$$j_0(z) = \frac{\sin z}{z} , \quad j_{-1}(z) = \frac{\cos z}{z} , \tag{9}$$

$$j_l(z) = z^l \left(-\frac{d}{zdz}\right)^l \frac{\sin z}{z} \quad (l \geq 1) . \tag{10}$$

Also, from Eq. (3) of Sec. 7.6, we deduce that

$$n_l(z) = (-)^{l+1} j_{-l-1}(z) . \tag{11}$$

From the relations between $H_\nu^{(1,2)}(z)$, $J_\nu(z)$ and $Y_\nu(z)$ [Eqs. (1) and (2) of Sec. 7.7], we have

$$h_l^{(1)}(z) = j_l(z) + in_l(z) \; , \tag{12}$$

$$h_l^{(2)}(z) = j_l(z) - in_l(z) \; . \tag{13}$$

(These two relations also hold for non-integral l.) In particular, we have

$$h_0^{(1)}(z) = \frac{e^{i\left(z-\frac{\pi}{2}\right)}}{z} \; , \tag{14}$$

$$h_0^{(2)}(z) = \frac{e^{-i\left(z-\frac{\pi}{2}\right)}}{z} \; . \tag{15}$$

7.10. Asymptotic Expansions for the Case $|z| \to \infty$

In Sec. 7.7, we have found the relations between the Hankel functions and the Whittaker functions [Eqs. (16) and (17) of Sec. 7.7]:

$$H_\nu^{(1)}(z) = \sqrt{\frac{2}{\pi z}} e^{-i\left(\frac{\nu\pi}{2}+\frac{\pi}{4}\right)} W_{0,\nu}(2e^{-\pi i/2}z) \; , \tag{1}$$

$$H_\nu^{(2)}(z) = \sqrt{\frac{2}{\pi z}} e^{i\left(\frac{\nu\pi}{2}+\frac{\pi}{4}\right)} W_{0,\nu}(2e^{\pi i/2}z) \; . \tag{2}$$

Therefore, we may utilize the asymptotic expansion of the Whittaker function [Eq. (2) of Sec. 6.6] to obtain

$$H_\nu^{(1)}(z) \sim \sqrt{\frac{2}{\pi z}} e^{i\left(z-\frac{\nu\pi}{2}-\frac{\pi}{4}\right)} \left[1 + \sum_{n=1}^{\infty} \frac{\left(\frac{1}{2}+\nu\right)_\nu \left(\frac{1}{2}-\nu\right)_n}{n!(2iz)^n}\right] \tag{3}$$
$$(-\pi < \arg z < 2\pi) \; ,$$

$$H_\nu^{(2)}(z) \sim \sqrt{\frac{2}{\pi z}} e^{-i\left(z-\frac{\nu\pi}{2}-\frac{\pi}{4}\right)} \left[1 + \sum_{n=1}^{\infty} (-)^n \frac{\left(\frac{1}{2}+\nu\right)_n \left(\frac{1}{2}-\nu\right)_n}{n!(2iz)^n}\right] \tag{4}$$
$$(-2\pi < \arg z < \pi) \; .$$

From these formulae, and using Eqs. (8) and (10) of Sec. 7.7, we have

$$J_\nu(z) \sim \sqrt{\frac{2}{\pi z}} \left[\cos\left(z - \frac{\nu\pi}{2} - \frac{\pi}{4}\right) \sum_{m=0}^{\infty} \frac{(-)^m (\nu, 2m)}{(2z)^{2m}} \right.$$

$$\left. - \sin\left(z - \frac{\nu\pi}{2} - \frac{\pi}{4}\right) \sum_{m=0}^{\infty} \frac{(-)^m (\nu, 2m+1)}{(2z)^{2m+1}} \right] \qquad (5)$$

$$(-\pi < \arg z < \pi)$$

$$Y_\nu(z) \sim \sqrt{\frac{2}{\pi z}} \left[\sin\left(z - \frac{\nu\pi}{2} - \frac{\pi}{4}\right) \sum_{m=0}^{\infty} \frac{(-)^m (\nu, 2m)}{(2z)^{2m}} \right.$$

$$\left. + \cos\left(z - \frac{\nu\pi}{2} - \frac{\pi}{4}\right) \sum_{m=0}^{\infty} \frac{(-)^m (\nu, 2m+1)}{(2z)^{2m+1}} \right] \qquad (6)$$

$$(-\pi < \arg z < \pi) \,,$$

where (ν, p) is a symbol commonly used; its definition is

$$(\nu, 0) = 1 \,,$$

$$(\nu, p) = (-)^p \frac{\left(\frac{1}{2} - \nu\right)_p \left(\frac{1}{2} + \nu\right)_p}{p!} = \frac{\Gamma\left(\frac{1}{2} + \nu + p\right)}{p!\,\Gamma\left(\frac{1}{2} + \nu - p\right)}$$

$$= \frac{\{4\nu^2 - 1\}\{4\nu^2 - 3^2\} \ldots \{4\nu^2 - (2p-1)^2\}}{2^{2p} p!} \qquad (7)$$

$$(p = 1, 2, \ldots) \,.$$

Note that

$$(\nu, p) = (-\nu, p) \,. \qquad (8)$$

From (5) we see that, if $\nu = n + \frac{1}{2}$ (half an odd integer), the series breaks up into a finite sum; the result is just Eq. (5) of Sec. 7.3 since

$$\left(n + \frac{1}{2}, 2m\right) = \frac{\Gamma(n + 2m + 1)}{(2m)!\,\Gamma(n - 2m + 1)} = \frac{(n + 2m)!}{(2m)!(n - 2m)!} \,,$$

$$\left(n + \frac{1}{2}, 2m + 1\right) = \frac{(n + 2m + 1)!}{(2m + 1)!(n - 2m - 1)!} \,.$$

For $\nu = -n - \frac{1}{2}$, if we take note of relation (8), we can also obtain Eq. (6) of Sec. 7.3 from Eq. (5).

In order to find the asymptotic expansion of the Bessel function $J_\nu(z)$ in other phases of z, for example, $0 < \arg z < 2\pi$, we may use Eq. (18) of Sec. 7.2. Thus,

$$
\begin{aligned}
J_\nu(z) &= e^{\nu\pi i} J_\nu(ze^{-\pi i}) \\
&\sim e^{(\nu+\frac{1}{2})\pi i}\sqrt{\frac{2}{\pi z}}\left[\cos\left(z+\frac{\nu\pi}{2}+\frac{\pi}{4}\right)\sum_{m=0}^{\infty}\frac{(-)^m(\nu,2m)}{(2z)^{2m}}\right.\\
&\quad\left.-\sin\left(z+\frac{\nu\pi}{2}+\frac{\pi}{4}\right)\sum_{m=0}^{\infty}\frac{(-)^m(\nu,2m+1)}{(2z)^{2m+1}}\right]
\end{aligned}
\tag{9}
$$

$$(0 < \arg z < 2\pi) .$$

In the range $0 < \arg z < \pi$ where both (5) and (9) are valid, the apparent difference of the two expressions is the Stokes phenomenon discussed in Sec. 6.8.

Also, by Eq. (7) of Sec. 7.8, or directly from Eq. (8) of Sec. 7.8, we obtain the asymptotic expansion for the Modified Bessel function of the second kind.

$$
K_\nu(z) \sim \sqrt{\frac{\pi}{2z}}e^{-z}\left[1+\sum_{n=1}^{\infty}\frac{(\nu,n)}{(2z)^n}\right] \qquad (|\arg z| < 3\pi/2) .
\tag{10}
$$

And, from the relation $I_\nu(z) = e^{\nu\pi i/2}J_\nu(ze^{-\pi i/2})$ and $I_\nu(z) = e^{-\nu\pi i/2}$ $\times J_\nu(ze^{\pi i/2})$, we obtain

$$
I_\nu(z) \sim \frac{e^z}{\sqrt{2\pi z}}\sum_{n=0}^{\infty}\frac{(-)^n(\nu,n)}{(2z)^n}+\frac{e^{-z+(\nu+\frac{1}{2})\pi i}}{\sqrt{2\pi z}}\sum_{n=0}^{\infty}\frac{(\nu,n)}{(2z)^n}
\tag{11}
$$

$$(-\pi/2 < \arg z < 3\pi/2) ,$$

$$
I_\nu(z) \sim \frac{e^z}{\sqrt{2\pi z}}\sum_{n=0}^{\infty}\frac{(-)^n(\nu,n)}{(2z)^n}+\frac{e^{-z-(\nu+\frac{1}{2})\pi i}}{\sqrt{2\pi z}}\sum_{n=0}^{\infty}\frac{(\nu,n)}{(2z)^n}
\tag{12}
$$

$$(-3\pi/2 < \arg z < \pi/2) .$$

The asymptotic expansions for other kinds of Bessel functions (for example, of the Thomson functions given in Sec. 7.8, of spherical Bessel functions in Sec. 7.9) can be derived from the above basic asymptotic expansion formulae according to the definitions of the various functions, and will not be enumerated here.

7.11. The Method of Steepest Descent

For the sake of deriving the asymptotic expansion of Bessel functions of order ν in the next section for the case when both $|\nu|$ and $|z|$ are large, we shall give a sketch of the important Method of Steepest Descent.

Suppose that the function to be expanded can be represented by the following integral:

$$f(z) = \int_a^b g(t) e^{zh(t)} \, dt \, , \tag{1}$$

where $g(t)$ and $h(t)$ are analytic functions of the complex variable t in a certain domain.

If $z, t, g(t), h(t)$ are all real and $h(t)$ has a maximum at a certain point, t_0 in the interval of integration $[a, b]$, then the function $\exp\{zh(t)\}$ will have a sharp (very steep) maximum at t_0 when $z \to +\infty$. Under this condition, so long as the variation of the function $g(t)$ does not match with that of $\exp\{zh(t)\}$, it is quite probable that the value of the integral in (1) comes mainly from that part of integration on the vicinity of t_0, and hence we can take the value of the integral on that part of the path as an approximation to the function $f(z)$ when $z \to +\infty$. Of course, a serious theory must give an estimation of the error thus introduced.

But, when $h(t)$ is a function of a complex variable, the situation becomes much involved. For, in the first place, the function $\text{Re}\{zh(t)\}$, which determines the absolute value of $\exp\{zh(t)\}$, being the real part of an analytic function, has no maximum. Secondly, when $|z|$ becomes large, $\exp\{i\text{Im}[zh(t)]\}$ is in general a very rapidly oscillating function, and hence it is rather difficult to look for an approximation to the integral as in the case of real function and real variable. We shall in the following show a way out of this dilemma. In the discussion, we shall assume that z is a positive real number, since, if it is complex, equal to $|z|e^{i\varphi}, \varphi = \arg z$, we can absorb the factor $e^{i\varphi}$ into $h(t)$.

According to Cauchy theorem, we can deform the path of integration in (1) so as to make it passing through the zero, say t_0, of $h'(t)$ and to keep, on the path,

$$\text{Im}[h(t)] = \text{Im}[h(t_0)] \, , \tag{2}$$

i.e., to keep the imaginary part of $h(t)$ unchanged on the path, at least being such on a small segment in the vicinity of the point t_0.

The curve expressed by (2) possesses the following two important properties: (i) Along it (or the segment of it near t_0), $\text{Im}[h(t)] = \text{constant}$, so that $\exp\{i\text{Im}[zh(t)]\}$ is no longer an oscillating function. (ii) Since $h(t)$ is an analytic function, $|h'(t)| = \sqrt{u_s^2 + v_s^2}$ has a definite value at every point of the curve, where u_s and v_s are respectively the directional derivatives of the real and imaginary parts of $h(t)$ along an arbitrary direction s. As $\text{Im}[h(t)]$ is unchanged when t varies along this curve, $v_s = 0$ and u_s takes its largest value, i.e., the value of $u = \text{Re}[h(t)]$ changes most rapidly along this curve, compared with any other curve passing through the same point t_0. Hence, the curve represented by (2) is called a *steepest path*.

In general, there are many paths (at least two, see below) passing through the zero t_0 of $h'(t)$ and satisfying the condition (2) — steepest paths. As the path of integration, we should choose among them the one along which the value of $\text{Re}[h(t)]$ decreases on both sides of t_0; such paths are called *Paths of Steepest Descent*. Then, as t varies along such path of steepest descent, $\exp\{\text{Re}[zh(t)]\}$ will exhibit a very steep maximum when $z \to +\infty$, and we can evaluate the approximation for the integral (1) when z is large (> 0) just as we would do in the case of real function and real variable described previously.

Let C denote a path of steepest descent. If the end points of C are a and b, and there are no singular points of the integrand in between C and the original path of integration, then

$$f(z) = \int_C g(t)e^{zh(t)}dt .\tag{3}$$

Expand $h(t)$ in the vicinity of the zero of $h'(t)$, t_0, we have

$$h(t) = h(t_0) + \frac{h''(t_0)}{2!}(t - t_0)^2 + \ldots ;\tag{4}$$

let $h''(t_0) = ae^{i\theta_0}, \theta_0 = \arg[h''(t_0)]$, and assume $a = |h''(t_0)| \neq 0$, then, according to (2), we have, along the steepest path in the vicinity of t_0,

$$\text{Im}\left[\frac{h''(t_0)}{2}(t - t_0)^2 + \ldots\right] = 0 .\tag{5}$$

Putting $t - t_0 = \rho e^{i\theta}, \theta = \arg(t - t_0)$, (5) becomes

$$\text{Im}\left[\frac{a}{2}\rho^2 e^{i(2\theta + \theta_0)} + \ldots\right] = \frac{a}{2}\rho^2 \sin(2\theta + \theta_0) + O(\rho^3) = 0 ,$$

and hence the direction θ of the steepest path, when it passes through the point t_0, satisfies the equation

$$\sin(2\theta + \theta_0) = 0 ,\tag{6}$$

i.e.,

$$2\theta + \theta_0 = n\pi \quad (n = 0, 1, 2, 3) . \tag{7}$$

Thus, there are two steepest paths: one corresponds to $n = 0$ (and $n = 2$); the other corresponds to $n = 1$ (and $n = 3$).

On the steepest path,

$$\begin{aligned} h(t) - h(t_0) &= \mathrm{Re}[h(t) - h(t_0)] \\ &= \frac{a}{2}\rho^2 \cos(2\theta + \theta_0) + O(\rho^3) . \end{aligned} \tag{8}$$

Therefore, the path of steepest descent, C, is the one corresponding to $n = 1$ (and $n = 3$); on C,

$$h(t) - h(t_0) = \mathrm{Re}[h(t) - h(t_0)] = -\frac{a}{2}\rho^2 + O(\rho^3) . \tag{9}$$

Substituting this result in (3) and taking the value of the integral for a small segment $C_\varepsilon(t_0)$ in the vicinity of t_0 as an approximation, we obtain

$$\begin{aligned} f(z) &\sim \int_{C_\varepsilon(t_0)} g(t) e^{z[h(t_0) - \frac{a}{2}\rho^2 + O(\rho^3)]} d(\rho e^{i\theta}) \\ &\sim g(t_0) e^{zh(t_0)} \left\{ e^{i(\pi - \theta_0)/2} \int_0^\varepsilon e^{-\frac{za}{2}\rho^2} d\rho + e^{i(3\pi - \theta_0)/2} \int_\varepsilon^0 e^{-\frac{za}{2}\rho^2} d\rho \right\} \\ &\sim g(t_0) e^{zh(t_0) - i\theta_0/2} 2i \int_0^\varepsilon e^{-\frac{za}{2}\rho^2} d\rho \quad (\varepsilon > 0) . \end{aligned}$$

Now,

$$\int_0^\varepsilon e^{-\lambda\rho^2} d\rho = \left(\int_0^\infty - \int_\varepsilon^\infty \right) e^{-\lambda\rho^2} d\rho \quad (\lambda > 0) ,$$

$$\int_0^\infty e^{-\lambda\rho^2} d\rho = \frac{1}{2}\sqrt{\frac{\pi}{\lambda}} \quad [\text{by} \quad \text{Eq. (9) of Sec. 3.8}] ,$$

$$\int_\varepsilon^\infty e^{-\lambda\rho^2} d\rho = \frac{1}{2}\frac{1}{\sqrt{\lambda}} \int_{\lambda\varepsilon^2}^\infty e^{-u} u^{-\frac{1}{2}} du \le \frac{1}{2\lambda\varepsilon} \int_{\lambda\varepsilon^2}^\infty e^{-u} du = \frac{e^{-\lambda\varepsilon^2}}{2\lambda\varepsilon} .$$

The value of the last integral, namely, $e^{-\lambda\varepsilon^2}/2\lambda\varepsilon$, decreases very rapidly as ε increases; hence,

$$f(z) \sim i\sqrt{\frac{2\pi}{az}} g(t_0) e^{zh(t_0) - i\theta_0/2} . \tag{10}$$

Of course, the derivation of formula (10) is rather crude, since we have neglected the value which the remaining part of C contributes to the integral

(3). Besides, the integrand has been approximated as we have discarded the terms $O(\rho^3)$ in the exponential function, and for $g(t)$, we only keep the first term $g(t_0)$ of its power series expansion in the vicinity of t_0.

To obtain a precise asymptotic expansion for the integral (1), it is necessary to put

$$
\begin{aligned}
\tau &= h(t_0) - h(t) \\
&= -\left[\frac{h''(t_0)}{2!}(t - t_0)^2 + \frac{h'''(t_0)}{3!}(t - t_0)^3 + \ldots \right] .
\end{aligned}
\tag{11}
$$

Substituting in (3) gives

$$
f(z) = e^{zh(t_0)} \int_{C'} e^{-z\tau} g(t(\tau)) \frac{dt}{d\tau} d\tau ,
\tag{12}
$$

where $t(\tau)$ can be found by the inversion of the series (11) with the Lagrange formula given in Eq. (9) of Sec. 1.4. The integral in (12) can often be evaluated by applying the formula for asymptotic expansion given in the Watson lemma stated in Sec. 1.9. For the actual calculation, see the following section.

If $h''(t_0) = h'''(t_0) = \ldots = h^{(m-1)}(t_0)$, but $h^{(m)}(t_0) = ae^{i\theta_0} \neq 0$, then (6) becomes

$$
\sin(m\theta + \theta_0) = 0 ,
\tag{13}
$$

so

$$
m\theta + \theta_0 = n\pi \quad (n = 0, 1, 2, \ldots, 2m - 1) .
\tag{14}
$$

There are then m steepest paths in total, among them those corresponding to $n = 1, 3, \ldots$ are paths of steepest descent. On the path of steepest descent,

$$
\begin{aligned}
h(t) - h(t_0) &= \text{Re}[h(t) - h(t_0)] \\
&= -\frac{a}{m!}\rho^m + O(\rho^{m+1}) .
\end{aligned}
\tag{15}
$$

7.12. Asymptotic Expansions of Bessel Functions of Order ν for Large $|\nu|$ and $|z|$

When z is fixed and only $|\nu|$ is very large, the asymptotic expansion formula for $J_\nu(z)$ can be easily found from its series expression [Eq. (7) of Sec. 7.2]. It is only necessary to apply Eq. (5) of Sec. 3.21 — the asymptotic expansion of the Gamma function. The result is

$$
J_\nu(z) \sim \exp\left\{ \nu + \nu \ln \frac{z}{2} - \left(\nu + \frac{1}{2} \right) \ln \nu \right\} \left[c_0 + \frac{c_1}{\nu} + \frac{c_2}{\nu^2} + \ldots \right] ,
\tag{1}
$$

where $c_0 = 1/\sqrt{2\pi}$.

When both $|\nu|$ and $|z|$ are large, it is more complicated and difficult to find the asymptotic expansions. The method of steepest descent introduced in the last section is an effective method in dealing with such cases.

To begin with, it is necessary to express the Bessel function in an integral form which is apt to the application of the method of steepest descent. From Eq. (18) of Sec. 7.4, Eqs. (14) and (15) of Sec. 7.7, it is known that the functions $J_\nu(z), H_\nu^{(1)}(z), H_\nu^{(2)}(z)$, etc. can be represented by the contour integrals of the following form:

$$\int_C e^{z\,\mathrm{sht}-\nu t}\,dt \ .$$

As we are going to investigate the case that z and ν are large at the same time, we put $z = \nu\lambda, \lambda$ being a constant. The above integral then takes the form

$$\int_C e^{\nu[\lambda\mathrm{sht}-t]}\,dt \ , \tag{2}$$

which is the expression desired [(1) of the last section].

Next, we are going to study the related paths of steepest descent for (2) and to see whether there is any path suitable for obtaining the asymptotic expansion of the integral when $\nu \to \infty$. We shall in the following discuss in detail[c] the case that *both ν and z are positive*; thus, x is written for z.

First, let us find the stationary points t_0 of the function $\lambda\mathrm{sht} - t$, which are the roots of the equation

$$\frac{d}{dt}[\lambda\mathrm{sht} - t] = \lambda\mathrm{cht} - 1 = 0 \ ,$$

i.e.,

$$\mathrm{sech}\, t_0 = \lambda = \frac{x}{\nu} \ . \tag{3}$$

It is necessary to distinguish three cases for discussion, namely, $x/\nu \gtreqless 1$.

1. $x/\nu < 1$. In this case, there exists a positive number α satisfying

$$\mathrm{sech}\,\alpha = \frac{x}{\nu} \ . \tag{4}$$

Hence,

$$t_0 = \pm\alpha + 2n\pi i \quad (n = 0, \pm 1, \pm 2, \dots) \ . \tag{5}$$

We only need to consider the paths of steepest descent which pass through the two points $t_0 = \pm\alpha$; for other stationary points, the paths are merely translated by distances $2n\pi$ parallel to the real axis of the t plane.

---

[c] For the general case, see Watson (1944), Sec. 8.6, p. 262.

According to the condition for steepest paths [(2) of the preceding section], remembering that α is a real number, we have

$$\text{Im}[\text{sech } \alpha \text{ sh} t - t] = \text{Im}[\text{sech } \alpha \text{ sh } (\pm \alpha) \mp \alpha] = 0 . \tag{6}$$

Let $t = u + iv$, u and v being real variables. (6) becomes

$$\text{ch } u \sin v - v \text{ ch } \alpha = 0 . \tag{7}$$

Therefore, the equation of the steepest path is $v = 0$ (the real axis) or

$$\text{ch } u = \frac{v \text{ ch } \alpha}{\sin v} . \tag{8}$$

The curve represented by (8) is symmetric with respect to both the u and the v axes. When v varies from 0 to π, the positive value of u increases monotonically from α to ∞, since then $du/dv = \text{ch } \alpha \cos v (\tan v - v)/\text{sh } u \sin^2 v \geq 0$. Thus, the curve is sketched as that shown in Fig. 31 denoted by C_+ and C_-.

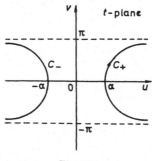

Fig. 31.

From the expression [Eq. (18) of Sec. 7.4]

$$J_\nu(x) = \frac{1}{2\pi i} \int_{\infty - \pi i}^{\infty + \pi i} e^{x \text{ sh} t - \nu t} dt$$

for $J_\nu(x)$, we see immediately that the path of integration can be deformed into the steepest path C_+ passing through the point $t_0 = \alpha$, i.e.,

$$J_\nu (\nu \text{ sech } \alpha) = \frac{1}{2\pi i} \int_{C_+} e^{\nu [\text{sech } \alpha \text{ sh} t - t]} dt . \tag{9}$$

Expanding the function sech α sh $t - t$ in the vicinity of $t = \alpha$ and putting $t - \alpha = \rho e^{i\theta}$, $\theta = \arg(t - \alpha)$, we have

$$\text{sech } \alpha \text{ sh } t - t = \text{th } \alpha - \alpha + \frac{\text{th } \alpha}{2}(t - \alpha)^2 + \dots$$

$$= \text{th } \alpha - \alpha + \frac{\text{th } \alpha}{2}\rho^2 e^{i2\theta} + O(\rho^3) \ . \tag{10}$$

We obtain thus two steepest paths through the point $t_0 = \alpha$; their directions are given by $\theta = 0, \pi$ and $\theta = \pm\pi/2$ [cf. (6) and (7) of the preceding section]. The one with $\theta = 0, \pi$ corresponds to the line $v = 0$; the other, with $\theta = \pm\pi/2$, corresponds to the curve C_+ represented by (8). So, C_+ is the path of steepest descent [cf. (8) and (9) of the preceding section].

If we take a small segment C_ϵ of C_+ in the vicinity of $t = \alpha$ and approximate the integral (9) by the value of the integral along C_ϵ, substituting (10) into it, neglecting terms of order $O(\rho^3)$, then we obtain, by (10) of the last section,

$$J_\nu(\nu \text{ sech } \alpha) \sim \frac{e^{\nu(\text{th } \alpha - \alpha)}}{\sqrt{2\nu\pi \text{ th } \alpha}} \quad (\nu \to +\infty) \ . \tag{11}$$

Of course, the derivation of this formula is crude, because we have discarded the contribution to the integral (9) by the remaining part of C_+ in the calculation without giving an estimation of the error thus introduced.

To show that (11) is indeed the correct asymptotic expression of $J_\nu(\nu$ $\times$sech α) and to look for the asymptotic expansion of the function, we need more precise calculations. For this purpose, let [see (11) of the last section]

$$\tau = \text{th } \alpha - \alpha - (\text{sech } \alpha \text{ sh } t - t) \ . \tag{12}$$

(9) becomes

$$J_\nu(\nu \text{ sech } \alpha) = \frac{e^{\nu(\text{th } \alpha - \alpha)}}{2\pi i} \int_{C'} e^{-\nu\tau} \frac{dt}{d\tau} \, d\tau \ , \tag{13}$$

where C' is the curve in the τ-plane corresponding to C_+. Since $\text{Im}[\text{sech } \alpha \text{ sh } t - t] = 0$ on C_+, τ is real and C' is the real axis of the τ-plane; the point corresponding to $t = \alpha$ is $\tau = 0$.

Expanding the r.h.s. of (12) in a power series of $t - \alpha$, we have

$$\tau = -\frac{\text{th } \alpha}{2!}(t - \alpha)^2 - \frac{1}{3!}(t - \alpha)^3 - \frac{\text{th } \alpha}{4!}(t - \alpha)^4 - \dots$$

$$= -(t - \alpha)^2[c_0 + c_1(t - \alpha) + c_2(t - \alpha)^2 + \dots] \ ,$$

where

$$c_{2m} = \frac{\mathrm{th}\,\alpha}{(2m+2)!} \;, \quad c_{2m+1} = \frac{1}{(2m+3)!} \;, \tag{14}$$

$$(m = 0, 1, 2, \ldots) \;.$$

Thus,

$$t - \alpha = \pm i\tau^{\frac{1}{2}}[c_0 + c_1 w + c_2 w^2 + \ldots]^{-\frac{1}{2}} \;, \tag{15}$$

where $w = t - \alpha$ and $\arg[c_0 + c_1 w \ldots] = 0$ when $w = 0$; thus the positive sign represents the upper part of C_+ $(v > 0)$ while the negative sign represents the lower part of $C_+ (v < 0)$.

With the Lagrange expansion formula [Eq. (9) of Sec. 1.4] we find from (15) the inversion which expresses $t - \alpha$ in terms of power of τ (in fact, of $\tau^{\frac{1}{2}}$):

$$t - \alpha = \sum_{n=1}^{\infty} \frac{(\pm i)^n}{n!} \tau^{\frac{n}{2}} \frac{d^{n-1}}{dw^{n-1}}[c_0 + c_1 w + \ldots]^{-\frac{n}{2}} \Big|_{w=0} \;. \tag{16}$$

Let

$$a_n = \frac{d^{n-1}}{dw^{n-1}}[c_0 + c_1 w + c_2 w^2 + \ldots]^{-\frac{n}{2}} \Big|_{w=0} \;.$$

We obtain

$$a_1 = c_0^{-\frac{1}{2}} = (2\,\mathrm{cth}\,\alpha)^{\frac{1}{2}} \;,$$

$$a_2 = -c_0^{-2} c_1 = -\frac{2}{3}(\mathrm{cth}\,\alpha)^2 \;,$$

$$a_3 = -c_0^{-\frac{7}{2}}\left(3c_0 c_2 - \frac{15}{4}c_1^2\right) = -(2\,\mathrm{cth}\,\alpha)^{\frac{1}{2}}\left(\frac{1}{4} - \frac{5}{12}\mathrm{cth}^2\,\alpha\right) \;, \tag{17}$$

$$a_4 = (2\,\mathrm{cth}\,\alpha)^3 \left(\frac{2}{5} - \frac{4}{9}\mathrm{cth}^2\,\alpha\right) \;,$$

$$a_5 = (2\,\mathrm{cth}\,\alpha)^{\frac{5}{2}}\left(\frac{9}{16} - \frac{77}{24}\mathrm{cth}^2\alpha + \frac{385}{144}\mathrm{cth}^4\,\alpha\right), \ldots \;.$$

Substituting (16) in integral (13), denoting by t_+ and t_- the inverse functions with the positive and negative sign respectively, and applying Watson lemma[d]

---

[d] Since $dt/d\tau = (1 - \mathrm{sech}\alpha\,\mathrm{ch}t)^{-1}$ and when $\tau \to \infty$, $t \to \infty \pm \pi i$, so $dt/d\tau \to 0$ and the condition of the lemma is fulfilled.

of Sec. 1.9 together with formula (3), we obtain

$$\int_{C_1} e^{-\nu\tau} \frac{dt}{d\tau} d\tau = \int_0^\infty e^{-\nu\tau} \frac{dt_+}{d\tau} d\tau + \int_\infty^0 e^{-\nu\tau} \frac{dt_-}{d\tau} d\tau$$

$$= \int_0^\infty e^{-\nu\tau} \frac{d(t_+ - t_-)}{d\tau} d\tau$$

$$= i \int_0^\infty e^{-\nu\tau} \sum_{m=0}^\infty \frac{(-)^m a_{2m+1}}{(2m)!} \tau^{m-\frac{1}{2}} d\tau$$

$$\sim i \sum_{m=0}^\infty \frac{(-)^m a_{2m+1}}{(2m)!} \Gamma\left(m + \frac{1}{2}\right) \nu^{-(m+\frac{1}{2})}$$

$$= i \sum_{m=0}^\infty \frac{(-)^m \Gamma\left(\frac{1}{2}\right) a_{2m+1}}{2^{2m} m!} \nu^{-(m+\frac{1}{2})}$$

[by Eq. (8) of Sec. 3.6] .

Hence,

$$J_\nu(\nu \operatorname{sech} \alpha) \sim \frac{e^{\nu(\operatorname{th}\alpha - \alpha)}}{\sqrt{2\nu\pi \operatorname{th}\alpha}} \sum_{m=0}^\infty \frac{D_m}{(\nu \operatorname{th}\alpha)^m} \quad (\nu \to +\infty) , \qquad (18)$$

where $D_m = (-)^m a_{2m+1}/\{2^m m! (2\operatorname{cth}\alpha)^{m+\frac{1}{2}}\}$; $D_0 = 1, D_1 = \frac{1}{8} - \frac{5}{24}\operatorname{cth}^2\alpha$, $D_2 = \frac{9}{128} - \frac{77}{192}\operatorname{cth}^2\alpha + \frac{385}{1152}\operatorname{cth}^4\alpha,\dots$. It is seen that the rough approximation (11) is precisely the first term of the exact asymptotic expansion.

2. $x/\nu > 1$. In this case, there exists a positive number $\beta < \pi/2$ satisfying

$$\sec\beta = \frac{x}{\nu} \qquad (19)$$

and the integral considered can be written as

$$\int_C e^{\nu(\sec\beta \operatorname{sht} - t)} dt . \qquad (20)$$

From (3), we find the stationary points of the function in the exponent to be

$$t_0 = \pm i\beta + 2n\pi i \quad (n = 0, \pm1, \pm2, \dots) . \qquad (21)$$

As before, only the points $t_0 = \pm i\beta$ need to be considered.

The equation of the path of steepest descent which passes through the point $t_0 = i\beta$ is

$$\operatorname{Im}[\sec\beta \operatorname{sh} t - t] = \operatorname{Im}[\sec\beta \operatorname{sh}(i\beta) - i\beta] = \tan\beta - \beta .$$

Let $t = u + iv$, u and v being real variables. The above equation becomes

$$\sec \beta \operatorname{ch} u \sin v - v = \tan \beta - \beta ,$$

or

$$\operatorname{ch} u = \frac{\tan \beta - \beta + v}{\sec \beta \sin v} . \qquad (22)$$

In $0 \leq v \leq \pi$, to every value of v, there corresponds a pair of values of u with equal magnitude but opposite signs. When $v = 0$ or π, $u = \pm\infty$. Now,

$$\frac{du}{dv} = \frac{\cos v}{\sec \beta} \frac{(\tan v - \tan \beta) - (v - \beta)}{\sin^2 v \operatorname{sh} u} .$$

Therefore, when $v < \beta$, du/dv and u are different in sign; while for $v > \beta$, they are of the same sign. Hence, the steepest paths passing through the point $i\beta$ are roughly of the form $C_+^{(1)}$ and $C_-^{(1)}$ drawn in Fig. 32.

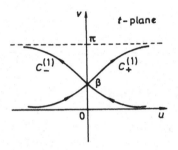

Fig. 32.

In the vicinity of $t = i\beta$, expanding the function $\sec \beta \operatorname{sh} t - t$ in a power series of $t - i\beta$, and putting $t - i\beta = \rho e^{i\theta}$, we have

$$\sec \beta \operatorname{sh} t - t = i \tan \beta - i\beta + i \frac{\tan \beta}{2!}(t - i\beta)^2 + \ldots$$

$$= i(\tan \beta - \beta) + \frac{\tan \beta}{2} \rho^2 e^{i(2\theta + \frac{\pi}{2})} + O(\rho^3) .$$

From which we find that the directions of the steepest paths passing through the point $i\beta$ are $\theta = -\pi/4, 3\pi/4$ and $\theta = \pi/4, 5\pi/4$; the first corresponds to $C_-^{(1)}$ and the latter to $C_+^{(1)}$, $C_+^{(1)}$ being the path of steepest descent.

By Eq. (14) of Sec. 7.7,

$$H_\nu^{(1)}(x) = \frac{1}{\pi i} \int_{-\infty}^{\infty + \pi i} e^{x \operatorname{sh} t - \nu t} dt ;$$

deforming the path of integration into $C_+^{(1)}$, we have

$$H_\nu^{(1)}(\nu \sec \beta) = \frac{1}{\pi i} \int_{C_+^{(1)}} e^{\nu(\sec \beta \, \text{sh} \, t - t)} dt \; . \tag{23}$$

As before, let

$$\tau = i(\tan \beta - \beta) - (\sec \beta \, \text{sh} \, t - t) \; . \tag{24}$$

When t varies along $C_+^{(1)}$, τ takes real values: from ∞ to 0, then from 0 to ∞.

Expanding the r.h.s. of (24) in a power series of $t - i\beta$, we obtain

$$\begin{aligned}
\tau &= -\frac{i}{2!} \tan \beta (t - i\beta)^2 - \frac{1}{3!}(t - i\beta)^3 - \frac{i}{4!} \tan \beta (t - i\beta)^4 - \ldots \\
&= -(t - i\beta)^2 [c_0 + c_1 w + c_2 w^2 + \ldots] \; ,
\end{aligned}$$

where $w = t - i\beta, c_{2m} = i \tan \beta / (2m + 2)!$, $c_{2m+1} = 1/(2m + 3)!, m = 0, 1, 2, \ldots$. Thus,

$$t - i\beta = \pm i \tau^{\frac{1}{2}} [c_0 + c_1 w + \ldots]^{-\frac{1}{2}} \; ,$$

where $\arg[c_0 + c_1 w + \ldots] = \pi/2$ when $w = 0$, since the inclination of $C_+^{(1)}$ when passing through the point $i\beta$ is $\pi/4$ or $5\pi/4$ (see above).

Proceed in the same way as in the case $x/\nu < 1$, we find the asymptotic expansion

$$H_\nu^{(1)}(\nu \sec \beta) \sim \frac{2e^{i[\nu(\tan \beta - \beta) - \frac{\pi}{4}]}}{\sqrt{2\nu\pi \tan \beta}} \sum_{m=0}^{\infty} \frac{D_m'}{(i\nu \tan \beta)^m} \; , \tag{25}$$

$$(\nu \to +\infty)$$

where D_m' can be obtained from D_m [see (18)] by merely replacing α with $i\beta$; for example, $D_0' = 1, D_1' = \frac{1}{8} + \frac{5}{24} \cot^2 \beta, D_2' = \frac{9}{128} + \frac{77}{192} \cot^2 \beta + \frac{385}{1152} \cot^4 \beta, \ldots$.

Similarly, we obtain the asymptotic expansion for $H_\nu^{(2)}(\nu \sec \beta)$ [replacing i in (25) by $-i$]:

$$H_\nu^{(2)}(\nu \sec \beta) \sim \frac{2e^{-i[\nu(\tan \beta - \beta) - \frac{\pi}{4}]}}{\sqrt{2\nu\pi \tan \beta}} \sum_{m=0}^{\infty} \frac{D_m'}{(-i\nu \tan \beta)^m} \tag{26}$$

$$(\nu \to +\infty) \; .$$

Note that in (25) and (26), β is an acute angle.

By Eqs. (8) and (10) of Sec. 7.7, we can deduce respectively from (25) and (26), the following asymptotic expansions:

$$J_\nu(\nu \sec \beta) \sim \left(\frac{2}{\nu \pi \tan \beta}\right)^{\frac{1}{2}} \left\{ \cos\left(\nu \tan \beta - \nu \beta - \frac{\pi}{4}\right) \sum_{m=0}^{\infty} \frac{(-)^m D'_{2m}}{(\nu \tan \beta)^{2m}} \right.$$

$$\left. + \sin\left(\nu \tan \beta - \nu \beta - \frac{\pi}{4}\right) \sum_{m=0}^{\infty} \frac{(-)^m D'_{2m+1}}{(\nu \tan \beta)^{2m+1}} \right\}, \tag{27}$$

$$(\nu \to +\infty)$$

$$Y_\nu(\nu \sec \beta) \sim \left(\frac{2}{\nu \pi \tan \beta}\right)^{\frac{1}{2}} \left\{ \sin\left(\nu \tan \beta - \nu \beta - \frac{\pi}{4}\right) \sum_{m=0}^{\infty} \frac{(-)^m D'_{2m}}{(\nu \tan \beta)^{2m}} \right.$$

$$\left. - \cos\left(\nu \tan \beta - \nu \beta - \frac{\pi}{4}\right) \sum_{m=0}^{\infty} \frac{(-)^m D'_{2m+1}}{(\nu \tan \beta)^{2m+1}} \right\} \tag{28}$$

$$(\nu \to +\infty) .$$

3. $x/\nu \approx 1$. In such cases, α or β is very small. The above formulae are evidently not good approximations, since they are all expressed in terms of powers of ν th α or $\nu \tan \beta$ in descending order. However, the method used in deriving the asymptotic expansions is still applicable provided that it is slightly modified.

Let us consider the expansion of $H_\nu^{(1)}(z)$ first, where ν and z could be complex numbers. Suppose that both $|z|$ and $|\nu|$ are very large, but $|z - \nu|$ is not (the proper order of magnitude of its value will be given below). Write

$$\nu = z(1 - \epsilon) . \tag{29}$$

Assuming $|\arg z| < \pi/2$, we have

$$H_\nu^{(1)}(z) = \frac{1}{\pi i} \int_{-\infty}^{\infty + \pi i} e^{z(\operatorname{sh}t - t) + z \epsilon t} dt . \tag{30}$$

Denoting $\arg z$ by φ, the last integral can be written as

$$\int_{-\infty}^{\infty + \pi i} e^{|z| h(t)} g(t) dt , \tag{31}$$

where $h(t) = e^{i\varphi}(\operatorname{sh} t - t), g(t) = e^{z \epsilon t}$. (31) is a typical integral for the application of the method of steepest descent [cf. (1) of the last section].

The stationary points of $h(t)$ are the roots of the equation $h'(t) = e^{i\varphi}(\operatorname{ch} t - 1) = 0$, namely, $t_0 = 0 + 2n\pi i (n = 0, \pm 1, \pm 2, \ldots)$. As before, we only need to consider the point $t_0 = 0$. The steepest paths passing through this point satisfy the equation

$$\operatorname{Im}[h(t)] = \operatorname{Im}[e^{i\varphi}(\operatorname{sh} t - t)] = \operatorname{Im}[e^{i\varphi}(\operatorname{sh} t_0 - t_0)] = 0 . \tag{32}$$

We shall find only the first two terms of the asymptotic expansion of (31) for very large $|z|$.

Expand $h(t) = e^{i\varphi}(\operatorname{sh} t - t)$ in the vicinity of $t_0 = 0$:

$$h(t) = e^{i\varphi}\left[\frac{t^3}{3!} + \frac{t^5}{5!} + \ldots\right] .$$

Let $t = \rho e^{i\theta}$. We have by (32)

$$\operatorname{Im}\left[\frac{1}{6}\rho^3 e^{i(3\theta+\varphi)} + \ldots\right] = \frac{1}{6}\rho^3 \sin(3\theta + \varphi) + O(\rho^5) = 0 .$$

Thus, the directions of the paths of steepest descent when passing through the point $t = 0$ are $\theta = (n\pi - \varphi)/3, n = 1, 3, 5$. Deforming the path of integration in (31) so that it approaches the point $t = 0$ along the path of steepest descent corresponding to $n = 3$ and then leaves $t = 0$ along the path corresponding to $n = 1$, we have, in the vicinity of $t = 0$,

$$h(t) = \operatorname{Re}[h(t)] \sim -\frac{1}{6}\rho^3 ,$$

and

$$g(t) = e^{z\varepsilon t} \sim 1 + z\varepsilon e^{i(3\pi-\varphi)/3}\rho \quad (t \text{ approaches } 0)$$
$$\sim 1 + z\varepsilon e^{i(\pi-\varphi)/3}\rho \quad (t \text{ leaves } 0) .$$

Hence,

$$\int_{-\infty}^{\infty+\pi i} e^{|z|h(t)}g(t)dt \sim \int_0^\delta e^{-\frac{|z|}{6}\rho^3}\left[1 + z\varepsilon e^{i(\pi-\varphi)/3}\rho\right]e^{i(\pi-\varphi)/3}d\rho$$
$$+ \int_\delta^0 e^{-\frac{|z|}{6}\rho^3}\left[1 + z\varepsilon e^{i(3\pi-\varphi)/3}\rho\right]e^{i(3\pi-\varphi)/3}d\rho . \tag{33}$$

Now,

$$\int_0^\delta e^{-\frac{|z|}{6}\rho^3}(1+\lambda\rho)d\rho = \left(\int_0^\infty - \int_\delta^\infty\right)e^{-\frac{|z|}{6}\rho^3}(1+\lambda\rho)d\rho$$

$$= \frac{1}{3}\left(\int_0^\infty - \int_{\delta^3}^\infty\right)e^{-\frac{|z|}{6}u}(1+\lambda u^{\frac{1}{3}})u^{-\frac{2}{3}}du,$$

$$\int_0^\infty e^{-\frac{|z|}{6}u}(1+\lambda u^{\frac{1}{3}})u^{-\frac{2}{3}}du = \left(\frac{|z|}{6}\right)^{-\frac{1}{3}}\Gamma\left(\frac{1}{3}\right) + \lambda\left(\frac{|z|}{6}\right)^{-\frac{2}{3}}\Gamma\left(\frac{2}{3}\right),$$

$$\left|\int_{\delta^3}^\infty e^{-\frac{|z|}{6}u}(1+\lambda u^{\frac{1}{3}})u^{-\frac{2}{3}}du\right| \le (\delta^{-2}+|\lambda|\delta^{-1})\frac{6}{|z|}e^{-\frac{|z|}{6}\delta^3}.$$

Substituting in (33), after little calculation, we obtain

$$\int_{-\infty}^{\infty+\pi i} e^{|z|h(t)}g(t)dt$$

$$\sim \frac{2}{3}\left[e^{\frac{\pi}{6}i}\sin\frac{\pi}{3}\Gamma\left(\frac{1}{3}\right)\left(\frac{z}{6}\right)^{-\frac{1}{3}} + e^{\frac{5\pi}{6}i}\sin\frac{2\pi}{3}\Gamma\left(\frac{2}{3}\right)(z\varepsilon)\left(\frac{z}{6}\right)^{-\frac{2}{3}}\right]$$

$$+ O\left(\frac{e^{-\gamma|z|}}{|z|}\right).$$

Therefore, when $|z|$ becomes very large, we have

$$H_\nu^{(1)}(z) \sim -\frac{2}{3\pi}\left[e^{\frac{2\pi i}{3}}\sin\frac{\pi}{3}\Gamma\left(\frac{1}{3}\right)\left(\frac{z}{6}\right)^{-\frac{1}{3}}\right.$$

$$\left. + e^{\frac{4\pi i}{3}}\sin\frac{2\pi}{3}\Gamma\left(\frac{2}{3}\right)(z\varepsilon)\left(\frac{z}{6}\right)^{-\frac{2}{3}}\right] \qquad (34)$$

$$(|\arg z| < \pi).$$

Although this result is derived in a rather crude way, it can be shown that the r.h.s. of it is actually the first two terms of the exact asymptotic expansion (for $|\arg z| < \pi$). The exact asymptotic expansion is not difficult to obtain by the method employed in the previous cases $x/\nu \gtrless 1$ [cf. Ex. 49 at the end of this chapter and Watson (1944), Sec. 8.42, p. 247]. Besides, we see from (34) that if $z\varepsilon z^{-1/3} = o(1)$, i.e., $z\varepsilon = z - \nu = o(z^{1/3})$, the second term is much smaller than the first.

(34) is proved under the condition $|\arg z| < \pi/2$. But the path of integration in (30) can be rotated as a whole around $t = 0$ by a proper angle η, $|\eta| < \pi/2$, so that the above result will be valid in the range $-\pi/2 + \eta < \arg z < \pi/2 + \eta$, and hence (34) holds in $|\arg z| < \pi$.

In a similar way, we obtain under the same condition the asymptotic expansion of $H_\nu^{(2)}(z)$:

$$H_\nu^{(2)}(z) \sim -\frac{2}{3\pi}\left[e^{-\frac{2\pi i}{3}}\sin\frac{\pi}{3}\Gamma\left(\frac{1}{3}\right)\left(\frac{z}{6}\right)^{-\frac{1}{3}}\right.$$
$$\left. +e^{-\frac{4\pi i}{3}}\sin\frac{2\pi}{3}\Gamma\left(\frac{2}{3}\right)(z\varepsilon)\left(\frac{z}{6}\right)^{-\frac{2}{3}}\right] \tag{35}$$
$$(|\arg z| < \pi) .$$

With (34) and (35), we find from Eqs. (8) and (10) of Sec. 7.7, respectively

$$J_\nu(z) \sim \frac{1}{3\pi}\left[\sin\frac{\pi}{3}\Gamma\left(\frac{1}{3}\right)\left(\frac{z}{6}\right)^{-\frac{1}{3}} + \sin\frac{2\pi}{3}\Gamma\left(\frac{2}{3}\right)(z\varepsilon)\left(\frac{z}{6}\right)^{-\frac{2}{3}}\right] ,$$
$$(|\arg z| < \pi) \tag{36}$$

$$Y_\nu(z) \sim \frac{-2}{3\pi}\left[\sin^2\frac{\pi}{3}\Gamma\left(\frac{1}{3}\right)\left(\frac{z}{6}\right)^{-\frac{1}{3}} - \sin^2\frac{2\pi}{3}\Gamma\left(\frac{2}{3}\right)(z\varepsilon)\left(\frac{z}{6}\right)^{-\frac{2}{3}}\right]$$
$$(|\arg z| < \pi) . \tag{37}$$

If $x/\nu \approx 1$, but $|x-\nu|$ (i.e., $|x\varepsilon|$) is very large, the above approximation formulae (34)–(37) are not applicable. There is no simple asymptotic expansion, each term of which is some elementary function. Watson had investigated the case of such transition range and obtained some important results [cf. Watson (1944), Sec. 8.43, p. 248]. Besides, Schöbe, Tricomi have also derived some formulae in this respect [cf. Erdélyi (1953), Vol. II, Sec. 7.4.3].

7.13. Addition Formulae

In Sec. 7.5, we have found the addition formulae

$$J_n(x + y) = \sum_{k=-\infty}^{\infty} J_k(x)J_{n-k}(y) \tag{1}$$

and

$$J_0(R) = \sum_{m=-\infty}^{\infty} J_m(r_1)J_m(r_2)e^{im\theta} \tag{2}$$

$$= \sum_{m=0}^{\infty} \varepsilon_m J_m(r_1)J_m(r_2)\cos m\theta , \tag{3}$$

where $R = [r_1^2 + r_2^2 - 2r_1 r_2 \cos \theta]^{1/2}, \varepsilon_0 = 1, \varepsilon_m = 2 \, (m \geq 1)$.

In the present section, we shall generalize these formulae and introduce other addition formulae.

1. Graf formula

$$J_\nu(\varpi) \left\{ \frac{x - ye^{-i\theta}}{x - ye^{i\theta}} \right\}^{\frac{\nu}{2}} = \sum_{m=-\infty}^{\infty} J_{\nu+m}(x) J_m(y) e^{im\theta} , \qquad (4)$$

where x and y are complex numbers, $|ye^{\pm i\theta}| < |x|, \varpi = [x^2 + y^2 - 2xy \cos \theta]^{1/2}$, and it is stipulated that $\varpi \to +x$ as $y \to 0$. Obviously, (4) is a generalization of (1) through (3) above. The proof is as follows.

From Eq. (14) of Sec. 7.4,

$$J_\nu(\varpi) = \frac{1}{2\pi i} \int_{-\infty e^{-i\alpha}}^{(0+)} e^{\frac{\varpi}{2}(t - t^{-1})} t^{-\nu-1} dt ,$$

where $\alpha = \arg \varpi, |\arg t| < \pi$. Write ϖ as $(x - ye^{i\theta})^{1/2}(x - ye^{-i\theta})^{1/2}$ and let $t = (x - ye^{-i\theta})^{1/2}(x - ye^{i\theta})^{-1/2} u$, then

$$J_\nu(\varpi) \left\{ \frac{x - ye^{-i\theta}}{x - ye^{i\theta}} \right\}^{\nu/2}$$

$$= \frac{1}{2\pi i} \int_{-\infty e^{-i\beta}}^{(0+)} e^{\frac{x}{2}(u - u^{-1})} e^{-\frac{y}{2}(ue^{-i\theta} - u^{-1}e^{i\theta})} u^{-\nu-1} du ,$$

where $\beta = \arg(x - ye^{-i\theta})$. Using Eq. (2) of Sec. 7.5, we have

$$e^{\frac{z}{2}(t - t^{-1})} = \sum_{n=-\infty}^{\infty} J_n(z) t^n = \sum_{n=-\infty}^{\infty} J_{-n}(z) t^{-n} = \sum_{n=-\infty}^{\infty} J_n(-z) t^{-n} .$$

Substituting in the integral on the r.h.s. of the last expression, we obtain

$$\frac{1}{2\pi i} \int_{-\infty e^{-i\beta}}^{(0+)} e^{\frac{x}{2}(u - u^{-1})} \sum_{m=-\infty}^{\infty} J_m(y) e^{im\theta} u^{-\nu-m-1} du .$$

Let $\gamma = \arg x$. If $|ye^{-i\theta}| < |x|$, then $|\gamma - \beta| < \pi/2$ and we can rotate the path of integration as a whole around $u = 0$ by an angle $(\beta - \gamma)$ so that

$$\int_{-\infty e^{-i\beta}}^{(0+)} = \int_{-\infty e^{-i\gamma}}^{(0+)} .$$

Interchanging the order of integration and summation [legitimate by Eq. (23) of Sec. 7.5], there results

$$
J_\nu(\varpi) \left\{ \frac{x - y e^{-i\theta}}{x - y e^{i\theta}} \right\}^{\nu/2}
$$

$$
= \frac{1}{2\pi i} \sum_{m=-\infty}^{\infty} J_m(y) e^{im\theta} \int_{-\infty e^{-i\eta}}^{(0+)} e^{\frac{x}{2}(u - u^{-1})} u^{-\nu - m - 1} du
$$

$$
= \sum_{m=-\infty}^{\infty} J_m(y) J_{\nu+m}(x) e^{im\theta} .
$$

2. Gegenbauer addition formula

In (4) putting $\nu = 0$, we obtain

$$
J_0(\varpi) = \sum_{m=-\infty}^{\infty} J_m(x) J_m(y) e^{im\theta} = \sum_{m=0}^{\infty} \varepsilon_m J_m(x) J_m(y) \cos m\theta . \tag{5}
$$

Differentiating with respect to $\cos\theta$ n times, as

$$
\frac{d}{d(\cos\theta)} = \frac{d\varpi}{d(\cos\theta)} \frac{d}{d\varpi} = -\frac{xy}{\varpi} \frac{d}{d\varpi} ,
$$

we have

$$
(-)^n (xy)^n \left(\frac{1}{\varpi} \frac{d}{d\varpi} \right)^n J_0(\varpi) = \sum_{m=0}^{\infty} \varepsilon_m J_m(x) J_m(y) \frac{d^n \cos m\theta}{d(\cos\theta)^n} .
$$

Now, $\cos m\theta$ can be expressed as a polynomial of $\cos\theta$ of degree m [cf. Eq. (2) of Sec. 4.11, noticing that $T_m(\cos\theta) = \cos m\theta$], so the derivatives in the last series vanish for $m < n$. Hence, applying Eq. (17) of Sec. 7.2 to the l.h.s. of the last equation, we obtain

$$
\frac{J_n(\varpi)}{\varpi^n} = \sum_{m=n}^{\infty} \varepsilon_m \frac{J_m(x)}{x^n} \frac{J_m(y)}{y^n} \frac{d^n \cos m\theta}{d(\cos\theta)^n}
$$

$$
= \sum_{m=0}^{\infty} \varepsilon_{m+n} \frac{J_{m+n}(x)}{x^n} \frac{J_{m+n}(y)}{y^n} \frac{d^n \cos(m+n)\theta}{d(\cos\theta)^n} , \tag{6}
$$

which is the Gegenbauer addition formula.

(6) can be generalized to the case of non-integral n [cf. Ex. 51 at the end of this chapter]:

$$\frac{J_\nu(\varpi)}{\varpi^\nu} = 2^\nu \Gamma(\nu) \sum_{m=0}^{\infty} (\nu + m) \frac{J_{\nu+m}(x)}{x^\nu} \frac{J_{\nu+m}(y)}{y^\nu} C_m^\nu(\cos\theta) , \qquad (7)$$

where x, y and θ can be any complex numbers: $\nu \neq 0, -1, -2, \ldots$; $C_m^\nu(x)$ are Gegenbauer polynomials [Sec. 5.23].

Also, we have

$$\frac{J_{-\nu}(\varpi)}{\varpi^\nu} = 2^\nu \Gamma(\nu) \sum_{m=0}^{\infty} (-)^m (\nu + m) \frac{J_{-\nu-m}(x)}{x^\nu} \frac{J_{\nu+m}(y)}{y^\nu} C_m^\nu(\cos\theta) , \qquad (8)$$

where $|ye^{\pm i\theta}| < |x|, \nu \neq 0, -1, -2, \ldots$.

When $\nu = n$, a positive integer, (7) reduces to (6) by Eq. (12) of Sec. 5.23.

An important special case of (7) is when $\nu = \frac{1}{2}$. Then, by Eq. (1) of Sec. 7.3, we have

$$\frac{\sin\varpi}{\varpi} = \pi \sum_{m=0}^{\infty} \left(m + \frac{1}{2}\right) \frac{J_{m+\frac{1}{2}}(x) J_{m+\frac{1}{2}}(y)}{\sqrt{x}\sqrt{y}} P_m(\cos\theta) . \qquad (9)$$

Similarly, by (8) and Eq. (2) of Sec. 7.3, we have

$$\frac{\cos\varpi}{\varpi} = \pi \sum_{m=0}^{\infty} (-)^m \left(m + \frac{1}{2}\right) \frac{J_{-m-\frac{1}{2}}(x) J_{m+\frac{1}{2}}(y)}{\sqrt{x}\sqrt{y}} P_m(\cos\theta) . \qquad (10)$$

$P_m(\cos\theta)$ are Legendre polynomials [cf. Sec. 5.23].

From (7) and (8), together with Eq. (3) of Sec. 7.6, it is found that

$$\frac{Y_\nu(\varpi)}{\varpi^\nu} = 2^\nu \Gamma(\nu) \sum_{m=0}^{\infty} (\nu + m) \frac{Y_{\nu+m}(x)}{x^\nu} \frac{J_{\nu+m}(y)}{y^\nu} C_m^\nu(\cos\theta) . \qquad (11)$$

Hence, by Eqs. (11) and (12) of Sec. 7.6, we have

$$\frac{Z_\nu(\varpi)}{\varpi^\nu} = 2^\nu \Gamma(\nu) \sum_{m=0}^{\infty} (\nu + m) \frac{Z_{\nu+m}(x)}{x^\nu} \frac{J_{\nu+m}(y)}{y^\nu} C_m^\nu(\cos\theta) , \qquad (12)$$

where $Z_\nu(z)$ is any cylinder function of arbitrary order ν.

From (12), let $\nu \to 0$ and note that $C_0^0(\cos\theta) = 1$, then, by the result of Ex. 56 at the end of Chap. 5, namely,

$$\lim_{\nu \to 0} [\Gamma(\nu)(\nu + m) C_m^\nu(\cos\theta)] = 2\cos m\theta \quad (m = 1, 2, \ldots)$$

we obtain

$$Z_0(\varpi) = \sum_{m=0}^{\infty} \varepsilon_m Z_m(x) J_m(y) \cos m\theta . \qquad (13)$$

We can derive further from (12) the following expansion formula:

$$e^{iy\cos\theta} = 2^\nu \Gamma(\nu) \sum_{m=0}^{\infty} (\nu + m) i^m \frac{J_{\nu+m}(y)}{y^\nu} C_m^\nu(\cos\theta) . \qquad (14)$$

The procedure is as follows. In (12), let $Z_\nu(\varpi)$ be $H_\nu^{(2)}(\varpi)$. Multiplying both sides with $x^{\nu+\frac{1}{2}}$, using the asymptotic expression of $H_\nu^{(2)}(\varpi)$ [Eq. (4) of Sec. 7.10], and noticing that $\varpi = \sqrt{x^2 + y^2 - 2xy\cos\theta} = x\left[1 - \frac{y}{x}\cos\theta + O(x^{-2})\right]$, we obtain (14) when we let $x \to \infty$.

An important special case of (14) is $\left(\nu = \frac{1}{2}\right)$

$$e^{iy\cos\theta} = \sqrt{\frac{\pi}{2y}} \sum_{n=0}^{\infty} (2n+1) i^n J_{n+\frac{1}{2}}(y) P_n(\cos\theta) , \qquad (15)$$

where $P_n(\cos\theta)$ are Legendre polynomials.

We can derive many other expansion formulae from (12) [cf. Ex. 52 at the end of this chapter].

7.14. Integrals Containing Bessel Functions. (1) Finite Integrals

In this section, we shall introduce several methods for the evaluation of integrals of this type.

The simplest method is to substitute the series expression of the Bessel function involved into the integrand and then integrate term by term. For example,

$$\int_0^{\frac{\pi}{2}} J_\mu(z\sin\theta)(\sin\theta)^{\mu+1}(\cos\theta)^{2\nu+1} d\theta \quad (\mathrm{Re}(\mu), \mathrm{Re}(\nu) > -1)$$

$$= \sum_{k=0}^{\infty} \frac{(-)^k}{k!} \frac{(z/2)^{\mu+2k}}{\Gamma(\mu+k+1)} \int_0^{\frac{\pi}{2}} (\sin\theta)^{2\mu+2k+1}(\cos\theta)^{2\nu+1} d\theta$$

$$= \sum_{k=0}^{\infty} \frac{(-)^k}{k!} \frac{(z/2)^{\mu+2k}}{\Gamma(\nu+k+1)} \frac{\Gamma(\mu+k+1)\Gamma(\nu+1)}{2\Gamma(\mu+\nu+k+2)} \quad \text{[by Eq. (7) of Sec. 3.8]}$$

$$= \frac{2^\nu \Gamma(\nu+1)}{z^{\nu+1}} J_{\mu+\nu+1}(z) . \qquad (1)$$

This is called *the First Sonine Finite Integral Formula.*

Another method is to use the integral representation of the Bessel function to transform the integral to be evaluated into a surface integral on a unit spherical surface and then simplify the calculation by coordinate transformation. Take again the above example. By Eq. (6) of Sec. 7.4, we have

$$\int_0^{\frac{\pi}{2}} J_\mu(z \sin \theta)(\sin \theta)^{\mu+1}(\cos \theta)^{2\nu+1} d\theta$$

$$= \frac{(z/2)^\mu}{\sqrt{\pi}\Gamma\left(\mu + \frac{1}{2}\right)} \int_0^{\frac{\pi}{2}} \int_0^\pi e^{iz \sin \theta \cos \varphi}(\sin \theta)^{2\mu+1}(\sin \varphi)^{2\mu}(\cos \theta)^{2\nu+1} d\varphi d\theta .$$

Let $l = \sin \theta \cos \varphi$, $m = \sin \theta \sin \varphi$, $n = \cos \theta$ be the direction cosines with respect to the coordinate system (x, y, z). The integral on the r.h.s. of the last equation can be written as

$$\frac{(z/2)^\mu}{\sqrt{\pi}\Gamma\left(\mu + \frac{1}{2}\right)} \iint_{n \geq 0, m \geq 0} e^{izl} m^{2\mu} n^{2\nu+1} d\omega , \quad d\omega = \sin \theta d\theta d\varphi ,$$

which is a surface integral on a fixed part $(n \geq 0, m \geq 0)$ of a unit spherical surface. Now, rotating the coordinate axes from (x, y, z) to (x', y', z') so that (l, m, n) change into (n', l', m') correspondingly, the above integral will become

$$\frac{(z/2)^\mu}{\sqrt{\pi}\Gamma\left(\mu + \frac{1}{2}\right)} \iint_{m' \geq 0, l' \geq 0} e^{izn'} l'^{2\mu} m'^{2\nu+1} d\omega'$$

$$= \frac{(z/2)^\mu}{\sqrt{\pi}\Gamma\left(\mu + \frac{1}{2}\right)} \int_0^{\frac{\pi}{2}} \int_0^\pi e^{iz \cos \theta'}(\sin \theta')^{2\mu+2\nu+1}(\cos \varphi')^{2\mu}$$

$$\times (\sin \varphi')^{2\nu+1} \sin \theta' d\theta' d\varphi'$$

$$= \frac{(z/2)^\mu}{\sqrt{\pi}\Gamma\left(\mu + \frac{1}{2}\right)} \int_0^{\frac{\pi}{2}} (\cos \varphi)^{2\mu}(\sin \varphi)^{2\nu+1} d\varphi \int_0^\pi e^{iz \cos \theta}(\sin \theta)^{2\mu+2\nu+2} d\theta$$

$$= \frac{(z/2)^\mu}{\sqrt{\pi}\Gamma\left(\mu + \frac{1}{2}\right)} \frac{\Gamma\left(\mu + \frac{1}{2}\right)\Gamma(\nu + 1)}{2\Gamma\left(\mu + \nu + \frac{3}{2}\right)} \frac{\sqrt{\pi}\Gamma\left(\mu + \nu + \frac{3}{2}\right)}{(z/2)^{\mu+\nu+1}} J_{\mu+\nu+1}(z)$$

$$= \frac{2^\nu \Gamma(\nu + 1)}{z^{\nu+1}} J_{\mu+\nu+1}(z) ,$$

where we have made use of Eq. (7) of Sec. 3.8 and Eq. (6) of Sec. 7.4. This proves (1).

With the same method, we obtain *the Second Sonine Finite Integral Formula:*

$$\int_0^{\frac{\pi}{2}} J_\mu(x \sin \theta) J_\nu(y \cos \theta)(\sin \theta)^{\mu+1}(\cos \theta)^{\nu+1} d\theta$$

$$= \frac{x^\mu y^\nu J_{\mu+\nu+1}(\sqrt{x^2 + y^2})}{(x^2 + y^2)^{(\mu+\nu+1)/2}} \qquad (\mathrm{Re}(\mu), \ \mathrm{Re}(\nu) > -1), \qquad (2)$$

[cf. Watson (1944), Sec. 12.13, p. 376].

Another important formula for the finite integral is

$$\int_0^\pi \frac{Z_\nu\left(\sqrt{x^2+y^2-2xy\cos\theta}\right)}{(x^2+y^2-2xy\cos\theta)^{\nu/2}} C_m^\nu(\cos\theta)(\sin\theta)^{2\nu}\,d\theta$$
$$= \frac{\pi\Gamma(2\nu+m)}{2^{\nu-1}m!\Gamma(\nu)} \frac{Z_{\nu+m}(x)}{x^\nu} \frac{J_{\nu+m}(y)}{y^\nu} \quad \left(\operatorname{Re}(\nu) > -\frac{1}{2}\right), \qquad (3)$$

which is called the *Sonine-Gegenbauer Formula*; $Z_\nu(z)$ is a cylinder function [cf. the end of Sec. 7.6] of any order ν and $C_m^\nu(\cos\theta)$ are Gegenbauer polynomials [Sec. 5.23].

The proof of (3) is simple. Multiplying both sides of Eq. (12) of Sec. 7.13 with $C_n^\nu(\cos\theta)(\sin\theta)^{2\nu}d\theta$ and integrating, by virtue of Eq. (11) of Sec. 5.23 — the orthonormal relations of Gegenbauer polynomials — we have (3). The legitimacy of the term by term integration is provided by the general theory of expansion in terms of an orthonormal set.

Besides the methods shown in the above examples, the formulae for indefinite integrals to be obtained in Exs. 15-19 at the end of this chapter may also be employed to calculate definite integrals.

7.15. Integrals Containing Bessel Functions. (2) Infinite Integrals

Similar to the case of finite integral, infinite integrals containing Bessel functions can also be evaluated by means of the following methods:

1. Use the series expression of the Bessel function and integrate term by term.

2. Use the various integral representations of Bessel functions and exchange the order of integrations.

3. When a product of Bessel functions occurs in the integrand, use an appropriate formula, for example, (3) of the last section or the Neumann formula given in Ex. 31 at the end of this chapter, to replace the product by an integral involving only one Bessel function and then exchange the order of integrations. We give in the following some fundamental integral formulae. For others, see the relating exercises at the end of this chapter, and Watson (1944), Chap. 13 and the bibliography supplied.

Hankel integral formula

$$\int_0^\infty e^{-at} J_\nu(bt) t^{\mu-1} dt = \frac{\Gamma(\mu+\nu)}{a^{\mu+\nu}\Gamma(\nu+1)} \left(\frac{b}{2}\right)^\nu$$
$$\times F\left(\frac{\mu+\nu}{2}, \frac{\mu+\nu+1}{2}, \nu+1, -\frac{b^2}{a^2}\right), \tag{1}$$

where $F(\alpha, \beta, \gamma, z)$ is the hypergeometric function, $\text{Re}(\mu + \nu) > 0$ and $\text{Re}(a \pm ib) > 0$ to ensure the convergence of the integral.

Suppose $\text{Re}(a) > 0$ and $|b/a| < 1$. Substituting the series expression of $J_\nu(z)$ [Eq. (7) of Sec. 7.2] in the integral and integrating term by term, we obtain

$$\int_0^\infty e^{-at} J_\nu(bt) t^{\mu-1} dt = \sum_{k=0}^\infty \frac{(-)^k}{k!} \frac{(b/2)^{\nu+2k}}{\Gamma(\nu+k+1)} \int_0^\infty e^{-at} t^{\mu+\nu+2k-1} dt$$
$$= \sum_{k=0}^\infty \frac{(-)^k}{k!} \frac{(b/2)^{\nu+2k}}{\Gamma(\nu+k+1)} a^{-\mu-\nu-2k} \Gamma(\mu+\nu+2k).$$

Using the duplication formula for the Γ-function [Eq. (8) of Sec. 3.6], we have

$$\Gamma(\mu+\nu+2k) = \Gamma(\mu+\nu) 2^{2k} \left(\frac{\mu+\nu}{2}\right)_k \left(\frac{\mu+\nu+1}{2}\right)_k,$$

and hence (1). The term by term integration is legitimate, since the series obtained is absolutely convergent under the condition $|b/a| < 1$ [Bromwich (1925), Sec. 176].

We have proved (1) above under the conditions $\text{Re}(a) > 0$ and $|a| > |b|$. But, by virtue of the asymptotic representation of $J_\nu(bt)$ for $t \to \infty$ [Eq. (5) of Sec. 7.10], we see that the integral on the l.h.s. of (1) is convergent (with respect to the upper limit) under the condition $\text{Re}(a \pm ib) > 0$, so it represents an analytic function of b for fixed a. Thus, (1) is valid under the conditions $\text{Re}(\mu + \nu) > 0$ and $\text{Re}(a \pm ib) > 0$ according to the principle of analytic continuation.

By applying separately Eqs. (8) and (9) of Sec. 4.3 to the r.h.s. of (1), we get

$$\int_0^\infty e^{-at} J_\nu(bt) t^{\mu-1} dt$$

$$= \frac{\Gamma(\mu+\nu)}{a^{\mu+\nu}\Gamma(\nu+1)} \left(\frac{b}{2}\right)^\nu \left(1+\frac{b^2}{a^2}\right)^{\frac{1}{2}-\mu}$$

$$F\left(\frac{\nu-\mu+2}{2}, \frac{\nu-\mu+1}{2}, \nu+1, -\frac{b^2}{a^2}\right)$$

$$= \frac{\Gamma(\mu+\nu)}{\Gamma(\nu+1)} \left(\frac{b}{2}\right)^\nu (a^2+b^2)^{-\frac{\mu+\nu}{2}} F\left(\frac{\nu+\mu}{2}, \frac{\nu-\mu+1}{2}, \nu+1, \frac{b^2}{a^2+b^2}\right)_{(2)}$$

With the relations of $Y_\nu(z), H_\nu^{(1,2)}(z), I_\nu(z), K_\nu(z)$ to $J_\nu(z)$, we can deduce from the above formulae (1) and (2) a series of formulae for infinite integrals involving these Bessel functions and modified Bessel functions [cf. the Exercises at the end of this chapter].

A simple special case of (1) is that $\nu = 0, \mu = 1$:

$$\int_0^\infty e^{-at} J_0(bt) dt = \frac{1}{a} F\left(\frac{1}{2}, 1, 1, -\frac{b^2}{a^2}\right) = \frac{1}{\sqrt{a^2+b^2}} \tag{3}$$

[by Eq. (10) of Sec. 4.2], where the square root takes the value a when $b = 0$.

Struve integral formula

$$\int_0^\infty \frac{J_\mu(t) J_\nu(t)}{t^{\mu+\nu}} dt = \frac{\Gamma\left(\frac{1}{2}\right) \Gamma(\mu+\nu)}{2^{\mu+\nu} \Gamma\left(\mu+\nu+\frac{1}{2}\right) \Gamma\left(\mu+\frac{1}{2}\right) \Gamma\left(\nu+\frac{1}{2}\right)} \tag{4}$$

$$(\operatorname{Re}(\mu+\nu) > 0) .$$

This is a special case of (11) below. But since the method of proving it is typical, we give it separately.

Assume first that $\operatorname{Re}(\mu)$ and $\operatorname{Re}(\nu)$ both be greater than $\frac{1}{2}$. Changing θ in Eq. (7) of Sec. 7.4 to $\pi/2 - \theta$, partial integrating once (in order to make the infinite integral of t below convergent) and substituting the result into the l.h.s. of (4), we obtain

$$\int_0^\infty \frac{J_\mu(t) J_\nu(t)}{t^{\mu+\nu}} dt = \frac{(2\mu-1)(2\nu-1)}{\pi 2^{\mu+\nu-2} \Gamma\left(\mu+\frac{1}{2}\right) \Gamma\left(\nu+\frac{1}{2}\right)}$$

$$\times \int_0^\infty \int_0^{\frac{\pi}{2}} \int_0^{\frac{\pi}{2}} \frac{\sin(t\sin\theta) \sin(t\sin\varphi)}{t^2}$$

$$\times (\cos\theta)^{2\mu-2} (\cos\varphi)^{2\nu-2} \sin\theta \sin\varphi \, d\theta \, d\varphi \, dt .$$

Interchanging the order of integrations (readily justified), and using the formula

$$\int_0^\infty \frac{\sin dt \sin \beta t}{t^2} dt = \frac{1}{2} \int_0^\infty \frac{\cos(\alpha - \beta)t - \cos(\alpha + \beta)t}{t^2} dt$$

$$= \begin{cases} \pi\beta/2 & (\alpha > \beta) \\ \pi\alpha/2 & (\alpha < \beta) \end{cases},$$

the triple integral above becomes

$$\frac{\pi}{2} \int_0^{\frac{\pi}{2}} \int_0^{\varphi} (\cos \theta)^{2\mu-2}(\cos \varphi)^{2\nu-2} \sin^2 \theta \sin \varphi d\theta d\varphi$$

$$+ \frac{\pi}{2} \int_0^{\frac{\pi}{2}} \int_0^{\theta} (\cos \theta)^{2\mu-2}(\cos \varphi)^{2\nu-2} \sin \theta \sin^2 \varphi d\varphi d\theta .$$

By Eq. (7) of Sec. 3.8, we have

$$\frac{\pi}{2} \int_0^{\frac{\pi}{2}} \int_0^{\varphi} (\cos \theta)^{2\mu-2}(\cos \varphi)^{2\nu-2} \sin^2 \theta \sin \varphi d\theta d\varphi$$

$$= \frac{\pi}{2} \int_0^{\frac{\pi}{2}} (\cos \theta)^{2\mu-2} \sin^2 \theta d\theta \int_\theta^{\frac{\pi}{2}} (\cos \varphi)^{2\nu-2} \sin \varphi d\varphi$$

$$= \frac{\pi}{2} \frac{1}{2\nu - 1} \int_0^{\frac{\pi}{2}} (\cos \theta)^{2\mu+2\nu-3} \sin^2 \theta d\theta$$

$$= \frac{\pi}{4} \frac{\Gamma(\mu + \nu - 1)\Gamma\left(\frac{3}{2}\right)}{(2\nu - 1)\Gamma\left(\mu + \nu + \frac{1}{2}\right)} .$$

Evaluating the second integral similarly (merely an exchange of μ and ν), we obtain

$$\int_0^\infty \frac{J_\mu(t)J_\nu(t)}{t^{\mu+\nu}} dt = \frac{\Gamma(\mu + \nu - 1)\Gamma(2\mu + 2\nu - 2)\Gamma\left(\frac{3}{2}\right)}{2^{\mu+\nu}\Gamma\left(\mu + \nu + \frac{1}{2}\right)\Gamma\left(\mu + \frac{1}{2}\right)\Gamma\left(\nu + \frac{1}{2}\right)} .$$

With the relation $z\Gamma(z) = \Gamma(z + 1)$ we see that the r.h.s. of this equation is the same as that of (4).

The condition $\mathrm{Re}(\mu + \nu) > 0$ ensures the convergence of the integral on the l.h.s. of (4) with respect to the upper limit. Therefore, the additional conditions $\mathrm{Re}(\mu) > \frac{1}{2}$ and $\mathrm{Re}(\nu) > \frac{1}{2}$ can be discarded according to the principle of analytic continuation.

Weber-Schafheitlin integral

Suppose that, in the integral

$$\int_0^\infty \frac{J_\mu(at) J_\nu(bt)}{t^\lambda} dt \, , \tag{5}$$

a and b are positive numbers and

$$\text{Re}(\mu + \nu + 1) > \text{Re}(\lambda) > -1 \quad (a \neq b) \, ,$$
$$\text{Re}(\mu + \nu + 1) > \text{Re}(\lambda) > 0 \quad (a = b) \, ,$$

then the integral is convergent at both the upper and lower limits. This can be shown respectively by using the asymptotic expression [Eq. (5) of Sec. 7.10] and the series expression [Eq. (7) of Sec. 7.2] of the Bessel functions.

We can evaluate (5) with formula (1) given above. For, under the conditions stated,

$$\int_0^\infty \frac{J_\mu(at) J_\nu(bt)}{t^\lambda} dt = \lim_{c \to +0} \int_0^\infty e^{-ct} \frac{J_\mu(at) J_\nu(bt)}{t^\lambda} dt \, . \tag{6}$$

For written simplicity, we introduce the following new constants:

$$\left. \begin{aligned} 2\alpha &= \nu + \mu - \lambda + 1 \, , \\ 2\beta &= \nu - \mu - \lambda + 1 \, , \\ \gamma &= \nu + 1 \, , \end{aligned} \right\} \qquad \left. \begin{aligned} \lambda &= \gamma - \alpha - \beta \, , \\ \mu &= \alpha - \beta \, , \\ \nu &= \gamma - 1 \, . \end{aligned} \right\}$$

When the constant c in the integral on the r.h.s. of (6) is an arbitrarily given positive number, by using the asymptotic expansion of $J_\nu(z)$, we can see that the integral is also convergent for complex values of b, so long as $|\text{Im}(b)| < c$. Replace b by z, then, in the domain $\text{Re}(z) > 0$ and $|\text{Im}(z)| < c$, the said integral is an analytic function of z.

By the series expression of the Bessel function and formula (2), we have

$$\int_0^\infty e^{-ct} \frac{J_{\alpha-\beta}(at) J_{\gamma-1}(zt)}{t^{\gamma-\alpha-\beta}} dt$$

$$= \int_0^\infty e^{-ct} J_{\alpha-\beta}(at) \sum_{k=0}^\infty \frac{(-)^k}{k!\Gamma(\gamma+k)} \left(\frac{z}{2}\right)^{\gamma+2k-1} t^{\alpha+\beta+2k-1} dt$$

$$= \sum_{k=0}^\infty \frac{(-)^k}{k!\Gamma(\gamma+k)} \left(\frac{z}{2}\right)^{\gamma+2k-1} \int_0^\infty e^{-ct} J_{\alpha-\beta}(at) t^{\alpha+\beta+2k-1} dt$$

$$= \sum_{k=0}^\infty \frac{(-)^k}{k!\Gamma(\gamma+k)} \left(\frac{z}{2}\right)^{\gamma+2k-1} \frac{\Gamma(2\alpha+2k)}{(a^2+c^2)^{\alpha+k}\Gamma(\alpha-\beta+1)}$$

$$\times \left(\frac{a}{2}\right)^{\alpha-\beta} F\left(\alpha+k, \frac{1}{2}-\beta-k, \alpha-\beta+1, \frac{a^2}{a^2+c^2}\right) \, .$$

When $|z| < c$, the interchange of the order of integration and summation is legitimate [cf. Watson (1944), p. 399].

It can be shown [Watson (1944), p. 400] that the last series is uniformly convergent in $0 \leq c \leq A$ for sufficiently small A. Therefore,

$$\lim_{c \to +0} \int_0^\infty e^{-ct} \frac{J_{\alpha-\beta}(at) J_{\gamma-1}(bt)}{t^{\gamma-\alpha-\beta}} dt$$

$$= \sum_{k=0}^\infty \frac{(-)^k \Gamma(2\alpha + 2k)}{k! \Gamma(\gamma+k) \Gamma(\alpha-\beta+1) a^{2\alpha+2k}} \left(\frac{b}{2}\right)^{\gamma+2k-1} \left(\frac{a}{2}\right)^{\alpha-\beta}$$

$$\times F\left(\alpha+k, \frac{1}{2} - \beta - k, \alpha - \beta + 1, 1\right) .$$

Then, by Eq. (4) of Sec. 4.7 and Eq. (8) of Sec. 3.6, we get from (6)

$$\int_0^\infty \frac{J_{\alpha-\beta}(at) J_{\gamma-1}(bt)}{t^{\gamma-\alpha-\beta}} dt = \frac{b^{\gamma-1} \Gamma(\alpha)}{2^{\gamma-\alpha-\beta} a^{\alpha+\beta} \Gamma(\gamma) \Gamma(1-\beta)} F\left(\alpha, \beta, \gamma, \frac{b^2}{a^2}\right) \quad (7)$$

or,

$$\int_0^\infty \frac{J_\mu(at) J_\nu(bt)}{t^\lambda} dt$$

$$= \frac{b^\nu \Gamma\left(\frac{\nu+\mu-\lambda+1}{2}\right)}{2^\lambda a^{\nu-\lambda+1} \Gamma(\nu+1) \Gamma\left(\frac{\mu-\nu+\lambda+1}{2}\right)}$$

$$\times F\left(\frac{\nu+\mu-\lambda+1}{2}, \frac{\nu-\mu-\lambda+1}{2}, \nu+1, \frac{b^2}{a^2}\right)$$

$$(0 < b < a) . \quad (8)$$

Exchanging a and b, μ and ν, we obtain separately from (7) and (8)

$$\int_0^\infty \frac{J_{\alpha-\beta}(at) J_{\gamma-1}(bt)}{t^{\gamma-\alpha-\beta}} dt$$

$$= \frac{a^{\alpha-\beta} \Gamma(\alpha)}{2^{\gamma-\alpha-\beta} b^{2\alpha-\gamma+1} \Gamma(\gamma-\alpha) \Gamma(\alpha-\beta+1)}$$

$$\times F\left(\alpha, \alpha-\gamma+1, \alpha-\beta+1, \frac{a^2}{b^2}\right) \quad (0 < a < b) \quad (9)$$

and the formula corresponding to (8).

With Eq. (8) of Sec. 4.8, it can be seen that the functions on the r.h.s. of (7) and (9) are not the analytic continuations of the same function. But when $b \to a$, we see from Eq. (4) of Sec. 4.7 that the r.h.s. of (7) and (8) approach to the same limit so long as $\text{Re}(\gamma - \alpha - \beta) > 0$. Thus, if integral (5) is continuous at the point $b = a$, then we have

$$\int_0^\infty \frac{J_{\alpha-\beta}(at)J_{\gamma-1}(at)}{t^{\gamma-\alpha-\beta}}\,dt = \frac{\Gamma(\alpha)\Gamma(\gamma-\alpha-\beta)}{2\Gamma(1-\beta)\Gamma(\gamma-\alpha)\Gamma(\gamma-\beta)}\left(\frac{a}{2}\right)^{\gamma-\alpha-\beta-1} \tag{10}$$
$$(\text{Re}(\alpha) > 0,\ \text{Re}(\gamma - \alpha - \beta) > 0)\,,$$

or,

$$\int_0^\infty \frac{J_\mu(at)J_\nu(at)}{t^\lambda}\,dt = \frac{\Gamma(\lambda)\Gamma\left(\frac{\mu+\nu-\lambda+1}{2}\right)\left(\frac{a}{2}\right)^{\lambda-1}}{2\Gamma\left(\frac{\mu-\nu+\lambda+1}{2}\right)\Gamma\left(\frac{\nu-\mu+\lambda+1}{2}\right)\Gamma\left(\frac{\mu+\nu+\lambda+1}{2}\right)} \tag{11}$$
$$(\text{Re}(\mu + \nu + 1) > \text{Re}(\lambda) > 0)\,.$$

For a rigorous proof of the continuity of the integral (5) at the point $b = a$, see Watson (1944), Sec. 13.41, p. 402.

By using the asymptotic expression of the Bessel function, it can be shown that when $\mu - \nu$ is an odd number, the integral on the l.h.s. of (11) also converges for $0 \geq \text{Re}(\lambda) > -1$. We shall deal with these cases now, especially when $\lambda = 0$.

Let $\mu = \sigma + p, \nu = \sigma - p - 1, p = 0, 1, 2, \ldots$ and $\mu - \nu = 2p + 1$ is an odd number. From (8) and (9), we have respectively

$$\int_0^\infty \frac{J_{\sigma+p}(at)J_{\sigma-p-1}(bt)}{t^\lambda}\,dt$$

$$= \frac{b^{\sigma-p-1}\Gamma\left(\sigma - \frac{\lambda}{2}\right)}{2^\lambda a^{\sigma-p-\lambda}\Gamma(\sigma-p)\Gamma\left(p + \frac{\lambda}{2} + 1\right)}F\left(\sigma - \frac{\lambda}{2},\ -p - \frac{\lambda}{2},\ \sigma - p,\ \frac{b^2}{a^2}\right) \tag{12}$$
$$(b < a)\,,$$

$$\int_0^\infty \frac{J_{\sigma+p}(at)J_{\sigma-p-1}(bt)}{t^\lambda}\,dt$$

$$= \frac{a^{\sigma+p}\Gamma\left(\sigma - \frac{\lambda}{2}\right)}{2^\lambda b^{\sigma+p-\lambda+1}\Gamma(\sigma+p+1)\Gamma\left(\frac{\lambda}{2} - p\right)}$$
$$\times F\left(\sigma - \frac{\lambda}{2},\ p + 1 - \frac{\lambda}{2},\ \sigma + p + 1,\ \frac{a^2}{b^2}\right)$$
$$(b > a)\,. \tag{13}$$

If $\lambda \neq 0$, then when $a \to b$, both the r.h.s. of (12) and (13) have no limits, since the respective hypergeometric series are divergent [cf. Sec. 4.7, before

(4)], whereas from (11), we have

$$\int_0^\infty \frac{J_{\sigma+p}(at) J_{\sigma-p-1}(at)}{t^\lambda} dt = \frac{a^{\lambda-1} \Gamma(\lambda) \Gamma\left(\sigma - \frac{\lambda}{2}\right)}{2^\lambda \Gamma\left(\frac{\lambda}{2} - p\right) \Gamma\left(p + \frac{\lambda}{2} + 1\right) \Gamma\left(\sigma + \frac{\lambda}{2}\right)}. \quad (14)$$

When $\lambda = 0$, the r.h.s. of (13) vanishes, since the factor $\Gamma\left(\frac{\lambda}{2} - p\right)$ in the denominator $\to \infty$, while the hypergeometric series on the r.h.s. of (12) reduces to a polynomial of degree p. Besides, when $\lambda \to 0$, the limit of the r.h.s. of (14) exists:

$$\lim_{\lambda \to 0} \frac{a^{\lambda-1} \Gamma(\lambda) \Gamma\left(\sigma - \frac{\lambda}{2}\right)}{2^\lambda \Gamma\left(\frac{\lambda}{2} - p\right) \Gamma\left(p + \frac{\lambda}{2} + 1\right) \Gamma\left(\sigma + \frac{\lambda}{2}\right)} = \frac{1}{a \times p!} \lim_{\lambda \to 0} \frac{\Gamma(\lambda)}{\Gamma\left(\frac{\lambda}{2} - p\right)}$$

$$= \frac{1}{a \times p!} \lim_{\lambda \to 0} \frac{\Gamma\left(1 + p - \frac{\lambda}{2}\right) \sin \pi \left(\frac{\lambda}{2} - p\right)}{\Gamma(1 - \lambda) \sin \pi\lambda} = \frac{1}{2a}(-)^p.$$

(For a rigorous calculation of the integral in (14) when $\lambda = 0$, see Watson (1944), p. 404.) Thus, summing up, we have the following formula:

$$\int_0^\infty J_{\sigma+p}(at) J_{\sigma-p-1}(bt) dt$$

$$= \begin{cases} \frac{\Gamma(\sigma)}{\Gamma(\sigma-p)p!b} \left(\frac{b}{a}\right)^{\sigma-p} F\left(\sigma, -p, \sigma-p, \frac{b^2}{a^2}\right), & (b < a) \\ (-)^p/(2a) & (b = a), \\ 0 & (b > a). \end{cases} \quad (15)$$

When $b \to a - 0$, by Eq. (2) of Sec. 4.10, we have $F(\sigma, -p, \sigma-p, 1) = (-p)_p/(\sigma - p)_p = (-)^p p! \Gamma(\sigma - p)/\Gamma(\sigma)$. Hence, the value of the integral in (15), when $b = a$, is the arithmetic mean of the limits for $b \to a - 0$ and $b \to a + 0$.

Sonine-Gegenbauer integral formula

$$\int_0^\infty J_\mu(bt) \frac{J_\nu\left(a\sqrt{t^2 + z^2}\right)}{(t^2 + z^2)^{\nu/2}} t^{\mu+1} dt$$

$$= \begin{cases} 0 & (a < b), \\ \frac{b^\mu}{a^\nu} \left(\frac{\sqrt{a^2 - b^2}}{z}\right)^{\nu-\mu-1} J_{\nu-\mu-1}\left(z\sqrt{a^2 - b^2}\right) & (a > b), \end{cases} \quad (16)$$

where z is any complex number, a and b are both positive numbers, $\mathrm{Re}(\nu) > \mathrm{Re}(\mu) > -1$ to ensure the convergence of the integral; if $a = b$, then it is necessary that $\mathrm{Re}(\nu) > \mathrm{Re}(\mu + 1) > 0$.

The proof of (16) is as follows. By Eq. (19) of Sec. 7.4, we have

$$\int_0^\infty J_\mu(bt) \frac{J_\nu\left(a\sqrt{t^2+z^2}\right)}{(t^2+z^2)^{\nu/2}} t^{\mu+1} dt$$

$$= \frac{1}{2\pi i} \int_0^\infty \int_{c-i\infty}^{c+i\infty} J_\mu(bt) t^{\mu+1} u^{-\nu-1} \exp\left\{\frac{a}{2}\left(u - \frac{t^2+z^2}{u}\right)\right\} du dt$$

$$= \frac{1}{2\pi i} \int_{c-i\infty}^{c+i\infty} u^{-\nu-1} \exp\left\{\frac{au}{2} - \frac{az^2}{2u}\right\} du \int_0^\infty e^{-\frac{a}{2u}t^2} J_\mu(bt) t^{\mu+1} dt .$$

The exchange of the order of integrations is justified since the integral concerned is absolutely convergent under the condition $\mathrm{Re}(\nu) > \mathrm{Re}(\mu+1) > 0$; the case for which the value of ν does not satisfy this condition may be considered from the view of analytic continuation.

With the series expression for $J_\mu(bt)$, we have

$$\int_0^\infty e^{-pt^2} J_\mu(bt) t^{\mu+1} dt \quad (\mathrm{Re}(p) > 0, \ \mathrm{Re}(\mu) > -1)$$

$$= \sum_{k=0}^\infty \frac{(-)^k (b/2)^{\mu+2k}}{k!\Gamma(\mu+k+1)} \int_0^\infty e^{-pt^2} t^{2\mu+2k+1} dt$$

$$= \sum_{k=0}^\infty \frac{(-)^k (b/2)^{\mu+2k}}{k!\Gamma(\mu+k+1)} (p)^{-\mu-k-1} \frac{1}{2} \int_0^\infty e^{-v} v^{\mu+k} dv$$

$$= \frac{(b/2)^\mu}{2p^{\mu+1}} \sum_{k=0}^\infty \frac{1}{k!} \left(\frac{-b^2}{4p}\right)^k = \frac{b^\mu}{(2p)^{\mu+1}} \exp\left\{-\frac{b^2}{4p}\right\} . \tag{17}$$

Hence,

$$\int_0^\infty J_\mu(bt) \frac{J_\nu\left(a\sqrt{t^2+z^2}\right)}{(t^2+z^2)^{\nu/2}} t^{\mu+1} dt$$

$$= \frac{b^\mu}{a^{\mu+1}} \frac{1}{2\pi i} \int_{c-i\infty}^{c+i\infty} u^{\mu-\nu} \exp\left\{\frac{(a^2-b^2)u}{2a} - \frac{az^2}{2u}\right\} du .$$

When $a < b$, the path of integration can be deformed into an infinite circular arc, with $u = 0$ as its center, to the right of the original path; the value of the integral is zero by Jordan's lemma. This gives a proof of the first part of formula (16).

When $a > b$, by Eq. (19) of Sec. 7.4, we obtain the second part of (16).

Another important integral formula [(21) below] is deduced from the consideration of the following contour integral:

$$\frac{1}{2\pi i} \int_C \frac{z^{\rho-1} H_\nu^{(1)}(az)}{(z^2 + k^2)^{\mu+1}} \, dz \,,$$

where C is the contour shown in Fig. 33 (assume $\mathrm{Re}(k) > 0$), $0 \le \arg z \le \pi$; when $z \to \infty$ along the positive real axis, $\arg(z^2 + k^2) \to 0$.

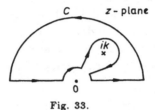

Fig. 33.

Assume $a > 0$. By the asymptotic property of $H_\nu^{(1)}(az)$ [Eq. (3) of Sec. 7.10] and Jordan's lemma, the value of the part of the integral along the large circular arc will approach zero when its radius tends to infinity, provided $\mathrm{Re}(\rho) < 2\mathrm{Re}(\mu) + 7/2$. Also if $|\mathrm{Re}(\nu)| < \mathrm{Re}(\rho)$, the integral over the small circular arc vanishes as the radius tends to zero. There is no singularity of the integrand inside the contour. Therefore, under the condition

$$|\mathrm{Re}(\nu)| < \mathrm{Re}(\rho) < 2\mathrm{Re}(\mu) + \frac{7}{2} \,,$$

we have

$$\frac{1}{2\pi i} \int_0^\infty \left[H_\nu^{(1)}(ax) - e^{(\rho-2\mu)\pi i} H_\nu^{(1)}(axe^{\pi i}) \right] \frac{x^{\rho-1} dx}{(x^2 + k^2)^{\mu+1}}$$
$$= \frac{1}{2\pi i} \int_0^{(ik+)} \frac{z^{\rho-1} H^{(1)}(az)}{(z^2 + k^2)^{\mu+1}} \, dz \,, \tag{18}$$

where $\arg(ik) = \pi/2 + \arg k$. $H_\nu^{(1)}(az)$ in the integral on the r.h.s. can be expanded in a power series of z in ascending order by Eq. (5) of Sec. 7.7.

Then, carrying out the integration term by term and observing that

$$\frac{1}{2\pi i} \int_0^{(ik+)} \frac{z^{\lambda-1} dz}{(z^2 + k^2)^{\mu+1}}$$

$$= \frac{e^{\lambda\pi i/2} k^\lambda}{e^{(\mu+1)\pi i} k^{2\mu+2}} \frac{1}{2\pi i} \int_0^{(1+)} \frac{z^{\lambda-1} dz}{(z^2 - 1)^{\mu+1}} \quad (|\arg(z^2 - 1)| < \pi)$$

$$= -\frac{1}{2\pi i} e^{(\frac{\lambda}{2} - \mu)\pi i} k^{\lambda - 2\mu - 2} \left(e^{\mu\pi i} - e^{-\mu\pi i} \right) \int_0^1 \frac{x^{\lambda-1} dx}{(1 - x^2)^{\mu+1}}$$

$$= -\frac{1}{\pi} e^{(\frac{\lambda}{2} - \mu)\pi i} k^{\lambda - 2\mu - 2} \sin \mu\pi \frac{\Gamma\left(\frac{\lambda}{2}\right) \Gamma(-\mu)}{2\Gamma\left(\frac{\lambda}{2} - \mu\right)} \quad [\text{cf. Sec. 3.8}]$$

$$= -e^{(\frac{\lambda}{2} - \mu)\pi i} k^{\lambda - 2\mu - 2} \frac{\Gamma\left(\frac{\lambda}{2}\right)}{2\Gamma\left(\frac{\lambda}{2} - \mu\right) \Gamma(1 + \mu)} , \tag{19}$$

we obtain

$$\frac{1}{2\pi i} \int_0^{(ik+)} \frac{z^{\rho-1} H_\nu^{(1)}(az) dz}{(z^2 + k^2)^{\mu+1}}$$

$$= \frac{-i}{2 \sin \nu\pi} \frac{e^{(\frac{\rho-\nu}{2} - \mu)\pi i} k^{\rho - 2\mu - 2}}{\Gamma(1 + \mu)}$$

$$\times \left\{ \left(\frac{ak}{2}\right)^\nu \frac{\Gamma\left(\frac{\rho+\nu}{2}\right)}{\Gamma\left(\frac{\rho+\nu}{2} - \mu\right) \Gamma(1 + \nu)} \right.$$

$$\times {}_1F_2\left(\frac{\rho+\nu}{2}; \frac{\rho+\nu}{2} - \mu, 1 + \nu; \frac{a^2 k^2}{4}\right)$$

$$- \left(\frac{ak}{2}\right)^{-\nu} \frac{\Gamma\left(\frac{\rho-\nu}{2}\right)}{\Gamma\left(\frac{\rho-\nu}{2} - \mu\right) \Gamma(1 - \nu)}$$

$$\left. \times {}_1F_2\left(\frac{\rho-\nu}{2}; \frac{\rho-\nu}{2} - \mu, 1 - \nu; \frac{a^2 k^2}{4}\right) \right\}, \tag{20}$$

where ${}_1F_2(\alpha; \gamma_1, \gamma_2; z)$ is the generalized hypergeometric series [Sec. 4.15].

From Eq. (18) of Sec. 7.2, Eq. (3) of Sec. 7.6 and Eq. (5) of Sec. 7.7, we have

$$H_\nu^{(1)}(z) - e^{(\rho-2\mu)\pi i} H_\nu^{(1)}(ze^{\pi i})$$

$$= 2e^{(\frac{\rho-\nu}{2} - \mu)\pi i} \left\{ \cos \pi \left(\frac{\rho - \nu}{2} - \mu\right) J_\nu(z) + \sin \pi \left(\frac{\rho - \nu}{2} - \mu\right) Y_\nu(z) \right\} .$$

Substituting these results in (18), we obtain the integral formula

$$
\int_0^\infty \left\{ \cos \pi \left(\frac{\rho - \nu}{2} - \mu \right) J_\nu(ax) \right.
$$

$$
\left. + \sin \pi \left(\frac{\rho - \nu}{2} - \mu \right) Y_\nu(ax) \right\} \frac{x^{\rho-1} dx}{(x^2 + k^2)^{\mu+1}}
$$

$$
= \frac{\pi}{2 \sin \nu \pi} \frac{k^{\rho-2\mu-2}}{\Gamma(1+\mu)} \left\{ \left(\frac{ak}{2} \right)^\nu \frac{\Gamma\left(\frac{\rho+\nu}{2} \right)}{\Gamma\left(\frac{\rho+\nu}{2} - \mu \right) \Gamma(1+\nu)} \right.
$$

$$
\times \, {}_1F_2 \left(\frac{\rho+\nu}{2}; \frac{\rho+\nu}{2} - \mu, \, 1+\nu; \frac{a^2 k^2}{4} \right)
$$

$$
- \left(\frac{ak}{2} \right)^{-\nu} \frac{\Gamma\left(\frac{\rho-\nu}{2} \right)}{\Gamma\left(\frac{\rho-\nu}{2} - \mu \right) \Gamma(1-\nu)}
$$

$$
\left. \times \, {}_1F_2 \left(\frac{\rho-\nu}{2}; \frac{\rho-\nu}{2} - \mu, \, 1-\nu; \frac{a^2 k^2}{4} \right) \right\} . \tag{21}
$$

7.16. Neumann Expansion

First we discuss the expansion of the function $(t - z)^{-1}$ in terms of Bessel functions. With this expansion and the Cauchy integral formula, we can derive the corresponding expansion for a general analytic function.

Assume $|z| < |t|$. From Eqs. (9) and (15) of Sec. 7.5, we have

$$
\frac{1}{t - z} = \frac{1}{t} + \sum \frac{z^s}{t^{s+1}}
$$

$$
= \frac{1}{t} \sum_{m=0}^\infty \varepsilon_{2m} J_{2m}(z) + \sum_{s=1}^\infty \frac{2^s}{t^{s+1}} \left\{ \sum_{m=0}^\infty \frac{(s + m - 1)!(s + 2m)}{m!} J_{s+2m}(z) \right\} .
$$

By Eq. (22) of Sec. 7.5, it can be shown that the double series here is absolutely convergent under the condition $|z| < |t|$ [cf. Watson (1944), p. 272], so that the order of integrations can be interchanged. Putting $n = s + 2m$, we obtain

$$
\frac{1}{t - z} = \frac{1}{t} \sum_{m=0}^\infty \varepsilon_{2m} J_{2m}(z) + \sum_{n=1}^\infty J_n(z) \sum_{m=0}^{\left[\frac{n-1}{2}\right]} \frac{2^{n-2m} n(n - m - 1)!}{m! t^{n-2m+1}}
$$

$$
= \frac{1}{t} J_0(z) + \sum_{n=1}^\infty \varepsilon_n J_n \sum_{m=0}^{\left[\frac{n}{2}\right]} \frac{2^{n-2m-1} n(n - m - 1)!}{m! t^{n-2m+1}} .
$$

Let

$$O_0(t) = \frac{1}{t} , \tag{1}$$

$$O_n(t) = \sum_{m=0}^{[\frac{n}{2}]} \frac{2^{n-2m-1} n (n-m-1)!}{m! t^{n-2m+1}} \quad (n \ge 1) . \tag{2}$$

We get the expansion

$$\frac{1}{t-z} = O_0(t) J_0(z) + 2O_1(t) J_1(z) + 2O_2(t) J_2(z) + \dots$$

$$= \sum_{n=0}^{\infty} \varepsilon_n O_n(t) J_n(z) . \tag{3}$$

It can be shown that the series is uniformly convergent in $|t| \ge R, |z| \le r$, where R and r are arbitrary positive numbers, $R > r$. $O_n(t)$ are called *Neumann polynomials*, being polynomials of $1/t$ of degrees $n+1$. The first six of them are:

$$O_0(t) = \frac{1}{t}, \quad O_1(t) = \frac{1}{t^2}, \quad O_2(t) = \frac{1}{t} + \frac{4}{t^3} ,$$

$$O_3(t) = \frac{3}{t^2} + \frac{24}{t^4}, \quad O_4(t) = \frac{1}{t} + \frac{16}{t^3} + \frac{192}{t^5} , \tag{4}$$

$$O_5(t) = \frac{5}{t^2} + \frac{120}{t^4} + \frac{1920}{t^6} .$$

There are recurrence relations satisfied by $O_n(t)$ which can be derived as follows. From the relation

$$\left(\frac{\partial}{\partial t} + \frac{\partial}{\partial z} \right) \frac{1}{t-z} = 0$$

and (3), we obtain, under the condition $|z| < |t|$,

$$\sum_{n=0}^{\infty} \varepsilon_n O_n'(t) J_n(z) + \sum_{n=0}^{\infty} \varepsilon_n O_n(t) J_n'(z) = 0 .$$

Using Eq. (14) of Sec. 7.2, we have

$$\sum_{n=0}^{\infty} \varepsilon_n O_n'(t) J_n(z)$$

$$= O_0(t) J_1(z) - \sum_{n=1}^{\infty} [J_{n-1}(z) - J_{n+1}(z)] O_n(t)$$

$$= -J_0(z) O_1(t) - \sum_{n=1}^{\infty} J_n(z) [O_{n+1}(t) - O_{n-1}(t)] ,$$

or,

$$J_0(z)[O_0'(t) + O_1(t)] + \sum_{n=1}^{\infty} J_n(z)[2O_n'(t) + O_{n+1}(t) - O_{n-1}(t)] = 0 .$$

For fixed t, if the coefficient of $J_0(z)$, namely, $O_0'(t) + O_1(t)$, is not equal to zero, then, since the series is uniformly convergent in the neighborhood of $z = 0$ [cf. Watson (1944), Sec. 9.11], the value of the first term in the above series will exceed the sum of the absolute values of all the remaining terms, provided $|z|$ is sufficiently small [$J_n(z) \approx z^n$ as $z \approx 0$]. Thus we must have

$$O_0'(t) + O_1(t) = 0 . \tag{5}$$

By similar arguments, we have

$$2O_n'(t) + O_{n+1}(t) - O_{n-1}(t) = 0 \quad (n \geq 1) . \tag{6}$$

Note that the recurrence relations (5) and (6) are the same as those for $J_n(z)$ [Eqs. (14) and (15) of Sec. 7.2].

Further, $O_n(t)$ also satisfies the following recurrence relation

$$(n - 1)O_{n+1}(t) + (n + 1)O_{n-1}(t) - \frac{2(n^2 - 1)}{t}O_n(t)$$
$$= \frac{2n\left(\sin \frac{n\pi}{2}\right)^2}{t} \quad (n \geq 1) . \tag{7}$$

[For proof, see Watson (1944), Sec. 9.11.]

Now we come to the Neumann expansion of an analytic function. Let $f(z)$ be an analytic function in the circle $|z| \leq R$. Denote by C the circumference of the circle. From Cauchy's integral formula, we have

$$f(z) = \frac{1}{2\pi i} \int_C \frac{f(t)}{t - z} dt .$$

By substituting the result in (3) into it, we obtain the Neumann expansion for $f(z)$:

$$f(z) = \sum_{n=0}^{\infty} a_n J_n(z) , \tag{8}$$

where

$$a_n = \frac{\varepsilon_n}{2\pi i} \int_C f(t)O_n(t)dt . \tag{9}$$

If the Taylor expansion of $f(z)$ in $|z| \leq R$ is $\sum_0^\infty b_s z^s$, then

$$a_n = \frac{\varepsilon_n}{2\pi i} \int_C O_n(t) \sum_{s=0}^\infty b_s t^s \, dt$$

$$= \frac{\varepsilon_n}{2\pi i} \sum_{s=0}^\infty b_s \int_C O_n(t) t^s \, dt .$$

From (1), (2) and the residue theorem, we find

$$a_0 = b_0 ,$$

$$a_n = n \times 2^n \sum_{m=0}^{[\frac{n}{2}]} \frac{(n-m-1)!}{m! 2^{2m}} b_{n-2m} \quad (n \geq 1) . \tag{10}$$

7.17. Kapteyn Expansion

Differed from the Neumann expansion, the general form of the Kapteyn expansion is

$$\sum_n a_n J_{\nu+n}\{(\nu+n)z\} ,$$

in which the argument of each Bessel function contains a factor $(\nu + n)$, the order of the corresponding function.

An example of the Kapteyn expansion — Consider the Fourier cosine expansion of the function $(1 - x\cos\theta)^{-1}, (|x| < 1)$:

$$(1 - x\cos\theta)^{-1} = \frac{A_0}{2} + \sum_{n=1}^\infty A_n \cos n\varphi , \tag{1}$$

where

$$\varphi = \theta - x\sin\theta . \tag{2}$$

(Equation (2) is the well known Kepler equation in dynamics, φ being the mean anomaly and θ the eccentric anomaly.)

According to the theory of Fourier series, the expansion coefficients in (1) are given by

$$A_n = \frac{2}{\pi} \int_0^\pi \frac{\cos n\varphi}{1 - x\cos\theta} d\varphi$$

$$= \frac{2}{\pi} \int_0^\pi \cos(n\theta - nx\sin\theta) d\theta = \begin{cases} 2 & (n=0) \\ 2J_n(nx) & (n \geq 1) \end{cases}$$

[cf. Eq. (20) of Sec. 7.5]. Hence,

$$\frac{1}{1 - x \cos \theta} = 1 + 2 \sum_{n=1}^{\infty} J_n(nx) \cos n\varphi , \tag{3}$$

the r.h.s. of which is just a Kapteyn series.

When $\theta = 0, \varphi = 0$ and we have

$$\frac{1}{1 - x} = 1 + 2 \sum_{n=1}^{\infty} J_n(nx) . \tag{4}$$

It can be shown [Watson (1944), Sec. 17.3] that for the complex variable z in the following domain

$$\omega(z) = \left| \frac{z \exp \sqrt{1 - z^2}}{1 + \sqrt{1 - z^2}} \right| < 1 . \tag{5}$$

(4) is also valid, i.e.,

$$\frac{1}{1 - z} = 1 + 2 \sum_{n=1}^{\infty} J_n(nz) . \tag{6}$$

Now, we use (6) to derive the Kapteyn expansion of z^n. To this end, we prove first the following theorem: If $f(z) = \sum_{m=1}^{\infty} a_m J_m(mz)$, then the sum $F(z) = \sum_{m=1}^{\infty} a_m J_m(mz) m^{-2}$ can be obtained from $f(z)$ by integrating twice with respect to z, so long as the Kapteyn series representing $f(z)$ is uniformly convergent. From the differential equation satisfied by the Bessel functions [Eq. (1) of Sec. 7.1], we have

$$z^2 F'' + z F' = \sum_{m=1}^{\infty} a_m \left[x^2 J_m''(mz) + \frac{z}{m} J_m'(mz) \right]$$

$$= (1 - z^2) \sum_{m=1}^{\infty} a_m J_m(mz) ,$$

where the prime $'$ expresses the differentiation with respect to the argument mz of J_m. Thus,

$$\left(z \frac{d}{dz} \right)^2 F(z) = (1 - z^2) f(z) , \tag{7}$$

and $F(z)$ can be obtained by integrating $f(z)$.

Changing z in (6) to $-z$ and using Eq. (8) of Sec. 7.2, we have

$$\frac{1}{1+z} = 1 + 2 \sum_{m=1}^{\infty} (-)^m J_m(mz) . \tag{8}$$

Adding (6) and (8), we find

$$\sum_{m=1}^{\infty} J_{2m}(2mz) = \frac{\frac{1}{2}z^2}{1-z^2} . \tag{9}$$

By (7), putting $f(z) = \frac{1}{2}z^2(1-z^2)^{-1}$, we then obtain

$$F(z) = \sum_{m=1}^{\infty} \frac{1}{4m^2} J_{2m}(2mz) = \frac{1}{8}z^2 + A \ln z + B .$$

Let $z \to 0$. We see that $A = B = 0$ and hence

$$z^2 = 2 \sum_{m=1}^{\infty} \frac{1}{m^2} J_{2m}(2mz) . \tag{10}$$

Further, by subtracting (8) from (6), we find

$$\sum_{m=1}^{\infty} J_{2m+1}\{(2m+1)z\} = \frac{\frac{1}{2}z}{1-z^2} . \tag{11}$$

In a similar way as that in obtaining (10), we have

$$z = 2 \sum_{m=0}^{\infty} \frac{1}{(2m+1)^2} J_{2m+1}\{(2m+1)z\} . \tag{12}$$

Now, suppose

$$z^n = \sum_{s=1}^{\infty} b_{n,s} J_s(sz) \quad (n \geq 1) .$$

By the above theorem, putting $f(z) = z^n$, we have

$$F(z) = \sum_{s=1}^{\infty} \frac{b_{n,s}}{s^2} J_s(sz) = \frac{z^n}{n^2} - \frac{z^{n+2}}{(n+2)^2} + C \ln z + D .$$

Letting $z \to 0$, we see that $C = D = 0$. Thus,

$$z^{n+2} = (n+2)^2 \left\{ \frac{z^n}{n^2} - \sum_{s=1}^{\infty} \frac{b_{n,s}}{s^2} J_s(sz) \right\}$$

$$= \frac{(n+2)^2}{n^2} \sum_{s=1}^{\infty} \left(1 - \frac{n^2}{s^2} \right) b_{n,s} J_s(sz)$$

$$= \sum_{s=1}^{\infty} b_{n+2,s} J_s(sz) \,,$$

where

$$b_{n+2,s} = \frac{(n+2)^2}{n^2} \left(1 - \frac{n^2}{s^2} \right) b_{n,s} \,. \tag{13}$$

If $n = 2(k-1)$, an even number, then, from (13), we have

$$b_{2k,s} = \frac{k^2}{(k-1)^2} \left[1 - \frac{(k-1)^2}{(s/2)^2} \right] b_{2(k-1),s}$$

$$= k^2 \left[1 - \frac{(k-1)^2}{(s/2)^2} \right] \left[1 - \frac{(k-2)^2}{(s/2)^2} \right] \cdots \left[1 - \frac{1}{(s/2)^2} \right] b_{2,s} \,.$$

But by (10), it is known that $b_{2,s} = 0$ when s is an odd number and $b_{2,2m} = 2/m^2$ when $s = 2m$, an even number. Therefore,

$$b_{2k,2m} = \frac{2k^2}{m^2} \left[1 - \frac{(k-1)^2}{m^2} \right] \left[1 - \frac{(k-2)^2}{m^2} \right] \cdots \left[1 - \frac{1}{m^2} \right]$$

$$= \frac{2k^2}{m^{2k}} \left[m - (k-1) \right] \left[m - (k-2) \right] \cdots \left[m - 1 \right]$$

$$\times \left[m + (k-1) \right] \left[(m + (k-2) \right] \cdots \left[m + 1 \right]$$

$$= \frac{2k^2 \Gamma(m+k)\Gamma(m)}{m^{2k}\Gamma(m+1)\Gamma(m-k+1)} = \frac{2k^2 \Gamma(m+k)}{m^{2k+1}\Gamma(m-k+1)} \,,$$

$$b_{2k,2m+1} = 0 \qquad (m = 0, 1, 2, \ldots) \,,$$

and it follows that

$$z^{2k} = 2k^2 \sum_{m=1}^{\infty} \frac{\Gamma(m+k)}{m^{2k+1}\Gamma(m-k+1)} J_{2m}(2mz)$$

$$= 2k^2 \sum_{m=k}^{\infty} \frac{\Gamma(m+k)}{m^{2k+1}\Gamma(m-k+1)} J_{2m}(2mz)$$

$$= 2k^2 \sum_{m=0}^{\infty} \frac{\Gamma(m+2k)}{(m+k)^{2k+1}m!} J_{2m+2k}\{(2m+2k)z\} \,.$$

Similarly, starting from (12), by (13), we obtain

$$z^{2k+1} = 2\left(k + \frac{1}{2}\right)^2 \sum_{m=0}^{\infty} \frac{\Gamma(m + 2k + 1)}{(m + k + \frac{1}{2})^{2k+2} \, m!} J_{2m+2k+1}\{(2m + 2k + 1)z\} \, .$$

By combining the two formulae obtained above, we may write

$$\left(\frac{z}{2}\right)^n = n^2 \sum_{m=0}^{\infty} \frac{\Gamma(n + m)}{m!(n + 2m)^{n+1}} J_{n+2m}\{(n + 2m)z\} \tag{14}$$

$$(n = 1, 2, \dots) \, ,$$

which is the Kapteyn expansion corresponding to the Neumann expansion of $(z/2)^n$ given in Eq. (15) of Sec. 7.5.

Kapteyn has shown that, as far as $z^2 - 1$ is not a positive number, the following inequality holds [cf. Watson (1944), Sec. 8.7]:

$$|J_n(nz)| \le \left| \frac{z^n \exp\{n\sqrt{1 - z^2}\}}{(1 + \sqrt{1 - z^2})^n} \right| \, . \tag{15}$$

With the asymptotic expression of $\Gamma(n+m)$ for $m \to \infty$ [Eq. (5) of Sec. 3.21], it can readily be shown that $\Gamma(n + m)/m!(n + 2m)^{n+1} = O(m^{-2})$ and hence the series $\sum_{m=0}^{\infty} \Gamma(n + m)/m!(n + 2m)^{n+1}$ is absolutely convergent. Therefore, the series in (14) is uniformly convergent in the closed domain

$$\omega(z) = \left| \frac{z \exp\sqrt{1 - z^2}}{1 + \sqrt{1 - z^2}} \right| \le 1 \, . \tag{16}$$

Having had the expansion (14) of the power function, we are ready to use the same method as that given in the preceding section to find the Kapteyn expansion of $(t - z)^{-1}$, namely,

$$\frac{1}{t - z} = \Theta_0(t) + 2 \sum_{n=1}^{\infty} \Theta_n(t) J_n(nz) \, , \tag{17}$$

where

$$\Theta_0(t) = \frac{1}{t} \, ,$$

$$\Theta_n(t) = \sum_{m=0}^{[n/2]} \frac{2^{n-2m-1}(n - 2m)^2(n - m - 1)!}{m!(nt)^{n-2m+1}} \tag{18}$$

and are called the *Kapteyn polynomials*. (17) holds in the following domain [see (16)]:

$$\omega(z) < \omega(t), \quad \omega(z) < \omega(1) \tag{19}$$

and the corresponding series is uniformly convergent in this domain [see Watson (1944), Sec. 17.34].

By (17) and the Cauchy integral formula, we arrive at once the theorem concerning the Kapteyn expansion of a general analytic function:

Let $f(z)$ be an analytic function in $\omega(z) \le a$ $(a \le 1)$, then

$$f(z) = \alpha_0 + 2 \sum_{n=1}^{\infty} \alpha_n J_n(nz) , \tag{20}$$

where

$$\alpha_n = \frac{1}{2\pi i} \int_C \Theta_n(t) f(t) dt , \tag{21}$$

the contour of integration C being $\omega(t) = a$.

If the Taylor expansion of $f(z)$ in the neighborhood of $z = 0$ is $\Sigma_{s=0}^{\infty} a_s z^s$, then, from (18) and (21), we have

$$\alpha_0 = a_0, \quad \alpha_n = \sum_{m=0}^{[\frac{n}{2}]} \frac{2^{n-2m-1}(n-2m)^2(n-m-1)!}{m!n^{n-2m+1}} a_{n-2m} . \tag{22}$$

7.18. The Zeros of Bessel Functions

We shall not go into details of the theory of zeros of Bessel functions, especially those formulae for finding the zeros. Only the most fundamental results are mentioned, which are results commonly used in the boundary-value problems in mathematical physics and are essential for the Fourier-Bessel expansion to be dealt in the following section. On the numerical values of the zeros, various function tables and literature may be consulted.[e]

We first show that *to any given real number ν, $J_\nu(z)$ has an infinite number of zeros of real values.*

Let us start with the case $\nu > -\frac{1}{2}$, then by Eq. (7) of Sec. 7.4, we have

$$J_\nu(x) = \frac{(x/2)^\nu}{\Gamma\left(\frac{1}{2}\right)\Gamma\left(\nu + \frac{1}{2}\right)} \int_0^\pi \cos(x\cos\theta) \sin^{2\nu}\theta\, d\theta$$

$$= \frac{2^{1-\nu}}{\Gamma\left(\frac{1}{2}\right)\Gamma\left(\nu + \frac{1}{2}\right) x^\nu} \int_0^x \frac{\cos t\, dt}{(x^2 - t^2)^{\frac{1}{2}-\nu}} .$$

---

[e] For example, Abramowitz & Stegun (1966), Chap. 9; in particular, Tables 9.5-9.7, pp. 409-415.

Setting $x = m\pi + \frac{1}{2}\pi\theta, 0 \leq \theta \leq 1$, it becomes

$$J_\nu\left(m\pi + \frac{1}{2}\pi\theta\right) = \frac{2(\pi/4)^\nu}{\Gamma\left(\frac{1}{2}\right)\Gamma\left(\nu + \frac{1}{2}\right)(2m + \theta)^\nu} \int_0^{2m+\theta} \frac{\cos(\pi t/2)dt}{[(2m + \theta)^2 - t^2]^{\frac{1}{2} - \nu}} .$$

Since ν is real,

$$\operatorname{sgn}J_\nu\left(m\pi + \frac{1}{2}\pi\theta\right) = \operatorname{sgn} \int_0^{2m+\theta} \frac{\cos(\pi t/2)dt}{[(2m + \theta)^2 - t^2]^{\frac{1}{2} - \nu}} .$$

In order to investigate the change of signs of the r.h.s. integral, we write it as

$$\sum_{r=1}^{m}(-)^r v_r + (-)^m v_m' ,$$

where

$$(-)^r v_r = \int_{2r-2}^{2r} \frac{\cos(\pi t/2)dt}{[(2m + \theta)^2 - t^2]^{\frac{1}{2} - \nu}} \quad (r = 1, 2, \ldots, m) ,$$

$$(-)^m v_m' = \int_{2m}^{2m+\theta} \frac{\cos(\pi t/2)dt}{[(2m + \theta)^2 - t^2]^{\frac{1}{2} - \nu}} .$$

Putting $t = 2r - 1 \pm s$, the first equation can be written as

$$v_r = \int_0^1 f_r(s)\sin(\pi s/2)ds ,$$

where

$$f_r(s) = [(2m + \theta)^2 - (2r - 1 + s)^2]^{\nu - \frac{1}{2}} - [(2m + \theta)^2 - (2r - 1 - s)^2]^{\nu - \frac{1}{2}} .$$

If we assume $\nu \leq \frac{1}{2}$, then by differentiation with respect to r, we see that $f_r(s)$ is a positive monotonic increasing function of r and so

$$0 \leq v_1 \leq v_2 \leq \ldots \leq v_m .$$

Again, let $t = 2m + s$, we see that $v_m' \geq 0$. Hence,

$$\operatorname{sgn}J_\nu\left(m\pi + \frac{1}{2}\pi\theta\right) = (-)^m\operatorname{sgn}\{v_m' + (v_m - v_{m-1}) + (v_{m-2} - v_{m-3}) + \ldots\}$$

$$= (-)^m ,$$

i.e., for $-\frac{1}{2} < \nu \le \frac{1}{2}$,

$$\text{sgn} J_\nu \left(m\pi + \frac{1}{2}\pi\theta \right) = \begin{cases} + & (m = 0, 2, 4, \ldots) \\ - & (m = 1, 3, 5, \ldots) \end{cases}.$$

Now, $J_\nu(x)$ is a continuous function of x, so in each interval of $(\pi/2, \pi)$, $(3\pi/2, 2\pi)$, ... there are an odd number of zeros of $J_\nu(x)$. Thus, there are an infinite number of real zeros.

By Eqs. (9) and (10) of Sec. 7.2, together with Roll's theorem, we infer immediately from the above results that for any real ν, $J_\nu(z)$ has an infinite number of real zeros.

Furthermore, all these zeros, except possibly $z = 0$ (if it is a zero), are simple, i.e., of the first order. This is because of the fact that the Bessel function has no singularities in a finite region except possibly the point $z = 0$. If $z = \alpha(\ne 0)$ is a zero of $J_\nu(z)$ of higher order, then both $J_\nu(\alpha)$ and $J_\nu'(\alpha)$ are equal to zero, and $J_\nu(z) \equiv 0$. This conclusion is obviously valid for the zeros of any solution of the Bessel equation.

Next, we shall show that, any real cylinder function [cf. end of Sec. 7.6] $Z_\nu(x) = \alpha J_\nu(x) + \beta Y_\nu(x)$, where α, β, ν are real and $x > 0$, has an infinite number of real positive zeros. For this purpose, let $Z_\nu^*(x) = \gamma J_\nu(x) + \delta Y_\nu(x)$ be another cylinder function of order ν linearly independent of $Z_\nu(x)$ (so, $\alpha\delta - \beta\gamma \ne 0$). By Eqs. (2) and (3) of Sec. 7.6, we have

$$Z_\nu(x) Z_\nu^{*'}(x) - Z_\nu'(x) Z_\nu^*(x) = \frac{2(\alpha\delta - \beta\gamma)}{\pi x}. \tag{1}$$

Take $Z_\nu^*(x) \equiv J_\nu(x)$, then, since the positive zeros of $J_\nu(x)$ are all simple, the signs of $J_\nu'(x)$ at two consecutive positive zeros must be opposite. Hence, according to (1), the signs of $Z_\nu(x)$ at the two said zeros must also be opposite, so there is at least one zero of $Z_\nu(x)$ lying between them. And this proves that $Z_\nu(x)$ has an infinite number of positive zeros.

By the same argument, we conclude that the respective zeros of any two linearly independent real cylinder functions, $Z_\nu(x)$ and $Z_\nu^*(x)$ of the same order, are interwoven, i.e., between any two consecutive positive zeros of $Z_\nu(x)$, there must be one and only one positive zero of $Z_\nu^*(x)$; the converse is also true.

Besides, with the recurrence relations of the cylinder functions [Eqs. (7) and (8) of Sec. 7.6], we can infer that the respective positive zeros of $Z_\nu(x)$ and $Z_{\nu-1}(x)$ are also interwoven.

Now, we come to show that: *when $\nu > -1$, all the zeros of $J_\nu(z)$ are real.*

First, let us prove the following important formula:

$$\int_0^x tJ_\nu(at)J_\nu(bt)dt = \frac{x}{a^2-b^2}\left[J_\nu(ax)\frac{dJ_\nu(bx)}{dx} - J_\nu(bx)\frac{dJ_\nu(ax)}{dx}\right], \qquad (2)$$

where $\nu > -1$ to ensure the convergence of the integral on the l.h.s. at the lower limit.

From the differential equation satisfied by $J_\nu(z)$ [Eq. (1) of Sec. 7.1], we have

$$\frac{1}{t}\frac{d}{dt}\left[t\frac{dJ_\nu(at)}{dt}\right] + \left(a^2 - \frac{\nu^2}{t^2}\right)J_\nu(at) = 0 ,$$

$$\frac{1}{t}\frac{d}{dt}\left[t\frac{dJ_\nu(bt)}{dt}\right] + \left(b^2 - \frac{\nu^2}{t^2}\right)J_\nu(bt) = 0 .$$

Multiplying respectively with $tJ_\nu(bt)$ and $tJ_\nu(at)$, subtracting and then integrating from 0 to x, it results

$$(a^2 - b^2)\int_0^x tJ_\nu(at)J_\nu(bt)dt = \left[tJ_\nu(at)\frac{dJ_\nu(bt)}{dt} - tJ_\nu(bt)\frac{dJ_\nu(at)}{dt}\right]_{t=0}^{t=x}$$

By using the series expression of $J_\nu(z)(\nu > -1)$ and noticing $\nu > -1$, it is seen that the value of the bracket on the r.h.s. is zero at $t = 0$, and hence (2).

Now, suppose that α is a complex zero of $J_\nu(z)$. α cannot be purely imaginary; otherwise we would have

$$J_\nu(\alpha) = \left(\frac{\alpha}{2}\right)^\nu \sum_{k=0}^\infty \frac{(-)^k(\alpha/2)^{2k}}{k!\Gamma(\nu+k+1)}$$

and $J_\nu(\alpha) \neq 0$, since the series is one of positive terms.

As the coefficients in the series expression of $J_\nu(z)$ are all real, $J_\nu(\overline{\alpha})$ also equals to zero, where $\overline{\alpha}$ is the complex conjugate of α. Since α is not purely imaginary, $\alpha^2 \neq \overline{\alpha}^2$. Thus, from (2), by putting $a = \alpha, b = \overline{\alpha}$, we have

$$\int_0^? tJ_\nu(\alpha t)J_\nu(\overline{\alpha}t)dt = 0 .$$

The integrand $t|J_\nu(\alpha t)|^2$ is, however, a non-negative continuous function, its integral cannot possibly be zero. So, α cannot be a complex number.

Concerning the numerical values of the real zeros of $J_\nu(x)$, we shall only point out that, when x is very large, from the asymptotic expansion of $J_\nu(x)$ [Eq. (5) of Sec. 7.10], the zeros are

$$x \sim \left(m + \frac{\nu}{2} - \frac{1}{4}\right)\pi . \qquad (3)$$

Having known the zeros of the Bessel functions, we can deduce the zeros of those special functions related to the Bessel functions; for instance, the zeros of the confluent hypergeometric function.[f] This is, of course, of primary importance in some boundary-value problems and eigen-value problems in mathematical physics.

7.19. Fourier-Bessel Expansion

This is a type of expansion in terms of an orthonormal function set, which is closely related to the eigenvalue problems in mathematical physics. We shall state some important conclusions without going into rigorous proofs and derivations. The rigorous theory is of the same nature as the theory of Fourier series.

Suppose that α_m and α_n are two different positive zeros of $J_\nu(x), \nu > -1$. From (2) of the last section, we have

$$\int_0^1 t J_\nu(\alpha_m t) J_\nu(\alpha_n t) dt = 0 . \tag{1}$$

This is often called the *orthogonality relation among the Bessel functions*, with weight t [cf. Eq. (1) of Sec. 1.10].

When $a = b$, applying l'Hospital rule to the r.h.s. of (2) in the last section, we find, as the limit for $a \to b$ [cf. Ex. 19 at the end of this chapter],

$$\int_0^1 t J_\nu^2(bt) dt = \frac{1}{2b^2} [b^2 \{J_\nu'(b)\}^2 + (b^2 - \nu^2) J_\nu^2(b)] , \tag{2}$$

from which we obtain

$$\int_0^1 t J_\nu^2(\alpha_m t) dt = \frac{1}{2} \{J_\nu'(\alpha_m)\}^2 , \tag{3}$$

α_m being the positive zeros of $J_\nu(x)$.

Now, let us consider the Fourier-Bessel expansion of the function of $f(x)$:

$$f(x) = \sum_m a_m J_\nu(\alpha_m x) . \tag{4}$$

Multiply both sides with $x J_\nu(\alpha_n x)$ and integrate from 0 to 1. Assume that the series on the r.h.s. can be integrated term by term, then, by the *orthonormal*

[f]Cf., for example, Slater (1960), Sec. 6.12, et seq.

relations (1) and (3), we find for the expansion coefficients in the expansion (4):

$$a_n = \frac{2}{\{J_\nu'(\alpha_n)\}^2} \int_0^1 tf(t) J_\nu(\alpha_n t) dt; \tag{5}$$

or, by Eq. (10) of Sec. 7.2, noting that $J_\nu(\alpha_n) = 0$, we have

$$a_n = \frac{2}{\{J_{\nu+1}(\alpha_n)\}^2} \int_0^1 tf(t) J_\nu(\alpha_n t) dt . \tag{6}$$

Of course, this deduction is only heuristic only. The rigorous theorem is:

Let $f(x)$ be a function defined on the interval $(0,1)$ and assume that $\int_0^1 t^{1/2}$ $\times f(t) dt$ exists; if the integral is an improper one, we assume that it is absolutely convergent. Let x be any point in the interval $(a,b), 0 < a < b < 1$, and $f(x)$ is of bounded variation in (a,b), then

$$\sum_m a_m J_\nu(\alpha_m x) = \frac{1}{2}\{f(x+0) + f(x-0)\} , \tag{7}$$

where a_m are given by (5) or (6), $\nu + \frac{1}{2} \geq 0; \alpha_m$ are the positive zeros of $J_\nu(x), \alpha_m \leq \alpha_{m+1}$. And if $f(x)$ is continuous in (a,b), then the series is uniformly convergent in $a + \Delta \leq x \leq b - \Delta (\Delta > 0)$ with its sum equal to $f(x)$. Regarding the proof of this fundamental expansion theorem, cf. Watson (1944), Secs. 18.24-18.25; or, Titchmarsh, Eigenfunction Expansions Associated with Second-Order Differential Equations, Sec. 4.9 (1946). The latter discusses the problem from the view of eigenfunction expansions.

Exercise 7

1. Show that

$$\frac{d^r}{dz^r} J_n(z) = \frac{1}{2^r} \sum_{k=0}^r (-)^k \binom{r}{k} J_{n-r+2k}(z) .$$

2. Show that

$$J_0\left(\sqrt{z^2 - t^2}\right) = \frac{1}{\pi} \int_0^\pi e^{t\cos\theta} \cos(z \sin\theta) d\theta .$$

[Hint: Write the integrand as $\exp\{t\cos\theta + iz\sin\theta\}$, expand in a series of $\sin\theta$ and $\cos\theta$, then integrate term by term.]

3. Show that

$$\cos(z \sin \theta) = J_0(z) + 2 \sum_{n=1}^{\infty} J_{2n}(z) \cos 2n\theta \ ,$$

$$\sin(z \sin \theta) = 2 \sum_{n=0}^{\infty} J_{2n+1}(z) \sin(2n+1)\theta \ ,$$

and thereupon, show that

$$z \sin z = 2 \left\{ 2^2 J_2(z) - 4^2 J_4(z) + 6^2 J_6(z) - \dots \right\} \ ,$$
$$z \cos z = 2 \left\{ 1^2 J_1(z) - 3^2 J_3(z) + 5^2 J_5(z) - \dots \right\} \ .$$

4. From the formula

$$\cos 2n\theta = \sum_{m=0}^{n} (-)^m \frac{2^{2m} n^2 [n^2 - 1] \dots [n^2 - (m-1)^2]}{(2m)!} \sin^{2m} \theta \ ,$$

which can be obtained from Eq. (6) of Sec. 4.11, prove by using Eq. (7) of Sec. 7.5 that

$$J_{2n}(z) = (-)^n \sum_{m=0}^{n} (-)^m \frac{2^m n^2 [n^2 - 1] \dots [n^2 - (m-1)^2]}{m!} \frac{J_m(z)}{z^m} \ .$$

5. Prove the *Jacobi Transformation Formula*:

$$\frac{d^{n-1}}{d\mu^{n-1}} \sin^{2n-1} \theta = (-)^{n-1} \frac{1 \cdot 3 \cdot 5 \dots (2n-1)}{n} \sin n\theta \quad (\mu = \cos \theta) \ .$$

[Hint: Write $\sin^{2n-1} \theta$ as $(1 - \mu)^{n-\frac{1}{2}} (1 + \mu)^{n-\frac{1}{2}}$.]
With this formula, derive the Bessel integral representation [Eq. (20) of Sec. 7.5] from the Poisson integral representation of $J_n(z)$ [Eq. (19) of Sec. 7.5].

6. Show that

$$J_n^2(z) = \frac{1}{\pi} \int_0^{\pi} J_{2n}(2z \cos \theta) d\theta = \frac{2}{\pi} \int_0^{\pi/2} J_{2n}(2z \sin \theta) d\theta$$
$$= \frac{1}{\pi} \int_0^{\pi} J_{2n}(2z \sin \theta) d\theta = \frac{1}{\pi} \int_0^{\pi} J_0(2z \sin \theta) \cos 2n\theta d\theta \ .$$

[Cf. Watson (1944), Sec. 2.6, p. 31; and Ex. 31 below.]

7. Using the result of the last problem, show that

$$J_n^2(z) = \sum_{m=0}^{\infty} \frac{(-)^m (2n+2m)!}{m!(2n+m)![(n+m)!]^2} \left(\frac{z}{2}\right)^{2n+2m} .$$

8. From Eq. (15) of Sec. 7.5, replacing z by $2z \sin \theta, m$ by $2m$, integrating with respect to θ from 0 to π, then using the result of Ex. 6, show that

$$\left(\frac{z}{2}\right)^{2m} = \frac{(m!)^2}{(2m)!} \sum_{n=0}^{\infty} \frac{(2m+n-1)!(2m+2n)}{n!} J_{m+n}^2(z)$$

and thereupon, prove also that

$$\left(\frac{z}{2}\right)^{2m-1} = \frac{m!(m-1)!}{(2m-1)!} \sum_{n=0}^{\infty} \frac{(2m+n-2)!(2m+2n-1)}{n!}$$
$$\times J_{m+n-1}(z) J_{m+n}(z) .$$

9. Show that

$$J_{\nu+n}(z) = \frac{(-i)^n \Gamma(2\nu) n! (z/2)^{\nu}}{\Gamma\left(\nu+\frac{1}{2}\right) \Gamma\left(\frac{1}{2}\right) \Gamma(2\nu+n)} \int_0^{\pi} e^{iz \cos \theta} \sin^{2\nu} \theta C_n^{\nu}(\cos \theta) d\theta ,$$

where $C_n^{\nu}(x)$ are the Gegenbauer polynomials [Sec. 5.23], n is an arbitrary non-negative integer, $\mathrm{Re}\left(\nu+\frac{1}{2}\right) > 0$. [Hint: Using Eq. (10) of Sec. 5.23 and Eq. (7) of Sec. 4.10 to obtain the derivative expression for $C_n^{\nu}(x)$, substitute into the r.h.s. and integrate.]
Equation (6) of Sec. 7.4 is a special case $(n = 0)$ of the above formula. When $\nu = \frac{1}{2}$, there is the formula

$$J_{n+\frac{1}{2}}(z) = (-i)^n \left(\frac{z}{2\pi}\right)^{1/2} \int_0^{\pi} e^{iz \cos \theta} P_n(\cos \theta) \sin \theta d\theta ,$$

where $P_n(\cos \theta)$ is the Legendre polynomial.

10. Show that

$$J_{\nu}(\varpi) = \frac{1}{\pi \Gamma(\nu)} \left(\frac{\varpi}{2}\right)^{\nu} \int_0^{\pi} \int_0^{\pi} \exp \{iz \cos \theta - iz_1 (\cos \theta \cos \theta_1$$
$$+ \sin \theta \sin \theta_1 \cos \varphi)\} \sin^{2\nu-1} \varphi \sin^{2\nu} \theta d\varphi d\theta ,$$

where $\varpi^2 = z^2 + z_1^2 - 2zz_1 \cos \theta_1$; z, z_1, θ_1 are any complex numbers. [Hint: Use the method of rotating of coordinate axes described in Sec. 7.14; cf. Watson (1944), p. 51.]

11. Show that

$$H_{-\nu}^{(1)}(z) = e^{\nu\pi i}H_{\nu}^{(1)}(z) \ , \quad H_{-\nu}^{(2)}(z) = e^{-\nu\pi i}H_{\nu}^{(2)}(z) \ .$$

12. Show that

$$Y_{\nu}(ze^{m\pi i}) = e^{-m\nu\pi i}Y_{\nu}(z) + 2i\sin m\nu\pi \cot \nu\pi J_{\nu}(z) \ ,$$

$$Y_{-\nu}(ze^{m\pi i}) = e^{-m\nu\pi i}Y_{-\nu}(z) + 2i\sin m\nu\pi \csc \nu\pi J_{\nu}(z) \ ,$$

$$H_{\nu}^{(1)}(ze^{m\pi i}) = e^{-m\nu\pi i}H_{\nu}^{(1)}(z) - 2e^{-\nu\pi i}\sin m\nu\pi \csc \nu\pi J_{\nu}(z)$$

$$= \sin(1-m)\nu\pi \csc \nu\pi H_{\nu}^{(1)}(z)$$

$$- e^{-\nu\pi i}\sin m\nu\pi \csc \nu\pi H_{\nu}^{(2)}(z) \ ,$$

$$H_{\nu}^{(2)}(ze^{m\pi i}) = e^{-m\nu\pi i}H_{\nu}^{(2)}(z) + 2e^{\nu\pi i}\sin m\nu\pi \csc \nu\pi J_{\nu}(z)$$

$$= \sin(1+m)\nu\pi \csc \nu\pi H_{\nu}^{(2)}(z)$$

$$+ e^{\nu\pi i}\sin m\nu\pi \csc \nu\pi H_{\nu}^{(1)}(z) \ .$$

13. Show that

$$I_{\nu}(ze^{m\pi i}) = e^{m\nu\pi i}I_{\nu}(z) \ ,$$

$$K_{\nu}(ze^{m\pi i}) = e^{-m\nu\pi i}K_{\nu}(z) - \pi i\sin m\nu\pi \csc \nu\pi I_{\nu}(z) \ .$$

14. Show that

$$I_{\nu-1} - I_{\nu+1} = \frac{2\nu}{z}I_{\nu} \ , \qquad\qquad I_{\nu-1} + I_{\nu+1} = 2I_{\nu}' \ ,$$

$$K_{\nu-1} - K_{\nu+1} = -\frac{2\nu}{z}K_{\nu} \ , \qquad K_{\nu-1} + K_{\nu+1} = -2K_{\nu}' \ ,$$

$$\left(\frac{d}{zdz}\right)^{m}(z^{\nu}I_{\nu}) = z^{\nu-m}I_{\nu-m} \ , \qquad \left(\frac{d}{zdz}\right)^{m}(z^{-\nu}I_{\nu}) = z^{-\nu-m}I_{\nu+m} \ ,$$

$$\left(\frac{d}{zdz}\right)^{m}(z^{\nu}K_{\nu}) = (-)^{m}z^{\nu-m}K_{\nu-m} \ , \quad \left(\frac{d}{zdz}\right)^{m}(z^{-\nu}K_{\nu}) = (-)^{m}z^{-\nu-m}K_{\nu+m} \ ,$$

$$I_{0}' = I_{1}, \quad I_{-n} = I_{n} \ ,$$

$$K_{0}' = -K_{1}, \quad K_{-\nu} = K_{\nu} \ .$$

15. By writing the equation satisfied by the cylinder function $Z_{\nu}(z)$ [end of Sec. 7.6] in the form $(zZ_{\nu}')' + (z - \nu^2 z^{-1})Z_{\nu} = 0$, prove the following reduction formula:

$$\int^{z} z^{\mu+1}Z_{\nu}(z)dz$$

$$= (\nu^2 - \mu^2)\int^{z} z^{\mu-1}Z_{\nu}(z)dz + \left[z^{\mu+1}Z_{\nu+1}(z) - (\nu - \mu)z^{\mu}Z_{\nu}(z)\right] \ .$$

16. $Z_\mu(z)$ and $\overline{Z}_\nu(z)$ are respectively cylinder functions of order μ and ν; show that

$$\int^z \left\{ (k^2 - l^2)z - \frac{\mu^2 - \nu^2}{z} \right\} Z_\mu(kz)\overline{Z}_\nu(lz)dz$$
$$= z\left\{ kZ_{\mu+1}(kz)\overline{Z}_\nu(lz) - lZ_\mu(kz)\overline{Z}_{\nu+1}(lz) \right\} - (\mu - \nu)Z_\mu(kz)\overline{Z}_\nu(lz) \ .$$

17. Using the result of the last problem, show that

$$\int^z zZ_\mu(kz)\overline{Z}_\mu(kz)dz$$
$$= -\frac{z}{2k}\left\{ kzZ_{\mu+1}(kz)\overline{Z}'_\mu(kz) - kzZ_\mu(kz)\overline{Z}_{\mu+1}(kz) - Z_\mu(kz)\overline{Z}_{\mu+1}(kz) \right\}$$
$$= \frac{z^2}{4}\left\{ 2Z_\mu(kz)\overline{Z}_\mu(kz) - Z_{\mu-1}(kz)\overline{Z}_{\mu+1}(kz) - Z_{\mu+1}(kz)\overline{Z}_{\mu-1}(kz) \right\} \ ,$$
$$\int^z Z_\mu(kz)\overline{Z}_\mu(kz)\frac{dz}{z}$$
$$= \frac{kz}{2\mu}\left\{ Z_{\mu+1}(kz)\frac{\partial}{\partial\mu}\overline{Z}_\mu(kz) - Z_\mu(kz)\frac{\partial}{\partial\mu}\overline{Z}_{\mu+1}(kz) \right\} + \frac{Z_\mu(kz)\overline{Z}_\mu(kz)}{2\mu}$$

18. Consider the derivatives $d\{z^\rho Z_\mu(z)\overline{Z}_\nu(z)\}/dz$ and $d\{z^\rho Z_{\mu+1}(z) \times \overline{Z}_{\mu+1}(z)\}/dz$. Show that

$$(\rho + \mu + \nu)\int^z z^{\rho-1}Z_\mu(z)\overline{Z}_\nu(z)dz + (\rho - \mu - \nu - 2)$$
$$\times \int^z z^{\rho-1}Z_{\mu+1}(z)\overline{Z}_{\nu+1}(z)dz$$
$$= z^\rho\{Z_\mu(z)\overline{Z}_\nu(z) + Z_{\mu+1}(z)\overline{Z}_{\nu+1}(z)\} \ .$$

19. With the result of the last problem, show that

$$\int^z z^{-\mu-\nu-1}Z_{\mu+1}(z)\overline{Z}_{\nu+1}(z)dz$$

$$= -\frac{z^{-\mu-\nu}}{2(\mu+\nu+1)}\left\{Z_\mu(z)\overline{Z}_\nu(z) + Z_{\mu+1}(z)\overline{Z}_{\nu+1}(z)\right\} ,$$

$$\int^z z^{\mu+\nu+1} Z_\mu(z)\overline{Z}_\nu(z)\,dz$$

$$= \frac{z^{\mu+\nu+2}}{2(\mu+\nu+1)}\left\{Z_\mu(z)\overline{Z}_\nu(z) + Z_{\mu+1}(z)\overline{Z}_{\nu+1}(z)\right\} ,$$

$$\int^z Z_n(z)\overline{Z}(z)\,dz/z$$

$$= -\frac{1}{2n}\left\{Z_0(z)\overline{Z}_0(z) + 2\sum_{m=1}^{n-1} Z_m(z)\overline{Z}_m(z) + Z_n(z)\overline{Z}_n(z)\right\}$$

$$(n = 1, 2, \dots) .$$

20. Assuming that μ is not a negative integer, show that

$$\left(\frac{z}{2}\right)^\mu = \sum_{n=0}^{\infty} \frac{(\mu+2n)\Gamma(\mu+n)}{n!} J_{\mu+2n}(z) ,$$

which is a generalization of Eq. (15) of Sec. 7.5. [Hint: Multiplying both sides with $(z/2)^{-\mu}$, show that the derivative of the series on the r.h.s. is equal to zero so that the series is a constant, then show that the constant is equal to 1.]

21. Show that

$$\left(\frac{z}{2}\right)^{\mu-\nu} J_\nu(z)$$

$$= \Gamma(\nu+1-\mu)\sum_{n=0}^{\infty} \frac{(\mu+2n)\Gamma(\mu+n)}{n!\Gamma(\nu+1-\mu-n)\Gamma(\nu+n+1)} J_{\mu+2n}(z) ,$$

where $\mu, \nu, \nu-\mu$ are not negative integers. [Hint: Using the result of the last problem expand each term of the power series expression of $(z/2)^{\mu-\nu} J_\nu(z)$ in terms of Bessel functions.]

22. Show that

$$\left(\frac{kz}{2}\right)^{\mu-\nu} J_\nu(kz)$$

$$= k^\mu \sum_{n=0}^{\infty} \frac{\Gamma(\mu+n)}{n!\Gamma(\nu+1)}\,{}_2F_1(\mu+n, -n; \nu+1; k^2)(\mu+2n)J_{\mu+2n}(z) .$$

23. Show that

$$(z+h)^{-\nu/2} J_\nu(\sqrt{z+h}) = \sum_{m=0}^{\infty} \frac{h^m}{m!} \frac{d^m}{dz^m} \{z^{-\nu/2} J_\nu(\sqrt{z})\}$$

$$= \sum_{m=0}^{\infty} \frac{(-h/2)^m}{m!} z^{-\frac{\nu \pm m}{2}} J_{\nu+m}(\sqrt{z}) ,$$

$$(z+h)^{\nu/2} J_\nu(\sqrt{z+h}) = \sum_{m=0}^{\infty} \frac{(h/2)^m}{m!} z^{\frac{\nu-m}{2}} J_{\nu-m}(\sqrt{z})$$

$$(|h| < |z|) .$$

24. Using the result of the last problem show that

$$\sqrt{\frac{2}{\pi z}} \cos \sqrt{z^2 - 2zt} = \sum_{m=0}^{\infty} \frac{t^m}{m!} J_{m-\frac{1}{2}}(z) ,$$

$$\sqrt{\frac{2}{\pi z}} \sin \sqrt{z^2 + 2zt} = \sum_{m=0}^{\infty} \frac{t^m}{m!} J_{\frac{1}{2}-m}(z) , \qquad \left(|t| < \frac{1}{2}|z| \right) .$$

The functions on the l.h.s. of these formulae may be regarded as the *generating functions of Bessel functions of order half an odd integer.*

25. From the result of Ex. 23, show that

$$J_\nu \left\{ z\sqrt{1+k} \right\} = (1+k)^{\nu/2} \sum_{m=0}^{\infty} \frac{(-)^m}{m!} \left(\frac{kz}{2} \right)^m J_{\nu+m}(z) ,$$

$$J_\nu \left\{ z\sqrt{1+k} \right\} = (1+k)^{-\nu/2} \sum_{m=0}^{\infty} \frac{1}{m!} \left(\frac{kz}{2} \right)^m J_{\nu-m}(z) \quad (|k| < 1) .$$

26. Applying the result of the last problem and the similar formulae for $Y_\nu \left(z\sqrt{1+k} \right)$, show that

$$Z_\nu(\lambda z) = \lambda^\nu \sum_{m=0}^{\infty} \frac{(-)^m (\lambda^2 - 1)^m}{m!} \left(\frac{z}{2} \right)^m Z_{\nu+m}(z) \quad (|\lambda^2 - 1| < 1) ,$$

where $Z_\nu(z)$ is any cylinder function [the end of Sec. 7.6].

27. Multiplying both sides of the first formula in Ex. 25 with $(1+k)^{-\nu/2}$, then letting $k \to -1$, we have

$$\frac{(z/2)^\nu}{\Gamma(\nu+1)} = \sum_{m=0}^{\infty} \frac{(z/2)^m}{m!} J_{\nu+m}(z) .$$

Prove, thereupon, that

$$J_\nu(z) = \frac{\Gamma(\mu+1)}{\Gamma(\nu-\mu)} \sum_{m=0}^{\infty} \frac{\Gamma(\nu-\mu+m)}{m! \Gamma(\nu+m+1)} \left(\frac{z}{2} \right)^{\nu-\mu+m} J_{\mu+m}(z) ,$$

where $\mu \neq \nu, \mu \neq$ negative integer.

28. Prove the following addition formula:

$$Z_\nu(z + t) = \sum_{m=-\infty}^{\infty} Z_{\nu-m}(t) J_m(z), \quad |z| < |t| ,$$

where $Z_\nu(z)$ is a cylinder function of arbitrary order ν. This formula is a generalization of Eq. (16) of Sec. 7.5. [Cf. Watson (1944), Sec. 5.3, p. 143].

29. Show that

$$J_\mu(az) J_\nu(bz) = \frac{1}{\Gamma(\nu + 1)} \left(\frac{az}{2}\right)^\mu \left(\frac{bz}{2}\right)^\nu$$

$$\times \sum_{m=0}^{\infty} \frac{(-)^m {}_2F_1(-m, -\mu - m; \nu + 1; b^2/a^2)}{m! \Gamma(\mu + m + 1)} \left(\frac{az}{2}\right)^{2m} .$$

30. By the result of the last problem, show that

$$e^{z \cos \theta} J_{\nu - \frac{1}{2}}(z \sin \theta) = \frac{\Gamma(\nu)}{\Gamma\left(\frac{1}{2}\right)} (2 \sin \theta)^{\nu - \frac{1}{2}} \sum_{n=0}^{\infty} \frac{z^{\nu + n - \frac{1}{2}}}{\Gamma(2\nu + n)} C_n^\nu(\cos \theta) ,$$

where $C_n^\nu(x)$ are Gegenbauer polynomials [Sec. 5.23]. [Hint: In the formula of the last problem, put $\mu = \pm \frac{1}{2}$ respectively.]

31. In Ex. 29, let $a = b = 1$, then apply the result of Ex. 16 at the end of Chap. 3. Show that

$$J_\mu(z) J_\nu(z) = \frac{2}{\pi} \int_0^{\frac{\pi}{2}} J_{\mu+\nu}(2z \cos \theta) \cos(\mu - \nu)\theta d\theta ,$$

where $\operatorname{Re}(\mu + \nu) > -1$. This is called the *Neumann integral formula*.

32. Show that

$$\left(\frac{z}{2}\right)^{\mu+\nu} = \frac{\Gamma(\mu + 1)\Gamma(\nu + 1)}{\Gamma(\mu + \nu + 1)}$$

$$\times \sum_{m=0}^{\infty} \frac{(\mu + \nu + 2m)\Gamma(\mu + \nu + m)}{m!} J_{\mu+m}(z) J_{\nu+m}(z) .$$

[Hint: Expand $(z \cos \theta)^{\mu+\nu}$ by the formula of Ex. 20, multiply with $\cos(\mu - \nu)\theta$, integrate, then apply the result of the last problem and the formula of Ex. 16, Chap. 3.]

33. Show that

$$\frac{z^2}{4} \left\{ J_{\nu-1}^2(z) - J_{\nu-2}(z) J_\nu(z) \right\} = \sum_{n=0}^{\infty} (\nu + 2n) J_{\nu+2n}^2(z) .$$

[Hint: Differentiating the series on the r.h.s. with respect to z, show that

$$\frac{1}{2} \int^z z J_{\nu-1}^2(z) dz = \sum_{n=0}^{\infty} (\nu + 2n) J_{\nu+2n}^2(z) ,$$

then apply the first formula of Ex. 17 to the integral on the l.h.s.]

34. Show that

$$J_\nu(z) = \lim_{\substack{\lambda \to \infty \\ \mu \to \infty}} \frac{(z/2)^\nu}{\Gamma(\nu+1)} {}_2F_1\left(\lambda, \mu; \nu+1; -\frac{z^2}{4\lambda\mu}\right) .$$

[Cf. Watson (1944), Sec. 5.7, p. 154.]

35. Show that

$$\lim_{n \to \infty} P_n\left(\cos\frac{\theta}{n}\right) = J_0(\theta) ,$$

$P_n(x)$ being the Legendre polynomial. [Cf. Watson (1944), Sec. 5.71, p. 155.]

36. Show that

$$J_\nu(z) = \frac{\Gamma\left(\frac{1}{2} - \nu\right)(z/2)^\nu}{\pi i \Gamma\left(\frac{1}{2}\right)} \int_0^{(1+)} (t^2 - 1)^{\nu-\frac{1}{2}} \cos(zt) dt ,$$

where it is stipulated that $\arg(t^2 - 1) = 0$ when t lies on the real axis to the right of $t = 1$.

37. Assuming $|\mathrm{Re}(\nu)| < \frac{1}{2}$, show that

$$H_\nu^{(1)}(z) = \frac{2}{i\Gamma\left(\frac{1}{2} - \nu\right)\Gamma\left(\frac{1}{2}\right)\left(\frac{z}{2}\right)^\nu} \int_1^\infty \frac{e^{izt} dt}{(t^2 - 1)^{\nu+\frac{1}{2}}}$$
$$(\mathrm{Im}(z) \geq 0) ,$$

$$H_\nu^{(2)}(z) = -\frac{2}{i\Gamma\left(\frac{1}{2} - \nu\right)\Gamma\left(\frac{1}{2}\right)\left(\frac{z}{2}\right)^\nu} \int_1^\infty \frac{e^{-izt} dt}{(t^2 - 1)^{\nu+\frac{1}{2}}}$$
$$(\mathrm{Im}(z) \leq 0) .$$

38. Show that

$$K_\nu(\alpha x) = \frac{\Gamma\left(\nu + \frac{1}{2}\right)(2\alpha)^\nu}{2\Gamma\left(\frac{1}{2}\right) x^\nu} \int_{-\infty}^\infty \frac{e^{-ixt} dt}{(t^2 + \alpha^2)^{\nu+\frac{1}{2}}} ,$$

where $\mathrm{Re}\left(\nu + \frac{1}{2}\right) > 0, x > 0, |\arg\alpha| < \pi/2.$

39. Show that

$$J_\nu(z) = \frac{e^{-\frac{1}{2}\left(\nu+\frac{1}{2}\right)\pi i}}{\pi\Gamma\left(\frac{1}{2}\right)} \left(\frac{z}{2}\right)^{\frac{1}{2}} \int_{\infty i\,\exp(-i\omega)}^{(-1+,1+)} e^{izt} Q_{\nu-\frac{1}{2}}(t)\,dt \ ,$$

where $Q_{\nu-\frac{1}{2}}(t)$ is the Legendre function of the second kind [Eq. (5) of Sec. 5.17], $-\pi/2 + \omega < \arg z < \pi/2 + \omega$, $|\omega| < \pi/2$, and it is stipulated that $\arg t = 0$ when t lies on the real axis to the right of $t = 1$. [Hint: Take a contour lying entirely outside the circle $|t| = 1$ and expand $Q_{\nu-\frac{1}{2}}(t)$ in descending powers of t.]

40. Using the result of the last problem and formula (i) of Ex. 36, Chap. 5, show that

$$H_\nu^{(2)}(z) = \frac{e^{\frac{1}{2}\left(\nu+\frac{1}{2}\right)\pi i}}{\pi\Gamma\left(\frac{1}{2}\right)\cos\nu\pi} \left(\frac{z}{2}\right)^{\frac{1}{2}} \int_{\infty i\,\exp(-i\omega)}^{(-1+,1+)} e^{izt} P_{\nu-\frac{1}{2}}(t)\,dt \ ,$$

where $P_{\nu-\frac{1}{2}}(t)$ is the Legendre function of the first kind [Ex. 31, Chap. 5]. Since $t = 1$ is not a singularity of $P_{\nu-\frac{1}{2}}(t)$ [cf. Eq. (10) of Sec. 5.16], the integration contour needs only to run around $t = -1$ once in the positive sense; $\arg(t + 1) = 0$ at the intersecting point of the contour and the real axis to the right of $t = -1$.

41. From the result of Ex. 39, show that

$$H_\nu^{(1)}(z) = \frac{e^{-\frac{1}{2}\left(\nu+\frac{1}{2}\right)\pi i}}{\pi\Gamma\left(\frac{1}{2}\right)} (2z)^{\frac{1}{2}} \int_{\infty i\,\exp(-i\omega)}^{(1+)} e^{izt} Q_{\nu-\frac{1}{2}}(t)\,dt \ ,$$

$$H_\nu^{(2)}(z) = \frac{e^{-\frac{1}{2}\left(\nu+\frac{1}{2}\right)\pi i}}{\pi\Gamma\left(\frac{1}{2}\right)} (2z)^{\frac{1}{2}} \int_{\infty i\,\exp(-i\omega)}^{(-1+)} e^{izt} Q_{\nu-\frac{1}{2}}(t)\,dt \ ;$$

in the first formula, $\arg(t - 1) = \arg(t + 1) = 0$ at the point of intersection of the contour and the real axis to the right of $t = 1$, in the second formula, $\arg(t + 1) = 0$, $\arg(t - 1) = \pi$ at the point of intersection of the contour and the real axis in between $t = -1$ and $t = 1$.

42. From Eqs. (14) and (15) of Sec. 7.7 show, respectively, that

$$H_\nu^{(1)}(z) = \frac{1}{\pi i} \int_0^{\infty\,\exp\pi i} u^{-\nu-1} \exp\left\{\frac{z}{2}(u - u^{-1})\right\} du \ ,$$

$$H_\nu^{(2)}(z) = -\frac{1}{\pi i} \int_0^{\infty\,\exp(-\pi i)} u^{-\nu-1} \exp\left\{\frac{z}{2}(u - u^{-1})\right\} du \ ,$$

where $|\arg z| < \pi/2$; the contour of integration is that shown in Fig. 34; when $u \to 0$ along the contour, $\arg u \to 0$ to ensure the convergence of the integral at the lower limit.

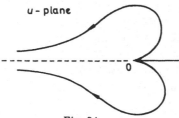

u - plane

Fig. 34.

If $|\arg z| > \pi/2$, assume $-\pi/2 + \omega < \arg z < \pi/2 + \omega, |\omega| < \pi$, then

$$H_\nu^{(1)}(z) = \frac{1}{\pi i} \int_{0 \exp i\omega}^{\infty \exp(\pi - \omega)i} u^{-\nu-1} \exp\left\{\frac{z}{2}(u - u^{-1})\right\} du ,$$

$$H_\nu^{(2)}(z) = -\frac{1}{\pi i} \int_{0 \exp i\omega}^{\infty \exp(-\pi - \omega)i} u^{-\nu-1} \exp\left\{\frac{z}{2}(u - u^{-1})\right\} du ;$$

the contours of integration are respectively that shown in Figs. 35a and 35b.

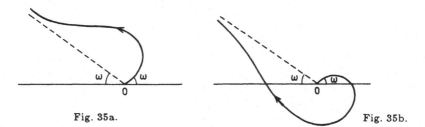

Fig. 35a.   Fig. 35b.

43. Assume that $x > 0, |\mathrm{Re}(\nu)| < 1$. With the result of the last Ex., show that

$$H_\nu^{(1)}(x) = \frac{e^{-\nu\pi i/2}}{\pi i} \int_{-\infty}^{\infty} e^{ix\,\mathrm{ch}\,t - \nu t} dt = \frac{2e^{-\nu\pi i/2}}{\pi i} \int_0^\infty e^{ix\,\mathrm{ch}\,t}\mathrm{ch}\,\nu t\, dt ,$$

$$H_\nu^{(2)}(x) = -\frac{e^{\nu\pi i/2}}{\pi i} \int_{-\infty}^{\infty} e^{-ix\,\mathrm{ch}\,t - \nu t} dt$$

$$= -\frac{2e^{\nu\pi i/2}}{\pi i} \int_0^\infty e^{-ix\,\mathrm{ch}\,t}\mathrm{ch}\,\nu t\, dt .$$

44. The *Airy Integral* is defined by

$$\int_0^\infty \cos(t^3 \pm xt)dt .$$

Consider a contour integral of the function $\exp\{it^3 \pm ixt\}$ in the t-plane; the contour starts from a point $t = -\rho(\rho > 0)$ on the real axis, running along the real axis to the point $t = \rho$, then to the point $\rho e^{\pi i/6}$ along a circular arc (with centre at $t = 0$ and radius equal to ρ), then to $t = 0$ along a straight line, then from $t = 0$ to $\rho e^{5\pi i/6}$ along a straight line, and finally returns to $-\rho$. Show that

$$\int_0^\infty \cos(t^3 \pm xt)dt = \frac{1}{2}\int_0^\infty \left\{ e^{\pi i/6}\exp(-\tau^3 \pm e^{2\pi i/3}x\tau) \right.$$
$$\left. + e^{-\pi i/6}\exp\left(-\tau^3 \pm e^{-2\pi i/3}x\tau\right) \right\}d\tau ,$$

and derive from it

$$\int_0^\infty \cos(t^3 - xt)dt = \frac{\pi}{3}\sqrt{\frac{x}{3}}\left\{ J_{-\frac{1}{3}}\left(\frac{2x\sqrt{x}}{3\sqrt{3}}\right) + J_{\frac{1}{3}}\left(\frac{2x\sqrt{x}}{3\sqrt{3}}\right) \right\} .$$
$$\int_0^\infty \cos(t^3 + xt)dt = \frac{\pi}{3}\sqrt{\frac{x}{3}}\left\{ I_{-\frac{1}{3}}\left(\frac{2x\sqrt{x}}{3\sqrt{3}}\right) - I_{\frac{1}{3}}\left(\frac{2x\sqrt{x}}{3\sqrt{3}}\right) \right\}$$
$$= \frac{\sqrt{x}}{3}K_{\frac{1}{3}}\left(\frac{2x\sqrt{x}}{3\sqrt{3}}\right) .$$

45. Show that

$$\frac{\mathrm{ber}(z)}{\mathrm{bei}(z)} = \frac{\exp\alpha(z)}{\sqrt{2\pi z}}\frac{\cos}{\sin}\beta(z) \quad (|\arg z| < \pi/4) ,$$

$$\frac{\mathrm{ker}(z)}{\mathrm{kei}(z)} = \frac{\exp\alpha(-z)}{\sqrt{2z/\pi}}\frac{\cos}{\sin}\beta(-z) \quad (|\arg z| < 5\pi/4) ,$$

where

$$\alpha(z) \sim \frac{z}{\sqrt{2}} + \frac{1}{8\sqrt{2}\,z} - \frac{25}{384\sqrt{2}\,z^3} - \frac{13}{128z^4} - \cdots ,$$
$$\beta(z) \sim \frac{z}{\sqrt{2}} - \frac{\pi}{8} - \frac{1}{8\sqrt{2}\,z} - \frac{1}{16z^2} - \frac{25}{384\sqrt{2}\,z^3} + \cdots .$$

46. Using the result of Ex. 29, show that

$$
\frac{\pi^3}{2 \sin \frac{1}{2}(\mu + \nu)\pi} \Bigg[\{J_\mu(z)J_\nu(z) + Y_\mu(z)Y_\nu(z)\}
$$

$$
- \cot \frac{1}{2}(\mu - \nu)\pi \{J_\mu(z)Y_\nu(z) - Y_\mu(z)J_\nu(z)\} \Bigg]
$$

$$
= \frac{1}{2\pi i} \int_{-\infty i}^{\infty i} \Gamma(2s+1)\Gamma\left(\frac{\mu+\nu}{2} - s\right) \Gamma\left(\frac{\mu-\nu}{2} - s\right)
$$

$$
\times \Gamma\left(\frac{\nu-\mu}{2} - s\right) \Gamma\left(-\frac{\mu+\nu}{2} - s\right) \cos s\pi \left(\frac{z}{2}\right)^{2s} ds
$$

$$
(\mu \pm \nu \neq \text{even numbers}) ,
$$

$$
\frac{\pi^3}{2 \cos \frac{1}{2}(\mu + \nu)\pi} \Bigg[\{J_\mu(z)J_\nu(z) + Y_\mu(z)Y_\nu(z)\}
$$

$$
+ \tan \frac{1}{2}(\mu - \nu)\pi \{J_\mu(z)Y_\nu(z) - Y_\mu(z)J_\nu(z)\} \Bigg]
$$

$$
= \frac{-1}{2\pi i} \int_{-\infty i}^{\infty i} \Gamma(2s+1)\Gamma\left(\frac{\mu+\nu}{2} - s\right) \Gamma\left(\frac{\mu-\nu}{2} - s\right)
$$

$$
\times \Gamma\left(\frac{\nu-\mu}{2} - s\right) \Gamma\left(-\frac{\mu+\nu}{2} - s\right) \sin s\pi \left(\frac{z}{2}\right)^{2s} ds
$$

$$
(\mu \pm \nu \neq \text{odd numbers}) ,
$$

where $|\arg z| < \pi$; $\Gamma(2s + 1)$ has its poles on the left of the integration contour, and the poles of the other Γ-functions lie on the right of the contour.

47. Using the result of the last problem, show that

$$
J_\mu(z)J_\nu(z) + Y_\mu(z)Y_\nu(z) - \cot \frac{1}{2}(\mu - \nu)\pi
$$

$$
\times \{J_\mu(z)Y_\nu(z) - J_\nu(z)Y_\mu(z)\} \sim \frac{\mu^2 - \nu^2}{\pi z^2 \sin \frac{1}{2}(\mu - \nu)\pi}
$$

$$
\times {}_4F_1\left(\frac{\mu+\nu}{2} + 1, \frac{\mu-\nu}{2} + 1 , \frac{\nu-\mu}{2} + 1, 1 - \frac{\mu+\nu}{2}; \frac{z}{2}; -\frac{1}{z^2}\right) ,
$$

$$J_\mu(z)J_\nu(z) + Y_\mu(z)Y_\nu(z) + \tan\frac{1}{2}(\mu-\nu)\pi$$

$$\times\{J_\mu(z)Y_\nu(z) - J_\nu(z)Y_\mu(z)\} \sim \frac{2}{\pi z \cos\frac{1}{2}(\mu-\nu)\pi}$$

$$\times\,_4F_1\left(\frac{\mu+\nu+1}{2},\ \frac{\mu-\nu+1}{2},\ \frac{\nu-\mu+1}{2},\ \frac{1-\mu-\nu}{2};\ \frac{1}{2};\ -\frac{1}{z^2}\right);$$

and,

$$J_\nu^2(z) + Y_\nu^2(z) \sim \frac{2}{\pi z}\sum_{m=0}^{\infty}\frac{(2m)!(\nu,m)}{2^{2m}m!z^{2m}}$$

when $\mu = \nu$.

48. Show that, when $\nu \to \infty$

$$Y_\nu(\nu\operatorname{sech}\alpha) \sim -\frac{e^{\nu(\alpha-\operatorname{th}\alpha)}}{\sqrt{\frac{1}{2}\nu\pi\operatorname{th}\alpha}}\sum_{m=0}^{\infty}\frac{\Gamma\left(m+\frac{1}{2}\right)}{\Gamma\left(\frac{1}{2}\right)}\frac{(-)^m D_m}{(\nu\operatorname{th}\alpha)^m},$$

where D_m are the coefficients in Eq. (18) of Sec. 7.12.

49. Show that, when $|z| \to \infty$, $|\nu| \to \infty$ while $|z| \approx |\nu|$,

$$H_\nu^{(1)}(z) \sim -\frac{2}{3\pi}\sum_{m=0}^{\infty}e^{\frac{2}{3}(m+1)\pi i}B_m(\varepsilon z)\sin\frac{1}{3}(m+1)\pi\frac{\Gamma\left(\frac{m+1}{3}\right)}{(z/6)^{\frac{1}{3}(m+1)}},$$

$$H_\nu^{(2)}(z) \sim -\frac{2}{3\pi}\sum_{m=0}^{\infty}e^{-\frac{2}{3}(m+1)\pi i}B_m(\varepsilon z)\sin\frac{1}{3}(m+1)\pi\frac{\Gamma\left(\frac{m+1}{3}\right)}{(z/6)^{\frac{1}{3}(m+1)}},$$

where $|\arg z| < \pi, \varepsilon z = z - \nu, B_0(\varepsilon z) = 1, B_1(\varepsilon z) = \varepsilon z, B_2(\varepsilon z) = \frac{1}{2}\varepsilon^2 z^2 - \frac{1}{20}, B_3(\varepsilon z) = \frac{1}{6}\varepsilon^3 z^3 - \frac{1}{15}\varepsilon z, B_4(\varepsilon z) = \frac{1}{24}\varepsilon^4 z^4 - \frac{1}{24}\varepsilon^2 z^2 + \frac{1}{280}, B_5(\varepsilon z) = \frac{1}{120}\varepsilon^5 z^5 - \frac{1}{60}\varepsilon^3 z^3 + \frac{43}{8400}\varepsilon z, \dots$.

50. By Eq. (4) of Sec. 7.13, show that

$$Z_\nu(\varpi)\frac{\cos}{\sin}\nu\psi = \sum_{m=-\infty}^{\infty}Z_{\nu+m}(x)J_m(y)\frac{\cos}{\sin}m\theta,$$

where Z_ν is the cylinder function of order ν [end of Sec. 7.6]; the definition of ψ is given by

$$\varpi\cos\psi = x - y\cos\theta, \quad \varpi\sin\psi = y\sin\theta$$

and $\psi \to 0$ when $y \to 0$.

51. In the first formula of Ex. 23, let $z = x^2 + y^2, h = -2xy\cos\theta$. We have

$$\frac{J_\nu(\varpi)}{\varpi^\nu} = \sum_{p=0}^{\infty} \frac{(xy\cos\theta)^p}{p!} \frac{J_{\nu+p}\left\{\sqrt{x^2+y^2}\right\}}{(x^2+y^2)^{\frac{1}{2}(\nu+p)}} .$$

Applying again the first formula of Ex. 23 to the last fractional factor in the sum, prove that

$$\frac{J_\nu(\varpi)}{\varpi^\nu} = 2^\nu \Gamma(\nu) \sum_{m=0}^{\infty} (\nu+m) \frac{J_{\nu+m}(x)}{x^\nu} \frac{J_{\nu+m}(y)}{y^\nu} C_m^\nu(\cos\theta)$$

$$(\nu \neq 0, -1, -2, \dots) ,$$

which is the addition formula (7) in Sec. 7.13.

Similarly, using the second formula in Ex. 23, prove that

$$\frac{J_{-\nu}(\varpi)}{\varpi^\nu} = 2^\nu \Gamma(\nu) \sum_{m=0}^{\infty} (-)^m (\nu+m) \frac{J_{-\nu-m}(x)}{x^\nu} \frac{J_{\nu+m}(y)}{y^\nu} C_m^\nu(\cos\theta) ,$$

$$(|ye^{\pm i\theta}| < |x|; \quad \nu \neq 0, -1, -2, \dots) ,$$

which is formula (8) in Sec. 7.13.

52. Let $R = \{r^2 + a^2 - 2ar\cos\theta\}^{1/2}$. Show that

$$\frac{\sin kR}{R} = \pi \sum_{m=0}^{\infty} \left(m + \frac{1}{2}\right) \frac{J_{m+\frac{1}{2}}(ka)}{\sqrt{a}} \frac{J_{m+\frac{1}{2}}(kr)}{\sqrt{r}} P_m(\cos\theta) ,$$

$$\frac{\cos kR}{R} = \pi \sum_{m=0}^{\infty} (-)^m \left(m + \frac{1}{2}\right) \frac{J_{-m-\frac{1}{2}}(ka)}{\sqrt{a}} \frac{J_{m+\frac{1}{2}}(kr)}{\sqrt{r}} P_m(\cos\theta) ,$$

$$\frac{e^{-kR}}{R} = \sum_{m=0}^{\infty} (2m+1) \frac{K_{m+\frac{1}{2}}(ka)}{\sqrt{a}} \frac{I_{m+\frac{1}{2}}(kr)}{\sqrt{r}} P_m(\cos\theta) .$$

53. By using the result of Ex. 44 of Chap. 5, show that

$$\int_0^\pi \exp\{iz(\cos\theta\cos\theta' + \sin\theta\sin\theta'\cos\varphi)\} (\sin\varphi)^{2\nu-1} d\varphi$$

$$= 2^{3\nu-1} [\Gamma(\nu)]^3 \sum_{m=0}^{\infty} \frac{i^m m!(\nu+m)}{\Gamma(2\nu+m)} \frac{J_{\nu+m}(z)}{z^\nu} C_m^\nu(\cos\theta) C_m^\nu(\cos\theta') .$$

54. Prove the *Bateman expansion formula:*

$$\frac{z}{2} J_\mu(z\cos\theta\cos\varphi) J_\nu(z\sin\theta\sin\varphi)$$

$$= \cos^\mu\theta\cos^\mu\varphi\sin^\nu\theta\sin^\nu\varphi$$

$$\times \sum_{n=0}^\infty (-)^n (\mu+\nu+2n+1) J_{\mu+\nu+2n+1}(z)$$

$$\times \frac{\Gamma(\mu+\nu+n+1)\Gamma(\nu+n+1)}{n!\,\Gamma(\mu+n+1)\{\Gamma(\nu+1)\}^2}$$

$$\times\ {}_2F_1\left(-n,\mu+\nu+n+1;\nu+1;\sin^2\theta\right)$$

$$\times\ {}_2F_1\left(-n,\mu+\nu+n+1,\nu+1;\sin^2\varphi\right)$$

$$(\mu,\nu \neq \text{negative integers})$$

[cf. Watson (1944), Sec. 11.6, p. 370].

55. Show that

$$\int_0^{\frac{\pi}{2}} J_\mu(z\sin\theta) I_\nu(z\cos\theta)(\tan\theta)^{\mu+1}\,d\theta = \frac{\Gamma\left(\frac{\nu-\mu}{2}\right)}{\Gamma\left(\frac{\nu+\mu}{2}+1\right)}\left(\frac{z}{2}\right)^\mu J_\nu(z)$$

$$(\operatorname{Re}(\nu) > \operatorname{Re}(\mu) > -1)\ .$$

[Hint: Expanding $I_\nu(z\cos\theta)$ in a power series, integrating term by term with formula (1) of Sec. 7.14, then use the result of Ex. 21.]

56. Show that

$$\int_0^\pi e^{iz\cos\theta\cos\varphi} J_{\nu-\frac{1}{2}}(z\sin\theta\sin\varphi) C_n^\nu(\cos\theta)(\sin\theta)^{\nu+\frac{1}{2}}\,d\theta$$

$$= \sqrt{\frac{2\pi}{z}}\, i^n (\sin\varphi)^{\nu-\frac{1}{2}} C_n^\nu(\cos\varphi) J_{\nu+n}(z)\ .$$

[Cf. Watson (1944), Sec. 12.14, p. 378.]

57. Using the Bateman formula in Ex. 54 and Eq. (13) of Sec. 4.10, show that

$$2\pi \int_0^{\frac{\pi}{2}} J_\mu(z\cos^2\theta) J_\nu(z\sin^2\theta)\sin\theta\cos\theta\,d\theta$$

$$= \int_0^z J_\mu(t) J_\nu(z-t)\,dt = 2\sum_{n=0}^\infty (-)^n J_{\mu+\nu+2n+1}(z)$$

$$(\operatorname{Re}(\mu), \operatorname{Re}(\nu) > -1)\ ,$$

and, thereupon, using the recurrence relations of the Bessel functions, prove the following formula:

$$\int_0^z J_\mu(t) J_\nu(z-t) \frac{dt}{t} = \frac{J_{\mu+\nu}(z)}{\mu}$$

$$(\text{Re}(\mu) > 0, \quad \text{Re}(\nu) > -1) .$$

58. By means of the first formula in the last problem and the result of Ex. 3, show that

$$\int_0^z J_\mu(t) J_{-\mu}(z-t) dt = \sin z \quad (|\text{Re}(\mu)| < 1) ,$$

$$\int_0^z J_\mu(t) J_{1-\mu}(z-t) dt = J_0(z) - \cos z \quad (-1 < \text{Re}(\mu) < 2) .$$

59. Show that

$$\int_0^z \cos(z-t) J_0(t) dt = z J_0(z) .$$

[Hint: Consider the differential equation satisfied by the integral, find its general solution, then determine the solution by its property when $z \to 0$.]

60. Show that

$$\int_0^z \sin(z-t) \frac{J_\mu(t)}{t} dt = \frac{2}{\mu} \sum_{n=0}^\infty (-)^n J_{\mu+2n+1}(z)$$

$$(\text{Re}(\mu) > 0) .$$

[Hint: Consider the differential equation satisfied by $v = \int_0^z J_0(z-t) \times J_\mu(t) dt$, find its general solution with the method of variation of parameters, then determine the solution by its property when $z \to 0$, together with the first formula of Ex. 57.]

61. Show that

$$\int_0^\infty \frac{J_\nu(bt) t^\nu dt}{e^{\pi t} - 1} = \frac{(2b)^\nu \Gamma\left(\nu + \frac{1}{2}\right)}{\sqrt{\pi}} \sum_{n=1}^\infty \frac{1}{(n^2\pi^2 + b^2)^{\nu+\frac{1}{2}}}$$

$$(\text{Re}(\nu) > 0, \quad |\text{Im}(b)| < \pi) .$$

62. Show that

$$\int_0^\infty e^{-t \, \text{ch} \, \alpha} I_\nu(t \, \text{sh} \, \alpha) t^\mu dt = \Gamma(\mu + \nu + 1) P_\mu^{-\nu}(\text{ch} \, \alpha)$$

$$(\text{Re}(\mu + \nu) > -1) ,$$

where $P_\mu^{-\nu}(z)$ is the associate Legendre function of the first kind [Sec. 5.16].

63. With the result of the last problem and the Whipple transformation given in Ex. 57, Chap. 5, show that

$$\int_0^\infty e^{-t\,\text{ch}\,\alpha} I_\nu(t) t^{\mu-1} dt = \frac{\cos\nu\pi}{\sin(\mu+\nu)\pi} \sqrt{\frac{2}{\pi}} \frac{Q_{\nu-\frac{1}{2}}^{\mu-\frac{1}{2}}(\text{ch}\,\alpha)}{(\text{sh}\,\alpha)^{\mu-\frac{1}{2}}} ,$$

where $\text{Re}(\mu+\nu) > 0, \text{Re}(\text{ch}\,\alpha) > 1$, $Q_{\nu-\frac{1}{2}}^{\mu-\frac{1}{2}}(z)$ is the associate Legendre function of the second kind [Sec. 5.17].

From the above formula, show that

$$\int_0^\infty e^{-t\,\text{ch}\,\alpha} K_\nu(t) t^{\mu-1} dt = \sqrt{\frac{\pi}{2}} \Gamma(\mu-\nu)\Gamma(\mu+\nu) \frac{P_{\nu-\frac{1}{2}}^{\frac{1}{2}-\mu}(\text{ch}\,\alpha)}{(\text{sh}\,\alpha)^{\mu-\frac{1}{2}}} ,$$

where $\text{Re}(\mu) > |\text{Re}(\nu)|, \text{Re}(\text{ch}\,\alpha) > -1$.

64. Show that

$$\int_0^\infty e^{-at} J_\nu(bt) J_\nu(ct) dt = \frac{1}{\pi\sqrt{bc}} Q_{\nu-\frac{1}{2}}\left(\frac{a^2+b^2+c^2}{2bc}\right) ,$$

where $\text{Re}(a \pm ib \pm ic)$ are all positive numbers, $\text{Re}(\mu+2\nu) > 0$.
[Hint: Use formula (3) in Sec. 7.14, putting it in $m = 0$.]

65. Show that

$$\int_0^\infty \frac{J_\nu(t) dt}{t^{\nu-\mu+1}} = \frac{\Gamma\left(\frac{\mu}{2}\right)}{2^{\nu-\mu+1}\Gamma\left(\nu-\frac{\mu}{2}+1\right)} ,$$

$$\left(0 < \text{Re}(\mu) < \text{Re}(\nu) + \frac{1}{2}\right) .$$

[Hint: Use the last formula of (2) in Sec. 7.15, letting $a \to 0$.]

66. Show that

$$\int_0^\infty e^{-p^2 t^2} J_\nu(at) t^{\mu-1} dt$$

$$= \frac{\Gamma\left(\frac{\mu+\nu}{2}\right)}{2p^\mu \Gamma(\nu+1)} \left(\frac{a}{2p}\right)^\nu {}_1F_1\left(\frac{\mu+\nu}{2}; \nu+1; -\frac{a^2}{4p^2}\right)$$

$$= \frac{\Gamma\left(\frac{\mu+\nu}{2}\right)}{2p^\mu \Gamma(\nu+1)} \left(\frac{a}{2p}\right)^\nu e^{-\frac{a^2}{4p^2}} {}_1F_1\left(\frac{\nu-\mu}{2}+1; \nu+1; \frac{a^2}{4p^2}\right)$$

$$(|\arg p| < \pi/4, \text{Re}(\mu+\nu) > 0) .$$

67. Show that

$$\int_0^\infty e^{-p^2 t^2} J_{2\nu}(at)\,dt = \frac{\sqrt{\pi}}{2p}\, e^{-\frac{a^2}{8p^2}}\, I_\nu\left(\frac{a^2}{8p^2}\right)$$

$$\left(|\arg p| < \pi/4,\ \text{Re}(\nu) > -\frac{1}{2}\right).$$

68. Show that

$$\int_0^\infty e^{-p^2 t^2} J_\nu(at)J_\nu(bt)t\,dt = \frac{1}{2p^2}\exp\left\{-\frac{a^2+b^2}{4p^2}\right\} I_\nu\left(\frac{ab}{2p^2}\right)$$

$$(\text{Re}(\nu) > -1, |\arg p| < \pi/4).$$

[Hint: Use Eq. (3) of Sec. 7.14, putting it in $m = 0$.]

69. Show that

$$\int_0^\infty e^{-p^2 t^2} J_\mu(at)J_\nu(at)t^{\lambda-1}\,dt = \frac{a^{\mu+\nu}}{2^{\mu+\nu}p^{\lambda+\mu+\nu}}\,\frac{\Gamma\left(\frac{\lambda+\mu+\nu}{2}\right)}{\Gamma(\mu+1)\Gamma(\nu+1)}$$

$$\times\,_3F_3\left(\frac{\mu+\nu+1}{2}\,;\,\frac{\mu+\nu+2}{2},\,\frac{\lambda+\mu+\nu}{2}\,;\right.$$

$$\left.\mu+1,\,\nu+1,\,\mu+\nu+1;\,-\frac{a^2}{p^2}\right)$$

$$(\text{Re}(\lambda+\mu+\nu) > 0,\ |\arg p| < \pi/4).$$

[Hint: Use the result of Ex. 29.]

70. Prove the following formulae of discontinuous integrals:

(i) $\displaystyle\int_0^\infty \frac{J_\mu(at)J_\mu(bt)}{t}\,dt = \begin{cases} \frac{1}{2\mu}\left(\frac{b}{a}\right)^\mu & (a \geq b) \\ \frac{1}{2\mu}\left(\frac{a}{b}\right)^\mu & (a \leq b) \end{cases}$ $(\text{Re}(\mu) > 0),$

(ii) $\displaystyle\int_0^\infty \frac{J_\mu(at)\sin bt}{t}\,dt = \begin{cases} \frac{1}{\mu}\sin\{\mu\,\text{arc sin}(b/a)\} & (a \geq b) \\ \frac{a^\mu \sin(\mu\pi/2)}{\mu\{b+\sqrt{b^2-a^2}\}^\mu} & (a \leq b) \end{cases}$

$(\text{Re}(\mu) > -1)$,

(iii) $\displaystyle\int_0^\infty \frac{J_\mu(at)\cos bt}{t}\,dt = \begin{cases} \frac{1}{\mu}\cos\{\mu\,\text{arc sin}(b/a)\} & (a \geq b) \\ \frac{a^\mu \cos(\mu\pi/2)}{\mu\{b+\sqrt{b^2-a^2}\}^\mu} & (a \leq b) \end{cases}$

$(\text{Re}(\mu) > 0)$,

(iv) $\displaystyle\int_0^\infty J_\mu(at)\sin bt \, dt = \begin{cases} \sin\{\mu\,\text{arc sin}(b/a)\}/\sqrt{a^2-b^2} & (a > b) \\ \frac{a^\mu \cos(\mu\pi/2)}{\sqrt{b^2-a^2}\{b+\sqrt{b^2-a^2}\}^\mu} & (a < b) \end{cases}$

$(\text{Re}(\mu) > -2)$,

(v) $\displaystyle\int_0^\infty J_\mu(at)\cos bt \, dt = \begin{cases} \cos\{\mu\,\text{arc sin}(b/a)\}/\sqrt{a^2-b^2} & (a > b) \\ -\frac{a^\mu \sin(\mu\pi/2)}{\sqrt{b^2-a^2}\{b+\sqrt{b^2-a^2}\}^\mu} & (a < b) \end{cases}$

$(\text{Re}(\mu) > -1)$.

71. Show that

$$\sum_{n=0}^\infty \varepsilon_n J_{\nu+n}^2(z) = 2\nu \int_0^z J_\nu^2(t)\frac{dt}{t} \quad (\text{Re}(\nu) > 0) .$$

Then, use (i) of the last Ex., when $\nu > 0, x > 0$, to prove

$$|J_\nu(x)| \leq 1 , \quad |J_{\nu+1}(x)| \leq \frac{1}{\sqrt{2}} .$$

72. Show that

$$\int_0^\infty \frac{K_\mu(at)J_\nu(bt)}{t^\lambda}\,dt = \frac{b^\nu \Gamma\left(\frac{\nu+\mu-\lambda+1}{2}\right)\Gamma\left(\frac{\nu-\mu-\lambda+1}{2}\right)}{2^{\lambda+1}a^{\nu-\lambda+1}\Gamma(\nu+1)}$$
$$\times\,{}_2F_1\left(\frac{\nu+\mu-\lambda+1}{2}, \frac{\nu-\mu-\lambda-1}{2}; \nu+1; -\frac{b^2}{a^2}\right)$$
$$(\text{Re}(a) > |\text{Im}(b)|, \ \text{Re}(\nu+1-\lambda) > |\text{Re}(\mu)|)$$

and its special case:

$$\int_0^\infty K_\mu(at) J_\nu(bt) t^{\mu+\nu+1} dt = \frac{(2a)^\mu (2b)^\nu \Gamma(\mu+\nu+1)}{(a^2+b^2)^{\mu+\nu+1}}$$

$$(\mathrm{Re}(a) > |\mathrm{Im}(b)|, \ \mathrm{Re}(\nu+1) > |\mathrm{Re}(\mu)|) \ .$$

[Hint: Use the second formula of Ex. 63, putting it in ch $\alpha = 0$.]

73. Show that

$$\int_0^\infty J_\mu(at) J_\nu(bt) J_\nu(ct) t^{1-\mu} dt = \frac{(bc)^\nu}{2^{\mu-1} a^\mu \Gamma(\mu-\nu) \Gamma\left(\nu+\frac{1}{2}\right) \Gamma\left(\frac{1}{2}\right)}$$

$$\times \int_0^A (a^2 - b^2 - c^2 + 2bc \cos\varphi)^{\mu-\nu-1} \sin^{2\nu} \varphi \, d\varphi \ ,$$

where both $\mathrm{Re}(\mu)$ and $\mathrm{Re}(\nu)$ are larger than $-\frac{1}{2}$ and

$$A = \begin{cases} 0 & (a^2 < (b-c)^2, (b+c)^2) \\ \arccos \frac{b^2+c^2-a^2}{2bc} & (a^2 \text{ lies between } (b-c)^2 \text{ and } (b+c)^2), \\ \pi & (a^2 > (b-c)^2, (b+c)^2) \ . \end{cases}$$

[Cf. Watson (1944), Sec. 13.46, p. 411.]

74. Show that

$$\int_0^\infty [J_\nu(at)]^4 \frac{dt}{t^{2\nu-1}} = \frac{a^{2\nu-2} \Gamma(2\nu) \Gamma(\nu)}{2\pi \Gamma(3\nu) \left[\Gamma\left(\nu+\frac{1}{2}\right)\right]^2}$$

$$(\mathrm{Re}(\nu) > 0, \ a > 0) \ .$$

[Cf. Watson (1944), p. 415.]

75. Show that

$$I_\nu(z) = \frac{(z/2)^\nu}{2\pi i} \int_{-\infty}^{(0+)} t^{-\nu-1} \exp\left\{t + \frac{z^2}{4t}\right\} dt \qquad (|\arg t| \le \pi)$$

$$= \frac{1}{2\pi i} \int_{-\infty}^{(0+)} u^{-\nu-1} \exp\left\{\frac{z}{2}\left(u - \frac{1}{u}\right)\right\} du \qquad (|\arg z| < \pi/2)$$

$$= \frac{1}{2\pi i} \int_{\infty-\pi i}^{\infty+\pi i} e^{z \operatorname{ch} w - \nu w} dw \qquad (|\arg z| < \pi/2) \ .$$

If $\mathrm{Re}(\nu) > 0$, then the last two formulae are also valid for $\arg z = \pm\pi/2$.

76. By the third formula of the last problem, show that

$$I_\nu(z) = \frac{1}{\pi} \int_0^\pi e^{z\cos\theta} \cos\nu\theta \, d\theta - \frac{\sin\nu\pi}{\pi} \int_0^\infty e^{-z\,\mathrm{ch}\,t - \nu t} dt \,,$$

and it follows

$$K_\nu(z) = \int_0^\infty e^{-z\,\mathrm{ch}\,t}\mathrm{ch}\,\nu t \, dt \qquad\qquad (|\arg z| < \pi/2)$$

$$= \frac{1}{2} \int_{-\infty}^\infty e^{-z\,\mathrm{ch}\,t - \nu t} dt \qquad\qquad (|\arg z| < \pi/2)$$

$$= \frac{1}{2} \int_{0\,\exp i\omega}^{\infty\,\exp(-i\omega)} u^{-\nu-1} \exp\left\{-\frac{z}{2}\left(u + \frac{1}{u}\right)\right\} du$$

$$(-\pi/2 + \omega < \arg z < \pi/2 + \omega, \ |\omega| < \pi) \,.$$

77. Use the last expression of $K_\nu(z)$ in the last problem to show that

$$\int_0^\infty J_\mu(bt) \frac{K_\nu\left\{a\sqrt{t^2 + z^2}\right\}}{(t^2 + z^2)^{\nu/2}} t^{\mu+1} dt = \frac{b^\mu}{a^\nu} \left\{\frac{\sqrt{a^2 + b^2}}{z}\right\}^{\nu-\mu-1}$$

$$\times K_{\nu-\mu-1}\left\{z\sqrt{a^2 + b^2}\right\} \,,$$

where a and b are positive numbers, $\mathrm{Re}(\mu) > -1, |\arg z| < \pi/2.$

78. With the result of the last Ex., show that

$$\int_0^\infty J_0(bt) \frac{\exp\left\{-a\sqrt{t^2 - y^2}\right\}}{\sqrt{t^2 - y^2}} t \, dt = \frac{\exp\left\{\mp iy\sqrt{a^2 + b^2}\right\}}{\sqrt{a^2 + b^2}} \qquad (y > 0) \,,$$

where the path of integration is to be curved around the singular point $t = y$; the exponential function on the r.h.s. takes the negative sign when the path is curved around $t = y$ by the upper half plane $(\mathrm{Im}(t) > 0)$, and takes the positive sign when the curved part of the path is in the lower half plane $(\mathrm{Im}(t) < 0)$.

79. From Eq. (16) of Sec. 7.15, dividing both sides by b^μ, letting $b \to 0$, obtain

$$\int_0^\infty \frac{J_\nu\left\{a\sqrt{t^2 + z^2}\right\}}{(t^2 + z^2)^{\nu/2}} t^{2\mu+1} dt = \frac{2^\mu \Gamma(\mu + 1)}{a^{\mu+1} z^{\nu-\mu-1}} J_{\nu-\mu-1}(az)$$

$$\left(a \geq 0, \ \mathrm{Re}\left(\frac{\nu}{2} - \frac{1}{4}\right) > \mathrm{Re}(\mu) > -1\right) \,.$$

Replacing in it ν by 2ν, a by $2\sin\theta$, then integrating from $\theta = 0$ to $\pi/2$, show, by using the Neumann formula given in Ex. 31, that

$$\int_0^\infty \frac{J_\nu^2\left\{\sqrt{t^2 + z^2}\right\}}{(t^2 + z^2)^\nu} t^{2\mu+1} dt = \frac{\Gamma(\mu+1)}{\pi z^{2\nu-\mu-1}} \int_0^{\frac{\pi}{2}} \frac{J_{2\nu-\mu-1}(2z\sin\theta)}{\sin^{\mu+1}\theta} d\theta$$

$$\left(\operatorname{Re}\left(\nu - \frac{1}{2}\right) > \operatorname{Re}(\mu) > -1\right).$$

80. In Eq. (16) of Sec. 7.15, write u for b, multiply both sides with $u^{\mu+1}$, then integrate from $u = 0$ to $u = b$. Show that, when $b > a$,

$$\int_0^\infty J_{\mu+1}(bt) \frac{J_\nu\left\{a\sqrt{t^2 + z^2}\right\}}{(t^2 + z^2)^{\nu/2}} t^\mu dt = \frac{2^\mu \Gamma(\mu+1)}{b^{\mu+1}} \frac{J_\nu(az)}{z^\nu}$$

$$(\operatorname{Re}(\nu+1) > \operatorname{Re}(\mu) > -1).$$

81. In Eq. (16) of Sec. 7.15, replacing a by u, dividing by $u^{\nu-1}$, then integrating from $u = a$ (assume $a \le b$) to $u = \infty$, show that

$$\int_0^\infty J_\mu(bt) \frac{J_{\nu-1}\left\{a\sqrt{t^2 + z^2}\right\}}{(t^2 + z^2)^{\frac{\nu}{2}+1}} t^{\mu+1} dt = \frac{a^{\nu-1} z^\mu}{2^{\nu-1}\Gamma(\nu)} K_\mu(bz)$$

$$(a < b, \ \operatorname{Re}(\nu+2) > \operatorname{Re}(\mu) > -1).$$

82. Using the Neumann formula given in Ex. 31 and the result of Ex. 80, show that

$$\int_0^\infty J_\mu(bt) \frac{J_\nu\left\{a\sqrt{t^2 + z^2}\right\} J_\lambda\left\{a\sqrt{t^2 + z^2}\right\}}{(t^2 + z^2)^{(\nu+\lambda)/2}} t^{\mu-1} dt$$

$$= \frac{2^{\mu-1}\Gamma(\mu)}{b^\mu} \frac{J_\nu(az) J_\lambda(az)}{z^{\nu+\lambda}}$$

$$\left(b > 2a, \ \operatorname{Re}\left(\nu+\lambda+\frac{5}{2}\right) > \operatorname{Re}(\mu) > 0\right).$$

83. Using Eq. (3) of Sec. 7.14 and the result of Ex. 80, show that

$$\int_0^\infty J_\mu(bt) \frac{J_\nu\left\{a\sqrt{t^2 + z^2}\right\} J_\nu\left\{c\sqrt{t^2 + z^2}\right\}}{(t^2 + z^2)^\lambda} t^{\mu-1} dt$$

$$= \frac{2^{\mu-1}\Gamma(\mu)}{b^\mu} \frac{J_\nu(az)}{z^\nu} \frac{J_\nu(cz)}{z^\nu}$$

$$\left(b > a + c, \ \operatorname{Re}\left(2\nu+\frac{5}{2}\right) > \operatorname{Re}(\mu) > 0\right).$$

Then, by induction, show that

$$\int_0^\infty J_\mu(bt) \frac{\prod\limits_a \left[J_\nu \left\{ a\sqrt{t^2 + z^2} \right\} \right]}{(t^2 + z^2)^{n\nu/2}} t^{\mu-1} dt = \frac{2^{\mu-1}\Gamma(\mu)}{b^\mu} \prod_a \left[\frac{J_\nu(az)}{z^\nu} \right]$$

$$\left(b > \sum a, \ \operatorname{Re}\left(n\nu + \frac{n}{2} + \frac{1}{2} \right) > \operatorname{Re}(\mu) > 0 \right) .$$

84. Show that

$$\int_{-\infty}^\infty \frac{J_\mu\{a(z+t)\} J_\nu\{b(\varsigma+t)\}}{(z+t)^\mu (\varsigma+1)^\nu} dt$$

$$= \frac{2(b/2)^\nu}{(2a)^\mu \Gamma\left(\mu + \frac{1}{2}\right) \Gamma\left(\nu + \frac{1}{2}\right)} \int_0^\pi (a^2 - b^2 \cos^2 \varphi)^{\mu - \frac{1}{2}}$$

$$\times \cos\left[b(z - \varsigma) \cos \varphi \right] \sin^{2\nu} \varphi \, d\varphi ,$$

where a and b are positive numbers; if $a \neq b$, assume $\operatorname{Re}(\mu + \nu) > -1$; if $a = b$, then assume $\operatorname{Re}(\mu + \nu) > 0$ to ensure the convergence of the integral. [Hint: Use Eq. (7) of Sec. 7.4 and Eq. (8) of Sec. 7.15.]

85. The integral on the r.h.s. of the formula of the last problem can be evaluated or simplified only when $\mu = \frac{1}{2}$, or when $a = b$. Show that

$$\int_{-\infty}^\infty \frac{\sin a(z+t)}{z+t} J_0(bt) dt = \pi J_0(bz) \quad (b \leq a) ,$$

$$\int_{-\infty}^\infty \frac{\sin a(z+t)}{z+t} J_0(bt) dt = 2 \int_0^a \frac{\cos uz \, du}{\sqrt{b^2 - u^2}} \quad (b \geq a);$$

and the *Hardy formula*:

$$\int_{-\infty}^\infty \frac{J_\mu\{a(z+t)\} J_\nu\{a(\varsigma+t)\}}{(z+t)^\mu (\varsigma+t)^\nu} dt$$

$$= \frac{\Gamma(\mu+\nu)\Gamma\left(\frac{1}{2}\right)}{\Gamma\left(\mu + \frac{1}{2}\right)\Gamma\left(\nu + \frac{1}{2}\right)} \left(\frac{2}{a}\right)^{\frac{1}{2}} \frac{J_{\mu+\nu-\frac{1}{2}}\{a(z-\varsigma)\}}{(z-\varsigma)^{\mu+\nu-\frac{1}{2}}}$$

$$(\operatorname{Re}(\mu+\nu) > 0) .$$

86. Show that

$$\frac{1}{2\pi i} \int_0^\infty \left[Z_\mu(bx) H_\nu^{(1)}(ax) - e^{\rho\pi i} Z_\mu(bxe^{\pi i}) H_\nu^{(1)}(axe^{\pi i}) \right] \frac{x^{\rho-1} dx}{(x^2 - r^2)^{m+1}}$$

$$= \frac{1}{2^{m+1} m!} \left(\frac{d}{r dr}\right)^m \left[r^{\rho-2} Z_\mu(br) H_\nu^{(1)}(ar) \right] ,$$

where $a \geq b > 0$; m is a positive integer; r is any complex number with an imaginary part greater than zero; $Z_\mu(z)$ is a cylinder function of any order μ [Sec. 7.6], $|\mathrm{Re}(\nu)| + |\mathrm{Re}(\mu)| < \mathrm{Re}(\rho) < 2m + 4$; if $a = b$, then, in the last inequality, $2m+4$ is to be replaced by $2m+3$. [Hint: Consider the contour integral

$$\frac{1}{2\pi i} \int_C z^{\rho-1} Z_\mu(bz) \frac{H_\nu^{(1)}(az)}{(z^2 - r^2)^{m+1}} \, dz \, ,$$

C being the contour shown in Fig. 32, Sec. 7.15.]

87. If, in the last problem, $\rho = 2m + 3$ and $a = b$, putting

$$Z_\mu(az) \equiv c_1 H_\mu^{(1)}(az) + c_2 H_\mu^{(2)}(az)$$

show that

$$\frac{1}{2\pi i} \int_0^\infty \left[Z_\mu(ax)H_\nu^{(1)}(ax) + Z_\mu(axe^{\pi i})H_\nu^{(1)}(axe^{\pi i}) \right] \frac{x^{2m+2}\,dx}{(x^2 - r^2)^{m+1}}$$
$$= \frac{1}{2^{m+1}m!} \left(\frac{d}{rdr} \right)^m \left[r^{2m+1} Z_\mu(ar)H_\nu^{(1)}(ar) \right] - \frac{c_2 e^{\frac{1}{2}(\mu-\nu)\pi i}}{\pi a} \, .$$

88. Show that

$$\int_0^\infty J_\nu(ax)J_\nu(bx)\frac{x\,dx}{x^2 - r^2} = \begin{cases} \frac{\pi i}{2} \, J_\nu(br)H_\nu^{(1)}(ar) & (a > b) \\ \frac{\pi i}{2} \, J_\nu(ar)H_\nu^{(1)}(br) & (a < b) \end{cases}$$
$$(\mathrm{Re}(\nu) > -1) \, .$$

89. Show that

$$\frac{1}{2\pi i} \int_0^\infty \frac{x^{\rho-1}}{(x^2 - r^2)^{m+1}} \frac{J_\mu\left\{ b\sqrt{x^2 + \varsigma^2} \right\}}{(x^2 + \varsigma^2)^{\mu/2}}$$
$$\times \left\{ H_\nu^{(1)}(ax) - e^{\rho\pi i} H_\nu^{(1)}(axe^{\pi i}) \right\} dx$$
$$= \frac{1}{2^{m+1}m!} \left(\frac{d}{rdr} \right)^m \left[r^{\rho-2} \frac{J_\mu\left\{ b\sqrt{r^2 + \varsigma^2} \right\}}{(r^2 + \varsigma^2)^{\mu/2}} H_\nu^{(1)}(ar) \right]$$
$$(|\mathrm{Re}(\nu)| < \mathrm{Re}(\rho) < 2m + 4 + \mathrm{Re}(\mu));$$

also its special case (use the result of Ex. 13):

$$\int_0^\infty \frac{x^{\nu+1}}{x^2+k^2} \frac{J_\mu\left\{b\sqrt{x^2+\varsigma^2}\right\}}{(x^2+\varsigma^2)^{\mu/2}} J_\nu(ax)dx$$

$$= \frac{J_\mu\left\{b\sqrt{\varsigma^2-k^2}\right\}}{(\varsigma^2-k^2)^{\mu/2}} k^\nu K_\nu(ak) .$$

90. With the series expression of $J_\nu(ax)$, it is easy to prove that

$$J_\nu(ax) = \frac{1}{2\pi i} \int_{-\infty i}^{\infty i} \frac{\Gamma(-s)}{\Gamma(\nu+s+1)} \left(\frac{ax}{2}\right)^{\nu+2s} ds ,$$

where $\text{Re}(\nu) > 0, a$ and x are both positive numbers. By this formula show that

$$\int_0^\infty \frac{x^{\rho-1} J_\nu(ax)}{(x^2+k^2)^{\mu+1}} dx$$

$$= \frac{a^\nu k^{\rho+\nu-2\mu-2}\Gamma\left(\frac{\rho+\nu}{2}\right)\Gamma\left(\mu+1-\frac{\rho+\nu}{2}\right)}{2^{\nu+1}\Gamma(\mu+1)\Gamma(\nu+1)}$$

$$\times {}_1F_2\left(\frac{\rho+\nu}{2}; \frac{\rho+\nu}{2}-\mu, \nu+1; \frac{a^2k^2}{4}\right)$$

$$+ \frac{a^{2\mu+2-\rho}\Gamma\left(\frac{\nu+\rho}{2}-\mu-1\right)}{2^{2\mu+3-\rho}\Gamma\left(\mu+2+\frac{\nu-\rho}{2}\right)}$$

$$\times {}_1F_2\left(\mu+1; \mu+2+\frac{\nu-\rho}{2}, \mu+2-\frac{\nu+\rho}{2}; \frac{a^2k^2}{4}\right)$$

$$\left(-\text{Re}(\nu) < \text{Re}(\rho) < 2\text{Re}(\mu)+\frac{7}{2}\right) .$$

91. Using the first integral expression of $K_\nu(z)$ in Ex. 76, show that

$$K_\mu(z)K_\nu(z) = 2\int_0^\infty K_{\mu+\nu}(2z\,\text{ch}\,t)\text{ch}(\mu-\nu)t\,dt$$

$$= 2\int_0^\infty K_{\mu-\nu}(2z\,\text{ch}\,t)\text{ch}(\mu+\nu)t\,dt$$

$$(|\arg z| < \pi/2) .$$

[Cf. Watson (1944), Sec. 13.72, p. 440.]

92. Show that the Neumann polynomial in Sec. 7.16 has the following integral expression:

$$O_n(z) = \int_0^\infty \frac{\left(t + \sqrt{t^2 + z^2}\right)^n + \left(t - \sqrt{t^2 + z^2}\right)^n}{2z^{n+1}} e^{-t} dt .$$

93. Show that

$$\frac{1}{t^2 - z^2} = J_0^2(z)\Omega_0(t) + 2J_1^2(z)\Omega_1(t) + 2J_2^2(z)\Omega_2(t) + \cdots$$

$$= \sum_{n=0}^{\infty} \varepsilon_n J_n^2(z)\Omega_n(t) ,$$

where

$$\Omega_0 = \frac{1}{t^2} ,$$

$$\Omega_n = \sum_{s=0}^{n} \frac{n(n+s-1)! 2^{2s}(s!)^2}{(n-s)! (2s)! t^{2s+2}} \qquad (n \geq 1)$$

are called *Neumann polynomials of the second kind*.

94. Show that the Neumann polynomials $\Omega_n(t)$ in the last Ex. satisfy the following recurrence relations:

$$\frac{2}{t}\Omega_0'(t) = -2\Omega_1(t) + 2\Omega_0(t) ,$$

$$\frac{2}{t}\Omega_1'(t) = \frac{1}{2}\Omega_0(t) - \frac{1}{2}\Omega_2(t) ,$$

$$\frac{2}{t}\Omega_n'(t) = \frac{\Omega_{n-1}(t)}{n-1} - \frac{\Omega_{n+1}(t)}{n+1} - \frac{2\Omega_0(t)}{n^2-1} \qquad (n \geq 2) .$$

[Hint: Cf. the derivation of the recurrence relations for $O_n(t)$ in Sec. 7.16, using, in addition, Eq. (10) of Sec. 7.5.]

95. Using the expansion formula in Ex. 20, prove that

$$\frac{z^\nu}{t-z} = \sum_{n=0}^{\infty} A_{n,\nu}(t) J_{\nu+n}(z) \quad (|z| < |t|) ,$$

where $\nu \neq 0, -1, -2, \ldots$, and

$$A_{n,\nu}(t) = \frac{2^{\nu+n}(\nu+n)}{t^{n+1}} \sum_{m=0}^{[n/2]} \frac{\Gamma(\nu+n-m)}{m!} \left(\frac{t}{2}\right)^{2m}$$

is a polynomial of $1/t$ of degree $n+1$.

96. Prove the following *Gegenbauer Expansion* (which is a generalization of the Neumann expansion in Sec. 7.16):

$$z^\nu f(z) = \sum_{n=0}^{\infty} a_n J_{\nu+n}(z) ,$$

where

$$a_n = \frac{1}{2\pi i} \int_C f(t) A_{n,\nu}(t) dt ;$$

C represents the circle $|z| = R$, $A_{n,\nu}(t)$ are the coefficients in the expansion of the last problem.

97 By the expansion formula in Ex. 32, show that

$$\frac{z^{\mu+\nu}}{t-z} = \sum_{n=0}^{\infty} B_{n;\mu,\nu}(t) J_{\mu+\frac{n}{2}}(z) J_{\nu+\frac{n}{2}}(z) \quad (|z| < |t|) ,$$

where

$$B_{n;\mu,\nu}(t) = \frac{2^{\mu+\nu+n}(\mu+\nu+n)}{t^{n+1}}$$

$$\times \sum_{m=0}^{[\frac{n}{2}]} \frac{\Gamma\left(\mu+\frac{n}{2}-m+1\right) \Gamma\left(\nu+\frac{n}{2}-m+1\right) \Gamma(\mu+\nu+n-m)}{m!\Gamma(\mu+\nu+n-2m+1)} \left(\frac{t}{2}\right)^{2m}$$

is a polynomial of $1/t$ of degree $n+1$.

98. Let $f(z)$ be analytic in $|z| \leq r$. Show that

$$z^{\mu+\nu} f(z) = \sum_{n=0}^{\infty} a_n J_{\mu+\frac{n}{2}}(z) J_{\nu+\frac{n}{2}}(z) ,$$

where

$$a_n = \frac{1}{2\pi i} \int_C f(t) B_{n;\mu,\nu}(t) dt \; ;$$

C represents the circle $|z| = r$ and $B_{n;\mu,\nu}(t)$ are the coefficients given in the last problem.

99. Show that, if $f(z)$ is an analytic function in $|z| \leq r$, and $f(-z) = f(z)$, then there is the expansion formula

$$f(z) = \sum_{n=0}^{\infty} a'_n J_n^2(z) \; ,$$

where

$$a'_n = \frac{\varepsilon_n}{2\pi i} \int t f(t) \Omega_n(t) dt \; ;$$

C represents the circle $|z| = r$, and $\Omega_n(t)$ are the coefficients of expansion given in Ex. 93.

100. Show that

$$\left(\frac{z}{2}\right)^{\nu} = \nu^2 \sum_{m=0}^{\infty} \frac{\Gamma(\nu + m)}{(\nu + 2m)^{\nu+1} m!} J_{\nu+2m}\{(\nu + 2m)z\} \; ,$$

where z satisfies Eq. (16) of Sec. 7.17. [Cf. Watson (1944), Sec. 17.5, p. 571.]

101. By the result of the last problem, show that

$$\frac{z^{\nu}}{t - z} = \sum_{n=0}^{\infty} \mathcal{A}_{n,\nu}(t) J_{\nu+n}\{(\nu + n)z\} \; ,$$

where

$$\mathcal{A}_{n,\nu}(t) = \frac{1}{2} \sum_{m=0}^{\left[\frac{n}{2}\right]} \frac{(\nu + n - 2m)^2 \Gamma(\nu + n - m)}{\left(\frac{1}{2}\nu + n\right)^{\nu+n-2m+1} m! t^{n-2m+1}} \; .$$

Then, from this expansion formula, obtain the general Kapteyn expansion [Sec. 7.17]:

$$z^{\nu} f(z) = \sum_{n=0}^{\infty} a_{n,\nu} J_{\nu+n}\{(\nu + n)z\} \; ,$$

$$a_{n,\nu} = \frac{1}{2\pi i} \int_C f(t) \mathcal{A}_{n,\nu}(t) dt \; ,$$

where $f(z)$ is an analytic function in the domain $\omega(z) \le a(a \le 1)$ given by Eq. (16) of Sec. 7.17; C is a contour lying in this domain and encircling $z = 0$ once in the positive sense.

102. From the expansion formula of Ex. 100, show that

$$\left(\frac{z}{2}\right)^{2\nu} = \frac{2\nu[\Gamma(\nu+1)]^2}{\Gamma(2\nu+1)} \sum_{m=0}^{\infty} \frac{\Gamma(2\nu+m)}{(\nu+m)^{2\nu+1}m!} J_{\nu+m}^2\{(\nu+m)z\} ,$$

and obtain a formula for the expansion of an analytic function in terms of $J_{\nu+m}^2\{(\nu+m)z\}$.

103. Suppose that the integral of $x^{1/2}f(x)$ in the integral $0 \le x \le 1$ is absolutely convergent, and k_m are the positive roots of the equation

$$k^{-\nu}\{kJ_\nu'(k) + HJ_\nu(k)\} = 0 , \qquad (*)$$

H being a real constant, $\nu \ge -\frac{1}{2}$. Show that, if $f(x)$ is of bounded variation in (a, b) and $0 \le a < b \le 1$, then, for an arbitrary point x in $a + \Delta \le x \le b - \Delta$ (Δ is any positive small number), there is the *Dini Expansion*:

$$\frac{1}{2}\{f(x+0) + f(x-0)\} = \sum_{m=1}^{\infty} A_m J_\nu(k_m x) ,$$

where

$$A_m = \frac{\int_0^1 x f(x) J_\nu(k_m x) dx}{\int_0^1 x\{J_\nu(k_m x)\}^2 dx} ;$$

if $f(x)$ is continuous in (a, b), then the series converges uniformly to $f(x)$.

When $H + \nu = 0$, Eq. (*) becomes $k^{1-\nu}J_{\nu+1}(k) = 0$ [cf. Eq. (12) of Sec. 7.2], and k is a root of it; an additional term $A_0 x^\nu$ is to be inserted into the above expansion of $f(x)$ with

$$A_0 = 2(\nu+1)\int_0^1 x^{\nu+1}f(x)dx .$$

104. Suppose that $f(z)$ is an analytic function in the ring-shape domain: within the concentric circles c and C. Show that

$$f(z) = \frac{1}{2}\alpha_0 J_0(z) + \alpha_1 J_1(z) + \alpha_2 J_2(z) + \dots$$
$$+ \frac{1}{2}\beta_0 O_0(z) + \beta_1 O_1(z) + \beta_2 O_2(z) + \dots ,$$

where

$$\alpha_n = \frac{1}{\pi i} \int_C f(t) O_n(t) dt \ , \quad \beta_n = \frac{1}{\pi i} \int_C f(t) J_n(t) dt \ ,$$

$O_n(t)$ being the Neumann polynomials [Sec. 7.16].

105. Show that

$$J_\nu(z) = \frac{1}{\Gamma(\nu + 1)} \left(\frac{z}{2}\right)^\nu {}_0F_1\left(\nu + 1; -\frac{z^2}{4}\right)$$

$$= \frac{1}{\Gamma(\nu + 1)} \left(\frac{z}{2}\right)^\nu e^{-iz} {}_1F_1\left(\nu + \frac{1}{2}, 2\nu + 1; 2iz\right) \ .$$

106. From Eq. (11) of Sec. 7.15, show that (when m and n are not negative integers, $\nu > -1$)

$$\int_0^\infty t^{-1} J_{\nu+2n+1}(t) J_{\nu+2m+1}(t) dt = \begin{cases} 0 & (m \neq n) \\ (2\nu + 4n + 2)^{-1} & (m = n) \ . \end{cases}$$

Then, when $x > 0$, we can formally obtain the expansion formula

$$f(x) = \sum_{n=0}^\infty a_n (2\nu + 4n + 2) J_{\nu+2n+1}(x) \ ,$$

where

$$a_n = \int_0^\infty t^{-1} f(t) J_{\nu+2n+1}(t) dt \ .$$

[As regards the theory of such expansions, cf. Wilkins, J. E., *Bull. Am. Math. Soc.* **54**, (1948) 232-234; *Trans. Am. Math. Soc.* **69** (1950) 55-65.]

Chapter 8
WEIERSTRASS ELLIPTIC FUNCTIONS

8.1. Elliptic Integrals and Elliptic Functions

In practical applications, elliptic functions are usually brought about from elliptic integrals. The general form of an elliptic integral is

$$\int R(x, y)\, dx \,, \tag{1}$$

where $R(x, y)$ is a rational function of x and y, and

$$y^2 = P(x) = ax^4 + bx^3 + cx^2 + dx + e \,. \tag{2}$$

When $a = 0$, the polynomial $P(x)$ reduces from quartic to cubic. Whatever it is (quartic or cubic), the relevant integral (1) belongs to elliptic integrals. But if $P(x)$ is a polynomial of degree higher than 4, the integral is named hyper-elliptic integral. The cubic case can be brought into a quartic one by the transformation $x = 1/t$:

$$P(x) = bx^3 + cx^2 + dx + e \,,$$
$$P\left(\frac{1}{t}\right) = \frac{1}{t^4} P_1(t), \quad P_1(t) = et^4 + dt^3 + ct^2 + bt \,, \tag{3}$$
$$y = \sqrt{P(x)} = t^{-2}\sqrt{P_1(t)} \,,$$

456

where $P_1(t)$ is a quartic polynomial. On the other hand, if one of the roots $x = x_1$ of the polynomial of degree 4 is known, the quartic case can be reduced to a cubic one by the transformation $\xi = 1/(x - x_1)$:

$$x = x_1 + \frac{1}{\xi} \; ,$$

$$P\left(x_1 + \frac{1}{\xi}\right) = a\left(x_1 + \frac{1}{\xi}\right)^4 + b\left(x_1 + \frac{1}{\xi}\right)^3$$

$$+ c\left(x_1 + \frac{1}{\xi}\right)^2 + d\left(x_1 + \frac{1}{\xi}\right) + e \tag{4}$$

$$= \frac{1}{\xi^4} P_1(\xi) \; ,$$

$$P_1(\xi) = (ax_1^4 + bx_1^3 + cx_1^2 + dx_1 + e)\xi^4$$
$$+ (4ax_1^3 + 3bx_1^2 + 2cx_1 + d)\xi^3$$
$$+ (6ax_1^2 + 3bx_1 + c)\xi^2 + (4ax_1 + b)\xi + a \; .$$

Since x_1 is a root of $P(x) = 0$, the coefficient of ξ^4 in $P_1(\xi)$ is zero, and $P_1(\xi)$ becomes a cubic polynomial. Thus, the corresponding elliptic integrals have the same properties whether $P(x)$ is quartic or cubic.

The general elliptical integral can be expressed as a combination of few fundamental elliptic integrals. We shall give some simple explanations here; details are left over until we have gone through the elliptic functions [cf. Sec. 10.8].

By (2) we can express $R(x, y)$ in the following form:

$$R(x, y) = R_1(x) + \frac{R_2(x)}{y} \; , \tag{5}$$

where $R_1(x)$ and $R_2(x)$ are rational functions of x. The integral $\int R_1(x)dx$ can be represented by elementary functions. Only the elliptic integral $\int R_2(x)y^{-1}dx$ is left. $R_2(x)$ can in turn be expressed as

$$R_2(x) = \sum_{m=0}^{n} a_m x^m + \sum_{p=1}^{q} \sum_{k=1}^{n_p} \frac{b_{pk}}{(x - h_p)^k} \; , \tag{6}$$

where a_m, b_{pk} are constants. Hence, it is seen that elliptic integrals can be reduced to two kinds:

$$I_m = \int \frac{x^m}{y} dx, \quad J_k = \int \frac{dx}{(x - h)^k y} \; . \tag{7}$$

When $P(x)$ is cubic, we shall show that I_m and J_k can be expressed in terms of three fundamental elliptic integrals I_0, I_1, J_1. By differentiation,

$$\frac{d}{dx}\left(x^m\sqrt{P(x)}\right) = mx^{m-1}\sqrt{P(x)} + \frac{x^m}{2}\frac{P'(x)}{\sqrt{P(x)}}$$

$$= \frac{1}{y}\left\{mx^{m-1}(ax^4 + bx^3 + cx^2 + dx + e)\right.$$

$$\left. + \frac{1}{2}x^m(4ax^3 + 3bx^2 + 2cx + d)\right\}$$

$$= \frac{1}{y}\left\{(m+2)ax^{m+3} + \left(m+\frac{3}{2}\right)bx^{m+2}\right.$$

$$\left. +(m+1)cx^{m+1} + \left(m+\frac{1}{2}\right)dx^m + mex^{m-1}\right\},$$

then integrating, we obtain

$$(m+2)aI_{m+3} + \left(m+\frac{3}{2}\right)bI_{m+2} + (m+1)cI_{m-1}$$

$$+ \left(m+\frac{1}{2}\right)dI_m + meI_{m-1} = x^m\sqrt{P(x)} + C, \qquad (8)$$

where C is an integration constant.

When $P(x)$ is cubic, $a = 0$, we obtain from (8) the expressions for representing I_m in terms of two fundamental elliptic integrals I_0, I_1 by putting $m = 0, 1, \ldots$ in turn.

Similarly, we have ($P(x)$ cubic)

$$\frac{d}{dx}\frac{\sqrt{P(x)}}{(x-h)^k} = \frac{1}{(x-h)^{k+1}y}\left\{\frac{x-h}{2}P'(x) - kP(x)\right\}$$

$$= \frac{1}{y}\left\{-\frac{kP(h)}{(x-h)^{k+1}} + \left(\frac{1}{2}-k\right)\frac{P'(h)}{(x-h)^k}\right.$$

$$\left. +\frac{(1-k)P''(h)}{2(x-h)^{k-1}} + \left(\frac{1}{4}-\frac{k}{6}\right)\frac{P'''(h)}{(x-h)^{k-2}}\right\}.$$

Integrating gives

$$-kP(h)J_{k+1} + \left(\frac{1}{2}-k\right)P'(h)J_k + \frac{1-k}{2}P''(h)J_{k-1}$$

$$+ \left(\frac{1}{4}-\frac{k}{6}\right)P'''(h)J_{k-2} = \frac{\sqrt{P(x)}}{(x-h)^k} + C. \qquad (9)$$

Evidently, $J_0 = I_0, J_{-1} = I_1 - hI_0$. Putting in turn $k = 1, 2, \ldots$ in (9), we find the expressions for J_k in terms of J_1 and I_0, I_1.

From (9) we see that if h is a root of $P(x) = 0$, the first term on the l.h.s. of (9) will be zero, so that J_1 is expressed in terms of I_0 and I_1 when $k = 1$, and hence also J_k.

The three fundamental elliptic integrals I_0, I_1 and J_1 are respectively called the elliptic integral of the first, second and third kind. The inversion of the elliptic integral of the first kind is an elliptic function. There are two standard forms, one of which is *Weierstrass' elliptic function* $\varsigma = \wp(z)$, in which the polynomial is cubic:

$$z = \int_\infty^\varsigma \frac{dx}{\sqrt{4x^3 - g_2 x - g_3}} \quad (\varsigma = \wp(z)), \tag{10}$$

where g_2 and g_3 are two constants. The other is *Jacobian elliptic function* $t = \operatorname{sn} u$, in which the polynomial is quartic:

$$u = \int_0^t \frac{dt}{\sqrt{(1 - t^2)(1 - k^2 t^2)}} \quad (t = \operatorname{sn} u), \tag{11}$$

where k is a positive constant less than 1. The reduction of a general elliptic integral to standard form will be treated later in Sec. 10.8.

The term elliptic function is originated from the problem of finding the circumference of the ellipse. Let the coordinates of an ellipse be

$$x = a\cos\varphi, \quad y = b\sin\varphi \quad (a > b) \tag{12}$$

and the length of an arc element be ds, then

$$ds = \sqrt{dx^2 + dy^2} = \sqrt{a^2 \sin^2\varphi + b^2 \cos^2\varphi}\, d\varphi$$
$$= a\sqrt{1 - e^2 \cos^2\varphi}\, d\varphi,$$

where $e^2 = (a^2 - b^2)/a^2$. Let $\cos\varphi = t$, then the arc length is

$$s = a\int \frac{\sqrt{1 - e^2 t^2}}{\sqrt{1 - t^2}}\, dt = a\int \frac{1 - e^2 t^2}{\sqrt{(1 - t^2)(1 - e^2 t^2)}}\, dt, \tag{13}$$

in which an elliptic integral appears.

8.2. The Periods of Elliptic Integrals

Consider the elliptic integral of the first kind

$$F(z) = \int_{z_0}^{z} \frac{dz}{\sqrt{P(z)}} \, , \tag{1}$$

where $P(z)$ is a cubic or a quartic polynomial. Since the quartic case can be reduced to a cubic one according to (4) of the last section, we shall consider only the cubic case. Let e_1, e_2, e_3 be the three roots of the cubic equation $P(z) = 0$. Draw three lines respectively from e_1, e_2, e_3 to ∞. Let L be a path leading directly from z_0 to z without intersecting any of the three lines just drawn (Fig. 36). The value of the integral (1) will be the same however complicated the path L taken from z_0 to z may be, so long as L does not come across the three lines mentioned above. Now, look at a path which encircles e_i once. Let

$$E_i = \int_{z_0}^{e_1} \frac{dz}{\sqrt{P(z)}} \, , \tag{2}$$

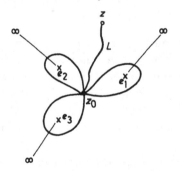

Fig. 36.

the path of integration being one from z_0 to e_i without crossing the three lines (casually, it may be the straight line $\overline{z_0 e_i}$), and suppose that $P(z_0) = y_0$ at the starting point. The simple contour around e_i may be made to consist of three parts: The first is $z_0 e_i$, the second, a small circle around e_i with infinitesimal radius, and the third, $e_i z_0$. After having encircled the point e_i once, $\sqrt{P(z)}$ changes sign, so the third part of the integral is equal to $-(-E_i) = E_i$. The second part approaches zero, and hence the value of the integral along the whole path is $2E_i$. Since the second part of the integral is zero, independent of the running direction, the value of the contour integral around e_i also does not depend on the running direction, but depends only upon the sign of the square root at the starting point. Had we defined $\sqrt{P(z_0)} = -y_0$, the value of the contour integral would be $-2E_i$.

Let I be the value of the integral (1) along the direct path L and the square root takes the value y_0 at the starting point. The general path of integration is one which encircles the three points e_1, e_2, e_3, each by several rounds, then goes to z along L. Consider the path which encircles first e_α by one round, then e_β (may be the same point e_α) also by one round, and then $e_\gamma, e_\delta, \ldots$ etc. The value of the integral for two consecutive encircling paths will be $2(E_\alpha - E_\beta)$. If the total number of encirclements is even, then the last value of the square root at the starting point will be y_0, and the value of the integral along L is I; if the total number of encirclements is odd, the last value of the square root at the starting point will be $-y_0$, and correspondingly the value of the integral along L will be $-I$. Thus, the value of the integral represented by $F(z)$ in (1) for the respective cases will be

$$F(z) = 2(E_\alpha - E_\beta) + 2(E_\gamma - E_\delta) + \ldots + 2(E_\kappa - E_\lambda) + I \,,$$
$$F(z) = 2(E_\alpha - E_\beta) + \ldots + 2(E_\kappa - E_\lambda) + 2E_\mu - I \,.$$

In order to simplify these two equations, let

$$\omega_1 = E_1 - E_3, \quad \omega_2 = E_2 - E_3 \,. \tag{3}$$

Since $E_\alpha - E_\beta = \omega_\alpha - \omega_\beta, E_\mu = \omega_\mu + E_3 (\omega_3 = 0)$, the above equations become

$$\left.\begin{array}{l} F(z) = I + 2m_1\omega_1 + 2m_2\omega_2 \,, \\ F(z) = 2E_3 - I + 2m_1\omega_1 + 2m_2\omega_2 \,, \end{array}\right\} \tag{4}$$

where m_1 and m_2 are integers, positive or negative. From (4), it is seen that $2\omega_1$ and $2\omega_2$ are *two periods of the elliptic integral (1)*; it means that, for different paths directing from z_0 to z, the values of the integral (1) are differed by integral multiples of $2\omega_1$ or $2\omega_2$.

Next, we shall show that the ratio ω_2/ω_1 of the two periods is not a real number, but a complex number. To make it simple, let $e_1 = a, e_2 = 1, e_3 = 0$, which can be realized by the transformation $z = e_3 + (e_2 - e_3)t$. Then

$$\omega_1 = \int_0^a \frac{dz}{\sqrt{z(z-1)(z-a)}} \,, \quad \omega_2 = \int_0^1 \frac{dz}{\sqrt{z(1-z)(a-z)}} \,. \tag{5}$$

If a is real, we assume $a > 1$. Then ω_2 is real while $\omega_1 - \omega_2$ is purely imaginary:

$$\omega_1 - \omega_2 = \int_1^a \frac{dz}{\sqrt{z(z-1)(z-a)}}$$
$$= \pm i \int_1^a \frac{dz}{\sqrt{z(z-1)(a-z)}} \,. \tag{6}$$

It is seen thus that the ratio of the periods, ω_2/ω_1, is not a real number.

If a is complex, we change the variable in the integral representing ω_1 by letting $z = at$, and obtain

$$\omega_1 = \int_0^1 \frac{dt}{\sqrt{t(1-t)(1-at)}} \ . \tag{7}$$

By which, it can be shown that if the imaginary part of a is positive, the imaginary part of ω_1 is positive too, while that of ω_2 is negative. It follows that the imaginary part of ω_2/ω_1 cannot vanish.

From the results obtain above, we see that the inversion of an elliptic integral, an elliptic function, is a doubly-periodic function, the periods of which are $2\omega_1$ and $2\omega_2$ and their ratio is not a real number.[a]

Weierstrass established his general theory of elliptic functions from the view of the doubly-periodic analytic functions. We shall deal first with his theory, then the Theta functions, and finally Jacobian elliptic functions and elliptic integrals.

8.3. The General Properties of Doubly-Periodic Functions and Elliptic Functions

Let $f(z)$ be a single-valued analytic function having two periods 2ω and $2\omega'$, and suppose that the imaginary part of the ratio ω'/ω is positive $(\text{Im}(\omega'/\omega) > 0)$:

$$f(z + 2\omega) = f(z), \quad f(z + 2\omega') = f(z) \ . \tag{1}$$

Let m and m' be arbitrary integers. (In the following, when we say integers, we always mean positive and negative integers including zero.) Denote

$$w = 2m\omega + 2m'\omega' \ , \tag{2}$$

the double periodicity can be expressed in general as

$$f(z + w) = f(z) \ . \tag{3}$$

Draw a straight line OA in the z-plane with the value of A equal to 2ω. Mark on the extension of OA the points $2m\omega$. Also, draw a straight line

---

[a] Jacobi had shown that a single-valued analytic function cannot have three periods, and it follows then that the periods of a doubly-periodic function cannot have their ratio equal to a real number. [Cf. Goursat, *Cours d'Analyse Mathématique*, II, p. 170.]

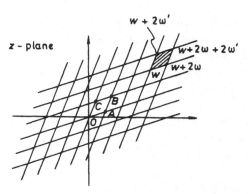

Fig. 37.

OC with the value of C equal to $2\omega'$ and mark on the extension of OC the points $2m'\omega'$. Through the points $2m\omega$ on OA draw straight lines parallel to OC, and through the points $2m'\omega'$ on OC, draw straight lines parallel to OA. These two sets of parallel lines divide the z-plane into parallelograms congruent to $OABC$ (Fig. 37).

The value of the point A is 2ω; that of point C is $2\omega'$, and that of B is $2\omega + 2\omega'$. When z runs along $OABC$ by one round, $z + w$ will pass successively through $w, w + 2\omega, w + 2\omega + 2\omega', w + 2\omega'$, at which the function has the same value. Each such parallelogram is called a *cell*. The parallelogram may not start from the origin O but from any point z_0, with the four points $z_0, z_0 + 2\omega$, $z_0 + 2\omega + 2\omega'$, $z_0 + 2\omega'$ as its vertices. Such a parallelogram is also called a *period parallelogram*.

The value of the doubly-periodic function at any point of any cell is equal to that at the corresponding point in $OABC$. Hence, we have the following theorem:

Theorem 1: Doubly-periodic function without singularities is a constant.

Proof: Since the doubly-periodic function $f(z)$ is assumed to have no singularities in the cell $OABC$, its absolute value must be less than a positive number M in the whole z-plane. Then, according to Liouville's theorem,[b] it has to be a constant.

Thus, a doubly-periodic single-valued function not equal to a constant must have singularities. *Only doubly-periodic functions with poles as its singularities (or doubly-periodic meromorphic functions) are nominated as elliptic functions.* The number of poles in a cell is called the *order of the elliptic function*; a pole of order p is counted as p poles. If the vertices of a cell are

---

[b]Cf. any text book on the theory of function of a complex variable.

poles, only one of them is counted. If there are poles on the boundary of the cell, only one of the two on the opposite sides is counted. To avoid a miscount of the number of poles, we might shift the period parallelogram slightly so that all the poles lie within the cell.

Obviously, *the derivative of an elliptic function is again an elliptic function.*

We have assumed that $\mathrm{Im}(\omega'/\omega) > 0$. If, on the contrary, $\mathrm{Im}(\omega'/\omega) < 0$, we may interchange ω and ω' so that the theorem holds. More generally, let p, p', q, q' be four integers satisfying the relation $pq' - p'q = \pm 1$ and let $\Omega = p\omega + q\omega', \Omega' = p'\omega + q'\omega'$. By inversion, we have $\omega = \pm(q'\Omega - q\Omega'), \omega' = \pm(p\Omega' - p'\Omega)$, so the periods $(2\omega, 2\omega')$ and $(2\Omega, 2\Omega')$ are equivalent. We can choose p, p', q, q' properly so that $\mathrm{Im}(\Omega'/\Omega) > 0$.

The following are some general properties of the elliptic functions.

Theorem 2: The sum of the residues at the poles of an elliptic function in its period parallelogram is equal to zero.

Proof: Choose the vertex $O(z = z_0)$ of the period parallelogram $OABC$ such that all the poles of the function lie within the cell. The contour integral $\int_{OABCO} f(z)dz$ along the sides is equal to $2\pi i$ times the sum of the residues. The integrals along two opposite sides exactly cancel each other. For example,

$$\int_{OA} f(z)dz = \int_{z_0}^{z_0+2\omega} f(z)dz \ , \quad \int_{BC} f(z)dz = \int_{z_0+2\omega+2\omega'}^{z_0+2\omega'} f(z)dz \ .$$

Putting $z = u + 2\omega'$ in the second integral and noticing the periodicity (1) of $f(z)$, we see that

$$\int_{BC} f(z)dz = \int_{z_0+2\omega}^{z_0} f(u)du = -\int_{z_0}^{z_0+2\omega} f(u)du = -\int_{OA} f(z)dz \ .$$

In a similar way, we can show that the integrals along AB and CO cancel each other too. So, the contour integral vanishes and the theorem is proved.

According to this theorem, an elliptic function cannot have only a single pole (note that a pole of order p is counted as p poles) in its period parallelogram. There are at least two, i.e., *elliptic functions are at least of order 2.*

Theorem 3: The number of zeros of an elliptic function in its period parallelogram is equal to its order (i.e., to its number of poles).

Proof: Let $f(z)$ be an elliptic function, then evidently $\varphi(z) = f'(z)/f(z)$ is also an elliptic function. The sum of the residues of $\varphi(z)$ in the period parallelogram is equal to the number of zeros of $f(z)$ minus its number of its poles [cf. Eq. (5) of Sec. 1.4]. Applying Theorem 2 to $\varphi(z)$, Theorem 3 is proved. Note that, similar to the counting of the number of poles, a zero of order p is counted as p zeros. In the proof, replacing $f(z)$ by $f(z) - C$, then, since $f(z) - C$ and $f(z)$ have the same poles, we conclude that:

The number of roots of $f(z) = C$ in the period parallelogram is equal to the order of $f(z)$.

Theorem 4: The sum of the zeros of an elliptic function $f(z)$ minus the sum of its poles in the parallelogram is equal to its period (period means w given in (2)).

Proof: Let s be the order of the elliptic function $f(z)$, α_k being its zeros and β_k its poles. By Eq. (1) of Sec. 1.4, we have

$$\sum_{k=1}^{s} \alpha_k - \sum_{k=1}^{s} \beta_k = \frac{1}{2\pi i} \int_{OABCO} z \frac{f'(z)}{f(z)} \, dz \, , \qquad (4)$$

where the vertices of $OABC$ are $z_0, z_0 + 2\omega, z_0 + 2\omega + 2\omega', z_0 + 2\omega'$, assuming that there are no zeros and poles of $f(z)$ on the boundary. Consider first the integrals along OA and BC. In the integral along BC, let $z = u + 2\omega'$:

$$\int_{OA} z \frac{f'(z)}{f(z)} \, dz = \int_{z_0}^{z_0+2\omega} z \frac{f'(z)}{f(z)} \, dz \, ,$$

$$\int_{BC} z \frac{f'(z)}{f(z)} \, dz = \int_{z_0+2\omega+2\omega'}^{z_0+2\omega'} z \frac{f'(z)}{f(z)} \, dz = \int_{z_0+2\omega}^{z_0} (u + 2\omega') \frac{f'(u)}{f(u)} \, du \, .$$

Thus,

$$\left(\int_{OA} + \int_{BC} \right) z \frac{f'(z)}{f(z)} \, dz = -2\omega' \int_{z_0}^{z_0+2\omega} \frac{f'(u)}{f(u)} \, du$$

$$= -2\omega' \left[\ln f(z) \right]_{z_0}^{z_0+2\omega} \, .$$

Since $f(z)$ has the same value at the two ends of OA, the variation of the value of $\ln f(z)$ is equal to the change of arg $f(z)$, which is equal to an integral multiple of $2\pi i$, i.e., $-2m'\pi i$. Hence, that part of the integral on the r.h.s. of (4) taken along OA and BC is equal to $2m'\omega'$. Similarly, the part of that integral taken along AB and CO has the value $2m\omega$. We thus obtain

$$\sum_{k=1}^{s} (\alpha_k - \beta_k) = 2m\omega + 2m'\omega' = w \, . \qquad (5)$$

Similar proof can be applied to the function $f(z) - C$; the α_k in (5) are then the roots of $f(z) = C$ in the period parallelogram.

8.4. The Function $\wp(z)$

In the corollary of Theorem 2 in the last section, we see that the order of an elliptic function is at least two. We shall now study the elliptic functions of order two. There are two different cases. One is that the poles in the period parallelogram are simple but different. This is the case of Jacobian elliptic functions which will be treated in details in Chap. 10. This is a rather complicated case, since at the different positions of the two poles, there may give rise to different situations. The second case is that there is only one pole of order two, which may be chosen to be $z = w$. Since the sum of the residues is zero, the principal part of the function at the pole is $A/(z-w)^2$. For simplicity, we may take $A = 1$. The function so constructed is the *Weierstrass elliptic function* $\wp(z)$:

$$\wp(z) = \frac{1}{z^2} + {\sum}' \left[\frac{1}{(z-w)^2} - \frac{1}{w^2} \right] \tag{1}$$
$$(w = 2mw + 2m'w') \,,$$

where Σ' means a summation over m and m' in which the term with $m = m' = 0$ is excluded. The term $-w^{-2}$ added is to make the series convergent. For, the general term of the series is

$$\frac{1}{(z-w)^2} - \frac{1}{w^2} = \frac{z(2w - z)}{(z-w)^2 w^2} \ .$$

The convergence of the series, when $|w| >> |z|$, depends upon the convergence of the series $\Sigma'|w|^{-3}$. We shall show that $\Sigma'|w|^{-\alpha}$ converges for $\alpha > 2$.

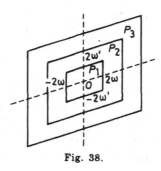

Fig. 38.

Let us arrange the terms of the series according to a sequence of parallelograms $P_k (k = 1, 2, \dots)$. The first parallelogram P_1 has the vertices $2w + 2w', -2w +$

$2\omega', -2\omega - 2\omega', 2\omega - 2\omega'$, and has eight w-points on its boundary, which include the four vertices and the four middle points of the sides, $\pm 2\omega, \pm 2\omega'$. The second parallelogram P_2 has the vertices $\pm 4\omega \pm 4\omega'$, and has $2 \times 8 = 16$ w-points on its boundary. In general, the parallelogram P_k has the vertices $\pm 2k\omega \pm 2k\omega'$, and has on its boundary $8k$ w-points. Let δ be the shortest distance from the center O to a w-point on the boundary of P_1, i.e., δ is the smallest among $|2\omega|, |2\omega'|$ and $|2\omega \pm 2\omega'|$. We have

$$\sum' \frac{1}{|w|^\alpha} < \sum_{k=1}^{\infty} \frac{8k}{(k\delta)^\alpha} = \frac{8}{\delta^\alpha} \sum_{k=1}^{\infty} \frac{1}{k^{\alpha-1}} \ .$$

The series on the r.h.s. is convergent for $\alpha > 2$, hence $\Sigma'|w|^{-\alpha}$ also converges for the same $\alpha > 2$.

After rearranging the terms of the series on the r.h.s. of (1), it can be shown that the function $\wp(z)$ has periods 2ω and $2\omega'$. The function $\wp(z)$ has the following properties:

1. $\wp(z)$ is a doubly-periodic function with w as its only pole;
2. In the neighborhood of the origin, its principal part is z^{-2};
3. $\wp(z) - z^{-2} = 0$ when $z \to 0$.

In the mean time, these three properties determine the function $\wp(z)$ completely. For, if there is another function $f(z)$ having the above mentioned three properties, then $f(z) - \wp(z)$ will be a doubly-periodic function without singularities. According to Theorem 1 of Sec. 8.3, it is a constant, and, as it is zero at $z = 0$, the constant has to be zero, and hence $f(z) \equiv \wp(z)$.

From (1) we see that $\wp(z)$ is an even function: $\wp(-z) = \wp(z)$. Differentiating (1) gives

$$\wp'(z) = -\frac{2}{z^3} - 2\sum' \frac{1}{(z-w)^3} = -2\sum \frac{1}{(z-w)^3} \ , \qquad (2)$$

where Σ denotes a sum over all m and m'. This is an elliptic function of order 3, and it is an odd function: $\wp'(-z) = -\wp(z)$. The periodicity of (2) is obvious: we have

$$\wp'(z + 2\omega) = \wp'(z) \ , \quad \wp'(z + 2\omega') = \wp'(z) \ .$$

Integrating gives

$$\wp(z + 2\omega) = \wp(z) + C, \quad \wp(z + 2\omega') = \wp(z) + C' \ .$$

Putting $z = -\omega$ and $-\omega'$, respectively in these two equations, we find $C = 0, C' = 0$ since $\wp(z)$ is an even function. This is another proof for the periodicity of $\wp(z)$.

In the neighborhood of the origin, $\wp(z) - z^{-2}$ can be expanded in a power series, whose radius of convergence is δ (δ being the shortest distance from the origin O to the w-point on the boundary of the parallelogram P_1). From

$$\frac{1}{(z-w)^2} - \frac{1}{w^2} = \frac{2z}{w^3} + \frac{2z^2}{w^4} + \ldots + \frac{(n+1)z^n}{w^{n+2}} + \ldots$$

we know that the odd-power terms of the series sought are all zero, since $\Sigma' w^{-2k-1} = 0$. Hence,

$$\wp(z) = \frac{1}{z^2} + c_2 z^2 + c_3 z^4 + \ldots + c_\lambda z^{2\lambda-2} + \ldots = \sum_{\lambda=0}^{\infty} c_\lambda z^{2\lambda-2} , \qquad (3)$$

where

$$c_0 = 1, \quad c_1 = 0, \quad c_2 = 3\sum\nolimits' w^{-4}, \quad c_3 = 5\sum\nolimits' w^{-6}, \ldots ,$$
$$c_\lambda = (2\lambda - 1)\sum\nolimits' w^{-2\lambda} \quad (\lambda \geq 2) . \qquad (4)$$

Differentiating (3), we obtain

$$\wp'(z) = -\frac{2}{z^3} + 2c_2 z^2 + 4c_3 z^3 + \ldots + (2\lambda - 2)c_\lambda z^{2\lambda-3} + \ldots . \qquad (5)$$

8.5. Algebraic Relation between $\wp(z)$ and $\wp'(z)$

It can be shown [see Ex. 1 at the end of this chapter] that between two elliptic functions with the same periods and the same poles there exists an algebraic relation.

To study the algebraic relation between $\wp(z)$ and $\wp'(z)$, consider the power series expansions in the neighborhood of the origin. By (3) and (5) of the preceding section, we have

$$\wp^3(z) = \frac{1}{z^6} + \frac{3c_2}{z^2} + 3c_3 + \ldots$$
$$\wp'^2(z) = \frac{4}{z^6} - \frac{8c_2}{z^2} - 16c_3 + \ldots ,$$

where the unwritten terms are zeros for $z = 0$. Combining these two expansions to cancel the z^{-6} terms, we find

$$\wp'^2(z) - 4\wp^3(z) = -\frac{20c_2}{z^2} - 28c_3 + \dots .$$

But the elliptic functions $\wp'^2(z) - 4\wp^3(z)$ and $-20c_2\wp(z) - 28c_3$ have the same periods and the same poles and principal parts, and their difference is zero at $z = 0$. Hence, they are identical, i.e.,

$$\wp'^2(z) = 4\wp^3(z) - g_2\wp(z) - g_3 , \tag{1}$$

where

$$g_2 = 20c_2 = 60 \sum{}' w^{-4}, \quad g_3 = 28c_3 = 140 \sum{}' w^{-6} . \tag{2}$$

(1) is a differential equation satisfied by $\wp(z)$; the constants g_2 and g_3 are called *invariants* (of Weierstrass elliptic function).

Differentiating (1) and dividing by $2\wp'(z)$, there results

$$\wp''(z) = 6\wp^2(z) - \frac{g_2}{2} . \tag{3}$$

On the other hand, by differentiating (5) of the last section, we obtain the expansion of $\wp''(z)$ in the neighborhood of the origin:

$$\wp''(z) = \frac{6}{z^4} + 2c_2 + 12c_3z^2 + \dots + (2\lambda - 2)(2\lambda - 3)c_\lambda z^{2\lambda - 4} + \dots . \tag{4}$$

Taking the square of $\wp(z)$ given by (3) of the last section and substituting into (3) above, then comparing with (4), we obtain

$$(2\lambda - 2)(2\lambda - 3)c_\lambda = 6 \sum_{\mu=0}^{\lambda} c_\mu c_{\lambda-\mu} \quad (\lambda \geq 3) .$$

Putting the values of c_0 and c_1 into it, the equation becomes

$$(2\lambda + 1)(\lambda - 3)c_\lambda = 3 \sum_{\mu=2}^{\lambda-2} c_\mu c_{\lambda-\mu} \quad (\lambda \geq 4) . \tag{5}$$

By this recurrence formula, all the c_λ with $\lambda \geq 4$ can be expressed in terms of c_2 and c_3, and in turn by (2), in terms of g_2 and g_3. The results are

$$c_2 = \frac{g_2}{2^2 \cdot 5}, \quad c_3 = \frac{g_3}{2^2 \cdot 7}, \quad c_4 = \frac{g_2^2}{2^4 \cdot 3 \cdot 5^2} , \tag{6}$$
$$c_5 = \frac{3g_2g_3}{2^4 \cdot 5 \cdot 7 \cdot 11}, \dots .$$

From (1), we know that the solutions of the following differential equation:

$$\left(\frac{d\varsigma}{dz}\right)^2 = 4\varsigma^3 - g_2\varsigma - g_3 \qquad (7)$$

are $\varsigma = \wp(\pm z + \alpha)$, α being an integration constant. Since $\wp(z)$ tends to ∞ when $z \to 0$, we find by integrating (7)

$$z = \int_\infty^\wp \frac{dt}{\sqrt{4t^3 - g_2 t - g_3}} , \qquad (8)$$

which is (10) of Sec. 8.1.

Let the three zeros of the cubic polynomial on the r.h.s. of (7) be e_1, e_2, e_3:

$$4\varsigma^3 - g_2\varsigma - g_3 = 4(\varsigma - e_1)(\varsigma - e_2)(\varsigma - e_3) . \qquad (9)$$

Comparing coefficients gives

$$e_1 + e_2 + e_3 = 0 , \qquad (10)$$

$$e_2 e_3 + e_3 e_1 + e_1 e_2 = -\frac{g_2}{4} , \quad e_1 e_2 e_3 = \frac{g_3}{4} . \qquad (11)$$

The discriminant of the polynomial (9) is

$$\Delta = 16(e_1 - e_2)^2(e_2 - e_3)^2(e_1 - e_3)^2 = g_2^3 - 27g_3^2 . \qquad (12)$$

To find the solution of a cubic equation, cf. App. I.

From (1), it is seen that the three roots e_k correspond to the three zeros, z_k, of $\wp'(z)$ in its period parallelogram. We shall now show that $z_k = \omega_k$, ω_k being the half periods: ω, ω' and $\omega + \omega'$. From the periodicity of $\wp'(z)$ and its being an odd function, we have

$$\wp'(z + 2\omega) = \wp'(z) = -\wp'(-z) .$$

Letting $z = -\omega$, we obtain

$$\wp'(\omega) = -\wp'(\omega) .$$

Now, ω is not a pole of $\wp'(z)$, so $\wp'(\omega)$ is finite and $\wp'(\omega) = 0$. In a similar way, it can be proved that $\wp'(\omega) = 0, \wp'(\omega + \omega') = 0$. These results show that ω_k are half periods, the correspondence relations between ω_k and e_k are

$$\wp(\omega_1) = e_1, \quad \wp(\omega_2) = e_2, \quad \wp(\omega_3) = e_3 . \qquad (13)$$

In order for ω_k to adapt to the condition (10) satisfied by e_k, we make a slightly different choice of ω_k, namely,

$$\omega_1 = \omega, \quad \omega_3 = \omega', \quad \omega_2 = -\omega - \omega', \tag{14}$$

so that

$$\omega_1 + \omega_2 + \omega_3 = 0. \tag{15}$$

But then $\omega_1, \omega_2, \omega_3$ will no longer fall in the same period parallelogram, while $\omega_1, \omega_3, \omega_2 + 2\omega + 2\omega'$ do. The reason that we choose ω' to be ω_3 instead of ω_2 is: By such choice, the conventional requirement $e_1 > e_2 > e_3$ is satisfied when the three roots e_k are all real. This point will become clear when we come to the theory of Theta functions. [Cf. the discussion after (5) of Sec. 9.7.]

8.6. The Function $\varsigma(z)$

For the representation of the integral of an elliptic function, we need the function $\varsigma(z)$ (this is not the Riemann's ς-function in Sec. 3.14). Integrating $\wp(z) - z^{-2}$ along an arbitrary path starting from the origin without passing through a pole, we obtain

$$\int_0^z \left\{ \wp(z) - \frac{1}{z^2} \right\} dz = -\sum{}' \left[\frac{1}{z-w} + \frac{1}{w} + \frac{z}{w^2} \right].$$

According to the discussion in Sec. 8.4, it can be proved that the series on the r.h.s. converges, and hence represents a meromorphic function with a simple pole, $z = w$. Define $\varsigma(z)$ by

$$\varsigma(z) = \frac{1}{z} + \sum{}' \left[\frac{1}{z-w} + \frac{1}{w} + \frac{z}{w^2} \right]. \tag{1}$$

Its relation to $\wp(z)$ is

$$\int_0^z \left\{ \wp(z) - \frac{1}{z^2} \right\} dz = -\varsigma(z) + \frac{1}{z}. \tag{2}$$

By differentiation, we have

$$\varsigma'(z) = -\wp(z). \tag{3}$$

It is evident from these equations that $\varsigma(z)$ is an odd function: $\varsigma(-z) = -\varsigma(z)$.

Substituting the series expression of $\wp(z)$ [Eq. (3) of Sec. 8.4] into (2) above and integrating, we obtain the expansion of $\varsigma(z)$ in the vicinity of the origin:

$$\varsigma(z) = \frac{1}{z} - \frac{c_2}{3}z^3 - \frac{c_3}{5}z^5 - \ldots - \frac{c_\lambda}{2\lambda - 1}z^{2\lambda-1} - \ldots . \tag{4}$$

Since $\varsigma(z)$ has only one simple pole in one period parallelogram, it cannot be an elliptic function and does not possess double periods. But $\varsigma(z + w)$ and $\varsigma(z)$ have the same derivative, $-\wp(z)$, they, therefore, can differ only by a constant. Let us find this constant. Introduce two fundamental constants η and η':

$$\varsigma(z + 2\omega) = \varsigma(z) + 2\eta, \quad \varsigma(z + 2\omega') = \varsigma(z) + 2\eta' . \tag{5}$$

Since $\varsigma(z)$ is an odd function, by putting $z = -\omega$ and $z = -\omega'$ respectively in the last two equations, we have

$$\varsigma(\omega) = \eta, \quad \varsigma(\omega') = \eta' . \tag{6}$$

From (5), we find the general relation

$$\varsigma(z + w) = \varsigma(z + 2m\omega + 2m'\omega') = \varsigma(z) + 2m\eta + 2m'\eta' . \tag{7}$$

There is a simple relation among the four constants $\eta, \eta', \omega, \omega'$, which can be obtained by integrating $\varsigma(z)$ along the period parallelogram. $\varsigma(z)$ has only one simple pole in the period parallelogram and, by (1), we know that the residue is 1. Therefore, if $\mathrm{Im}(\omega'/\omega) > 0$, we have

$$2\pi i = \int_{OABCO} \varsigma(z)dz .$$

The sum of the integrals along the lines OA and BC is

$$\int_{z_0}^{z_0+2\omega} [\varsigma(z) - \varsigma(z + 2\omega')] \, dz = -\int_{z_0}^{z_0+2\omega} 2\eta'dz = -4\omega\eta' ;$$

that along AB and CO can also be shown, in the same way, to be $4\omega'\eta$. Hence, we have $4\eta\omega' - 4\omega\eta' = 2\pi i$, or

$$\eta\omega' - \omega\eta' = \frac{\pi i}{2} . \tag{8}$$

This result depends on the assumption $\mathrm{Im}(\omega'/\omega) > 0$. If $\mathrm{Im}(\omega'/\omega) < 0$, the path $OABCO$ has a reverse direction, and we have to change i to $-i$ on the r.h.s. of (8).

Corresponding to ω_k, we also have η_k, whose relations to ω_k are

$$\varsigma(\omega_k) = \eta_k \quad (k = 1, 2, 3) . \tag{9}$$

From (6) and (7), it is seen that the relations corresponding to (15) of the last section are

$$\eta_1 \equiv \eta, \quad \eta_3 = \eta', \quad \eta_2 = -\eta - \eta' . \tag{10}$$

Hence, (8) can be generalized to

$$\omega_1 \eta_2 - \omega_2 \eta_1 = \omega_2 \eta_3 - \omega_3 \eta_2 = \omega_3 \eta_1 - \omega_1 \eta_3 = \frac{\pi i}{2} . \tag{11}$$

8.7. The Function $\sigma(z)$

The function $\sigma(z)$ is obtained from the integral of $\varsigma(z)$. Integrating $\varsigma(z) - z^{-1}$ along an arbitrary path starting from the origin without passing through any poles, we find

$$\int_0^z \left\{ \varsigma(z) - \frac{1}{z} \right\} dz = \sum{}' \left[\ln \left(1 - \frac{z}{w} \right) + \frac{z}{w} + \frac{z^2}{2w^2} \right] . \tag{1}$$

In order to avoid the occurrence of the multi-valued logarithmic function, we put the r.h.s. of (1) equal to $\ln(\sigma(z)/z)$:

$$\int_0^z \left\{ \varsigma(z) - \frac{1}{z} \right\} dz = \ln \frac{\sigma(z)}{z} . \tag{2}$$

Substituting in (1) gives

$$\sigma(z) = z \prod{}' \left\{ \left(1 - \frac{z}{w} \right) \exp \left(\frac{z}{w} + \frac{z^2}{2w^2} \right) \right\} , \tag{3}$$

where Π' denotes a product extending over all integers m and m', except $m = m' = 0$. The convergence of the infinite product (3) is the same as that of the infinite series (1) [cf. Sec. 1.6, Theorems 2 and 5]; when $|w| >> |z|$, by expanding the logarithmic function on the r.h.s. of (1) and using the discussion in Sec. 8.4, the series on the r.h.s. of (1) is seen to be convergent.

Differentiating (2), we obtain

$$\frac{\sigma'(z)}{\sigma(z)} = \varsigma(z) . \tag{4}$$

From (3) we see that $\sigma(z)$ is an odd function: $\sigma(-z) = -\sigma(z)$. Also, from (3), we find that $\sigma(z)$ is an integral function which has no singularities in the finite domain, and has $z = w$ as its only zero.

Substituting the series expansion of $\varsigma(z)$ [(4) of the last section] into (2) and integrating, we have

$$\ln \frac{\sigma(z)}{z} = -\frac{c_2}{3 \cdot 4} z^4 - \frac{c_3}{5 \cdot 6} z^6 - \cdots - \frac{c_\lambda}{2\lambda(2\lambda - 1)} z^{2\lambda} - \cdots . \tag{5}$$

Removing the logarithm and applying (6) of Sec. 8.5, we obtain the power series expansion of $\sigma(z)$:

$$\sigma(z) = z - \frac{g_2 z^5}{2^4 \cdot 3 \cdot 5} - \frac{g_3 z^7}{2^3 \cdot 3 \cdot 5 \cdot 7} \\ - \frac{g_2^2 z^9}{2^9 \cdot 3^2 \cdot 5 \cdot 7} - \frac{g_2 g_3 z^{11}}{2^7 \cdot 3^2 \cdot 5^2 \cdot 7 \cdot 11} - \cdots . \tag{6}$$

From (4) and Eq. (5) of Sec. 8.6, we have

$$\frac{\sigma'(z + 2w)}{\sigma(z + 2w)} - \frac{\sigma'(z)}{\sigma(z)} = \varsigma(z + 2w) - \varsigma(z) = 2\eta .$$

Integrating gives

$$\ln \frac{\sigma(z + 2w)}{\sigma(z)} = 2\eta z + \ln C ,$$

i.e.,

$$\sigma(z + 2w) = C e^{2\eta z} \sigma(z) ,$$

C being an integration constant. Put $z = -w$. Since $\sigma(z)$ is an odd function, we have

$$\sigma(w) = -C \sigma(w) e^{-2\eta w} ,$$

so $C = -e^{2\eta w}$, and

$$\sigma(z + 2w) = -e^{2\eta(z+w)} \sigma(z) . \tag{7}$$

In the same way, we can show that

$$\sigma(z + 2w') = -e^{2\eta'(z+w')} \sigma(z) . \tag{8}$$

In (7), changing z to $z - w$, we obtain the more symmetric form:

$$e^{-\eta z} \sigma(w + z) = e^{\eta z} \sigma(w - z) . \tag{9}$$

Similarly, from (8), we obtain

$$e^{-\eta' z}\sigma(\omega' + z) = e^{\eta' z}\sigma(\omega' - z) . \tag{10}$$

In (8), changing z to $z + 2\omega$, and using (7), we obtain

$$\sigma(z + 2\omega + 2\omega') = e^{2\eta'(z+2\omega+\omega')+2\eta(z+\omega)}\sigma(z) .$$

Applying (8) of the last section, this becomes

$$\sigma(z + 2\omega + 2\omega') = -e^{2(\eta+\eta')(z+\omega+\omega')}\sigma(z) . \tag{11}$$

Replacing z by $z - \omega - \omega'$, we again arrive at the more symmetric form:

$$e^{-(\eta+\eta')z}\sigma(\omega + \omega' + z) = e^{(\eta+\eta')z}\sigma(\omega + \omega' - z) . \tag{12}$$

The three formulae (9), (10) and (12) can be combined into one formula, namely

$$e^{-\eta_k z}\sigma(\omega_k + z) = e^{\eta_k z}\sigma(\omega_k - z) \quad (k = 1, 2, 3) . \tag{13}$$

Introducing the functions $\sigma_k(z)$:

$$\sigma_k(z) = e^{-\eta_k z}\frac{\sigma(\omega_k + z)}{\sigma(\omega_k)} = e^{\eta_k z}\frac{\sigma(\omega_k - z)}{\sigma(\omega_k)} . \tag{14}$$

It is obvious that these functions are even functions: $\sigma_k(-z) = \sigma_k(z)$, and $\sigma_k(0) = 1, \sigma_k(\omega_k) = 0; z = w + \omega_k$ are zeros. The formulae corresponding to (7) and (8) can easily be shown to be

$$\sigma_k(z + 2\omega_k) = -e^{2\eta_k(z+\omega_k)}\sigma_k(z) , \tag{15}$$

$$\sigma_k(z + 2\omega_l) = e^{2\eta_l(z+\omega_l)}\sigma_k(z) \quad (l \neq k) . \tag{16}$$

It can be proved [Ex. 2 at the end of this chapter] that

$$\ln \sigma_k(z) = -\sum_{n=1}^{\infty} \wp^{(2n-2)}(\omega_k)\frac{z^{2n}}{(2n)!} . \tag{17}$$

8.8. Homogeneity of the Weierstrass Elliptic Function

In order to expose the periods, or equivalently the invariants g_2 and g_3 [Eq. (1) of Sec. 8.5], of the elliptic functions, we sometimes write

$$\wp(z) = \wp(z|\omega,\omega') = \wp(z; g_2, g_3) ,$$
$$\varsigma(z) = \varsigma(z|\omega,\omega') = \varsigma(z; g_2, g_3) ,$$
$$\sigma(z) = \sigma(z|\omega,\omega') = \sigma(z; g_2, g_3) .$$

From the definitions of these three functions [Eq. (1) of Sec. 8.4, Eq. (1) of Sec. 8.6, and Eq. (3) of Sec. 8.7], we see immediately that for any number λ, there are the homogeneity relations:

$$\wp(\lambda z|\lambda\omega, \lambda\omega') = \lambda^{-2}\wp(z|\omega,\omega') ,$$
$$\varsigma(\lambda z|\lambda\omega, \lambda\omega') = \lambda^{-1}\varsigma(z|\omega,\omega') ,$$
$$\sigma(\lambda z|\lambda\omega, \lambda\omega') = \lambda\sigma(z|\omega,\omega') .$$

From Eq. (2) of Sec. 8.5, the corresponding homogeneity relations relevant to g_2 and g_3 are

$$\wp(\lambda z; \lambda^{-4}g_2, \lambda^{-6}g_3) = \lambda^{-2}\wp(z; g_2, g_3) ,$$
$$\varsigma(\lambda z; \lambda^{-4}g_2, \lambda^{-6}g_3) = \lambda^{-1}\varsigma(z; g_2, g_3) ,$$
$$\sigma(\lambda z; \lambda^{-4}g_2, \lambda^{-6}g_3) = \lambda\sigma(z; g_2, g_3) .$$

From the homogeneity relations, it is seen that the Weierstrass elliptic function depends essentially upon two parameters; for example, it can be expressed as a function of z/ω and ω'/ω. This will be realized when we come to Theta functions to be discussed later.

8.9. Representations of a General Elliptic Function

Any elliptic function can be expressed in terms of $\sigma(z)$, or $\varsigma(z)$, or $\wp(z)$. We shall show these one by one.

1. Expressed in terms of $\sigma(z)$

Let $f(z)$ be an elliptic function with periods $(2\omega, 2\omega')$ and of order $s; \alpha_r$ and $\beta_r (r = 1, 2, \ldots, s)$ being respectively the s zeros and s poles in the period parallelogram, which are counted according to their multiplicities. By Theorem 4, Eq. (5) of Sec. 8.3, we have

$$\sum_{r=1}^{s} \alpha_r - \sum_{r=1}^{s} \beta_r = 2\Omega , \tag{1}$$

where 2Ω is a certain period. Construct the following function:

$$\varphi(z) = \frac{\sigma(z - \alpha_1)\ldots\sigma(z - \alpha_s)}{\sigma(z - \beta_1)\ldots\sigma(z - \beta_{s-1})\sigma(z - \beta_s - 2\Omega)}$$

This function has the same zeros and poles as $f(z)$, since, by Eq. (3) of Sec. 8.7, we know that 0 is a simple zero of $\sigma(z)$. Let us examine the double periodicity of $\varphi(z)$. Changing z to $z + 2\omega$, then according to Eq. (7) of Sec. 8.7, the numerator and the denominator of $\varphi(z)$ will be multiplied respectively by the following factors

$$(-)^s \exp\{2\eta(sz + s\omega - \alpha_1 - \ldots - \alpha_s)\}\ ,$$
$$(-)^s \exp\{2\eta(sz + s\omega - \beta_1 - \ldots - \beta_s - 2\Omega)\}\ .$$

By (1), we see that these two factors are equal to each other, and hence $\varphi(z)$ has the period 2ω. Similarly, we can show that $\varphi(z)$ also has the period $2\omega'$. Therefore, $\varphi(z)$ and $f(z)$ are elliptic functions which have the same periods, the same zero and poles. Their ratio $f(z)/\varphi(z)$ is a doubly-periodic function without singularities, and hence, by Theorem 1 of Sec. 8.3, it has to be a constant. So,

$$f(z) = C\frac{\sigma(z - \alpha_1)\ldots\sigma(z - \alpha_s)}{\sigma(z - \beta_1)\ldots\sigma(z - \beta_{s-1})\sigma(z - \beta_s - 2\Omega)}\ . \tag{2}$$

The constant C can be determined by the value of $f(z)$ at a certain point which is neither a zero nor a pole.

From the fact that the zeros of $\sigma(z)$ have the period w, we see that it is not necessary to choose the zeros α_r and the poles β_r all lying inside one period parallelogram. We can choose arbitrarily s zeros α'_r and s poles β'_r, if only they embrace all the zeros and poles after adding to them the period w so as to make

$$\sum_{r=1}^{s}\alpha'_r = \sum_{r=1}^{s}\beta'_r\ . \tag{3}$$

Then, the expression for $f(z)$ is

$$f(z) = C\prod_{r=1}^{s}\frac{\sigma(z - \alpha'_r)}{\sigma(z - \beta'_r)}\ . \tag{4}$$

2. Expressed in terms of $\varsigma(z)$

Suppose that the k poles β_r of the elliptic function $f(z)$ with periods $(2\omega, 2\omega')$ are in general of order p_r and the principal part in the neighborhood of β_r is known to be

$$\frac{B_{r,1}}{z - \beta_r} + \frac{B_{r,2}}{(z - \beta_r)^2} + \cdots + \frac{B_{r,p_r}}{(z - \beta_r)^{p_r}} . \tag{5}$$

Construct the following function:

$$\varphi(z) = \sum_{r=1}^{k} \left\{ B_{r,1}\varsigma(z - \beta_r) - B_{r,2}\varsigma'(z - \beta_r) + \cdots \right.$$
$$\left. + (-)^{p_r-1} \frac{B_{r,p_r}}{(p_r - 1)!} \varsigma^{(p_r-1)}(z - \beta_r) \right\} .$$

Since $z = 0$ is a simple pole of $\varsigma(z)$ [Eq. (1) of Sec. 8.6], this function has the same poles and the same principal parts as $f(z)$. Let us examine the double periodicity of $\varphi(z)$. Replacing z by $z + 2\omega$. $\varphi(z)$ will be increased by $2\eta\Sigma_{r=1}^{k} B_{r,1}$ [cf. Eq. (5) of Sec. 8.6]. But $\Sigma_{r=1}^{k} B_{r,1}$ is the sum of the residues of $f(z)$ at the poles in a period parallelogram (note that β_r need not all lie in one period parallelogram), so it must be zero according to Theorem 2 of Sec. 8.3. Hence, $\varphi(z)$ has the period 2ω. Similarly, $\varphi(z)$ also has the period $2\omega'$, and the difference between $f(z)$ and $\varphi(z)$ is a doubly-periodic function without singularities. Then, by Theorem 1 in Sec. 8.3, the difference is a constant and

$$f(z) = C + \sum_{r=1}^{k} \sum_{q=1}^{p_r} \frac{(-)^{q-1} B_{r,q}}{(q-1)!} \varsigma^{(q-1)}(z - \beta_r) . \tag{6}$$

The order of the elliptic function $f(z)$ is $s = \sum_{r=1}^{k} p_r$.

The form (6) is most convenient for evaluating the integral of an elliptic function. For, integrating (6), we have

$$\int f(z)dz = Cz + C' + \sum_{r=1}^{k} B_{r,1} \ln \sigma(z - \beta_r)$$
$$+ \sum_{r=1}^{k} \sum_{q=1}^{p_r-1} \frac{(-)^q B_{r,q+1}}{q!} \varsigma^{(q-1)}(z - \beta_r) . \tag{7}$$

3. Expressed in terms of $\wp(z)$

As $\wp(z)$ is an even function, we have to consider even and odd functions separately. First, let $f_1(z)$ be an even elliptic function. In order to distinguish between even and odd functions, we choose the vertices $ABCD$ of the period parallelogram to be $\omega + \omega', \omega' - \omega, -\omega - \omega', \omega - \omega'$. Inside $ABCD$, the zeros, except possibly $z = 0$, all appear in pairs: $\pm\alpha_1, \pm\alpha_2, \ldots, \pm\alpha_k$, each having its number of occurrence equal to its order. The poles inside $ABCD$, except possibly $z = 0$, also appear in pairs: $\pm\beta_1, \pm\beta_2, \ldots, \pm\beta_l$, each again having its number of occurrence equal to its order. According to Theorem 3 of Sec. 8.3, the number of zeros in $ABCD$ is equal to the number of poles in it. Therefore, if $k < l, z = 0$ is a zero of order $2(l - k)$; if $k > l, z = 0$ is a pole of order $2(k - l)$. The order s of the elliptic function $f_1(z)$ is equal to the larger of $2k$ and $2l$. Applying the same reasoning as in the proof of (2), we obtain

$$f_1(z) = C \frac{[\wp(z) - \wp(\alpha_1)][\wp(z) - \wp(\alpha_2)]\ldots[\wp(z) - \wp(\alpha_k)]}{[\wp(z) - \wp(\beta_1)][\wp(z) - \wp(\beta_2)]\ldots[\wp(z) - \wp(\beta_l)]} . \tag{8}$$

If a certain zero, say α_r, is a half period, then α_r and $-\alpha_r$ both lie on the boundary of $ABCD$, and their difference $2\alpha_r$ is equal to a whole period. Thus, the order of α_r can only be counted by one half. If the order of this zero is $2p$, it will appear only p times in the product (8), not the same as those zeros lying inside $ABCD$, which will appear $2p$ times. The same conclusion is effected for poles equal to a half period.

If a zero of an even function is a half period, or a whole period, its order can only be an even number; the same holds for a pole. For, otherwise, let the half period ω of the even function $f(z)$ be a zero of odd order, $2n + 1$, then the $(2n+1)$th derivative $f^{(2n+1)}(z)$ would not be zero at $z = \omega$. From $f^{(2n+1)}(z + 2\omega) = f^{(2n+1)}(z)$, putting $z = -\omega$, we obtain $f^{(2n+1)}(\omega) = f^{(2n+1)}(-\omega)$. But $f^{(2n+1)}(z)$ is an odd function, so $f^{(2n+1)}(-\omega) = -f^{(2n+1)}(\omega)$. From these two equations, we see that, since ω is not a singular point, $f^{(2n+1)}(\omega)$ must be zero, which is contradictory to the conclusion $f^{(2n+1)}(\omega) \neq 0$. Therefore, the order of the half-period zero of an even function can only be even. Consider the poles; since a pole of $f(z)$ is a zero of $1/f(z)$, the order of the half-period pole of an even function can only be even. The same reasoning will show that the order of a half-period zero or pole of an odd function can only be odd. As twice a period is also a period, the above assertions about evenness and oddness are also valid when the zeros and poles are whole-period, i.e., in the case when $z = 0$ is a zero or a pole.

Next, let us consider an odd elliptic function $f_2(z)$. Evidently, $f_2(z)/\wp'(z)$

is an even function, and Eq. (8) applies. Thus,

$$f(z) = C' \frac{[\wp(z) - \wp(\alpha_1')] \dots [\wp(z) - \wp(\alpha_{k'}')]}{[\wp(z) - \wp(\beta_1')] \dots [\wp(z) - \wp(\beta_{l'}')]} \, \wp'(z) \, . \tag{9}$$

A general elliptic function $f(z)$ can be written as a sum of an even function and an odd function as follows:

$$f(z) = \frac{1}{2} \{f(z) + f(-z)\} + \frac{1}{2} \{f(z) - f(-z)\} = f_1(z) + f_2(z) \, . \tag{10}$$

Applying (8) and (9), we can express $f(z)$ in the form

$$f(z) = R_1(\wp(z)) + R_2(\wp(z))\wp'(z) \, , \tag{11}$$

where $R_1(\wp)$ and $R_2(\wp)$ are rational functions of $\wp$.

Using (11), we can evaluate the integral of $f(z)$:

$$\int f(z)dz = \int R_1(\wp)dz + \int R_2(\wp)d\wp \, .$$

The second integral on the r.h.s. can be expressed as an elementary function of $\wp(z)$. As regards the first integral, by the same discussion as in Sec. 8.1, expressing $R_1(\wp)$ in the form (6) of Sec. 8.1, we see that it can be reduced to two types of integrals, namely,

$$I_m = \int [\wp(z)]^m dz \, , \quad J_k = \int \frac{dz}{[\wp(z) - \wp(\beta)]^k} \, . \tag{12}$$

For I_m, there are recurrence relations same as (8) of Sec. 8.1 (making use of (1) and (3) of Sec. 8.5):

$$\frac{d}{dz} \{[\wp(z)]^m \wp'(z)\} = (4m+6)[\wp(z)]^{m+2} - \left(m + \frac{1}{2}\right) g_2[\wp(z)]^m - mg_3[\wp(z)]^{m-1}, \tag{13}$$

$$[\wp(z)]^m \wp'(z) = (4m + 6)I_{m+2} - \left(m + \frac{1}{2}\right) g_2 I_m - mg_3 I_{m-1} \, . \tag{14}$$

In the last equation, putting successively $m = 0, 1, 2, \dots$, we can express $I_m (m \geq 2)$ in terms of $I_0 = z$ and $I_1 = -\varsigma(z)$.

For J_k, similar to Eq. (9) of Sec. 8.1, taking $\wp$ as the independent variable, and using Eq. (1) of Sec. 8.5, we have

$$\frac{d}{dz}\frac{\wp'(z)}{[\wp(z)-\wp(\beta)]^k} = -\frac{k\wp'^2(\beta)}{[\wp(z)-\wp(\beta)]^{k+1}} + \frac{(1-2k)\wp''(\beta)}{[\wp(z)-\wp(\beta)]^k}$$
$$+ \frac{12(1-k)\wp(\beta)}{[\wp(z)-\wp(\beta)]^{k-1}} + \frac{6-4k}{[\wp(z)-\wp(\beta)]^{k-2}} , \tag{15}$$

$$\frac{\wp'(z)}{[\wp(z)-\wp(\beta)]^k} = -k\wp'^2(\beta)J_{k+1} + (1-2k)\wp''(\beta)J_k$$
$$+ 12(1-k)\wp(\beta)J_{k-1} + (6-4k)J_{k-2} . \tag{16}$$

Putting successively $k = 1, 2, \ldots$ in this last equation, we obtain $J_k(k > 1)$ in terms of J_1 and I_0, I_1. When β is a half period, $\wp'(\beta) = 0, J_1$ itself can also be expressed in terms of I_0 and I_1. The evaluation of J_1 will be shown at the end of the next section [Eq. (19)].

8.10. Addition Formulae

Addition formula means the expression for $f(z + u), f$ being any one of the functions $\wp, \varsigma, \sigma$. As the formulae are rather complicated and not easy to obtain directly, we shall derive them in some indirect manner. Firstly, we express $\wp(z) - \wp(u)$ in terms of the σ functions. As a function of $z, \wp(z) - \wp(u)$ has two zeros, $z = \pm u; z = 0$ is still its pole. By (2) of the last section, we have

$$\wp(z) - \wp(u) = C\frac{\sigma(z+u)\sigma(z-u)}{\sigma^2(z)} .$$

Multiplying by $\sigma^2(z)$, then let $z \to 0$, we find $1 = -C\sigma^2(u)$ and hence

$$\wp(z) - \wp(u) = -\frac{\sigma(z+u)\sigma(z-u)}{\sigma^2(z)\sigma^2(u)} , \tag{1}$$

which is the addition formula for the σ function.

Taking the logarithmic derivative of (1) with respect to z, we have

$$\frac{\wp'(z)}{\wp(z)-\wp(u)} = \varsigma(z+u) + \varsigma(z-u) - 2\varsigma(z) . \tag{2}$$

Interchanging z and u, it becomes

$$-\frac{\wp'(u)}{\wp(z)-\wp(u)} = \varsigma(z+u) - \varsigma(z-u) - 2\varsigma(u) . \tag{3}$$

Adding these two equations, we have

$$\varsigma(z + u) - \varsigma(z) - \varsigma(u) = \frac{1}{2} \frac{\wp'(z) - \wp'(u)}{\wp(z) - \wp(u)}, \tag{4}$$

which is the addition formula for the ς function.

From (4), by differentiation with respect to z, then expressing $\wp''(z)$ in terms of $\wp(z)$ by (3) of Sec. 8.5, we shall arrive at the addition formula for the $\wp$ function. But the calculation is somewhat tedious. So, we shall derive it indirectly as follows. First, we show that

$$\wp(z + u) + \wp(z) + \wp(u) = [\varsigma(z + u) - \varsigma(z) - \varsigma(u)]^2 . \tag{5}$$

Both sides of this equation are elliptic functions of z with $z = 0$ and $z = -u$ as their poles of the second order. In the vicinity of the origin, we have

$$\varsigma(z + u) - \varsigma(z) - \varsigma(u) = -\frac{1}{z} + z\varsigma'(u) + \alpha z^2 + \dots ,$$

$$[\varsigma(z + u) - \varsigma(z) - \varsigma(u)]^2 = \frac{1}{z^2} - 2\varsigma'(u) - 2\alpha z + \dots ,$$

with the principal part z^{-2}, same as that of the l.h.s. of (5). In the neighborhood of $z = -u$, let $z = v - u$, then we have

$$\varsigma(z + u) - \varsigma(z) - \varsigma(u)$$
$$= \varsigma(v) - \varsigma(v - u) - \varsigma(u)$$
$$= \frac{1}{v} - v\varsigma'(u) + \beta v^2 + \dots ,$$

$$[\varsigma(z + u) - \varsigma(z) - \varsigma(u)]^2 = \frac{1}{v^2} - 2\varsigma'(u) + 2\beta v + \dots ,$$

with the principal part $v^{-2} = (z + u)^{-2}$, again same as that of the l.h.s. of (5). Now, the two sides of (5) have the same principal part, so they can differ at most by a constant [Theorem 1 of Sec. 8.3]. But the constant term in the expansion of the l.h.s. in the vicinity of the origin is $2\wp(u) = -2\varsigma'(u)$, same as that of the r.h.s., hence (5) holds.

Substituting (4) in (5), we obtain the addition formula for the $\wp$ function:

$$\wp(z + u) + \wp(z) + \wp(u) = \frac{1}{4} \left[\frac{\wp'(z) - \wp'(u)}{\wp(z) - \wp(u)} \right]^2 . \tag{6}$$

Putting $u = z$, we have

$$\wp(2z) + 2\wp(z) = \frac{1}{4}\left[\frac{\wp''(z)}{\wp'(z)}\right]^2 = \frac{\left[6\wp^2(z) - \frac{1}{2}g_2\right]^2}{4\left\{4\wp^3(z) - g_2\wp(z) - g_3\right\}} , \tag{7}$$

which is the *duplication formula* for the $\wp$ function.

We shall now give another form of the addition formula for $\wp(z)$. Assume

$$\wp'(z) = A\wp(z) + B , \quad \wp'(u) = A\wp(u) + B , \tag{8}$$

which determine A and B with

$$A = \frac{\wp'(z) - \wp'(u)}{\wp(z) - \wp(u)} . \tag{9}$$

Since $\wp'(v)$ is a function of order 3, the equation $\wp'(v) - A\wp(v) - B = 0$ should have three roots, of which the sum has to be equal to the period [Theorem 4 of Sec. 8.3; the pole is now $v = 0$], and hence the roots are $v = z, v = u, v = -z - u$. Thus, we have

$$-\wp'(z + u) = A\wp(z + u) + B . \tag{10}$$

Eliminating A and B from (8) and (10), we arrive at

$$\begin{vmatrix} \wp(z) & \wp'(z) & 1 \\ \wp(u) & \wp'(u) & 1 \\ \wp(z + u) & -\wp'(z + u) & 1 \end{vmatrix} = 0 , \tag{11}$$

which is another form of the addition formula for the function $\wp$.

Another derivation of the addition formula (6) is as follows. By squaring (8) and (10), we have

$$\wp'^2(v) - [A\wp(v) + B]^2 = 0 . \tag{12}$$

This equation, as an equation of v, should have 6 roots: $v = \pm z, \pm u, \pm(z + u)$. Let $t = \wp(v)$, then, by (1) of Sec. 8.5, Eq. (12) reduces to

$$4t^3 - A^2t^2 - (2AB + g_2)t - (B^2 + g_3) = 0 . \tag{13}$$

This equation, as an equation of t, has three roots: $t = \wp(z), \wp(u), \wp(z + u)$, of which the sum is $A^2/4$. Using (9), we have (6).

In (6), putting $u = \omega_1$ and using Eqs. (9) and (13) of Sec. 8.5, we obtain

$$
\begin{aligned}
\wp(z + \omega_1) + \wp(z) + e_1 \\
= \frac{1}{4} \frac{\wp'^2(z)}{[\wp(z) - e_1]^2} = \frac{[\wp(z) - e_2][\wp(z) - e_3]}{\wp(z) - e_1} \\
= \wp(z) - e_1 + 2e_1 - e_2 - e_3 + \frac{(e_1 - e_2)(e_1 - e_3)}{\wp(z) - e_1} \, .
\end{aligned}
$$

Using Eq. (10) of Sec. 8.5, we have

$$
\wp(z + \omega_1) = e_1 + \frac{(e_1 - e_2)(e_1 - e_3)}{\wp(z) - e_1} , \tag{14}
$$

which gives the increment of the $\wp$ function when its argument is increased by a half period.

In (1), putting $u = \omega_1$ and using Eq. (14) of Sec. 8.7, we have by virtue that σ is an odd function,

$$
\wp(z) - e_1 = -\frac{\sigma(z + \omega_1)\sigma(z - \omega_1)}{[\sigma(z)\sigma(\omega_1)]^2} = \left[\frac{\sigma_1(z)}{\sigma(z)}\right]^2 . \tag{15}
$$

Extracting the root, there results

$$
\sqrt{\wp(z) - e_1} = \frac{\sigma_1(z)}{\sigma(z)} = e^{-\eta_1 z} \frac{\sigma(z + \omega_1)}{\sigma(\omega_1)\sigma(z)} . \tag{16}
$$

Here, by tradition, only one sign of the square root is taken. The same formula holds for e_2 and e_3, hence

$$
\sqrt{\wp(z) - e_k} = \frac{\sigma_k(z)}{\sigma(z)} \quad (k = 1, 2, 3) . \tag{17}
$$

In (14) and (16), putting $z = -\omega_1/2$, we have

$$
\wp\left(\frac{\omega_1}{2}\right) - e_1 = \pm\sqrt{(e_1 - e_2)(e_1 - e_3)} = \frac{e^{\eta_1 \omega_1}}{\sigma^2(\omega_1)} . \tag{18}
$$

Finally, we come to the evaluation of the integral J_1 of the last section. By integrating (3), we obtain

$$
J_1 = \int \frac{dz}{\wp(z) - \wp(u)} = \frac{1}{\wp'(u)} \left\{ \ln \frac{\sigma(z - u)}{\sigma(z + u)} + 2z\varsigma(u) \right\} . \tag{19}
$$

8.11. Expressing the Coordinates of a Cubic Curve by Elliptic Functions

Let us consider first the special case of a plane cubic curve, whose coordinates x, y satisfy the following relation:

$$y^2 = 4x^3 - g_2 x - g_3 \ . \tag{1}$$

If the discriminant $\Delta = g_2^3 - 27 g_3^2 \neq 0$, then the cubic curve has no double point. According to Eq. (1) of Sec. 8.5, the coordinates of this curve can be expressed by elliptic functions:

$$x = \wp(u), \quad y = \wp'(u) \ . \tag{2}$$

As $\wp(u)$ and $\wp'(u)$ are single-valued functions, to each value of u there can only be one pair of values (x, y). Conversely, it can be shown that, corresponding to a pair of values (x, y), there is only one value of u in the period parallelogram. For, $x = \wp(u)$ can only have two roots: u_1 and $u_2 = -u_1$, and since $\wp'(u)$ is an odd function, the values of y corresponding to these two values of u are different. Therefore, to one pair of values (x, y), there corresponds only one value of u.

Next, let us consider the following cubic curve:

$$y^2 = b_0 x^3 + 3 b_1 x^2 + 3 b_2 x + b_3 \ . \tag{3}$$

Performing the transformation

$$x = \frac{4}{b_0} x' - \frac{b_1}{b_0}, \quad y = \frac{4}{b_0} y' \ , \tag{4}$$

we obtain

$$y'^2 = 4x'^3 - g_2 x' - g_3 \ ,$$
$$g_2 = \frac{3}{4}(b_1^2 - b_0 b_2), \quad g_3 = \frac{1}{16}(3 b_0 b_1 b_2 - 2 b_1^3 - b_0^2 b_3) \ . \tag{5}$$

Then x and y can be expressed as (2) and substituted into (4) to give

$$x = \frac{4}{b_0} \wp(u) - \frac{b_1}{b_0}, \quad y = \frac{4}{b_0} \wp'(u) \ . \tag{6}$$

Now we can deal with the general cubic curve C_3. Assume that there is a point (α, β) which is known on C_3, and the tangent line at (α, β) intersects

$C_{3'}$ at another point (α', β'). Take the point (α', β') as the origin and write the equation of C_3 in the following form:

$$\varphi_3(x, y) + \varphi_2(x, y) + \varphi_1(x, y) = 0 , \tag{7}$$

where $\varphi_k(x, y)$ are homogeneous functions of degree k. As C_3 passes through the origin, there is no constant term in (7). An arbitrary straight line $y = tx$ will meet C_3 at three points. The x coordinates of these intersecting points besides the origin are determined by the equation

$$x^2\varphi_3(1, t) + x\varphi_2(1, t) + \varphi_1(1, t) = 0 ,$$

from which, we obtain

$$x = \frac{-\varphi_2(1, t) \pm \sqrt{P(t)}}{2\varphi_3(1, t)} , \quad y = tx , \tag{8}$$

where

$$P(t) = [\varphi_2(1, t)]^2 - 4\varphi_1(1, t)\varphi_3(1, t) . \tag{9}$$

This is a quartic polynomial in t. The root of this polynomial corresponds to a coincidence of two values of x, i.e., coincidence of two points, and hence it is the slope of a tangent to C_3 (since $t = y/x$). But we have known the slope of a tangent, which is the one mentioned at the beginning, namely, the slope of the tangent passing through the points (α, β). Let t_0 denote the slope, which is a root of $P(t) = 0$. Using Eq. (4) of Sec. 8.1, i.e., $t - t_0 = 1/\xi$, we obtain $\sqrt{P(t)} = \xi^{-2}\sqrt{P_1(\xi)}$. $P_1(\xi)$ being a cubic polynomial. Let $\eta^2 = P_1(\xi)$; ξ and η can be expressed by elliptic functions, as x and y in (6). Then, by (8), the coordinates (x, y) of C_3 will be expressed in terms of elliptic functions.

8.12. The Problem of a Quartic Polynomial

Suppose that $P(x)$ is a quartic polynomial and (x, y) are the coordinates of a curve C_4:

$$y^2 = P(x) = a_0 x^4 + 4a_1 x^3 + 6a_2 x^2 + 4a_3 x + a_4 . \tag{1}$$

If one root of the polynomial, α, is known, i.e., $P(\alpha) = 0$, then, by the transformation formula (4) of Sec. 8.1: $x - \alpha = 1/x'$, we have $y = \sqrt{P(x)} = x'^{-2}\sqrt{P_1(x')}$, where $P_1(x')$ is a cubic polynomial. Performing further the transformation $y = y'/x'^2$, we obtain $y'^2 = P_1(x')$. According to (6) of the last section, x' and y' can be expressed as $x' = b\wp(u) + c, y' = b\wp'(u)$. Hence, we have

$$x = \alpha + \frac{1}{b\wp(u) + c}, \quad y = \frac{b\wp'(u)}{[b\wp(u) + c]^2}, \tag{2}$$

where $b = 4/b_0 = 4/P'(\alpha); c = -b_1/b_0 = -P''(\alpha)/6P'(\alpha)$, b_0 and b_1 being the coefficients in $P_1(x')$ [see (3) of the last section]. From (2), we have $dx = -ydu$, so

$$u = -\int \frac{dx}{\sqrt{P(x)}}. \tag{3}$$

If none of the roots of $P(x) = 0$ is known, we do not have to solve the quartic equation (concerning the solution of such equation, see App. II), but the following method might be used. Express $P(t)$ in the following form:

$$\begin{aligned}
\eta^2 = P(t) &= a_0 t^4 + 4a_1 t^3 + 6a_2 t^2 + 4a_3 t + a_4 \\
&= [\varphi_2(t)]^2 - \varphi_1(t)\varphi_3(t),
\end{aligned} \tag{4}$$

where $\varphi_k(t)$ is a polynomial of degree k. Obviously, there are infinitely many possible ways in doing this. One option for $\varphi_k(t)$ is: Let (α, β) be the coordinates of a point on C_4. With arbitrarily chosen α, we have $\beta = \sqrt{P(\alpha)}$. Now, choose $\varphi_2(t)$ such that $\varphi_2(\alpha) = \beta$. There are evidently infinitely many ways of attaining this. Then, $t = \alpha$ will be a root of the equation $P(t) - [\varphi_2(t)]^2 = 0$, and we can set $\varphi_1(t) = t - \alpha$ to determine $\varphi_3(t)$. When $\varphi_k(t)$ have thus been fixed, construct an auxilliary cubic curve C_3 whose equation is

$$x^3\varphi_3\left(\frac{y}{x}\right) + 2x^2\varphi_2\left(\frac{y}{x}\right) + x\varphi_1\left(\frac{y}{x}\right) = 0. \tag{5}$$

This curve passes through the origin and its intersections with the straight line $y = tx$, besides the origin, are the two points satisfying the quadratic equation

$$x^2\varphi_3(t) + 2x\varphi_2(t) + \varphi_1(t) = 0,$$

from which we find

$$x = \frac{-\varphi_2(t) + \eta}{\varphi_3(t)}, \tag{6}$$

where $\eta^2 = P(t)$ as in (4). In the last section, we have obtained the formula for expressing the coordinates (x, y) of a cubic curve C_3 in terms of elliptic functions. Thus, through $y = tx$ and (6), we can express t and η in terms of elliptic functions. This procedure solves the problem of expressing the coordinates (t, η) of C_4 by elliptic functions.

Now, we shall find the explicit forms of $\varphi_k(t)$. In (1), perform the transformation $x = t - a_1/a_0$ to eliminate the term containing x^3. Then put $a_1 = 0$ in (4). We obtain

$$a_0 \eta^2 = a_0 P(t) = (a_0 t^2)^2 + 6a_0 a_2 t^2 + 4a_0 a_3 t + a_0 a_4 . \tag{7}$$

Let

$$\varphi_1(t) = -1, \quad \varphi_2(t) = a_0 t^2 ,$$
$$\varphi_3(t) = 6a_0 a_2 t^2 + 4a_0 a_3 t + a_0 a_4 . \tag{8}$$

Equation (5) of the auxilliary cubic curve C_3 reduces to

$$6a_0 a_2 xy^2 + 4a_0 a_3 x^2 y + a_0 a_4 x^3 + 2a_0 y^2 - x = 0 . \tag{9}$$

Putting $y = tx$, we have

$$\frac{1}{x^2} - \frac{2a_0 t^2}{x} - (6a_0 a_2 t^2 + 4a_0 a_3 t + a_0 a_4) = 0$$

of which the solution is

$$\frac{1}{x} = a_0 t^2 + \sqrt{a_0 P(t)} .$$

From this, we can express t and $\eta = \sqrt{P(t)}$ in terms of the coordinates (x, y) of the cubic curve C_3 as follows:

$$t = \frac{y}{x}, \quad \sqrt{a_0 P(t)} = \frac{1}{x} - a_0 \left(\frac{y}{x}\right)^2 . \tag{10}$$

On the other hand, from (9) we have

$$y = \frac{-2a_0 a_3 x^2 + \sqrt{4a_0^2 a_3^2 x^4 - x(a_0 a_4 x^2 - 1)(6a_0 a_2 x + 2a_0)}}{6a_0 a_2 x + 2a_0} . \tag{11}$$

One of the roots of the quartic polynomial under the radical sign is $x = 0$. We can put $x = 1/\xi$ [see Eq. (4) of Sec. 8.1] and obtain

$$y = \frac{-2a_0 a_3 + \sqrt{4a_0^2 a_3^2 - (a_0 a_4 - \xi^2)(6a_0 a_2 + 2a_0 \xi)}}{2a_0 \xi(3a_2 + \xi)} . \tag{12}$$

Further, let $\xi = 2a_0 x' - a_2$, the polynomial under the radical sign becomes $4a_0^4(4x'^3 - g_2 x' - g_3)$, where

$$g_2 = \frac{a_0 a_4 + 3a_2^2}{a_0^2}, \quad g_3 = \frac{a_0 a_2 a_4 - a_2^3 - a_0 a_3^2}{a_0^3} . \tag{13}$$

Hence,

$$x = \frac{1}{2a_0 \wp(u) - a_2}, \quad y = \frac{a_0 \wp'(u) - a_3}{[2a_0 \wp(u) - a_2][2a_0 \wp(u) + 2a_2]} . \quad (14)$$

In order to simplify the formula, parameter v is introduced such that

$$\wp(v) = -\frac{a_2}{a_0}, \quad \wp'(v) = \frac{a_3}{a_0} . \quad (15)$$

It is easy to verify that, with g_2 and g_3 given by (13), $\wp(v)$ and $\wp'(v)$ in (15) satisfy the relations (1) and (2) of the last section, and hence the equations in (15) hold simultaneously. After the introducing of v, by substituting (14) in (10), we obtain

$$t = \frac{1}{2} \frac{\wp'(u) - \wp'(v)}{\wp(u) - \wp(v)} , \quad (16)$$

$$\eta = \sqrt{P(t)} = \sqrt{a_0} \left\{ 2\wp(u) + \wp(v) - \frac{1}{4} \left[\frac{\wp'(u) - \wp'(v)}{\wp(u) - \wp(v)} \right]^2 \right\} . \quad (17)$$

Applying Eq. (6) of Sec. 8.10, the latter is simplified to

$$\eta = \sqrt{P(t)} = \sqrt{a_0} \{ \wp(u) - \wp(u + v) \} . \quad (18)$$

By differentiating Eq. (4) of Sec. 8.10, we obtain

$$\frac{1}{2} \frac{d}{du} \frac{\wp'(u) - \wp'(v)}{\wp(u) - \wp(v)} = \wp(u) - \wp(u + v) , \quad (19)$$

from which we have $dt/du = \sqrt{P(t)/a_0}$, and by integrating,

$$u = \sqrt{a_0} \int \frac{dt}{\sqrt{P(t)}} . \quad (20)$$

8.13. Curves of Genus (Deficiency) 1

The equation of a curve C_n of degree n, can be written as

$$F(x, y) = \sum_{m=0}^{n} \varphi_m(x, y) = 0 , \quad (1)$$

where $\varphi_m(x, y)$ is a homogeneous polynomial of degree m, which contains $m + 1$ coefficients. Since (1) represents the same curve when multiplied by an arbitrary constant, the total number of independent coefficients is $\Sigma_{m=0}^n (m + 1) - 1 = \Sigma_{m=1}^n (m + 1) = \frac{1}{2}n(n + 3)$, and this is just the number of points necessary to determine the curve C_n.

We shall show that an irresolvable curve C_n has at most $\frac{1}{2}(n - 1)(n - 2)$ double points. "Irresolvable" means the equation of the curve $F(x, y) = 0$ cannot be resolved into $F(x, y) = F_1(x, y) \times F_2(x, y) = 0$; i.e., not resolvable into two curves: $F_1(x, y) = 0$ and $F_2(x, y) = 0$. Let d be the number of double points of C_n. Draw a curve C_{n-1} passing through these d points, and, in addition, it can pass through another $N - d$ assigned points on C_n, N being the number of points needed to fix C_{n-1}, $N = \frac{1}{2}(n - 1)(n + 2)$. As a double point is counted as two points, the number of intersecting points is equal to $N - d + 2d = N + d$. But C_n can intersect a curve C_m of degree m at most at nm points,[c] therefore, the number $N + d$ cannot be larger than $n(n - 1)$, and

$$d \le n(n - 1) - N = \frac{1}{2}(n - 1)(n - 2) .$$

This shows that the curve C_n can have at most $\frac{1}{2}(n - 1)(n - 2)$ double points.

Let d be the number of double points of C_n, and

$$p = \frac{1}{2}(n - 1)(n - 2) - d \tag{2}$$

p is called the *genus*, or *deficiency*, of the curve C_n. A curve of zero genus has a maximum number of double points, and is called a *unicursal* curve, the coordinates of which can be expressed by rational functions of a certain parameter. Also, every curve which can be so expressed is unicursal. A curve C_n of degree n and genus 1 has $d = \frac{1}{2}(n - 1)(n - 2) - 1 = \frac{1}{2}n(n - 3)$ double points. When $n = 3, d = 0$, so a cubic curve without double points is of genus 1.

We shall show that the coordinates of a curve C_n of genus 1 can be expressed by elliptic functions.

Draw a curve C_{n-2} of degree $n - 2$ passing through all the $\frac{1}{2}n(n - 3)$ double points of C_n. To determine C_{n-2} completely, we need a further $\frac{1}{2}(n - 2)(n + 1) - \frac{1}{2}n(n - 3) = n - 1$ points. Now, for these $n - 1$ points, we choose $n - 3$ more points on C_n, then there remains two unspecified points which we describe by two parameters λ and μ. Thus, the equation of the curve C_{n-2} has the form:

$$\lambda f_1(x, y) + \mu f_2(x, y) + f_3(x, y) = 0 . \tag{3}$$

---

[c] This is Bezout's theorem, cf. Fricke, *Algebra*, I, p. 123.

Since a double point is counted as two points, the number of intersecting points of C_{n-2} and C_n already determined is $n(n-3) + n - 3 = n(n-2) - 3$. But the maximum number of intersecting points is $n(n-2)$, so, besides these already determined points, there can be at most 3 more.

Using the polynomials in (3), introduce new variables x' and y':

$$x' = \frac{f_1(x,y)}{f_3(x,y)}, \quad y' = \frac{f_2(x,y)}{f_3(x,y)} . \tag{4}$$

When the point (x,y) varies along the curve C_n, the point (x',y') will run along a corresponding curve C' whose equation is obtained from (1) and (4) by eliminating x and y. We see from (4) that, corresponding to one point (x,y) on C_n, there can only be one point (x',y') on C'. The converse is also true. For, if to a point (x',y') on C', there corresponds on C_n two points (a,b) and (a',b'), then we have from (4)

$$\frac{f_1(a',b')}{f_1(a,b)} = \frac{f_2(a',b')}{f_2(a,b)} = \frac{f_3(a',b')}{f_3(a,b)} .$$

By (3) we see that, if we let C_{n-2} pass through the point (a,b), it must also pass through (a',b'). Suppose that (a,b) and (a',b') do not belong to those points determining the family of curves C_{n-2} [Eq. (3)], then, after (a,b) has been fixed, one of the parameters λ and μ, say λ, is still arbitrary, and the family of curves C_{n-2} is simplified to a one parameter family of curves, which can still come across the curve C_n at a variable point (x,y). When this variable point is also fixed, the value of the parameter λ will follow. Conversely, when λ is given, the variable point (x,y) is also fixed. Hence, the coordinates of the variable point have to be rational functions of the parameter λ. Since the variable point can vary continuously along the curve C_n, we are led to the conclusion that the coordinates of C_n can be expressed by rational functions with one parameter. However, only unicursal curves can be represented by one-parameter rational functions. Since our C_n is not a unicursal curve, the supposition that there are two points (a,b) and (a',b') cannot be true, and corresponding to one point on C' there can only be one point on C_n. Thus, it is always possible to represent the coordinates of C_n by rational functions of the coordinates of C':

$$x = g_1(x',y'), \quad y = g_2(x',y') . \tag{5}$$

These two relations should have been obtained by solving (4). Every transformation for which both (4) and (5) hold is called a *double rational transformation*.

Now let us find the degree of the curve C', which is equal to the number of intersecting points of it with the straight line $ax' + by' + c = 0$. And, this number also equals to the number of intersecting points of C_n and the curve

$$af_1(x,y) + bf_2(x,y) + cf_3(x,y) = 0 ,$$

since there is one point on C' corresponding to one point on C_n. This curve belongs to the family of curves C_{n-2}, which can intersect with C_n at most at three points varying with a, b, c. Therefore, C' is a cubic curve. We have already known that the coordinates (x', y') of a cubic curve can be expressed in terms of elliptic functions. By substituting them into (5), we obtain the coordinates of C_n represented by elliptic functions. Thus, we have shown that the coordinates of curves of genus 1 can be represented by elliptic functions.

From above, we have solved the problem of representing the coordinates of a curve of genus 1 by elliptic functions only in principle. The practical calculations in the determination of the family of curves C_{n-2} is rather cumbersome. It is better to use the double rational transformation to reduce the degree of the curve in question step by step. In this way, we need only to determine one double point at a time, not to determine all the double points at one time. For example, suppose that we have a curve of degree 12:

$$y^6 = A(x - a)^3(x - b)^4(x - c)^5 . \tag{6}$$

The three points $(a, 0), (b, 0), (c, 0)$ on this curve are multiple points. Take as the first step the transformation:

$$x = x, \quad y = (x - b)(x - c)/\xi .$$

We obtain

$$(x - b)^2(x - c) = A(x - a)^3 \xi^6 ,$$

which is a curve of degree 9 in coordinates (x, ξ). Next, as the second step, use the transformation: $\xi = \xi, x = a + 1/\eta$. There results

$$A\xi^6 = [(a - b)\eta + 1]^2[(a - c)\eta + 1] ,$$

which is a curve of degree 6. The third step is to perform the transformation: $\xi = \xi, (a - b)\eta + 1 = \xi^3/\varsigma$. The result is

$$(a - c)\xi^3 - (b - c)\varsigma = A(a - b)\varsigma^3 ,$$

which is already a cubic curve. For the fourth step, use the transformation: $\xi = t/v, \varsigma = 1/v$. We obtain

$$(b - c)v^2 = (a - c)t^3 - A(a - b) \tag{7}$$

which is already close to the standard form. Finally, then, with the transformation

$$t = \frac{4(b - c)}{a - c} t', \quad v = \frac{4(b - c)}{a - c} v',$$

we get $\tag{8}$

$$v'^2 = 4t'^3 - g_3; \quad g_2 = 0, \quad g_3 = \frac{A(a - b)(a - c)^2}{16(b - c)^3}$$

which is of the standard form. The relations between (x, y) and (t, v) are

$$x = \frac{at^3 - bv^2}{t^3 - v^2}, \quad y = \frac{(a - b)[(a - c)t^3 - (b - c)v^2]t^2 v}{(t^3 - v^2)^2};$$

$$t = \frac{y^2}{(x - a)(x - b)(x - c)^2}, \quad v = \frac{y^3}{(x - a)(x - b)^2(x - c)^3}. \tag{9}$$

Expressed by elliptic functions, they are

$$t = \frac{4(b - c)}{a - c} \wp(u), \quad v = \frac{4(b - c)}{a - c} \wp'(u). \tag{10}$$

This method can be used to solve the differential equation

$$\left(\frac{dz}{dx}\right)^6 = A(x - a)^3(x - b)^4(x - c)^5. \tag{11}$$

Following the above procedure, let $y = dz/dx$, we have

$$z = \int y\, dx$$

and this integral can be expressed in terms of elliptic functions.

Exercise 8

1. Show that there is an algebraic relation between two elliptic functions $x = f_1(z)$ and $y = f_2(z)$ with the same periods and the same poles. [Hint:

Choose the coefficients in the polynomial $F(x, y)$ such that the principal part at each pole is zero.]

2. Multiplying the two equations in Eq. (14) of Sec. 8.7, and expanding $\ln \sigma(\omega_k \pm z)$ in Taylor series, then adding, show that [Eq. (17) of Sec. 8.7],

$$\ln \sigma_k(z) = -\sum_{n=1}^{\infty} \wp^{(2n-2)}(\omega_k) \frac{z^{2n}}{(2n)!}$$

$$= -e_k \frac{z^2}{2!} - (6e_k^2 - e_1^2 - e_2^2 - e_3^2) \frac{z^4}{4!}$$

$$- 12e_k(6e_k^2 - e_1^2 - e_2^2 - e_3^2) \frac{z^6}{6!}$$

$$- 36(6e_k^2 - e_1^2 - e_2^2 - e_3^2)(10e_k^2 - e_1^2 - e_2^2 - e_3^2) \frac{z^8}{8!}$$

$$- \cdots .$$

3. From Eqs. (6) and (11) of Sec. 8.5, show that

$$c_2 = \frac{1}{5}(e_3^2 - e_1 e_2), \quad c_3 = \frac{e_1 e_2 e_3}{7}, \quad c_4 = \frac{(e_3^2 - e_1 e_2)^2}{75} .$$

4. From Exs. 2, 3 and Eq. (5) of Sec. 8.7, show that

$$\ln \frac{\sigma(z)}{z\sigma_3(z)} = e_3 \frac{z^2}{2} + (2e_3^2 + 3e_1 e_2) \frac{z^4}{30} + e_3 \left(e_3^2 + \frac{3}{7} e_1 e_2\right) \frac{z^6}{15}$$

$$+ \left[(2e_3^2 + e_1 e_2)(4e_3^2 + e_1 e_2) - \frac{(e_3^2 - e_1 e_2)^2}{15}\right] \frac{z^8}{280}$$

$$+ \cdots ,$$

$$\ln \frac{\sigma_1(z)}{\sigma_3(z)} = (e_1 - e_3) \left\{-\frac{z^2}{2} + e_2 \frac{z^4}{4} - (e_2^2 - e_1 e_3) \frac{z^6}{15}\right.$$

$$\left. + e_2(7e_2^2 - 22e_1 e_3) \frac{z^8}{280} - \cdots\right\} ,$$

$$\ln \frac{\sigma_2(z)}{\sigma_3(z)} = (e_2 - e_3) \left\{-\frac{z^2}{2} + e_1 \frac{z^4}{4} - (e_1^2 - e_2 e_3) \frac{z^6}{15}\right.$$

$$\left. + e_1(7e_1^2 - 22e_2 e_3) \frac{z^8}{280} - \cdots\right\} .$$

5. From Eqs. (1) and (7) of Sec. 8.10, derive the following:

$$\wp'(z) = -\frac{\sigma(2z)}{\sigma^4(z)}, \quad \frac{\sigma(3z)}{\sigma^9(z)} = 3\wp(z)\wp'^2(z) - \frac{1}{4}\wp''^2(z) .$$

6. From Eq. (4) of Sec. 8.10, show that

$$\varsigma(2z) - 2\varsigma(z) = \frac{\wp''(z)}{2\wp'(z)},$$

$$\varsigma(z + \omega_1) - \varsigma(z) - \eta_1 = \frac{1}{2} \frac{\wp'(z)}{\wp(z) - e_1};$$

then by integration derive that

$$\sqrt{\wp(z) - e_1} = \frac{\sigma_1(z)}{\sigma(z)}.$$

7. By Eq. (14) of Sec. 8.10, show that

$$(e_1 - e_2)(e_1 - e_3) \int \frac{dz}{\wp(z) - e_1} = -\varsigma(z + \omega_1) - e_1 z + C.$$

8. From Eqs. (16) and (17) of Sec. 8.10, show that

$$\sqrt{e_1 - e_2} = -\frac{e^{-\omega_1 \eta_3} \sigma(\omega_3)}{\sigma(\omega_1)\sigma(\omega_2)},$$

$$\sqrt{e_1 - e_3} = -\frac{e^{-\omega_1 \eta_3} \sigma(\omega_2)}{\sigma(\omega_1)\sigma(\omega_3)},$$

$$\sqrt{e_2 - e_3} = -\frac{e^{-\omega_2 \eta_3} \sigma(\omega_1)}{\sigma(\omega_2)\sigma(\omega_3)}.$$

9. Show that $x = \wp\left(\frac{w}{5}\right) + \wp\left(\frac{2w}{5}\right)$, where $w = 2m\omega + 2m'\omega'$, is the root of the equation of the sixth degree:

$$x^6 - 5g_2 x^4 - 40g_3 x^3 - 5g_2^2 x^2 - 8g_2 g_3 x - 5g_3^2 = 0,$$

and find the expressions for the six roots.

10. Show that a cubic curve has nine points of inflection whose coordinates can be expressed in terms of a parameter u, $u = \frac{1}{3}(2m\omega + 2m'\omega')$, which is the independent variable of elliptic functions; m and m' can take any one of the values $(0,1,2)$.

11. By the transformation $x = x' - a_1/a_0$, $P(x) = a_0 x^4 + 4a_1 x^3 + 6a_2 x^2 + 4a_3 x + a_4$ becomes

$$P\left(x' - \frac{a_1}{a_0}\right) = a_0 x'^4 + 6a_2' x'^2 + 4a_3' x' + a_4' ,$$

$$a_2' = a_2 - \frac{a_1^2}{a_0}, \quad a_3' = a_3 - \frac{3a_1 a_2}{a_0} + \frac{2a_1^3}{a_0^2} ,$$

$$a_4' = a_4 - \frac{4a_1 a_3}{a_0} + \frac{6a_1^2 a_2}{a_0^2} - \frac{3a_1^4}{a_0^3} .$$

From which, show that, when $a_1 \neq 0$, (13) - (18) of Sec. 8.12 become respectively

$$g_2 = \frac{a_0 a_4 - 4a_1 a_3 + 3a_2^2}{a_0^2} ,$$

$$g_3 = \frac{a_0 a_2 a_4 + 2a_1 a_2 a_3 - a_1^2 a_4 - a_0 a_3^2 - a_2^3}{a_0^3} ,$$

$$\wp(v) = \frac{a_1^2}{a_0^2} - \frac{a_2}{a_0}, \quad \wp'(v) = \frac{a_3}{a_0} - \frac{3a_1 a_2}{a_0^2} + \frac{2a_1^3}{a_0^3} ,$$

$$x = \frac{1}{2} \frac{\wp'(u) - \wp'(v)}{\wp(u) - \wp(v)} - \frac{a_1}{a_0} ,$$

$$y = \sqrt{P(x)} = \sqrt{a_0} \{\wp(u) - \wp(u+v)\} .$$

12. Show that, in Ex. 11 if we change g_2, g_3 to G_2, G_3:

$$G_2 = a_0^2 g_2 = a_0 a_4 - 4a_1 a_3 + 3a_2^2 ,$$

$$G_3 = a_0^3 g_3 = a_0 a_2 a_4 - a_1^2 a_4 + 2a_1 a_2 a_3 - a_0 a_3^2 - a_2^3 ,$$

then u will be changed to $U = a_0^{-\frac{1}{2}} u$, and

$$\wp(V) = \frac{a_1^2}{a_0} - a_2 , \quad \wp'(V) = \sqrt{a_0}\left(a_3 - \frac{3a_1 a_2}{a_0} + \frac{2a_1^3}{a_0^2}\right) ,$$

$$x = \frac{1}{2\sqrt{a_0}} \frac{\wp'(U) - \wp'(V)}{\wp(U) - \wp(V)} - \frac{a_1}{a_0} ,$$

$$y = \sqrt{P(x)} = \frac{1}{\sqrt{a_0}} \{\wp(U) - \wp(U+V)\} .$$

Also show that the G_2, G_3 here are the same as the invariants in Eq. (2) of Sec. 8.12, and the relation between u there and U is $u = U + V/2$. [Hint: Applying Eq. (9) of Sec. 8.9.]

13. Express the coordinates of the following curve in terms of elliptic functions:

$$y^5 + Axy^4 + x^4 \left(Bx - \frac{4^4}{5^5} \frac{A^5}{4B} \right)^2 = 0 .$$

14. Express the coordinates of the following curve in terms of elliptic functions

$$y^5 + Axy^4 + \left(Bx^5 - \frac{4^4}{5^5} \frac{A^5}{4B} \right)^2 = 0 .$$

15. From Eq. (17) of Sec. 8.10 derive that

$$\sigma_k^2(z) - \sigma_l^2(z) + (e_k - e_l)\sigma^2(z) = 0 \quad (k, l = 1, 2, 3) ,$$
$$(e_2 - e_3)\sigma_1^2(z) + (e_3 - e_1)\sigma_2^2(z) + (e_1 - e_2)\sigma_3^2(z) = 0 .$$

Chapter 9
THETA FUNCTIONS

9.1. The Theta Function $\theta(v)$

For practical purpose it is necessary to choose one of the periods of the elliptic function to be real. This may be realized by putting $v = z/2\omega$. At the end of Sec. 8.8, we have mentioned that the functional relation of an elliptic function can be expressed in terms of the variables z/ω and ω'/ω. Let

$$v = \frac{z}{2\omega} , \quad \tau = \frac{\omega'}{\omega} , \tag{1}$$

then, corresponding to the two periods 2ω and $2\omega'$ in z are the two periods 1 and τ in v. We still assume that $\text{Im}(\tau) > 0$. Since all elliptic functions can be expressed by $\sigma(z)$, whereas $\sigma(z)$ is an integral function, all of its zeros being known, $\sigma(z)$ is a suitable choice. In order to make the characteristics of the real period outstanding, we multiply $\sigma(z)$ with the factor e^{az^2+bz} and properly choose a and b so that the product has the period 2ω. It is easy to show that $a = -\eta/2\omega, b = \pi i/2\omega$. Let

$$\varphi(z) = e^{-\frac{\eta z^2}{2\omega}+i\frac{\pi z}{2\omega}}\sigma(z) . \tag{2}$$

Applying Eqs. (7) and (8) of Sec. 8.7 and Eq. (8) of Sec. 8.6, we have

$$\varphi(z + 2\omega) = \varphi(z) , \quad \varphi(z + 2\omega') = -e^{-\frac{i\pi z}{\omega}}\varphi(z) . \tag{3}$$

A periodic function can be expanded in a Fourier series, thus [see Sec. 12.5]

$$\varphi(z) = \sum_{n=-\infty}^{\infty} c_n e^{2n\pi v i} = \sum_{n=-\infty}^{\infty} c_n x^{2n} , \tag{4}$$

where $x = e^{i\pi v}, z = 2\omega v$. In (4), changing z to $z + 2\omega'$, v to $v + \tau$, and putting

$$x = e^{i\pi v} , \quad q = e^{i\pi \tau} , \tag{5}$$

we obtain

$$\varphi(z + 2\omega') = \sum_{n=-\infty}^{\infty} c_n q^{2n} x^{2n} .$$

Substituting in the second equation of (3) gives

$$\sum_{n=-\infty}^{\infty} c_n q^{2n} x^{2n} = - \sum_{n=-\infty}^{\infty} c_n x^{2n-2} = - \sum_{n=-\infty}^{\infty} c_{n+1} x^{2n} .$$

The coefficients of x^{2n} on the two sides must be equal, so

$$c_{n+1} = -q^{2n} c_n ,$$

which can be reduced to

$$(-)^{n+1} q^{-\left(n+\frac{1}{2}\right)^2} c_{n+1} = (-)^n q^{-\left(n-\frac{1}{2}\right)^2} c_n = C ,$$

C being a constant. Hence, we have

$$\varphi(z) = C \sum_{n=-\infty}^{\infty} (-)^n q^{\left(n-\frac{1}{2}\right)^2} x^{2n} . \tag{6}$$

To preserve the oddness of $\sigma(z)$, we employ, instead of $\varphi(z)$, $\varphi(z)/x$ and define the θ-function:

$$\theta(v) = i \sum_{n=-\infty}^{\infty} (-)^n q^{\left(n-\frac{1}{2}\right)^2} e^{(2n-1)\pi v i} . \tag{7}$$

The factor i is introduced such that, when v is real and $\theta(v)$ is expressed in terms of $\sin(2n+1)\pi v, i$ does not occur in the coefficients [see (8) below]. In the r.h.s. series of (7), replacing n by $-n$ for that part of the sum from 0 to $-\infty$ and replacing n by $n+1$ in the part from $n=1$ to ∞, we obtain

$$\theta(v) = 2 \sum_{n=0}^{\infty} (-)^n q^{\left(n+\frac{1}{2}\right)^2} \sin(2n+1)\pi v$$

$$= 2q^{\frac{1}{4}} \sin \pi v - 2q^{\frac{9}{4}} \sin 3\pi v + 2q^{\frac{25}{4}} \sin 5\pi v - \dots . \qquad (8)$$

This expression clearly discloses that $\theta(v)$ is an odd function.

From (2), (6), (7), we have

$$\sigma(z) = -iCe^{\frac{\eta z^2}{2\omega}} \theta\left(\frac{z}{2\omega}\right) .$$

The constant C can be determined by letting $z \to 0$ (after both sides have been divided by z). The result is $1 = -iC\theta'(0)/2\omega$, and hence

$$\sigma(z) = \frac{2\omega}{\theta'} e^{\frac{\eta z^2}{2\omega}} \theta\left(\frac{z}{2\omega}\right) , \qquad (9)$$

where θ' is an abbreviation for $\theta'(0)$.

One advantage of the θ function is that it has a real period. It has the further merit that the series expansion (8) converges very rapidly. The quick convergence is due to the fact that $|q| < 1$ when $\text{Im}(\tau) > 0$ and that the degrees of q in the consecutive terms differ by a rather large number.

Noting the relation between $\wp(z)$ and $\theta(v)$, we have from (3)

$$\theta(v+1) = -\theta(v) , \quad \theta(v+\tau) = -q^{-1}e^{-2\pi vi}\theta(v) , \qquad (10)$$

which can also be proved directly from the definition (7). The zeros of $\theta(v)$ are the same as those of $\sigma(z)$. Therefore, $v = m + m'\tau$ are the only zeros, m and m' being arbitrary integers. As regards the zeros of $\theta(v)$, it is not necessary to use the result obtained for $\sigma(z)$. We only have to calculate directly the value of the contour integral $\int \theta'(v)/\theta(v)dv$ along the sides of the period parallelogram, which is found to be $2\pi i$ and hence the assertion is proved.

9.2. The Functions $\vartheta_k(v)$

Theta functions ϑ_k were introduced by Jacobi. There are four of them: $k = 1, 2, 3, 4; \vartheta_1(v) = \theta(v)$. The other three ϑ functions correspond to the three σ functions [Eq. (14) of Sec. 8.7], $\sigma_k(z), k = 1, 2, 3$. Jordan uses the symbols $\theta(v)$ and $\theta_k(v)$ with $\theta_k(v)$ corresponding to $\sigma_k(z)$. More commonly used are

Jacobi' symbols, in which $\vartheta_{k+1}(v)$ correspond to $\sigma_k(z)$. The definitions of $\vartheta_k(v)$ are

$$\vartheta_1(v) = \theta(v) = i \sum_{n=-\infty}^{\infty} (-)^n q^{\left(n-\frac{1}{2}\right)^2} e^{(2n-1)\pi v i}$$

$$= 2 \sum_{n=0}^{\infty} (-)^n q^{\left(n+\frac{1}{2}\right)^2} \sin(2n+1)\pi v$$

$$= 2q^{\frac{1}{4}} \sin \pi v - 2q^{\frac{9}{4}} \sin 3\pi v + 2q^{\frac{25}{4}} \sin 5\pi v - \dots . \tag{1}$$

$$\vartheta_2(v) = \theta_1(v) = \vartheta_1\left(v + \frac{1}{2}\right) = \theta\left(v + \frac{1}{2}\right)$$

$$= \sum_{n=-\infty}^{\infty} q^{\left(n-\frac{1}{2}\right)^2} e^{(2n-1)\pi v i} = 2 \sum_{n=0}^{\infty} q^{\left(n+\frac{1}{2}\right)^2} \cos(2n+1)\pi v$$

$$= 2q^{\frac{1}{4}} \cos \pi v + 2q^{\frac{9}{4}} \cos 3\pi v + 2q^{\frac{25}{4}} \cos 5\pi v + \dots . \tag{2}$$

$$\vartheta_3(v) = \theta_2(v) = q^{\frac{1}{4}} e^{i\pi v} \vartheta_1\left(v + \frac{1}{2} + \frac{\tau}{2}\right) = q^{\frac{1}{4}} e^{i\pi v} \vartheta_2\left(v + \frac{\tau}{2}\right)$$

$$= \sum_{n=-\infty}^{\infty} q^{n^2} e^{2n\pi v i} = 1 + 2 \sum_{n=1}^{\infty} q^{2n} \cos 2n\pi v$$

$$= 1 + 2q \cos 2\pi v + 2q^4 \cos 4\pi v + 2q^9 \cos 6\pi v + \dots . \tag{3}$$

$$\vartheta_4(v) = \theta_3(v) = \vartheta_3\left(v + \frac{1}{2}\right) = -iq^{\frac{1}{4}} e^{i\pi v} \vartheta_1\left(v + \frac{\tau}{2}\right)$$

$$= iq^{\frac{1}{4}} e^{-i\pi v} \vartheta_1\left(v - \frac{\tau}{2}\right)$$

$$= \sum_{n=-\infty}^{\infty} (-)^n q^{n^2} e^{2n\pi v i} = 1 + 2 \sum_{n=1}^{\infty} (-)^n q^{n^2} \cos 2n\pi v$$

$$= 1 - 2q \cos 2\pi v + 2q^4 \cos 4\pi v - 2q^9 \cos 6\pi v + \dots . \tag{4}$$

All $\vartheta_k(v)(k = 2, 3, 4)$ are even functions: $\vartheta_k(-v) = \vartheta_k(v)$: only $\vartheta_1(v)$ is an odd function. Formula (1) is a repeat of (7) and (8) of the last section.

The connection among these functions and the functions $\sigma(z), \sigma_k(z)$ of Sec. 8.7 can be found from their definitions:

$$\sigma(z) = \frac{2\omega}{\vartheta_1'} e^{\frac{\eta z^2}{2\omega}} \vartheta_1\left(\frac{z}{2\omega}\right) , \tag{5}$$

$$\sigma_k(z) = \frac{e^{\frac{\eta z^2}{2\omega}}}{\vartheta_{k+1}} \vartheta_{k+1}\left(\frac{z}{2\omega}\right) \quad (k = 1, 2, 3) , \tag{6}$$

where ϑ_k are written for $\vartheta_k(0)(k = 2, 3, 4)$. Formula (5) is a repeat of (9) of the last section.

The periodicities of these functions, according to (10) of the last section and the definitions (1) – (4) above, are presented by

$$\begin{aligned}
\vartheta_1(v + 1) &= -\vartheta_1(v) , & \vartheta_1(v + \tau) &= -q^{-1}e^{-2\pi vi}\vartheta_1(v) , \\
\vartheta_2(v + 1) &= -\vartheta_2(v) , & \vartheta_2(v + \tau) &= q^{-1}e^{-2\pi vi}\vartheta_2(v) , \\
\vartheta_3(v + 1) &= \vartheta_3(v) , & \vartheta_3(v + \tau) &= q^{-1}e^{-2\pi vi}\vartheta_3(v) , \\
\vartheta_4(v + 1) &= \vartheta_4(v) , & \vartheta_4(v + \tau) &= -q^{-1}e^{-2\pi vi}\vartheta_4(v) .
\end{aligned} \tag{7}$$

By the results obtained at the end of the last section, we see that the zeros of $\vartheta_k(v)$ are

$$m + m'\tau + v_k ; \quad v_1 = 0 , \quad v_2 = \frac{1}{2} , \quad v_3 = \frac{1}{2} + \frac{\tau}{2} , \quad v_4 = \frac{\tau}{2} , \tag{8}$$

all being simple. From the definitions, we find the values of $\vartheta_k(v)$ at their respective half periods to be

$$\begin{aligned}
\vartheta_1\left(\frac{1}{2}\right) &= \vartheta_2 , & \vartheta_1\left(\frac{\tau}{2}\right) &= iq^{-\frac{1}{4}}\vartheta_4 , & \vartheta_1\left(\frac{1}{2} + \frac{\tau}{2}\right) &= q^{-\frac{1}{4}}\vartheta_3 ; \\
\vartheta_2\left(\frac{1}{2}\right) &= 0 , & \vartheta_2\left(\frac{\tau}{2}\right) &= q^{-\frac{1}{4}}\vartheta_3 , & \vartheta_2\left(\frac{1}{2} + \frac{\tau}{2}\right) &= -iq^{-\frac{1}{4}}\vartheta_4 ; \\
\vartheta_3\left(\frac{1}{2}\right) &= \vartheta_4 , & \vartheta_3\left(\frac{\tau}{2}\right) &= q^{-\frac{1}{4}}\vartheta_2 , & \vartheta_3\left(\frac{1}{2} + \frac{\tau}{2}\right) &= 0 ; \\
\vartheta_4\left(\frac{1}{2}\right) &= \vartheta_3 , & \vartheta_4\left(\frac{\tau}{2}\right) &= 0 , & \vartheta_4\left(\frac{1}{2} + \frac{\tau}{2}\right) &= q^{-\frac{1}{4}}\vartheta_2 .
\end{aligned} \tag{9}$$

9.3. Elliptic Functions Represented by Theta Functions

From the relation between $\sigma(z)$ and $\vartheta_1(v)$ [Eq. (5) of Sec. 9.2], we can express Eq. (2) or (4) of Sec. 8.9 in terms of $\vartheta_1(v)$. Suppose that the periods of an elliptic function $f(z)$ are 2ω and $2\omega'$, and $\text{Im}(\omega'/\omega) > 0, \omega'/\omega = \tau$. Assume that its zeros α_r and poles $\beta_r(r = 1, 2, \ldots, s)$ satisfy $\sum_{r=1}^{s}(\alpha_r - \beta_r) = 0$. Then

$$f(z) = C \prod_{r=1}^{s} \left\{ \vartheta_1\left(\frac{z - \alpha_r}{2\omega}\right) \Big/ \vartheta_1\left(\frac{z - \beta_r}{2\omega}\right) \right\} . \tag{1}$$

Besides, Eq. (6) of Sec. 8.9 can also be expressed in terms of $\vartheta_1(v)$. From Eq. (4) of Sec. 8.7 and Eq. (5) of Sec. 9.2, we have

$$\varsigma(z) = \frac{\sigma'(z)}{\sigma(z)} = \frac{\eta z}{\omega} + \frac{1}{2\omega} \frac{\vartheta_1'(z/2\omega)}{\vartheta_1(z/2\omega)} \ . \tag{2}$$

Substituting in Eq. (6) of Sec. 8.9 leads to

$$f(z) = C + \sum_{r=1}^{s} \sum_{q=1}^{p_r} \frac{(-)^{q-1} B_{r,q}}{(q-1)!} \frac{d^q}{dz^q} \ln \vartheta_1\left(\frac{z-\beta_r}{2\omega}\right) \ . \tag{3}$$

This formula differs from Eq. (6) of Sec. 8.9 by terms corresponding to the first term $\eta z/\omega$ on the r.h.s. of (2), which can be combined into the constant C in (3); here, we have made use of the fact that the sum of the residues $\Sigma_r B_{r,1} = 0$.

From Eqs. (15) and (17) of Sec. 8.10 we obtain $\wp(z)$ expressed in terms of $\vartheta_k(v)$:

$$\wp(z) - e_1 = \left[\frac{\sigma_1(z)}{\sigma(z)}\right]^2 = \left[\frac{\vartheta_1'}{2\omega\vartheta_2} \frac{\vartheta_2(z/2\omega)}{\vartheta_1(z/2\omega)}\right]^2 \ , \tag{4}$$

$$\wp(z) - e_k = \left[\frac{\vartheta_1'}{2\omega\vartheta_{k+1}} \frac{\vartheta_{k+1}(z/2\omega)}{\vartheta_1(z/2\omega)}\right]^2 \quad (k = 1, 2, 3) \ . \tag{5}$$

By multiplying the three equations of (17) in Sec. 8.10, we obtain $\wp'(z)$ expressed in terms of $\sigma_k(z)$:

$$\wp'(z) = -\frac{2\sigma_1(z)\sigma_2(z)\sigma_3(z)}{\sigma^3(z)} \ ; \tag{6}$$

the negative sign on the r.h.s. is determined by the principal part in the vicinity of the origin. Expressing in terms of $\vartheta_k(v)$, we have

$$\wp'(z) = \frac{-\vartheta_1'^3}{4\omega^3\vartheta_2\vartheta_3\vartheta_4} \frac{\vartheta_2(v)\vartheta_3(v)\vartheta_4(v)}{\vartheta_1^3(v)} \quad \left(v = \frac{z}{2\omega}\right) \ . \tag{7}$$

9.4. Relations among the Squares of $\vartheta_k(v)$

Consider the function

$$f(v) = \frac{a\vartheta_1^2(v) + b\vartheta_4^2(v)}{\vartheta_2^2(v)} \ ,$$

which is a doubly-periodic function with periods 1 and τ, as can be shown by Eq. (7) of Sec. 9.2. If we choose a and b so as to cancel one common zero of the denominator and the numerator, then $f(v)$ will be an elliptic function with at most one pole, and, according to Theorem 2 and Theorem 1 in Sec. 8.3, must be a constant. Since both a and b can be multiplied with an arbitrary constant, we may put the constant $f(v) = 1$ and obtain

$$\vartheta_2^2(v) = a\vartheta_1^2(v) + b\vartheta_4^2(v) \ .$$

To determine a and b, we put $v = 0$ and $v = \tau/2$ successively. Utilizing Eq. (9) of Sec. 9.2, we obtain $b = \vartheta_2^2/\vartheta_4^2$, $a = -\vartheta_3^2/\vartheta_4^2$, so

$$\vartheta_4^2\vartheta_2^2(v) = \vartheta_2^2\vartheta_4^2(v) - \vartheta_3^2\vartheta_1^2(v) \ . \tag{1}$$

Similarly, we can show that

$$\vartheta_4^2\vartheta_3^2(v) = \vartheta_3^2\vartheta_4^2(v) - \vartheta_2^2\vartheta_1^2(v) \ . \tag{2}$$

Changing v to $v + \frac{1}{2}$ in (1) and (2) and using Eqs. (2) and (4) of Sec. 9.2, we obtain

$$\vartheta_4^2\vartheta_1^2(v) = \vartheta_2^2\vartheta_3^2(v) - \vartheta_3^2\vartheta_2^2(v) \ , \tag{3}$$

$$\vartheta_4^2\vartheta_4^2(v) = \vartheta_3^2\vartheta_3^2(v) - \vartheta_2^2\vartheta_2^2(v) \ . \tag{4}$$

By putting $v = 0$ in (4), we have

$$\vartheta_3^4 = \vartheta_2^4 + \vartheta_4^4 \ . \tag{5}$$

With this relation, we see that (3) and (4) can be derived from (1) and (2) by eliminating respectively $\vartheta_4^2(v)$ and $\vartheta_1^2(v)$.

It is worthwhile to note that the relations obtained in this section can easily be deduced from the relations between $\vartheta_k(v)$ and $\sigma_k(z)$ after making use of (5) in the last section [see Ex. 15 of Chap. 8].

9.5. Addition Formulae

Consider the function

$$f(v) = \frac{a\vartheta_1^2(v) + b\vartheta_3^2(v)}{\vartheta_3(v+w)\vartheta_3(v-w)} \ .$$

By the same reasoning as in the previous section, we can suitably choose a and b such that $f(v)$ is an elliptic function with at most one pole, and hence must be a constant. Thus we have

$$\vartheta_3(v+w)\vartheta_3(v-w) = a\vartheta_1^2(v) + b\vartheta_3^2(v) .$$

Put successively $v = 0$ and $v + \frac{1}{2} + \frac{\tau}{2}$. Using Eqs. (9) and (3) of Sec. 9.2, we obtain $b = \vartheta_3^2(w)/\vartheta_3^2, a = \vartheta_1^2(w)/\vartheta_3^2$, so

$$\vartheta_3^2\vartheta_3(v+w)\vartheta_3(v-w) = \vartheta_1^2(v)\vartheta_1^2(w) + \vartheta_3^2(v)\vartheta_3^2(w) . \tag{1}$$

Changing v to $v + \frac{1}{2}$, it becomes

$$\vartheta_3^2\vartheta_4(v+w)\vartheta_4(v-w) = \vartheta_2^2(v)\vartheta_1^2(w) + \vartheta_4^2(v)\vartheta_3^2(v) . \tag{2}$$

Changing further v to $v + \frac{\tau}{2}$ and $v + \frac{1}{2} + \frac{\tau}{2}$ successively, we obtain the expressions for $\vartheta_3^2\vartheta_k(v+w)\vartheta_k(v-w)(k=1,2)$. In the same way, we can show that

$$\vartheta_2^2\vartheta_3(v+w)\vartheta_3(v-w) = \vartheta_1^2(v)\vartheta_4^2(w) + \vartheta_2^2(v)\vartheta_3^2(w) . \tag{3}$$

From this formula, we can also obtain the expression for $\vartheta_2^2\vartheta_k(v+w)\vartheta_k(v-w)$.

In (2) putting $v = 0$, we obtain (2) of the last section. Thus, the various relations obtained in the last section are special cases of those derived here. Another formula is

$$\vartheta_4^2\vartheta_1(v+w)\vartheta_1(v-w) = \vartheta_1^2(v)\vartheta_4^2(w) - \vartheta_4^2(v)\vartheta_1^2(w) . \tag{4}$$

Changing v to $v + \frac{1}{2}$ and w to $w + \frac{1}{2}$ leads to

$$\vartheta_4^2\vartheta_1(v+w)\vartheta_1(v-w) = \vartheta_3^2(v)\vartheta_2^2(w) - \vartheta_2^2(v)\vartheta_3^2(w) . \tag{5}$$

In general, each $\vartheta_r^2\vartheta_k(v+w)\vartheta_k(v-w)$ has two representations such as (4) and (5).

In the same way, it can be shown that

$$\vartheta_2\vartheta_3\vartheta_1(v+w)\vartheta_4(v-w)$$
$$= \vartheta_1(v)\vartheta_4(v)\vartheta_2(w)\vartheta_3(w) + \vartheta_2(v)\vartheta_3(v)\vartheta_1(w)\vartheta_4(w) . \tag{6}$$

In (1), (2), (3), (6), putting $v = w$, we obtain the duplication formulae:

$$\begin{aligned}
\vartheta_3^3\vartheta_3(2v) &= \vartheta_1^4(v) + \vartheta_3^4(v) , \\
\vartheta_3^2\vartheta_4\vartheta_4(2v) &= \vartheta_1^2(v)\vartheta_2^2(v) + \vartheta_3^2(v)\vartheta_4^2(v) , \\
\vartheta_2^2\vartheta_3\vartheta_3(2v) &= \vartheta_1^2(v)\vartheta_4^2(v) + \vartheta_2^2(v)\vartheta_3^2(v) , \\
\vartheta_2\vartheta_3\vartheta_4\vartheta_1(2v) &= 2\vartheta_1(v)\vartheta_2(v)\vartheta_3(v)\vartheta_4(v) .
\end{aligned} \tag{7}$$

9.6. Differential Equations Satisfied by Theta Functions

We shall now show that each function $\vartheta_k(v)$ satisfies the partial differential equation

$$\frac{\partial^2 \vartheta_k(v)}{\partial v^2} = 4\pi i \cdot \frac{\partial \vartheta_k(v)}{\partial \tau} \quad (k = 1, 2, 3, 4) . \tag{1}$$

Take a definite k, say $k = 3$. By the definition given in Eq. (3) of Sec. 9.2, we have

$$\frac{\partial^2 \vartheta_3(v)}{\partial v^2} = -4\pi^2 \sum_{n=-\infty}^{\infty} n^2 q^{n^2} e^{2n\pi v i} ,$$

$$\frac{\partial \vartheta_3(v)}{\partial \tau} = i\pi \sum_{n=-\infty}^{\infty} n^2 q^{n^2} e^{2n\pi v i} ,$$

from which we see that (1) is satisfied. The same holds for $k = 1, 2, 4$.

Next, we shall show that the following ordinary differential equation is satisfied:

$$\frac{d}{dv} \left\{ \frac{\vartheta_1(v)}{\vartheta_4(v)} \right\} = \pi \vartheta_4^2 \frac{\vartheta_2(v)\vartheta_3(v)}{\vartheta_4^2(v)} . \tag{2}$$

To this end, consider the function

$$f(v) = \frac{\vartheta_1'(v)}{\vartheta_1(v)} - \frac{\vartheta_4'(v)}{\vartheta_4(v)} .$$

From Eqs. (7) and (8) of Sec. 9.2, it is easy to see that $f(v)$ is a doubly-periodic function with periods 1 and τ, with 0 and $-\tau/2$ as its poles and $\frac{1}{2}, -\frac{1}{2} - \frac{\tau}{2}$ its zeros. By Eq. (1) of Sec. 9.3, we obtain

$$f(v) = C \frac{\vartheta_1\left(v - \frac{1}{2}\right)\vartheta_1\left(v + \frac{1}{2} + \frac{\tau}{2}\right)}{\vartheta_1(v)\vartheta_1\left(v + \frac{\tau}{2}\right)} = iC \frac{\vartheta_2(v)\vartheta_3(v)}{\vartheta_1(v)\vartheta_4(v)} .$$

Since $vf(v) \to 1$ when $v \to 0$, we have

$$1 = \frac{iC\vartheta_2\vartheta_3}{\vartheta_1'\vartheta_4} .$$

Using then the result to be proved below, namely, $\vartheta_1' = \pi\vartheta_2\vartheta_3\vartheta_4$, we arrive at (2).

(2) can also be obtained simply by applying Eqs. (5) and (7) of Sec. 9.3. Let $k = 3$ in Eq. (5) of Sec. 9.3 and take the derivative with respect to z to obtain $\wp'(z)$. Then, by canceling $\wp'(z)$ with Eq. (7) of Sec. 9.3, we have (2).

Now we show that

$$\vartheta'_1 = \pi \vartheta_2 \vartheta_3 \vartheta_4 . \tag{3}$$

Expand the r.h.s. of Eq. (5) of Sec. 9.3 in power series of z in the vicinity of the origin:

$$\wp(z) - e_k = \left[\frac{\vartheta'_1}{2\omega \vartheta_{k+1}} \frac{\vartheta_{k+1} + \frac{z^2}{8\omega^2} \vartheta''_{k+1} + \ldots}{\frac{z}{2\omega} \vartheta'_1 + \frac{z^3}{48\omega^3} \vartheta'''_1 + \ldots} \right]^2 ,$$

where we have made use of the fact that $\vartheta_1(v)$ is an odd function and $\vartheta_{k+1}(v)$ even functions; $\vartheta''_k, \vartheta'''_k$ are abbreviations for $\vartheta''_k(0), \vartheta'''_k(0)$. Further calculations show that the r.h.s. is

$$\frac{1}{z^2} \left[1 + \frac{z^2}{8\omega^2} \left(\frac{\vartheta''_{k+1}}{\vartheta_{k+1}} - \frac{\vartheta'''_1}{3\vartheta'_1} \right) + \ldots \right]^2 = \frac{1}{z^2} + \frac{1}{4\omega^2} \left(\frac{\vartheta''_{k+1}}{\vartheta_{k+1}} - \frac{\vartheta'''_1}{3\vartheta'_1} \right) + \ldots .$$

Expanding $\wp(z)$ on the l.h.s. according to Eq. (3) of Sec. 8.4 and comparing the constant terms on the two sides, we obtain

$$4\omega^2 e_k = \frac{\vartheta'''_1}{3\vartheta'_1} - \frac{\vartheta''_{k+1}}{\vartheta_{k+1}} \quad (k = 1, 2, 3) . \tag{4}$$

The condition $e_1 + e_2 + e_3 = 0$ [Eq. (10) of Sec. 8.5] gives

$$\frac{\vartheta'''_1}{\vartheta'_1} = \frac{\vartheta''_2}{\vartheta_2} + \frac{\vartheta''_3}{\vartheta_3} + \frac{\vartheta''_4}{\vartheta_4} . \tag{5}$$

On the other hand, by putting $v = 0$ in the differential equation (1), we obtain

$$\vartheta''_k = 4\pi i \frac{\partial \vartheta_k}{\partial \tau} \quad (k = 2, 3, 4) . \tag{6}$$

For $k = 1$, we have to differentiate (1) once more with respect to v and then put $v = 0$. The result is

$$\vartheta'''_1 = 4\pi i \frac{\partial \vartheta'_1}{\partial \tau} . \tag{7}$$

Substituting (6) and (7) in (5) gives

$$\frac{1}{\vartheta'_1} \frac{\partial \vartheta'_1}{\partial \tau} = \frac{1}{\vartheta_2} \frac{\partial \vartheta_2}{\partial \tau} + \frac{1}{\vartheta_3} \frac{\partial \vartheta_3}{\partial \tau} + \frac{1}{\vartheta_4} \frac{\partial \vartheta_4}{\partial \tau} . \tag{8}$$

Integrating, we have

$$\vartheta'_1 = C \vartheta_2 \vartheta_3 \vartheta_4 .$$

The constant of integration is determined by letting $q \to 0$ and using Eqs. (1) – (4) of Sec. 9.2. It is found that $C = \pi$, which proves (3).

9.7. The Values of Some Constants

In the definitions of $\vartheta_k(v)$ [Sec. 9.2], putting $v = 0$, we have

$$\vartheta_2 = 2 \sum_{n=0}^{\infty} q^{\left(n+\frac{1}{2}\right)^2} = 2q^{\frac{1}{4}} + 2q^{\frac{9}{4}} + 2q^{\frac{25}{4}} + 2q^{\frac{49}{4}} + \ldots$$

$$= 2q^{\frac{1}{4}} \left(1 + q^2 + q^6 + q^{12} + \ldots\right),$$

$$\vartheta_3 = 1 + 2 \sum_{n=1}^{\infty} q^{n^2} = 1 + 2q + 2q^4 + 2q^9 + \ldots, \tag{1}$$

$$\vartheta_4 = 1 + 2 \sum_{n=1}^{\infty} (-)^n q^{n^2} = 1 - 2q + 2q^4 - 2q^9 + \ldots.$$

These three constants are connected by the relation [Eq. (5) of Sec. 9.4]:

$$\vartheta_3^4 = \vartheta_2^4 + \vartheta_4^4.$$

From Eq. (17) of Sec. 8.10 and Eq. (5) of Sec. 9.3, we have

$$\sqrt{\wp(z) - e_k} = \frac{\vartheta_1'}{2\omega\vartheta_{k+1}} \frac{\vartheta_{k+1}(v)}{\vartheta_1(v)} \quad \left(v = \frac{z}{2\omega}\right). \tag{2}$$

Putting $k = 2, z = \omega$, we obtain, by means of Eq. (9) of Sec. 9.2 and Eq. (3) of Sec. 9.6,

$$\sqrt{e_1 - e_2} = \frac{\pi\vartheta_4^2}{2\omega}. \tag{3}$$

Putting $k = 3, z = \omega$, we obtain

$$\sqrt{e_1 - e_3} = \frac{\pi\vartheta_3^2}{2\omega}. \tag{4}$$

Putting $k = 3, z = \omega + \omega'$, we obtain

$$\sqrt{e_2 - e_3} = \frac{\pi\vartheta_2^2}{2\omega}. \tag{5}$$

From these formulae, it is seen that, for real e_k, we can choose real ω and q so that $e_1 > e_2 > e_3$. If, in Sec. 8.5, we had chosen $\omega' = \omega_2$ instead of $\omega' = \omega_3$, then e_2 and e_3 would be interchanged and not satisfy the requirement $e_2 > e_3$.

Putting $k = 1, z = \omega + \omega'$, we have

$$\sqrt{e_2 - e_1} = -i\sqrt{e_1 - e_2} . \tag{6}$$

Similarly,

$$\sqrt{e_3 - e_1} = -i\sqrt{e_1 - e_3} , \quad \sqrt{e_3 - e_2} = -i\sqrt{e_2 - e_3} . \tag{7}$$

In these formulae, e_2 corresponds to $z = \omega + \omega' = -\omega_2$. If we take $z = \omega_2$, all the square roots with e_2 as its first term should change sign, and i is to be replaced by $-i$.

By squaring (3) and (4), adding the results and applying the relation $e_1 + e_2 + e_3 = 0$, we have

$$12\omega^2 e_1 = \pi^2(\vartheta_3^4 + \vartheta_4^4) . \tag{8}$$

Similarly, we obtain

$$12\omega^2 e_2 = \pi^2(\vartheta_2^4 - \vartheta_4^4) , \quad 12\omega^2 e_3 = -\pi^2(\vartheta_2^4 + \vartheta_3^4) . \tag{9}$$

From the following relation

$$(e_2 - e_3)^2 + (e_3 - e_1)^2 + (e_1 - e_2)^2 - 2(e_1 + e_2 + e_3)^2$$
$$= -6(e_2 e_3 + e_3 e_1 + e_1 e_2) = \frac{3}{2} g_2$$

and $g_3 = 4e_1 e_2 e_3$ [see Eq. (11) of Sec. 8.5], we obtain

$$g_2 = \frac{2}{3} \left(\frac{\pi}{2\omega}\right)^4 (\vartheta_2^8 + \vartheta_3^8 + \vartheta_4^8) , \tag{10}$$

$$g_3 = \frac{4}{27} \left(\frac{\pi}{2\omega}\right)^6 (\vartheta_2^4 + \vartheta_3^4)(\vartheta_3^4 + \vartheta_4^4)(\vartheta_4^4 - \vartheta_2^4) . \tag{11}$$

The quartic (4th) root of the discriminant Δ [Eq. (12) of Sec. 8.5] is

$$\Delta^{\frac{1}{4}} = 2 \left(\frac{\pi}{2\omega}\right)^3 (\vartheta_2 \vartheta_3 \vartheta_4)^2 = \frac{\pi \vartheta_1'^2}{4\omega^3} . \tag{12}$$

Now, let us find the expressions for η and η'. Taking the logarithmic derivative of Eq. (2) of Sec. 9.3, we have

$$\wp(z) = -\varsigma'(z) = -\frac{\eta}{\omega} - \frac{1}{(2\omega)^2} \frac{d}{dv} \frac{\vartheta_1'(v)}{\vartheta_1(v)} . \tag{13}$$

The expansion of $\vartheta_1(v)$ at the origin is

$$\vartheta_1(v) = \vartheta_1' v + \frac{\vartheta_1'''}{6} v^3 + \dots, \quad \vartheta_1'(v) = \vartheta_1' + \frac{\vartheta_1'''}{2} v^2 + \dots,$$

$$\frac{\vartheta_1'(v)}{\vartheta_1(v)} = \frac{1}{v} + \frac{\vartheta_1'''}{3\vartheta_1'} v + \dots.$$

Substituting in (13), comparing with the expansion of $\wp(z)$ [Eq. (3) of Sec. 8.4], and putting the constant term equal to zero, we obtain

$$\eta = -\frac{\vartheta_1'''}{12\omega\vartheta_1'}. \tag{14}$$

Using Eq. (8) of Sec. 8.6, we have

$$\eta' = -\frac{\pi i}{2\omega} - \frac{\tau\vartheta_1'''}{12\omega\vartheta_1'}. \tag{15}$$

The expansions of ϑ_1' and ϑ_1''' can be obtained from Eq. (1) of Sec. 9.2 by differentiations:

$$\vartheta_1' = 2\pi q^{\frac{1}{4}} \left(1 - 3q^2 + 5q^6 - 7q^{12} + \dots\right), \tag{16}$$

$$\vartheta_1''' = -2\pi^3 q^{\frac{1}{4}} \left(1 - 27q^2 + 125q^6 - 343q^{12} + \dots\right). \tag{17}$$

9.8. Legendre's Elliptic Integral of the First Kind

In Eq. (2) of Sec. 9.6, let

$$t = \frac{\vartheta_3\vartheta_1(v)}{\vartheta_2\vartheta_4(v)}, \quad k = \frac{\vartheta_2^2}{\vartheta_3^2}, \quad u = \pi\vartheta_3^2 v. \tag{1}$$

Squaring and applying Eqs. (1) and (2) of Sec. 9.4, we obtain

$$\left(\frac{dt}{du}\right)^2 = (1 - t^2)(1 - k^2 t^2), \tag{2}$$

from which we have

$$u = \int_0^t \frac{dt}{\sqrt{(1 - t^2)(1 - k^2 t^2)}}. \tag{3}$$

This is conforming to the requirement that $t \to 0$ and $dt/du \to 1$ when $u \to 0$ as posed by (1).

Integral (3) is called Legendre's elliptic integral of the first kind; k the *modulus*. Ordinarily, it is assumed that $k < 1$, and hence $k' = \sqrt{1 - k^2}$ is a real number also less than 1; k' is called the *complementary modulus*. By (1) and Eq. (5) of Sec. 9.4, we have

$$k' = \frac{\vartheta_4^2}{\vartheta_3^2} \, . \tag{4}$$

Complete Elliptic Integrals K and K' are defined by

$$K = \int_0^1 \frac{dt}{\sqrt{(1 - t^2)(1 - k^2 t^2)}} \, ,$$

$$K' = \int_0^1 \frac{dt}{\sqrt{(1 - t^2)(1 - k'^2 t^2)}} \, . \tag{5}$$

According to Eq. (7) of Sec. 9.2, we know that t is a doubly-periodic function; the periods are $(2, \tau)$ with respect to v and $(2\pi\vartheta_3^2, \pi\tau\vartheta_3^2)$ with respect to u. This function is the Jacobian elliptic function $t = \operatorname{sn} u$ [Eq. (11) of Sec. 8.1], which will be discussed in detail in the following chapter. When $v = \frac{1}{2}, t = 1$ by Eq. (9) of Sec. 9.2. Hence, from (5) we know that the corresponding u is K, and by (1), we have

$$K = \frac{\pi}{2}\vartheta_3^2 \, . \tag{6}$$

Applying (1) of the last section, we obtain

$$\left(\frac{2K}{\pi}\right)^{\frac{1}{2}} = 1 + 2\sum_{n=1}^{\infty} q^{n^2} = 1 + 2q + 2q^4 + 2q^9 + \dots \, . \tag{7}$$

When $v = \frac{1}{2} + \frac{\tau}{2}$, from (1), applying Eq. (9) of Sec. 9.2, we have $t = 1/k$. In the meantime, we have

$$\pi\vartheta_3^2 \cdot \left(\frac{1}{2} + \frac{\tau}{2}\right) = u = \int_0^{1/k} \frac{dt}{\sqrt{(1 - t^2)(1 - k^2 t^2)}}$$

$$= \int_0^1 \frac{dt}{\sqrt{(1 - t^2)(1 - k^2 t^2)}} + i \int_1^{1/k} \frac{dt}{\sqrt{(t^2 - 1)(1 - k^2 t^2)}} \, ,$$

where $\operatorname{Im}(t) > 0$ when the path of integration passes by and around $t = 1$. In the second integral, making the transformation $k^2 t^2 = 1 - k'^2 t'^2$, we obtain

$$\int_0^{1/k} \frac{dt}{\sqrt{(1 - t^2)(1 - k^2 t^2)}} = \int_0^1 \frac{dt'}{\sqrt{(1 - t'^2)(1 - k'^2 t'^2)}} = K' \, .$$

Hence,

$$\frac{\pi}{2}\vartheta_3^2 \cdot (1+\tau) = K + iK' .$$

Using (6), we obtain

$$iK' = \frac{\pi}{2}\vartheta_3^2\tau , \quad \tau = \frac{iK'}{K} . \tag{8}$$

So,

$$\frac{\pi K'}{K} = -i\pi\tau = \ln\frac{1}{q} . \tag{9}$$

By (1), (4), (7), (9), we can calculate the values of k, K, K' from a given value of q. Conversely, we can also calculate q for a given value of k by (4). The procedure is as follows. Let

$$2\varepsilon = \frac{1 - \sqrt{k'}}{1 + \sqrt{k'}} . \tag{10}$$

By (4) and Eq. (1) of Sec. 9.7, we have

$$2\varepsilon = \frac{\vartheta_3 - \vartheta_4}{\vartheta_3 + \vartheta_4} = \frac{2(q + q^9 + q^{25} + \ldots)}{1 + 2q^4 + 2q^{16} + \ldots} . \tag{11}$$

Conversely, we find

$$q = \varepsilon + 2\varepsilon^5 + 15\varepsilon^9 + 150\varepsilon^{13} + 1707\varepsilon^{17} + 20910\varepsilon^{21} + \ldots . \tag{12}$$

From (10) we see that $\varepsilon < \frac{1}{2}$. The series (12) converges rapidly and is suitable for calculating q from a given value of k. Even when k is as large as $\sqrt{0.8704} = 0.933$, so ε is $1/8$, $2\varepsilon^5 = 0.0000609$, $15\varepsilon^9 = 0.0000002$, the first three terms of the series will furnish an approximation with an accuracy to the 7th decimal place. The smaller the k, the smaller the q also, and the more rapid the convergence. If $k' < k$, we can exchange k and k', which corresponds to a sort of transformation to be discussed in detail in the next section.

From (3) − (5) of the last section and (1), (4) of the present section, we find the relations among k and e_1, e_2, e_3:

$$k^2 = \frac{e_2 - e_3}{e_1 - e_3} , \quad k'^2 = \frac{e_1 - e_2}{e_1 - e_3} . \tag{13}$$

Under the condition that e_r are real numbers and satisfy the relation $e_1 > e_2 > e_3$, we may use the above mentioned method to calculate q. The integral

$$z = \int_x^\infty \frac{dx}{\sqrt{4(x - e_1)(x - e_2)(x - e_3)}} , \tag{14}$$

after the transformation

$$x = e_3 + \frac{e_1 - e_3}{t^2}, \quad z = \frac{u}{\sqrt{e_1 - e_3}}, \tag{15}$$

becomes Legendre's standard form (3), where k is determined by (13). Then, $\omega\sqrt{e_1 - e_3} = K$. $\omega'\sqrt{e_1 - e_3} = iK'$ [cf. Ex. 10 at the end of this chapter].

When only one of the e_r is real, let

$$e_1 = \alpha + i\beta, \quad e_2 = -2\alpha, \quad e_3 = \alpha - i\beta \quad (\beta > 0) . \tag{16}$$

To reduce (14) to Legendre's standard form, we first make the transformation $x - \alpha = \beta/\xi$ to express the polynomial in the square root in a quartic form. The result is

$$z = \int_x^\infty \frac{dx}{\sqrt{4[(x-\alpha)^2 + \beta^2](x + 2\alpha)}}$$

$$= \int_0^\xi \frac{d\xi}{2\sqrt{(1 + \xi^2)(\beta + 3\alpha\xi)\xi}} . \tag{17}$$

In order for the integrand to be a function of squares of its argument, we need the following transformation

$$\xi = \frac{\eta - \lambda}{\lambda\eta + 1}, \quad \eta = \frac{\lambda + \xi}{1 - \lambda\xi},$$

where the parameter λ is not yet determined. We choose it so that the product $(\beta + 3\alpha\xi)\xi$ becomes $\frac{\eta^2 - \lambda^2}{(\lambda\eta + 1)^2} \cdot C(\lambda)$ where $C(\lambda)$ is some function of λ. This requires that $\eta = -\lambda$ when $\xi = -\beta/3\alpha$, and hence $-\beta/3\alpha = 2\lambda/(\lambda^2 - 1)$. The solution is

$$\lambda = \frac{\sigma^2 - 3\alpha}{\beta}, \quad \sigma^4 = 9\alpha^2 + \beta^2 .$$

After the transformation, (17) reduces to

$$z = \sqrt{\frac{\lambda}{2\beta}} \int_\lambda^\eta \frac{d\eta}{\sqrt{(\eta^2 + 1)(\eta^2 - \lambda^2)}} . \tag{18}$$

Making further transformation

$$\eta^2 = \frac{\lambda^2}{1 - t^2}, \quad k^2 = \frac{1}{1 + \lambda^2} = \frac{1}{2} + \frac{3\alpha}{2\sigma^2},$$

we obtain   (19)

$$2\sigma z = \int_0^t \frac{dt}{\sqrt{(1-t^2)(1-k^2t^2)}} = u \, ,$$

which is then of Legendre's standard form. By the above transformation formulae, we can express $\omega_1, \omega_2, \omega_3$ in terms of K and K' [see Ex. 11 at the end of this chapter]. The relations between t and x are

$$t = \frac{2\sigma\sqrt{x+2\alpha}}{x^2+2\alpha+\sigma^2} \, , \quad x+2\alpha+\sigma^2 = \frac{2\sigma^2}{t^2}(1+\sqrt{1-t^2}) \, . \tag{20}$$

9.9. Jacobi's Imaginary Transformation

In Weierstrass' elliptic functions, the interchange of ω and ω' does not cause any alteration. In order to preserve $\mathrm{Im}(\omega'/\omega) > 0$, it is necessary to replace ω and ω' respectively by ω' and $-\omega$. In the Theta functions, however, since the two periods are not symmetric, the corresponding transformation is to replace v by $v' = z/2\omega' = v/\tau$ and, at the same time, τ by $\tau' = -1/\tau$. As τ is imaginary, this transformation is called an imaginary transformation. To indicate the period τ explicitly, we write $\vartheta_k(v|\tau)$ for $\vartheta_k(v)$. In the transformation from (ω, ω') to $(\omega', -\omega)$, $\sigma(z)$ is kept unaltered, whereas $e_1 \to e_3, e_3 \to e_1, e_2$ unchanged. From Eq. (5) of Sec. 9.2, we have

$$\sigma(z) = \frac{2\omega}{\vartheta_1'(0|\tau)} e^{\frac{\eta z^2}{2\omega}} \vartheta_1(v|\tau) = \frac{2\omega'}{\vartheta_1'(0|\tau')} e^{\frac{\eta' z^2}{2\omega'}} \vartheta_1\left(\frac{v}{\tau}\Big|-\frac{1}{\tau}\right) \, .$$

Using Eq. (8) of Sec. 8.6, and $z = 2\omega v$, we obtain

$$\left(\frac{\eta}{2\omega} - \frac{\eta'}{2\omega'}\right) z^2 = \frac{2\omega v^2}{\omega'}(\eta\omega' - \eta'\omega) = \frac{i\pi v^2}{\tau} \, .$$

Further, as e_k are merely interchanged among themselves, the discriminant Δ is invariant. Hence, by Eq. (12) of Sec. 9.7, we have

$$\frac{\vartheta_1'(0|\tau')}{\omega'^{3/2}} = \varepsilon \frac{\vartheta_1'(0|\tau)}{\omega^{3/2}} \, ,$$

where ε is a factor independent of τ. To determine this factor, let $\tau = \tau' = i$, then we obtain $\xi = i^{-3/2}$. Thus, we have

$$\vartheta_1'\left(0\Big|-\frac{1}{\tau}\right) = (-i\tau)^{3/2}\vartheta_1'(0|\tau) \tag{1}$$

$$\vartheta_1\left(\frac{v}{\tau}\Big|-\frac{1}{\tau}\right) = -i(-i\tau)^{1/2}e^{i\pi v^2/\tau}\vartheta_1(v|\tau) \, . \tag{2}$$

Replacing v successively by $v + \tau/2, v + \tau/2 - 1/2, v - 1/2$, with the definitions of Eqs. (2) – (4) in Sec. 9.2, we obtain

$$\vartheta_2 \left(\frac{v}{\tau} \Big| -\frac{1}{\tau} \right) = (-i\tau)^{1/2} e^{i\pi v^2/\tau} \vartheta_4(v|\tau) , \tag{3}$$

$$\vartheta_3 \left(\frac{v}{\tau} \Big| -\frac{1}{\tau} \right) = (-i\tau)^{1/2} e^{i\pi v^2/\tau} \vartheta_3(v|\tau) , \tag{4}$$

$$\vartheta_4 \left(\frac{v}{\tau} \Big| -\frac{1}{\tau} \right) = (-i\tau)^{1/2} e^{i\pi v^2/\tau} \vartheta_2(v|\tau) . \tag{5}$$

This transformation is important in numerical computations, since it transforms a series with comparatively slow convergence (in case $|\tau|$ is small) into a series which converges quickly (the case when $\tau' = -1/\tau$ has a comparatively larger absolute value).

Let $\tau = i\sigma$ and use Eq. (3) of Sec. 9.2. We can write the above Eq. (4) as

$$\sum_{n=-\infty}^{\infty} e^{-\pi(n+v)^2/\sigma} = \sigma^{1/2} \sum_{n=-\infty}^{\infty} e^{-\pi\sigma n^2 + 2n\pi v i} . \tag{6}$$

This formula can also be justified by the method of Fourier expansion. The l.h.s. of it is evidently a periodic function of v with period 1. Therefore, it can be expanded in a Fourier series $\sum_{-\infty}^{\infty} c_n e^{2n\pi v i}$; the coefficients c_n are given by

$$c_n = \int_0^1 \sum_{m=-\infty}^{\infty} e^{-\pi(m+v)^2/\sigma} e^{-2n\pi v i} dv$$

$$= \sum_{m=-\infty}^{\infty} \int_m^{m+1} e^{-\pi t^2/\sigma} e^{-2n\pi t i} dt = \int_{-\infty}^{\infty} e^{(-\pi/\sigma)t^2 - 2n\pi t i} dt = \sigma^{1/2} e^{-\pi\sigma n^2} ,$$

which proves (6).

In (3), (4), (5), putting $v = 0$, we have

$$\vartheta_2 \left(0 \Big| -\frac{1}{\tau} \right) = (-i\tau)^{1/2} \vartheta_4(0|\tau) ,$$

$$\vartheta_3 \left(0 \Big| -\frac{1}{\tau} \right) = (-i\tau)^{1/2} \vartheta_3(0|\tau) , \tag{7}$$

which give the transformation relations of the constants ϑ_k.

From the expressions of k and k' given by (1) and (4) of Sec. 9.8, it is seen that the above relations (7) show: the transformation from τ to τ' leads to changing k into k'. Therefore, if we carry out the transformation τ to τ' in Eq. (6) of Sec. 9.8, we have

$$K' = \frac{\pi}{2}\left[\vartheta_3\left(0\Big|-\frac{1}{\tau}\right)\right]^2 = -\frac{i\tau\pi}{2}[\vartheta_3(0|\tau)]^2 = -i\tau K ,$$

i.e., $\tau = iK'/K$. This is the same as Eq. (8) of Sec. 9.8.

The more general transformation [see Sec. 8.3, where p, q, p', q' are integers]:

$$\Omega = p\omega + q\omega' , \quad \Omega' = p'\omega + q'\omega' , \quad pq' - p'q = 1 \tag{8}$$

can be reduced to a combined operation of two successive fundamental transformations. One of the fundamental transformation is Jacobi's imaginary transformation, corresponding to $p = 0, q = 1, p' = -1, q' = 0$; the other is

$$\Omega = \omega , \quad \Omega' = \omega + \omega' , \tag{9}$$

corresponding to $(v, \tau) \to (v, \tau_2)$:

$$\tau_2 = \tau + 1 . \tag{10}$$

To this transformation, we have, by the definitions in Sec. 9.2,

$$\begin{aligned}
\vartheta_1(v|\tau + 1) &= e^{i\pi/4}\vartheta_1(v|\tau) , & \vartheta_2(v|\tau + 1) &= e^{i\pi/4}\vartheta_2(v|\tau) , \\
\vartheta_3(v|\tau + 1) &= \vartheta_4(v|\tau) , & \vartheta_4(v|\tau + 1) &= \vartheta_3(v|\tau) .
\end{aligned} \tag{11}$$

9.10. Transformation of Landen-Type

The transformation $v' = 2v, \tau' = 2\tau$ is of Landen-type. We now show that

$$\frac{\vartheta_3(v|\tau)\vartheta_4(v|\tau)}{\vartheta_4(2v|2\tau)} = \frac{\vartheta_3(0|\tau)\vartheta_4(0|\tau)}{\vartheta_4(0|2\tau)} . \tag{1}$$

The l.h.s. of this equation has two periods, $v = \frac{1}{2}$ and $v = \tau$; the zero of the numerator is $v = \frac{1}{2}\tau$ (the other zero, $\frac{1}{2} + \frac{1}{2}\tau$ is a duplication since $\frac{1}{2}$ is a period) and that of the denominator is the same, so they cancel each other. By Theorem 1 of Sec. 8.3, the l.h.s. of (1) must be a constant, the value of which is determined by putting $v = 0$ and is found to be the value given on the r.h.s. of (1). Thus (1) holds.

Replacing v by $v + \frac{1}{2}\tau$ and using the formulae given in Sec. 9.2, (1) reduces to

$$\frac{\vartheta_1(v|\tau)\vartheta_2(v|\tau)}{\vartheta_1(2v|2\tau)} = \frac{\vartheta_3(0|\tau)\vartheta_4(0|\tau)}{\vartheta_4(0|2\tau)} . \tag{2}$$

Similarly, it can be shown that

$$\frac{[\vartheta_3(v|\tau)]^2 + [\vartheta_4(v|\tau)]^2}{\vartheta_3(2v|2\tau)} = \frac{\vartheta_3^2(0|\tau) + \vartheta_4^2(0|\tau)}{\vartheta_3(0|2\tau)} , \tag{3}$$

$$\frac{[\vartheta_1(v|\tau)]^2 + [\vartheta_2(v|\tau)]^2}{\vartheta_3(2v|2\tau)} = \frac{\vartheta_2^2(0|\tau)}{\vartheta_3(0|2\tau)} . \tag{4}$$

Changing v to $v + \frac{1}{2}\tau$ gives

$$\frac{[\vartheta_2(v|\tau)]^2 - [\vartheta_1(v|\tau)]^2}{\vartheta_2(2v|2\tau)} = \frac{\vartheta_3^2(0|\tau) + \vartheta_4^2(0|\tau)}{\vartheta_3(0|2\tau)} , \tag{5}$$

$$\frac{[\vartheta_3^2(v|\tau)]^2 - [\vartheta_4(v|\tau)]^2}{\vartheta_2(2v|2\tau)} = \frac{\vartheta_2^2(0|\tau)}{\vartheta_3(0|2\tau)} . \tag{6}$$

Further, by the definitions given in Sec. 9.2, we have

$$\vartheta_3(v|\tau) + \vartheta_4(v|\tau) = 2 \sum_{n=-\infty}^{\infty} q^{4n^2} e^{4n\pi v i} = 2\vartheta_3(2v|4\tau) , \tag{7}$$

$$\vartheta_3(v|\tau) - \vartheta_4(v|\tau) = 2 \sum_{n=-\infty}^{\infty} q^{4\left(n-\frac{1}{2}\right)^2} e^{2(2n-1)\pi v i} = 2\vartheta_2(2v|4\tau) . \tag{8}$$

9.11. Representation of Theta Functions by Infinite Product

Consider first the representation of $\vartheta_4(v)$ by an infinite product. From Eq. (8) of Sec. 9.2, we know that the zeros of $\vartheta_4(v)$ are $m + m'\tau + \frac{1}{2}\tau$. Writing $n - 1$ for m' and using the periodicity of $e^{2\pi v i}$, these zeros will depend only on one integer, n, and contribute the factors $(e^{2\pi v i} - e^{(2n-1)\tau\pi i})$. To ensure the convergence of the infinite product, the coefficient $2n - 1$ of τ must be positive (since $\text{Im}(\tau) > 0$), so we have

$$\vartheta_4(v) = G \prod_{n=1}^{\infty} \left\{ (1 - q^{2n-1} e^{2\pi v i})(1 - q^{2n-1} e^{-2\pi v i}) \right\} . \tag{1}$$

Whether the factor G depends on v or not can be judged by the periodicity of $\vartheta_4(v)$ with respect to v [Eq. (7) of Sec. 9.2]. When v changes to $v + 1$, the product on the r.h.s. of (1) is unaltered; when v changes to $v + \tau$, the product becomes

$$\prod_{n=1}^{\infty} (1 - q^{2n+1} e^{2\pi vi})(1 - q^{2n-3} e^{-2\pi vi})$$

$$= \prod_{n=1}^{\infty} (1 - q^{2n-1} e^{2\pi vi})(1 - q^{2n-1} e^{-2\pi vi})(1 - q^{-1} e^{-2\pi vi})/(1 - q e^{2\pi vi})$$

$$= -q^{-1} e^{-2\pi vi} \prod_{n=1}^{\infty} (1 - q^{2n-1} e^{2\pi vi})(1 - q^{2n-1} e^{-2\pi vi}) .$$

The multiplying factors in the product are the same as those in the expression for $\vartheta_4(v)$. Therefore, the ratio of this product to $\vartheta_4(v)$ is a doubly-periodic function with no singularities. So, according to Theorem 1 of Sec. 8.3, it must be a constant, and hence G is independent of v. The value of G will be determined later. This infinite product is simpler than that of $\sigma(z)$ in Sec. 8.7, since only one integer, n, occurs here.

Replacing v successively by $v + \frac{1}{2}, v + \frac{1}{2}\tau, v + \frac{1}{2} + \frac{1}{2}\tau$, we obtain

$$\vartheta_3(v) = G \prod_{n=1}^{\infty} (1 + q^{2n-1} e^{2\pi vi})(1 + q^{2n-1} e^{-2\pi vi}) , \tag{2}$$

$$\vartheta_1(v) = 2Gq^{1/4} \sin \pi v \prod_{n=1}^{\infty} (1 - q^{2n} e^{2\pi vi})(1 - q^{2n} e^{-2\pi vi}) , \tag{3}$$

$$\vartheta_2(v) = 2Gq^{1/4} \cos \pi v \prod_{n=1}^{\infty} (1 + q^{2n} e^{2\pi vi})(1 + q^{2n} e^{-2\pi vi}) . \tag{4}$$

By multiplying up the pair factors therein, these become

$$\vartheta_1(v) = 2Gq^{1/4} \sin \pi v \prod_{n=1}^{\infty} (1 - 2q^{2n} \cos 2\pi v + q^{4n}) , \tag{5}$$

$$\vartheta_2(v) = 2Gq^{1/4} \cos \pi v \prod_{n=1}^{\infty} (1 + 2q^{2n} \cos 2\pi v + q^{4n}) , \tag{6}$$

$$\vartheta_3(v) = G \prod_{n=1}^{\infty} (1 + 2q^{2n-1} \cos 2\pi v + q^{4n-2}) , \tag{7}$$

$$\vartheta_4(v) = G \prod_{n=1}^{\infty} (1 - 2q^{2n-1} \cos 2\pi v + q^{4n-2}) . \tag{8}$$

To determine the factor G, we use Eq. (3) of Sec. 9.6, i.e., $\vartheta_1' = \pi\vartheta_2\vartheta_3\vartheta_4$. From (1) – (4), we find

$$\vartheta_1' = 2\pi G q^{1/4} \prod_{n=1}^{\infty} (1 - q^{2n})^2 , \tag{9}$$

$$\vartheta_2 = 2G q^{1/4} \prod_{n=1}^{\infty} (1 + q^{2n})^2 , \tag{10}$$

$$\vartheta_3 = G \prod_{n=1}^{\infty} (1 + q^{2n-1})^2 , \tag{11}$$

$$\vartheta_4 = G \prod_{n=1}^{\infty} (1 - q^{2n-1})^2 . \tag{12}$$

Putting them in $\vartheta_1' = \pi\vartheta_2\vartheta_3\vartheta_4$, we obtain

$$\prod_{n=1}^{\infty} (1 - q^{2n})^2 = G^2 \prod_{n=1}^{\infty} [(1 + q^{2n})(1 + q^{2n-1})]^2 (1 - q^{2n-1})^2$$

$$= G^2 \prod_{n=1}^{\infty} (1 + q^n)^2 (1 - q^{2n-1})^2$$

$$= G^2 \prod_{n=1}^{\infty} (1 + q^n)^2 [(1 - q^n)/(1 - q^{2n})]^2 = G^2 .$$

Taking its square root, on account of $\vartheta_4 \to 1$ when $q \to 0$ [see Eq. (1) of Sec. 9.7], we have

$$G = \prod_{n=1}^{\infty} (1 - q^{2n}) . \tag{13}$$

Applying these formulae, it is found that

$$\vartheta_3(0|\tau)\vartheta_4(0|\tau) = \prod_{n=1}^{\infty} (1 - q^{2n})^2 (1 + q^{2n-1})^2 (1 - q^{2n-1})^2$$

$$= \prod_{n=1}^{\infty} (1 - q^{2n})^2 (1 - q^{4n-2})^2$$

$$= \left[\prod_{n=1}^{\infty} (1 - q^{4n})(1 - q^{4n-2})(1 - q^{4n-2}) \right]^2 ,$$

i.e.,

$$\vartheta_3(0|\tau)\vartheta_4(0|\tau) = [\vartheta_4(0|2\tau)]^2 . \tag{14}$$

By multiplying up (7) and (8) of the last section, then putting $v = 0$, we obtain

$$\vartheta_3^2(0|\tau) - \vartheta_4^2(0|\tau) = 4\vartheta_2(0|4\tau)\vartheta_3(0|4\tau)$$

$$= 8q \prod_{n=1}^{\infty} (1 - q^{8n})^2 (1 + q^{8n})^2 (1 + q^{8n-4})^2$$

$$= 8q \prod_{n=1}^{\infty} (1 - q^{8n})^2 (1 + q^{4n})^2$$

$$= 8q \prod_{n=1}^{\infty} (1 - q^{4n})^2 (1 + q^{4n})^4 ,$$

i.e.,

$$\vartheta_3^2(0|\tau) - \vartheta_4^2(0|\tau) = 2\vartheta_2^2(0|2\tau) . \tag{15}$$

From (14) and (15), we have

$$[\vartheta_3^2(0|\tau) + \vartheta_4^2(0|\tau)]^2 = [\vartheta_3^2(0|\tau) - \vartheta_4^2(0|\tau)]^2 + 4\vartheta_3^2(0|\tau)\vartheta_4^2(0|\tau)$$

$$= 4[\vartheta_2^4(0|2\tau) + \vartheta_4^4(0|2\tau)] .$$

Using Eq. (5) of Sec. 9.4 and noticing the property when $q \to 0$, we have

$$\vartheta_3^2(0|\tau) + \vartheta_4^2(0|\tau) = 2\vartheta_3^2(0|2\tau) . \tag{16}$$

Multiplying up (15) and (16) and applying Eq. (5) of Sec. 9.4, we obtain, after extracting the square root,

$$\vartheta_2^2(0|\tau) = 2\vartheta_2(0|2\tau)\vartheta_3(0|2\tau) . \tag{17}$$

With these formulae, the transformation of Landen-type in the last section can be simplified into

$$\vartheta_1(2v|2\tau) = \frac{\vartheta_1(v|\tau)\vartheta_2(v|\tau)}{\vartheta_4(0|2\tau)} , \tag{18}$$

$$\vartheta_2(2v|2\tau) = \frac{\vartheta_3^2(v|\tau) - \vartheta_4^2(v|\tau)}{2\vartheta_2(0|2\tau)}$$

$$= \frac{\vartheta_2^2(v|\tau) - \vartheta_1^2(v|\tau)}{2\vartheta_3(0|2\tau)} , \tag{19}$$

$$\vartheta_3(2v|2\tau) = \frac{\vartheta_2^2(v|\tau) + \vartheta_1^2(v|\tau)}{2\vartheta_2(0|2\tau)}$$

$$= \frac{\vartheta_3^2(v|\tau) + \vartheta_4^2(v|\tau)}{2\vartheta_3(0|2\tau)} , \tag{20}$$

$$\vartheta_4(2v|2\tau) = \frac{\vartheta_3(v|\tau)\vartheta_4(v|\tau)}{\vartheta_4(0|2\tau)} . \tag{21}$$

Formulae (18), (21) and the first of (19) can be proved by using (1) − (4) of the present section and (7), (8) of the last section. The second formula of (19) can be proved by using Eqs. (2), (4) of Sec. 9.4 and (16), (17) of the present section. (20) can be derived from (19) by replacing v with $v + \frac{1}{2}\tau$.

9.12. Fourier Expansion of the Logarithmic Derivatives of Theta Functions

The logarithmic derivatives of the Theta functions also have the period $v = 1$ and thus can be expanded in Fourier series. Consider first the expansion of $\vartheta_4'(v)/\vartheta_4(v)$. It is an odd function, and hence its Fourier expansion is a sine series:

$$\frac{\vartheta_4'(v)}{\vartheta_4(v)} = \sum_{n=1}^{\infty} c_n \sin 2n\pi v ,$$

where

$$c_n = 2 \int_{-\frac{1}{2}}^{\frac{1}{2}} \frac{\vartheta_4'(v)}{\vartheta_4(v)} \sin 2n\pi v\, dv .$$

Owing to the periodicity of $\vartheta_4(v)$, this integral can be evaluated by the method of contour integration. The contour is chosen to be the rectangle $ABCD$ with vertices $-\frac{1}{2}, \frac{1}{2}, \frac{1}{2} + \tau, -\frac{1}{2} + \tau$ in succession; the integrand is taken to be $e^{2n\pi v i}\vartheta_4'(v)/\vartheta_4(v)$, which has the unique pole $v = \frac{1}{2}\tau$ in $ABCD$ with a residue equal to

$$e^{n\pi\tau i}\vartheta_4'\left(\frac{\tau}{2}\right)\Big/\vartheta_4\left(\frac{\tau}{2}\right) = q^n .$$

The values of the integrals along BC and DA cancel each other. Those along AB and CD are respectively

$$\int_{AB} \frac{\vartheta_4'(v)}{\vartheta_4(v)} e^{2n\pi v i}\, dv = \int_{-\frac{1}{2}}^{\frac{1}{2}} \frac{\vartheta_4'(v)}{\vartheta_4(v)} e^{2n\pi v i}\, dv$$

$$= i \int_{-\frac{1}{2}}^{\frac{1}{2}} \frac{\vartheta_4'(v)}{\vartheta_4(v)} \sin 2n\pi v\, dv ,$$

$$\int_{CD} \frac{\vartheta_4'(v)}{\vartheta_4(v)} e^{2n\pi v i}\, dv = -\int_{-\frac{1}{2}}^{\frac{1}{2}} \frac{\vartheta_4'(v+\tau)}{\vartheta_4(v+\tau)} e^{2n\pi(v+\tau)i}\, dv$$

$$= -\int_{-\frac{1}{2}}^{\frac{1}{2}} \left[\frac{\vartheta_4'(v)}{\vartheta_4(v)} - 2\pi i\right] q^{2n} e^{2n\pi v i}\, dv$$

$$= -i q^{2n} \int_{-\frac{1}{2}}^{\frac{1}{2}} \frac{\vartheta_4'(v)}{\vartheta_4(v)} \sin 2n\pi v\, dv .$$

Adding them together and dividing by $2\pi i$, the result should be equal to the residue q^n. Hence, we have $q^n = (1 - q^{2n})c_n/4\pi$, so

$$\frac{\vartheta_4'(v)}{\vartheta_4(v)} = 4\pi \sum_{n=1}^{\infty} \frac{q^n \sin 2n\pi v}{1 - q^{2n}} . \tag{1}$$

Changing v to $v + \frac{1}{2}$ and using Eq. (4) of Sec. 9.2, we obtain

$$\frac{\vartheta_3'(v)}{\vartheta_3(v)} = 4\pi \sum_{n=1}^{\infty} \frac{(-)^n q^n \sin 2n\pi v}{1 - q^{2n}} . \tag{2}$$

In (1), replacing v by $v - \frac{1}{2}\tau$ and expressing the sine functions on the r.h.s. by exponential functions, we find, by Eq. (4) of Sec. 9.2 and assuming $\text{Im}(v) > 0$,

$$\frac{\vartheta_1'(v)}{\vartheta_1(v)} = \pi \cot \pi v + 4\pi \sum_{n=1}^{\infty} \frac{q^{2n} \sin 2n\pi v}{1 - q^{2n}} . \tag{3}$$

Changing v to $v + \frac{1}{2}$, we obtain

$$\frac{\vartheta_2'(v)}{\vartheta_2(v)} = -\pi \tan \pi v + 4\pi \sum_{n=1}^{\infty} \frac{(-)^n q^{2n} \sin 2n\pi v}{1 - q^{2n}} . \tag{4}$$

The series in (3) and (4) converge for real values of v, so the assumption that $\text{Im}(v) > 0$ can be retracted. The condition for the convergence of series (1) and (2) is that $|\text{Im}(v)| < \frac{1}{2}\text{Im}(\tau)$, and that for (3) and (4) is $|\text{Im}(v)| < \text{Im}(\tau)$. (Note that $q = e^{i\pi\tau}$.)

9.13. The Functions $\Theta(u)$ and $H(u)$

The earlier notations for Theta functions as introduced by Jacobi were $\Theta(u)$ and $H(u)$. Their relations to $\vartheta_k(v)$ are

$$\Theta(u) = \vartheta_4(v) , \quad H(u) = \vartheta_1(v) , \quad u = 2Kv . \tag{1}$$

Corresponding to the periods 1 and τ with respect to v are the two periods $2K$ and $2K'i$ with respect to u, and

$$\tau = \frac{iK'}{K} , \quad q = e^{-\pi K'/K} . \tag{2}$$

The relation between $H(u)$ and $\Theta(u)$ is, according to Eq. (4) of Sec. 9.2,

$$H(u) = -iq^{1/4} e^{i\pi u/2K} \Theta(u + iK') . \tag{3}$$

Besides, the following function $Z(u)$ was also introduced:

$$Z(u) = \frac{\Theta'(u)}{\Theta(u)} \; . \tag{4}$$

These notations still appear in the literature sometimes. We shall discuss them further in the next chapter when we deal with Jacobian elliptic functions.

Exercise 9

1. Let $M = q^{1/4} e^{i\pi v}$ Show that

$$\vartheta_1(v) = -\vartheta_2\left(v + \frac{1}{2}\right) = -iM\vartheta_3\left(v + \frac{1}{2} + \frac{\tau}{2}\right)$$
$$= -iM\vartheta_4\left(v + \frac{\tau}{2}\right) \; ,$$

$$\vartheta_2(v) = M\vartheta_3\left(v + \frac{\tau}{2}\right) = M\vartheta_4\left(v + \frac{1}{2} + \frac{\tau}{2}\right)$$
$$= \vartheta_1\left(v + \frac{1}{2}\right) \; ,$$

$$\vartheta_3(v) = \vartheta_4\left(v + \frac{1}{2}\right) = M\vartheta_1\left(v + \frac{1}{2} + \frac{\tau}{2}\right)$$
$$= M\vartheta_2\left(v + \frac{\tau}{2}\right) \; ,$$

$$\vartheta_4(v) = -iM\vartheta_1\left(v + \frac{\tau}{2}\right) = iM\vartheta_2\left(v + \frac{1}{2} + \frac{\tau}{2}\right)$$
$$= \vartheta_3\left(v + \frac{1}{2}\right) \; .$$

2. Show that

$$\vartheta_3^2\vartheta_1(v+w)\vartheta_1(v-w) = \vartheta_1^2(v)\vartheta_3^2(w) - \vartheta_3^2(v)\vartheta_1^2(w)$$
$$= \vartheta_4^2(v)\vartheta_2^2(w) - \vartheta_2^2(v)\vartheta_4^2(w) \; ,$$
$$\vartheta_3^2\vartheta_2(v+w)\vartheta_2(v-w) = \vartheta_2^2(v)\vartheta_3^2(w) - \vartheta_4^2(v)\vartheta_1^2(w)$$
$$= \vartheta_3^2(v)\vartheta_2^2(w) - \vartheta_1^2(v)\vartheta_4^2(w) \; ,$$
$$\vartheta_4^2\vartheta_2(v+w)\vartheta_2(v-w) = \vartheta_2^2(v)\vartheta_4^2(w) - \vartheta_3^2(v)\vartheta_1^2(w)$$
$$= \vartheta_4^2(v)\vartheta_2^2(w) - \vartheta_1^2(v)\vartheta_3^2(w) \; ,$$

$$\vartheta_4^2\vartheta_3(v+w)\vartheta_3(v-w) = \vartheta_3^2(v)\vartheta_4^2(w) - \vartheta_2^2(v)\vartheta_1^2(w)$$
$$= \vartheta_4^2(v)\vartheta_3^2(w) - \vartheta_1^2(v)\vartheta_2^2(w) \,,$$
$$\vartheta_4^2\vartheta_4(v+w)\vartheta_4(v-w) = \vartheta_4^2(v)\vartheta_4^2(w) - \vartheta_1^2(v)\vartheta_1^2(w)$$
$$= \vartheta_3^2(v)\vartheta_3^2(w) - \vartheta_2^2(v)\vartheta_2^2(w) \,,$$
$$\vartheta_2^2\vartheta_4(v+w)\vartheta_4(v-w) = \vartheta_4^2(v)\vartheta_2^2(w) - \vartheta_3^2(v)\vartheta_1^2(w)$$
$$= \vartheta_2^2(v)\vartheta_4^2(w) - v_1^2(v)\vartheta_3^2(w) \,.$$

3. Show that

$$\vartheta_2\vartheta_3\vartheta_2(v+w)\vartheta_3(v-w)$$
$$= \vartheta_2(v)\vartheta_3(v)\vartheta_2(w)\vartheta_3(w) - \vartheta_1(v)\vartheta_4(v)\vartheta_1(w)\vartheta_4(w) \,,$$
$$\vartheta_2\vartheta_4\vartheta_1(v+w)\vartheta_3(v-w)$$
$$= \vartheta_1(v)\vartheta_3(v)\vartheta_2(w)\vartheta_4(w) + \vartheta_2(v)\vartheta_4(v)\vartheta_1(w)\vartheta_3(w) \,,$$
$$\vartheta_2\vartheta_4\vartheta_2(v+w)\vartheta_4(v-w)$$
$$= \vartheta_2(v)\vartheta_4(v)\vartheta_2(w)\vartheta_4(w) - \vartheta_1(v)\vartheta_3(v)\vartheta_1(w)\vartheta_3(w) \,,$$
$$\vartheta_3\vartheta_4\vartheta_1(v+w)\vartheta_2(v-w)$$
$$= \vartheta_1(v)\vartheta_2(v)\vartheta_3(w)\vartheta_4(w) + \vartheta_3(v)\vartheta_4(v)\vartheta_1(w)\vartheta_2(w) \,,$$
$$\vartheta_3\vartheta_4\vartheta_3(v+w)\vartheta_4(v-w)$$
$$= \vartheta_3(v)\vartheta_4(v)\vartheta_3(w)\vartheta_4(w) - \vartheta_1(v)\vartheta_2(v)\vartheta_1(w)\vartheta_2(w) \,.$$

4. Show that

$$\vartheta_3^2\vartheta_3(v+w)\vartheta_3(v-w) = \vartheta_2^2(v)\vartheta_2^2(w) + \vartheta_4^2(v)\vartheta_4^2(w) \,,$$

then

$$\vartheta_3^3\vartheta_3(2v) = \vartheta_2^4(v) + \vartheta_4^4(v)$$

and from it

$$\vartheta_1^4(v) + \vartheta_3^4(v) = \vartheta_2^4(v) + \vartheta_4^4(v) \,.$$

5. By the results obtained in Ex. 2, show that

$$\vartheta_2\vartheta_4^2\vartheta_2(2v) = \vartheta_2^2(v)\vartheta_4^2(v) - \vartheta_1^2(v)\vartheta_3^2(v) \,,$$
$$\vartheta_3\vartheta_4^2\vartheta_3(2v) = \vartheta_3^2(v)\vartheta_4^2(v) - \vartheta_1^2(v)\vartheta_2^2(v) \,,$$
$$\vartheta_4^3\vartheta_4(2v) = \vartheta_3^4(v) - \vartheta_2^4(v) = \vartheta_4^4(v) - \vartheta_1^4(v) \,.$$

6. Suppose that (w', x', y', z') and (w, x, y, z) are related by

$$2w' = -w + x + y + z \qquad 2x' = w - x + y + z \,,$$
$$2y' = w + x - y + z \,, \qquad 2z' = w + x + y - z \,.$$

Use the simplifying symbols $[r]$ to represent $\vartheta_r(w)\vartheta_r(x)\vartheta_r(y)\vartheta_r(z)$, $[pqrs]$ to represent $\vartheta_p(w)\vartheta_q(x)\vartheta_r(y)\vartheta_s(z)$, $[r]'$ to represent $\vartheta_r(w')\vartheta_r(x')$ $\times\vartheta_r(y')\vartheta_r(z')$, $[pqrs]'$ to represent $\vartheta_p(w')\vartheta_q(x')\vartheta_r(y')\vartheta_s(z')$. Show that

$$2[3] = -[1]' + [2]' + [3]' + [4]' ,$$
$$2[4] = [1]' - [2]' + [3]' + [4]' ,$$
$$2[1] = [1]' + [2]' - [3]' + [4]' ,$$
$$2[2] = [1]' + [2]' + [3]' - [4]' ,$$

$$[1] + [2] = [1]' + [2]' , \qquad [1] + [3] = [2]' + [4]' ,$$
$$[1] + [4] = [1]' + [4]' , \qquad [2] + [3] = [2]' + [3]' ,$$
$$[2] + [4] = [1]' + [3]' , \qquad [3] + [4] = [3]' + [4]' ,$$
$$[3344] + [2211] = [4433]' + [1122]',$$

$$2[1234] = [3412]' + [2143]' - [1234]' + [4321]' ,$$
$$2[1122] = [1122]' + [2211]' - [4433]' + [3344]' ,$$
$$2[1133] = [1133]' + [3311]' - [4422]' + [2244]' ,$$
$$2[1144] = [1144]' + [4411]' - [3322]' + [2233]' ,$$
$$2[2233] = [2233]' + [3322]' - [4411]' + [1144]' ,$$
$$2[2244] = [2244]' + [4422]' - [3311]' + [1133]' ,$$
$$2[3344] = [3344]' + [4433]' - [2211]' + [1122]' .$$

7. Show that

$$\frac{\vartheta_1'(v)}{\vartheta_1(v)} - \frac{\vartheta_2'(v)}{\vartheta_2(v)} = \pi\vartheta_2^2 \frac{\vartheta_3(v)\vartheta_4(v)}{\vartheta_1(v)\vartheta_2(v)} ,$$

$$\frac{\vartheta_1'(v)}{\vartheta_1(v)} - \frac{\vartheta_3'(v)}{\vartheta_3(v)} = \pi\vartheta_3^2 \frac{\vartheta_2(v)\vartheta_4(v)}{\vartheta_1(v)\vartheta_3(v)} ,$$

$$\frac{\vartheta_2'(v)}{\vartheta_2(v)} - \frac{\vartheta_3'(v)}{\vartheta_3(v)} = -\pi\vartheta_4^2 \frac{\vartheta_1(v)\vartheta_4(v)}{\vartheta_2(v)\vartheta_3(v)} ,$$

$$\frac{\vartheta_2'(v)}{\vartheta_2(v)} - \frac{\vartheta_4'(v)}{\vartheta_4(v)} = -\pi\vartheta_3^2 \frac{\vartheta_1(v)\vartheta_3(v)}{\vartheta_2(v)\vartheta_4(v)} ,$$

$$\frac{\vartheta_3'(v)}{\vartheta_3(v)} - \frac{\vartheta_4'(v)}{\vartheta_4(v)} = -\pi\vartheta_2^2 \frac{\vartheta_1(v)\vartheta_2(v)}{\vartheta_3(v)\vartheta_4(v)} .$$

[Hint: Using the method for proving Eq. (2) of Sec. 9.6 and the formulae given in Sec. 9.4, or Eqs. (5) and (6) of Sec. 9.3.]

8. By Eq. (2) of Sec. 9.6, and Eqs. (3) – (5) of Sec. 9.7, show that

$$\frac{\vartheta_4''}{\vartheta_4} - \frac{\vartheta_3''}{\vartheta_3} = 4\omega^2(e_2 - e_3) = \pi^2\vartheta_2^4 \;,$$

$$\frac{\vartheta_4''}{\vartheta_4} - \frac{\vartheta_2''}{\vartheta_2} = 4\omega^2(e_1 - e_3) = \pi^2\vartheta_3^4 \;,$$

$$\frac{\vartheta_3''}{\vartheta_3} - \frac{\vartheta_2''}{\vartheta_2} = 4\omega^2(e_1 - e_2) = \pi^2\vartheta_4^4 \;.$$

9. By Eqs. (1), (4), (6) of Sec. 9.8 and Eq. (1) of Sec. 9.7, show that

$$\left(\frac{2kK}{\pi}\right)^{1/2} = \vartheta_2 = 2\sum_{n=0}^{\infty} q^{\left(n+\frac{1}{2}\right)^2}$$

$$= 2q^{1/4}(1 + q^2 + q^6 + q^{12} + q^{20} + \cdots) \;,$$

$$\left(\frac{2k'K}{\pi}\right)^{1/2} = \vartheta_4 = 1 + 2\sum_{n=1}^{\infty}(-)^n q^{n^2}$$

$$= 1 - 2q + 2q^4 - 2q^9 + \cdots \;.$$

10. When the real roots satisfy the condition $e_1 > e_2 > e_3$, we have

$$2\omega = \int_{e_1}^{+\infty} \frac{dx}{\sqrt{(x-e_1)(x-e_2)(x-e_3)}} \;,$$

$$2\omega' = \int_{e_3}^{+\infty} \frac{dx}{\sqrt{(x-e_1)(x-e_2)(x-e_3)}} \;;$$

the second path of integration lies in the upper half plane when it passes around the points e_2 and e_1. By the transformation

$$x = e_2 - \frac{(e_1 - e_2)(e_2 - e_3)}{\varsigma - e_2} \;,$$

which transforms the four points (e_3, e_2, e_1, ∞) respectively to (e_1, ∞, e_3, e_2), show that

$$2\omega = \int_{e_1}^{\infty} \frac{dx}{\sqrt{(x-e_1)(x-e_2)(x-e_3)}}$$

$$= \int_{e_3}^{e_2} \frac{d\varsigma}{\sqrt{(\varsigma-e_1)(\varsigma-e_2)(\varsigma-e_3)}} \;,$$

and obtain from which

$$2\omega' = i \int_{e_2}^{e_1} \frac{dx}{\sqrt{(e_1 - x)(x - e_2)(x - e_3)}}$$

By a further transformation: $\varsigma = e_3 + (e_2 - e_3)t^2$, show that $\omega\sqrt{e_1 - e_3} = K$, and $k^2 = (e_2 - e_3)/(e_1 - e_3)$.

In ω', use the transformation

$$x - e_2 = \frac{(e_1 - e_2)(e_2 - e_3)\tau^2}{(e_1 - e_3) - (e_1 - e_2)\tau^2}$$

to show that $\omega'\sqrt{e_1 - e_3} = iK'$.

11. When there are imaginary roots and the roots are taken in accord with Eq. (16) of Sec. 9.8, show that, by properly choosing the paths of integration,

$$2\sigma\omega_1 = -K - iK' , \quad 2\sigma\omega_2 = 2K , \quad 2\sigma\omega_3 = -K + iK' ;$$

the corresponding modulus is

$$k = (1 + \lambda^2)^{-1/2} .$$

[The paths of integration in the x plane are: from $a + i\infty$ to e_1 for ω_1; from e_2 to $+\infty$ for ω_2; and from $a - i\infty$ to e_3 for ω_3. In the η plane, the paths of integration are: from $-i$ to $-i\infty$, then from $+\infty$ to λ, for ω_1; from λ to $+\infty$, then from $-\infty$ to $-\lambda$, for ω_2; from i to $i\infty$, then from $+\infty$ to λ, for ω_3.]

12. With Eqs. (10) – (12) of Sec. 9.11, show that

$$k = 4q^{\frac{1}{2}} \prod_{n=1}^{\infty} \left(\frac{1 + q^{2n}}{1 + q^{2n-1}} \right)^4 , \quad k' = \prod_{n=1}^{\infty} \left(\frac{1 - q^{2n-1}}{1 + q^{2n-1}} \right)^4 .$$

13. With Eq. (9) of Sec. 9.1 and Eq. (5) of Sec. 9.11, show that

$$\sigma(z) = \frac{2\omega}{\pi} \exp\left(\frac{\eta z^2}{2\omega} \right) \sin\left(\frac{\pi z}{2\omega} \right)$$

$$\times \prod_{n=1}^{\infty} \left\{ \left(1 - 2q^{2n} \cos\frac{\pi z}{\omega} + q^{4n} \right) (1 - q^{2n})^{-2} \right\} .$$

14. By Eq. (3) of Sec. 9.12, show that

$$\varsigma(z) = \frac{\eta z}{\pi} + \frac{\pi}{2\omega} \cot \frac{\pi z}{2\omega} + \frac{2\pi}{\omega} \sum_{n=1}^{\infty} \frac{q^{2n}}{1 - q^{2n}} \sin \frac{n\pi z}{\omega} ,$$

$$\wp(z) = -\frac{\eta}{\omega} + \left(\frac{\pi}{2\omega}\right)^2 \csc^2\left(\frac{\pi z}{2\omega}\right) - 2\left(\frac{\pi}{\omega}\right)^2 \sum_{n=1}^{\infty} \frac{n q^{2n}}{1 - q^{2n}} \cos \frac{n\pi z}{\omega} .$$

15. Show that, when $|\mathrm{Im}(v)| < \mathrm{Im}(\tau)$ and $|\mathrm{Im}(w)| < \mathrm{Im}(\tau)$,

$$\frac{\vartheta_1' \vartheta_1(v + w)}{\vartheta_1(v)\vartheta_1(w)} = \pi \cot \pi v + \pi \cot \pi w$$

$$+ 4\pi \sum_{m,n=1}^{\infty} q^{2mn} \sin(2m\pi v + 2n\pi w) .$$

16. Show that

$$\frac{K k^{1/2} \vartheta_4}{\pi \vartheta_4(v)} = \frac{1}{2} a_0 + \sum_{n=1}^{\infty} a_n \cos 2n\pi v ,$$

$$a_n = 2 \sum_{m=0}^{\infty} (-)^m q^{\left(m+\frac{1}{2}\right)\left(2n+m+\frac{1}{2}\right)} .$$

17. From Eqs. (5) – (8) of Sec. 9.11, by differentiation, show that

$$\frac{\vartheta_1'(v)}{\vartheta_1(v)} = \pi \cot \pi v + 4\pi \sum_{n=1}^{\infty} \frac{q^{2n} \sin 2\pi v}{1 - 2q^{2n} \cos 2\pi v + q^{4n}} ,$$

$$\frac{\vartheta_2'(v)}{\vartheta_2(v)} = -\pi \tan \pi v - 4\pi \sum_{n=1}^{\infty} \frac{q^{2n} \sin 2\pi v}{1 + 2q^{2n} \cos 2\pi v + q^{4n}} ,$$

$$\frac{\vartheta_3'(v)}{\vartheta_3(v)} = -4\pi \sum_{n=1}^{\infty} \frac{q^{2n-1} \sin 2\pi v}{1 + 2q^{2n-1} \cos 2\pi v + q^{4n-2}} ,$$

$$\frac{\vartheta_4'(v)}{\vartheta_4(v)} = 4\pi \sum_{n=1}^{\infty} \frac{q^{2n-1} \sin 2\pi v}{1 - 2q^{2n-1} \cos 2\pi v + q^{4n-2}} .$$

18. By Eq. (7) of Sec. 9.6, and Eq. (9) of Sec. 9.11, together with the result of Ex. 17, show that

$$\frac{\vartheta_1'''}{\vartheta_1'} = -\pi^2 \left\{ 1 - 24 \sum_{n=1}^{\infty} \frac{nq^{2n}}{1 - q^{2n}} \right\}$$

$$= -\pi^2 \left\{ 1 - 24 \sum_{n=1}^{\infty} \frac{q^{2n}}{(1 - q^{2n})^2} \right\} .$$

19. In the result of Ex. 17, show, by letting $v \to 0$, that

$$\frac{\vartheta_2''}{\vartheta_2} = -\pi^2 \left\{ 1 + 8 \sum_{n=1}^{\infty} \frac{q^{2n}}{(1 + q^{2n})^2} \right\} ,$$

$$\frac{\vartheta_3''}{\vartheta_3} = -8\pi^2 \sum_{n=1}^{\infty} \frac{q^{2n-1}}{(1 + q^{2n-1})^2} ,$$

$$\frac{\vartheta_4''}{\vartheta_4} = 8\pi^2 \sum_{n=1}^{\infty} \frac{q^{2n-1}}{(1 - q^{2n-1})^2} .$$

Verify also Eq. (5) of Sec. 9.6 with the results of Ex. 18 and of the present exercise.

20. Show that

$$\frac{\vartheta_4'(v)}{\vartheta_4(v)} + \frac{\vartheta_4'(w)}{\vartheta_4(w)} - \frac{\vartheta_4'(v + w)}{\vartheta_4(v + w)} = \pi \vartheta_2 \vartheta_3 \frac{\vartheta_1(v)\vartheta_1(w)\vartheta_1(v + w)}{\vartheta_4(v)\vartheta_4(w)\vartheta_4(v + w)} .$$

Chapter 10

JACOBIAN ELLIPTIC FUNCTIONS

10.1. Jacobian Elliptic Functions sn u, cn u, dn u

Jacobian elliptic functions are functions which possess two simple poles in
the period parallelogram with one of the periods real and the other purely
imaginary. Owing to these characteristics, Jacobian elliptic functions are
closely connected with the Theta functions discussed in the preceding chapter
and their theory can be built up entirely on the theory of Theta functions. On
the other hand, Jacobian elliptic functions are quite similar to the trigonometric functions, so it is direct and easy to understand the theory from the
point of view of the inversion of Legendre's elliptic integral of the first kind.
In the present chapter, we shall deal with both approaches, first from the theory of the inversion of the elliptic integral and then from the theory of Theta
functions.

The function sn u is defined to be the inversion of Legendre's elliptic integral of the first kind:

$$t = \operatorname{sn} u , \quad u = \int_0^t \frac{dt}{\sqrt{(1 - t^2)(1 - k^2 t^2)}} . \tag{1}$$

It is common to introduce the variable φ such that

$$t = \sin \varphi , \quad u = \int_0^{\varphi} \frac{d\varphi}{\sqrt{1 - k^2 \sin^2 \varphi}} . \tag{2}$$

The inversion of this integral is usually written as

$$\varphi = \text{am } u . \tag{3}$$

So, we have

$$t = \sin \text{ am } u = \text{sn } u . \tag{4}$$

The functions cn u and dn u are defined by

$$\text{cn } u = \sqrt{1 - \text{sn}^2 u} , \quad \text{dn } u = \sqrt{1 - k^2 \text{sn}^2 u} , \tag{5}$$

where the sign of the square root is so chosen that cn $u \to 1$, dn $u \to 1$ when $u \to 0$. Sometimes, these two functions are also written as

$$\text{cn } u = \cos \text{ am } u , \quad \text{dn } u = \Delta \text{am } u . \tag{6}$$

The relations (5) can also be represented by the equations

$$\text{sn}^2 u + \text{cn}^2 u = 1 , \quad \text{dn}^2 u + k^2 \text{sn}^2 u = 1 . \tag{7}$$

Differentiating (1), we obtain

$$\frac{dt}{du} = \sqrt{(1 - t^2)(1 - k^2 t^2)} ,$$

from which we have

$$\frac{d}{du} \text{sn } u = \text{cn } u \text{ dn } u ; \tag{8}$$

the sign of it is determined by $u \to 0$.

Differentiating (7) and using (8), we have

$$\frac{d}{du} \text{cn } u = -\text{sn } u \text{ dn } u \tag{9}$$

$$\frac{d}{du} \text{dn } u = -k^2 \text{sn } u \text{ cn } u . \tag{10}$$

In more complicated formulae, the following Glaisher's symbols are often employed:

$$\text{ns } u = \frac{1}{\text{sn } u}, \quad \text{nc } u = \frac{1}{\text{cn } u}, \quad \text{nd } u = \frac{1}{\text{dn } u},$$

$$\text{sc } u = \frac{\text{sn } u}{\text{cn } u}, \quad \text{sd } u = \frac{\text{sn } u}{\text{dn } u}, \quad \text{cd } u = \frac{\text{cn } u}{\text{dn } u}, \tag{11}$$

$$\text{cs } u = \frac{\text{cn } u}{\text{sn } u}, \quad \text{ds } u = \frac{\text{dn } u}{\text{sn } u}, \quad \text{dc } u = \frac{\text{dn } u}{\text{cn } u}.$$

In the formulae of the last two rows, the first letter denotes the numerator and the second letter denotes the denominator.

From integral (1) we see that sn u is an odd function: $\text{sn}(-u) = -\text{sn } u$; while from the definitions (5), cn u and dn u are even functions: $\text{cn}(-u) = \text{cn } u, \text{dn}(-u) = \text{dn } u$.

10.2. Geometric Representations of Jacobian Elliptic Functions
1. The first geometric representation

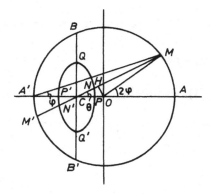

Fig. 39.

There are two kinds of geometric representations for Jacobian elliptic functions. For the first kind, draw a circle with centre O and radius equal to 1. [See Fig. 39.] On the diameter $A'OA$, take a point C such that $CO = \varepsilon$,

$$\varepsilon = \frac{1 - k'}{1 + k'} \quad \left(k' = \sqrt{1 - k^2}\right). \tag{1}$$

Through C an arbitrary chord $M'M$ is drawn. Take two points N and N' on $M'M$ such that $CN = N'C \propto (M'M)^{-1/2}$. Draw a line from O perpendicular to $M'M$ at the point H. Let $HM = m$, then $M'M = 2m$. Let $\angle OCN = \theta$, $\angle AA'M = \varphi$, then $\angle AOM = 2\varphi$. $\angle OMC = 2\varphi - \theta$. Since $OM = 1$, we have

$$m = \cos(2\varphi - \theta) , \tag{2}$$

$$\varepsilon \sin \theta = \sin(2\varphi - \theta) . \tag{3}$$

From (3), we obtain the relation between θ and φ:

$$\tan \theta = \frac{\sin 2\varphi}{\varepsilon + \cos 2\varphi} . \tag{4}$$

Choosing the ratio of CN to $(M'M)^{-1/2}$ to be $\lambda\sqrt{2(1 + \varepsilon)}$, λ being an arbitrary constant factor, we have

$$r = CN = \lambda\sqrt{\frac{1 + \varepsilon}{m}} . \tag{5}$$

By (2) and (3), we obtain

$$m^2 = 1 - \varepsilon^2 \sin^2 \theta . \tag{6}$$

Substituting in (5) gives

$$r^4(1 - \varepsilon^2 \sin^2 \theta) = \lambda^4(1 + \varepsilon)^2 . \tag{7}$$

This is the equation of the locus of N in polar coordinates, which is a quartic curve somewhat like an ellipse. Let P be the point of intersection of this curve with the line $A'A$ and $[PCN]$ represent a fan-shaped area. We have

$$[PCN] = \frac{1}{2} \int_0^\theta r^2 d\theta = \frac{\lambda^2(1 + \varepsilon)}{2} \int_0^\theta \frac{d\theta}{m} .$$

From (2) and (4) we find

$$m = \frac{1 + \varepsilon \cos 2\varphi}{\sqrt{1 + \varepsilon^2 + 2\varepsilon \cos 2\varphi}} ,$$

$$\frac{d\theta}{d\varphi} = \frac{2(1 + \varepsilon \cos 2\varphi)}{1 + \varepsilon^2 + 2\varepsilon \cos 2\varphi} .$$

By relation (1) between ε and k, we have

$$1 + \varepsilon^2 + 2\varepsilon \cos 2\varphi = (1 + \varepsilon)^2 - 4\varepsilon \sin^2 \varphi$$
$$= (1 + \varepsilon)^2 (1 - k^2 \sin^2 \varphi) \ .$$

Therefore,

$$[PCN] = \lambda^2 \int_0^\varphi \frac{d\varphi}{\sqrt{1 - k^2 \sin^2 \varphi}} = \lambda^2 u \ . \tag{8}$$

This equation exhibits the geometric meaning of u. It is the area of the fan-shaped region PCN (λ is arbitrary. In the figure shown, we have chosen $\lambda < \varepsilon < \frac{1}{2}$. If we choose $\lambda = 1$, then $[PCN] = u$).

Now, we have determined $\varphi = \text{am } u$ by (8), so

$$\text{sn } u = \sin \varphi = \frac{AM}{A'A} = \frac{1}{2} AM \ ,$$

$$\text{cn } u = \cos \varphi = \frac{A'M}{A'A} = \frac{1}{2} A'M \ , \tag{9}$$

$$\text{dn } u = \frac{CM}{CA} = \frac{CM}{1 + \varepsilon} \ .$$

The proof of the last formula is as follows:

$$CM^2 = CA^2 + AM^2 - 2CA \cdot AM \cos \angle OAM$$
$$= (1 + \varepsilon)^2 + 4\sin^2 \varphi - 4(1 + \varepsilon) \sin \varphi \cos \left(\frac{\pi}{2} - \varphi \right)$$
$$= (1 + \varepsilon)^2 - 4\varepsilon \sin^2 \varphi = (1 + \varepsilon)^2 (1 - k^2 \sin^2 \varphi)$$
$$= (1 + \varepsilon)^2 (1 - k^2 \text{sn}^2 \ u) = (1 + \varepsilon)^2 \text{dn}^2 \ u \ .$$

2. The second geometric representation

For the second geometric representation, draw a sphere of radius 1. Take $\frac{\pi}{2} - \varphi$ as the latitude and θ the longitude, then the coordinates of a point on the sphere are

$$x = \sin \varphi \cos \theta \ , \quad y = \sin \varphi \sin \theta \ , \quad z = \cos \varphi \ . \tag{10}$$

The φ and θ here are just in reversal to the traditional convention. It is made for the purpose of ensuring the relation $\varphi = \text{am } u$. Let

$$t = \sin \varphi \tag{11}$$

then (t, θ, z) are the cylindrical coordinates. On the sphere, the square of a line element is

$$ds^2 = d\varphi^2 + \sin^2 \varphi d\theta^2 = \frac{dt^2}{1 - t^2} + t^2 d\theta^2 \ . \tag{12}$$

On the sphere a spiral is drawn, the equation of which is

$$\theta = ks \ , \tag{13}$$

where s is the length of the arc measured from the polar axis $\varphi = 0$ $(t = 0)$. Substituting (13) in (12) gives

$$(1 - k^2 \sin^2 \varphi)ds^2 = d\varphi^2 \ .$$

Thus $u = s$, i.e., u is equal to the length of the curve. The cylindrical coordinates of the curve are

$$t = \operatorname{sn} s \ , \quad \theta = ks \ , \quad z = \operatorname{cn} s \quad (s = u) \ . \tag{14}$$

The cosine of its intersecting angle with the meridian is given by

$$\frac{d\varphi}{ds} = \sqrt{1 - k^2 \sin^2 \varphi} = \operatorname{dn} s \ . \tag{15}$$

10.3. Complete Elliptic Integrals

The complete elliptic integral K is defined to be

$$K = \int_0^1 \frac{dt}{\sqrt{(1 - t^2)(1 - k^2 t^2)}} = \int_0^{\pi/2} \frac{d\varphi}{\sqrt{1 - k^2 \sin^2 \varphi}} \ , \tag{1}$$

from which we have

$$\operatorname{sn} K = 1 \ , \quad \operatorname{cn} K = 0 \ , \quad \operatorname{dn} K = k' \ . \tag{2}$$

When $k < 1$, by expanding the integral in power series of k^2 and putting $t^2 = \xi$, we obtain

$$K = \sum_{r=0}^{\infty} \binom{-\frac{1}{2}}{r} (-k^2)^r \int_0^1 \frac{t^{2r} dt}{\sqrt{1 - t^2}}$$

$$= \frac{1}{2} \sum_{r=0}^{\infty} \left(\frac{1}{2}\right)_r \frac{k^{2r}}{r!} \int_0^1 \xi^{r-\frac{1}{2}} (1 - \xi)^{-\frac{1}{2}} d\xi$$

$$= \frac{1}{2} \sum_{r=0}^{\infty} \left(\frac{1}{2}\right)_r \frac{k^{2r}}{r!} \frac{\Gamma\left(\frac{1}{2}\right) \Gamma\left(r + \frac{1}{2}\right)}{\Gamma(r + 1)}$$

$$= \frac{\pi}{2} \sum_{r=0}^{\infty} \left[\left(\frac{1}{2}\right)_r\right]^2 \frac{k^{2r}}{(r!)^2} \ ,$$

i.e.,

$$K = \frac{\pi}{2} F \left(\frac{1}{2}, \frac{1}{2}, 1, k^2 \right) . \tag{3}$$

The complete elliptic integral corresponding to the complementary modulus k' is

$$K' = \int_0^1 \frac{dt}{\sqrt{(1 - t^2)(1 - k'^2 t^2)}} = \frac{\pi}{2} F \left(\frac{1}{2}, \frac{1}{2}, 1, k'^2 \right)$$

$$= \frac{\pi}{2} F \left(\frac{1}{2}, \frac{1}{2}, 1, 1 - k^2 \right) . \tag{4}$$

In Sec. 9.8, we have obtained, by the transformation $k^2 t^2 = 1 - k'^2 t'^2$, the following formula:

$$\int_0^{1/k} \frac{dt}{\sqrt{(1 - t^2)(1 - k^2 t^2)}} = \int_0^1 \frac{dt}{\sqrt{(1 - t^2)(1 - k^2 t^2)}}$$

$$+ i \int_1^{1/k} \frac{dt}{\sqrt{(t^2 - 1)(1 - k^2 t^2)}} = K + i K' .$$

Hence, we have

$$\operatorname{sn}(K + iK') = \frac{1}{k} , \quad \operatorname{cn}(K + iK') = -\frac{ik'}{k} ,$$

$$\operatorname{dn}(K + iK') = 0 . \tag{5}$$

The sign in the second formula is determined by the stipulation that the path of integration is curved around $t = 1$ by the upper half plane ($\operatorname{Im}(t) > 0$) [cf. the paragraph below Eq. (7) of Sec. 9.8].

Consider the integral from $t = 0$ to $t = \infty$, the path of integration being curved around the points $t = 1$ and $t = 1/k$ by the upper half plane. We have

$$\int_0^\infty \frac{dt}{\sqrt{(1 - t^2)(1 - k^2 t^2)}} = K + iK' - \int_{1/k}^\infty \frac{dt}{\sqrt{(t^2 - 1)(k^2 t^2 - 1)}} .$$

Putting $kt = 1/\xi$ in the last integral, we obtain

$$\int_{1/k}^\infty \frac{dt}{\sqrt{(t^2 - 1)(k^2 t^2 - 1)}} = \int_0^1 \frac{d\xi}{\sqrt{(1 - \xi^2)(1 - k^2 \xi^2)}} = K ,$$

and hence

$$\int_0^\infty \frac{dt}{\sqrt{(1 - t^2)(1 - k^2 t^2)}} = iK' .$$

It follows,

$$\operatorname{sn}(iK') = \infty . \tag{6}$$

By the definitions of cn u and dn u given in Eq. (5) of Sec. 10.1, we have

$$\operatorname{cn}(iK') = \infty , \quad \operatorname{dn}(iK') = \infty . \tag{7}$$

Thus, $u = iK'$ is a pole of these three functions.

10.4. Addition Formulae

Addition formula means the expression of $\operatorname{sn}(u+v)$, etc. For the sake of written simplicity, we use the following notations:

$$\begin{aligned}
s_1 &= \operatorname{sn} u , & s_2 &= \operatorname{sn} v ; \\
c_1 &= \operatorname{cn} u , & c_2 &= \operatorname{cn} v ; \\
d_1 &= \operatorname{dn} u , & d_2 &= \operatorname{dn} v .
\end{aligned} \tag{1}$$

The derivation of the addition formula is a little bit tricky. We assume that $u + v = C$ (a constant), which is a solution of the differential equation $du + dv = 0$. Use a dot to express the differentiation with respect to u, then, since $\dot{v} = -1$, we have, by Eq. (8) of Sec. 10.1,

$$\dot{s}_1^2 = (1 - s_1^2)(1 - k^2 s_1^2) , \quad \dot{s}_2^2 = (1 - s_2^2)(1 - k^2 s_2^2) .$$

Differentiating again, there results

$$\ddot{s}_1 = -(1 + k^2)s_1 + 2k^2 s_1^3 , \quad \ddot{s}_2 = -(1 + k^2)s_2 + 2k^2 s_2^3 .$$

So, we have

$$\frac{\ddot{s}_1 s_2 - \ddot{s}_2 s_1}{\dot{s}_1^2 s_2^2 - \dot{s}_2^2 s_1^2} = \frac{2k^2 s_1 s_2 (s_1^2 - s_2^2)}{(s_2^2 - s_1^2)(1 - k^2 s_1^2 s_2^2)} = -\frac{2k^2 s_1 s_2}{1 - k^2 s_1^2 s_2^2}$$

and hence

$$\frac{1}{\dot{s}_1 s_2 - \dot{s}_2 s_1} \frac{d}{du}(\dot{s}_1 s_2 - \dot{s}_2 s_1) - \frac{1}{1 - k^2 s_1^2 s_2^2} \frac{d}{du}(1 - k^2 s_1^2 s_2^2) = 0 ,$$

which is a differential equation also of the form $du + dv = 0$. An integral of this equation is

$$\frac{\dot{s}_1 s_2 - \dot{s}_2 s_1}{1 - k^2 s_1^2 s_2^2} = C' \quad \text{(const.)} .$$

Applying Eq. (8) of Sec. 10.1, it becomes

$$\frac{c_1 d_1 s_2 + c_2 d_2 s_1}{1 - k^2 s_1^2 s_2^2} = C' .$$

This result must be in agreement with the fact that $u + v = C$ is a solution of the same equation, and therefore, there must exist a functional relation between the two constants C and C', i.e.,

$$\frac{c_1 d_1 s_2 + c_2 d_2 s_1}{1 - k^2 s_1^2 s_2^2} = f(u + v) .$$

Putting $v = 0$, we have $f(u) = \text{sn}\, u$, and hence

$$\text{sn}(u + v) = \frac{s_1 c_2 d_2 + s_2 c_1 d_1}{1 - k^2 s_1^2 s_2^2} = \frac{\text{sn}\, u\, \text{cn}\, v\, \text{dn}\, v + \text{sn}\, v\, \text{cn}\, u\, \text{dn}\, u}{1 - k^2 \text{sn}^2\, u\, \text{sn}^2\, v} \qquad (2)$$

which is the addition formula sought.

Starting from this formula, we can derive the expressions for $\text{cn}(u+v)$ and $\text{dn}(u + v)$:

$$\begin{aligned}
(1 - k^2 s_1^2 s_2^2)^2 \text{cn}^2(u + v) &= (1 - k^2 s_1^2 s_2^2)^2 \{1 - \text{sn}^2(u + v)\} \\
&= (1 - k^2 s_1^2 s_2^2)^2 - (s_1 c_2 d_2 + s_2 c_1 d_1)^2 \\
&= 1 - 2k^2 s_1^2 s_2^2 + k^4 s_1^4 s_2^4 - s_1^2 (1 - s_2^2)(1 - k^2 s_2^2) \\
&\quad - s_2^2 (1 - s_1^2)(1 - k^2 s_1^2) - 2 s_1 s_2 c_1 c_2 d_1 d_2 \\
&= (1 - s_1^2)(1 - s_2^2) + s_1^2 s_2^2 (1 - k^2 s_1^2)(1 - k^2 s_2^2) - 2 s_1 s_2 c_1 c_2 d_1 d_2 \\
&= (c_1 c_2 - s_1 s_2 d_1 d_2)^2 .
\end{aligned}$$

Taking the square root and determining the sign by setting $v = 0$, we obtain

$$\text{cn}(u + v) = \frac{c_1 c_2 - s_1 s_2 d_1 d_2}{1 - k^2 s_1^2 s_2^2} = \frac{\text{cn}\, u\, \text{cn}\, v - \text{sn}\, u\, \text{sn}\, v\, \text{dn}\, u\, \text{dn}\, v}{1 - k^2 \text{sn}^2\, u\, \text{sn}^2\, v} . \qquad (3)$$

Similarly, from

$$\begin{aligned}
(1 - k^2 s_1^2 s_2^2)^2 \text{dn}^2(u + v) &= (1 - k^2 s_1^2 s_2^2)^2 - k^2 (s_1 c_2 d_2 + s_2 c_1 d_1)^2 \\
&= (1 - k^2 s_1^2)(1 - k^2 s_2^2) + k^4 s_1^2 s_2^2 (1 - s_1^2)(1 - s_2^2) - 2k^2 s_1 s_2 c_1 c_2 d_1 d_2
\end{aligned}$$

we obtain

$$\text{dn}(u + v) = \frac{d_1 d_2 - k^2 s_1 s_2 c_1 c_2}{1 - k^2 s_1^2 s_2^2} = \frac{\text{dn}\, u\, \text{dn}\, v - k^2 \text{sn}\, u\, \text{sn}\, v\, \text{cn}\, u\, \text{cn}\, v}{1 - k^2 \text{sn}^2\, u\, \text{sn}^2\, v} . \qquad (4)$$

Putting $u = v$ in the addition formulae, and introducing the abbreviations:

$$s = \text{sn } u , \qquad c = \text{cn } u , \qquad d = \text{dn } u ,$$
$$S = \text{sn } (2u) , \quad C = \text{cn } (2u) \quad D = \text{dn } (2u) , \tag{5}$$

we arrive at the duplication formulae:

$$S = \frac{2scd}{1 - k^2 s^4} , \quad C = \frac{1 - 2s^2 + k^2 s^4}{1 - k^2 s^4} , \quad D = \frac{1 - 2k^2 s^2 + k^2 s^4}{1 - k^2 s^4} , \tag{6}$$

and their inversions:

$$s^2 = \frac{1 - C}{1 + D} , \quad c^2 = \frac{D + C}{1 + D} , \quad d^2 = \frac{D + C}{1 + C} . \tag{7}$$

10.5. The Periodicity of Jacobian Elliptic Functions

From the addition formula derived in the last section and Eq. (2) of Sec. 10.3, we have

$$\text{sn}(u + K) = \frac{\text{cn } u \text{ dn } u}{1 - k^2 \text{sn}^2 u} = \frac{\text{cn } u}{\text{dn } u} = \text{cd } u . \tag{1}$$

Similarly, we have

$$\acute{\text{c}}\text{n}(u + K) = -k'\text{sd } u , \quad \text{dn}(u + K) = k'\text{nd } u . \tag{2}$$

Adding one more K, we have

$$\text{sn}(u + 2K) = -\text{sn } u , \quad \text{cn}(u + 2K) = -\text{cn } u ,$$
$$\text{dn}(u + 2K) = \text{dn } u . \tag{3}$$

Adding $2K$ again:

$$\text{sn}(u + 4K) = \text{sn } u , \quad \text{cn}(u + 4K) = \text{cn } u . \tag{4}$$

Thus, we see that the periods of sn u and cn u are all $4K$, while that of dn u is $2K$.

Similarly, from the addition formulae and Eq. (5) of Sec. 10.3, we obtain

$$\text{sn}(u + K + iK') = k^{-1}\text{dc } u , \quad \text{cn}(u + K + iK') = -ik'k^{-1}\text{nc } u ,$$
$$\text{dn}(u + K + iK') = ik' \text{ sc } u . \tag{5}$$

Adding again $K + K'i$ to the argument, we have

$$\text{sn}(u + 2K + 2K'i) = -\text{sn } u , \quad \text{cn}(u + 2K + 2K'i) = \text{cn } u ,$$
$$\text{dn}(u + 2K + 2K'i) = -\text{dn } u . \tag{6}$$

Thus, $4K + 4K'i$ is the period of both sn u and dn u, and $2K + 2K'i$ is that of cn u.

When $K'i$ is added to the argument, we are not able to calculate the function with Eq. (6) of Sec. 10.3, and have to resort to the following method:

$$\text{sn}(u + K'i) = \text{sn}(u - K + K + K'i)$$
$$= k^{-1} \text{ dc}(u - K) = k^{-1} \text{ ns } u . \tag{7}$$

Similarly,

$$\text{cn}(u + K'i) = -ik^{-1} \text{ ds } u , \quad \text{dn}(u + K'i) = -i \text{ cs } u . \tag{8}$$

Adding $K'i$ once more:

$$\text{sn}(u + 2K'i) = \text{sn } u , \quad \text{cn}(u + 2K'i) = -\text{cn } u ,$$
$$\text{d}(u + 2K'i) = -\text{dn } u . \tag{9}$$

Thus, $2K'i$ is the period of sn u, while $4K'i$ is the period of both cn u and dn u.

In summary, since $4K'i = 2(2K + 2K'i) - 4K$, we arrive at the following conclusion:

$(4K, 2K'i)$ are the periods of sn u; $(4K, 2K + 2K'i)$ are the periods of cn u; $(2K, 4K'i)$ are those of dn u.

10.6. The Zeros and Poles of Jacobian Elliptic Functions

From the evaluation of the last integral in Sec. 10.3, we see that the pole u of sn u is

$$u = \int_0^\infty \frac{dt}{\sqrt{(1 - t^2)(1 - k^2 t^2)}} = K \pm iK' \pm K .$$

The positive or negative sign depends upon the path of integration: whether it is curved around $t = 1$ and $t = 1/k$ by the upper half or by the lower half plane. From the definition Eq. (5) of Sec. 10.1 for cn u and dn u, we note that the poles of sn u are also poles of cn u and dn u. According to the specific values of the periods for these functions given in the last section, we know that each function has only two poles in the period parallelogram. Thus,

In the respective period parallelogram of the functions, sn u and cn u have the two poles: $K'i$ and $2K + K'i$; while dn u has the two poles: $K'i$ and $3K'i$.

According to the general theory in Sec. 8.3, these functions also have only two zeros in the period parallelogram. From the definitions of sn u given in Sec. 10.1, it is seen that $u = 0$ is a zero of sn u; the other zero of it is $u = 2K$, as may be seen from (3) of the last section. By (2) of the last section, it can be seen that $u = K$ and $u = 3K$ are the two zeros of cn u and, by (5) of the last section, the two zeros of dn u are $u = K + K'i$ and $u = -K - K'i$ or $u = K + K'i$ and $u = K + 3K'i$; the latter is equal to $2K + 4K'i - (K + K'i)$. To sum up,

In the respective period parallelogram, (0, 2K) are the two zeros of sn u, (K, 3K) are those of cn u, and (K + K'i, K + 3K'i) are those of dn u.

Now, let us find the expansions at the zeros and at the poles. First, consider the power series expansion of sn u at $u = 0$, i.e., its Taylor expansion. Using Eqs. (8) – (10) of Sec. 10.1, we can find the derivatives of sn u, so, by Taylor's expansion formula, we obtain

$$\text{sn } u = u - \frac{1+k^2}{6}u^3 + (1 + 14k^2 + k^4)\frac{u^5}{5!} - \dots . \tag{1}$$

Substituting in the definitions in Eq. (5) of Sec. 10.1, we obtain the expansions of cn u and dn u at the origin:

$$\text{cn } u = 1 - \frac{u^2}{2} + (1 + 4k^2)\frac{u^4}{4!} - (1 + 44k^2 + 16k^4)\frac{u^6}{6!} + \dots , \tag{2}$$

$$\text{dn } u = 1 - k^2\frac{u^2}{2} + k^2(4 + k^2)\frac{u^4}{4!} - k^2(16 + 44k^2 + k^4)\frac{u^6}{6!} + \dots . \tag{3}$$

Substituting in Eq. (2) of Sec. 10.5, we obtain the expansion of cn u at its zero K:

$$\text{cn}(u + K) = -k'u + \frac{k'(1 - 2k^2)}{6}u^3 - \dots . \tag{4}$$

Substituting in Eq. (5) of Sec. 10.5, we obtain the expansion of dn u at its zero $K + K'i$:

$$\text{dn}(u + K + K'i) = ik'u + \frac{ik'(2 - k^2)}{6}u^3 + \dots . \tag{5}$$

Substituting in Eqs. (7) and (8) of Sec. 10.5, we obtain the expansions at the pole $K'i$:

$$\text{sn}(u + K'i) = \frac{1}{ku} + \frac{1 + k^2}{6k}u + \frac{7 - 22k^2 + 7k^4}{360k}u^3 + \ldots , \tag{6}$$

$$\text{cn}(u + K'i) = -\frac{i}{ku} + \frac{i(2k^2 - 1)}{6k}u + \ldots , \tag{7}$$

$$\text{dn}(u + K'i) = -\frac{i}{u} + \frac{i(2 - k^2)}{6}u + \ldots . \tag{8}$$

From these expressions, we see that the residues of sn u, cn u, and dn u at the pole $K'i$ are respectively $1/k, -i/k, -i$.

10.7. Transformations of Elliptic Functions

We do not contemplate to discuss the transformations of elliptic functions to the full. Only three of them will be dealt. The first is Jacobi's imaginary transformation; the second, Landen's transformation; and the third, imaginary modulus transformation.

Since the modulus k is also changed in the transformation, we have to include k in the expression of the function. For example, we write $\text{sn}(u, k)$ instead of simply sn u.

1. Jacobi's imaginary transformation

Jacobi's imaginary transformation is to find $\text{sn}(iu, k)$. When u changes to iu, let the upper limit of the integral be replaced by iy (which defines y):

$$\begin{aligned} iu &= \int_0^{iy} \frac{dt}{\sqrt{(1 - t^2)(1 - k^2t^2)}} \\ &= i\int_0^y \frac{dy}{\sqrt{(1 + y^2)(1 + k^2y^2)}} . \end{aligned} \tag{1}$$

Making the transformation $y^2 = \eta^2/(1 - \eta^2)$, we have

$$u = \int_0^\eta \frac{d\eta}{\sqrt{(1 - \eta^2)(1 - k'^2\eta^2)}} .$$

Hence, $\eta = \text{sn}(u, k')$ and

$$t = \text{sn}(iu, k) = iy = i\eta(1 - \eta^2)^{-\frac{1}{2}} = i\,\text{sc}(u, k') \tag{2}$$

$$\text{cn}(iu, k) = (1 + y^2)^{\frac{1}{2}} = (1 - \eta^2)^{-\frac{1}{2}} = \text{nc}(u, k') \tag{3}$$

$$\text{dn}(iu, k) = (1 + k^2y^2)^{\frac{1}{2}} = (1 - k'^2\eta^2)(1 - \eta^2)^{-\frac{1}{2}} = \text{dc}(u, k') . \tag{4}$$

2. *Transformation of Landen's type*

We have dealt with Landen's type of transformation in the theory of the Theta function [Sec. 9.10]. The transformation is from τ to $\tau_1 = 2\tau$; the corresponding modulus k is transformed to k_1, the value of which can be computed by Eqs. (15) and (16) of Sec. 9.11. [For k, see Eq. (1) of Sec. 9.8.]:

$$k_1 = \frac{\vartheta_2^2(0|2\tau)}{\vartheta_3^2(0|2\tau)} = \frac{\vartheta_3^2(0|\tau) - \vartheta_4^2(0|\tau)}{\vartheta_3^2(0|\tau) + \vartheta_4^2(0|\tau)} = \frac{1 - k'}{1 + k'} .$$

Therefore, the Landen transformation is to find $\mathrm{sn}(u_1, k_1)$:

$$t_1 = \mathrm{sn}(u_1, k_1) , \quad u_1 = \int_0^{t_1} \frac{dt_1}{\sqrt{(1 - t_1^2)(1 - k_1^2 t_1^2)}} ,$$

$$k_1 = \frac{1 - k'}{1 + k'} . \tag{5}$$

Making the transformation

$$t_1 = \frac{(1 + k')t(1 - t^2)^{\frac{1}{2}}}{(1 - k^2 t^2)^{\frac{1}{2}}} , \tag{6}$$

we obtain

$$u_1 = (1 + k') \int_0^t \frac{dt}{\sqrt{(1 - t^2)(1 - k^2 t^2)}} = (1 + k')u . \tag{7}$$

Hence,

$$\mathrm{sn}(u_1, k_1) = (1 + k')\mathrm{sn}(u, k)\mathrm{cd}(u, k) , \tag{8}$$

$$\mathrm{cn}(u_1, k_1) = \{1 - (1 + k')\mathrm{sn}^2(u, k)\}\mathrm{nd}(u, k) , \tag{9}$$

$$\mathrm{dn}(u_1, k_1) = \{1 - (1 - k')\mathrm{sn}^2(u, k)\}\mathrm{nd}(n, k) \tag{10}$$

$$(u_1 = (1 + k')u , \quad k_1 = (1 - k')/(1 + k')) .$$

It is not easy to find the transformation (6) directly, but it is convenient to derive it from the theory of Theta function.

The periods $2K_1$ and $2K_1' i$ can be obtained from the zeros and poles of $\mathrm{sn}\, u_1$. From (8), we see that $\mathrm{sn}\, u_1$ has the zeros $u_1 = 0$ and $u_1 = 2K_1$, corresponding to the zeros $u = 0$ of $\mathrm{sn}\, u$ and $u = K$ of $\mathrm{cn}\, u$. Hence, we obtain from (7)

$$2K_1 = (1 + k')K . \tag{11}$$

Also by (8), we know that the pole of sn u_1, namely $u_1 = K_1'i$, is just the pole $u = K'i$ of sn u. Hence, from (7), we have

$$K_1' = (1 + k')K' . \tag{12}$$

Combining (11) and (12), we have $K_1'/K_1 = 2K'/K$, i.e., $\tau_1 = 2\tau$. (11) and (12) can also be proved by using the integral (7). As t varies from 0 to $(1 + k')^{-1/2}$, t_1 varies from 0 to 1; and when t varies from $(1 + k')^{-1/2}$ to 1, t_1 varies from 1 back to 0. Thus, the value of the integral for t from 0 to 1 is equal to $2K_1$, and this is (11). When t varies from 1 to $1/k$, t_1 varies from 0 to ∞ along the negative imaginary axis. But, $\sqrt{1 - t_1^2} = -1$ at $t_1 = 0$. Thus, we have (12).

3. Imaginary modulus transformation

This corresponds to the transformation from τ to $\tau_2 = \tau + 1$ in the Theta function. The corresponding modulus k changes to k_2 [cf. Eq. (11) of Sec. 9.9]:

$$k_2 = \frac{\vartheta_2^2(0|\tau + 1)}{\vartheta_3^2(0|\tau + 1)} = \frac{i\vartheta_2^2(0|\tau)}{\vartheta_4^2(0|\tau)} = \frac{ik}{k'} , \tag{13}$$

which is purely imaginary. This transformation is used to find $\mathrm{sn}(u_2, k_2)$:

$$t_2 = \mathrm{sn}(u_2, k_2) , \quad u_2 = \int_0^{t_2} \frac{dt_2}{\sqrt{(1 - t_2^2)(1 - k_2^2 t_2^2)}} , \quad k_2 = \frac{ik}{k'} . \tag{14}$$

Make the transformation

$$t_2 = \frac{k't}{\sqrt{1 - k^2 t^2}} , \tag{15}$$

then

$$u_2 = \int_0^t \frac{k' \, dt}{\sqrt{(1 - t^2)(1 - k^2 t^2)}} = k'u . \tag{16}$$

Hence,

$$\mathrm{sn}(u_2, k_2) = \mathrm{sn}(k'u, ik/k') = \frac{k't}{\sqrt{1 - k^2 t^2}} = k' \, \mathrm{sd}(u, k) , \tag{17}$$

$$\mathrm{cn}(k'u, ik/k') = \mathrm{cd}(u, k) ,$$

$$\mathrm{dn}(k'u, ik/k') = \mathrm{nd}(u, k) . \tag{18}$$

From the zeros of cn u_2 and the poles of dn u_2 we can find the periods $2K_2$ and $2K_2'i$:

$$K_2 = k'K , \quad iK_2' = k'(K + K'i) , \tag{19}$$

from which we obtain $iK_2'/K_2 = 1 + K'i/K = \tau + 1$. We are not able to determine off hand from the poles of dn u whether the coefficient of K in the second equation of (19) be positive or negative. But it is rather easy to make a decision directly from the integral (16). When t varies from 0 to 1, (15) shows that t_2 varies correspondingly also from 0 to 1. Hence (16) leads to the first equation of (19). When t_2 varies from 1 to $1/k_2 = -ik'/k$, t varies from 1 to ∞, and the square root in (15) has to be positive imaginary for $t > 1/k$. Using the method of calculation shown at the end of Sec. 10.3, we obtain the second equation of (19).

10.8. Reductions of Elliptic Integrals

The general elliptic integral is $\int R(x,y)dx$, where $R(x,y)$ is a rational function of x and y, while

$$y^2 = P(x) = a_0 x^4 + 4a_1 x^3 + 6a_2 x^2 + 4a_3 x + a_4 . \tag{1}$$

In Sec. 8.1 we have shown that the case of a cubic polynomial $P(x)$ $(a_0 = 0)$ can be transformed into the case of a quartic polynomial, while the case of a quartic polynomial, if one of its roots is known, can be transformed into a cubic case. In the case of a cubic polynomial, it has been shown that the general elliptic integral can be reduced to three fundamental elliptic integrals, I_0, I_1 and J_1.

In Sec. 8.11, we have shown how to write an arbitrary cubic polynomial in the standard form of Weierstrass:

$$y^2 = 4x^3 - g_2 x - g_3 . \tag{2}$$

In Sec. 8.9, we have discussed the reduction of an elliptic integral in the standard form of Weierstrass to the three fundamental integrals I_0, I_1 and J_1, and obtained the result: $I_0 = z, I_1 = -\varsigma(z)$. The expression for J_1 was derived in Sec. 8.10.

In Sec. 8.12, we have dealt with the case when no root of the quartic polynomial is known and shown how to represent it in the standard form of Weierstrass, which is equivalent to reducing it to the case of a cubic polynomial.

Therefore, theoretically, the problem of elliptic integrals has been completely solved.

However, in practice, especially when numerical calculations are concerned, it is advantageous for one of the periods to be real. Hence, Legendre's standard form of the elliptic integral is more convenient. Although we have shown in Sec. 9.8 how to transform Weierstrass' standard form to Legendre's, namely,

$$y^2 = L(t) = (1 - t^2)(1 - k^2 t^2) , \quad k < 1 , \tag{3}$$

it is more desirable for practical importance to reduce a quartic polynomial (1) to Legendre's standard form (3) without the intrusion of Weierstrass' form (2).

Now we shall show how to reduce a general quartic polynomial (1) into Legendre's standard form (3). We assume that the roots of $P(x) = 0$ have been known [cf. App. II for the solution of the quartic equation], and $P(x)$ is represented by

$$P(x) = S_1 S_2 , \quad S_l = p_l x^2 + 2 q_l x + r_l , \quad l = 1, 2, \tag{4}$$

or, with the four roots a, b, c, d written out explicitly,

$$S_1 = p_1(x - a)(x - b) , \quad S_2 = p_2(x - c)(x - d) . \tag{5}$$

Choose λ such that $S_1 - \lambda S_2$ becomes a perfect square:

$$\begin{aligned} S_1 - \lambda S_2 &= (p_1 - \lambda p_2)x^2 + 2(q_1 - \lambda q_2)x + (r_1 - \lambda r_2) \\ &= (p_1 - \lambda p_2)(x - \alpha)^2 . \end{aligned} \tag{6}$$

In order that (6) holds, λ has to satisfy the following condition:

$$(q_1 - \lambda q_2)^2 - (p_1 - \lambda p_2)(r_1 - \lambda r_2) = 0 , \tag{7}$$

which furnishes two roots, λ and μ. Comparing the coefficients of $2x$ in (6), we obtain

$$(p_1 - \lambda p_2)\alpha = -(q_1 - \lambda q_2) . \tag{8}$$

Solving for λ gives

$$\lambda = \frac{p_1 \alpha + q_1}{p_2 \alpha + q_2} . \tag{9}$$

Substituting in (7), we have

$$(p_1 q_2 - p_2 q_1)\alpha^2 + (p_1 r_2 - p_2 r_1)\alpha + (q_1 r_2 - q_2 r_1) = 0 . \tag{10}$$

The roots α and β of this equation are in correspondence with the roots λ and μ of Eq. (7).

For real coefficients in $P(x)$, if there are complex roots among a, b, c, d, they must occur in pairs. So, we can make all the p_l, q_l, r_l real and let the two roots of Eq. (10) to be real. The condition for this is

$$D = (p_1 r_2 - p_2 r_1)^2 - 4(p_1 q_2 - p_2 q_1)(q_1 r_2 - q_2 r_1) > 0 . \tag{11}$$

This relation can be expressed in terms of the roots of $P(x) = 0$. Comparing the coefficients of (4) and (5), we obtain

$$p_1(a + b) = -2q_1 , \quad p_1 ab = r_1;$$
$$p_2(c + d) = -2q_2 , \quad p_2 cd = r_2 . \tag{12}$$

Substituting in (10) gives

$$(a + b - c - d)\alpha^2 + 2(cd - ab)\alpha + ab(c + d) - cd(a + b) = 0 . \tag{13}$$

Substituting in (11) gives

$$D = (p_1 p_2)^2 \{(cd - ab)^2 - (a + b - c - d)[ab(c + d) - cd(a + b)]\}$$
$$= a_0^2(a - c)(a - d)(b - c)(b - d) > 0 . \tag{14}$$

If all the four roots are real, we may put $a > b > c > d$ so that (14) is satisfied. In general, so far as the intervals (a, b) and (c, d) do not overlap, (14) can be made satisfied. If there is one pair of complex roots (c, d), then c and d are complex conjugate to each other; if four roots are complex, a and b must be complex conjugate to each other also; (14) obviously holds in both cases.

Assume that $p_1 q_2 - p_2 q_1 \neq 0$. Then, Eq. (10) has two real roots α and β. Let λ and μ be the two roots of (7) corresponding to α and β. By (6), we have

$$S_1 - \lambda S_2 = (p_1 - \lambda p_2)(x - \alpha)^2 ,$$
$$S_1 - \mu S_2 = (p_1 - \mu p_2)(x - \beta)^2 .$$

Solving for S_1 and S_2, we obtain

$$S_1 = b_1(x - \alpha)^2 + c_1(x - \beta)^2 ,$$
$$S_2 = b_2(x - \alpha)^2 + c_2(x - \beta)^2 , \tag{15}$$
$$b_1 = \frac{\mu p_1 - \lambda \mu p_2}{\mu - \lambda} = \frac{p_1 \beta + q_1}{\alpha - \beta} , \quad c_1 = -\frac{\lambda p_1 - \lambda \mu p_2}{\mu - \lambda} = -\frac{p_1 \alpha + q_1}{\alpha - \beta} ,$$
$$b_2 = \frac{p_1 - \lambda p_2}{\mu - \lambda} = \frac{p_2 \beta + q_2}{\alpha - \beta} , \quad c_2 = -\frac{p_1 - \mu p_2}{\mu - \lambda} = -\frac{p_2 \alpha + q_2}{\alpha - \beta} .$$

By the transformation

$$s = \frac{x - \alpha}{x - \beta} ,$$

$$P(x) = (x - \beta)^4 Q(s) = (x - \beta)^4 (b_1 s^2 + c_1)(b_2 s^2 + c_2) , \tag{16}$$

we obtain

$$\frac{dx}{\sqrt{P(x)}} = \frac{1}{\alpha - \beta} \frac{ds}{\sqrt{(b_1 s^2 + c_1)(b_2 s^2 + c_2)}} . \tag{17}$$

If $p_1 q_2 - p_2 q_1 = 0$, Eq. (10) only has one finite root: $\alpha = -q_1/p_1 = -q_2/p_2$; the other root is $\beta = \infty$. Then, evidently,

$$S_1 = p_1(x - \alpha)^2 + r_1 - p_1\alpha^2 , \quad S_2 = p_2(x - \alpha)^2 + r_2 - p_2\alpha^2 . \tag{18}$$

By the transformation $x - \alpha = s$, we can make $P(s + \alpha)$ containing only even powers of s so as to take the form of $Q(s)$ in (16). Note that, from (13), such case occurs when $a + b = c + d$. Thus, it can occur only when there are complex roots; then, $\alpha = \frac{1}{2}(a + b) = -q_1/p_1$.

The following elliptic integrals

$$I_m = \int \frac{s^{2m} ds}{\sqrt{Q(s)}} , \quad J_n = \int \frac{ds}{(s^2 - k)^n \sqrt{Q(s)}} \tag{19}$$

can be reduced to the three fundamental integrals I_0, I_1, J_1. The procedure of proving this is similar to that in Sec. 8.1, based on the following relations:

$$\frac{d}{ds} \left\{ s^{2m-1} \sqrt{Q(s)} \right\}$$

$$= \frac{1}{\sqrt{Q(s)}} \left\{ (2m + 1)b_1 b_2 s^{2m+2} + 2m(b_1 c_2 + b_2 c_1)s^{2m} \right.$$

$$\left. + (2m - 1)c_1 c_2 s^{2m-2} \right\} , \tag{20}$$

$$\frac{d}{ds} \left\{ \frac{s\sqrt{Q(s)}}{(s^2 - h)^n} \right\} = \frac{1}{(s^2 - h)^{n+1}\sqrt{Q(s)}}$$

$$\times \left\{ (3 - 2n)b_1 b_2 (s^2 - h)^3 + (2 - 2n)[b_1 c_2 + b_2 c_1 + 3b_1 b_2 h](s^2 - h)^2 \right.$$

$$+ (1 - 2n)[c_1 c_2 + 2(b_1 c_2 + b_2 c_1)h + 3b_1 b_2 h^2](s^2 - h)$$

$$\left. - 2nh(c_1 + hb_1)(c_2 + hb_2) \right\} . \tag{21}$$

Now, we shall discuss further the procedure of reducing $Q(s)$ to Legendre's standard form $L(t)$. There are different cases. We shall deal with them one by one.

1. $Q(s) = (a^2 - s^2)(b^2 - s^2)$, $s < a < b$.
Let $s = at$. We obtain $Q(s) = a^2b^2L(t)$, $k = a/b$,

$$\frac{ds}{\sqrt{Q(s)}} = \frac{1}{b}\frac{dt}{\sqrt{L(t)}} = \frac{du}{b}.$$

2. $Q(s) = (s^2 - a^2)(s^2 - b^2)$, $a < b < s$.
Let $s = b/t$. We obtain $Q(s) = (b/t)^4 L(t)$, $k = a/b$,

$$\frac{ds}{\sqrt{Q(s)}} = -\frac{dt}{b\sqrt{L(t)}}.$$

3. $Q(s) = (s^2 - a^2)(b^2 - s^2)$, $a < s < b$.
Let $s^2 = a^2/(1 - k^2t^2)$, $k^2 = 1 - a^2/b^2$. We obtain

$$Q(s) = \frac{a^2b^2k^4t^2(1 - t^2)}{(1 - k^2t^2)^2}, \quad \frac{ds}{\sqrt{Q(s)}} = \frac{dt}{b\sqrt{L(t)}}.$$

Or, by the transformation $s^2 = b^2(1 - k^2\tau^2)$, $k^2 = 1 - a^2/b^2$. We obtain

$$Q(s) = b^4k^4\tau^2(1 - \tau^2), \quad \frac{ds}{\sqrt{Q(s)}} = -\frac{d\tau}{b\sqrt{L(\tau)}}.$$

The relation between the variables τ and t is $\tau^2 = (1 - t^2)/(1 - k^2t^2)$, or $1 - k^2\tau^2 = (1 - k^2)/(1 - k^2t^2)$, which exchanges 0 and 1.

4. $Q(s) = (a^2 - s^2)(s^2 + b^2)$, $s < a$.
Let

$$s^2 = \frac{a^2(1 - k^2)t^2}{1 - k^2t^2} = \frac{b^2}{1 - k^2t^2} - b^2, \quad k^2 = \frac{a^2}{a^2 + b^2},$$

or, $s^2 = a^2(1 - \tau^2)$. We obtain

$$\frac{ds}{\sqrt{Q(s)}} = \frac{dt}{\sqrt{a^2 + b^2}\sqrt{L(t)}} = -\frac{d\tau}{\sqrt{a^2 + b^2}\sqrt{L(\tau)}}.$$

5. $Q(s) = (s^2 - a^2)(s^2 + b^2)$, $s > a$.
Let $s^2 = a^2/(1 - t^2)$, $k^2 = b^2/(a^2 + b^2)$. We obtain

$$\frac{ds}{\sqrt{Q(s)}} = \frac{dt}{\sqrt{a^2 + b^2}\sqrt{L(t)}}.$$

6. $Q(s) = (s^2 + a^2)(s^2 + b^2)$, $a < b$.

Let $s^2 = a^2 t^2/(1 - t^2), k^2 = 1 - a^2/b^2$. We obtain

$$\frac{ds}{\sqrt{Q(s)}} = \frac{dt}{b\sqrt{L(t)}} .$$

Example. $P(x) = x^4 + 1$.

The four roots are $a = e^{\pi i/4}, b = e^{\frac{\pi i}{4} + \frac{3\pi i}{2}} = e^{\frac{\pi i}{4} - \frac{\pi i}{2}} = e^{-\pi i/4}, c = e^{\frac{\pi i}{4} + \frac{\pi i}{2}} = e^{\frac{3\pi i}{4}}, d = e^{\frac{\pi i}{4} + \pi i} = e^{\frac{\pi i}{4} - \pi i} = e^{-\frac{3\pi i}{4}}$. Substituting in (13) gives $\alpha^2 = 1$, i.e., $\alpha = 1, \beta = -1$. Let

$$s = \frac{x - 1}{x + 1} , \quad x = \frac{1 + s}{1 - s} = \frac{2}{1 - s} - 1 ,$$

then

$$P(x) = \frac{(1 + s)^4 + (1 - s)^4}{(1 - s)^4} = \frac{2(s^4 + 6s^2 + 1)}{(1 - s)^4} ,$$

$$\frac{dx}{\sqrt{P(x)}} = \frac{\sqrt{2}ds}{\sqrt{s^4 + 6s^2 + 1}} .$$

Now $Q(s) = s^4 + 6s^2 + 1 = (s^2 + 3)^2 - 8 = (s^2 + 3 - \sqrt{8})(s^2 + 3 + \sqrt{8})$, which belongs to 6 above. Hence, put

$$s^2 = \frac{(3 - \sqrt{8})t^2}{1 - t^2} , \quad k^2 = 1 - \frac{3 - \sqrt{8}}{3 + \sqrt{8}} = \frac{2\sqrt{8}}{3 + \sqrt{8}} .$$

We have

$$k' = \sqrt{\frac{3 - \sqrt{8}}{3 + \sqrt{8}}} = \frac{1}{3 + \sqrt{8}} = 3 - \sqrt{8} ,$$

$$\frac{dx}{\sqrt{x^4 + 1}} = \sqrt{\frac{2}{3 + \sqrt{8}}} \frac{dt}{\sqrt{(1 - t^2)(1 - k^2 t^2)}} .$$

When all the four roots are real, the intermediate step from x to s is unnecessary, and the transformation may be carried out directly from x to t. The procedure is as follows.

7. $P(x) = (x - a)(x - b)(x - c)(x - d)$, $a > b > c > d$.

There are two cases in which $P(x) \geq 0$:

A. $x \geq a$ or $x \leq d$. Let

$$\frac{x-a}{x-b} = \frac{a-d}{b-d}t^2 , \qquad k^2 = \frac{(a-d)(b-c)}{(a-c)(b-d)} .$$

We obtain

$$\frac{dx}{\sqrt{P(x)}} = \frac{2}{\sqrt{(a-c)(b-d)}} \frac{dt}{\sqrt{L(t)}} .$$

From $(a-b)(c-d) + (a-d)(b-c) = (a-c)(b-d)$ we know $k^2 < 1$.

B. $b \geq x \geq c$. Let

$$\frac{x-c}{x-d} = \frac{b-c}{b-d}t^2 , \qquad k^2 = \frac{(a-d)(b-c)}{(a-c)(b-d)} .$$

We obtain

$$\frac{dx}{\sqrt{P(x)}} = \frac{2}{\sqrt{(a-c)(b-d)}} \frac{dt}{\sqrt{L(t)}} .$$

8. $P(x) = -(x-a)(x-b)(x-c)(x-d)$, $a > b > c > d$.
Again, there are two cases in which $P(x) \geq 0$:

A. $a \geq x \geq b$. Let

$$\frac{x-b}{x-c} = \frac{a-b}{a-c}t^2 , \qquad k^2 = \frac{(a-b)(c-d)}{(a-c)(b-d)} .$$

We obtain

$$\frac{dx}{\sqrt{P(x)}} = \frac{2}{\sqrt{(a-c)(b-d)}} \frac{dt}{\sqrt{L(t)}} .$$

B. $c \geq x \geq d$. Let

$$\frac{x-d}{x-a} = \frac{c-d}{c-a}t^2 , \qquad k^2 = \frac{(a-b)(c-d)}{(a-c)(b-d)} .$$

We obtain

$$\frac{dx}{\sqrt{P(x)}} = \frac{2}{\sqrt{(a-c)(b-d)}} \frac{dt}{\sqrt{L(t)}} .$$

Case 7 corresponds to 3; 8A to 1; 8B to 2; for, by (15), it can be shown that a^2 and b^2 in 1, 2, 3 are equal respectively to $-c_1/b_1$, $-c_2/b_2$. After some calculations, it is seen that the modulus k in 1, 2, 3 is different from the k in 7 and 8; they are connected by transformation of Landen's type [Eq. (5) of Sec. 10.7].

10.9. Elliptic Integral of the Second Kind

The standard form of Legendre's elliptic integral of the second kind is

$$E(u) = \int_0^t \frac{(1 - k^2 t^2) dt}{\sqrt{(1 - t^2)(1 - k^2 t^2)}} = \int_0^t \sqrt{\frac{1 - k^2 t^2}{1 - t^2}}\, dt$$

$$= \int_0^\varphi \sqrt{1 - k^2 \sin^2 \varphi}\, d\varphi \quad (t = \operatorname{sn} u = \sin \varphi) . \tag{1}$$

It can also be represented as

$$E(u) = \int_0^u \operatorname{dn}^2 u\, du . \tag{2}$$

The complete elliptic integral of the second kind is [cf. Eq. (3) of Sec. 10.3]

$$E = E(K) = \int_0^1 \sqrt{\frac{1 - k^2 t^2}{1 - t^2}}\, dt = \sum_{r=0}^\infty \binom{\frac{1}{2}}{r} (-k^2)^r \int_0^1 \frac{t^{2r} dt}{\sqrt{1 - t^2}}$$

$$= \frac{1}{2} \sum_{r=0}^\infty \left(-\frac{1}{2}\right)_r \frac{k^{2r}}{r!} \frac{\Gamma\left(\frac{1}{2}\right) \Gamma\left(r + \frac{1}{2}\right)}{\Gamma(r+1)}$$

$$= \frac{\pi}{2} \sum_{r=0}^\infty \left(-\frac{1}{2}\right)_r \left(\frac{1}{2}\right)_r \frac{k^{2r}}{(r!)^2} ,$$

i.e.,

$$E = \frac{\pi}{2} F\left(-\frac{1}{2}, \frac{1}{2}, 1, k^2\right) . \tag{3}$$

To evaluate (2), we apply Eq. (3) of Sec. 9.3, the representation of the Theta function. Since the periods of $\operatorname{dn}^2 u$ are $2K$ and $2Ki$ [Sec. 10.5] and the pole of it is $K'i$ with the principal part equal to $-(u - K'i)^{-2}$ [Sec. 10.6], we have

$$\operatorname{dn}^2 u = \frac{d^2}{du^2} \ln \vartheta_1 \left(\frac{u - K'i}{2K}\right) + C = \frac{d^2}{du^2} \ln \Theta(u) + C ,$$

where

$$\Theta(u) = iq^{\frac{1}{4}} e^{-\frac{i\pi u}{2K}} \vartheta_1 \left(\frac{u - K'i}{2K}\right) = iq^{\frac{1}{4}} e^{-\frac{i\pi u}{2K}} \vartheta_1 \left(\frac{u}{2K} - \frac{\tau}{2}\right) = \vartheta_4 \left(\frac{u}{2K}\right) .$$

This symbol agrees with that in Sec. 9.13. Substituting in (2) and integrating, we have

$$E(u) = \frac{\Theta'(u)}{\Theta'(u)} + Cu .$$

Now, in $\Theta'(u+2K) = \Theta'(u)$, put $u = -K$. Since $\Theta'(u)$ is an odd function of u, we see that $\Theta'(K) = 0$. Hence, the last equation gives

$$E = CK \ ,$$

which determines the constant C, and we have

$$E(u) = Z(u) + \frac{E}{K}u \ , \tag{4}$$

where

$$Z(u) = \frac{\Theta'(u)}{\Theta(u)} = \frac{d}{du}\ln\vartheta_4\left(\frac{u}{2K}\right) \ . \tag{5}$$

10.10. Elliptic Integral of the Third Kind

The definition now commonly used for Legendre's elliptic integral of the third kind is

$$\Pi(u,c) = \int_0^u \frac{du}{1 + c\,\text{sn}^2\,u} \ . \tag{1}$$

Let $t = \text{sn}\,u = \sin\varphi$, (1) can also be written as

$$\Pi(u,c) = \int_0^t \frac{dt}{(1+ct^2)\sqrt{(1-t^2)(1-k^2t^2)}}$$

$$= \int_0^\varphi \frac{d\varphi}{(1 + c\sin^2\varphi)\sqrt{1 - k^2\sin^2\varphi}} \ . \tag{2}$$

Putting $t = 1$, or $u = K$, we obtain the complete elliptic integral of the third kind:

$$\Pi_1(c) = \int_0^K \frac{du}{1 + c\,\text{sn}^2\,u} = \int_0^1 \frac{dt}{(1+ct^2)\sqrt{(1-t^2)(1-k^2t^2)}} \ . \tag{3}$$

In order to represent the elliptic integral of the third kind by means of elliptic functions, we introduce the constant a so as to make

$$c = -\frac{1}{\text{sn}^2\,a} = -\text{ns}^2\,a \ . \tag{4}$$

The integrand in (1) has the periods $2K$ and $2K'i$; $u = \pm a$ are its poles with residues $\mp \text{sn}\,a/2\,\text{cn}\,a\,\text{dn}\,a$ respectively. By Eq. (1) of Sec. 9.8, namely $\text{sn}\,u = \vartheta_3\vartheta_1(v)/\vartheta_2\vartheta_4(v), u = 2Kv = \pi\vartheta_3^2v$, we can express the residues as

$$\mp \left[2\frac{d}{da}\left\{\ln\vartheta_1\left(\frac{a}{2K}\right) - \ln\vartheta_4\left(\frac{a}{2K}\right)\right\}\right]^{-1} \ .$$

Applying Eq. (3) of Sec. 9.3, we obtain

$$\frac{1}{1+c\,\text{sn}^2 u} = \frac{\text{sn}\,a}{\text{cn}\,a\,\text{dn}\,a}\left\{\frac{1}{2}\frac{d}{du}\left[\ln\vartheta_1\left(\frac{u+a}{2K}\right)-\ln\vartheta_1\left(\frac{u-a}{2K}\right)\right]+C\right\}.$$

Let $u \to 0$. The value of C is determined to be

$$C = -\frac{d}{da}\ln\vartheta_4\left(\frac{a}{2K}\right) = -Z(a).$$

Thus, we obtain for the elliptic integral of the third kind the expression

$$\Pi(u,c) = \frac{\text{sn}\,a}{\text{cn}\,a\,\text{dn}\,a}\left\{\frac{1}{2}\ln\frac{\vartheta_1\left(\frac{u+a}{2K}\right)}{\vartheta_1\left(\frac{u-a}{2K}\right)}-u\frac{d}{da}\ln\vartheta_4\left(\frac{a}{2K}\right)\right\}. \tag{5}$$

Putting $u = K$, we have the complete elliptic integral of the third kind expressed as

$$\Pi_1(c) = \frac{\text{sn}\,a}{\text{cn}\,a\,\text{dn}\,a}\{Ea - KE(a)\}, \tag{6}$$

from which we see that the complete elliptic integral of the third kind can be expressed in terms of elliptic integrals of the first and second kind.

As regards the values of a, we see that, from the definition (4), if $c < -1$, then $0 < a < K$. If, then, $u < a, \vartheta_1\left(\frac{u-a}{2K}\right)$ in (5) should change sign, which can be seen from the derivation of (5). Hence, (5) should be changed into

$$\Pi(u,c) = \frac{\text{sn}\,a}{\text{cn}\,a\,\text{dn}\,a}\left\{\frac{1}{2}\ln\frac{\vartheta_1\left(\frac{u+a}{2K}\right)}{\vartheta_1\left(\frac{a-u}{2K}\right)}-u\frac{d}{da}\ln\vartheta_4\left(\frac{a}{2K}\right)\right\}. \tag{7}$$

If $-1 < c < -k^2$, then the value of a has to be replaced by $K + ia'$, and, by Eq. (1) of Sec. 10.5 and Eqs. (3) and (4) of Sec. 10.7, we have

$$c = -dn^2(a', k') \quad (a = K + ia'). \tag{8}$$

In the mean time, (7) reduces to

$$\Pi(u,c) = \frac{i\,\text{dn}\,(a',k')}{k'^2\,\text{sn}(a',k')\,\text{cn}(a',k')}\left\{\frac{1}{2}\ln\frac{\vartheta_2\left(\frac{u+ia'}{2K}\right)}{\vartheta_2\left(\frac{u-ia'}{2K}\right)}+iu\frac{d}{da'}\ln\vartheta_3\left(\frac{ia'}{2K}\right)\right\}. \tag{9}$$

In the interval $-k^2 < c < 0$, we have to use the transformation $a = b + K'i$ so that

$$c = -k^2\,\text{sn}^2 b \quad (a = b + K'i), \tag{10}$$

$$\Pi(u,c) = -\frac{\operatorname{sn} b}{\operatorname{cn} b \operatorname{dn} b} \left\{ \frac{1}{2} \ln \frac{\vartheta_4 \left(\frac{u+b}{2K}\right)}{\vartheta_4 \left(\frac{u-b}{2K}\right)} - u \frac{d}{db} \ln \vartheta_1 \left(\frac{b}{2K}\right) \right\} . \tag{11}$$

If $c > 0$, we can use the imaginary transformation of Sec. 10.7; let $b = ib'$ and obtain

$$c = k^2 \operatorname{sc}^2(b', k') = k^2 \operatorname{sn}^2(b', k')/\{1 - \operatorname{sn}^2(b', k')\} . \tag{12}$$

a is now equal to $(b' + K')i$, and (11) becomes

$$\Pi(u,c) = -\frac{i \operatorname{sn}(b', k') \operatorname{cn}(b', k')}{\operatorname{dn}(b', k')}$$

$$\times \left\{ \frac{1}{2} \ln \frac{\vartheta_4 \left(\frac{u+ib'}{2K}\right)}{\vartheta_4 \left(\frac{u-ib'}{2K}\right)} + iu \frac{d}{db'} \ln \vartheta_1 \left(\frac{ib'}{2K}\right) \right\} . \tag{13}$$

In the different regions, the corresponding expressions for the complete elliptic integrals of the third kind are:

$$\Pi_1(-\operatorname{dn}^2(a', k')) = K + \frac{\operatorname{dc}(a', k')}{k'^2 \operatorname{sn}(a', k')} \left[\frac{\pi}{2} - (E - K)a' - K E(a', k') \right] , \tag{14}$$

$$\Pi_1(-k^2 \operatorname{sn}^2 b) = K + \frac{\operatorname{sc} b}{\operatorname{dn} b} [K E(b) - Eb] , \tag{15}$$

$$\Pi_1(k^2 \operatorname{sc}^2(b', k')) = K \operatorname{cn}^2(b', k')$$

$$+ \operatorname{sn}(b', k')\operatorname{cd}(b', k')[(E - K)b' + K E(b', k')] . \tag{16}$$

10.11. Properties of the Function $E(u)$

The function $E(u)$ introduced in Sec. 10.9 to represent the elliptic integral of the second kind is an odd function since $Z(u)$ is. It has, in the period parallelogram, a zero, $u = 0$, and a pole, $u = K'i$ with residue 1. This can easily be proved, as the residue is equal to $\Theta'(K'i)/\left[\frac{d}{du}\Theta(u)\right]_{u=K'i} = 1$.

$E(u)$ is not a periodic function. From the definition in Eq. (2) of Sec. 10.9, we have

$$E(u + 2K) = \int_0^{u+2K} \operatorname{dn}^2 u \, du = \int_0^{2K} \operatorname{dn}^2 u \, du + \int_{2K}^{u+2K} \operatorname{dn}^2 u \, du .$$

The first integral can be reduced to

$$\int_0^K \operatorname{dn}^2 u \, du + \int_K^{2K} \operatorname{dn}^2 u \, du = \int_0^K [\operatorname{dn}^2 u + \operatorname{dn}^2(2K - u)]du$$

$$= 2 \int_0^K \operatorname{dn}^2 u \, du = 2E ,$$

i.e.,

$$E(2K) = 2E(K) = 2E .\tag{1}$$

The second integral can be reduced to

$$\int_0^u \mathrm{dn}^2(u + 2K)du = \int_0^u \mathrm{dn}^2 u \, du = E(u) .$$

Hence,

$$E(u + 2K) = E(u) + 2E .\tag{2}$$

Similarly,

$$E(u + 2K'i) = E(u) + E(2K'i) .\tag{3}$$

To evaluate $E(2K'i)$, consider the imaginary transformation for $E(u)$, writing $E(u, k)$ for $E(u)$. When we discussed the imaginary transformation $u' = iu$ in Sec. 10.7, we replaced t by it. By Eq. (4) of Sec. 10.7, we have

$$E(iu, k) = i \int_0^u \mathrm{dn}^2(iu, k)du = i \int_0^u \mathrm{dc}^2(u, k')du .$$

Using the formula given in Sec. 10.1, we have

$$\frac{d}{du}(u + \mathrm{dn}\ u\ \mathrm{sc}\ u)$$

$$= 1 + \mathrm{dn}\ u\ \mathrm{sc}\ u\left(-\frac{k^2\ \mathrm{sn}\ u\ \mathrm{cn}\ u}{\mathrm{dn}\ u} + \frac{\mathrm{cn}\ u\ \mathrm{dn}\ u}{\mathrm{sn}\ u} + \frac{\mathrm{sn}\ u\ \mathrm{dn}\ u}{\mathrm{cn}\ u}\right)$$

$$= 1 + \mathrm{dn}\ u\ \mathrm{sc}\ u\left(-\frac{k^2\ \mathrm{sn}\ u\ \mathrm{cn}\ u}{\mathrm{dn}\ u} + \frac{\mathrm{dn}\ u}{\mathrm{sn}\ u\ \mathrm{cn}\ u}\right)$$

$$= 1 - k^2\ \mathrm{sn}^2\ u + \frac{\mathrm{dn}^2 u}{\mathrm{cn}^2 u} = \mathrm{dn}^2 u + \mathrm{dc}^2 u ,$$

from which we obtain the imaginary transformation for $E(u)$:

$$E(iu, k) = iu + i\ \mathrm{dn}(u, k')\ \mathrm{sc}(u, k') - iE(u, k') .\tag{4}$$

Let $u = 2K'$ and apply (1) to the case in which the modulus is equal to k'. We have $E(2K', k') = 2E'$, so

$$E(2K'i, k) = 2(K' - E')i .\tag{5}$$

Substituting in (3) gives

$$E(u + 2K'i) = E(u) + 2(K' - E')i .\tag{6}$$

Consider a rectangle with vertices $-K, K, K+2K'i, -K+2K'i$. It includes a pole $K'i$ of $E(u)$. The integral $\int E(u)du$ along the sides of the rectangle will be equal to $2\pi i$ times the residue of $E(u)$. Thus,

$$2\pi i = \int_{-K}^{K} [E(u) - E(u + 2K'i)]du + \int_{K}^{K+2K'i} [E(u) - E(u - 2K)]du$$
$$= -4K(K' - E')i + 4K'Ei ,$$

i.e.,

$$KE' + K'E - KK' = \frac{\pi}{2} . \tag{7}$$

This is called the *Legendre relation*.

The periodic properties of $Z(u)$ can be deduced from those of the Theta function. From Eq. (7) of Sec. 9.2, we have

$$\frac{\vartheta_4'(v + 1)}{\vartheta_4(v + 1)} = \frac{\vartheta_4'(v)}{\vartheta_4(v)} , \quad \frac{\vartheta_4'(v + \tau)}{\vartheta_4(v + \tau)} = \frac{\vartheta_4'(v)}{\vartheta_4(v)} - 2\pi i .$$

Hence,

$$Z(u + 2K) = Z(u) , \quad Z(u + 2K'i) = Z(u) - \frac{\pi i}{K} . \tag{8}$$

The first equation of (8) follows immediately from Eq. (4) of Sec. 10.9 and Eq. (2) of the present section. By Eq. (4) of Sec. 10.9 and Eq. (6) above, we obtain

$$Z(u + 2K'i) = Z(u) + 2\left(K' - E' - \frac{K'E}{K}\right)i .$$

Comparing this result with the second equation of (8), we have (7); this is another proof for (7).

By Eq. (4) of Sec. 10.9 and (4) above, we have

$$Z(iu, k) = iu + i\,\mathrm{dn}(u, k')\,\mathrm{sc}(u, k') - iZ(u, k') - iu\left(\frac{E}{K} + \frac{E'}{K'}\right) .$$

Applying (7), we find

$$Z(iu, k) = i\,\mathrm{dn}(u, k')\,\mathrm{sc}(u, k') - iZ(u, k') - \frac{i\pi u}{2KK'} . \tag{9}$$

This is the imaginary transformation for the function $Z(u)$.

We now prove the following addition formula for $E(u)$:

$$E(u) + E(v) - E(u + v) = Z(u) + Z(v) - Z(u + v)$$
$$= k^2\,\mathrm{sn}\,u\,\mathrm{sn}\,v\,\mathrm{sn}(u + v) . \tag{10}$$

According to (2) and (6), we know that the l.h.s. of this equation is a doubly periodic function of u with periods of $2K$ and $2K'i$; $K'i$ and $K'i - v$ are its poles, the residues being 1 and -1 respectively. The same is for the r.h.s. Moreover, both sides equal to zero when $u = 0$. Hence, Eq. (10) holds. By this formula, together with Eq. (2) of Sec. 10.4, we can also derive Eq. (11) of Sec. 10.10. In (10), putting $v = u$, we obtain the duplication formula:

$$2E(u) - E(2u) = 2Z(u) - Z(2u) = k^2 \operatorname{sn}^2 u \operatorname{sn}(2u) . \tag{11}$$

10.12. Differential Equations Satisfied by K and E with Respect to k and Expansions of K and E with Respect to k

First,

$$\frac{dK}{dk} = \frac{d}{dk} \int_0^1 \frac{dt}{\sqrt{(1-t^2)(1-k^2t^2)}} = \int_0^1 \frac{kt^2 \, dt}{(1-t^2)^{\frac{1}{2}}(1-k^2t^2)^{\frac{3}{2}}}$$

$$= \frac{1}{k} \int_0^1 \frac{dt}{(1-t^2)^{\frac{1}{2}}(1-k^2t^2)^{\frac{3}{2}}} - \frac{1}{k} \int_0^1 \frac{dt}{\sqrt{(1-t^2)(1-k^2t^2)}} .$$

From

$$\frac{d}{dt} \left\{ k^2 t \sqrt{\frac{1-t^2}{1-k^2t^2}} \right\}$$

$$= k^2 \sqrt{\frac{1-t^2}{1-k^2t^2}} - \frac{k^2 t^2}{\sqrt{(1-t^2)(1-k^2t^2)}} + \frac{k^4 t^2 \sqrt{1-t^2}}{(1-k^2t^2)^{\frac{3}{2}}}$$

$$= \frac{k^2 \sqrt{1-t^2}}{(1-k^2t^2)^{\frac{3}{2}}} - \frac{k^2 t^2}{\sqrt{(1-t^2)(1-k^2t^2)}}$$

$$= \frac{k^2(1-t^2)}{(1-t^2)^{\frac{1}{2}}(1-k^2t^2)^{\frac{3}{2}}} - \frac{k^2 t^2}{\sqrt{(1-t^2)(1-k^2t^2)}}$$

$$= \sqrt{\frac{1-k^2t^2}{1-t^2}} - \frac{1-k^2}{(1-t^2)^{\frac{1}{2}}(1-k^2t^2)^{\frac{1}{2}}} ,$$

by integration, we have

$$\int_0^1 \frac{k'^2 \, dt}{(1-t^2)^{\frac{1}{2}}(1-k^2t^2)^{\frac{3}{2}}} = \int_0^1 \sqrt{\frac{1-k^2t^2}{1-t^2}} \, dt = E .$$

Hence,

$$\frac{dK}{dk} = \frac{E}{kk'^2} - \frac{K}{k} . \tag{1}$$

Next,

$$\frac{dE}{dk} = \frac{d}{dk} \int_0^1 \sqrt{\frac{1 - k^2 t^2}{1 - t^2}}\, dt = -\int_0^1 \frac{k t^2\, dt}{\sqrt{(1 - t^2)(1 - k^2 t^2)}}$$

$$= \frac{1}{k} \int_0^1 \frac{(1 - k^2 t^2 - 1)\, dt}{\sqrt{(1 - t^2)(1 - k^2 t^2)}} \, ,$$

i.e.,

$$\frac{dE}{dk} = \frac{E - K}{k} . \tag{2}$$

When $k < 1$, the expansion of K has been given in Eq. (3) of Sec. 10.3:

$$K = \frac{\pi}{2} \left\{ 1 + \frac{k^2}{4} + \frac{9k^4}{64} + \dots \right\} . \tag{3}$$

The expansion of E has been given in Eq. (3) of Sec. 10.9:

$$E = \frac{\pi}{2} \left\{ 1 - \frac{k^2}{4} - \frac{3k^4}{64} - \dots \right\} . \tag{4}$$

As regards K' and E', using the transformation of the hypergeometric functions [Eq. (8) of Sec. 4.9], we obtain

$$K' = \frac{\pi}{2} F\left(\frac{1}{2}, \frac{1}{2}, 1, 1 - k^2 \right)$$

$$= -\frac{1}{2} F\left(\frac{1}{2}, \frac{1}{2}, 1, k^2 \right) \ln k^2$$

$$- \sum_{r=0}^{\infty} \left[\left(\frac{1}{2} \right)_r \right]^2 \frac{k^{2r}}{(r!)^2} \left[\psi\left(\frac{1}{2} + r \right) - \psi(1 + r) \right]$$

$$= \left\{ 1 + \frac{k^2}{4} + \frac{9k^4}{64} + \dots \right\} \ln \frac{1}{k}$$

$$+ \left[\psi(1) - \psi\left(\frac{1}{2} \right) \right] + \left[\psi(2) - \psi\left(\frac{3}{2} \right) \right] \frac{k^2}{4} + \dots ; \tag{5}$$

with Eq. (9) of Sec. 4.9, we obtain

$$
\begin{aligned}
E' &= \frac{\pi}{2} F\left(-\frac{1}{2}, \frac{1}{2}, 1, 1 - k^2\right) \\
&= 1 - \frac{k^2}{4} F\left(\frac{3}{2}, \frac{1}{2}, 2, k^2\right) \ln k^2 - \frac{k^2}{4} \sum_{r=0}^{\infty} \frac{\left(\frac{3}{2}\right)_r \left(\frac{1}{2}\right)_r k^{2r}}{r!(r+1)!} \\
&\quad \times \left[\psi\left(\frac{3}{2} + r\right) + \psi\left(\frac{1}{2} + r\right) - \psi(1 + r) - \psi(2 + r)\right] \\
&= 1 - \frac{k^2}{2}\left(1 + \frac{3}{8} k^2 + \ldots\right) \ln k \\
&\quad - \frac{k^2}{4}\left\{\left[\psi\left(\frac{1}{2}\right) + \psi\left(\frac{3}{2}\right) - \psi(1) - \psi(2)\right]\right. \\
&\quad \left. + \frac{3k^2}{8}\left[\psi\left(\frac{3}{2}\right) + \psi\left(\frac{5}{2}\right) - \psi(2) - \psi(3)\right] + \ldots\right\} .
\end{aligned}
\tag{6}
$$

By Eq. (2) of Sec. 3.11, we have

$$
\begin{aligned}
\psi(2 + r) - \psi\left(\frac{3}{2} + r\right) &= \psi(1 + r) - \psi\left(\frac{1}{2} + r\right) + \frac{1}{1+r} - \frac{1}{\frac{1}{2} + r} \\
&= \psi(1 + r) - \psi\left(\frac{1}{2} + r\right) - \frac{1}{(1+r)(1+2r)} , \\
\psi(1 + r) - \psi\left(\frac{1}{2} + r\right) &= \psi(1) - \psi\left(\frac{1}{2}\right) + \sum_{p=1}^{r}\left(\frac{1}{p} - \frac{1}{p - \frac{1}{2}}\right) \\
&= \ln 4 - \sum_{p=1}^{r} \frac{1}{p(2p - 1)} .
\end{aligned}
$$

Hence,

$$
K' = F\left(\frac{1}{2}, \frac{1}{2}, 1, k^2\right) \ln \frac{4}{k} - \sum_{r=1}^{\infty}\left[\left(\frac{1}{2}\right)_r\right]^2 \frac{k^{2r}}{(r!)^2} \sum_{p=1}^{r} \frac{1}{p(2p - 1)} ,
\tag{7}
$$

$$
\begin{aligned}
E' &= 1 + \frac{k^2}{2} F\left(\frac{3}{2}, \frac{1}{2}, 2, k^2\right) \ln \frac{4}{k} \\
&\quad - \frac{k^2}{4} \sum_{r=0}^{\infty} \left(\frac{3}{2}\right)_r \left(\frac{1}{2}\right)_r \frac{k^{2r}}{r!(r+1)!} \\
&\quad \times \left[\sum_{p=0}^{r} \frac{2}{(p+1)(2p+1)} - \frac{1}{(r+1)(2r+1)}\right] .
\end{aligned}
\tag{8}
$$

10.13. Relations between Jacobian Elliptic Functions and Theta Functions

We have already made use of the Theta functions in dealing with elliptic integrals. We shall now see that theory of Jacobian elliptic functions can be built up entirely on the theory of Theta functions. This is done by defining the Jacobian elliptic functions by the Theta functions.

In Eq. (1) of Sec. 9.8, we have found the relation between sn u and the Theta function to be

$$\text{sn } u = \frac{\vartheta_3 \vartheta_1(v)}{\vartheta_2 \vartheta_4(v)} \ , u = 2Kv \ , k = \frac{\vartheta_2^2}{\vartheta_3^2} \ , \tag{1}$$

$$2K = \pi \vartheta_3^2 \ , K'i = K\tau \ . \tag{2}$$

Now we take (1) as the definition of sn u. Similarly, cn u and dn u are defined by

$$\text{cn } u = \frac{\vartheta_4 \vartheta_2(v)}{\vartheta_2 \vartheta_4(v)} \ , \quad \text{dn } u = \frac{\vartheta_4 \vartheta_3(v)}{\vartheta_3 \vartheta_4(v)} \ . \tag{3}$$

By Eq. (7) of Sec. 9.2, we see that sn u has the periods $(4K, 2K'i)$; cn u has the periods $(4K, 2K + 2K'i)$; dn u has the periods $(2K, 4K'i)$; the same as in Sec. 10.5. From Eq. (8) of Sec. 9.2, we see that the zeros and poles of sn u, cn u, dn u are the same as those shown in Sec. 10.6. By Eq. (1) of Sec. 9.4, we have $\text{sn}^2 u + \text{cn}^2 u = 1$; by Eq. (2) of Sec. 9.4, we have $k^2 \text{sn}^2 u + \text{dn}^2 u = 1$; and also the conditions cn $0=1$, dn $0 = 1$ are conformed.

Starting from the known zeros and poles of sn u and using Eq. (1) of Sec. 9.3, we can arrive at (1) above. Since the periods of $\text{sn}^2 u$ are $(2K, 2K'i)$, its zero being 0 and its poles being $\pm K'i$, Eq. (1) of Sec. 9.3 then gives

$$\text{sn}^2 u = C \frac{\left[\vartheta_1 \left(\frac{u}{2K} \right) \right]^2}{\vartheta_1 \left(\frac{u - K'i}{2K} \right) \vartheta_1 \left(\frac{u + K'i}{2K} \right)} = C q^{\frac{1}{2}} \left[\frac{\vartheta_1 \left(\frac{u}{2K} \right)}{\vartheta_4 \left(\frac{u}{2K} \right)} \right]^2 \ .$$

Taking the square root, we have

$$\text{sn } u = C' \frac{\vartheta_1(v)}{\vartheta_4(v)} \ .$$

To determine the constant C', dividing by u and let $u \to 0$, we obtain

$$1 = \frac{C'}{2K} \frac{\vartheta_1'}{\vartheta_4} = \frac{C'}{\pi \vartheta_3^2} \frac{\pi \vartheta_2 \vartheta_3 \vartheta_4}{\vartheta_4} = \frac{C' \vartheta_2}{\vartheta_3} \ ,$$

and hence (1).

If it is desirable to express sn u in terms of $\sigma(u)$, we use Eq. (2) of Sec. 8.9 and obtain

$$\text{sn}^2 u = C \frac{[\sigma(u)]^2}{\sigma(u - K'i)\sigma(u + K'i)}$$

$$= -C \left[\frac{\sigma(u)}{\sigma_3(u)\sigma(K'i)} \right]^2 .$$

In the last step, Eq. (14) of Sec. 8.7 and the relation $\omega_3 = \omega' = K'i$ have been employed. Taking the square root, we have

$$\text{sn } u = C' \frac{\sigma(u)}{\sigma_3(u)} .$$

To determine C', dividing by u and let $u \to 0$. We find $C' = 1$. Thus

$$\text{sn } u = \frac{\sigma(u)}{\sigma_3(u)} \quad (\omega = K, \omega' = K'i) . \tag{4}$$

This result can also be derived from Eqs. (5) and (6) of Sec. 9.2. These formulae give at the same time

$$\text{cn } u = \frac{\sigma_1(u)}{\sigma_3(u)} , \quad \text{dn } u = \frac{\sigma_2(u)}{\sigma_3(u)} . \tag{5}$$

For more general formulae, without assuming $\omega = K, \omega' = K'i$, see Ex. 8 at the end of this chapter.

The series expansions given in Sec. 10.6 can be obtained from (4) and (5) by utilizing Eq. (6) of Sec. 8.7.

The derivative of sn u [Eq. (8) of Sec. 10.1] can be obtained from the differential Eq. (2) of Sec. 9.6.

The addition formulae in Sec. 10.4 can be obtained from their counterparts (2) and (6) of Sec. 9.5.

The formulae in Sec. 10.5 can be derived from the definition of the Theta function given in Sec. 9.2.

The imaginary transformation of Jacobi and the transformation of Landen type in Sec. 10.7 can be obtained respectively by the formulae given in Sec. 9.9 and those given in Sec. 9.10 as follows: Dividing Eqs. (1) and (2) of Sec. 9.10, we have

$$\frac{\vartheta_1(2v|2\tau)}{\vartheta_4(2v|2\tau)} = \frac{\vartheta_1(v|\tau)\vartheta_2(v|\tau)}{\vartheta_3(v|\tau)\vartheta_4(v|\tau)} ,$$

i.e.,

$$k_1^{\frac{1}{2}} \operatorname{sn}(4K_1 v, k_1) = k \operatorname{sn}(2Kv, k)\, \operatorname{cd}(2Kv, k) .$$

It was shown in Sec. 10.7, by applying the theory of Theta functions, that $k_1 = (1 - k')/(1 + k') = [k/(1 + k')]^2$. So we have

$$\operatorname{sn}(4K_1 v, k_1) = (1 + k')\operatorname{sn}(2Kv, k)\, \operatorname{cd}(2Kv, k) .$$

This is Eq. (8) of Sec. 10.7. We need to prove further that $2K_1 = (1 + k')K$, i.e., Eq. (11) of Sec. 10.7. This is obtained by using Eq. (16) of Sec. 9.11:

$$2K_1 = \pi \vartheta_3^2(0|2\tau) = \frac{\pi}{2}[\vartheta_3^2(0|\tau) + \vartheta_4^2(0|\tau)] = (1 + k')K .$$

The same result can be obtained from the above expression for $\operatorname{sn}(4K_1 v, k_1)$ by first dividing it with v and then let $v \to 0$.

As having mentioned in Sec. 10.7, the imaginary transformation in Sec. 10.7 corresponds to the transformation in Eq. (11) of Sec. 9.9. By means of the latter, we have

$$\frac{\vartheta_1(v|\tau + 1)}{\vartheta_4(v|\tau + 1)} = e^{i\pi/4}\frac{\vartheta_1(v|\tau)}{\vartheta_3(v|\tau)} ,$$

i.e.,

$$k_2^{\frac{1}{2}} \operatorname{sn}(2K_2 v, k_2) = e^{i\pi/4}(kk')^{\frac{1}{2}}\operatorname{sd}(2Kv, k) .$$

It has been shown in Sec. 10.7 by the theory of Theta functions that $k_2 = ik/k'$. Hence,

$$\operatorname{sn}(2K_2 v, k_2) = k'\, \operatorname{sd}(2Kv, k) .$$

Dividing both sides by v and let $v \to 0$, we find $K_2 = k'K$, which can also be proved by Eq. (11) of Sec. 9.9 as follows:

$$2K_2/\pi = \vartheta_3^2(0|\tau + 1) = \vartheta_4^2(0|\tau) = k'\vartheta_3^2(0|\tau) = 2k'K/\pi .$$

From $\tau_2 = \tau + 1$, we arrive immediately at

$$iK_2' = (\tau + 1)K_2 = (\tau + 1)k'K = k'(K + K'i) ,$$

which is the second equation of (19) in Sec. 10.7.

In Sec. 10.9, we have expressed the elliptic integral of the second kind, $E(u)$, in terms of the Theta function $\vartheta_4(v)$. The constant E can be determined from the equation

$$\operatorname{dn}^2 u = \left[\frac{\vartheta_4\vartheta_3(v)}{\vartheta_3\vartheta_4(v)}\right]^2 = \frac{d^2}{du^2}\ln\vartheta_4(v) + \frac{E}{K}$$

by putting in it $u = 2Kv \to 0$. When $v \to 0$, we have

$$E - K - \frac{1}{4K}\frac{\vartheta_4''}{\vartheta_4} = \frac{\pi}{2}\vartheta_3^2 - \frac{\vartheta_4''}{2\pi\vartheta_3^2\vartheta_4} \ . \tag{6}$$

From Eqs. (4) of Secs. 9.6 and 9.7, we have

$$4\omega^2(e_1 - e_3) = \frac{\vartheta_4''}{\vartheta_4} - \frac{\vartheta_2''}{\vartheta_2} = \pi^2\vartheta_3^4 \ . \tag{7}$$

Substituting in (6) leads to

$$E = -\frac{\vartheta_2''}{2\pi\vartheta_3^2\vartheta_2} \ . \tag{8}$$

By Eq. (6) of Sec. 9.6 and Eq. (1) of Sec. 9.7, we have

$$\vartheta_2'' = -8\pi^2\sum_{n=0}^{\infty}\left(n + \frac{1}{2}\right)^2 q^{\left(n + \frac{1}{2}\right)^2}$$
$$= -2\pi^2 q^{\frac{1}{4}}\left(1 + 9q^2 + 25q^6 + 49q^{12} + \ldots\right) , \tag{9}$$

$$\vartheta_3'' = -8\pi^2\sum_{n=1}^{\infty}n^2 q^{n^2}$$
$$= -8\pi^2\left(q + 4q^4 + 9q^9 + 16q^{16} + \ldots\right) , \tag{10}$$

$$\vartheta_4'' = -8\pi^2\sum_{n=1}^{\infty}(-)^n n^2 q^{n^2}$$
$$= 8\pi^2\left(q - 4q^4 + 9q^9 - 16q^{16} + \ldots\right) . \tag{11}$$

Substituting (9) and Eq. (1) of Sec. 9.7 in (8), we obtain the power series expansion of E in q.

Applying the imaginary transformation of Sec. 9.9, we can show that the relation between K' and k' is the same as that between K and k. This is manifested by

$$2K' = -i\tau\pi\vartheta_3^2(0|\tau) = \pi\vartheta_3^2(0|\tau') \quad (\tau' = -1/\tau) \ .$$

In the imaginary transformation, the relation between k and k' is given by

$$k'^2 = \frac{\vartheta_2^4(0|\tau')}{\vartheta_3^4(0|\tau')} = \frac{\vartheta_4^4(0|\tau)}{\vartheta_3^4(0|\tau)} = 1 - \frac{\vartheta_2^4(0|\tau)}{\vartheta_3^4(0|\tau)} = 1 - k^2 \ .$$

The imaginary transformation of $Z(u)$ given by Eq. (9) of Sec. 10.11 can be obtained from Eq. (5) of Sec. 9.9. For, from the latter, putting $v' = v/\tau, \tau' = -1/\tau$, we have

$$\frac{1}{\tau}\frac{\vartheta_4'(v'|\tau')}{\vartheta_4(v'|\tau')} = \frac{2\pi vi}{\tau} + \frac{\vartheta_2'(v|\tau)}{\vartheta_2(v|\tau)}$$

$$= \frac{2\pi vi}{\tau} + \frac{d}{dv} \ln \frac{\vartheta_2(v|\tau)}{\vartheta_4(v|\tau)} + \frac{\vartheta_4'(v|\tau)}{\vartheta_4(v|\tau)} \ .$$

Multiplying with $\tau/2K' = i/2K$ and putting $u = 2Kv$, we obtain

$$Z\left(\frac{2K'v}{\tau}, k'\right) = \frac{\pi vi}{K'} + i\frac{d}{du} \ln \ \mathrm{cn}(n, k) + iZ(u, k) \ ,$$

i.e.,

$$Z(-iu, k') = \frac{\pi ui}{2KK'} - i \ \mathrm{dn}(u, k) \ \mathrm{sc}(u, k) + iZ(u, k) \ .$$

Replacing i by $-i$, the result is just Eq. (9) of Sec. 10.11.

We have obtained $\vartheta_4'(v'|\tau')$ from Eq. (5) of Sec. 9.9. Differentiating once more, we have

$$\frac{\vartheta_4''(v'|\tau')}{\vartheta_4'(v'|\tau')} - \left[\frac{\vartheta_4'(v'|\tau')}{\vartheta_4(v'|\tau')}\right]^2$$

$$= \tau^2 \left\{ \frac{\vartheta_2''(v|\tau)}{\vartheta_2(v|\tau)} - \left[\frac{\vartheta_2'(v|\tau)}{\vartheta_2(v|\tau)}\right]^2 \right\} + 2\pi\tau i \ ,$$

where $v' = v/\tau$. Let $v \to 0$, we obtain

$$\frac{\vartheta_4''(0|\tau')}{\vartheta_4(0|\tau')} = \tau^2 \frac{\vartheta_2''(0|\tau)}{\vartheta_2(0|\tau)} + 2\pi\tau i \ . \tag{12}$$

Dividing by τ^2, then replacing τ by τ', it gives

$$\frac{\vartheta_2''(0|\tau')}{\vartheta_2(0|\tau')} = \tau^2 \frac{\vartheta_4''(0|\tau)}{\vartheta_4(0|\tau)} + 2\pi\tau i \ . \tag{13}$$

In (8), change τ to τ'. The result is

$$E' = -\frac{\vartheta_2''(0|\tau')}{4K'\vartheta_2(0|\tau')} = -\frac{\tau^2\vartheta_4''(0|\tau)}{4K'\vartheta_4(0|\tau)} - \frac{\pi\tau i}{2K'}$$

$$= \frac{\tau^2 K(E - K)}{K'} + \frac{\pi}{2K} = -\frac{K'(E - K)}{K} + \frac{\pi}{2K} \ ,$$

from which we arrive again at Legendre's relation [Eq. (7) of Sec. 10.11]:

$$KE' + K'E - KK' = \frac{\pi}{2} \; .$$

This can also be written as

$$\frac{E'}{K'} + \frac{E}{K} = 1 + \frac{\pi}{2KK'} \; . \tag{14}$$

For the power series expansion of K' in k, we can make use of Eq. (9) of Sec. 9.8, whereas the relation between k and q can be obtained from Eq. (1) of Sec. 9.8 and Eq. (1) of Sec. 9.7; the result is

$$k^2 = 16q(1 - 8q + 44q^2 - 192q^3 + \dots)$$
$$q = \left(\frac{k}{4}\right)^2 \left(1 + \frac{k^2}{2} + \dots\right) \; . \tag{15}$$

10.14. Expressing Jacobian Elliptic Functions in Infinite Products and Fourier Series

From (1) and (3) of the last section and the expressions of Theta functions in infinite products, Eqs. (5) – (8) of Sec. 9.11, we have

$$\operatorname{sn}(2Kv) = 2q^{\frac{1}{4}} k^{-\frac{1}{2}} \sin \pi v \prod_{n=1}^{\infty} \frac{1 - 2q^{2n} \cos 2\pi v - q^{4n}}{1 - 2q^{2n-1} \cos 2\pi v + q^{4n-2}} \; , \tag{1}$$

$$\operatorname{cn}(2Kv) = 2q^{\frac{1}{4}} \left(\frac{k'}{k}\right)^{\frac{1}{2}} \cos \pi v \prod_{n=1}^{\infty} \frac{1 + 2q^{2n} \cos 2\pi v + q^{4n}}{1 - 2q^{2n-1} \cos 2\pi v + q^{4n-2}} \; , \tag{2}$$

$$\operatorname{dn}(2Kv) = k'^{\frac{1}{2}} \prod_{n=1}^{\infty} \frac{1 + 2q^{2n-1} \cos 2\pi v + q^{4n-2}}{1 - 2q^{2n-1} \cos 2\pi v + q^{4n-2}} \; . \tag{3}$$

Taking the logarithm of (1), we obtain

$$\ln \operatorname{sn}(2Kv) = \ln(2q^{\frac{1}{4}}) - \frac{1}{2} \ln k + \ln \sin \pi v$$
$$+ \sum_{n=1}^{\infty} \{\ln(1 - q^{2n} e^{2\pi vi})(1 - q^{2n} e^{-2\pi vi})$$
$$- \ln(1 - q^{2n-1} e^{2\pi vi})(1 - q^{2n-1} e^{-2\pi vi})\} \; .$$

Expanding the logarithms in powers of $e^{\pm 2\pi vi}$, we arrive at

$$\ln \operatorname{sn}(2Kv) = \frac{1}{4} \ln \frac{16q}{k^2} + \ln \sin \pi v + \sum_{m=1}^{\infty} \frac{2q^m \cos(2m\pi v)}{m(1 + q^m)} \; . \tag{4}$$

Similarly, from (2) and (3), we find

$$\ln \ \text{cn}(2Kv) = \frac{1}{4}\ln\frac{16qk'^2}{k^2} + \ln\cos\pi v + \sum_{m=1}^{\infty}\frac{2q^m\cos(2m\pi v)}{m[1+(-q)^m]} \ , \tag{5}$$

$$\ln \ \text{dn}(2Kv) = \frac{1}{2}\ln k' + \sum_{m=0}^{\infty}\frac{4q^{2m+1}\cos(2m+1)2\pi v}{(2m+1)(1-q^{4m+2})} \ . \tag{6}$$

These are the Fourier expansions; the condition for convergence is

$$|\text{Im}(v)| < \frac{1}{2}\text{Im}(\tau) \ .$$

sn u itself can also be expanded in a Fourier series. Since sn u is an odd function, we may assume that

$$\text{sn}(2Kv) = \sum_{m=1}^{\infty} b_m\sin(m\pi v) \ .$$

Here, corresponding to the period $4K$ of sn u, the period with respect to v is 2, and the coefficient b_m is given by the following integral:

$$b_m = \int_{-1}^{1}\text{sn}(2Kv)\sin(m\pi v)dv \ .$$

This integral can be evaluated by the same method depicted in Sec. 9.12, i.e., by evaluating the contour integral $\int \text{sn}(2Kv)e^{m\pi vi}dv$ along the sides of the parallelogram with vertices at $-1, +1, 2+\tau, \tau$, within which $\text{sn}(2Kv)$ has two poles, $v = \tau/2$ and $1+\tau/2$. From Eq. (6) of Sec. 10.6, we know that the residues at these two poles of the integrand are respectively

$$\frac{e^{\frac{m\pi\tau i}{2}}}{2Kk} = \frac{q^{\frac{m}{2}}}{2Kk} \ , \qquad -\frac{e^{m\pi\left(1+\frac{\tau}{2}\right)i}}{2Kk} = -\frac{(-)^m q^{\frac{m}{2}}}{2Kk} \ .$$

Hence,

$$[1-(-)^m]\frac{q^{\frac{m}{2}}}{2Kk} = \frac{1}{2\pi i}\left\{\int_{-1}^{1} - \int_{\tau}^{2+\tau}\right\}\text{sn}(2Kv)e^{m\pi vi}dv$$

$$= \frac{1}{2\pi i}\int_{-1}^{1}\left\{\text{sn}(2Kv)e^{m\pi vi} - \text{sn}\ 2K(v+1+\tau)\cdot e^{m\pi(v+1+\tau)i}\right\}dv$$

$$= \frac{1}{2\pi i}\int_{-1}^{1}\{1+(-)^m q^m\}\ \text{sn}(2Kv)e^{m\pi vi}dv$$

$$= \frac{1}{2\pi}\{1+(-)^m q^m\}\int_{-1}^{1}\text{sn}(2Kv)\sin(m\pi v)dv \ ,$$

and so

$$b_m = \frac{\pi[1 - (-)^m]q^{\frac{m}{2}}}{Kk[1 + (-q)^m]} .$$

Only for $m = 2n+1$, an odd integer, is b_m non-vanishing. Thus, the expansion of sn u is

$$\mathrm{sn}(2Kv) = \frac{2\pi}{Kk} \sum_{n=0}^{\infty} \frac{q^{n+\frac{1}{2}} \sin(2n + 1)\pi v}{1 - q^{2n+1}} . \tag{7}$$

Similarly, we find

$$\mathrm{cn}(2Kv) = \frac{2\pi}{Kk} \sum_{n=0}^{\infty} \frac{q^{n+\frac{1}{2}} \cos(2n + 1)\pi v}{1 + q^{2n+1}} , \tag{8}$$

$$\mathrm{dn}(2Kv) = \frac{\pi}{2K} + \frac{2\pi}{K} \sum_{n=1}^{\infty} \frac{q^n \cos 2n\pi v}{1 + q^{2n}} . \tag{9}$$

From (9), by integrating, we have

$$\mathrm{am}(2Kv) = \int_0^{2Kv} \mathrm{dn}\, u \, du = \pi v + \sum_{n=1}^{\infty} \frac{2q^n \sin 2n\pi v}{n(1 + q^{2n})} . \tag{10}$$

The Fourier expansion for the reciprocal of sn u(ns u) can be derived from Eq. (7) of Sec. 10.5 as follows:

$$\mathrm{ns}(2Kv) = k\, \mathrm{sn}\left[2K\left(v + \frac{\tau}{2}\right)\right]$$

$$= \frac{2\pi}{K} \sum_{n=0}^{\infty} \frac{q^{n+\frac{1}{2}} \sin(2n + 1)\pi\left(v + \frac{\tau}{2}\right)}{1 - q^{2n+1}}$$

$$= \frac{\pi}{Ki} \sum_{n=0}^{\infty} \frac{q^{n+\frac{1}{2}}}{1 - q^{2n+1}} \left\{ q^{n+\frac{1}{2}} e^{(2n+1)\pi vi} - q^{-n-\frac{1}{2}} e^{-(2n+1)\pi vi} \right\}$$

$$= \frac{\pi}{Ki} \sum_{n=0}^{\infty} \frac{1}{1 - q^{2n+1}} \left\{ q^{2n+1} \left[e^{(2n+1)\pi vi} - e^{-(2n+1)\pi vi} \right] \right.$$

$$\left. - (1 - q^{2n+1})e^{-(2n+1)\pi vi} \right\} .$$

The condition for the convergence of this series is $-\mathrm{Im}(\tau) < \mathrm{Im}(v) < 0$. Under this condition, we have

$$\sum_{n=0}^{\infty} e^{-(2n+1)\pi vi} = \frac{e^{-\pi vi}}{1 - e^{-2\pi vi}} = \frac{1}{2i \sin \pi v} .$$

Hence,

$$\text{ns}(2Kv) = \frac{\pi}{2K \sin \pi v} + \frac{2\pi}{K} \sum_{n=0}^{\infty} \frac{q^{2n+1} \sin(2n+1)\pi v}{1 - q^{2n+1}} . \tag{11}$$

Since the series on the r.h.s. converges when $|\text{Im}(v)| < \text{Im}(\tau)$, the superfluous condition $\text{Im}(v) < 0$ stated in the proof can be retracted.

In a similar way, we find

$$\text{nc}(2Kv) = \frac{\pi}{2Kk' \cos \pi v} - \frac{2\pi}{Kk'} \sum_{n=0}^{\infty} \frac{(-)^n q^{2n+1} \cos(2n+1)\pi v}{1 + q^{2n+1}} , \tag{12}$$

$$\text{nd}(2Kv) = \frac{\pi}{2Kk'} + \frac{2\pi}{Kk'} \sum_{n=1}^{\infty} \frac{(-)^n q^n \cos 2n\pi v}{1 + q^{2n}} . \tag{13}$$

Exercise 10

1. Using the symbols in Eq. (5) of Sec. 10.4, show that

$$s = \frac{\sqrt{1+S} - \sqrt{1-S}}{\sqrt{1+kS} + \sqrt{1-kS}} ,$$

$$s^2 = \frac{1-C}{1+D} = \frac{1-D}{k^2(1+C)}$$

$$= \frac{D - k^2C - k'^2}{k^2(D-C)} = \frac{D-C}{k'^2 + D - k^2C} ,$$

$$c^2 = \frac{D+C}{1+D} = \frac{D + k^2C - k'^2}{k^2(1+C)}$$

$$= \frac{k'^2(1-D)}{k^2(D-C)} = \frac{k'^2(1+C)}{k'^2 + D - k^2C} ,$$

$$d^2 = \frac{k'^2 + D + k^2C}{1+D} = \frac{D+C}{1+C}$$

$$= \frac{k'^2(1-C)}{D-C} = \frac{k'^2(1+D)}{k'^2 + D - k^2C} .$$

2. Show that

$$\text{sn}\frac{K}{2} = (1+k')^{-\frac{1}{2}} , \quad \text{cn}\frac{K}{2} = \sqrt{\frac{k'}{1+k'}} , \quad \text{dn}\frac{K}{2} = k'^{\frac{1}{2}} .$$

3. Derive the addition formula for $Z(u)$ given by Eq. (10) of Sec. 10.11, from the result of Ex. 20 at the end of Chap. 9.

4. Show that

$$\text{sn}\frac{iK'}{2} = ik^{-\frac{1}{2}}, \quad \text{cn}\frac{iK'}{2} = \sqrt{1+\frac{1}{k}}, \quad \text{dn}\frac{iK'}{2} = \sqrt{1+k},$$

$$\text{sn}\frac{K+iK'}{2} = \frac{\sqrt{1+k}+i\sqrt{1-k}}{\sqrt{2k}},$$

$$\text{cn}\frac{K+iK'}{2} = \frac{(1-i)\sqrt{k'}}{\sqrt{2k}},$$

$$\text{dn}\frac{K+iK'}{2} = \frac{\sqrt{k'}(\sqrt{1+k'}-i\sqrt{1-k'})}{\sqrt{2}}.$$

5. Show that

$$\int \text{sn } u \, du = \frac{1}{2k}\ln\frac{1-k\,\text{cd } u}{1+k\,\text{cd } u},$$

$$\int \text{cn } u \, du = k^{-1}\arctan(k\,\text{sd } u),$$

$$\int \text{dn } u \, du = \text{am } u,$$

$$\int \text{sc } u \, du = \frac{1}{2k'}\ln\frac{\text{dn } u+k'}{\text{dn } u-k'},$$

$$\int \text{ds } u \, du = \frac{1}{2}\ln\frac{1-\text{cn } u}{1+\text{cn } u},$$

$$\int \text{dc } u \, du = \frac{1}{2}\ln\frac{1+\text{sn } u}{1-\text{sn } u}.$$

By changing u to $u + K$, obtain another six formulae.

6. Show that

$$\text{sn}\left(u+\frac{K}{2}\right) = \frac{1}{\sqrt{1+k'}}\frac{k'\,\text{sn } u+\text{cn } u\,\text{dn } u}{1-(1-k')\text{sn}^2 u},$$

$$\text{sn}\left(u+\frac{iK'}{2}\right) = \frac{1}{\sqrt{k}}\frac{(1+k)\,\text{sn } u+i\,\text{cn } u\,\text{dn } u}{1+k\,\text{sn}^2 u}.$$

7. Show that

$$\int_0^K \cos\frac{\pi u}{K}\ln\text{sn } u \, du = \frac{K}{2}\text{th}\frac{i\pi\tau}{2}.$$

8. From Eq. (15) of Sec. 9.8, obtain

$$\wp(z) = e_3 + \frac{e_1-e_3}{\text{sn}^2(z\sqrt{e_1-e_3})},$$

with which show that

$$K = \omega\sqrt{e_1 - e_3} \ , \quad iK' = \omega'\sqrt{e_1 - e_3} \ ,$$

$$\text{sn } u = \sqrt{e_1 - e_3} \ \frac{\sigma(u/\sqrt{e_1 - e_3})}{\sigma_3(u/\sqrt{e_1 - e_3})} \ ,$$

$$\text{cn } u = \frac{\sigma_1(u/\sqrt{e_1 - e_3})}{\sigma_3(u/\sqrt{e_1 - e_3})} \ , \quad \text{dn } u = \frac{\sigma_2(u/\sqrt{e_1 - e_3})}{\sigma_3(u/\sqrt{e_1 - e_3})} \ .$$

9. Show that when $K = \omega$,

$$e_1 - e_3 = 1 \ , \quad e_1 = \frac{2 - k^2}{3} \ , \quad e_2 = -\frac{1 - 2k^2}{3} \ , \quad e_3 = -\frac{1 + k^2}{3} \ .$$

Hence, derive the formulae (1), (2), (3) of Sec. 10.6 from the result of Ex. 4 at the end of Chap. 8.

10. By Eq. (8) of Sec. 10.13 and Eq. (4) of Sec. 9.6, show that

$$E = \frac{2\omega^2 e_1}{\pi\vartheta_3^2} - \frac{\vartheta_1'''}{6\pi\vartheta_3^2\vartheta_1'} = \frac{\pi(\vartheta_3^4 + \vartheta_4^4)}{6\vartheta_3^2} - \frac{\vartheta_1'''}{6\pi\vartheta_3^2\vartheta_1'}$$

$$= \frac{K}{3}(2 - k^2) - \frac{\vartheta_1'''}{12K\vartheta_1'} \ .$$

11. Show that

$$EK = \eta\omega + \omega^2 e_1 = \eta\omega + \frac{\pi^2}{12}(\vartheta_3^4 + \vartheta_4^4) \ ,$$

$$\frac{E}{K} = \frac{2 - k^2}{3} - \frac{i\eta\omega'}{KK'} \ , \quad \frac{E'}{K'} = \frac{2 - k'^2}{3} + \frac{i\eta'\omega}{KK'} \ .$$

12. Show that

$$Z(u) = \frac{\omega}{K}\left\{\varsigma\left(\frac{\omega u}{K} + \omega'\right) - \eta' - \frac{\eta}{K}u\right\} \ ,$$

$$E(u) = \frac{\omega}{K}\left\{\varsigma\left(\frac{\omega u}{K} + \omega'\right) - \eta' + \frac{\omega e_1}{K}u\right\} \ .$$

13. **Show that**

$$\text{sc}(2Kv) = \frac{\pi}{2Kk'} \tan \pi v + \frac{2\pi}{Kk'} \sum_{n=1}^{\infty} \frac{(-)^n q^{2n} \sin 2n\pi v}{1 + q^{2n}} \,,$$

$$\text{sd}(2Kv) = \frac{2\pi}{Kkk'} \sum_{n=0}^{\infty} \frac{(-)^n q^{n+\frac{1}{2}} \sin(2n+1)\pi v}{1 + q^{2n+1}} \,,$$

$$\text{cs}(2Kv) = \frac{\pi}{2K} \cot \pi v - \frac{2\pi}{K} \sum_{n=1}^{\infty} \frac{q^{2n} \sin 2n\pi v}{1 + q^{2n}} \,,$$

$$\text{cd}(2Kv) = \frac{2\pi}{Kk} \sum_{n=0}^{\infty} \frac{(-)^n q^{n+\frac{1}{2}} \cos(2n+1)\pi v}{1 - q^{2n+1}} \,,$$

$$\text{ds}(2Kv) = \frac{\pi}{2K} \csc \pi v - \frac{2\pi}{K} \sum_{n=0}^{\infty} \frac{q^{2n+1} \sin(2n+1)\pi v}{1 + q^{2n+1}} \,,$$

$$\text{dc}(2Kv) = \frac{\pi}{2K} \sec \pi v + \frac{2\pi}{K} \sum_{n=0}^{\infty} \frac{(-)^n q^{2n+1} \cos(2n+1)\pi v}{1 - q^{2n+1}} \,.$$

14. **Show that**

$$E(u + K) = E(u) - k^2 \text{ sn } u \text{ cd } u + E \,,$$
$$E(u + iK') = E(u) + \text{cn } u \text{ ds } u + i(K' - E') \,,$$
$$E(u + K + iK') = E(u) - \text{sn } u \text{ dc } u + E + i(K' - E') \,.$$

15. **From Eq. (8) of Sec. 10.13, show that after Landen's type of transformation** E **changes to**

$$E_1 = \frac{E + k'K}{1 + k'} \,.$$

16. **Prove the following relations among the elliptic integrals of the third kind** [Sec. 10.10]:

$$\Pi(iu, c, k) = \frac{ic}{c+1} \prod(u, -c-1, k') + \frac{iu}{c+1} \,,$$
$$\Pi(u, c) + \Pi(v, c) - \Pi(u+v, c)$$
$$= \frac{\text{sn } b}{2\text{cn } b \text{ dn } b} \ln \frac{1 - k^2 \text{ sn } b \text{ sn } v \text{ sn}(u+v-b)}{1 + k^2 \text{ sn } b \text{ sn } u \text{ sn } v \text{ sn}(u+v+b)} \,,$$

where $b = a - K'i$. [Hint: Apply the result of Ex. 6 in Chap. 9, namely, the relation $[1] + [4] = [1]' + [4]'$, taking $x : y : z : w = u : v : \pm b : u + v \pm b$.]

17. By using case 4 of Sec. 10.8 show that

$$\int_0^1 \frac{ds}{\sqrt{1-s^4}} = \frac{1}{\sqrt{2}} \int_0^1 \frac{d\tau}{\sqrt{(1-\tau^2)(1-k^2\tau^2)}} ,$$

$$s^2 = 1 - \tau^2 , \quad k^2 = \frac{1}{2} ,$$

and also

$$q = e^{-\pi} , \quad K = \frac{1}{4\sqrt{\pi}} \left[\Gamma\left(\frac{1}{4}\right) \right]^2 ,$$

$$E = \frac{1}{8\sqrt{\pi}} \left[\Gamma\left(\frac{1}{4}\right) \right]^2 + \pi^{\frac{3}{2}} \left[\Gamma\left(\frac{1}{4}\right) \right]^{-2} .$$

18. Consider three special values of k: (1) $k = \sqrt{2} - 1$, show that $K' = K\sqrt{2}$; (2) $k = \sin\frac{\pi}{12}$, show that $K' = K\sqrt{3}$; (3) $k = \tan^2\frac{\pi}{8}$, show that $K' = 2K$. [(1) corresponds to $k_1 = k' = \sqrt{2} - 1$ in Landen's transformation; (2) corresponds to $P(x) = x^3 - 1$; with Eq. (16) of Sec. 9.8, we have $\alpha = -\frac{1}{2}, \beta = \frac{\sqrt{3}}{2} = \cos\frac{\pi}{6}$; (3) corresponds to $k = \frac{1}{\sqrt{2}}, k_1 = (\sqrt{2}-1)/(\sqrt{2}+1) = \tan^2\frac{\pi}{8}$.]

19. For $k = \sin\frac{\pi}{12}$, show that

$$3^{-\frac{1}{4}}2K = 3^{-\frac{3}{4}}2K'$$

$$= 3^{-\frac{1}{2}} \int_0^1 t^{-\frac{2}{3}}(1-t)^{-\frac{1}{2}}dt = \frac{\sqrt{\pi}}{3}\Gamma\left(\frac{1}{6}\right)/\Gamma\left(\frac{2}{3}\right)$$

$$\frac{\pi}{4\sqrt{3}} = K\left\{E - \frac{\sqrt{3}+1}{2\sqrt{3}}K\right\} ,$$

$$\frac{\pi\sqrt{3}}{4} = K'\left\{E' - \frac{\sqrt{3}-1}{2\sqrt{3}}K'\right\} .$$

20. *Another derivation for the addition formula.* Consider the three functions: sn u sn$(u + v)$, cn u cn$(u + v)$, dn u dn$(u + v)$. With respect to u they are all of periods $2K$ and $2K'i$; all have the two simple poles, $K'i$ and $K'i - v$. Therefore, a pole can be cancelled by a linear combination of any two functions; and, according to Theorem 2 of Sec. 8.3, the second pole must have been cancelled also, so that the linear combination is equal to a constant. In this way, we can have two linear combinations, equal to constants, each containing two constants which may be determined by

setting $u = 0$ in the linear combinations themselves and in the derivatives (with respect to u) of them. The results are:

$$\text{cn } u \text{ cn}(u + v) + \text{dn } v \text{ sn } u \text{ sn}(u + v) = \text{cn } v \ ,$$
$$\text{dn } u \text{ dn}(u + v) + k^2 \text{ cn } v \text{ sn } u \text{ sn}(u + v) = \text{dn } v \ .$$

Changing u to $-u$ and v to $u + v$, we obtain

$$\text{cn } u \text{ cn } v - \text{dn}(u + v) \text{ sn } u \text{ sn } v = \text{cn}(u + v) \ ,$$
$$\text{dn } u \text{ dn } v - k^2 \text{cn}(u + v) \text{ sn } u \text{ sn } v = \text{dn}(u + v) \ .$$

From these equations, we solve for $\text{cn}(u + v)$ and $\text{dn}(u + v)$. Then, by substituting them in the previous equations, we have the addition formula for $\text{sn}(u + v)$; and so on.

Chapter 11

LAMÉ FUNCTIONS

11.1. Ellipsoidal Coordinates

Lamé functions are given rise from solving Laplace equation in ellipsoidal coordinates. In Cartesian coordinates (x, y, z), the equation of an ellipsoid is

$$\frac{x^2}{a^2} + \frac{y^2}{b^2} + \frac{z^2}{c^2} = 1 \,, \tag{1}$$

where a, b, c are the semi-axes of the ellipsoid. We shall choose the order of the coordinates so that $a > b > c$, assuming for the time being that none of them are equal.

The quadrics confocal to (1) are

$$\frac{x^2}{a^2 + \theta} + \frac{y^2}{b^2 + \theta} + \frac{z^2}{c^2 + \theta} = 1 \,. \tag{2}$$

When $\theta > -c^2$, (2) represents an ellipsoid; when $-b^2 < \theta < -c^2$, it represents a single-sheet hyperboloid; when $-a^2 < \theta < -b^2$, it represents a double-sheet hyperboloid. Corresponding to three different values of θ, there are three curved surfaces whose intersecting point determines uniquely a group of values (x^2, y^2, z^2) representing eight points of the space. Conversely, passing through any point (x, y, z), there correspond three curved surfaces (2) with three values of θ. To investigate these values of θ, let

$$\varphi(\theta) = (a^2 + \theta)(b^2 + \theta)(c^2 + \theta) , \tag{3}$$

then (2) can be written as

$$
\begin{aligned}
f(\theta) &\equiv \varphi(\theta) \left\{ 1 - \frac{x^2}{a^2 + \theta} - \frac{y^2}{b^2 + \theta} - \frac{z^2}{c^2 + \theta} \right\} \\
&\equiv (a^2 + \theta)(b^2 + \theta)(c^2 + \theta) \\
&\quad - x^2(b^2 + \theta)(c^2 + \theta) - y^2(c^2 + \theta)(a^2 + \theta) - z^2(a^2 + \theta)(b^2 + \theta) \\
&= 0 .
\end{aligned}
\tag{4}
$$

$f(\theta)$ is a cubic polynomial of θ, having three roots λ, μ, ν:

$$f(\theta) = (\theta - \lambda)(\theta - \mu)(\theta - \nu) . \tag{5}$$

When θ takes in turn the values $-\infty, -a^2, -b^2, -c^2, +\infty$, $f(\theta)$, by (4), will assume the corresponding values $-\infty, -x^2(a^2 - b^2)(a^2 - c^2), y^2(a^2 - b^2)(b^2 - c^2), -z^2(a^2 - c^2)(b^2 - c^2), +\infty$. Thus, we see that the three roots of $f(\theta)$ are situating separately in the intervals $(-a^2, -b^2), (-b^2, -c^2), (-c^2, +\infty)$, and hence are three nonequal real roots. Arranging them in the order $\lambda > \mu > \nu$, we have

$$\lambda > -c^2 > \mu > -b^2 > \nu > -a^2 ,$$

which shows that the three curved surfaces (2) passing through a point in the space are: an ellipsoid corresponding to $\theta = \lambda$; a single-sheet hyperboloid corresponding to $\theta = \mu$; and a double-sheet hyperboloid corresponding to $\theta = \nu$.

We may employ λ, μ, ν as the coordinates, the ellipsoidal coordinates; x, y, z can be expressed in terms of λ, μ, ν. In (4) and (5), putting in turn $\theta = -a^2, -b^2, -c^2$, we obtain

$$
\begin{aligned}
x^2 &= \frac{(a^2 + \lambda)(a^2 + \mu)(a^2 + \nu)}{(a^2 - b^2)(a^2 - c^2)} , \\
y^2 &= \frac{(b^2 + \lambda)(b^2 + \mu)(b^2 + \nu)}{(b^2 - c^2)(b^2 - a^2)} , \\
z^2 &= \frac{(c^2 + \lambda)(c^2 + \mu)(c^2 + \nu)}{(c^2 - a^2)(c^2 - b^2)} .
\end{aligned}
\tag{6}
$$

We shall now show that the three curved surfaces passing through a point are orthogonal to each other, and hence λ, μ, ν are orthogonal curvilinear coordinates. The direction cosines (l, m, n) of the normal at a point (x, y, z) on the surface $F(x, y, z) = 0$ are proportional to $\left(\frac{\partial F}{\partial x}, \frac{\partial F}{\partial y}, \frac{\partial F}{\partial z}\right)$. Therefore, the direction cosines $(l_\lambda, m_\lambda, n_\lambda)$ of the normal to the surface $\theta = \lambda$ are in the proportion:

$$l_\lambda : m_\lambda : n_\lambda = \frac{x}{a^2 + \lambda} : \frac{y}{b^2 + \lambda} : \frac{z}{c^2 + \lambda} . \tag{7}$$

Subtracting Eq. (2) of the surfaces $\theta = \lambda$ and $\theta = \mu$ to eliminate the factor $\lambda - \mu$, we obtain

$$\frac{x^2}{(a^2 + \lambda)(a^2 + \mu)} + \frac{y^2}{(b^2 + \lambda)(b^2 + \mu)} + \frac{z^2}{(c^2 + \lambda)(c^2 + \mu)} = 0 , \tag{8}$$

which leads to $l_\lambda l_\mu + m_\lambda m_\mu + n_\lambda n_\mu = 0$, i.e., the surfaces $\theta = \lambda$ and $\theta = \mu$ are orthogonal to each other. Similarly, we can show that the surfaces $\theta = \lambda$ and $\theta = \nu$ are orthogonal, and also $\theta = \mu$ and $\theta = \nu$.

Now, let us find the relation between dx and $d\lambda$. From (6), taking the logarithmic derivative, we have

$$2\frac{dx}{x} = \frac{d\lambda}{a^2 + \lambda} + \frac{d\mu}{a^2 + \mu} + \frac{d\nu}{a^2 + \nu} . \tag{9}$$

Replacing a in turn by b and c, we obtain similar formulae for dy and dz. With these formulae, we can write the orthogonality condition (8) as

$$\frac{\partial x}{\partial \lambda}\frac{\partial x}{\partial \mu} + \frac{\partial y}{\partial \lambda}\frac{\partial y}{\partial \mu} + \frac{\partial z}{\partial \lambda}\frac{\partial z}{\partial \mu} = 0 . \tag{10}$$

The square of the line element is

$$ds^2 = dx^2 + dy^2 + dz^2 = H_1^2(d\lambda)^2 + H_2^2(d\mu)^2 + H_3^2(d\nu)^2 , \tag{11}$$

in which no cross terms $d\lambda d\mu, d\mu d\nu, d\nu d\lambda$ occur in virtue of the orthogonality condition (10). For the coefficient of $(d\lambda)^2$, we have

$$H_1^2 = \left(\frac{\partial x}{\partial \lambda}\right)^2 + \left(\frac{\partial y}{\partial \lambda}\right)^2 + \left(\frac{\partial z}{\partial \lambda}\right)^2$$

$$= \frac{1}{4}\left\{\frac{x^2}{(a^2 + \lambda)^2} + \frac{y^2}{(b^2 + \lambda)^2} + \frac{z^2}{(c^2 + \lambda)^2}\right\} .$$

Applying (6), we have

$$4H_1^2 = \frac{(a^2 + \mu)(a^2 + \nu)}{(a^2 - b^2)(a^2 - c^2)(a^2 + \lambda)} + \frac{(b^2 + \mu)(b^2 + \nu)}{(b^2 - c^2)(b^2 - a^2)(b^2 + \lambda)}$$
$$+ \frac{(c^2 + \mu)(c^2 + \nu)}{(c^2 - a^2)(c^2 - b^2)(c^2 + \lambda)}$$
$$= \frac{(\lambda - \mu)(\lambda - \nu)}{(a^2 + \lambda)(b^2 + \lambda)(c^2 + \lambda)} . \tag{12}$$

The last step can be justified by partial fraction. The coefficients H_2 and H_3 may be obtained from H_1 by cyclically permuting λ, μ, ν. By multiplying and extracting square root, we have

$$H = H_1 H_2 H_3 = \frac{(\lambda - \mu)(\lambda - \nu)(\mu - \nu)}{8\sqrt{-\varphi(\lambda)\varphi(\mu)\varphi(\nu)}} , \tag{13}$$

where the function φ is determined by (3).

In ellipsoidal coordinates, the expression of Laplace equation [see App. III] is

$$\nabla^2 \Psi \equiv \frac{4}{(\lambda - \mu)(\lambda - \nu)(\mu - \nu)}$$
$$\times \left\{ (\mu - \nu)\sqrt{\psi(\lambda)} \frac{\partial}{\partial \lambda} \left(\sqrt{\varphi(\lambda)} \frac{\partial \Psi}{\partial \lambda} \right) \right.$$
$$+ (\lambda - \nu)\sqrt{-\varphi(\mu)} \frac{\partial}{\partial \mu} \left(\sqrt{-\varphi(\mu)} \frac{\partial \Psi}{\partial \mu} \right)$$
$$\left. + (\lambda - \mu)\sqrt{\varphi(\nu)} \frac{\partial}{\partial \nu} \left(\sqrt{\varphi(\nu)} \frac{\partial \Psi}{\partial \nu} \right) \right\} = 0 . \tag{14}$$

11.2. Representing the Coordinates with Elliptic Functions

The transformations from Cartesian coordinates to ellipsoidal coordinates in (6) of the last section are not single valued. We may settle the correspondence relations single valued by employing elliptic functions. Express the ellipsoidal coordinates in terms of elliptic functions:

$$\wp(u) = -\lambda - \frac{1}{3}(a^2 + b^2 + c^2) ,$$
$$\wp(v) = -\mu - \frac{1}{3}(a^2 + b^2 + c^2) , \tag{1}$$
$$\wp(w) = -\nu - \frac{1}{3}(a^2 + b^2 + c^2) .$$

The invariants g_2 and g_3 are determined by the following equation:

$$4\wp^3(u) - g_2\wp(u) - g_3 = -4(a^2 + \lambda)(b^2 + \lambda)(c^2 + \lambda) . \tag{2}$$

Then the three roots will be [cf. Eq. (9) of Sec. 8.5]

$$e_1 = a^2 - \frac{1}{3}(a^2 + b^2 + c^2), \quad e_2 = b^2 - \frac{1}{3}(a^2 + b^2 + c^2) ,$$

$$e_3 = c^2 - \frac{1}{3}(a^2 + b^2 + c^2) , \tag{3}$$

which satisfy the condition $e_1 > e_2 > e_3$.

Substituting (1) in (6) of the last section gives

$$x^2 = \frac{-\{\wp(u) - e_1\}\{\wp(v) - e_1\}\{\wp(w) - e_1\}}{(e_1 - e_2)(e_1 - e_3)} .$$

Applying Eqs. (17) and (18) of Sec. 8.10, we have

$$x^2 = -\left[\frac{e^{-\eta_1\omega_1}\sigma_1(u)\sigma_1(v)\sigma_1(w)}{\sigma(u)\sigma(v)\sigma(w)}\sigma^2(\omega_1)\right]^2 .$$

Taking the square root, we obtain

$$x = ie^{-\eta_1\omega_1}\sigma^2(\omega_1)\frac{\sigma_1(u)\sigma_1(v)\sigma_1(w)}{\sigma(u)\sigma(v)\sigma(w)} . \tag{4}$$

Similarly, we have

$$y = ie^{-\eta_2\omega_2}\sigma^2(\omega_2)\frac{\sigma_2(u)\sigma_2(v)\sigma_2(w)}{\sigma(u)\sigma(v)\sigma(w)} ,$$

$$z = ie^{-\eta_3\omega_3}\sigma^2(\omega_3)\frac{\sigma_3(u)\sigma_3(v)\sigma_3(w)}{\sigma(u)\sigma(v)\sigma(w)} . \tag{5}$$

These formulae give the cartesian coordinates in terms of single valued functions of u, v, w.

With Eqs. (5) and (6) of Sec. 9.2, we can also express the coordinates in terms of Theta functions:

$$x = \frac{i\vartheta_1'}{2\omega_1\vartheta_2}\frac{\vartheta_2\left(\frac{u}{2\omega_1}\right)\vartheta_2\left(\frac{v}{2\omega_1}\right)\vartheta_2\left(\frac{w}{2\omega_1}\right)}{\vartheta_1\left(\frac{u}{2\omega_1}\right)\vartheta_1\left(\frac{v}{2\omega_1}\right)\vartheta_1\left(\frac{w}{2\omega_1}\right)} ,$$

$$y = \frac{i\vartheta_1'}{2\omega_1\vartheta_3}\frac{\vartheta_3\left(\frac{u}{2\omega_1}\right)\vartheta_3\left(\frac{v}{2\omega_1}\right)\vartheta_3\left(\frac{w}{2\omega_1}\right)}{\vartheta_1\left(\frac{u}{2\omega_1}\right)\vartheta_1\left(\frac{v}{2\omega_1}\right)\vartheta_1\left(\frac{w}{2\omega_1}\right)} , \tag{6}$$

$$z = \frac{i\vartheta_1'}{2\omega_1\vartheta_4}\frac{\vartheta_4\left(\frac{u}{2\omega_1}\right)\vartheta_4\left(\frac{v}{2\omega_1}\right)\vartheta_4\left(\frac{w}{2\omega_1}\right)}{\vartheta_1\left(\frac{u}{2\omega_1}\right)\vartheta_1\left(\frac{v}{2\omega_1}\right)\vartheta_1\left(\frac{w}{2\omega_1}\right)} .$$

The coordinates can also be represented by Jacobian elliptic functions, which are of advantage in numerical computations. Let

$$a^2 + \lambda = (a^2 - b^2)t^2 , \quad k^2 = \frac{a^2 - b^2}{a^2 - c^2} . \tag{7}$$

We obtain

$$\varphi(\lambda) = (a^2 - b^2)^2(a^2 - c^2)t^2(1 - t^2)(1 - k^2t^2) , \tag{8}$$

$$K = \int_0^1 \frac{dt}{\sqrt{(1 - t^2)(1 - k^2t^2)}} = \frac{\sqrt{a^2 - c^2}}{2} \int_{-a^2}^{-b^2} \frac{d\lambda}{\sqrt{\varphi(\lambda)}} . \tag{9}$$

Putting $t = \operatorname{sn} \alpha$, and letting $\operatorname{sn} \beta$, $\operatorname{sn} \gamma$ correspond to μ and ν respectively, by substituting (7) in (6) of the last section, taking square root, the result is

$$x = k^2\sqrt{a^2 - c^2} \operatorname{sn} \alpha \operatorname{sn} \beta \operatorname{sn} \gamma ,$$

$$y = -\frac{k^2}{k'}\sqrt{a^2 - c^2} \operatorname{cn} \alpha \operatorname{cn} \beta \operatorname{cn} \gamma , \tag{10}$$

$$z = \frac{i}{k'}\sqrt{a^2 - c^2} \operatorname{dn} \alpha \operatorname{dn} \beta \operatorname{dn} \gamma .$$

From (7) and Eqs. (5) and (6) of Sec. 10.3, we see that, when λ varies from $-c^2$ to $+\infty$, $\pm\alpha$ vary from $K + K'i$ to $K'i$, or α varies from $-K + K'i$ to $K'i$, then from $K'i$ to $K + K'i$; $\alpha = \pm K + K'i$ correspond to a repeat of the degenerate ellipsoid in the plane $z = 0$ containing the foci. When μ varies from $-b^2$ to $-c^2$, $\pm\beta$ vary from K to $K + K'i$: $\beta = \pm K$ correspond to the region inside the degenerate hyperboloid, including the origin, in the plane $y = 0$, whereas $\beta = \pm K + K'i$ correspond to the region exterior to the degenerate ellipsoid in the plane $z = 0$ containing the foci. When ν varies from $-a^2$ to $-b^2$, $\pm\gamma$ vary from 0 to K, or γ from $-K$ through 0 to K; $\gamma = 0$ corresponds to the coincidence of the degenerate double-sheet hyperboloid in the plane $x = 0$; $\gamma = \pm K$ correspond to the region, not including the origin, of the degenerate hyperboloid in the plane $y = 0$.

11.3. Lamé Equation

The Laplace equation in ellipsoidal coordinates [Eq. (14) of Sec. 11.1] can be solved by the separation of variables. By assuming $\Psi = \Lambda(\lambda)M(\mu)N(\nu)$, we obtain

$$\frac{\mu - \nu}{\Lambda}\sqrt{\varphi(\lambda)}\frac{d}{d\lambda}\left(\sqrt{\varphi(\lambda)}\frac{d\Lambda}{d\lambda}\right) + \frac{\lambda - \nu}{M}\sqrt{-\varphi(\mu)}\frac{d}{d\mu}\left(\sqrt{-\varphi(\mu)}\frac{dM}{d\mu}\right)$$

$$+ \frac{\lambda - \mu}{N}\sqrt{\varphi(\nu)}\frac{d}{d\nu}\left(\sqrt{\varphi(\nu)}\frac{dN}{d\nu}\right) = 0 .$$

Comparing with the identity

$$(\mu - \nu)(K\lambda + C) + (\nu - \lambda)(K\mu + C) + (\lambda - \mu)(K\nu + C) \equiv 0 ,$$

where K and C are constants, we have

$$4\sqrt{\varphi(\lambda)}\frac{d}{d\lambda}\left(\sqrt{\varphi(\lambda)}\frac{d\Lambda}{d\lambda}\right) = (K\lambda + C)\Lambda . \tag{1}$$

This is the *Lamé equation*. The constant K is often written as $n(n+1)$:

$$4\sqrt{\varphi(\lambda)}\frac{d}{d\lambda}\left(\sqrt{\varphi(\lambda)}\frac{d\Lambda}{d\lambda}\right) = \{n(n+1)\lambda + C\}\,\Lambda . \tag{2}$$

When n is a positive integer, the solutions of this differential equation are called *Lamé functions*.

From $M(\mu)$ and $N(\nu)$ we have the same equations as (1):

$$-4\sqrt{-\varphi(\mu)}\frac{d}{d\mu}\left(\sqrt{-\varphi(\mu)}\frac{dM}{d\mu}\right) = \{n(n+1)\mu + C\}\,M , \tag{3}$$

$$4\sqrt{\varphi(\nu)}\frac{d}{d\nu}\left(\sqrt{\varphi(\nu)}\frac{dN}{d\nu}\right) = \{n(n+1)\nu + C\}\,N . \tag{4}$$

Written out in full, (2) takes the following form:

$$\frac{d^2\Lambda}{d\lambda^2} + \frac{1}{2}\left(\frac{1}{a^2 + \lambda} + \frac{1}{b^2 + \lambda} + \frac{1}{c^2 + \lambda}\right)\frac{d\Lambda}{d\lambda}$$
$$- \frac{n(n+1)\lambda + C}{4(a^2 + \lambda)(b^2 + \lambda)(c^2 + \lambda)}\Lambda = 0 . \tag{5}$$

This is a differential equation of the Fuchsian type [Sec. 2.7] with four regular singular points: $\lambda = -a^2, -b^2, -c^2, \infty$; the exponents at the first three are all 0 and $\frac{1}{2}$, and those at ∞ are $-\frac{1}{2}n$ and $(n+1)/2$.

By the transformations (1) and (2) of the preceding section:

$$\lambda = -p - \frac{1}{3}(a^2 + b^2 + c^2) , \tag{6}$$

$$e_1 = a^2 - \frac{1}{3}(a^2 + b^2 + c^2) , \quad e_2 = b^2 - \frac{1}{3}(a^2 + b^2 + c^2) ,$$
$$e_3 = c^2 - \frac{1}{3}(a^2 + b^2 + c^2) , \tag{7}$$

(5) becomes

$$\frac{d^2\Lambda}{dp^2} + \frac{1}{2}\left(\frac{1}{p-e_1} + \frac{1}{p-e_2} + \frac{1}{p-e_3}\right)\frac{d\Lambda}{d\lambda}$$
$$- \frac{n(n+1)p+B}{4(p-e_1)(p-e_2)(p-e_3)}\Lambda = 0 , \tag{8}$$

where B and C are connected by

$$B + C = \frac{1}{3}n(n+1)(a^2 + b^2 + c^2) . \tag{9}$$

Making further transformation $p = \wp(u)$, Lamé equation reduces to

$$\frac{d^2\Lambda}{du^2} = \{n(n+1)\wp(u) + B\}\Lambda . \tag{10}$$

This is the Lamé equation in Weierstrass' form.

Lamé equation can be expressed in Jacobian form by using Jacobian elliptic functions. Applying the transformation (7) of the last section:

$$a^2 + \lambda = (a^2 + b^2)\mathrm{sn}^2\alpha , \quad k^2 = \frac{a^2 - b^2}{a^2 - c^2} , \tag{11}$$

we have

$$\frac{d^2\Lambda}{d\alpha^2} = \{n(n+1)k^2 \mathrm{sn}^2\alpha + A\}\Lambda , \tag{12}$$

where A and C are connected by

$$C - n(n+1)a^2 = (a^2 - c^2)A . \tag{13}$$

In studying the properties of Lamé equation, it is better not to be confined to one specific form, but to choose a form apt to a definite purpose. For computational practice, Jacobian form (12) is more convenient. For the investigation of the general properties of the solutions for the equation, the algebraic form (8) is better.

If we take $\mathrm{sn}^2\alpha = s$ as the independent variable, Lamé equation becomes

$$\frac{d^2\Lambda}{ds^2} + \frac{1}{2}\left(\frac{1}{s} + \frac{1}{s-1} + \frac{1}{s-h}\right)\frac{d\Lambda}{ds}$$
$$- \frac{n(n+1)s+H}{4s(s-1)(s-h)}\Lambda = 0 , \tag{14}$$

where $h = k^{-2} = (a^2 - c^2)/(a^2 - b^2), H = hA$, and $h > 1$.

Another form is to use trigonometric functions. Let sn $\alpha = \cos \varsigma$, then we have

$$(1 - k^2 \cos^2 \varsigma)\frac{d^2 \Lambda}{d\varsigma^2} + k^2 \cos \varsigma \sin \varsigma \frac{d\Lambda}{d\varsigma}$$
$$- \{n(n+1)k^2 \cos^2 \varsigma + A\}\Lambda = 0 . \tag{15}$$

11.4. Four Types of Lamé Functions

We employ now the algebraic form of Lamé equation to investigate its solutions. The three algebraic forms, (5), (8), (14) of the last section, are equivalent. We shall use (14), which may be written as

$$4[s^3 - (1+h)s^2 + hs]\frac{d^2 \Lambda}{ds^2} + 2\left[3s^2 - 2(1+h)s + h\right]\frac{d\Lambda}{ds}$$
$$- [n(n+1)s + H]\Lambda = 0 . \tag{1}$$

This is an equation of Fuchsian type with the four regular singularities: $s = 0, 1, h, \infty$. The first three, namely $0, 1, h$, have the property that the corresponding exponents are all the same: 0 and $\frac{1}{2}$. The exponents at ∞ are $-n/2$ and $(n+1)/2$.

Assume the solution of (1) to be

$$\Lambda = \sum_{\nu=0}^{\infty} a_\nu s^{\rho+\nu} \quad (a_0 \neq 0) . \tag{2}$$

Substituting in (1), we find the equations for determining the coefficients a_ν:

$$2h\rho(2\rho - 1)a_0 = 0 ,$$
$$2h(\rho + 1)(2\rho + 1)a_1 - [4(1+h)\rho^2 + H]a_0 = 0 ,$$
$$2h(\rho + 2)(2\rho + 3)a_2 - [4(1+h)(\rho + 1)^2 + H]a_1$$
$$- (n - 2\rho)(n + 2\rho + 1)a_0 = 0 ,$$

$$\cdots$$

$$2h(\rho + \nu + 2)(2\rho + 2\nu + 3)a_{\nu+2} - [4(1+h)(\rho + \nu + 1)^2 + H]a_{\nu+1}$$
$$- (n - 2\rho - 2\nu)(n + 2\rho + 2\nu + 1)a_\nu = 0 , \tag{3}$$

$$\cdots \cdot$$

From the first one, we obtain the values of the exponents $\rho, 0$ and $\frac{1}{2}$. The succeeding equations determine $a_1/a_0, a_2/a_0, \ldots, a_\nu/a_0, \ldots$. It is easy to see that a_ν/a_0 is a polynomial of H of degree ν. In general, the solutions corresponding to $\rho = 0$ and $\rho = \frac{1}{2}$ are infinite series.

When n is a positive integer, we can properly choose the value of H such that the solution is not an infinite series but a polynomial. Thus, when $\nu = m$, which satisfies $2\rho + 2m = n$, and H is a root of $a_{m+1} = 0, a_{m+2} = 0$ and all $a_\nu = 0$ for $\nu > m$; the solution is a polynomial of degree m. For even n, the solution corresponding to $\rho = 0$ is a polynomial of degree $m = n/2$. For odd n, the solution corresponding to $\rho = \frac{1}{2}$ is a polynomial of degree $m = (n-1)/2$, while the highest degree in Λ, namely $m + \rho$, is still $n/2$.

These polynomials are called *Lamé functions of the first type* (also of the first kind; cf. Sec. 11.8). Since H is a root of the equation $a_{m+1} = 0$ of degree $m + 1$, there are $m + 1$ Lamé functions of the first type. The solutions for $n = 0, 1, 2, 3, 4, 5$ are:

$$n = 0 : \Lambda = 1 , \quad H = 0 .$$

$$n = 1 : \Lambda = s^{\frac{1}{2}} , \quad H = -1 - h .$$

$$n = 2 : \Lambda = s + \frac{2h}{H} , \quad H^2 + 4(1-h)H + 12h = 0 ;$$

the two roots are:

$$H_1 = -2(1+h) - 2\sqrt{1 - h + h^2} ,$$
$$H_2 = -2(1+h) + 2\sqrt{1 - h + h^2} .$$

$$n = 3 : \Lambda = s^{\frac{3}{2}} + \frac{6h}{1 + h + H}s^{\frac{1}{2}} ,$$
$$H^2 + 10(1+h)H + 9(1+h)^2 + 60h = 0 ;$$

the two roots are:

$$H_1 = -5(1+h) - 2\sqrt{4(1-h)^2 + h} ,$$
$$H_2 = -5(1+h) + 2\sqrt{4(1-h)^2 + h} .$$

$$n = 4 : \Lambda = s^2 + \frac{12hH}{H^2 + 4(1+h)H + 40h}s$$
$$+ \frac{24h^2}{H^2 + 4(1+h)H + 40h} ,$$
$$H^3 + 20(1+h)H^2 + 16(4 + 21h + 4h^2)H$$
$$+ 640(1+h)h = 0 .$$

$$n = 5: \Lambda = s^{\frac{5}{2}} + \frac{20h(H+1+h)}{H^2 + 10(1+h)H + 9(1+h)^2 + 168h}s^{\frac{3}{2}}$$

$$+ \frac{120h^2}{H^2 + 10(1+h)H + 9(1+h)^2 + 168h}s^{\frac{1}{2}} , \qquad (4)$$

$$H^3 + 35(1+h)H^2 + [259(1+h)^2 + 528h]H$$

$$+ 225(1+h)^3 + 4560(1+h)h = 0 .$$

Besides Lamé functions of the first type, there are three further types of Lamé functions, which correspond to the series solutions at $s = 1$ and $s = h$ with the exponent $\frac{1}{2}$.

Lamé functions of the second type are $\Lambda = (s-1)^{\frac{1}{2}}\Phi, \Phi = \Sigma_{\nu=0}^{\infty} b_\nu s^{\rho+\nu}$. Substituting in (1), we obtain the equation satisfied by Φ:

$$4[s^3 - (1+h)s^2 + hs]\frac{d^2\Phi}{ds^2} + 2[5s^2 - 2(1+2h)s + h]\frac{d\Phi}{ds}$$

$$- [(n-1)(n+2)s + H + h]\Phi = 0 . \qquad (5)$$

The coefficients b_ν are determined by the equations:

$$2h(\rho + \nu + 2)(\rho + 2\nu + 3)b_{\nu+2}$$

$$- \{4(\rho + \nu + 1)[(1+h)(\rho + \nu + 1 + h)] + H + h\} b_{\nu+1}$$

$$- (n - 2\rho - 2\nu - 1)(n + 2\rho + 2\nu + 2)b_\nu = 0 \qquad (6)$$

$$(\nu = 0, 1, 2, \dots) .$$

The values of the exponent ρ are still 0 and $\frac{1}{2}$; b_ν/b_0 is a polynomial of H of degree ν. When n is a positive integer, by properly choosing H, the solution could be made a polynomial instead of an infinite series. When $\nu = m$ satisfies $2\rho + 2m + 1 = n$, and H is a root of $b_{m+1} = 0$, the solution will be a polynomial. The exponent $\rho = 0$ corresponds to odd n and $\rho = \frac{1}{2}$ corresponds to even n with n not less than 1.

Lamé functions of the third type are $\Lambda = (s - h)^{\frac{1}{2}}\Psi, \Psi = \Sigma_{\nu=0}^{\infty} c_\nu s^{\rho+\nu}$. Substituting in (1), we obtain the equation satisfied by Ψ:

$$4[s^3 - (1+h)s^2 + hs]\frac{d^2\Psi}{ds^2} + 2\left[5s^2 - 2(2+h)s + h\right]\frac{d\Psi}{ds}$$

$$- [(n-1)(n+2)s + H + 1]\Psi = 0 . \qquad (7)$$

The equations for the determination of the coefficients c_ν are

$$2h(\rho + \nu + 2)(2\rho + 2\nu + 3)c_{\nu+2}$$

$$- \{4(\rho + \nu + 1)[(1+h)(\rho + \nu + 1) + 1] + H + 1\} c_{\nu+1}$$

$$- (n - 2\rho - 2\nu - 1)(n + 2\rho + 2\nu + 2)c_\nu = 0 \quad (\nu = 0, 1, 2, \dots) .$$

$$(8)$$

The properties of this type of Lamé functions are similar to those of the second type.

Lamé functions of the fourth type are $\Lambda = (s-1)^{\frac{1}{2}}(s-h)^{\frac{1}{2}}\Omega, \Omega = \Sigma_{\nu=0}^{\infty}d_\nu$ $\times s^{\rho+\nu}$; or $\Phi = (s-h)^{\frac{1}{2}}\Omega$. By putting this form of Φ in (5), we obtain

$$4\left[s^3 - (1+h)s^2 + hs\right]\frac{d^2\Omega}{ds^2} + 2\left[7s^2 - 4(1+h)s + h\right]\frac{d\Omega}{ds} \tag{9}$$
$$- \left[(n-2)(n+3)s + H + 1 + h\right]\Omega = 0 .$$

The equations for the determination of the coefficients d_ν, are

$$2h(\rho + \nu + 2)(2\rho + 2\nu + 3)d_{\nu+2}$$
$$- \left[4(1+h)(\rho + \nu + 1)(\rho + \nu + 2) + H + 1 + h\right]d_{\nu+1} \tag{10}$$
$$- (n - 2\rho - 2\nu - 2)(n + 2\rho + 2\nu + 3)d_\nu = 0 \quad (\nu = 0, 1, 2, \ldots) .$$

Just as the previous types of Lamé functions, when n is a positive integer, we can choose H such that $d_{m+1} = 0, 2\rho + 2m + 2 = n$ to obtain a polynomial solution of degree m for $n \geq 2$.

To summarize, for all the four types of Lamé functions, when n is an even number, there are $\frac{1}{2}n + 1$ Lamé functions of the first type, and $\frac{1}{2}n$ Lamé functions for other types; totally, there are $2n + 1$ functions. When n is an odd number, there are $(n + 1)/2$ functions for each of the first, second and third types and $(n - 1)/2$ functions for the fourth type; totally, there are again $2n + 1$ functions. These four types of Lamé functions are denoted by the symbol $E_n^m(s)$, in which m assumes the values $-n, -n+1, \ldots, n-1, n$.[a]

For $n = 0, 1, 2, 3$, the four types of Lamé functions are of the following forms (the superscript m assumes the values from $-n$ to n, and they are arranged according to the magnitudes of the corresponding eigenvalues H_n^m

---

[a] In the literature, m is usually assumed to take the values from 1 to $2n + 1$.

in ascending order):

$$n = 0 : \quad E_0 = 1, \quad H_0 = 0 .$$

$$n = 1 : \quad E_1^{-1} = s^{\frac{1}{2}}, \quad E_1^0 = (s-1)^{\frac{1}{2}}, \quad E_1^1 = (s-h)^{\frac{1}{2}} ;$$
$$\qquad\qquad H_1^{-1} = -1 - h, \quad H_1^0 = -h, \quad H_1^1 = -1 .$$

$$n = 2 : \quad E_2^{-2} = s + \frac{2h}{H_1}, \quad E_2^2 = s + \frac{2h}{H_2} ,$$
$$\qquad E_2^{-1} = s^{\frac{1}{2}}(s-1)^{\frac{1}{2}}, \quad E_2^0 = s^{\frac{1}{2}}(s-h)^{\frac{1}{2}} ,$$
$$\qquad E_2^1 = (s-1)^{\frac{1}{2}}(s-h)^{\frac{1}{2}} ;$$
$$\qquad H_2^{-2} = H_1 = -2(1+h) - 2\sqrt{1-h+h^2} ,$$
$$\qquad H_2^2 = H_2 = -2(1+h) + 2\sqrt{1-h+h^2} ,$$
$$\qquad H_2^{-1} = -1 - 4h, \ , H_2^0 = -4 - h, \quad H_2^1 = -1 - h .$$

$$n = 3 : \quad E_3^{-3} = s^{\frac{1}{2}} \left(s + \frac{6h}{1+h+H_1} \right) ,$$
$$\qquad E_3^1 = s^{\frac{1}{2}} \left(s + \frac{6h}{1+h+H_2} \right) ,$$
$$\qquad E_3^{-2} = (s-1)^{\frac{1}{2}} \left(s + \frac{2h}{H_3+h} \right) , \qquad\qquad (11)$$
$$\qquad E_3^2 = (s-1)^{\frac{1}{2}} \left(s + \frac{2h}{H_4+h} \right) ,$$
$$\qquad E_3^{-1} = (s-h)^{\frac{1}{2}} \left(s + \frac{2h}{H_5+1} \right) ,$$
$$\qquad E_3^3 = (s-h)^{\frac{1}{2}} \left(s + \frac{2h}{H_6+1} \right) ,$$
$$\qquad E_3^0 = s^{\frac{1}{2}}(s-1)^{\frac{1}{2}}(s-h)^{\frac{1}{2}} ;$$
$$\qquad H_3^{-3} = H_1 = -5(1+h) - 2\sqrt{4(1-h)^2+h} ,$$
$$\qquad H_3^1 = H_2 = -5(1+h) + 2\sqrt{4(1-h)^2+h} ,$$
$$\qquad H_3^{-2} = H_3 = -(2+5h) - 2\sqrt{1-h+4h^2} ,$$
$$\qquad H_3^2 = H_4 = -(2+5h) + 2\sqrt{1-h+4h^2} ,$$
$$\qquad H_3^{-1} = H_5 = -(5+2h) - 2\sqrt{4-h+h^2} ,$$
$$\qquad H_3^3 = H_6 = -(5+2h) + 2\sqrt{4-h+h^2} ,$$
$$\qquad H_3^0 = -4(1+h) .$$

It can be shown that all the eigenvalues H_n^m are real numbers. The first method to show this is to determine the coefficients in the series $\Sigma_{\nu=0}^{\infty} a_\nu (p$

$-e_2)^{\rho+\nu}$ by Eq. (8) of Sec. 11.3. The $m+1$ roots of the equation $a_{m+1}=0$ are all different and real. This is owing to the fact that a_ν, qua polynomial of B of degree ν, form a Sturm's sequence of functions (see Ex. 2 at the end of this chapter). Another method is to consider the orthogonality of the solutions (eigenfunctions) corresponding to different eigenvalues [cf. Hobson (1931), p. 465]. Let Λ_1 and Λ_2 be the eigenfunctions corresponding to the two eigenvalues H_1 and H_2, which satisfy respectively the equations [(14) of the preceding section]:

$$\frac{d^2\Lambda_r}{ds^2} + \frac{1}{2}\left(\frac{1}{s} + \frac{1}{s-1} + \frac{1}{s-h}\right)\frac{d\Lambda_r}{ds}$$

$$- \frac{n(n+1)s + H}{4s(s-1)(s-h)}\Lambda_r = 0 \quad (r=1,2) .$$

Multiplying the equation for $r=1$ by Λ_2 and that for $r=2$ by Λ_1, then subtracting, we have

$$\Lambda_2\frac{d^2\Lambda_1}{ds^2} - \Lambda_1\frac{d^2\Lambda_2}{ds^2} + \frac{1}{2}\left(\frac{1}{s} + \frac{1}{s-1} + \frac{1}{s-h}\right)\left(\Lambda_2\frac{d\Lambda_1}{ds} - \Lambda_1\frac{d\Lambda_2}{ds}\right)$$

$$- \frac{(H_1 - H_2)\Lambda_1\Lambda_2}{4s(s-1)(s-h)} = 0 ,$$

i.e.,

$$\sqrt{4s(s-1)(s-h)}\frac{d}{ds}\left\{\sqrt{4s(s-1)(s-h)}\left(\Lambda_2\frac{d\Lambda_1}{ds} - \Lambda_1\frac{d\Lambda_2}{ds}\right)\right\}$$

$$= (H_1 - H_2)\Lambda_1\Lambda_2 .$$

By integrating from $s=0$ to 1, we find

$$(H_1 - H_2)\int_0^1 \frac{\Lambda_1\Lambda_2 ds}{\sqrt{4s(s-1)(s-h)}}$$

$$= \left[\sqrt{4s(s-1)(s-h)}\left(\Lambda_2\frac{d\Lambda_1}{ds} - \Lambda_1\frac{d\Lambda_2}{ds}\right)\right]_{s=0}^{s=1} . \tag{12}$$

If Λ_1 and Λ_2 are Lamé functions of the same type, then the r.h.s. on (12) is equal to zero for all four types of functions, and hence

$$\int_0^1 \frac{\Lambda_1\Lambda_2 ds}{\sqrt{4s(s-1)(s-h)}} = 0 , \tag{13}$$

i.e., the eigenfunctions associate to different eigenvalues are orthogonal to each other. If H_1 is a complex number $P + iQ$, there must be a complex conjugate $H_2 = P - iQ$ with corresponding eigenfunctions $\Lambda_1 = I + iJ$ and $\Lambda_2 = I - iJ$, I and J being real. Then (13) gives

$$\int_0^1 \frac{(I^2 + J^2)ds}{\sqrt{4s(s-1)(s-h)}} = 0 \ . \tag{14}$$

The integrand here is positive definite, so the integral cannot vanish. Therefore, the eigenvalue H cannot possibly be a complex number and must be real.

From the orthogonality relation, it can be shown that the Lamé functions are linearly independent to each other. For, if there exists a linear relation $\Sigma_m a_m E_n^m(s) = 0$, we have, by multiplying it with $E_n^{m'}(s)ds/\sqrt{4s(s-1)(s-h)}$ and integrating, $a_{m'} = 0$ for all m'.

From the method of series solution of the differential equation, it can be proved that the zeros of Lamé functions are all simple (of order 1). Let the m zeros of a Lamé function be s_r, which are not singularities but ordinary points of the differential equation, then they are different from either of 0, 1, h. From the series solution in the vicinity of s_r [Sec. 2.2], we see that, if the coefficients of $(s - s_r)^0$ and $(s - s_r)^1$ are zeros, the series itself must be identically zero and thus cannot possess the factor $(s - s_r)^\nu (\nu > 1)$. For the singular points 0, 1, h, since their corresponding exponents are 0 and $\frac{1}{2}$, they cannot be the zeros of the solution.

11.5. Ellipsoidal Harmonics

If instead of s we use λ as the independent variable and denote the m distinct roots of a polynomial $F(\lambda)$ of degree m by θ_r, then by writing

$$F(\lambda) = \prod_{r=1}^m (\lambda - \theta_r) \ , \tag{1}$$

the four types of Lamé functions in the last section can be reclassified into the following four groups:

(i) $F(\lambda)$;

(ii) $\sqrt{\lambda + a^2} F(\lambda)$, $\sqrt{\lambda + b^2} F(\lambda)$, $\sqrt{\lambda + c^2} F(\lambda)$;

(iii) $\sqrt{(\lambda + b^2)(\lambda + c^2)} F(\lambda)$, $\sqrt{(\lambda + c^2)(\lambda + a^2)} F(\lambda)$,
     $\sqrt{(\lambda + a^2)(\lambda + b^2)} F(\lambda)$;

(iv) $\sqrt{(\lambda + a^2)(\lambda + b^2)(\lambda + c^2)}F(\lambda)$.   (2)

Here, the first group (i) of functions $F(\lambda)$ corresponds to $n = 2m$; the second group (ii) of functions corresponds to $n = 2m + 1$; the third group (iii) of functions corresponds to $n = 2m + 2$; the fourth group (iv) of functions corresponds to $n = 2m + 3$. The polynomials $F(\lambda)$ in each group and those $F(\lambda)$ in different groups all have different roots θ_r, and hence are all different. This can be seen from the equations determining the coefficients in $F(\lambda)$ [(3), (6), (8), (10) of the last section]. However, for brevity, we use the same symbol $F(\lambda)$ for these polynomials.

From the four groups of Lamé functions above, we can construct *four species of ellipsoidal harmonic functions* Ψ which satisfy the Laplace equation $\nabla^2\Psi = 0$. According to Eqs. (2) – (4) of Sec. 11.3, the constant C in them is the same for the three variables λ, μ, ν, and hence the three polynomials $F(\lambda), F(\mu), F(\nu)$ have the same roots θ_r. Each root θ_r contributes to Ψ three multiplying factors $(\lambda - \theta_r)(\mu - \theta_r)(\nu - \theta_r)$. By Eqs. (4) and (5) of Sec. 11.1, this product is equal to

$$-f(\theta_r) = \varphi(\theta_r)\Theta_r \; ,$$

where $\varphi(\theta_r) = (a^2 + \theta_r)(b^2 + \theta_r)(c^2 + \theta_r)$ is a constant and

$$\Theta_r = \frac{x^2}{a^2 + \theta_r} + \frac{y^2}{b^2 + \theta_r} + \frac{z^2}{c^2 + \theta_r} - 1 \; .   (3)$$

Therefore, $F(\lambda)F(\mu)F(\nu)$ and $\Pi_{r=1}^{m}\Theta_r$ differ only by a constant factor.

Now, we consider the factors outside $F(\lambda)$ in the four groups of Lamé functions, the factor in the first group is 1, which leads to the *first species of ellipsoidal harmonics* $\Psi = \Pi_{r=1}^{m}\Theta_r$.

There are three different kinds of factors in the second group. The first is

$$\sqrt{(\lambda + a^2)(\mu + a^2)(\nu + a^2)} \; .$$

Comparing with Eq. (6) of Sec. 11.1, we see that it differs from x only by a constant factor. Hence, this leads to the first case of the *second species of ellipsoidal harmonics*, namely $x\Pi_{r=1}^{m}\Theta_r$. It is easy to see that the second and the third cases of the ellipsoidal harmonics of the second species are respectively $y\Pi_{r=1}^{m}\Theta_r$ and $z\Pi_{r=1}^{m}\Theta_r$.

There are also three cases for the factors in the third group. It is easy to show that this leads to three cases for the *ellipsoidal harmonics of the third species*, namely, $yz\Pi_{r=1}^{m}\Theta_r$, $zx\Pi_{r=1}^{m}\Theta_r$ and $xy\Pi_{r=1}^{m}\Theta_r$.

The factor in the fourth group gives xyz, and correspondingly the *ellipsoidal harmonics of the fourth species* is $xyz\Pi^m_{r=1}\Theta_r$.

In summary, the four species of ellipsoidal harmonics can be written in the following simple scheme

$$\Psi = \left\{ \begin{matrix} x, & yz, \\ 1, & y, & zx, & xyz \\ & z, & xy, \end{matrix} \right\} \prod_{r=1}^{m} \left\{ \frac{x^2}{a^2 + \theta_r} + \frac{y^2}{b^2 + \theta_r} + \frac{z^2}{c^2 + \theta_r} - 1 \right\}. \quad (4)$$

It is seen from this scheme that the first species of ellipsoidal harmonics is a polynomial of x, y, z of degree $2m$; the second, third and fourth species of ellipsoidal harmonics are respectively polynomials of x, y, z of degrees $(2m + 1), (2m + 2)$ and $(2m + 3)$.

11.6. Niven's Representation

Niven has found a formula for representing the ellipsoidal harmonics by homogeneous harmonic functions. Let $G_n(x, y, z)$ be an ellipsoidal harmonics of degree n, and $H_n(x, y, z)$ a homogeneous ellipsoidal harmonics also of degree n of the same species:

$$G_n(x, y, z) = \left\{ \begin{matrix} x, & yz, \\ 1, & y, & zx, & xyz \\ & z, & xy, \end{matrix} \right\} \prod_{r=1}^{m} \left\{ \frac{x^2}{a^2 + \theta_r} + \frac{y^2}{b^2 + \theta_r} + \frac{z^2}{c^2 + \theta_r} - 1 \right\}, \quad (1)$$

$$H_n(x, y, z) = \left\{ \begin{matrix} x, & yz, \\ 1, & y, & zx, & xyz \\ & z, & xy, \end{matrix} \right\} \prod_{r=1}^{m} \left\{ \frac{x^2}{a^2 + \theta_r} + \frac{y^2}{b^2 + \theta_r} + \frac{z^2}{c^2 + \theta_r} \right\}, \quad (2)$$

where n and m are connected by the following relations for the four species respectively:

$$\text{(i) } n = 2m, \qquad \text{(ii) } n = 2m + 1,$$
$$\text{(iii) } n = 2m + 2, \quad \text{(iv) } n = 2m + 3. \quad (3)$$

Niven's representation is

$$G_n(x, y, z)$$
$$= \left\{ 1 - \frac{D}{2(2n - 1)} + \frac{D^2}{2 \times 4(2n - 1)(2n - 3)} - \cdots \right.$$
$$\left. + (-)^p \frac{D^p}{2^p p!(2n - 1) \ldots (2n - 2p + 1)} \right\} \times H_n(x, y, z), \quad (4)$$

where

$$D \equiv a^2 \frac{\partial^2}{\partial x^2} + b^2 \frac{\partial^2}{\partial y^2} + c^2 \frac{\partial^2}{\partial z^2} \; .$$

We shall give a proof of (4) only for the first species. Let

$$K_p = \frac{x^2}{a^2 + \theta_p} + \frac{y^2}{b^2 + \theta_p} + \frac{z^2}{c^2 + \theta_p} \; , \tag{5}$$

then, $(n = 2m)$

$$G_{2m}(x, y, z) = \prod_{p=1}^{m} (K_p - 1) = \sum_{r=0}^{m} (-)^r S_{2m-2r} \; , \tag{6}$$

where S_{2m-2r} is a homogeneous function of degree $2m - 2r$ in the form of a sum of the products of $K_1, K_2, \ldots , K_m$. Each term of the sum consists of the product of $m - r$ different factors K_p, and

$$S_{2m} = H_{2m}(x, y, z) = \prod_{p=1}^{m} K_p \; . \tag{7}$$

By differentiating, we obtain

$$\frac{\partial S_{2m-2r}}{\partial x} = \sum_{p=1}^{m} \frac{\partial S_{2m-2r}}{\partial K_p} \frac{\partial K_p}{\partial x} = \sum_{p=1}^{m} \frac{\partial S_{2m-2r}}{\partial K_p} \frac{2x}{a^2 + \theta_p} \; ,$$

$$\frac{\partial^2 S_{2m-2r}}{\partial x^2} = \sum_{p=1}^{m} \frac{\partial S_{2m-2r}}{\partial K_p} \frac{2}{a^2 + \theta_p} + \sum_{p<q} \frac{\partial^2 S_{2m-2r}}{\partial K_p \partial K_q} \frac{8x^2}{(a^2 + \theta_p)(a^2 + \theta_q)} \; ,$$

from which, we have

$$DS_{2m-2r} = \sum_{p=1}^{m} \frac{\partial S_{2m-2r}}{2K_p} \left(\frac{2a^2}{a^2 + \theta_p} + \frac{2b^2}{b^2 + \theta_p} + \frac{2c^2}{c^2 + \theta_p} \right)$$

$$+ \sum_{p<q} \frac{\partial^2 S_{2m-2r}}{\partial K_p \partial K_q} \left\{ \frac{8a^2 x^2}{(a^2 + \theta_p)(a^2 + \theta_q)} \right.$$

$$+ \frac{8b^2 y^2}{(b^2 + \theta_p)(b^2 + \theta_q)} + \frac{8c^2 z^2}{(c^2 + \theta_p)(c^2 + \theta_q)} \left. \right\} \; . \tag{8}$$

On the other hand, θ_p are the zeros of $\Lambda(\lambda)$, i.e.,

$$\Lambda(\lambda) = \prod_{p=1}^{m} (\lambda - \theta_p) \; .$$

Substituting in the Lamé equation satisfied by Λ [Eq. (5) of Sec. 11.3], putting $\lambda = \theta_p$ and note that

$$\frac{\Lambda''(\theta_p)}{\Lambda'(\theta_p)} = 2\sum_{q=1}^{m}{}' (\theta_p - \theta_q)^{-1} ,$$

we obtain

$$\frac{1}{a^2 + \theta_p} + \frac{1}{b^2 + \theta_p} + \frac{1}{c^2 + \theta_p} + \sum_{q=1}^{m}{}' \frac{4}{\theta_p - \theta_q} = 0 , \qquad (9)$$

where the primed symbol Σ' means that $q \neq p$ in the summation.

Now, applying (9), we find for the expression in the first sum on the r.h.s. of (8),

$$\frac{a^2}{a^2 + \theta_p} + \frac{b^2}{b^2 + \theta_p} + \frac{c^2}{c^2 + \theta_p}$$

$$= 3 - \theta_p \left(\frac{1}{a^2 + \theta_p} + \frac{1}{b^2 + \theta_p} + \frac{1}{c^2 + \theta_p} \right)$$

$$= 3 + \theta_p \sum_{q=1}^{m}{}' \frac{4}{\theta_p - \theta_q} ,$$

and, for the expression in the second sum on the r.h.s. of (8),

$$\frac{a^2 x^2}{(a^2 + \theta_p)(a^2 + \theta_q)} + \frac{b^2 y^2}{(b^2 + \theta_p)(b^2 + \theta_q)} + \frac{c^2 z^2}{(c^2 + \theta_p)(c^2 + \theta_q)} = \frac{\theta_p K_p - \theta_q K_q}{\theta_p - \theta_q} .$$

Let S^*_{2m-2r} be a sum of the products of $K_1, K_2, \ldots, K_m$, which is also a homogeneous function as S_{2m-2r}; each term of the sum consists of a product of $m - r$ different factors $K_s (s \neq p, q)$, but lacking the factors K_p and K_q which occur in S_{2m-2r}. Then, evidently,

$$S_{2m-2r} = S^*_{2m-2r} + K_p S^*_{2m-2r-2} + K_q S^*_{2m-2r-2} + K_p K_q S^*_{2m-2r-4} , \qquad (10)$$

from which we obtain

$$\frac{\partial S_{2m-2}}{\partial K_p} - K_q \frac{\partial^2 S_{2m-2r}}{\partial K_p \partial K_q} = \frac{\partial S_{2m-2r}}{\partial K_q} - K_p \frac{\partial^2 S_{2m-2r}}{\partial K_p \partial K_q}$$

and

$$K_p \frac{\partial S_{2m-2r}}{\partial K_p} - K_q \frac{\partial S_{2m-2r}}{\partial K_q} = (K_p - K_q) S^*_{2m-2r-2} \; .$$

By the first relation, we find

$$\frac{\partial^2 S_{2m-2r}}{\partial K_p \partial K_q} = -\frac{1}{K_p - K_q} \left\{ \frac{\partial S_{2m-2r}}{\partial K_p} - \frac{\partial S_{2m-2r}}{\partial K_q} \right\} \; ,$$

whence (8) reduces to

$$
\begin{aligned}
DS_{2m-2r} &= \sum_{p=1}^{m} \frac{\partial S_{2m-2r}}{\partial K_p} \left\{ 6 + \theta_p \sum_{q=1}^{m}{}' \frac{8}{\theta_p - \theta_q} \right\} \\
&\quad + 8 \sum_{p<q} \frac{\partial^2 S_{2m-2r}}{\partial K_p \partial K_q} \frac{\theta_p K_p - \theta_q K_q}{\theta_p - \theta_q} \\
&= \sum_{p=1}^{m} \frac{\partial S_{2m-2r}}{\partial K_p} \left\{ 6 + \theta_p \sum_{q=1}^{m}{}' \frac{8}{\theta_p - \theta_q} - 8 \sum_{q=1}^{m}{}' \frac{\theta_p K_p - \theta_q K_q}{(\theta_p - \theta_q)(K_p - K_q)} \right\} \\
&= \sum_{p=1}^{m} \frac{\partial S_{2m-2r}}{\partial K_p} \left\{ 6 - 8 \sum_{q=1}^{m}{}' \frac{K_q}{K_p - K_q} \right\} \\
&= \sum_{p=1}^{m} \frac{\partial S_{2m-2r}}{\partial K_p} \left\{ 6 + 8(m-1) - 8 \sum_{q=1}^{m}{}' \frac{K_p}{K_p - K_q} \right\} \\
&= (8m-2) \sum_{p=1}^{m} \frac{\partial S_{2m-2r}}{\partial K_p} - 8 \sum_{p<q} \frac{1}{K_p - K_q} \\
&\quad \times \left\{ K_p \frac{\partial S_{2m-2r}}{\partial K_p} - K_q \frac{\partial S_{2m-2r}}{\partial K_q} \right\}^{\iota} \\
&= (8m-2) \sum_{p=1}^{m} \frac{\partial S_{2m-2r}}{\partial K_p} - 8 \sum_{p<q} S^*_{2m-2r-2} \; .
\end{aligned}
$$

Obviously, the r.h.s. is a homogeneous function of $K_1, K_2, \ldots, K_m$. Being a sum of products of $m - r - 1$ different factors, it must be proportional to $S_{2m-2r-2}$. To find the constant of proportion, we compare the number of terms on the r.h.s. with that of $S_{2m-2r-2}$. The number of terms of $S_{2m-2r-2}$ is $\binom{m}{r+1}$, and that of $\partial S_{2m-2r}/\partial K_p$ is $\binom{m-1}{r}$. The number of terms of $S^*_{2m-2r-2}$ is $\binom{m-2}{r-1}$. Hence, the proportional constant sought is equal to

$$\left[(8m-2)m\binom{m-1}{r} - 8\binom{m}{2}\binom{m-2}{r-1}\right]\bigg/\binom{m}{r+1}$$
$$= (2r+2)(4m-2r-1) = (2r+2)(2n-2r-1)\,,$$

which gives

$$DS_{2m-2r} = (2r+2)(2n-2r-1)S_{2m-2r-2} \quad (n=2m)\,. \tag{11}$$

Thus, we have

$$D^p S_{2m-2r} = (2r+2)(2r+4)\ldots(2r+2p)$$
$$\times (2n-2r-1)(2n-2r-3)\ldots(2n-2r-2p+1)S_{2m-2r-2p}\,. \tag{12}$$

Putting $r = 0$, we obtain

$$D^p S_{2m} = 2\times 4\ldots 2p(2n-1)(2n-3)\ldots(2n-2p+1)S_{2m-2p}\,.$$

Substituting into (6) gives

$$G_{2m}(x,y,z) = \sum_{r=0}^{m} \frac{(-)^r D^r H_{2m}(x,y,z)}{2\times 4\ldots 2r(2n-1)(2n-3)\ldots(2n-2r+1)}\,,$$

which is (4) for the ellipsoidal harmonics of the first species.

Formulae similar to (4) for the second, third and fourth species of ellipsoidal harmonics can be proved as above.

11.7. On the Zeros of Lamé Polynomials

The four types of Lamé polynomials introduced in Sec. 11.4 can be written in the general form

$$\Lambda(\lambda) = (\lambda+a^2)^\rho(\lambda+b^2)^\sigma(\lambda+c^2)^\tau \prod_{p=1}^{m}(\lambda-\theta_p)\,, \tag{1}$$

where ρ,σ,τ are equal to 0 or $\frac{1}{2}$, and the relation between n, the degree of the Lamé function, and m is

$$\frac{n}{2} = m+\rho+\sigma+\tau\,. \tag{2}$$

When $\rho = \sigma = \tau = 0$, we have Lamé functions of the first type; when one of ρ, σ, τ equals to $\frac{1}{2}$ and the other two equal to 0, we have Lamé functions of the second type; when two of ρ, σ, τ equal to $\frac{1}{2}$ and the remaining one equal to 0, we have Lamé functions of the third type; when all ρ, σ, τ equal to $\frac{1}{2}$, we have Lamé functions of the fourth type. To each type of Lamé functions (with given values of m, ρ, σ, τ), there are $m + 1$ linearly independent functions corresponding to $m + 1$ eigenvalues H or C. Each function has m zeros θ_p, which are distinct from each other, and also not equal to $-a^2, -b^2, -c^2$ (see the last paragraph of Sec. 11.4).

Substituting the expression (1) in Eq. (5) of Sec. 11.3, putting $\lambda = \theta_p$ we obtain

$$\frac{\rho + \frac{1}{4}}{a^2 + \theta_p} + \frac{\sigma + \frac{1}{4}}{b^2 + \theta_p} + \frac{\tau + \frac{1}{4}}{c^2 + \theta_p} + \sum_{q=1}^{m}{}' \frac{1}{\theta_p - \theta_q} = 0 , \tag{3}$$

which is the condition satisfied by the m roots θ_p, a generalization of (9) of the last section. Every equation (3) is an equation of degree $m + 1$ in θ_p, giving $m + 1$ values of θ_p. For example, when $m = 1$, Eq. (3) reduces to

$$\frac{\rho + \frac{1}{4}}{a^2 + \theta} + \frac{\sigma + \frac{1}{4}}{b^2 + \theta} + \frac{\tau + \frac{1}{4}}{c^2 + \theta} = 0 ,$$

i.e.,

$$(4\rho + 1)(b^2 + \theta)(c^2 + \theta) + (4\sigma + 1)(c^2 + \theta)(a^2 + \theta)$$
$$+ (4\tau + 1)(a^2 + \theta)(b^2 + \theta) = 0 , \tag{4}$$

which gives two values of θ corresponding to two different eigenvalues H.

Stieltjes has proved that the $m + 1$ Lamé functions can be arranged in such a way that the $r - 1$ zeros of the rth function lie in the interval $-a^2$ and $-b^2$, and the remaining $m - r + 1$ zeros lie between $-b^2$ and $-c^2$. To show this, let $\varphi_1, \varphi_2, \ldots, \varphi_m$ be real numbers satisfying the following conditions:

$$-a^2 \le \varphi_p \le -b^2 \quad (p = 1, 2, \ldots, r - 1) ,$$
$$-b^2 \le \varphi_p \le -c^2 \quad (p = r, r + 1, \ldots, m) .$$

Build up the product

$$\Phi = \prod_{p=1}^{m} |\varphi_p + a^2|^{p + \frac{1}{4}} |\varphi_p + b^2|^{\sigma + \frac{1}{4}} |\varphi_p + c^2|^{\tau + \frac{1}{4}} \prod_{p<q} |\varphi_p - \varphi_q| .$$

The value of Φ will be zero when φ_p attains the end points: $-a^2, -b^2, -c^3$. As φ_p is finite, so is Φ. Since the values of Φ are positive, finite and continuous, apart from being equal to zero for φ_p at the end points, there must be a set of values of φ_p for which Φ assumes its maxima. These values of φ_p satisfy the conditions

$$\frac{\partial}{\partial \varphi_p} \ln \Phi = 0 \quad (p = 1, 2, \ldots, m) ,$$

i.e.,

$$\frac{\rho + \frac{1}{4}}{\varphi_p + a^2} + \frac{\sigma + \frac{1}{4}}{\varphi_p + b^2} + \frac{\tau + \frac{1}{4}}{\varphi_p + c^2} + \sum_{q=1}^{m} {}' \frac{1}{\varphi_p - \varphi_q} = 0 .$$

This is the same as (3), so the solution is $\varphi_p = \theta_p$, and the above assertion about the zeros of Lamé functions is proved.

11.8. Lamé Functions of the Second Kind

When n is a positive integer, if one of the solution of Lamé equation is a polynomial, then the other solution must be an infinite series. The polynomial solution is called *Lamé function of the first kind*, denoted by $E_n^m(p)$; the series solution is *Lamé function of the second kind*, denoted by $F_n^m(p)$, where $p = \wp(u) = -\lambda - \frac{1}{3}(a^2 + b^2 + c^2)$ [see Eq. (6) of Sec. 11.3].

By the method employed in proving Eq. (12) of Sec. 11.4, noticing that both kinds of Lamé functions belong to the same eigenvalue, we obtain from Eq. (10) of Sec. 11.3,

$$\frac{d}{du} \left(E_n^m \frac{dF_n^m}{du} - F_n^m \frac{dE_n^m}{du} \right) = 0 .$$

Integrating gives

$$E_n^m \frac{dF_n^m}{du} - F_n^m \frac{dE_n^m}{du} = C .$$

The constant of integration C depends on the choice of the constant factor in Lamé function of the second kind. Heine has chosen $C = 2n + 1$.

Integrating once again, we have

$$F_n^m(p) = (2n + 1)E_n^m(p) \int_0^u \frac{du}{[E_n^m(p)]^2} . \tag{1}$$

In the vicinity of $u = 0$, since $E_n^m(p)$ is a polynomial p of degree $n/2$, and $p = \wp(u) = u^{-2}[1 + O(u^4)]$ [See Eq. (3) of Sec. 8.4], we have

$$E_n^m(p) = u^{-n}[1 + O(u)] ,$$

$$F_n^m(p) = (2n + 1)u^{-n}[1 + O(u)] \int_0^u u^{2n}[1 + O(u)]du$$

$$= u^{n+1}[1 + O(u)] .$$

This explains the previous choice of $C = 2n + 1$ — to make the coefficient of the lowest term u^{n+1} in the expansion of $F_n^m(p)$ in the vicinity of $u = 0$ equal to 1.

The integral (1) can be expressed by elliptic functions. We shall show this for the first group of Lamé functions. For the first group of Lamé functions E_n^m, n is even $(2k)$ and E_n^m has $n/2$ zeros, p_s. Corresponding to each zero p_s, there are two values of u (in the period parallelogram). So, in total, there are n values $u_r, r = 1, 2, \ldots, n$. Now, suppose that u_r are so arranged that $u_{n-r} = -u_{r+1}$.

Let the expansion of $E_n^m(p)$ in the vicinity of u_r be

$$E_n^m(p) = k_1(u - u_r) + k_2(u - u_r)^2 + k_3(u - u_r)^3 + \ldots \quad (k_1 \neq 0) .$$

Substituting in Lamé equation [Eq. (10) of Sec. 11.3], we obtain $k_2 = 0$. Whence, the principal part of $[E_n^m(p)]^{-2}$ at u_r is $[k_1(u - u_r)]^{-2}$. Therefore, we can find constants A_r such that [see Eq. (6) of Sec. 8.9]

$$\frac{1}{[E_n^m(p)]^2} = A + \sum_{r=1}^{n} A_r \wp(u - u_r) , \tag{2}$$

where $A_{n-r} = A_{r+1}$. Integrating gives

$$\int_0^u \frac{du}{[E_n^m(p)]^2} = Au - \sum_{r=1}^{n} A_r \left\{ \varsigma(u - u_r) + \varsigma(u_r) \right\}$$

$$= Au - \sum_{r=1}^{n/2} A_r \left\{ \varsigma(u - u_r) + \varsigma(u + u_r) \right\} ,$$

where we have used the relations $u_{n-r} = -u_{r+1}, A_{n-r} = A_{r+1}$. Applying Eq. (2) of Sec. 8.10, we obtain

$$\int_0^u \frac{du}{[E_n^m(p)]^2} = Au - 2\varsigma(u) \sum_{r=1}^{n/2} A_r - \sum_{r=1}^{n/2} \frac{A_r \wp'(u)}{\wp(u) - \wp(u_r)} . \tag{3}$$

Substituting in (1) gives

$$F_n^m(p) = (2n + 1) \left\{ Au - 2\varsigma(u) \sum_{r=1}^{n/2} A_r \right\} E_n^m(p) + \wp'(u) W_{\frac{n}{2} - 1}(p) , \tag{4}$$

where $W_{\frac{n}{2} - 1}(p)$ is a polynomial in p of degree $\frac{1}{2} n - 1$.

For Lamé polynomials of the other groups, we can also reduce (1) into the form (4). Lamé functions of the second kind, $F_n^m(p)$, are suitable for the exterior problems of an ellipsoid.

11.9. Generalized Lamé Functions

When n is not a positive integer, the solutions of Lamé equation are called the *generalized Lamé functions*. Ince and Erdélyi[b] have investigated the Lamé equation expressed in terms of Jacobian elliptic functions [Eq. (12) of Sec. 11.3]:

$$\frac{d^2 \Lambda}{dz^2} + \left\{ h - n(n+1)(k \operatorname{sn} z)^2 \right\} \Lambda = 0 . \tag{1}$$

Under the assumptions of $k < 1$ and $n(n+1)$ being real, they studied the singly-periodic solutions of Lamé equation. These singly-periodic solutions (note that, by Eq. (11) of Sec. 11.3, Lamé polynomials are doubly-periodic functions of z) are called *periodic Lamé functions*.

We shall deal only with functions of real periods. The problem of imaginary period can be solved by Jacobi's imaginary transformation described in Sec. 10.7, which reduces the problem to one of real period. For details, see Erdélyi (1955), Vol. III, p. 63.

Since the real period of $\operatorname{sn}^2 z$ is $2K$ [Eq. (3) of Sec. 10.5], the real period of Lamé function must be $P = 2pK, p = 1, 2, \dots$. Now, $\operatorname{sn}^2 z$ is an even function of $z - K$, so, when $\Lambda(z)$ is a periodic Lamé function, $\Lambda(2K - z)$ must also be a periodic Lamé function. Hence, $\Lambda(z) \pm \Lambda(2K - z)$ is a periodic Lamé function and we can study separately the even and odd functions of $z - K$. Let $\operatorname{Ec}_n(z)$ or $\operatorname{Ec}_n(z, k^2)$ denote a Lamé function with a real period, which is an even function of $z - K$; $\operatorname{Es}_n(z)$ or $\operatorname{Es}_n(z, k^2)$ denotes an odd Lamé function of real period. More specifically, let $\operatorname{Ec}_n^m(z, k^2)$ and $\operatorname{Es}_n^m(z, k^2)$ denote functions with period $P = 2pK$ and with exactly pm zeros in the interval $0 \le z \le 2pK$. The eigenvalues h in (1) corresponding to Ec_n^m and Es_n^m are denoted by $a_n^m(k^2)$ and $b_n^m(k^2)$, or simply a_n^m and b_n^m, respectively.

Consider now the periods $2K$ and $4K$ (i.e., $p = 1, 2$). Then, $\operatorname{Ec}(z)$ is an even function of $z - K$, and also of $z + K$. As an even function of $z - K$, it is required that $\operatorname{Ec}(z) = \operatorname{Ec}(2K - z)$; and, as an even function of $z + K$, it is required that $\operatorname{Ec}(z) = \operatorname{Ec}(-2K - z)$. Thus we have the boundary condition

$$\operatorname{Ec}'(-K) = \operatorname{Ec}'(K) = 0 . \tag{2}$$

---

[b]Erdélyi (1955), Vol. III, p. 63.

Similarly, we have

$$\text{Es}(-K) = \text{Es}(K) = 0 \ . \tag{3}$$

Although these functions can be expressed, according to the method given in Sec. 11.4, as an infinite power series of $s = \text{sn}^2 z$, Ince has found, by Eq. (15) of Sec. 11.3, an expansion in terms of trigonometric functions, which converges more rapidly. Let $\text{sn } z = \cos \varsigma$, $H = 2h - n(n + 1)k^2$, Eq. (1) reduces [see Eq. (15) of Sec. 11.3] to

$$(2 - k^2 - k^2 \cos 2\varsigma) \frac{d^2 \Lambda}{d\varsigma^2} + k^2 \sin 2\varsigma \frac{d\Lambda}{d\varsigma}$$
$$+ \left\{ H - n(n + 1)k^2 \cos 2\varsigma \right\} \Lambda = 0 \ . \tag{4}$$

Assume the expansions in trigonometric functions to be

$$\text{Ec}_n^{2m}(z) = \frac{1}{2}A_0 + \sum_{r=1}^{\infty} A_{2r} \cos(2r\varsigma)$$
$$= \text{dn } z \left[\frac{1}{2}C_0 + \sum_{r=1}^{\infty} C_{2r} \cos(2r\varsigma) \right] \ , \tag{5}$$

$$\text{Ec}_n^{2m+1}(z) = \sum_{r=0}^{\infty} A_{2r+1} \cos\left[(2r + 1)\varsigma \right]$$
$$= \text{dn } z \sum_{r=0}^{\infty} C_{2r+1} \cos\left[(2r + 1)\varsigma \right] \ , \tag{6}$$

$$\text{Es}_n^{2m}(z) = \sum_{r=1}^{\infty} B_{2r} \sin(2r\varsigma) = \text{dn } z \sum_{r=1}^{\infty} D_{2r} \sin(2r\varsigma) \ , \tag{7}$$

$$\text{Es}_n^{2m+1}(z) = \sum_{r=0}^{\infty} B_{2r+1} \sin\left[(2r + 1)\varsigma \right]$$
$$= \text{dn } z \sum_{r=0}^{\infty} D_{2r+1} \sin\left[(2r + 1)\varsigma \right] \ . \tag{8}$$

Substituting in (4), we obtain the equations which determine the coefficients:

$$- HA_0 + (n - 1)(n + 2)k^2 A_2 = 0 \ ,$$
$$\frac{1}{2}(n - 2r + 2)(n + 2r - 1)k^2 A_{2r-2} - \left[H - 4r^2(2 - k^2) \right] A_{2r}$$
$$+ \frac{1}{2}(n - 2r - 1)(n + 2r + 2)k^2 A_{2r+2} = 0 \ . \tag{9}$$

$$- HC_0 + n(n+1)k^2 C_2 = 0 \,,$$

$$\frac{1}{2}(n-2r+1)(n+2r)k^2 C_{2r-2} - [H - 4r^2(2-k^2)]C_{2r} \tag{10}$$

$$+ \frac{1}{2}(n-2r)(n+2r+1)k^2 C_{2r+2} = 0 \,.$$

$$- \left[H - 2 + k^2 - \frac{1}{2}n(n+1)k^2 \right] A_1$$

$$+ \frac{1}{2}(n-2)(n+3)k^2 A_3 = 0 \,,$$

$$\frac{1}{2}(n-2r+1)(n+2r)k^2 A_{2r-1} - \left[H - (2r+1)^2(2-k^2) \right] A_{2r+1} \tag{11}$$

$$+ \frac{1}{2}(n-2r-2)(n+2r+3)k^2 A_{2r+3} = 0 \,.$$

$$- \left[H - 2 + k^2 - \frac{1}{2}n(n+1)k^2 \right] C_1 + \frac{1}{2}(n-1)(n+2)k^2 C_3 = 0 \,,$$

$$\frac{1}{2}(n-2r)(n+2r+1)k^2 C_{2r-1} - \left[H - (2r+1)^2(2-k^2) \right] C_{2r+1} \tag{12}$$

$$+ \frac{1}{2}(n-2r-1)(n+2r+2)k^2 C_{2r+3} = 0 \,.$$

$$- (H - 8 + 4k^2)B_2 + \frac{1}{2}(n-3)(n+4)k^2 B_4 = 0 \,,$$

$$\frac{1}{2}(n-2r)(n+2r+1)k^2 B_{2r} - \left[H - (2r+2)^2(2-k^2) \right] B_{2r+2} \tag{13}$$

$$+ \frac{1}{2}(n-2r-3)(n+2r+4)k^2 B_{2r+4} = 0 \,.$$

$$- (H - 8 + 4k^2)D_2 + \frac{1}{2}(n-2)(n+3)k^2 D_4 = 0 \,,$$

$$\frac{1}{2}(n-2r-1)(n+2r+2)k^2 D_{2r} - \left[H - (2r+2)^2(2-k^2) \right] D_{2r+2} \tag{14}$$

$$+ \frac{1}{2}(n-2r-2)(n+2r+3)K^2 D_{2r+4} = 0 \,.$$

$$-\left[H - 2 + k^2 - \frac{1}{2}n(n+1)k^2\right]B_1 + \frac{1}{2}(n-2)(n+3)k^2 B_3 = 0 \,,$$

$$\frac{1}{2}(n - 2r + 1)(n + 2r)k^2 B_{2r-1} - \left[H - (2r+1)^2(2-k^2)\right]B_{2r+1} \qquad (15)$$

$$+ \frac{1}{2}(n - 2r - 2)(n + 2r + 3)k^2 B_{2r+3} = 0 \,.$$

$$-\left[H - 2 + k^2 + \frac{1}{2}n(n+1)k^2\right]D_1 + \frac{1}{2}(n-1)(n+2)k^2 D_3 = 0 \,,$$

$$\frac{1}{2}(n - 2r)(n + 2r + 1)k^2 D_{2r-1} - \left[H - (2r+1)^2(2-k^2)\right]D_{2r+1} \qquad (16)$$

$$+ \frac{1}{2}(n - 2r - 1)(n + 2r + 2)k^2 D_{2r+3} = 0 \,.$$

11.10. Integral Equations of Lamé Functions[c]

Let $\Lambda(\alpha)$ be a Lamé function satisfying the Lamé equation , Eq. (12) of Sec. 11.3:

$$\frac{d^2\Lambda}{d\alpha^2} - \left\{n(n+1)k^2 \operatorname{sn}^2\alpha + A\right\}\Lambda = 0 \,. \qquad (1)$$

Let $N(\alpha, \theta)$ be a function which satisfies the following partial differential equation

$$\frac{\partial^2 N}{\partial\alpha^2} - n(n+1)k^2 \operatorname{sn}^2\alpha \cdot N = \frac{\partial^2 N}{\partial\theta^2} - n(n+1)k^2 \operatorname{sn}^2\theta \cdot N \,. \qquad (2)$$

Integrating by parts, we have

$$\left\{\frac{d^2}{d\alpha^2} - n(n+1)k^2 \operatorname{sn}^2\alpha - A\right\}\int_a^b N(\alpha, \theta)\Lambda(\theta)d\theta$$

$$= \int_a^b \left\{\frac{\partial^2 N}{\partial\theta^2} - \left[n(n+1)k^2 \operatorname{sn}^2\theta + A\right]N\right\}\Lambda(\theta)d\theta$$

$$= \left[\frac{\partial N}{\partial\theta}\Lambda(\theta) - N(\alpha, \theta)\frac{d\Lambda}{d\theta}\right]_a^b$$

$$+ \int_a^b N(\alpha, \theta)\left\{\frac{d^2\Lambda}{d\theta^2} - \left[n(n+1)k^2 \operatorname{sn}^2\theta + A\right]\Lambda\right\}d\theta \,.$$

By properly choosing $N(\alpha, \theta)$ such that

$$\left[\frac{\partial N}{\partial\theta}\Lambda(\theta) - N(\alpha, \theta)\frac{d\Lambda}{d\theta}\right]_a^b = 0 \,, \qquad (3)$$

---

[c] Cf. Sec. 12.14, the discussion of the same problem for Mathieu functions; also, Sec. 2.12.

$\int_a^b N(\alpha, \theta) \Lambda(\theta) d\theta$ will be a solution of the Lamé equation (1), with the same eigenvalue A as for $\Lambda(\alpha)$. In many cases, this solution behaves in the same ways as $\Lambda(\alpha)$ and hence is proportional to $\Lambda(\alpha)$. This leads to the integral equation for $\Lambda(\alpha) = E_n^m(\alpha)$:

$$E_n^m(\alpha) = \lambda \int_{-2K}^{2K} N(\alpha, \theta) E_n^m(\theta) d\theta \ , \tag{4}$$

where the limits of integration are chosen to be $\pm 2K$ in order to meet the case that $N(\alpha, \theta)$ has a period equal to $4K$ and satisfies (3) when $E_n^m(\theta)$ has the period $4K$. As regards the specific forms of $N(\alpha, \theta)$, we shall discuss them in the following for different cases.

For Lamé functions of the first type [Sec. 11.4],

$$N(\alpha, \theta) = P_n(k \text{ sn } \alpha \text{ sn } \theta) \ , \tag{5}$$

where $P_n(\mu)$ is Legendre's function. The proof is as follows. Let $\mu = k \text{ sn}\alpha \text{ sn}\theta$, we have

$$\left\{ \frac{\partial^2}{\partial \alpha^2} - \frac{\partial^2}{\partial \theta^2} \right\} P_n(k \text{ sn } \alpha \text{ sn } \theta)$$
$$= k^2 \left\{ \text{cn}^2\alpha \text{ dn}^2\alpha \text{ sn}^2\theta - \text{cn}^2\theta \text{ dn}^2\theta \text{ sn}^2\alpha \right\} P_n''(\mu)$$
$$+ 2k^3 \text{ sn } \alpha \text{ sn } \theta (\text{sn}^2\alpha - \text{sn}^2\theta) P_n'(\mu)$$
$$= k^2 (\text{sn}^2\alpha - \text{sn}^2\theta) \left[(\mu^2 - 1) P_n''(\mu) + 2\mu P_n'(\mu) \right]$$
$$= k^2 (\text{sn}^2\alpha - \text{sn}^2\theta) n(n + 1) P_n(\mu) \ ,$$

which shows that $P_n(\mu)$ satisfies the partial differential equation (2). Besides, since $P_n(\mu)$ is a polynomial of degree n in μ, the r.h.s. of (4) will be a polynomial of degree n in sn α and hence behaves exactly as $E_n^m(\alpha)$ (Lamé function of the first type) and must be proportional to $E_n^m(\alpha)$. This proves (4), λ being the proportional constant.

For Lamé functions of the second type which contain the factor cn α, we have, if n is an odd number,

$$N(\alpha, \theta) = P_n \left(\frac{ik}{k'} \text{cn } \alpha \text{ cn } \theta \right) \ , \tag{6}$$

which is also valid for Lamé functions of the first type with even n.

For Lamé functions of the third type, which contain the factor dn α, we have, if n is an odd number,

$$N(\alpha, \theta) = P_n \left(\frac{1}{k'} \text{dn } \alpha \text{ dn } \theta \right) \ . \tag{7}$$

This type of Lamé functions belongs to the third of the second group [Sec. 11.5]. (7) holds also for Lamé functions of the first type with even n (i.e., Lamé functions of the first group).

For Lamé functions of the second type which contain the factor snα $\times$ cnα (which is the third case of the third group Lamé function with even n), we have

$$N(\alpha, \theta) = \text{sn}\alpha \ \text{cn}\alpha \ \text{sn}\theta \ \text{cn}\theta \ P_n'' \left(\frac{1}{k'}\text{dn}\alpha \ \text{dn}\theta \right) . \tag{8}$$

For Lamé functions of the third type which contain the factor snα dnα (the second case of the third group with even n), we have

$$N(\alpha, \theta) = \text{sn}\alpha \ \text{dn}\alpha \ \text{sn}\theta \ \text{dn}\theta P_n'' \left(\frac{ik}{k'}\text{cn}\alpha \ \text{cn}\theta \right) . \tag{9}$$

For Lamé functions of the fourth type which contain the factor cnα dnα (the first case of the third group with even n), we have

$$N(\alpha, \theta) = \text{cn}\alpha \ \text{dn}\alpha \ \text{cn}\theta \ \text{dn}\theta P_n''(k \ \text{sn}\alpha \ \text{sn}\theta) . \tag{10}$$

This is also valid for Lamé functions of the fourth type containing the factor snα cnα dnα; then n is odd (Lamé functions of the fourth group). The kernels (8) and (9) also apply to Lamé functions of the fourth group, since they have the factor snα cnα dnα when n is odd.

11.11. The Integral Representation of Ellipsoidal Harmonics

The homogeneous ellipsoidal harmonics $H_n^m(x, y, z)$ introduced in Sec. 11.6 can be expressed in the following form of an integral:

$$H_n^m(x, y, z) = \int_{-\pi}^{\pi} (x \cos t + y \sin t + iz)^n f(t) dt , \tag{1}$$

where $f(t)$ is a periodic function of t with period 2π to be determined. Applying on it the operator D introduced in Sec. 11.6, the result is

$$D(x \cos t + y \sin t + iz)^n$$
$$= n(n-1)(a^2 \cos^2 t + b^2 \sin^2 t - c^2)(x \cos t + y \sin t + iz)^{n-2} . \tag{2}$$

Thus it is seen that the r.h.s. of (1) satisfies the Laplace equation ($a = b = c$). Obviously, it is homogeneous and hence a homogeneous harmonic function of degree n. By properly choosing $f(t)$, it can be made equal to $H_n^m(x, y, z)$.

By making use of (2), we can write Niven's expression [Eq. (4) of Sec 11.6] as

$$G_n^m(x,y,z)$$
$$= \int_{-\pi}^{\pi} \left\{ p^n - \frac{n(n-1)}{2(2n-1)} p^{n-2}q^2 + \frac{n(n-1)(n-2)(n-3)}{2 \times 4(2n-1)(2n-3)} p^{n-4}q^4 - \ldots \right\}$$
$$\times f(t)dt ,$$

where

$$p = x\cos t + y\sin t + iz , \qquad q = \sqrt{a^2\cos^2 t + b^2\sin^2 t - c^2} . \tag{3}$$

Using the first equation of Eq. (7) of Sec. 5.2, we obtain

$$G_n^m(x,y,z) = \frac{2^n(n!)^2}{(2n)!} \int_{-\pi}^{\pi} P_n\left(\frac{p}{q}\right) q^n f(t)dt . \tag{4}$$

By the transformation

$$\sin t = \mathrm{cd}\,\theta , \qquad k^2 = \frac{a^2 - b^2}{a^2 - c^2} , \qquad \frac{2^n(n!)^2}{(2n)!} q^n f(t) = \varphi(\theta)d\theta ,$$

we obtain

$$G_n^m(x,y,z) = \int_{-2K}^{2K} P_n\left(\frac{k'x\,\mathrm{sn}\,\theta + y\,\mathrm{cn}\,\theta + iz\,\mathrm{dn}\,\theta}{\sqrt{b^2 - c^2}}\right) \varphi(\theta)d\theta , \tag{5}$$

where the original limits of integration K and $-3K$ have been replaced by $2K$ and $-2K$ in virtue of the periodicity of the integrand.

To determine $\varphi(\theta)$, we express x, y, z in terms of α, β, γ [see Eq. (10) of Sec. 11.2] and represent the ellipsoidal harmonics $G_n^m(x,y,z)$ by a product of Lamé functions [see Sec. 11.5]. Then (5) becomes

$$E_n^m(\alpha)E_n^m(\beta)E_n^m(\gamma) = C\int_{-2K}^{2K} P_n(\mu)\varphi(\theta)d\theta , \tag{6}$$

where C is a constant factor, and

$$\mu = k^2\,\mathrm{sn}\alpha\,\mathrm{sn}\beta\,\mathrm{sn}\gamma\,\mathrm{sn}\theta - \frac{k^2}{k'^2}\mathrm{cn}\alpha\,\mathrm{cn}\beta\,\mathrm{cn}\theta$$
$$- \frac{1}{k'^2}\mathrm{dn}\alpha\,\mathrm{dn}\beta\,\mathrm{dn}\gamma\,\mathrm{dn}\theta . \tag{7}$$

If the ellipsoidal harmonics is of the first species, or the first case of the second species (i.e., Lamé function of the first type), let $\beta = K, \gamma = K + K'i$, then $\mu = k \operatorname{sn} \alpha \operatorname{sn} \theta$ and (6) shows that

$$\int_{-2K}^{2K} P_n(k \operatorname{sn}\alpha \operatorname{sn}\theta)\varphi(\theta)d\theta$$

is a solution of the Lamé equation and is proportional to $E_n^m(\alpha)$. By (4) and (5) of the last section, it is seen that $\varphi(\theta) \propto E_n^m(\theta)$. Hence,

$$G_n^m(x, y, z) = \lambda \int_{-2K}^{2K} P_n\left(\frac{k'x \operatorname{sn}\theta + y \operatorname{cn}\theta + iz \operatorname{dn}\theta}{\sqrt{b^2 - c^2}}\right) E_n^m(\theta)d\theta . \qquad (8)$$

If the ellipsoidal harmonics is the second case of the second species, putting $\beta = 0, \gamma = K + K'i$, then $\mu = ik \operatorname{cn}\alpha \operatorname{cn}\theta/k'$ and, by (6) of the last section, we again arrive at (8) above.

If the ellipsoidal harmonics is the third case of the second species, putting $\beta = 0, \gamma = K$, then $\mu = -\operatorname{dn}\alpha \operatorname{dn}\theta/k'$ and, according to (7) of the last section, we again have (8).

If the ellipsoidal harmonics is of the third or fourth species, assuming that $E_n^m(\alpha)$ possesses the factor $\operatorname{cn}\alpha \operatorname{dn}\alpha$, we can differentiate (6) first with respect to β, then with respect to γ and put $\beta = K, \gamma = K + K'i$; we find

$$E_n^m(\alpha)\left[\frac{d}{d\beta}E_n^m(\beta)\right]_{\beta=K}\left[\frac{d}{d\gamma}E_n^m(\gamma)\right]_{\gamma=K+K'i}$$

$$= C\int_{-2K}^{2K}\left[\frac{\partial^2 P_n(\mu)}{\partial\beta\partial\gamma}\right]_{\substack{\beta=K\\\gamma=K+K'i}}\varphi(\theta)d\theta .$$

Further, from

$$\left[\frac{\partial P_n(\mu)}{\partial\gamma}\right]_{\gamma=K+K'i} = -\frac{i}{k'}\operatorname{dn}\alpha \operatorname{dn}\beta \operatorname{dn}\theta P_n'(\mu) ,$$

$$\left[\frac{\partial^2 P_n(\mu)}{\partial\beta\partial\gamma}\right]_{\substack{\beta=K\\\gamma=K+K'i}} = -k \operatorname{cn}\alpha \operatorname{dn}\alpha \operatorname{cn}\theta \operatorname{dn}\theta P_n''(k \operatorname{sn}\alpha \operatorname{sn}\theta)$$

and (10) of the last section, we again obtain (8).

Thus, (8) holds for any species of the ellipsoidal harmonics. It is easy to see that the corresponding homogeneous function satisfies the integral formula

$$H_n^m(x, y, z)$$

$$= \lambda\frac{(2n)!}{2^n(n!)^2(b^2 - c^2)^{n/2}}\int_{-2K}^{2K}(k'x \operatorname{sn}\theta + y \operatorname{cn}\theta + iz \operatorname{dn}\theta)^n E_n^m(\theta)d\theta . \qquad (9)$$

Exercise 11

1. Show that the ellipsoidal coordinates satisfy the following relation

$$x^2 + y^2 + z^2 = a^2 + b^2 + c^2 + \lambda + \mu + \nu \ .$$

2. Taking Eq. (8) of Sec. 11.3 as the Lamé equation, express the four types of Lamé functions in the following series forms:

(i) $\Lambda = \Sigma_r b_r (p - e_2)^{\frac{n}{2} - r}$,

(ii) $\Lambda = (p - e_1)^{\frac{1}{2}} \Sigma_r b'_r (p - e_2)^{\frac{n}{2} - \frac{1}{2} - r}$,

(iii) $\Lambda = (p - e_3)^{\frac{1}{2}} \Sigma_r b''_r (p - e_2)^{\frac{n}{2} - \frac{1}{2} - r}$,

(iv) $\Lambda = (p - e_1)^{\frac{1}{2}} (p - e_3)^{\frac{1}{2}} \Sigma_r b'''_r (p - e_2)^{\frac{n}{2} - 1 - r}$.

Show that the equations for the determination of the coefficients are

$$r \left(n - r + \frac{1}{2} \right) b_r$$

$$= \left\{ 3e_2 \left(\frac{n}{2} - r + 1 \right)^2 - \frac{1}{4} n(n+1) e_2 - \frac{1}{4} B \right\} b_{r-1}$$

$$- (e_1 - e_2)(e_2 - e_3) \left(\frac{n}{2} - r + 2 \right) \left(\frac{n}{2} - r + \frac{3}{2} \right) b_{r-2} \ ,$$

$$r \left(n - r + \frac{1}{2} \right) b'_r$$

$$= \left\{ 3e_2 \left(\frac{n}{2} - r + \frac{1}{2} \right)^2 + (e_2 - e_3) \left(\frac{n}{2} - r + \frac{3}{4} \right) \right.$$

$$\left. - \frac{1}{4} n(n+1) e_2 - \frac{1}{4} B \right\} b'_{r-1}$$

$$- (e_1 - e_2)(e_2 - e_3) \left(\frac{n}{2} - r + \frac{3}{2} \right) \left(\frac{n}{2} - r + 1 \right) b'_{r-2} \ ,$$

$$r\left(n - r + \frac{1}{2}\right) b_r''$$

$$= \left\{3e_2 \left(\frac{n}{2} - r + \frac{1}{2}\right)^2 - (e_1 - e_2)\left(\frac{n}{2} - r + \frac{3}{4}\right)\right.$$

$$\left. - \frac{1}{4}n(n+1)e_2 - \frac{1}{4}B\right\} b_{r-1}''$$

$$- (e_1 - e_2)(e_2 - e_3)\left(\frac{n}{2} - r + \frac{3}{2}\right)\left(\frac{n}{2} - r + 1\right) b_{r-2}'',$$

$$r\left(n - r + \frac{1}{2}\right) b_r'''$$

$$= \left\{3e_2 \left(\frac{n}{2} - r + \frac{1}{2}\right)^2 - \frac{1}{4}e_2(n^2 + n + 1) - \frac{1}{4}B\right\} b_{r-1}'''$$

$$- (e_1 - e_2)(e_2 - e_3)\left(\frac{n}{2} - r + 1\right)\left(\frac{n}{2} - r + \frac{1}{2}\right) b_{r-2}'''.$$

Show that b_r is a polynomial of degree r in B and the coefficient of B^r in it is

$$(-)^r/[2 \times 4 \ldots 2r(2n-1)(2n-1)\ldots(2n-2r+1)];$$

when $r < \frac{1}{2}(n+3)$, $r < n$, if $b_{r-1} = 0$, then b_r and b_{r-2} are of opposite signs, and hence $b_0, b_1, \ldots, b_r$ form a Sturm's sequence of functions. Thus, when n is even, $b_{\frac{n}{2}+1} = 0$; when n is odd, $b_{\frac{1}{2}(n+1)} = 0$, and all the roots are real and distinct [cf. Whittaker and Watson, (1927), p. 556].

3. Show that the values of θ given in Eq. (4) of Sec. 11.7 are in conformity with the roots s given by Eqs. (4) and (11) of Sec. 11.4.

4. Complete the proofs of formula (4) in Sec. 11.6 for the second, third and fourth species of ellipsoidal harmonics.

5. Study the Heun equation

$$z(z-1)(z-a)\frac{d^2y}{dz^2}$$

$$+ \left\{(\alpha + \beta + 1)z^2 - [\alpha + \beta - \delta + 1 + (\gamma - \delta)a]z + \alpha\gamma\right\}\frac{dy}{dz}$$

$$+ \alpha\beta(z - q)y = 0.$$

Derive the following series solution:

$$y = 1 + \alpha\beta \sum_{n=1}^{\infty} \frac{G_n(q)}{n!(\gamma)_n} \left(\frac{z}{a}\right)^n ,$$

where

$$G_1(q) = q ,$$
$$G_2(q) = \alpha\beta q^2 + [(\alpha + \beta - \delta + 1) + (\gamma + \delta)a]q - \alpha\gamma ,$$
$$\ldots$$
$$G_{n+1}(q) = \{n[(\alpha + \beta - \delta + n) + (\gamma + \delta + n - 1)a] + \alpha\beta q\} G_n(q)$$
$$- (\alpha + n - 1)(\beta + n - 1)(\gamma + n - 1)naG_{n-1}(q) .$$

Chapter 12

MATHIEU FUNCTIONS

12.1. Mathieu Equation

Mathieu equation is

$$\frac{d^2y}{dz^2} + (\lambda - 2q\cos 2z)y = 0 \ , \tag{1}$$

where λ and q are parameters. This is an equation with periodic-function coefficient.

The solutions of (1) are called *Mathieu functions*. However, the name Mathieu function is often specifically used to denote those solutions with period π or 2π. It is necessary to point out that a differential equation, even with single-valued periodic functions as coefficients, need not have periodic solutions. For example, the equation

$$\frac{dy}{dx} + (a + b\cos 2x)y = 0$$

does not have a periodic solution unless $a = 0$. Moreover, even if the equation does have a periodic solution, the period of the solution may not be the same as that of the coefficient. We shall see presently that only when the parameters

λ and q satisfy certain conditions, then there can be solutions with period π or 2π for (1).

To find out such conditions is one of the central problems of the theory of Mathieu function. This is on account of the fact that the physical problems leading to Mathieu equation often require solutions with periodicity; for example, the problems of vibrations and waves (see below).

Example for the origin of Mathieu equation

Among the various sources that lead to Mathieu equation, one of the most important is that derived from the solution of Helmholtz equation

$$\nabla^2 u + k^2 u = 0 \tag{2}$$

by the method of separation of variables in elliptic cylinder coordinates.

The relations between the elliptic cylinder coordinates ξ, η, z and the cartesian coordinates are [cf. App. III, 4]

$$x = a \operatorname{ch} \xi \, \cos \eta \,, \quad y = a \operatorname{sh} \xi \, \sin \eta \,, \quad z = z \, ; \tag{3}$$

if it is stipulated that $\xi \geq 0, -\pi \leq \eta \leq \pi$, then (x, y) and (ξ, η) are in one-to-one correspondence.

Let the solution of (2) be $u = F(\xi)G(\eta)Z(z)$. We have

$$Z''(z) + m^2 Z = 0 \,, \tag{4}$$

$$\frac{d^2 F}{d\xi^2} + (\alpha^2 \operatorname{ch}^2 \xi - \mu)F(\xi) = 0 \,, \tag{5}$$

$$\frac{d^2 G}{d\eta^2} + (\mu - \alpha^2 \cos^2 \eta)G(\eta) = 0 \,, \tag{6}$$

where $\alpha^2 = a^2(k^2 - m^2)$, μ and m are parameters introduced during the separation of variables. In (5), putting $\xi = i\varsigma$, it happens to be a Mathieu equation with $\lambda = \mu - \frac{1}{2}\alpha^2, q = \frac{1}{4}\alpha^2$; it is the same for (6).

From (1), it is seen that the coefficient of Mathieu equation is an analytic function of z in the whole z-plane; the only singularity is at $z = \infty$, which is an irregular singularity. Hence, the solutions of Mathieu equation must be integral (entire) functions.

Let $t = \cos^2 z$, then we arrive at another form of Mathieu equation whose coefficients are algebraic functions:

$$4t(1-t)\frac{d^2y}{dt^2} + 2(1-2t)\frac{dy}{dt} + (\lambda + 2q - 4qt)y = 0 \ . \tag{7}$$

[cf. Ex. 3 at the end of Chap. 2.] This equation has two regular singularities: $t = 0$ and $t = 1$; the other singularity $t = \infty$ is irregular.

Mathieu equation can be generalized to the following form:

$$\frac{d^2u}{dz^2} + \left\{ \theta_0 + 2\sum_{n=1}^{\infty} \theta_n \cos 2nz \right\} u = 0 \ , \tag{8}$$

where $\theta_0, \theta_1, \ldots$ are constants. Equation (8) is called *Hill equation* and will be discussed in Sec. 12.11.

12.2. General Properties of the Solution. Fundamental Solutions

In the present and next sections, we shall discuss a more general equation of the form:

$$\frac{d^2y}{dz^2} + \{\lambda - \varphi(z)\}\, y = 0 \ , \tag{1}$$

where $\varphi(z)$ is a periodic function with period ω. Equation (1) not only encompasses Mathieu equation and Hill equation as its special cases, but also Lamé equation [cf. Eqs. (10) and (11) of Sec. 11.3], for example.

Fundamental solutions

Suppose that $f(z)$ and $g(z)$ are solutions satisfying the following initial conditions:

$$\begin{aligned} f(0) &= 1 \ , \quad f'(0) = 0 \ ; \\ g(0) &= 0 \ , \quad g'(0) = 1 \ . \end{aligned} \tag{2}$$

By (1) and the initial conditions for $f(z)$ and $g(z)$, we have

$$f(z)\, g'(z) - f'(z)\, g(z) = C = 1 \ . \tag{3}$$

Therefore, $f(z)$ and $g(z)$ are two linearly independent solutions and can be taken as the fundamental solutions.

As $\varphi(z)$ is a periodic function with period ω, $f(z \pm \omega)$ and $g(z \pm \omega)$ must also be solutions of (1), so we have

$$f(z \pm \omega) = A_\pm f(z) + B_\pm g(z) \ .$$

Using the initial conditions for $f(z)$ and $g(z)$, we can determine $A_\pm$ and $B_\pm$. It is thus found that

$$f(z \pm \omega) = f(\pm\omega)f(z) + f'(\pm\omega)g(z) . \tag{4}$$

Similarly,

$$g(z \pm \omega) = g(\pm\omega)f(z) + g'(\pm\omega)g(z) . \tag{5}$$

If $\varphi(z)$ is an even function, then $f(-z)$ and $g(-z)$ must also be solutions of (1), so

$$f(-z) = Af(z) + Bg(z) .$$

Again by (2), we find $A = 1, B = 0$, so

$$f(-z) = f(z) . \tag{6}$$

Similarly,

$$g(-z) = -g(z) . \tag{7}$$

That is, *when $\varphi(z)$ is an even function, the fundamental solutions determined by the initial conditions (2) are such that $f(z)$ is an even function and $g(z)$ an odd function.* In virtue of this result, (1) cannot have two linearly independent solutions both being even or odd.

In case (6) and (7) hold, (4) and (5) reduce respectively to

$$f(z \pm \omega) = f(\omega)f(z) \pm f'(\omega)g(z) \tag{8}$$
$$g(z \pm \omega) = \pm g(\omega)f(z) + g'(\omega)g(z) . \tag{9}$$

Taking the lower signs in (8) and (9), and putting $z = \omega$, with due regard to the conditions (2), we have

$$1 = f^2(\omega) - f'(\omega)g(\omega) , \tag{10}$$
$$0 = g(\omega)\{g'(\omega) - f(\omega)\} . \tag{11}$$

If $g(\omega) \neq 0$, we obtain from (11)

$$f(\omega) = g'(\omega) . \tag{12}$$

If $g(\omega) = 0$, then we obtain $f(\omega) = \pm 1$ from (10). Substituting this result in (3), we have $g'(\omega) = 1/f(\omega) = \pm 1$, and hence (12) holds too.

12.3. Floquet Solution

If the solution $y(z)$ of the equation

$$\frac{d^2 y}{dz^2} + \{\lambda - \varphi(z)\}\, y = 0 \tag{1}$$

with periodic coefficient $\varphi(z)$ has the following property:

$$y(z + \omega) = \sigma y(z) \,, \tag{2}$$

where σ is a constant independent of z and ω is the period of $\varphi(z)$, then $y(z)$ is called a *Floquet solution.*

Since any solution of (1) is determined by certain initial values, condition (2) is equivalent to

$$y(\omega) = \sigma y(0) \,, \quad y'(\omega) = \sigma y'(0) \,. \tag{3}$$

Now let us investigate the condition for which a solution $y(z)$ satisfying (3), hence also satisfying (2), exists. We assume

$$y(z) = A f(z) + B g(z) \,, \tag{4}$$

where $f(z)$ and $g(z)$ are the fundamental solutions introduced in the last section.

From (3) and the initial conditions for f and g [(2) of the preceding section], we have

$$y(\omega) = A f(\omega) + B g(\omega) = \sigma A \,,$$
$$y'(\omega) = A f'(\omega) + B g'(\omega) = \sigma B \,,$$

or

$$\{f(\omega) - \sigma\}\, A + g(\omega) B = 0 \,,$$
$$f'(\omega) A + \{g'(\omega) - \sigma\}\, B = 0 \,. \tag{5}$$

In order that A and B are not both zero, σ has to satisfy the determinant equation

$$\begin{vmatrix} f(\omega) - \sigma & g(\omega) \\ f'(\omega) & g'(\omega) - \sigma \end{vmatrix} = 0 \,, \tag{6}$$

or, by (3) of the last section

$$\sigma^2 - \{f(\omega) + g'(\omega)\}\, \sigma + 1 = 0 \,. \tag{7}$$

Putting

$$\sigma = e^{i\nu\omega} ,\tag{8}$$

we obtain from (7)

$$\cos \nu\omega = \frac{1}{2}\{f(\omega) + g'(\omega)\} .\tag{9}$$

If $\varphi(z) = \varphi(-z)$, i.e., $\varphi(z)$ is an even function, then, according to (10) of the last section, $f(\omega) = g'(\omega)$, and (9) becomes

$$\cos \nu\omega = f(\omega) .\tag{10}$$

The values of ν determined by (9) or (10) are called *characteristic exponents*. By putting the values of σ corresponding to the characteristic exponents into (5) and solving for A and B, we obtain Floquet solutions from (4).

We can always write the Floquet solution in the form

$$y(z) = e^{i\nu z}u(z) ,\tag{11}$$

where $u(z)$ is a periodic function with period ω. For, from (2) and (8), we have

$$u(z + \omega) = e^{-i\nu(z+\omega)}y(z + \omega) = e^{-i\nu z}y(z) = u(z) .$$

When $\omega = \pi$, we see from (9) and (10) that if ν_0 is a characteristic exponent, then $-\nu_0$ and $\pm\nu_0 + 2k(k = \pm 1, \pm 2, \ldots)$ are also such. And, if ν_0 is not an integer, then $e^{i\nu_0 z}u_1(z)$ and $e^{-i\nu_0 z}u_2(z)$ $(u_i(z + \pi) = u_i(z), i = 1, 2)$ are two linearly independent Floquet solutions, since the ratio $e^{2i\nu_0 z}u_1(z)/u_2(z)$ cannot be a constant, or $u_1(z)/u_2(z)$ would not be a periodic function with period π.

If ν_0 is an integer, then the Floquet solutions $e^{\pm i\nu_0 z}u(z)$ are periodic functions with period π or 2π; when ν_0 is an even number, the period is π and the solution is said to be *whole-period*; when ν_0 is an odd number, the period is 2π and the solution, *half-period*.

If ν_0 is a rational number $r/s, r$ and s being non-zero integers without a common factor, then the Floquet solutions $e^{\pm i\nu_0 z}u(z)$ are periodic functions with period $s\pi$ or $2s\pi$.

12.4. Periodic Solutions of Mathieu Equation

The problems on the solutions of Mathieu equation can be classified, in general, into two classes. One is that the parameters λ and q in the equation are constants to be determined by the periodic condition in the form of a functional relation between them. This class of problems, known as double-parameter eigenvalue problem, is the subject of the present section.

The second class of problem is to find the solutions for given values of λ and q, which will be treated in Sec. 12.11.

Let us investigate the periodic solutions of Mathieu equation

$$\frac{d^2y}{dz^2} + (\lambda - 2q\cos 2z)y = 0 . \tag{1}$$

We shall see that, only when the parameters λ and q satisfy certain functional relation [see (8) below], the equation has periodic solution with period π or 2π. The function $\lambda(q)$ expressing this relation is then called the *eigenvalue* of (1), q being a parameter. Corresponding to each eigenvalue, there is only one solution with period π or 2π, unless $q = 0, \lambda = m^2 (m = 1, 2, \ldots)$. In the latter cases, the solutions of (1) are the well-known periodic functions $\cos mz$ and $\sin mz$.

Again, let $f(z)$ and $g(z)$ be the two fundamental solutions of (1), which satisfy the initial conditions of Eq. (2) of Sec. 12.2. Since the coefficient of (1) is an even periodic function with period π, we have, from the results obtained in Sec. 12.2,

$$f(z) = f(-z) , \quad g(z) = -g(-z) , \tag{2}$$
$$f(z \pm \pi) = f(\pi)f(z) \pm f'(\pi)g(z) , \tag{3}$$
$$g(z \pm \pi) = \pm g(\pi)f(z) + g'(\pi)g(z) , \tag{4}$$
$$1 = f^2(\pi) - f'(\pi)g(\pi) , \tag{5}$$
$$f(\pi) = g'(\pi) . \tag{6}$$

Besides, according to Eq. (10) of Sec. 12.3, the characteristic exponents ν satisfy the equation

$$\cos \nu\pi = f(\pi; \lambda, q) , \tag{7}$$

where, the dependence of the solution on the parameters λ and q is expressed explicitly by $f(z; \lambda, q)$.

From (7), we see that only when λ and q satisfy the relation

$$f(\pi; \lambda, q) = \pm 1 \tag{8}$$

will ν be an integer and the equation has solutions with period π or 2π. [See the last paragraph of the preceding section.]

Now we come to the proof that, when λ and q satisfy (8), at least one of the two fundamental solutions, $f(z)$ and $g(z)$, is a periodic function with period π or 2π. When (8) is satisfied, we know from (5) that $f'(\pi)g(\pi) = 0$, i.e., either $f'(\pi) = 0$ or $g(\pi) = 0$, or $f'(\pi) = g(\pi) = 0$.

If $f'(\pi) = 0$ while $g(\pi) \neq 0$, then, from (3) and (4), by virtue of (6), we obtain

$$f(z \pm \pi) = f(\pi)f(z) , \tag{9}$$

$$g(z \pm \pi) = \pm g(\pi)f(z) + f(\pi)g(z) . \tag{10}$$

Using (8) again, we see immediately that, in this case, $f(z)$ is a function with period π (when $f(\pi) = 1$) or with period 2π (when $f(\pi) = -1$), while $g(z)$ is not a periodic function.

If $g(\pi) = 0$ while $f'(\pi) \neq 0$, then $g(z)$ is a function with period π (when $f(\pi) = 1$) or with period 2π (when $f(\pi) = -1$). The proof is similar.

Only when $f'(\pi)$ and $g(\pi)$ equal to zero simultaneously, $f(z)$ and $g(z)$ are both solutions with period π or 2π. We shall show in the following section that this occurs only for $q = 0$.

12.5. Fourier Expansion of the Floquet Solution

Let

$$y(z) = e^{i\nu z}u(z) \tag{1}$$

be a Floquet solution [Sec. 12.3] of Mathieu equation, where $u(z)$ is a function with period π. $u(z)$ is analytic in the whole z-plane, since Mathieu equation has no singularity in the finite region. By the transformation $z = (\ln t)/2i$, the function

$$v(t) = u\left(\frac{1}{2i}\ln t\right) ,$$

which is single-valued and analytic in the whole t-plane except for the point $t = 0$, can be expanded in a Laurent series

$$v(t) = \sum_{k=-\infty}^{\infty} c_k t^k , \quad 0 < a \leq |t| \leq b < \infty .$$

Returning to the variable z, we have

$$y(z) = \sum_{k=-\infty}^{\infty} c_k e^{i(\nu+2k)z} , \tag{2}$$

which is the Fourier expansion of the Floquet solution; the series is absolutely and uniformly convergent in an arbitrary strip domain parallel to the real axis of the z-plane, since the Laurent expansion converges absolutely and uniformly in $a \leq |t| \leq b$, while $|t| = |e^{i2z}| = e^{-2y}, y = \text{Im}(z)$.

When $k \to \pm\infty$,

$$\lim_{k \to \pm\infty} |c_k|^{\frac{1}{|k|}} = 0 , \tag{3}$$

since the ring-shape domain for the above Laurent expansion can have its inner radius arbitrarily small (approaching zero) and its outer radius arbitrarily large (approaching infinity).

Let us determine the expansion coefficients c_k. Substituting the series (2) into Mathieu equation, we have

$$\sum_{k=-\infty}^{\infty} \left\{ \left[\lambda - (\nu + 2k)^2\right] c_k - q(c_{k-1} + c_{k+1}) \right\} e^{i(\nu+2k)z} = 0 ,$$

from which we obtain the recurrence relation among the consecutive coefficients:

$$\left[\lambda - (\nu + 2k)^2\right] c_k - q(c_{k-1} + c_{k+1}) = 0 . \tag{4}$$

If $q \neq 0$, (4) can be written as

$$c_{k+1} - D_k c_k + c_{k-1} = 0 , \tag{5}$$

where

$$D_k = \left[\lambda - (\nu + 2k)^2\right]/q \quad (k = 0, \pm 1, \pm 2, \dots) . \tag{6}$$

(4) or (5) is a 3-term recurrence relation. Reader can refer to Sec. 1.8 of Meixner and Schäfke (1954) for its solution.

We now show that, since $c_k \to 0$ when $k \to \pm\infty$ [see (3)], the solution of (5), except for a constant factor, is unique for given ν, λ and q. Suppose that there are two solutions c_k and $c'_k (k = 0, \pm 1, \pm 2, \dots)$; c'_k satisfy

$$c'_{k+1} - D_k c'_k + c'_{k-1} = 0 . \tag{7}$$

Multiplying (5) and (7) respectively by c'_k and c_k, and subtracting, we have

$$c'_k c_{k+1} - c_k c'_{k+1} = c'_{k-1} c_k - c_{k-1} c'_k ,$$

i.e.,

$$\delta(c_k, c'_k) \equiv c'_k c_{k+1} - c_k c'_{k+1} = \text{const. (independent of } k) . \tag{8}$$

Now, we know that $c_k \rightarrow 0$ (when $k \rightarrow \pm\infty$). Therefore, this constant has to be zero and

$$\frac{c_{k+1}}{c'_{k+1}} = \frac{c_k}{c'_k} \quad (k = 0, \pm 1, \pm 2, \ldots)$$

which proves the said uniqueness.

When ν is an integer, $y(z)$ of (1) is a periodic function. If ν is an even number, then $y(z)$ is a whole-period solution with period π; if ν is odd, then $y(z)$ is a half-period solution with period 2π. From the uniqueness property showed above, we know that if $q \neq 0$, then, for a definite ν (we may assume $\nu = 0$ or 1, since the addition of an arbitrary even number does not affect the result), corresponding to an eigenvalue $\lambda(q)$ satisfying (8) of the last section, the values of the coefficients c_k $(k = 0, \pm 1, \pm 2, \ldots)$ are unique, except for a constant factor, and hence the solution is also unique. Thus, there can only be one periodic solution. But if $q = 0$, then (4) determining the coefficients becomes

$$[\lambda - (\nu + 2k)^2]c_k = 0 ,$$

from which we see that, in this case in general, all the coefficients c_k must vanish and we have a trivial (zero) solution, unless $\lambda = m^2 (m = 1, 2, \ldots)$. In the latter case, we obtain from

$$[m^2 - (\nu + 2k)^2]c_k = 0$$

two non-zero coefficients c_k, where $2k = -\nu \pm m$; the corresponding periodic solutions are $e^{\pm imz}$.

Continued fraction of the consecutive coefficients

From (5), we have

$$\frac{c_{k-1}}{c_k} = D_k - \frac{c_{k+1}}{c_k} ,$$

of which the reciprocal is

$$\frac{c_k}{c_{k-1}} = \cfrac{1}{D_k - \cfrac{c_{k+1}}{c_k}} = \cfrac{1}{D_k - \cfrac{1}{D_{k+1} - \cfrac{c_{k+2}}{c_{k+1}}}} = \cdots$$

$$= \cfrac{1}{D_k - \cfrac{1}{D_{k+1} - \cfrac{1}{D_{k+2} - \cdots}}} \tag{9}$$

It can be shown that this infinite continued fraction is convergent [cf. Meixner and Schäfke (1954) Theorem 4 of Sec. 1.8].

For large k, we obtain from (9) and (6)

$$
\frac{c_k}{c_{k-1}} = \frac{1}{D_k + qO(k^{-2})} = \frac{1}{D_k}\left\{1 + q^2 O(k^{-4})\right\}
$$
$$
= \frac{q}{\lambda - (\nu + 2k)^2}\left\{1 + q^2 O(k^{-4})\right\} . \tag{10}
$$

When k is a negative integer, proceed as above, we obtain

$$
\frac{c_k}{c_{k+1}} = \cfrac{1}{D_k - \cfrac{1}{D_{k-1} - \cfrac{1}{D_{k-2} - \dots}}} . \tag{11}
$$

For large k, we have

$$
\frac{c_k}{c_{k+1}} = \frac{1}{D_k}\left\{1 + q^2 O(k^{-4})\right\} = \frac{q}{\lambda - (\nu + 2k)^2}\left\{1 + q^2 O(k^{-4})\right\} . \tag{12}
$$

It is observed from (10) and (12) that $c_k \neq 0$ when $|k|$ is sufficiently large.

12.6. Formulae for Computing Eigenvalues $\lambda(q)$

In the relation [Eq. (7) of Sec. 12.4],

$$
\cos\nu\pi = f(\pi; \lambda, q)
$$

among λ, q, ν, there appears the fundamental solution $f(z; \lambda, q)$. Although it is not difficult to obtain the power series representation of $f(z; \lambda, q)$, it is not convenient for the purpose of evaluating the eigenvalues $\lambda(q)$. In this section, we shall derive some formulae in the form of continued fractions which are important in numerical calculations.

Assuming $q \neq 0$, by (9) and (6) of the preceding section, we have

$$
\frac{c_k}{c_{k-1}} = \cfrac{q}{\lambda - (\nu + 2k)^2 - \cfrac{q^2}{\lambda - (\nu + 2k + 2)^2 - \cfrac{q^2}{\lambda - (\nu + 2k + 4)^2 - \dots}}} . \tag{1}
$$

Similarly, by (11) of the same section, we have

$$
\frac{c_{k-1}}{c_k} = \cfrac{q}{\lambda - (\nu + 2k - 2)^2 - \cfrac{q^2}{\lambda - (\nu + 2k - 4)^2 - \cfrac{q^2}{\lambda - (\nu + 2k - 6)^2 - \dots}}} \tag{2}
$$

Comparing these two expressions, we see that, for any value of $k(k = 0, \pm 1, \pm 2, \ldots)$, there is the relation

$$\lambda - (\nu + 2k)^2 - \cfrac{q^2}{\lambda - (\nu + 2k + 2)^2 - \cfrac{q^2}{\lambda - (\nu + 2k + 4)^2 - \ldots}}$$
$$= \cfrac{q^2}{\lambda - (\nu + 2k - 2)^2 - \cfrac{q^2}{\lambda - (\nu + 2k - 4)^2 - \cfrac{q^2}{\lambda - (\nu + 2k - 6)^2 - \ldots}}} .$$

$$(3)$$

Let us now look for the formulae for computing the eigenvalues $\lambda(q)$.

(i) *Whole-period solution*; period $= \pi, \nu$ even.

In (3), put $\nu = 0, k = 0$. If both sides of the resulting equation are finite (the case $c_0 \neq 0$), we have

$$\lambda = - \cfrac{2q^2}{4 - \lambda - \cfrac{q^2}{16 - \lambda - \cfrac{q^2}{36 - \lambda - \ldots}}} ; \qquad (4)$$

if both sides are infinite (this happens when $c_0 = 0$; cf. (1) and the following section), then we have

$$\lambda - 4 = - \cfrac{q^2}{16 - \lambda - \cfrac{q^2}{36 - \lambda - \ldots}} . \qquad (5)$$

Both (4) and (5) hold for any even number ν. For, if $\nu = 2n$, then, by choosing $k = -n$, we obtain again these equations from (3).

(4) and (5) are formulae for computing the eigenvalues corresponding to the whole-period solutions. From them, we find a sequence of $\lambda(q)$. We shall denote the eigenvalues determined from (4) by $a_{2n}(q)$, which satisfy the condition $a_{2n}(0) = (2n)^2$, and denote the eigenvalues determined from (5) by $b_{2n+2}(q)$, which satisfy the condition $b_{2n+2}(0) = (2n + 2)^2$.

From (4), we find the formula for computing $\lambda = a_{2n}(q)(n = 1, 2, \ldots)$ to be

$$\lambda - (2n)^2 - \cfrac{q^2}{\lambda - (2n-2)^2 - \ldots \cfrac{}{\ldots - \cfrac{q^2}{\lambda - 4 - \cfrac{2q^2}{\lambda}}}}$$

$$= -\cfrac{q^2}{(2n+2)^2 - \lambda - \cfrac{q^2}{(2n+4)^2 - \lambda - \ldots}}, \tag{6}$$

and from (5), we find that for computing $\lambda = b_{2n+2}(q)$:

$$\lambda - (2n+2)^2 - \cfrac{q^2}{\lambda - (2n)^2 - \ldots \cfrac{}{\ldots - \cfrac{q^2}{\lambda - 4}}}$$

$$= -\cfrac{q^2}{(2n+4)^2 - \lambda - \cfrac{q^2}{(2n+6)^2 - \lambda - \ldots}}. \tag{7}$$

We shall give a proof of (6) below; that of (7) is similar.

From (4), we have

$$-\frac{\lambda}{2q^2} = \cfrac{1}{4 - \lambda - \cfrac{q^2}{16 - \lambda - \cfrac{q^2}{36 - \lambda - \ldots}}},$$

hence,

$$4 - \lambda - \cfrac{q^2}{16 - \lambda - \cfrac{q^2}{36 - \lambda - \ldots}} = -\frac{2q^2}{\lambda},$$

which is exactly (6) for $n = 1$. Then, by induction we have (6) in general.

(ii) *Half-period solution*: period $= 2\pi, \nu$ odd.

By putting $\nu = 1, k = 0$ in (3), we obtain

$$\lambda - 1 - \cfrac{q^2}{\lambda - 9 - \cfrac{q^2}{\lambda - 25 - \ldots}}$$

$$= \cfrac{q^2}{\lambda - 1 - \cfrac{q^2}{\lambda - 9 - \cfrac{q^2}{\lambda - 25 - \ldots}}} = \pm q .$$

As the first equality is of the form $\alpha = q^2/\alpha$, we have $\alpha = \pm q$.

Taking the positive sign, we have

$$\lambda = 1 + q - \cfrac{q^2}{q - \lambda - \cfrac{q^2}{25 - \lambda - \ldots}} \ . \tag{8}$$

Taking the negative sign, we have

$$\lambda = 1 - q - \cfrac{q^2}{q - \lambda - \cfrac{q^2}{25 - \lambda - \ldots}} \ . \tag{9}$$

These are the formulae for computing the eigenvalues corresponding to the half-period solutions. They hold for any odd ν, since the same result is obtained from (3) by putting $k = -n$ when $\nu = 2n + 1$. We shall denote the eigenvalues determined from (8) and (9) respectively by $a_{2n+1}(q)$ and $b_{2n+1}(q)$, which satisfy the conditions $a_{2n+1}(0) = b_{2n+1}(0) = (2n + 1)^2$.

Besides, since (9) can be obtained from (8) by changing q to $-q$, we have

$$a_{2n+1}(q) = b_{2n+1}(-q) \ . \tag{10}$$

We can also derive from (8) and (9) formulae convenient for computing $a_{2n+1}(q)$ and $b_{2n+1}(q)$ $(n = 1, 2, \ldots)$:

$$\lambda - (2n+1)^2 - \cfrac{q^2}{\lambda - (2n-1)^2 - \ldots} $$
$$\ldots - \cfrac{q^2}{\lambda - q - \cfrac{q^2}{\lambda - 1 - q}}$$
$$= - \cfrac{q^2}{(2n+3)^2 - \lambda - \cfrac{q^2}{(2n+5)^2 - \lambda - \ldots}} \ , \tag{11}$$

$$\lambda - (2n+1)^2 - \cfrac{q^2}{\lambda - (2n-1)^2 - \ldots} $$
$$\ldots - \cfrac{q^2}{\lambda - q - \cfrac{q^2}{\lambda - 1 - q}}$$
$$= - \cfrac{q^2}{(2n+3)^2 - \lambda - \cfrac{q^2}{(2n+5)^2 - \lambda - \ldots}} \ . \tag{12}$$

All the formulae (4)–(12) given above are of importance in numerical calculations.

12.7. Mathieu Functions $ce_m(z)$, $m = 0,1,2,\ldots$ and $se_m(z)$, $m = 1,2,\ldots$

According to the results obtained in the last section, the periodic solution of Mathieu equation

$$\frac{d^2y}{dz^2} + (\lambda - 2q\cos 2z)y = 0$$

can be classified into four kinds: the whole-period solutions $ce_{2n}(z, q)$ and $se_{2n+2}(z, q)$ corresponding respectively to the eigenvalues $a_{2n}(q)$ and $b_{2n+2}(q)$; the half-period solutions $ce_{2n+1}(z, q)$ and $se_{2n+1}(z, q)$ corresponding to the eigenvalues $a_{2n+1}(q)$ and $b_{2n+1}(q)$, respectively; $n = 0, 1, 2, \ldots$. The symbols denoting these various functions manifest the forms of the respective Fourier expansions (see below for the proof):

$$ce_{2n}(z, q) = \sum_{r=0}^{\infty} A_{2r} \cos 2rz , \tag{1}$$

$$se_{2n+2}(z, q) = \sum_{r=0}^{\infty} B_{2r+2} \sin[(2r + 2)z] , \tag{2}$$

$$ce_{2n+1}(z, q) = \sum_{r=0}^{\infty} A_{2r+1} \cos[(2r + 1)z] , \tag{3}$$

$$se_{2n+1}(z, q) = \sum_{r=0}^{\infty} B_{2r+1} \sin[(2r + 1)z] , \tag{4}$$

where $A_{2r}, A_{2r+1}, B_{2r+1}, B_{2r+2}$ are functions of q.

There are two sorts of normalization for $ce_m(z, q)$ and $se_m(z, q)$. One is to specify that the coefficient A_m of $\cos mz$ in the expansions (1) and (3) of $ce_m(z, q)$ and that of B_m of $\sin mz$ in the expansions (2) and (4) of $se_m(z, q)$ are both equal to 1. The other requires that

$$\frac{1}{\pi} \int_0^{2\pi} ce_m^2(z + q)dz = \frac{1}{\pi} \int_0^{2\pi} se_m^2(z, q)dz = 1 , \tag{5}$$

i.e.,

$$2A_0^2 + \sum_{r=1}^{\infty} A_{2r}^2 = \sum_{r=0}^{\infty} A_{2r+1}^2 = \sum_{r=0}^{\infty} B_{2r+1}^2 = \sum_{r=0}^{\infty} B_{2r+2}^2 \tag{6}$$

$$= 1 .$$

Now we give the proofs of (1) – (4) based on the results obtained in the last section.

(i) *Whole-period solutions* $ce_{2n}(z)$, $se_{2n+2}(z)$.

If the eigenvalue $\lambda(q)$ is equal to $a_0(q)$ determined by (4) of the last section, then by (1) of that section, putting $\nu = 0, k = 0$, we have

$$\frac{c_0}{c_{-1}} = \cfrac{q}{\lambda - \cfrac{q^2}{\lambda - 4 - \dots}} . \tag{7}$$

Hence, if $q \neq 0$, then, $c_0 \neq 0$.

Now, putting $\nu = 0, k = 1$ in (2) of the last section, we have

$$\frac{c_0}{c_1} = \cfrac{q}{\lambda - \cfrac{q^2}{\lambda - 4 - \dots}} .$$

The r.h.s. of it is the same as the r.h.s. of (7). Thus, $c_{-1} = c_1$. By the recurrence relation Eq. (4) of Sec. 12.5, putting $\nu = 0$ in it and let $k = \pm 1$, we obtain

$$(\lambda - 4)c_1 - q(c_0 + c_2) = 0$$
$$(\lambda - 4)c_{-1} - q(c_{-2} + c_0) = 0 .$$

Hence we have $c_{-2} = c_2$. By induction, we can show that $c_{-k} = c_k, k = 0, 1, 2, \dots$. Denoting the corresponding solution by $ce_0(z)$, we obtain (1) above for the case $n = 0$ by virtue of Eq. (2) of Sec. 12.5.

To the eigenvalues $a_{2n}(q)(n = 1, 2, \dots)$ there are similar results. Putting in (1) of the preceding section $\nu = 2n, k = -n$, we have

$$\frac{c_{-n}}{c_{-n-1}} = \cfrac{q}{\lambda - \cfrac{q^2}{\lambda - 4 - \dots}} .$$

Thus $c_{-n} \neq 0$. Again, by (2) of the preceding section, let $\nu = 2n, k = -n+1$, we obtain

$$\frac{c_{-n}}{c_{-n+1}} = \cfrac{q}{\lambda - \cfrac{q^2}{\lambda - 4 - \dots}} ,$$

hence, $c_{-n-1} = c_{-n+1}$. Using the recurrence relation as before, we have $c_{-n-k} = c_{-n+k}, k = 0, 1, 2, \dots$. Denoting the corresponding solutions by $ce_{2n}(z)$ [Eq. (2) of Sec. 12.5], we have (1) for the general case.

$ce_{2n}(z)$ are even functions; $ce_{2n}(z) = ce_{2n}(-z), n = 0, 1, 2, \dots$.

If the eigenvalues $\lambda(q)$ are those $b_{2n+2}(q)$ determined by (5) of the last section, assuming $\nu = 2n$, then by (2) of that section, setting $k = -n$, we see that $c_{-n} = 0$. Thus, from the recurrence relation between the coefficients, we obtain $c_{-n+k} = -c_{-n-k}, k = 0, 1, 2, \ldots$. Denoting the corresponding solutions by $\mathrm{se}_{2n+2}(z)$ [Eq. (2) of Sec. 12.5], we have (2) above.

$\mathrm{se}_{2n+2}(z)$ are odd functions; $\mathrm{se}_{2n+2}(z) = -\mathrm{se}_{2n+2}(-z), n = 0, 1, 2, \ldots$.

(ii) *Half-period solutions*: $\mathrm{ce}_{2n+1}(z), \mathrm{se}_{2n+1}(z)$.

If $\lambda(q)$ are the eigenvalues $a_{2n+1}(q)$ determined by (8) of the last section, then by (1) of that section, putting $\nu = 2n + 1, k = -n$, we have $c_{-n} = c_{-n-1}$. Using again the recurrence relation, we obtain $c_{k-n} = c_{-k-n-1}, k = 0, 1, 2, \ldots$. Denoting the corresponding solutions by $\mathrm{ce}_{2n+1}(z)$ [Eq. (2) of Sec. 12.5], we have

$$
\begin{aligned}
\mathrm{ce}_{2n+1}(z) &= \sum_{k=-\infty}^{\infty} c_k e^{i(2n+1+2k)z} = \sum_{s=-\infty}^{\infty} c_{s-n} e^{i(2s+1)z} \\
&= \sum_{s=0}^{\infty} c_{s-n} e^{i(2s+1)z} + \sum_{s=-1}^{-\infty} c_{s-n} e^{i(2s+1)z} \\
&= \sum_{s=0}^{\infty} c_{s-n} e^{i(2s+1)z} + \sum_{s=0}^{\infty} c_{-s-n-1} e^{-i(2s+1)z} \\
&= 2 \sum_{s=0}^{\infty} c_{s-n} \cos\left[(2s+1)z\right] ,
\end{aligned}
$$

which proves (3).

$\mathrm{ce}_{2n+1}(z)$ are even functions; $\mathrm{ce}_{2n+1}(z) = \mathrm{ce}_{2n+1}(-z), n = 0, 1, 2, \ldots$.

If $\lambda(q)$ are those eigenvalues $b_{2n+1}(q)$ determined by (9) of the last section, then from (2) of that section, putting $\nu = 2n + 1, k = -n$, we obtain $c_{-n-1} = -c_{-n}$. Using the recurrence relation, we have $c_{-k-n-1} = -c_{k-n}, k = 0, 1, 2, \ldots$. Denoting the corresponding solutions by $\mathrm{se}_{2n+1}(z)$ [Eq. (2) of Sec. 12.5], we have

$$
\begin{aligned}
\mathrm{se}_{2n+1}(z) &= \sum_{k=-\infty}^{\infty} c_k e^{i(2n+1+2k)z} = \sum_{s=-\infty}^{\infty} c_{s-n} e^{i(2s+1)z} \\
&= 2i \sum_{s=0}^{\infty} c_{s-n} \sin\left[(2s+1)z\right] ,
\end{aligned}
$$

which proves (4).

$\mathrm{se}_{2n+1}(z)$ are odd functions; $\mathrm{se}_{2n+1}(z) = -\mathrm{se}_{2n+1}(-z), n = 0, 1, 2, \ldots$.

Orthogonality relations

From the Fourier expansions (1) – (4) of Mathieu functions $ce_m(z, q)$ and $se_m(z, q)$, we obtain immediately the following orthogonality relations:

$$\int_0^{2\pi} ce_m(z, q) ce_{m'}(z, q) dz = 0 \quad (m \neq m') , \tag{8}$$

$$\int_0^{2\pi} se_m(z, q) se_{m'}(z, q) dz = 0 \quad (m \neq m') , \tag{9}$$

$$\int_0^{2\pi} ce_m(z, q) se_{m'}(z, q) dz = 0 . \tag{10}$$

The coefficients A_s, B_s in the Fourier expansions can be found from the recurrence relations similar to Eq. (4) of Sec. 12.5. We shall discuss this problem in detail in Sec. 12.9.

12.8. Expansion of $\lambda_\nu(q)$ in Powers of q

For small enough q, the parameter λ in the Mathieu equation can be expanded in a power series of q. We shall use the notation $\lambda_\nu(q)$ to express explicitly the dependence of λ on q and ν.

In Eq. (3) of Sec. 12.6, putting $k = 0$, we have

$$\lambda_\nu(q) = \nu^2 + \cfrac{q^2}{\lambda - (\nu+2)^2 - \cfrac{q^2}{\lambda - (\nu+4)^2 - \dots}}$$
$$+ \cfrac{q^2}{\lambda - (\nu-2)^2 - \cfrac{q^2}{\lambda - (\nu-4)^2 - \dots}} . \tag{1}$$

For sufficiently small q, if $\nu \neq integer$ (except 0), we see from (1) that

$$\lambda_\nu(q) = \nu^2 + O(q^2) .$$

Substituting this result in the r.h.s. of (1), we obtain the second order approximation

$$\begin{aligned}
\lambda_\nu(q) &= \nu^2 + \frac{q^2}{\nu^2 - (\nu+2)^2 + O(q^2)} + \frac{q^2}{\nu^2 - (\nu-2)^2 + O(q^2)} + O(q^4) \\
&= \nu^2 + \frac{q^2}{-4(\nu+1)} + \frac{q^2}{4(\nu-1)} + O(q^4) \\
&= \nu^2 + \frac{q^2}{2(\nu^2 - 1)} + O(q^4) .
\end{aligned} \tag{2}$$

By iteration, we can attain approximation to any order. The approximation to the fourth order is

$$\lambda_\nu(q) = \nu^2 + \frac{1}{2(\nu^2 - 1)}q^2 + \frac{5\nu^2 + 7}{32(\nu^2 - 1)^3(\nu^2 - 4)}q^4$$
$$+ \frac{9\nu^4 + 58\nu^2 + 29}{64(\nu^2 - 1)^5(\nu^2 - 4)(\nu^2 - 9)}q^6 + O(q^8) . \tag{3}$$

As the order of approximation increased, the computations become more tedious.

When ν is an integer m (which is most important), the above formulae fail in general. We have to expand the eigenvalue $\lambda_m(q)$ in powers of q by Eqs. (4) – (12) of Sec. 12.6. Using the symbols $a_m(q)$ and $b_m(q)$ in the last section to denote these eigenvalues and arranging them in ascending order (assume $q > 0$), we have the following:

$$a_0 = -\frac{1}{2}q^2 + \frac{7}{128}q^4 - \frac{29}{2304}q^6 + \frac{68687}{188\,74368}q^8 + O(q^{10}) ,$$

$$b_1 = 1 - q - \frac{1}{8}q^2 + \frac{1}{64}q^3 - \frac{1}{1536}q^4 - \frac{11}{36864}q^5 + \frac{49}{5\,89824}q^6$$
$$- \frac{55}{94\,37184}q^7 - \frac{83}{353\,89440}q^8 + O(q^9) ,$$

$a_1 =$ the last expression for b_1 by changing q into $-q$,

$$b_2 = 4 - \frac{1}{12}q^2 + \frac{5}{13824}q^4 - \frac{289}{796\,26240}q^6$$
$$+ \frac{213941}{45\,86471\,42400}q^8 + O(q^{10}) ,$$

$$a_2 = 4 + \frac{5}{12}q^2 - \frac{763}{13824}q^4 + \frac{10\,02401}{796\,26240}q^6 - \frac{16690\,68401}{45\,86471\,42400}q^8 + O(q^{10}) ,$$

$$b_3 = q + \frac{1}{16}q^2 - \frac{1}{64}q^3 + \frac{13}{20480}q^4 + \frac{5}{16384}q^5 - \frac{1961}{235\,92960}q^6$$
$$+ \frac{609}{1048\,57600}q^7 + O(q^8) ,$$

a_3 = the last expression for b_3 by changing q into $-q$,

$$b_4 = 16 + \frac{1}{30}q^2 - \frac{317}{8\,64000}q^4 + \frac{10049}{27216\,00000}q^6 + O(q^8) ,$$

$$a_4 = 16 + \frac{1}{30}q^2 + \frac{433}{8\,64000}q^4 - \frac{5701}{27216\,00000}q^6 + O(q^8) ,$$

$$b_5 = 25 + \frac{1}{48}q^2 + \frac{11}{7\,74144}q^4 - \frac{1}{1\,47456}q^5 + \frac{37}{8918\,13888}q^6 + O(q^7) , \quad (4)$$

a_5 = the last expression for b_5 by changing q into $-q$,

$$b_6 = 36 + \frac{1}{70}q^2 + \frac{187}{439\,04000}q^4 - \frac{58\,61633}{9293\,59872\,00000}q^6 + O(q^8) ,$$

$$a_6 = 36 + \frac{1}{70}q^2 + \frac{187}{439\,04000}q^4 + \frac{67\,43617}{9293\,59872\,00000}q^6 + O(q^8) .$$

These results, except $a_0(q)$, cannot be obtained from (3). However, it can be shown that, *when* $m \geq 7$, we may get results corresponding to (3):

$$a_m, b_m = m^2 + \frac{1}{2(m^2-1)}q^2 + \frac{5m^2+7}{32(m^2-1)^3(m^2-4)}q^4$$
$$+ \frac{9m^4+58m^2+29}{64(m^2-1)^5(m^2-4)(m^2-9)}q^6 + O(q^7) . \quad (5)$$

The difference lies on that the order of the remainder is $O(q^7)$, not $O(q^8)$. But if $m \geq 8$, then the remainder is $O(q^8)$. Note further that (5) does not mean $a_m = b_m$; only that their difference is $O(q^7)$.

As the general term of the above power series expansion is difficult to obtain, it is not easy to determine the radius of convergence of the series precisely. Nevertheless, the following results concerning the region of convergence are given in Meixner and Schäfke (1954), Sec. 2.22:

The radius of convergence of the power series expansion of $\lambda_m(q)$ in q is always larger than ρ_m;

$$\rho_0 = 1 , \quad \rho_1 = 2 , \quad \rho_m = m - 1 \quad (m \geq 2) . \quad (6)$$

The radii of convergence of $a_0(q)$ and $a_2(q)$ are smaller than $5/2$.

12.9. Fourier Expansions of ce$_m(z)$ and se$_m(z)$ for Small q

There are commonly two approaches to this problem. One is the method of perturbation, which will be sketched in the following. In the Mathieu equation

$$\frac{d^2y}{dz^2} + (\lambda - 2q\cos 2z)y = 0 \tag{1}$$

assume

$$\lambda = m^2 + \sum_{r=1}^{\infty} \alpha_r q^r \tag{2}$$

$$y = \beta_0 + \sum_{r=1}^{\infty} \beta_r(z)q^r , \tag{3}$$

where $\beta_r(z)(r = 1, 2, \ldots)$ are periodic functions independent of q, and each of them is assumed not to contain a constant term: β_0 is not zero only for ce$_{2n}(z)$ [see Eqs. (1) – (4) of Sec. 12.7]. Substituting (2) and (3) in (1) and putting the coefficients of the different powers of q equal to zero, we obtain a sequence of inhomogeneous differential equations with constant coefficients for $\beta_r(z)$ and the parameters α_r to be determined. We then solve in succession the periodic solutions $\beta_r(z)$ of these equations and determine at the same time the values of α_r.

The other approach is to substitute directly the Fourier expansions of ce$_m(z)$, se$_m(z)$ in (1) and determine the coefficients of the expansions. We shall show it in detail [cf. Sec. 12.5].

Take ce$_0(z)$ as an example. Substituting the Fourier expansion

$$\text{ce}_0(z) = \sum_{s=0}^{\infty} A_{2s} \cos 2sz , \tag{4}$$

in (1) ($\lambda = a_0(q)$), we obtain

$$\sum_{s=0}^{\infty} \left[a_0 - (2s)^2\right] A_{2s} \cos 2sz$$

$$- q\left\{\sum_{s=1}^{\infty} A_{2s} \cos\left[(2s - 2)z\right] + \sum_{s=0}^{\infty} A_{2s} \cos\left[(2s + 2)z\right]\right\} = 0$$

from which there results the recurrence relations

$$a_0 A_0 - q A_2 = 0 , \tag{5}$$

$$\left[a_0 - (2s)^2\right] A_{2s} - q(A_{2s+2} + A_{2s-2}) = 0 \quad (s \geq 1) . \tag{6}$$

Let

$$v_s = A_{2s+2}/A_{2s} .$$ (7)

We have from (5) and (6)

$$v_0 = \frac{a_0}{q} , \quad v_s + \frac{1}{v_{s-1}} = \frac{a_0 - (2s)^2}{q} \quad (s \geq 1) ,$$ (8)

by which we can express the coefficients $A_{2s} (s = 1, 2, \dots)$ in terms of A_0; q may take any non-zero values.

When q is small enough, we can obtain from (8) and Eq. (4) of Sec. 12.8 the expansions of the coefficients A_{2s} in terms of powers of q as follows.

The second equation of (8) can be written as

$$v_{s-1} = \cfrac{q}{a_0 - (2s)^2 - qv_s}$$

$$= \cfrac{q}{a_0 - (2s)^2 - \cfrac{q^2}{a_0 - (2s+2)^2 - qv_{s+1}}} = \dots$$

$$= \cfrac{q}{a_0 - (2s)^2 - \cfrac{q^2}{a_0 - (2s+2)^2 - \dots}} .$$ (9)

For any fixed value of q, the last continued fraction is convergent.

If q is sufficiently small, taking the two terms written in (9) as an approximation and using the expansion of a_0 in Eq. (4) of Sec. 12.8, we have

$$v_{s-1} = \cfrac{q}{-(2s)^2 - \frac{q^2}{2} + \frac{q^2}{(2s+2)^2} + O(q^4)}$$

$$= -\frac{q}{(2s)^2} \left[1 - \frac{(2s+2)^2 - 2}{2(2s)^2(2s+2)^2} q^2 + O(q^4) \right] ,$$ (10)

by which we obtain

$$\frac{A_{2s}}{A_0} = \frac{A_{2s}}{A_{2s-2}} \frac{A_{2s-2}}{A_{2s-4}} \dots \frac{A_4}{A_2} \frac{A_2}{A_0} = v_{s-1} v_{s-2} \dots v_1 v_0$$

$$= \prod_{k=2}^{s} \left[-\frac{q}{(2k)^2} \right] \left[1 - \sum_{k=2}^{s} \frac{2k^2 + 4k + 1}{k^2(k+1)^2} \left(\frac{q}{4} \right)^2 + O(q^4) \right]$$

$$\times \left[-\frac{q}{2} + \frac{7}{128} q^3 + O(q^5) \right]$$

$$= (-)^s \frac{2}{(s!)^2} \left(\frac{q}{4} \right)^s \left[1 - \frac{7}{64} q^2 - \sum_{k=2}^{s} \frac{2k^2 + 4k + 1}{k^2(k+1)^2} \left(\frac{q}{4} \right)^2 + O(q^4) \right]$$

$$= (-)^s \frac{2}{(s!)^2} \left(\frac{q}{4} \right)^s \left[1 - \sum_{k=1}^{s} \frac{2k^2 + 4k + 1}{k^2(k+1)^2} \left(\frac{q}{4} \right)^2 + O(q^4) \right] .$$

Now

$$\sum_{k=1}^{s} \frac{2k^2 + 4k + 1}{k^2(k+1)^2} = 2\sum_{k=1}^{s} \frac{1}{k(k+1)} + \sum_{k=1}^{s} \frac{2k+1}{k^2(k+1)^2}$$

$$= 2\sum_{k=1}^{s} \left(\frac{1}{k} - \frac{1}{k+1}\right) + \sum_{k=1}^{s} \left[\frac{1}{k^2} - \frac{1}{(k+1)^2}\right]$$

$$= 2\left(1 - \frac{1}{s+1}\right) + \left[1 - \frac{1}{(s+1)^2}\right] = \frac{s(3s+4)}{(s+1)^2}.$$

Therefore,

$$\frac{A_{2s}}{A_0} = (-)^s \frac{2}{(s!)^2}\left(\frac{q}{4}\right)^s - (-)^s \frac{2s(3s+4)}{[(s+1)!]^2}\left(\frac{q}{4}\right)^{s+2} + O(q^{s+4}). \tag{11}$$

This is the desired formula. Approximations of higher orders can be derived similarly. The calculations, however, become more laborious.

The Fourier expansions of other $ce_m(z)$ and $se_m(z)$ can be obtained in a similar way [cf. Exs. 1 and 2 at the end of this chapter].

12.10. Infinite Determinant

For the sake of discussing the Hill equation in the following section, we shall give some ideas about the theory of infinite determinant needed. Let

$$D_m = \det|A_{ik}| \tag{1}$$

represents the determinant formed of the elements $A_{ik}(i, k = -m, -m +1, \ldots, m-1, m; m = \text{integer})$. If D_m approaches a definite limit D when $m \to \infty$, we say that the infinite determinant converges to D; if D does not exist, then the determinant is said to be divergent. $A_{ii}(i = 0, \pm 1, \pm 2, \ldots)$ are called the principal diagonal elements of D : A_{00}, the origin of the determinant and $A_{ik}(i \neq k)$ the off-diagonal elements.

Theorem: If both the product of the principal diagonal elements and the sum of the off-diagonal elements converge absolutely, then the infinite determinant is convergent.

Proof: Write the principal diagonal elements of the infinite determinant D as $1 + a_{ii}$ and the off-diagonal elements as $a_{ik}(i \neq k)$. Then according to the assumptions, $\Sigma_{i,k=-\infty}^{\infty}|a_{ik}|$ is convergent, and hence the infinite product

$$\overline{P} = \prod_{i=-\infty}^{\infty}\left(1 + \sum_{k=-\infty}^{\infty}|a_{ik}|\right) \tag{2}$$

is convergent [Theorem 4 of Sec. 1.6]. Let

$$P_m = \prod_{i=-m}^{m} \left(1 + \sum_{k=-m}^{m} a_{ik}\right), \quad \overline{P}_m = \prod_{i=-m}^{m} \left(1 + \sum_{k=-m}^{m} |a_{ik}|\right).$$

It is easily seen that every term of the development of the determinant D_m does occur in the product P_m, at most differing by a negative sign. Hence, to every term of the development of D_m, there corresponds a term in the expansion of $\overline{P}_m$, and the modulus of the corresponding term in $\overline{P}_m$ is equal to or greater than the modulus of the corresponding term in D_m.

Now, let us look at the difference $D_{m+p} - D_m$, p being an arbitrary positive integer. To each term of this difference, there always corresponds a term in the expansion of $\overline{P}_{m+p} - \overline{P}_m$. Moreover, the modulus of the latter is equal to or greater than that of the former. Hence,

$$|D_{m+p} - D_m| \le \overline{P}_{m+p} - \overline{P}_m.$$

As the limit of $\overline{P}_m$ [(2)] exists when $m \to \infty$, the limit of D_m also exists and the theorem is proved.

If a_{ik} are functions of a complex variable z, the corresponding infinite determinant is uniformly convergent provided that the infinite product (2) converges uniformly.

12.11. Hill Equation

We have discussed above the periodic solutions of Mathieu equation as solutions of an eigenvalue problem. Another type of problem about Mathieu equation is to find solutions for given values of λ and q. Of course, the solutions in such case are, in general, not periodic.

The equation we are to deal with is the more general Hill equation:

$$\frac{d^2u}{dz^2} + J(z)u = 0 \tag{1}$$

$J(z)$ is an even function with period equal to π, which can be expanded in a Fourier cosine series:

$$J(z) = \theta_0 + 2\theta_1 \cos 2z + 2\theta_2 \cos 4z + \ldots \tag{2}$$

with known coefficients θ_n. Besides, $\Sigma_{n=0}^{\infty} \theta_n$ is assumed to be absolutely convergent. If z is a complex variable, we assume $J(z)$ to be an analytic function in a strip region parallel to the real axis and containing the axis in it. Hence, it can also be expanded in a Fourier series of the form (2), and $\Sigma\theta_n$ also converges absolutely [cf. Sec. 12.5 on the Fourier expansion of functions of a complex variable].

After the discussion in Sec. 12.3, we may assume that the solution of (1) takes the form

$$u = e^{i\nu z} \sum_{n=-\infty}^{\infty} b_n e^{i2nz} = \sum_{n=-\infty}^{\infty} b_n e^{i(\nu+2n)z} . \tag{3}$$

Substituting in (1) and stipulating that $\theta_{-n} = \theta_n$, we have

$$-\sum_{n=-\infty}^{\infty} b_n(\nu + 2n)^2 e^{i(\nu+2n)z} + \sum_{k=-\infty}^{\infty} \theta_k e^{i2kz} \sum_{n=-\infty}^{\infty} b_n e^{i(\nu+2n)z} = 0 ,$$

from which we obtain a system of simultaneous equations for the coefficients b_n:

$$-(\nu + 2n)^2 b_n + \sum_{k=-\infty}^{\infty} \theta_k b_{n-k} = 0 \quad (n = 0, \pm 1, \dots) . \tag{4}$$

Dividing (4) by $\theta_0 - (\nu + 2n)^2$ gives

$$\sum_{m=-\infty}^{\infty} B_{nm} b_m = 0 \quad (n = 0, \pm 1, \dots) , \tag{5}$$

where

$$B_{nn} = 1$$

$$B_{nm} = \frac{\theta_{n-m}}{\theta_0 - (\nu + 2n)^2} \quad (m \neq n) . \tag{6}$$

The condition of the existence of solutions for (5) is that $\det |B_{nm}| = 0$, i.e.,

$$\Delta(\nu) = \begin{vmatrix} \cdots 1 & \frac{\theta_1}{\theta_0-(\nu-4)^2} & \frac{\theta_2}{\theta_0-(\nu-4)^2} & \frac{\theta_3}{\theta_0-(\nu-4)^2} & \frac{\theta_4}{\theta_0(\nu-4)^2} & \cdots \\ \cdots \frac{\theta_1}{\theta_0-(\nu-2)^2} & 1 & \frac{\theta_1}{\theta_0-(\nu-2)^2} & \frac{\theta_2}{\theta_0-(\nu-2)^2} & \frac{\theta_3}{\theta_0-(\nu-2)^2} & \cdots \\ \cdots \frac{\theta_2}{\theta_0-\nu^2} & \frac{\theta_1}{\theta_0-\nu^2} & 1 & \frac{\theta_1}{\theta_0-\nu^2} & \frac{\theta_2}{\theta_0-\nu^2} & \cdots \\ \cdots \frac{\theta_3}{\theta_0-(\nu+2)^2} & \frac{\theta_2}{\theta_0-(\nu+2)^2} & \frac{\theta_1}{\theta_0-(\nu+2)^2} & 1 & \frac{\theta_1}{\theta_0-(\nu+2)^2} & \cdots \\ \cdots \frac{\theta_4}{\theta_0-(\nu+4)^2} & \frac{\theta_3}{\theta_0-(\nu+4)^2} & \frac{\theta_2}{\theta_0-(\nu+4)^2} & \frac{\theta_1}{\theta_0-(\nu+4)^2} & 1 & \cdots \end{vmatrix} = 0$$

$$\tag{7}$$

which is the equation for the determination of the characteristic exponents ν of the Floquet solutions (3).

We shall now show that, from (7), we can derive a very simple formula for the determination of the characteristic exponents ν, namely,

$$\sin^2\left(\frac{\nu\pi}{2}\right) = \Delta(0)\sin^2\left(\frac{\sqrt{\theta_0}\,\pi}{2}\right) . \tag{8}$$

From (7), it is seen that $\Delta(\nu)$ is an even function of ν with period 2. Further, since the determinant is absolutely and uniformly convergent except for the points $\nu = \pm\sqrt{\theta_0} - 2n(n = 0, \pm 1, \ldots)$ [see the theorem in the last section], $\Delta(\nu)$ is a meromorphic function. And, when Im $(\nu) \to \pm\infty$, $\Delta(\nu) \to 1$.

The even function $\cot\frac{\pi}{2}(\nu - \sqrt{\theta_0}) - \cot\frac{\pi}{2}(\nu + \sqrt{\theta_0})$ possesses the same poles as $\Delta(\nu)$ and also has the period 2, together with the property that it is bounded when $\text{Im}(\nu) \to \pm\infty$. Hence, there exists a constant K such that

$$D(\nu) \equiv \Delta(\nu) - K\left[\cot\frac{\pi}{2}(\nu - \sqrt{\theta_0}) - \cot\frac{\pi}{2}(\nu + \sqrt{\theta_0})\right]$$

has no singularities in the whole ν-plane. In addition, because of its periodicity, $D(\nu)$ is bounded when $|\nu| \to \infty$. Then, according to Liouville's theorem, $D(\nu)$ has to be a constant. Let $\text{Im}(\nu) \to +\infty$, we see immediately that $D(\nu) = 1$, and

$$\Delta(\nu) = 1 + K\left[\cot\frac{\pi}{2}\left(\nu - \sqrt{\theta_0}\right) - \cot\frac{\pi}{2}\left(\nu + \sqrt{\theta_0}\right)\right] .$$

Putting $\nu = 0$, we obtain

$$K = \frac{1 - \Delta(0)}{2\cot\frac{\pi}{2}\sqrt{\theta_0}} .$$

Substituting this result into the preceding equation, after some manipulation, we see that the equation $\Delta(\nu) = 0$ for the characteristic exponents reduces to (8).

From (8), it is seen that the main problem in determining the characteristic exponents is to evaluate the infinite determinant

$$
\Delta(0) = \begin{vmatrix}
\cdots 1 & \frac{\theta_1}{\theta_0-16} & \frac{\theta_2}{\theta_0-16} & \frac{\theta_3}{\theta_0-16} & \frac{\theta_4}{\theta_0-16} & \cdots \\
\cdots \frac{\theta_1}{\theta_0-4} & 1 & \frac{\theta_1}{\theta_0-4} & \frac{\theta_2}{\theta_0-4} & \frac{\theta_3}{\theta_0-4} & \cdots \\
\cdots \frac{\theta_2}{\theta_0-0} & \frac{\theta_1}{\theta_0-0} & 1 & \frac{\theta_1}{\theta_0-0} & \frac{\theta_2}{\theta_0-0} & \cdots \\
\cdots \frac{\theta_3}{\theta_0-4} & \frac{\theta_2}{\theta_0-4} & \frac{\theta_1}{\theta_0-4} & 1 & \frac{\theta_1}{\theta_0-4} & \cdots \\
\cdots \frac{\theta_4}{\theta_0-16} & \frac{\theta_3}{\theta_0-16} & \frac{\theta_2}{\theta_0-16} & \frac{\theta_1}{\theta_0-16} & 1 & \cdots
\end{vmatrix}. \tag{9}
$$

When the values of $\theta_1, \theta_2, \ldots$ are sufficiently small, the most crude approximation is furnished by the determinant of order 3 with B_{00} as the center element:

$$
\Delta(0) \simeq 1 + \frac{2\theta_1^2}{\theta_0(4-\theta_0)} + \frac{2\theta_1^2\theta_2}{\theta_0(4-\theta_0)^2} - \frac{\theta_2^2}{(4-\theta_0)^2}. \tag{10}
$$

A better approximation is obtained by preserving in the development of the infinite determinant the terms involving $\theta_n^2 (n = 1, 2, \ldots)$; the result will be

$$
\Delta(0) \simeq 1 + \frac{\pi \cot\left(\pi\sqrt{\theta_0/2}\right)}{4\sqrt{\theta_0}} \left[\frac{\theta_1^2}{1^2 - \theta_0} + \frac{\theta_2^2}{2^2 - \theta_0} + \frac{\theta_3^2}{3^2 - \theta_0} + \cdots \right], \tag{11}
$$

which is proved as follows. First, look at the terms containing θ_1^2. The two 2×2 determinants marked in (9) with squares represent the typical contribution of them to $\Delta(0)$, namely,

$$
-2\theta_1^2 \sum_{r=0}^{\infty} \frac{1}{\theta_0 - (2r)^2} \frac{1}{\theta_0 - (2r+2)^2}
$$

$$
= -\frac{\theta_1^2}{2} \sum_{r=0}^{\infty} \frac{1}{2r+1} \left[\frac{1}{\theta_0 - (2r+2)^2} - \frac{1}{\theta_0 - (2r)^2} \right]
$$

$$
= -\frac{\theta_1^2}{2} \left\{ \sum_{r=1}^{\infty} \left(\frac{1}{2r-1} - \frac{1}{2r+1} \right) \frac{1}{\theta_0 - (2r)^2} - \frac{1}{\theta_0} \right\}
$$

$$
= -\frac{\theta_1^2}{2} \left\{ \sum_{r=1}^{\infty} \frac{2}{(2r)^2 - 1} \frac{1}{\theta_0 - (2r)^2} - \frac{1}{\theta_0} \right\}
$$

$$
= -\frac{\theta_1^2}{2} \left\{ 2\sum_{r=1}^{\infty} \frac{1}{\theta_0 - 1} \left[\frac{1}{(2r)^2 - 1} - \frac{1}{\theta_0 - (2r)^2} \right] - \frac{1}{\theta_0} \right\}.
$$

Now

$$\sum_{r=1}^{\infty} \frac{1}{(2r)^2 - 1} = \frac{1}{2}\sum_{r=1}^{\infty}\left[\frac{1}{2r-1} - \frac{1}{2r+1}\right] = \frac{1}{2} ;$$

therefore, the preceding expression is equal to

$$-\frac{\theta_1^2}{2(\theta_0 - 1)}\left\{\frac{1}{\theta_0} + \sum_{r=1}^{\infty}\frac{2}{\theta_0 - (2r)^2}\right\} = \frac{\theta_1^2}{1-\theta_0}\frac{\pi\cot\left(\pi\sqrt{\theta_0}/2\right)}{4\sqrt{\theta_0}} ,$$

the rational fraction expansion of $\cot z, z = \pi\sqrt{\theta_0}/2$ [Eq. (9) of Sec. 1.5], having been used in the last step.

The contributions to $\Delta(0)$ of terms involving θ_2^2 can be found from the two typical 3×3 determinants marked with squares in (9) as

$$-\theta_2^2\left\{\frac{1}{(\theta_0 - 4)^2} + 2\sum_{r=1}^{\infty}\frac{1}{\theta_0 - (2r+2)^2}\frac{1}{\theta_0 - (2r-2)^2}\right\} .$$

By similar calculation as above, it can be shown that this expression is equal to

$$\frac{\theta_2^2}{4 - \theta_0}\frac{\pi\cot\left(\pi\sqrt{\theta_0}/2\right)}{4\sqrt{\theta_0}} .$$

The contributions of terms involving θ_n^2 can be calculated in a similar way and is found to be

$$\frac{\theta_n^2}{n^2 - \theta_0}\frac{\pi\cot\left(\pi\sqrt{\theta_0}/2\right)}{4\sqrt{\theta_0}} .$$

Thus, (11) is proved.

After the characteristic exponents have been found, we can, in principle, determine the Fourier expansion coefficients b_n of the Floquet solution (3) from (5) or (4). Of course, this method of calculation is practical only when $\theta_1, \theta_2, \ldots$ are comparatively small.

For the Mathieu equation, $\theta_0 = \lambda, \theta_1 = -q, \theta_n = 0(n \geq 2)$, the system of equations (4) reduces to a three-term recurrence relation, and can be solved by the method of continued fractions described in Sec. 12.9.

12.12. Stable and Unstable Solutions of Mathieu Equation. Stable Region and Unstable Region

The solution of the Mathieu equation, which is bounded in the interval $-\infty < z < +\infty$, is called a *stable solution*; if the solution approaches infinity as $z \to +\infty$ or $z \to -\infty$, it is *unstable*.

From the form of the Floquet solution, $e^{i\nu z}u(z)$, where $u(z + \pi) = u(z)$, we see that the solution is stable only for real characteristic exponent ν. According to Eq. (7) of Sec. 12.4, this requires

$$|\cos \nu\pi| = |f(\pi; \lambda, q)| \leq 1 . \tag{1}$$

If ν is not an integer, the inequality holds; $e^{i\nu z}u(z)$ and $e^{-i\nu z}u(-z)$ are two linearly independent stable solutions. In the $\lambda - q$ plane, the region satisfying (1) is called a *stable region*; all solutions are finite in the interval $-\infty < z < +\infty$ for every pair of values (λ, q) lying in this stable region.

If

$$|\cos \nu\pi| = |f(\pi; \lambda, q)| > 1 , \tag{2}$$

ν must be a complex number with non-zero imaginary part. Then, $e^{\pm i\nu z}u(\pm z)$ are two linearly independent unstable solutions. The corresponding region in the $\lambda - q$ plane is called an *unstable region*.

When $\lambda(q)$ is an eigenvalue,

$$\cos \nu\pi = f(\pi; \lambda, q) = \pm 1 \tag{3}$$

[Eq. (8) of Sec. 12.4]. Then, there is a unique periodic solution (unless $q = 0, \lambda = 1, 4, 9, \ldots, m^2, \ldots$; [see Sec. 12.5]), which is stable. The other solution must be unstable as can be shown as follows.

We have proved in Sec. 12.4 that when (3) is satisfied, one of the fundamental solutions $f(z)$ and $g(z)$ is a periodic function. Let it be $f(z)$, then $f(z + \pi) = f(z)$. For the other solution $g(z)$, we have

$$g(z + \pi) = g(\pi)f(z) + g(z)$$

[Eq. (10) of Sec. 12.4; $f(\pi) = 1$]. Hence,

$$\frac{g(z + \pi)}{f(z + \pi)} = g(\pi) + \frac{g(z)}{f(z)} ,$$

from which we see that

$$\varphi(z) \equiv \frac{g(z)}{f(z)} - \frac{g(\pi)}{\pi}z$$

is a periodic function with period π, since

$$\varphi(z + \pi) = \frac{g(z + \pi)}{f(z + \pi)} - \frac{g(\pi)}{\pi}(z + \pi)$$

$$= \frac{g(z)}{f(z)} - \frac{g(\pi)}{\pi}z = \varphi(z) .$$

Thus, we have

$$g(z) = \frac{g(\pi)}{\pi} z f(z) + \varphi(z) f(z) .$$

Obviously, this other solution $g(z)$ cannot be stable. For, when $z \to \pm\infty$, $|g(z)| \to \infty$, unless $g(\pi) = 0$, which occurs only when $q = 0$ and $\lambda = 1, 4, 9, \ldots, m^2, \ldots$ [see Sec. 12.5].

If $g(z)$ is periodic, there is a similar result. Besides, the above conclusion holds also when the period is 2π instead of π.

We see in Fig. 40 that the curves $f(\pi; \lambda, q) = \pm 1$ are the border lines separating the stable and unstable regions in the $\lambda - q$ plane.

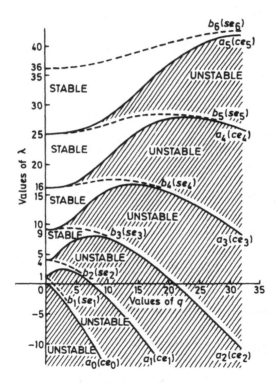

Fig. 40. Curves a_{2n}, b_{2n} are symmetric about the λ–axis; a_{2n+1}, b_{2n+1} are not, but $a_{2n+1}(-q) = b_{2n+1}(q)$. [After Mclachlan (1947), p. 40.]

12.13. Approximate Solutions of Mathieu Equation for $\lambda \gg q > 0$

Under the condition $\lambda \gg q > 0$, we may employ the following method (usually called the *WKBJ method*) to find approximate solutions of the Mathieu equation.

In the Mathieu equation

$$\frac{d^2 y}{dz^2} + (\lambda - q \cos 2z) y = 0 \tag{1}$$

let

$$y = e^{\beta \int^z w(z) dz} , \quad \beta = \sqrt{\lambda} . \tag{2}$$

We obtain the equation for w:

$$\frac{1}{\beta} \frac{dw}{dz} + w^2 + \rho^2 = 0 , \tag{3}$$

where

$$\rho^2 = 1 - \frac{2q}{\lambda} \cos 2z . \tag{4}$$

If $\beta = \sqrt{\lambda} \gg 1$, by discarding the first term of (3), we have

$$w = \pm i \rho = \pm i \left(1 - \frac{2q}{\lambda} \cos 2z \right)^{\frac{1}{2}} \simeq \pm i \left(1 - \frac{q}{\lambda} \cos 2z \right) ,$$

and obtain the approximate solutions:

$$y = e^{\beta \int_0^z w(z) dz} \simeq e^{\pm i \sqrt{\lambda} \left(z - \frac{q}{2\lambda} \sin 2z \right)} .$$

Take the two linearly independent solutions

$$y_1(z) \simeq \frac{1}{2} \left[e^{i\sqrt{\lambda} \left(z - \frac{q}{2\lambda} \sin 2z \right)} + e^{-i\sqrt{\lambda} \left(z - \frac{q}{2\lambda} \sin 2z \right)} \right]$$

$$= \cos \left[\sqrt{\lambda} \left(z - \frac{q}{2\lambda} \sin 2z \right) \right] , \tag{5}$$

$$y_2(z) \simeq \frac{1}{2i} \left[e^{i\sqrt{\lambda} \left(z - \frac{q}{2\lambda} \sin 2z \right)} - e^{-i\sqrt{\lambda} \left(z - \frac{q}{2\lambda} \sin 2z \right)} \right]$$

$$= \sin \left[\sqrt{\lambda} \left(z - \frac{q}{2\lambda} \sin 2z \right) \right] . \tag{6}$$

Let $q/2\lambda = h$. Using Eqs. (7) and (8) of Sec. 7.5, we obtain

$$y_1(z) \simeq \cos\sqrt{\lambda}z\,\cos(h\sin 2z) + \sin\sqrt{\lambda}z\,\sin(h\sin 2z)$$

$$= \cos\sqrt{\lambda}z$$

$$\times \sum_{n=0}^{\infty} \varepsilon_{2n}J_{2n}(h)\cos 4nz + \sin\sqrt{\lambda}z\sum_{n=0}^{\infty}\varepsilon_{2n+1}J_{2n+1}(h)\sin(4\pi+2)z$$

$$= \frac{1}{2}\sum_{n=0}^{\infty}\varepsilon_{2n}J_{2n}(h)\left[\cos\left(\sqrt{\lambda}-4n\right)z + \cos\left(\sqrt{\lambda}+4n\right)z\right]$$

$$+ \frac{1}{2}\sum_{n=0}^{\infty}\varepsilon_{2n+1}J_{2n+1}(h)\left[\cos\left(\sqrt{\lambda}-4n-2\right)z - \cos\left(\sqrt{\lambda}+4n+2\right)z\right]$$

$$= \frac{1}{2}\sum_{n=0}^{\infty}\varepsilon_n J_n(h)\cos\left(\sqrt{\lambda}-2n\right)z + \frac{1}{2}\sum_{n=-\infty}^{0}\varepsilon_n J_n(h)\cos\left(\sqrt{\lambda}-2n\right)z$$

$$= \sum_{n=-\infty}^{\infty}J_n(h)\cos\left(\sqrt{\lambda}-2n\right)z \ . \tag{7}$$

Similarly, we have

$$y_2(z) \simeq \sum_{n=-\infty}^{\infty}J_n(h)\sin\left(\sqrt{\lambda}-2n\right)z \ . \tag{8}$$

Since the value of $J_n(h)$ drops steeply as n increases [cf. Eq. (1) of Sec. 7.12], the series in (7) and (8) converge rapidly. Thus, they are of much advantageous, under the condition $\lambda >> q > 0$, in comparison with the Fourier expansions given in Sec. 12.9.

To obtain more precise approximations, we may set

$$w = w_0 + \frac{1}{\beta}w_1 + \frac{1}{\beta^2}w_2 + \dots \ . \tag{9}$$

Substituting into (3) and expanding in descending powers of β, we have

$$w_0^2 + \rho^2 + \frac{1}{\beta}(w_0' + 2w_0w_1)$$

$$+ \frac{1}{\beta^2}(w_1' + w_1^2 + 2w_0w_2) + \dots = 0 \ ,$$

from which we obtain a system of equations for approximations of different orders:

$$w_0^2 + \rho^2 = 0 \ ,$$
$$w_0' + 2w_0w_1 = 0 \ ,$$
$$w_1' + w_1^2 + 2w_0w_2 = 0, \ \dots \ . \tag{10}$$

The first equation of it gives $w_0 = \pm i\rho$, which is just the zeroth approximation already found. From the second equation of (10), we have

$$w_1 = -\frac{1}{2}\frac{w_0'}{w_0} \ ,$$

hence,

$$\int^z w_1 dz = \ln(A\rho^{-\frac{1}{2}}) \mp \frac{\pi i}{4} \ .$$

From the third equation of (10), we have

$$w_2 = \mp \frac{i}{8\rho^3}(2\rho\rho'' - 3\rho'^2) \ ,$$

$$\int^z w_2 dz = \mp \frac{i}{8}\int^z \frac{2\rho\rho'' - 3\rho'^2}{\rho^3} dz \ .$$

Substituting these results in (9) and (2), we obtain the second order approximation solutions:

$$y(z) = e^{\beta \int^z w\,dz} \simeq \text{constant} \times \rho^{-\frac{1}{2}} e^{\pm i\sqrt{\lambda}\int^z \left(\rho - \frac{2\rho\rho'' - 3\rho'^2}{8\lambda\rho^3}\right)dz} \ , \tag{11}$$

or,

$$\begin{matrix} y_1(z) \\ y_2(z) \end{matrix} \simeq \text{constant} \times \rho^{-\frac{1}{2}} \begin{matrix} \cos \\ \sin \end{matrix} \left[\sqrt{\lambda}\int_0^z \varphi(z)dz\right] \ , \tag{12}$$

where

$$\varphi(z) \equiv \rho - (2\rho\rho'' - 3\rho'^2)/8\lambda\rho^3 \ , \tag{13}$$

ρ being function of z [(4)].

If we discard the term w_2, we shall have the first-order approximation:

$$\sqrt{\lambda}\int_0^z \varphi(z)dz \simeq \int_0^z (\lambda - 2q\cos 2z)^{\frac{1}{2}} dz$$

$$= (\lambda + 2q)^{\frac{1}{2}} \int_0^z (1 - k^2\cos^2 z)^{\frac{1}{2}} dz$$

$$= (\lambda + 2q)^{\frac{1}{2}} E(z, k) \quad \left(k^2 = \frac{4q}{\lambda + 2q}\right) \ .$$

$E(z, k)$ being the elliptic integral of the second kind [Eq. (1) of Sec. 10.9], with the modulus

$$k = 2\left(\frac{q}{\lambda + 2q}\right)^{\frac{1}{2}} < 1 \ .$$

The corresponding solutions are:

$$\begin{matrix} y_1(z) \\ y_2(z) \end{matrix} \simeq \text{constant} \times (\lambda - 2q \cos 2z)^{-\frac{1}{4}} \begin{matrix} \cos \\ \sin \end{matrix} \left[(\lambda + 2q)^{\frac{1}{2}} E(z, k) \right] . \tag{14}$$

12.14. Integral Equations for Mathieu Functions

Let the integral solution of the Mathieu equation

$$\frac{d^2 y}{dz^2} + (\lambda - 2q \cos 2z)y = 0 \tag{1}$$

be

$$y(z) = \int_C K(z, \varsigma) v(\varsigma) d\varsigma \tag{2}$$

[cf. Sec. 2.12]. Denote $L \equiv \partial^2 / \partial z^2 + \lambda - 2q \cos 2z$, then

$$L[y] = \int_C \left[\frac{\partial^2 K}{\partial z^2} + (\lambda - 2q \cos 2z)K \right] v(\varsigma) d\varsigma .$$

Assuming that $K(z, \varsigma)$ satisfies the partial differential equation

$$\frac{\partial^2 K}{\partial z^2} - 2q \cos 2z \times K = \frac{\partial^2 K}{\partial \varsigma^2} - 2q \cos 2\varsigma \times K , \tag{3}$$

we have

$$\begin{aligned} L[y] &= \int_C \left[\frac{\partial^2 K}{\partial \varsigma^2} + (\lambda - 2q \cos 2\varsigma)K \right] v(\varsigma) d\varsigma \\ &= \left[v(\varsigma) \frac{\partial K}{\partial \varsigma} - K \frac{\partial v}{\partial \varsigma} \right]_C + \int_C \left[\frac{d^2 v}{d\varsigma^2} + (\lambda - 2q \cos 2\varsigma)v \right] K d\varsigma . \end{aligned}$$

Hence, if we choose the path of integration C such that

$$\left[v \frac{\partial K}{\partial \varsigma} - K \frac{\partial v}{\partial \varsigma} \right]_C = 0 \tag{4}$$

and $v(\varsigma)$ to be a solution of the Mathieu equation

$$\frac{d^2 v}{d\varsigma^2} + (\lambda - 2q \cos 2\varsigma)v = 0 \tag{5}$$

then the integral will be a solution of the Mathieu equation, at least in a formal way.

Let us now find the kernel $K(z, \varsigma)$ for the integral solution. In (3), putting $\varsigma = i\xi$, we obtain

$$\frac{\partial^2 K}{\partial \xi^2} + \frac{\partial^2 K}{\partial z^2} + 2q(\text{ch } 2\xi - \cos 2z)K = 0 .\tag{6}$$

Let

$$x = a \text{ ch } \xi \, \cos z , \quad y = a \text{ sh } \xi \, \sin z$$

[cf. Sec. 12.1]. Equation (6) reduces to the Helmholtz equation in cartesian coordinates:

$$\frac{\partial^2 K}{\partial x^2} + \frac{\partial^2 K}{\partial y^2} + k^2 K = 0 \quad \left(k^2 = \frac{4q}{a^2} \right) .\tag{7}$$

A particular solution of (7) is

$$K(x, y) = e^{ik(\alpha x + \beta y)} \quad (\alpha^2 + \beta^2 = 1) .\tag{8}$$

Choosing $\alpha = 1, \beta = 0$ and returning to the variables z, ς, we have

$$K(z, \varsigma) = e^{ika \, \text{ch } \xi \, \cos z} = e^{ika \cos \varsigma \cos z} .\tag{9}$$

Take $v(\varsigma)$ to be the periodic solution of the Mathieu equation (5) and C a segment of a straight line, namely $-\pi \le \varsigma \le \pi$, then (4) is satisfied and we have

$$y(z) = \int_{-\pi}^{\pi} e^{ika \cos z \cos \varsigma} v(\varsigma) d\varsigma \tag{10}$$

which is a periodic solution of (1).

But it is known that, to the same eigenvalue $\lambda(q), q \ne 0$, Mathieu equation can only have one periodic solution [see the paragraph below Eq. (8) of Sec. 12.5]. Hence, $v(\varsigma)$ and $y(\varsigma)$ can be differed by at most a constant multiple γ, and (10) thus becomes

$$y(z) = \gamma \int_{-\pi}^{\pi} e^{ika \cos z \cos \varsigma} y(\varsigma) d\varsigma \quad (ka = \sqrt{4q}) .\tag{11}$$

This is an integral equation for Mathieu functions.

From (8), choosing different values of α and β, or, choosing another solution of (7) as the kernel $K(z, \varsigma)$, we arrive at other integral equations; all their kernels are symmetric, $K(z, \varsigma) = K(\varsigma, z)$, and hence the equations have non-trivial solutions only when γ are eigenvalues.

With the integral equation, it is often possible to obtain, from one form of series expansion for the Mathieu function, another form of series expansion. Let us take as an example the function $y(z) = \text{ce}_{2n}(z)$.

We first find the eigenvalue γ. From (11), we have

$$\text{ce}_{2n}(z) = \gamma \int_{-\pi}^{\pi} e^{ika \cos z \cos \varsigma} \text{ce}_{2n}(\varsigma) d\varsigma \ .$$

Putting $z = \pi/2$ and using the Fourier expansion of $\text{ce}_{2n}(\varsigma)$ [Eq. (5) of Sec. 12.7], we have

$$\text{ce}_{2n}\left(\frac{\pi}{2}\right) = \gamma \int_{-\pi}^{\pi} \sum_{r=0}^{\infty} A_{2r} \cos 2r\varsigma \times d\varsigma = \gamma A_0 2\pi \ .$$

Thus

$$\gamma = \text{ce}_{2n}(\pi/2)/2\pi A_0 \ ,$$

and by Eq. (16) of Sec. 7.4, we obtain

$$\text{ce}_{2n}(z) = \frac{1}{2\pi A_0} \text{ce}_{2n}\left(\frac{\pi}{2}\right) \int_{-\pi}^{\pi} e^{ika \cos z \cos \varsigma} \sum_{r=0}^{\infty} A_{2r} \cos 2r\varsigma d\varsigma$$

$$= \frac{1}{A_0} \text{ce}_{2n}\left(\frac{\pi}{2}\right) \sum_{r=0}^{\infty} (-)^r A_{2r} J_{2r}(ka \cos z) \ , \tag{12}$$

where $A_{2r}(r = 0, 1, 2, \ldots)$ are the Fourier coefficients of $\text{ce}_{2n}(z)$.

Exercise 12

1. Prove the following formulae on the Fourier coefficients of Mathieu functions, where $t = q/4$; $A_{2r}^{2n}, A_{2r+1}^{2n+1}, B_{2r+2}^{2n+2}, B_{2r+1}^{2n+1}$ are the Fourier coefficients respectively of $\text{ce}_{2n}(z), \text{ce}_{2n+1}(z), \text{se}_{2n+2}(z), \text{se}_{2n+1}(z) (r \geq 1)$:

(i) $A_{2r+1}^1 = (-)^r \left[\dfrac{1}{r!(r+1)!} t^r + \dfrac{r}{[(r+1)!]^2} t^{r+1} \right.$

$\left. + \dfrac{1}{4(r-1)!(r+2)!} t^{r+2} + O(t^{r+3}) \right] A_1^1 \ ;$

(ii) $A_0^2 = \left[t - \frac{5}{3}t^3 + \frac{1363}{216}t^5 + O(t^7) \right] A_2^2 ,$

$$A_{2r+2}^2 = (-)^r \left[\frac{2}{r!(r+2)!}t^r + \frac{r(47r^2 + 222r + 247)}{18(r+2)!(r+3)!}t^{r+2} \right.$$
$$\left. + O(t^{r+4}) \right] A_2^2 ;$$

(iii) $B_{2r+1}^1 = (-)^r \left[\frac{1}{r!(r+1)!}t^r - \frac{r}{[(r+1)!]^2}t^{r+1} \right.$

$$\left. + \frac{1}{4(r-1)!(r+2)!}t^{r+2} + O(t^{r+3}) \right] B_1^1 ;$$

(iv) $B_{2r+2}^2 = (-)^r \left[\frac{2}{r!(r+2)!}t^r - \frac{r(r+1)(7r+23)}{18(r+2)!(r+3)!}t^{r+2} \right.$

$$\left. + O(t^{r+4}) \right] B_2^2 ;$$

(v) $A_{m-2r}^m = \left[\frac{(m-r-1)!}{r!(m-1)!}t^r + O(t^{r+1}) \right] A_m^m$

$$(r > 0, m - 2r \geq 0) ,$$

$$A_{m+2r}^m = \left[(-)^r \frac{m!}{r!(r+m)!}t^r + O(t^{r+1}) \right] A_m^m$$

$$(r > 0, m > 0) ;$$

(vi) $B_{m-2r}^m = \left[\frac{(m-r-1)!}{r!(m-1)!}t^r + O(t^{r+1}) \right] B_m^m$

$$(r > 0, m - 2r \geq 0) ,$$

$$B_{m+2r}^m = \left[(-)^r \frac{m!}{r!(r+m)!}t^r + O(t^{r+1}) \right] B_m^m$$

$$(r > 0, m > 0) .$$

$O(t^{r+1})$ can be replaced by the more precise $O(t^{r+2})$ in the second formula of (v) when $m \geq 3$, in the first formulae of (v) and (vi) when $m - 2r \geq 3$, and in all cases when m is an even number.

2. Prove the following expansion formulae for Mathieu functions, normalized

according to Eq. (6) of Sec. 12.7 when q is small:

$$\sqrt{2}\mathrm{ce}_0(z,q) = 1 - q \times \frac{1}{2}\cos 2z + q^2 \left[\frac{1}{32}\cos 4z - \frac{1}{16}\right]$$
$$- q^3 \left[\frac{1}{1152}\cos 6z - \frac{11}{128}\cos 2z\right] + O(q^4),$$

$$\mathrm{ce}_1(z,q) = \cos z - q \times \frac{1}{8}\cos 3z$$
$$+ q^2 \left[\frac{1}{192}\cos 5z - \frac{1}{64}\cos 3z - \frac{1}{128}\cos z\right]$$
$$- q^3 \left[\frac{1}{9 \times 2^{10}}\cos 7z - \frac{1}{9 \times 2^7}\cos 5z - \frac{1}{3 \times 2^{10}}\cos 3z\right.$$
$$\left.+ \frac{1}{2^9}\cos z\right] + O(q^4),$$

$$\mathrm{se}_1(z,q) = \sin z - q \times \frac{1}{8}\sin 3z$$
$$+ q^2 \left[\frac{1}{192}\sin 5z + \frac{1}{64}\sin 3z - \frac{1}{128}\sin z\right]$$
$$- q^3 \left[\frac{1}{9 \times 2^{10}}\sin 7z + \frac{1}{9 \times 2^7}\sin 5z - \frac{1}{3 \times 2^{10}}\sin 3z\right.$$
$$\left.- \frac{1}{2^9}\sin z\right] + O(q^4),$$

$$\mathrm{ce}_2(z,q) = \cos 2z - q \left[\frac{1}{12}\cos 4z - \frac{1}{4}\right]$$
$$+ q^2 \left[\frac{1}{384}\cos 6z - \frac{19}{288}\cos 2z\right] + O(q^3),$$

$$\mathrm{se}_2(z,q) = \sin 2z - q \times \frac{1}{12}\sin 4z$$
$$+ q^2 \left[\frac{1}{384}\sin 6z - \frac{1}{288}\sin 2z\right] + O(q^3),$$

$$\mathrm{ce}_3(z,q) = \cos 3z - q \left[\frac{1}{16}\cos 5z - \frac{1}{8}\cos z\right]$$
$$+ q^2 \left[\frac{1}{640}\cos 7z - \frac{5}{512}\cos 3z + \frac{1}{64}\cos z\right] + O(q^3),$$

$$\mathrm{se}_3(z,q) = \sin 3z - q \left[\frac{1}{16}\sin 5z - \frac{1}{8}\sin z\right]$$
$$+ q^2 \left[\frac{1}{640}\sin 7z - \frac{5}{512}\sin 3z - \frac{1}{64}\sin z\right] + O(q^3);$$

when $m \geq 4$,

$$\mathrm{ce}_m(z, q) = \cos mz - q\left[\frac{1}{4(m+1)}\cos(m+2)z - \frac{1}{4(m-1)}\cos(m-2)z\right]$$

$$+ q^2\left\{\frac{1}{32(m+1)(m+2)}\cos(m+4)z\right.$$

$$+ \frac{1}{32(m-1)(m-2)}\cos(m-4)z$$

$$\left.- \frac{1}{32}\left[\frac{1}{(m+1)^2} + \frac{1}{(m-1)^2}\right]\cos mz\right\} + O(q^3) \,,$$

$$\mathrm{se}_m(z, q) = \sin mz - q\left[\frac{1}{4(m+1)}\sin(m+2)z - \frac{1}{4(m-1)}\sin(m-2)z\right]$$

$$+ q^2\left\{\frac{1}{32(m+1)(m+2)}\sin(m+4)z\right.$$

$$+ \frac{1}{32(m-1)(m-2)}\sin(m-4)z$$

$$\left.- \frac{1}{32}\left[\frac{1}{(m+1)^2} + \frac{1}{(m-1)^2}\right]\sin mz\right\} + O(q^3) \,.$$

3. $f(z; \lambda, q)$ is a fundamental solution of the Mathieu equation $y'' + (\lambda - 2q\cos 2z)y = 0$; $f(0; \lambda, q) = 1, \partial f/\partial z|_{z=0} = 0$ [see Eq. (2) of Sec. 12.2]. Assume the expansion of $f(z; \lambda, q)$ in ascending powers of q to be

$$f(z; \lambda, q) = \cos\sqrt{\lambda}z + q\Phi_1(z; \lambda) + q^2\Phi_2(z; \lambda) + \ldots$$

then

$$\frac{d^2\Phi_i}{dz^2} + \lambda\Phi_i = 2\cos 2z \times \Phi_{i-1} \quad (i = 1, 2, \ldots) \,,$$

$$\Phi_0(z; \lambda) = \cos\sqrt{\lambda}z \,,$$

$$\Phi_i(0) = \frac{\partial\Phi_i}{\partial z}\bigg|_{z=0} = 0 \quad (i = 1, 2, \ldots) \,.$$

Show that

$$\Phi_{2k+1}(\pi; \lambda) = 0 \quad (k = 0, 1, 2, \ldots)$$

and the equation for the characteristic exponents is [Eq. (7) of Sec. 12.4]

$$\cos \nu\pi = \cos \sqrt{\lambda}\pi + \sum_{k=1}^{\infty} \Phi_{2k}(\pi; \lambda) q^{2k}$$

$$= \cos \sqrt{\lambda}\pi + q^2 \frac{\pi \sin \sqrt{\lambda}\pi}{4\sqrt{\lambda}(\lambda-1)}$$

$$+ q^4 \left[\frac{15\lambda^2 - 35\lambda + 8}{64(\lambda-1)^3(\lambda-4)\lambda^{3/2}} \pi \sin \sqrt{\lambda}\pi - \frac{\pi^2 \cos \sqrt{\lambda}\pi}{32\lambda(\lambda-1)^2} \right]$$

$$+ q^6 \left[\frac{105\lambda^5 - 1155\lambda^4 + 3815\lambda^3 - 4705\lambda^2 + 1652\lambda - 288}{256(\lambda-1)^5(\lambda-4)^2(\lambda-9)\lambda^{5/2}} \right.$$

$$\times \pi \sin \sqrt{\lambda}\pi - \frac{\pi^3 \sin \sqrt{\lambda}\pi}{384(\lambda-1)^3\lambda^{3/2}} - \frac{15\lambda^2 - 35\lambda + 8}{256\lambda^2(\lambda-1)^4(\lambda-4)} \pi^2$$

$$\left. \times \cos \sqrt{\lambda}\pi \right] + \dots .$$

When $\lambda \approx 0$,

$$\cos \nu\pi = \left(1 - \frac{\lambda\pi^2}{2} + \frac{\lambda^2\pi^4}{24} + \dots \right)$$

$$- q^2 \frac{\pi^2}{4} \left[1 + \lambda \left(1 - \frac{\pi^2}{6} \right) + \dots \right]$$

$$+ q^4 \left[-\frac{25\pi^2}{256} + \frac{\pi^4}{96} + \dots \right] + \dots .$$

Similarly, find the expansions in the vicinities of $\lambda = 1, 4, 9, \dots, m^2, \dots$.

4. $f(z)$ and $g(z)$ are the fundamental solutions of the Mathieu equation;

$$f(0) = 1, \quad f'(0) = 0;$$
$$g(0) = 0, \quad g'(0) = 1.$$

Show that

$$f(\pi) + 1 = 2f\left(\frac{\pi}{2}\right) g'\left(\frac{\pi}{2}\right),$$
$$f(\pi) - 1 = 2g\left(\frac{\pi}{2}\right) f'\left(\frac{\pi}{2}\right),$$
$$f'(\pi) = 2f\left(\frac{\pi}{2}\right) f'\left(\frac{\pi}{2}\right),$$
$$g'(\pi) = 2g\left(\frac{\pi}{2}\right) g'\left(\frac{\pi}{2}\right).$$

And, it follows that, for

$$\text{even whole-period solution} - f'\left(\frac{\pi}{2};\lambda,q\right) = 0 \ ,$$

$$\text{odd whole-period solution} - f\left(\frac{\pi}{2};\lambda,q\right) = 0 \ ,$$

$$\text{even half-period solution} - g\left(\frac{\pi}{2};\lambda,q\right) = 0 \ ,$$

$$\text{odd half-period solution} - g'\left(\frac{\pi}{2};\lambda,q\right) = 0 \ ,$$

where $'$ denotes the derivative with respect to z [cf. Secs. 12.3 and 12.4].

5. Show that $\text{ce}_m(z)$ can be expanded in series of the form

$$\sum_{n=0}^{\infty} A_n \cos^{2n} z \quad \text{or} \quad \sum_{n=0}^{\infty} B_n \cos^{2n+1} z \ ,$$

depending on whether m is even or odd; the series converge in $|\cos z| < 1$.

6. Show that the eigenvalue γ_m of the integral equation

$$\text{ce}_m(z) = \gamma_m \int_{-\pi}^{\pi} e^{\alpha \cos z \cos \theta} \text{ce}_m(\theta) d\theta$$

is determined by the following series:

$$A_0 = 2\pi\gamma_m \sum_{n=0}^{\infty} \frac{(2n)!}{2^{2n}(n!)^2} A_n$$

or

$$B_0 = 2\pi\gamma_m \alpha \sum_{n=0}^{\infty} \frac{(2n+1)!}{2^{2n+1}n!(n+1)!} B_n \ ,$$

depending on whether m is even or odd; A_n, B_n are the expansion coefficients in Ex. 5 above.

7. Show that

$$ce_{2n}(z,q) = \frac{p_{2n}}{A_0} \sum_{r=0}^{\infty} (-)^r A_{2r} J_r\left(\sqrt{q}e^{iz}\right) J_r\left(\sqrt{q}e^{-iz}\right) ,$$

$$ce_{2n+1}(z,q) = \frac{p_{2n+1}}{A_1} \sum_{r=0}^{\infty} (-)^r A_{2r+1} \left[J_r\left(\sqrt{q}e^{iz}\right) J_{r+1}\left(\sqrt{q}e^{-iz}\right) \right.$$
$$\left. + J_{r+1}\left(\sqrt{q}e^{iz}\right) J_r\left(\sqrt{q}e^{-iz}\right) \right] ,$$

$$se_{2n+1}(z,q) = -\frac{s_{2n+1}}{iB_1} \sum_{r=0}^{\infty} (-)^r B_{2r+1} \left[J_r\left(\sqrt{q}e^{iz}\right) J_{r+1}\left(\sqrt{q}e^{-iz}\right) \right.$$
$$\left. - J_{r+1}\left(\sqrt{q}e^{iz}\right) J_r\left(\sqrt{q}e^{-iz}\right) \right] ,$$

$$se_{2n+2}(z,q) = \frac{s_{2n+2}}{iB_2} \sum_{r=0}^{\infty} (-)^r B_{2r+2} \left[J_r\left(\sqrt{q}e^{iz}\right) J_{r+2}\left(\sqrt{q}e^{-iz}\right) \right.$$
$$\left. - J_{r+2}\left(\sqrt{q}e^{iz}\right) J_r\left(\sqrt{q}e^{-iz}\right) \right] ,$$

where $A_{2r}, A_{2r+1}, B_{2r+1}, B_{2r+2}(r = 0, 1, \ldots)$ are respectively the Fourier coefficients of the corresponding functions [Eqs. (1) – (4) of Sec. 12.7], and p_m, s_m are given respectively by the following equations:

$$A_0 p_{2n} = ce_{2n}(0)ce_{2n}\left(\frac{\pi}{2}\right) ,$$

$$\sqrt{q}A_1 p_{2n+1} = -ce_{2n+1}(0)ce'_{2n+1}\left(\frac{\pi}{2}\right) ,$$

$$\sqrt{q}B_1 s_{2n+1} = se'_{2n+1}(0)se_{2n+1}\left(\frac{\pi}{2}\right) ,$$

$$qB_2 s_{2n+2} = se'_{2n+2}(0)se'_{2n+2}\left(\frac{\pi}{2}\right) ,$$

where ' denotes the derivative with respect to z. [cf McLachlan (1947), p. 193.]

8. *Modified Mathieu Functions* $Ce_m(z,q)$, $Se_m(z,q)$ are defined by

$$Ce_m(z,q) = ce_m(iz,q), \quad Se_m(z,q) = -i\, se_m(iz,q) .$$

Prove the following asymptotic formulae for $\mathrm{Re}(z) \to \infty$:

$$\mathrm{Ce}_{2n}(z, q) \sim p_{2n} \left(\frac{2}{\pi}\right)^{\frac{1}{2}} q^{-\frac{1}{4}} e^{-\frac{z}{2}} \cos\left(\sqrt{q}\, e^z - \frac{\pi}{4}\right),$$

$$\mathrm{Ce}_{2n+1}(z, q) \sim p_{2n+1} \left(\frac{2}{\pi}\right)^{\frac{1}{2}} q^{-\frac{1}{4}} e^{-\frac{z}{2}} \cos\left(\sqrt{q}\, e^z - \frac{3\pi}{4}\right),$$

$$\mathrm{Se}_{2n+1}(z, q) \sim s_{2n+1} \left(\frac{2}{\pi}\right)^{\frac{1}{2}} q^{-\frac{1}{4}} e^{-\frac{z}{2}} \cos\left(\sqrt{q}\, e^z - \frac{3\pi}{4}\right),$$

$$\mathrm{Se}_{2n+2}(z, q) \sim s_{2n+2} \left(\frac{2}{\pi}\right)^{\frac{1}{2}} q^{-\frac{1}{4}} e^{-\frac{z}{2}} \cos\left(\sqrt{q}\, e^z - \frac{\pi}{4}\right),$$

$$\left(|\tfrac{1}{2} \arg q + \mathrm{Im}(z)| < \pi\right),$$

where p_m and s_m are the constants given in the last exercise. [Hint: Utilize the expansion of the Mathieu function in terms of the Bessel functions, for example, Eq. (12) of Sec. 12.14, or the results of the last Exercise, and the asymptotic expressions of the Bessel functions, Eq. (5) of Sec. 7.10.]

9. Show that, when $q \to \infty$, the eigenvalues of the Mathieu equation $y'' + (\lambda - 2q \cos 2z) y = 0$ can be represented by

$$a_m(q) \sim b_{m+1}(q)$$
$$\sim -2q + 2(2m + 1)\sqrt{q} - \frac{1}{4}(2m^2 + 2m + 1);$$

the asymptotic expressions of the corresponding eigenfunctions are:

$$\mathrm{ce}_m(z, q) \sim C_m (\cos z)^{-m-1} \left\{ \left[\cos\left(\frac{z}{2} + \frac{\pi}{4}\right)\right]^{2m+1} e^{2\sqrt{q}\sin z} \right.$$

$$\left. + \left[\sin\left(\frac{z}{2} + \frac{\pi}{4}\right)\right]^{2m+1} e^{-2\sqrt{q}\sin z} \right\},$$

$$\mathrm{se}_{m+1}(z, q) \sim S_{m+1} (\cos z)^{-m-1} \left\{ \left[\cos\left(\frac{z}{2} + \frac{\pi}{4}\right)\right]^{2m+1} e^{2\sqrt{q}\sin z} \right.$$

$$\left. - \left[\sin\left(\frac{z}{2} + \frac{\pi}{4}\right)\right]^{2m+1} e^{-2\sqrt{q}\sin z} \right\}$$

$$\left(|z| < \frac{\pi}{2}\right).$$

Also,

$$\mathrm{Ce}_m(z,q) \sim C_m 2^{\frac{1}{2}-m}(\mathrm{ch}\ z)^{-\frac{1}{2}}$$
$$\times \cos\left[2\sqrt{q}\,\mathrm{sh}\ z - (2m+1)\tan^{-1}\left(\mathrm{th}\frac{z}{2}\right)\right]\ ,$$
$$\mathrm{Se}_m(z,q) \sim S_{m+1} 2^{\frac{1}{2}-m}(\mathrm{ch}\ z)^{-\frac{1}{2}}$$
$$\times \sin\left[2\sqrt{q}\,\mathrm{sh}\ z - (2m+1)\tan^{-1}\left(\mathrm{th}\frac{z}{2}\right)\right]$$
$$(z>0)\ .$$

The constant appear in the formula

$$C_m = (-)^{[m/2]} 2^{m-\frac{1}{2}} q^{-\frac{1}{4}} \pi^{-\frac{1}{2}} p_m\ ,$$
$$S_m = (-)^{[m/2]} 2^{m-\frac{3}{2}} q^{-\frac{1}{4}} \pi^{-\frac{1}{2}} s_m\ ;$$

p_m and s_m are the constants given in Ex. 7 above.

APPENDICES

Appendix I. Roots of a Cubic Equation

Let the cubic equation be

$$x^3 + bx^2 + cx + d = 0 . \tag{1}$$

By the transformation $x = y - b/3$, the equation becomes

$$y^3 + py + q = 0 , \tag{2}$$

where

$$p = c - \frac{b^2}{3} , \quad q = d - \frac{bc}{3} + \frac{2b^3}{27} . \tag{3}$$

Making another transformation $y = z - p/3z$, we have

$$z^3 - \frac{p^3}{37z^3} + q = 0 , \tag{4}$$

which is a quadratic equation of z^3 with the solutions

$$z^3 = -\frac{q}{2} \pm \sqrt{R} , \tag{5}$$

where

$$R = \frac{p^3}{27} + \frac{q^2}{4} . \tag{6}$$

Putting

$$A = \left(-\frac{q}{2} + \sqrt{R}\right)^{\frac{1}{3}} , \quad B = -\frac{p}{3A} = \left(-\frac{q}{2} - \sqrt{R}\right)^{\frac{1}{3}} , \tag{7}$$

$$\omega = -\frac{1}{2} + \frac{i\sqrt{3}}{2} \quad (\omega^3 = 1) , \tag{8}$$

we obtain the three roots for z to be $z_1 = A, z_2 = \omega A, z_3 = \omega^2 A$; the corresponding roots for y are

$$y_1 = A + B , \quad y_2 = \omega A + \omega^2 B , \quad y_3 = \omega^2 A + \omega B . \tag{9}$$

This is *Cardan's formula*.

The discriminant of the cubic equation is

$$\begin{aligned}
\Delta &\equiv (x_1 - x_2)^2 (x_1 - x_3)^2 (x_2 - x_3)^2 \\
&= (y_1 - y_2)^2 (y_1 - y_3)^2 (y_2 - y_3)^2 \\
&= -108R = -4p^3 - 27q^2 .
\end{aligned} \tag{10}$$

Substituting the p and q of (3) in (10) gives

$$\Delta = 18bcd - 4b^3 d + b^2 c^2 - 4c^3 - 27d^2 . \tag{11}$$

When $R < 0$, the discriminant $\Delta > 0$ and all the three roots are real; formula (9) fails to hold. We can solve the problem by employing trigonometric functions. Consider the following trigonometric formula:

$$4\cos^3 u - 3\cos u = \cos 3u . \tag{12}$$

Let $y = n\cos u$. Substituting in (12) and multiplying with $n^3/4$, we obtain

$$y^3 - \frac{3n^2}{4}y - \frac{n^3}{4}\cos 3u = 0 .$$

Comparing with Eq. (2), we get

$$p = -\frac{3n^2}{4} , \quad q = -\frac{n^3}{4}\cos 3u .$$

It follows,

$$n = \left(-\frac{4p}{3}\right)^{\frac{1}{2}} , \quad \cos 3u = -\frac{q}{2}\left(-\frac{p}{3}\right)^{-\frac{3}{2}} . \tag{13}$$

Since $R < 0$, we see from (6) that $p < 0$, and hence the u given by (13) is real (take the positive root), so that $|\cos 3u| < 1$ and we can find the angle $3u$. Suppose that we have found the angle u from (13), then the three roots of Eq. (2) are

$$y_1 = n \cos u, \quad y_2 = n \cos \left(u + \frac{2\pi}{3} \right), \quad y_3 = n \cos \left(u + \frac{4\pi}{3} \right) . \tag{14}$$

Appendix II. Roots of the Quartic Equation

Let the quartic equation be

$$x^4 + bx^3 + cx^2 + dx + e = 0 . \tag{1}$$

By the transformation $x = z - b/4$, it becomes

$$z^4 + qz^2 + rz + s = 0 , \tag{2}$$

where

$$q = c - \frac{3b^2}{8}, \quad r = d - \frac{bc}{2} + \frac{b^3}{8} ,$$
$$s = e - \frac{bd}{4} + \frac{b^2 c}{16} - \frac{3b^4}{256} . \tag{3}$$

Resolve (2) into two factors:

$$z^4 + qz^2 + rz + s \equiv (z^2 + kz + l)(z^2 - kz + m) = 0 , \tag{4}$$

where

$$2l = q + k^2 - \frac{r}{k}, \quad 2m = q + k^2 + \frac{r}{k}, \quad lm = s . \tag{5}$$

Eliminating l and m gives

$$k^6 + 2qk^4 + (q^2 - 4s)k^2 - r^2 = 0 \tag{6}$$

which is a cubic equation in k^2, the roots of which can be obtained by the method described in App. I.

Let the three roots of (6) be k_1^2, k_2^2, k_3^2, then

$$k_1^2 k_2^2 k_3^2 = r^2 .$$

Now, choose the signs of k_1, k_2, k_3 so that

$$k_1 k_2 k_3 = -r . \tag{7}$$

Among the two roots $\pm k_i$ of each k_i^2 we need only to take the $+k_i$, since a change from $+k_i$ to $-k_i$ effects only an exchange of the two factors in (4).

Let us first find the roots of $z^2 - kz + m = 0$. Let $k = k_1$, then

$$z = \frac{1}{2}\left(k_1 \pm \sqrt{k_1^2 - 4m}\right) .$$

From (6) we have

$$k_1^2 + k_2^2 + k_3^2 = -2q ,$$

then by (5)

$$k_1^2 - 4m = k_1^2 - 2q - 2k_1^2 - \frac{2r}{k_1}$$
$$= k_2^2 + k_3^2 + 2k_2k_3 = (k_2 + k_3)^2 ,$$

from which we obtain two roots:

$$z_1 = \frac{1}{2}(k_1 + k_2 + k_3) , \tag{8}$$

$$z_2 = \frac{1}{2}(k_1 - k_2 - k_3) . \tag{9}$$

The other two roots $k = k_2$ and $k = k_3$ give

$$z_3 = \frac{1}{2}(-k_1 + k_2 - k_3) , \tag{10}$$

$$z_4 = \frac{1}{2}(-k_1 - k_2 + k_3) . \tag{11}$$

It can be shown that the first factor in (4), $z^2 + kz + l = 0$, also furnishes the same roots z_1, z_2, z_3, z_4; therefore, these are the four roots of the quartic equation.

The discriminant of the quartic equation is

$$\Delta = (x_1 - x_2)^2 (x_1 - x_3)^2 (x_1 - x_4)^2 (x_2 - x_3)^2 (x_2 - x_4)^2 (x_3 - x_4)^2$$
$$= (z_1 - z_2)^2 (z_1 - z_3)^2 (z_1 - z_4)^2 (z_2 - z_3)^2 (z_2 - z_4)^2 (z_3 - z_4)^2$$
$$= (k_1^2 - k_2^2)^2 (k_1^2 - k_3^2)^2 (k_2^2 - k_3^2)^2 .$$

From which, we see that the discriminant of the quartic equation is exactly the same as the discriminant of the auxiliary cubic equation (6). Applying Eq. (10) of App. I, we obtain

$$\Delta = 4\left(\frac{q^2}{3} + 4s\right)^3 - 27\left(\frac{2q^3}{27} - \frac{8qs}{3} + r^2\right)^2$$
$$= 4\left(\frac{c^2}{3} - bd + 4e\right)^3 - 27\left(\frac{2c^3}{27} + d^2 + b^2e - \frac{bcd}{3} - \frac{8ce}{3}\right) . \tag{12}$$

Appendix III. Orthogonal Curvilinear Coordinate Systems

1. *General formulae*

Let ξ_1, ξ_2, ξ_3 be a system of curvilinear coordinates whose relations to the rectangular coordinates x, y, z are

$$x = x(\xi_1, \xi_2, \xi_3), \quad y = y(\xi_1, \xi_2, \xi_3),$$
$$z = z(\xi_1, \xi_2, \xi_3). \tag{1}$$

The square of the line element is

$$ds^2 = dx^2 + dy^2 + dz^2 = \sum_{i,j=1}^{3} g_{ij} d\xi_i \xi_j, \tag{2}$$

where

$$g_{ij} = \frac{\partial x}{\partial \xi_i} \frac{\partial x}{\partial \xi_j} + \frac{\partial y}{\partial \xi_i} \frac{\partial y}{\partial \xi_j} + \frac{\partial z}{\partial \xi_i} \frac{\partial z}{\partial \xi_j}. \tag{3}$$

Let the inverse of the matrix (g_{ij}) be (g^{ij}), i.e.,

$$\sum_{j=1}^{3} g_{ij} g^{ik} = \delta_i^k = \begin{cases} 1, & i = k. \\ 0, & i \neq k. \end{cases} \tag{4}$$

Let $g = \det |g_{ij}|$ and G^{ij} be the cofactor of g_{ij} in the determinant, then

$$G^{ij} = g g^{ij}. \tag{5}$$

In the text book on Tensor Analysis, it is proved that the expression of the Laplacian operator in curvilinear coordinates is

$$\nabla^2 \Phi \equiv \frac{1}{\sqrt{g}} \sum_{i,j=1}^{3} \frac{\partial}{\partial \xi_i} \left(\sqrt{g}\, g^{ij} \frac{\partial \Phi}{\partial \xi_j} \right). \tag{6}$$

When the curvilinear coordinates are orthogonal, we have $g_{ij} = 0$ for $i \neq j$, and (2) reduces to

$$ds^2 = H_1^2 d\xi_1^2 + H_2^2 d\xi_2^2 + H_3^2 d\xi_3^2, \tag{7}$$

i.e.,

$$g_{ii} = H_i^2 = \frac{1}{g^{ii}}, \quad g = H_1^2 H_2^2 H_3^2 = H^2 \tag{8}$$

and (6) becomes

$$\nabla^2 \Phi = \frac{1}{H} \sum_{i=1}^{3} \frac{\partial}{\partial \xi_i} \left(\frac{H}{H_i^2} \frac{\partial \Phi}{\partial \xi_i} \right) . \tag{9}$$

The direction cosines of the normal to the surface $\xi_i = $ constant are

$$H_i \frac{\partial \xi_i}{\partial x}, \quad H_i \frac{\partial \xi_i}{\partial y}, \quad H_i \frac{\partial \xi_i}{\partial z} , \tag{10}$$

which are also equal to

$$\frac{1}{H_i} \frac{\partial x}{\partial \xi_i}, \quad \frac{1}{H_i} \frac{\partial y}{\partial \xi_i}, \quad \frac{1}{H_i} \frac{\partial z}{\partial \xi_i} . \tag{11}$$

The projection, or components, (u_1, u_2, u_3) of a vector **v** in the curvilinear coordinate system are related to its cartesian components (v_x, v_y, v_z) by

$$u_i = \frac{v_x}{H_i} \frac{\partial x}{\partial \xi_i} + \frac{v_y}{H_i} \frac{\partial y}{\partial \xi_i} + \frac{v_z}{H_i} \frac{\partial z}{\partial \xi_i}$$

$$= H_i \left(v_x \frac{\partial \xi_i}{\partial x} + v_y \frac{\partial \xi_i}{\partial y} + v_z \frac{\partial \xi_i}{\partial z} \right) . \tag{12}$$

The gradient, divergence and Curl (or, rotation) are defined respectively by

$$(\nabla \Phi)_i = \frac{1}{H_i} \frac{\partial \Phi}{\partial \xi_i} , \tag{13}$$

$$\nabla \cdot \mathbf{v} = \frac{1}{H} \sum_i \frac{\partial}{\partial \xi_i} \left(\frac{H}{H_i} u_i \right) , \tag{14}$$

$$(\nabla \times \mathbf{v})_{ij} = \frac{1}{H_i H_j} \left\{ \frac{\partial (H_j u_j)}{\partial \xi_i} - \frac{\partial (H_i u_i)}{\partial \xi_j} \right\} . \tag{15}$$

2. Cylindrical coordinates ρ, φ, z

$$x = \rho \cos \varphi, \quad y = \rho \sin \varphi, \quad z = z . \tag{16}$$

$$ds^2 = d\rho^2 + \rho^2 d\varphi^2 + dz^2 . \tag{17}$$

$$H_\rho = 1, \quad H_\varphi = \rho, \quad H_z = 1; \quad H = \rho . \tag{18}$$

$$\nabla^2 \Phi \equiv \frac{1}{\rho} \frac{\partial}{\partial \rho} \left(\rho \frac{\partial \Phi}{\partial \rho} \right) + \frac{1}{\rho^2} \frac{\partial^2 \Phi}{\partial \varphi^2} + \frac{\partial^2 \Phi}{\partial z^2} . \tag{19}$$

$$u_\rho = v_x \cos \varphi + v_y \sin \varphi, \quad u_\varphi = -v_x \sin \varphi + v_y \cos \varphi \ ,$$

$$u_z = v_z \ . \tag{20}$$

3. Spherical polar coordinates r, θ, φ

$$x = r \sin \theta \cos \varphi, \quad y = r \sin \theta \sin \varphi, \quad z = r \cos \theta \ . \tag{21}$$

$$ds^2 = dr^2 + r^2 d\theta^2 + r^2 \sin^2 \theta d\varphi^2 \ . \tag{22}$$

$$H_r = 1, \quad H_\theta = r, \quad H_\varphi = r \sin \theta; \quad H = r^2 \sin \theta \ . \tag{23}$$

$$\nabla^2 \Phi \equiv \frac{1}{r^2} \frac{\partial}{\partial r} \left(r^2 \frac{\partial \Phi}{\partial r} \right) + \frac{1}{r^2 \sin \theta} \frac{\partial}{\partial \theta} \left(\sin \theta \frac{\partial \Phi}{\partial \theta} \right) + \frac{1}{r^2 \sin^2 \theta} \frac{\partial^2 \Phi}{\partial \varphi^2} \ . \tag{24}$$

$$u_r = v_x \sin \theta \cos \varphi + v_y \sin \theta \sin \varphi + v_z \cos \theta \ ,$$

$$u_\theta = v_x \cos \theta \cos \varphi + v_z \cos \theta \sin \varphi - v_z \sin \theta \ , \tag{25}$$

$$u_\varphi = -v_x \sin \varphi + v_y \cos \varphi \ ,$$

OQ being the projection of OP on the $x - y$ plane (Fig. 41).

4. Elliptical cylinder coordinates ξ, η, z

Let the coordinates of the foci A and B of the ellipse be $(a, 0)$ and $(-a, 0)$ respectively, and the distances from an arbitrary point P to these foci be r_A and r_B (Fig. 42). The elliptical cylinder coordinates ξ, η, z are defined by

$$\xi = \frac{r_A + r_B}{2a}, \quad \eta = \frac{r_B - r_A}{2a}, \quad z = z \ . \tag{26}$$

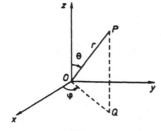
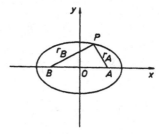

Fig. 41.   Fig. 42.

But

$$r_A^2 = (x - a)^2 + y^2, \quad r_B^2 = (x + a)^2 + y^2 \ ,$$

therefore,

$$x = a\xi\eta, \quad y = a\sqrt{(\xi^2 - 1)(1 - \eta^2)}, \quad z = z \ . \tag{27}$$

$$ds^2 = a^2 \left\{ \frac{\xi^2 - \eta^2}{\xi^2 - 1} d\xi^2 + \frac{\xi^2 - \eta^2}{1 - \eta^2} d\eta^2 \right\} + dz^2 . \tag{28}$$

$$H_\xi^2 = \frac{a^2(\xi^2 - \eta^2)}{\xi^2 - 1}, \quad H_\eta^2 = \frac{a^2(\xi^2 - \eta^2)}{1 - \eta^2}, \quad H_z = 1 . \tag{29}$$

$$\nabla^2 \Phi \equiv \frac{1}{a^2(\xi^2 - \eta^2)} \left\{ \sqrt{\xi^2 - 1} \frac{\partial}{\partial \xi} \left(\sqrt{\xi^2 - 1} \frac{\partial \Phi}{\partial \xi} \right) \right.$$
$$\left. + \sqrt{1 - \eta^2} \frac{\partial}{\partial \eta} \left(\sqrt{1 - \eta^2} \frac{\partial \Phi}{\partial \eta} \right) \right\} + \frac{\partial^2 \Phi}{\partial z^2} . \tag{30}$$

From (27), we obtain

$$\frac{x^2}{\xi^2} + \frac{y^2}{\xi^2 - 1} = a^2 , \quad \frac{x^2}{\eta^2} - \frac{y^2}{1 - \eta^2} = a^2 . \tag{31}$$

It is seen, by (26), that the range of variation of ξ is $(1, \infty)$ and that of η is $(-1, 1)$. From (26) or (31), we observe that $\xi = $ constant represents an ellipse and $\eta = $ constant represents a hyperbola; these ellipses and hyperbolas all have the same fixed foci A and B.

Another kind of elliptical cylinder coordinates are u, v, z; their relations to ξ and η are

$$\xi = \text{ch } u, \quad \eta = \cos v . \tag{32}$$

$$ds^2 = a^2(\text{ch}^2 u - \cos^2 v)(du^2 + dv^2) + dz^2 . \tag{33}$$

$$H_u^2 = H_v^2 = H = a^2(\text{ch}^2 u - \cos^2 v), \quad H_z = 1 . \tag{34}$$

$$\nabla^2 \Phi \equiv \frac{1}{a^2(\text{ch}^2 u - \cos^2 v)} \left\{ \frac{\partial^2 \Phi}{\partial u^2} + \frac{\partial^2 \Phi}{\partial v^2} \right\} + \frac{\partial^2 \Phi}{\partial z^2} . \tag{35}$$

5. Parabolic cylinder coordinates λ, μ, z

$$x = \frac{1}{2}(\lambda - \mu), \quad y = \sqrt{\lambda \mu}, \quad z = z . \tag{36}$$

$$ds^2 = \frac{\lambda + \mu}{4} \left\{ \frac{d\lambda^2}{\lambda} + \frac{d\mu^2}{\mu} \right\} + dz^2 . \tag{37}$$

$$H_\lambda^2 = \frac{\lambda + \mu}{4\lambda}, \quad H_\mu^2 = \frac{\lambda + \mu}{4\mu}, \quad H_z = 1 . \tag{38}$$

$$\nabla^2 \Phi \equiv \frac{4}{\lambda + \mu} \left\{ \sqrt{\lambda} \frac{\partial}{\partial \lambda} \left(\sqrt{\lambda} \frac{\partial \Phi}{\partial \lambda} \right) + \sqrt{\mu} \frac{\partial}{\partial \mu} \left(\sqrt{\mu} \frac{\partial \Phi}{\partial \mu} \right) \right\} + \frac{\partial^2 \Phi}{\partial z^2} . \tag{39}$$

From (36), we have

$$y^2 + 2\lambda x - \lambda^2 = 0, \quad y^2 - 2\mu x - \mu^2 = 0 . \tag{40}$$

When λ varies from 0 to ∞ and μ varies from 0 to ∞, Eq. (40) represents two sets of parabolas mutually orthogonal; the focus of these parabolas is at the origin.

Another kind of parabolic cylinder coordinates are ξ, η, z; their relations to λ, μ are

$$\lambda = \xi^2, \quad \mu = \eta^2 . \tag{41}$$

$$ds^2 = (\xi^2 + \eta^2)(d\xi^2 + d\eta^2) + dz^2 . \tag{42}$$

$$H_\xi^2 = H_\eta^2 = H = \xi^2 + \eta^2, \quad H_z = 1 . \tag{43}$$

$$\nabla^2 \Phi \equiv \frac{1}{\xi^2 + \eta^2} \left\{ \frac{\partial^2 \Phi}{\partial \xi^2} + \frac{\partial^2 \Phi}{\partial \eta^2} \right\} + \frac{\partial^2 \Phi}{\partial z^2} . \tag{44}$$

6. Conical coordinates r, λ, μ

The conical coordinates r, λ, μ are defined by

$$r = \sqrt{x^2 + y^2 + z^2} , \tag{45}$$

$$\begin{aligned} \frac{x^2}{\lambda - \alpha^2} + \frac{y^2}{\lambda + \beta^2} + \frac{z^2}{\lambda} &= 0 , \\ \frac{x^2}{\mu + \alpha^2} + \frac{y^2}{\mu - \beta^2} + \frac{z^2}{\mu} &= 0 , \end{aligned} \qquad (\alpha^2 + \beta^2 = 1) . \tag{46}$$

$$\begin{aligned} x = \frac{r}{\alpha}\sqrt{(\alpha^2 - \lambda)(\alpha^2 + \mu)}, \quad y &= \frac{r}{\beta}\sqrt{(\beta^2 + \lambda)(\beta^2 - \mu)} , \\ z = \frac{r\sqrt{\lambda\mu}}{\alpha\beta}, \quad (\alpha^2 + \beta^2 &= 1) . \end{aligned} \tag{47}$$

The ranges of variation of λ and μ are $0 \leq \lambda \leq \alpha^2$, $0 \leq \mu \leq \beta^2$.

$$ds^2 = dr^2 + \frac{r^2(\lambda + \mu)}{4} \left\{ \frac{d\lambda^2}{\lambda(\alpha^2 - \lambda)(\beta^2 + \lambda)} + \frac{d\mu^2}{\mu(\alpha^2 + \mu)(\beta^2 - \mu)} \right\} \tag{48}$$

$$\begin{aligned} H_r = 1 , \quad H_\lambda^2 &= \frac{r^2(\lambda + \mu)}{4\lambda(\alpha^2 - \lambda)(\beta^2 + \lambda)} , \\ H_\mu^2 &= \frac{r^2(\lambda + \mu)}{4\mu(\alpha^2 + \mu)(\beta^2 - \mu)} . \end{aligned} \tag{49}$$

$$\nabla^2 \Phi \equiv \frac{1}{r^2}\frac{\partial}{\partial r}\left(r^2\frac{\partial \Phi}{\partial r}\right)$$
$$+ \frac{4}{r^2(\lambda + \mu)}\left\{\sqrt{\lambda(\alpha^2 - \lambda)(\beta^2 + \lambda)}\frac{\partial}{\partial \lambda}\left(\sqrt{\lambda(\alpha^2 - \lambda)(\beta^2 + \lambda)}\frac{\partial \Phi}{\partial \lambda}\right)\right.$$
$$\left. + \sqrt{\mu(\alpha^2 + \mu)(\beta^2 - \mu)}\frac{\partial}{\partial \mu}\left(\sqrt{\mu(\alpha^2 + \mu)(\beta^2 - \mu)}\frac{\partial \Phi}{\partial \mu}\right)\right\} .$$

$$(50)$$

Another kind of conical coordinates are r, ξ, η; their relations to λ, μ are

$$\lambda = \xi^2 , \quad \mu = \eta^2 . \tag{51}$$

$$ds^2 = dr^2 + r^2(\xi^2 + \eta^2)\left\{\frac{d\xi^2}{(\alpha^2 - \xi^2)(\beta^2 + \xi^2)} + \frac{d\eta^2}{(\alpha^2 + \eta^2)(\beta^2 - \eta^2)}\right\} . \tag{52}$$

$$H_r = 1, \quad H_\xi^2 = \frac{r^2(\xi^2 + \eta^2)}{(\alpha^2 - \xi^2)(\beta^2 + \xi^2)}, \quad H_\eta^2 = \frac{r^2(\xi^2 + \eta^2)}{(\alpha^2 + \eta^2)(\beta^2 - \eta^2)} . \tag{53}$$

$$\nabla^2 \Phi \equiv \frac{1}{r^2}\frac{\partial}{\partial r}\left(r^2\frac{\partial \Phi}{\partial r}\right)$$
$$+ \frac{1}{r^2(\xi^2 + \eta^2)}\left\{\sqrt{(\alpha^2 - \xi^2)(\beta^2 + \xi^2)}\frac{\partial}{\partial \xi}\left(\sqrt{(\alpha^2 - \xi^2)(\beta^2 + \xi^2)}\frac{\partial \Phi}{\partial \xi}\right)\right.$$
$$\left. + \sqrt{(\alpha^2 + \eta^2)(\beta^2 - \eta^2)}\frac{\partial}{\partial \eta}\left(\sqrt{(\alpha^2 + \eta^2)(\beta^2 - \eta^2)}\frac{\partial \Phi}{\partial \eta}\right)\right\} .$$

$$(54)$$

Another kind of conical coordinates are r, u, v; their relations to ξ, η, are

$$\xi = \alpha \text{ cn }(u, \alpha), \quad \eta = \beta \text{ cn }(v, \beta) \tag{55}$$

$$x = r \text{ dn }(u, \alpha) \text{ sn }(v, \beta), \quad y = r \text{ sn }(u, \alpha) \text{ dn }(v, \beta) ,$$
$$z = r \text{ cn }(u, \alpha) \text{ cn }(v, \beta) . \tag{56}$$

$$ds^2 = dr^2 + r^2(\alpha^2 \text{ cn}^2 u + \beta^2 \text{ cn}^2 v)(du^2 + dv^2) . \tag{57}$$

$$H_r = 1, \quad H_u^2 = H_v^2 = H = r^2(\alpha^2 \text{ cn}^2 u + \beta^2 \text{ cn}^2 v) . \tag{58}$$

$$\nabla^2 \Phi \equiv \frac{1}{r^2}\frac{\partial}{\partial r}\left(r^2\frac{\partial \Phi}{\partial r}\right) + \frac{1}{r^2[\alpha^2 \text{ cn}^2 (u, \alpha) + \beta^2 \text{ cn}^2 (v, \beta)]}$$
$$\times \left\{\frac{\partial^2 \Phi}{\partial u^2} + \frac{\partial^2 \Phi}{\partial v^2}\right\} . \tag{59}$$

7. *Ellipsoidal coordinates* λ, μ, ν

In Sec. 11.1, we have discussed in detail the ellipsoidal coordinates.

$$x^2 = \frac{(a^2 + \lambda)(a^2 + \mu)(a^2 + \nu)}{(a^2 - b^2)(a^2 - c^2)}, \quad y^2 = \frac{(b^2 + \lambda)(b^2 + \mu)(b^2 + \nu)}{(b^2 - c^2)(b^2 - a^2)},$$

$$z^2 = \frac{(c^2 + \lambda)(c^2 + \mu)(c^2 + \nu)}{(c^2 - a^2)(c^2 - b^2)}. \tag{60}$$

The range of variation of λ, μ, ν is

$$\lambda > -c^2 > \mu > -b^2 > \nu > -a^2.$$

$$ds^2 = \frac{(\lambda - \mu)(\lambda - \nu)}{4\varphi(\lambda)} d\lambda^2 + \frac{(\mu - \nu)(\mu - \lambda)}{4\varphi(\mu)} d\mu^2$$

$$+ \frac{(\nu - \lambda)(\nu - \mu)}{4\varphi(\nu)} d\nu^2, \tag{61}$$

where

$$\varphi(\theta) = (a^2 + \theta)(b^2 + \theta)(c^2 + \theta). \tag{62}$$

$$H_\lambda^2 = \frac{(\lambda - \mu)(\lambda - \nu)}{4\varphi(\lambda)}, \quad H_\mu^2 = \frac{(\mu - \lambda)(\mu - \nu)}{4\varphi(\mu)},$$

$$H_\nu^2 = \frac{(\nu - \lambda)(\nu - \mu)}{4\varphi(\nu)}, \quad H = \frac{(\lambda - \mu)(\lambda - \nu)(\mu - \nu)}{8\sqrt{-\varphi(\lambda)\varphi(\mu)\varphi(\nu)}}. \tag{63}$$

$$\nabla^2 \Phi \equiv \frac{4}{(\lambda - \mu)(\lambda - \nu)(\mu - \nu)}$$

$$\times \left\{ (\mu - \nu)\sqrt{\varphi(\lambda)} \frac{\partial}{\partial \lambda} \left(\sqrt{\varphi(\lambda)} \frac{\partial \Phi}{\partial \lambda} \right) \right.$$

$$+ (\lambda - \nu)\sqrt{-\varphi(\mu)} \frac{\partial}{\partial \mu} \left(\sqrt{-\varphi(\mu)} \frac{\partial \Phi}{\partial \mu} \right)$$

$$\left. + (\lambda - \mu)\sqrt{\varphi(\nu)} \frac{\partial}{\partial \nu} \left(\sqrt{\varphi(\nu)} \frac{\partial \Phi}{\partial \nu} \right) \right\}. \tag{64}$$

Another type of ellipsoidal coordinates are u, v, w; their relations to λ, μ, ν are [Sec. 11.2]

$$\wp(u) = -\lambda - \frac{1}{3}(a^2 + b^2 + c^2),$$

$$\wp(v) = -\mu - \frac{1}{3}(a^2 + b^2 + c^2), \tag{65}$$

$$\wp(w) = -\nu - \frac{1}{3}(a^2 + b^2 + c^2).$$

$$x = ie^{-\eta_1\omega_1}\sigma^2(\omega_1)\frac{\sigma_1(u)\sigma_1(v)\sigma_1(w)}{\sigma(u)\sigma(v)\sigma(w)}$$

$$= \frac{i\vartheta_1'}{2\omega_1\vartheta_2}\frac{\vartheta_2\left(\dfrac{u}{2\omega_1}\right)\vartheta_2\left(\dfrac{v}{2\omega_1}\right)\vartheta_2\left(\dfrac{w}{2\omega_1}\right)}{\vartheta_1\left(\dfrac{u}{2\omega_1}\right)\vartheta_1\left(\dfrac{v}{2\omega_1}\right)\vartheta_1\left(\dfrac{w}{2\omega_1}\right)} \, ,$$

$$y = ie^{-\eta_2\omega_2}\sigma^2(\omega_2)\frac{\sigma_2(u)\sigma_2(v)\sigma_2(w)}{\sigma(u)\sigma(v)\sigma(w)}$$

$$= \frac{i\vartheta_1'}{2\omega_1\vartheta_3}\frac{\vartheta_3\left(\dfrac{u}{2\omega_1}\right)\vartheta_3\left(\dfrac{v}{2\omega_1}\right)\vartheta_3\left(\dfrac{w}{2\omega_1}\right)}{\vartheta_1\left(\dfrac{u}{2\omega_1}\right)\vartheta_1\left(\dfrac{v}{2\omega_1}\right)\vartheta_1\left(\dfrac{w}{2\omega_1}\right)} \, , \qquad (66)$$

$$z = ie^{-\eta_3\omega_3}\sigma^2(\omega_3)\frac{\sigma_3(u)\sigma_3(v)\sigma_3(w)}{\sigma(u)\sigma(v)\sigma(w)}$$

$$= \frac{i\vartheta_1'}{2\omega_1\vartheta_4}\frac{\vartheta_4\left(\dfrac{u}{2\omega_1}\right)\vartheta_4\left(\dfrac{v}{2\omega_1}\right)\vartheta_4\left(\dfrac{w}{2\omega_1}\right)}{\vartheta_1\left(\dfrac{u}{2\omega_1}\right)\vartheta_1\left(\dfrac{v}{2\omega_1}\right)\vartheta_1\left(\dfrac{w}{2\omega_1}\right)} \, .$$

$$\begin{aligned}
ds^2 = & \,[\wp(w) - \wp(u)][\wp(u) - \wp(v)]du^2 \\
& + [\wp(u) - \wp(v)][\wp(v) - \wp(w)]dv^2 \\
& + [\wp(v) - \wp(w)][\wp(w) - \wp(u)]dw^2 \, .
\end{aligned} \qquad (67)$$

$$\begin{aligned}
H_u^2 &= [\wp(w) - \wp(u)][\wp(u) - \wp(v)] \, , \\
H_v^2 &= [\wp(u) - \wp(v)][\wp(v) - \wp(w)] \, , \\
H_w^2 &= [\wp(v) - \wp(w)][\wp(w) - \wp(u)] \, , \\
H &= [\wp(u) - \wp(v)][\wp(v) - \wp(w)][\wp(w) - \wp(u)] \, .
\end{aligned} \qquad (68)$$

$$\begin{aligned}
\nabla^2\Phi \equiv \frac{1}{H}\Bigg\{ &[\wp(v) - \wp(w)]\frac{\partial^2\Phi}{\partial u^2} + [\wp(w) - \wp(u)]\frac{\partial^2\Phi}{\partial v^2} \\
&+ [\wp(u) - \wp(v)]\frac{\partial^2\Phi}{\partial w^2}\Bigg\} \, .
\end{aligned} \qquad (69)$$

Another kind of ellipsoidal coordinates are α, β, γ; their relations to λ, μ, ν are [see Sec. 11.2]

$$a^2 + \lambda = (a^2 - b^2)\,\mathrm{sn}^2\,\alpha, \quad a^2 + \mu = (a^2 - b^2)\,\mathrm{sn}^2\,\beta \, , \qquad (70)$$
$$a^2 + \nu = (a^2 - b^2)\,\mathrm{sn}^2\gamma \, .$$

$$x = k^2 \sqrt{a^2 - c^2} \; \text{sn}\alpha \; \text{sn}\beta \; \text{sn}\gamma \; ,$$

$$y = -\frac{k^2}{k'} \sqrt{a^2 - c^2} \; \text{cn}\alpha \; \text{cn}\beta \; \text{cn}\gamma \; , \qquad (71)$$

$$z = \frac{i}{k'} \sqrt{a^2 - c^2} \; \text{dn}\alpha \; \text{dn}\beta \; \text{dn}\gamma \; , \qquad \left(k^2 = \frac{a^2 - b^2}{a^2 - c^2} \right) \; .$$

$$ds^2 = (a^2 - c^2)k^4 \left\{ (\text{sn}^2\alpha - \text{sn}^2\beta)(\text{sn}^2\alpha - \text{sn}^2\gamma)d\alpha^2 \right.$$
$$+ (\text{sn}^2\beta - \text{sn}^2\alpha)(\text{sn}^2\beta - \text{sn}^2\gamma)d\beta^2 \qquad (72)$$
$$\left. + (\text{sn}^2\gamma - \text{sn}^2\alpha)(\text{sn}^2\gamma - \text{sn}^2\beta)d\gamma^2 \right\} \; .$$

$$\nabla^2 \Phi \equiv \left\{ (a^2 - c^2)k^4(\text{sn}^2\alpha - \text{sn}^2\beta)(\text{sn}^2\alpha - \text{sn}^2\gamma)(\text{sn}^2\beta - \text{sn}^2\gamma) \right\}^{-1}$$
$$\times \left\{ (\text{sn}^2\beta - \text{sn}^2\gamma)\frac{\partial^2\Phi}{\partial\alpha^2} + (\text{sn}^2\gamma - \text{sn}^2\alpha)\frac{\partial^2\Phi}{\partial\beta^2} + (\text{sn}^2\alpha - \text{sn}^2\beta)\frac{\partial^2\Phi}{\partial\gamma^2} \right\} \; . \qquad (73)$$

8. Rotational prolate spheroidal coordinates ξ, η, φ

The ξ, η here are equivalent to that of the elliptic cylinder coordinates; φ is the angle of rotation around the z axis.

$$x = a\sqrt{(\xi^2 - 1)(1 - \eta^2)} \cos\varphi \; ,$$
$$y = a\sqrt{(\xi^2 - 1)(1 - \eta^2)} \sin\varphi \; , \qquad (74)$$
$$z = a\xi\eta \; .$$

The ranges of variation are: $(1, \infty)$ for ξ, $(-1, 1)$ for η, and $(0, 2\pi)$ for φ. From (74) we find

$$\frac{x^2 + y^2}{\xi^2 - 1} + \frac{z^2}{\xi^2} = a^2 \; , \qquad -\frac{x^2 + y^2}{1 - \eta^2} + \frac{z^2}{\eta^2} = a^2 \; . \qquad (75)$$

Thus, $\xi = $ constant represents a rotational ellipsoid with z in the direction of its major axis: $\eta = $ constant represents rotational hyperboloid of two sheets.

$$ds^2 = a^2 \left\{ \frac{\xi^2 - \eta^2}{\xi^2 - 1}d\xi^2 + \frac{\xi^2 - \eta^2}{1 - \eta^2}d\eta^2 + (\xi^2 - 1)(1 - \eta^2)d\varphi^2 \right\} \; . \qquad (76)$$

$$H_\xi^2 = \frac{a^2(\xi^2 - \eta^2)}{\xi^2 - 1}, \qquad H_\eta^2 = \frac{a^2(\xi^2 - \eta^2)}{1 - \eta^2} \; , \qquad (77)$$
$$H_\varphi^2 = a^2(\xi^2 - 1)(1 - \eta^2), \qquad H = a^3(\xi^2 - \eta^2) \; .$$

$$\nabla^2 \Phi \equiv \frac{1}{a^2(\xi^2 - \eta^2)} \left\{ \frac{\partial}{\partial \xi} \left[(\xi^2 - 1) \frac{\partial \Phi}{\partial \xi} \right] + \frac{\partial}{\partial \eta} \left[(1 - \eta^2) \frac{\partial \Phi}{\partial \eta} \right] \right.$$

$$\left. + \left[\frac{1}{\xi^2 - 1} + \frac{1}{1 - \eta^2} \right] \frac{\partial^2 \Phi}{\partial \varphi^2} \right\} . \tag{78}$$

Another kind of rotational prolate spheroidal coordinates are λ, μ, φ; their relations to ξ, η are $\lambda = \xi^2$, $\mu = \eta^2$.

$$ds^2 = \frac{a^2}{4} \left\{ \frac{\lambda - \mu}{\lambda(\lambda - 1)} d\lambda^2 + \frac{\lambda - \mu}{\mu(1 - \mu)} d\mu^2 \right.$$

$$\left. + 4(\lambda - 1)(1 - \mu) d\varphi^2 \right\} . \tag{79}$$

$$\nabla^2 \Phi \equiv \frac{4}{a^2(\lambda - \mu)} \left\{ \sqrt{\lambda} \frac{\partial}{\partial \lambda} \left[(\lambda - 1) \sqrt{\lambda} \frac{\partial \Phi}{\partial \lambda} \right] \right.$$

$$+ \sqrt{\mu} \frac{\partial}{\partial \mu} \left[(1 - \mu) \sqrt{\mu} \frac{\partial \Phi}{\partial \mu} \right]$$

$$\left. + \frac{1}{4} \left(\frac{1}{\lambda - 1} + \frac{1}{1 - \mu} \right) \frac{\partial^2 \Phi}{\partial \varphi^2} \right\} . \tag{80}$$

A further kind of rotational prolate spheroidal coordinates are u, v, φ; their relations to ξ, η are $\xi = \mathrm{ch}\, u, \eta = \cos v$.

$$ds^2 = a^2 \left\{ (\mathrm{ch}^2 u - \cos^2 v)(du^2 + dv^2) + \mathrm{sh}^2 u \sin^2 v \, d\varphi^2 \right\} . \tag{81}$$

$$\nabla^2 \Phi \equiv \frac{1}{a^2(\mathrm{ch}^2 u - \cos^2 v)} \left\{ \frac{1}{\mathrm{sh}\, u} \frac{\partial}{\partial u} \left(\mathrm{sh}\, u \frac{\partial \Phi}{\partial u} \right) + \frac{1}{\sin v} \frac{\partial}{\partial v} \left(\sin v \frac{\partial \Phi}{\partial v} \right) \right.$$

$$\left. + \left(\frac{1}{\mathrm{sh}^2 u} + \frac{1}{\sin^2 v} \right) \frac{\partial^2 \Phi}{\partial \varphi^2} \right\} . \tag{82}$$

9. Rotational oblate spheroidal coordinates ξ, η, φ

The ξ, η here are equivalent to that in the elliptic cylinder coordinates: φ is the angle of rotation around the z axis.

$$x = a\xi\eta\cos\varphi, \quad y = a\xi\eta\sin\varphi, \quad z = a\sqrt{(\xi^2 - 1)(1 - \eta^2)} \, . \qquad (83)$$

The ranges of variation are: $(1, \infty)$ for ξ, $(-1, 1)$ for η, $(0, 2\pi)$ for φ. From (8) we obtain

$$\frac{x^2 + y^2}{\xi^2} + \frac{z^2}{\xi^2 - 1} = a^2, \quad \frac{x^2 + y^2}{\eta^2} - \frac{z^2}{1 - \eta^2} = a^2 \, . \qquad (84)$$

$\xi = $ constant represents a rotational ellipsoid with its minor axis in the z direction: $\eta = $ constant represents a rotational hyperboloid of one sheet.

$$ds^2 = a^2 \left\{ \frac{\xi^2 - \eta^2}{\xi^2 - 1} d\xi^2 + \frac{\xi^2 - \eta^2}{1 - \eta^2} d\eta^2 + \xi^2\eta^2 d\varphi^2 \right\} \, , \qquad (85)$$

$$\nabla^2\Phi \equiv \frac{1}{a^2(\xi^2 - \eta^2)} \left\{ \frac{\sqrt{\xi^2 - 1}}{\xi} \frac{\partial}{\partial\xi} \left(\xi\sqrt{\xi^2 - 1} \frac{\partial\Phi}{\partial\xi} \right) \right.$$
$$+ \frac{\sqrt{1 - \eta^2}}{\eta} \frac{\partial}{\partial\eta} \left(\eta\sqrt{1 - \eta^2} \frac{\partial\Phi}{\partial\eta} \right) \qquad (86)$$
$$\left. + \left(\frac{1}{\eta^2} - \frac{1}{\xi^2} \right) \frac{\partial^2\Phi}{\partial\varphi^2} \right\} \, .$$

If the square roots are undesirable in $\nabla^2\Phi$, we shall make the following transformation

$$\xi^2 = \lambda^2 + 1 \, , \quad \eta^2 = 1 - \mu^2 \, . \qquad (87)$$

Then,

$$ds^2 = a^2 \left\{ \frac{\lambda^2 + \mu^2}{\lambda^2 + 1} d\lambda^2 + \frac{\lambda^2 + \mu^2}{1 - \mu^2} d\mu^2 + (\lambda^2 + 1)(1 - \mu^2) d\varphi^2 \right\} \, . \qquad (88)$$

$$\nabla^2\Phi \equiv \frac{1}{a^2(\lambda^2 + \mu^2)} \left\{ \frac{\partial}{\partial\lambda} \left[(\lambda^2 + 1) \frac{\partial\Phi}{\partial\lambda} \right] + \frac{\partial}{\partial\mu} \left[(1 - \mu^2) \frac{\partial\Phi}{\partial\mu} \right] \right.$$
$$\left. + \left(\frac{1}{1 - \mu^2} - \frac{1}{\lambda^2 + 1} \right) \frac{\partial^2\Phi}{\partial\varphi^2} \right\} \, . \qquad (89)$$

Another kind of rotational oblate spheroidal coordinates are u, v, φ; their relations to ξ, η are

$$\xi = \mathrm{ch}\, u, \quad \eta = \cos v; \quad \text{also} \quad \lambda = \mathrm{sh}\, u, \quad \mu = \sin v . \tag{90}$$

$$ds^2 = a^2 \left\{ (\mathrm{ch}^2 u - \cos^2 v)(du^2 + dv^2) + \mathrm{ch}^2 u \cos^2 v \, d\varphi^2 \right\} . \tag{91}$$

$$
\nabla^2 \Phi \equiv \frac{1}{a^2(\mathrm{ch}^2 u - \cos^2 v)} \left\{ \frac{1}{\mathrm{ch}\, u} \frac{\partial}{\partial u} \left(\mathrm{ch}\, u \frac{\partial \Phi}{\partial u} \right) \right.
$$
$$
\left. + \frac{1}{\cos v} \frac{\partial}{\partial v} \left(\cos v \frac{\partial \Phi}{\partial v} \right) + \left(\frac{1}{\cos^2 v} - \frac{1}{\mathrm{ch}^2 u} \right) \frac{\partial^2 \Phi}{\partial \varphi^2} \right\} . \tag{92}
$$

10. Rotational paraboloidal coordinates λ, μ, φ

The λ, μ here are equivalent to that in the parabolic cylinder coordinates; φ is the angle of rotation around the z axis.

$$x = \sqrt{\lambda\mu}\cos\varphi, \quad y = \sqrt{\lambda\mu}\sin\varphi, \quad z = \frac{1}{2}(\lambda - \mu) . \tag{93}$$

$$ds^2 = \frac{\lambda + \mu}{4} \left(\frac{d\lambda^2}{\lambda} + \frac{d\mu^2}{\mu} \right) + \lambda\mu \, d\varphi^2 . \tag{94}$$

$$\nabla^2 \Phi \equiv \frac{4}{\lambda + \mu} \left\{ \frac{\partial}{\partial \lambda} \left(\lambda \frac{\partial \Phi}{\partial \lambda} \right) + \frac{\partial}{\partial \mu} \left(\mu \frac{\partial \Phi}{\partial \mu} \right) + \frac{1}{4} \left(\frac{1}{\lambda} + \frac{1}{\mu} \right) \frac{\partial^2 \Phi}{\partial \varphi^2} \right\} . \tag{95}$$

Another kind of rotational paraboloidal coordinates are ξ, η, φ; their relations to λ, μ are $\lambda = \xi^2, \mu = \eta^2$.

$$ds^2 = (\xi^2 + \eta^2)(d\xi^2 + d\eta^2) + \xi^2\eta^2 d\varphi^2 . \tag{96}$$

$$
\nabla^2 \Phi \equiv \frac{1}{\xi^2 + \eta^2} \left\{ \frac{1}{\xi} \frac{\partial}{\partial \xi} \left(\xi \frac{\partial \Phi}{\partial \xi} \right) + \frac{1}{\eta} \frac{\partial}{\partial \eta} \left(\eta \frac{\partial \Phi}{\partial \eta} \right) \right.
$$
$$
\left. + \left(\frac{1}{\xi^2} + \frac{1}{\eta^2} \right)^2 \frac{\partial^2 \Phi}{\partial \varphi^2} \right\} . \tag{97}
$$

11. Paraboloidal coordinates λ, μ, ν

Let the equation of the paraboloid be

$$\frac{x^2}{a^2 + \theta} + \frac{y^2}{b^2 + \theta} = 2z + \theta . \tag{98}$$

Through an arbitrary point (x, y, z) in the space there pass three surfaces corresponding to three values of θ, namely, λ, μ, ν, which satisfy the conditions

$$\lambda > -b^2 > \mu > -a^2 > \nu . \tag{99}$$

In $f(\theta) = (a^2 + \theta)(b^2 + \theta)(2z + \theta) - (b^2 + \theta)x^2 - (a^2 + \theta)y^2 = \theta - \lambda)(\theta - \mu)(\theta - \nu)$, put successively $\theta = -a^2, -b^2, 0$. We obtain

$$
\begin{aligned}
x^2 &= -\frac{(a^2 + \lambda)(a^2 + \mu)(a^2 + \nu)}{a^2 - b^2} , \\
y^2 &= \frac{(b^2 + \lambda)(b^2 + \mu)(b^2 + \nu)}{a^2 - b^2} , \\
2z &= -\lambda - \mu - \nu - a^2 - b^2 .
\end{aligned}
\tag{100}
$$

The x-y intercept of the paraboloid $\lambda =$ constant is an ellipse with the z axis as the positive direction, satisfying $z > -\frac{1}{2}\lambda$. The x-y intercept of the paraboloid $\mu =$ constant is a hyperbola, which degenerates into two intersecting straight lines lying in the intercepting plane $z = -\frac{1}{2}\mu$. The x-y intercept of the paraboloid $\nu =$ constant is again an ellipse with the z axis in the negative direction, satisfying $z < -\frac{1}{2}\nu$.

$$
\begin{aligned}
ds^2 &= \frac{(\lambda - \mu)(\lambda - \nu)}{4(a^2 + \lambda)(b^2 + \lambda)} d\lambda^2 + \frac{(\mu - \lambda)(\mu - \nu)}{4(a^2 + \mu)(b^2 + \mu)} d\mu^2 \\
&+ \frac{(\nu - \lambda)(\nu - \mu)}{4(a^2 + \nu)(b^2 + \nu)} d\nu^2 .
\end{aligned}
\tag{101}
$$

$$
\begin{aligned}
\nabla^2 \Phi &\equiv \frac{4}{(\lambda - \mu)(\lambda - \nu)(\mu - \nu)} \\
&\times \left\{ (\mu - \nu)\sqrt{(a^2 + \lambda)(b^2 + \lambda)}\frac{\partial}{\partial \lambda}\left[\sqrt{(a^2 + \lambda)(b^2 + \lambda)}\frac{\partial \Phi}{\partial \lambda} \right] \right. \\
&+ (\lambda - \nu)\sqrt{-(a^2 + \mu)(b^2 + \mu)}\frac{\partial}{\partial \mu}\left[\sqrt{-(a^2 + \mu)(b^2 + \mu)}\frac{\partial \Phi}{\partial \mu} \right] \\
&+ \left. (\lambda - \mu)\sqrt{(a^2 + \nu)(b^2 + \nu)}\frac{\partial}{\partial \nu}\left[\sqrt{(a^2 + \nu)(b^2 + \nu)}\frac{\partial \Phi}{\partial \nu} \right] \right\} .
\end{aligned}
\tag{102}
$$

Another kind of paraboloidal coordinates are u, v, w; their relations to λ, μ, ν are

$$
\lambda = \frac{a^2 - b^2}{2}\operatorname{ch} u - \frac{a^2 + b^2}{2}, \quad \mu = \frac{a^2 - b^2}{2}\cos v - \frac{a^2 + b^2}{2} ,
$$
$$
\nu = -\frac{a^2 - b^2}{2}\operatorname{ch} w - \frac{a^2 + b^2}{2} .
\tag{103}
$$

$$x = (a^2 - b^2)\operatorname{ch}\frac{u}{2}\cos\frac{v}{2}\operatorname{sh}\frac{w}{2}, \quad y = (a^2 - b^2)\operatorname{sh}\frac{u}{2}\sin\frac{v}{2}\operatorname{ch}\frac{w}{2},$$

$$2z = -(a^2 - b^2)\left[\left(\operatorname{ch}\frac{u}{2}\right)^2 + \left(\cos\frac{v}{2}\right)^2 - \left(\operatorname{ch}\frac{w}{2}\right)^2\right] + a^2, \tag{104}$$

$$ds^2 = \left(\frac{a^2 - b^2}{4}\right)^2 \left\{ (\operatorname{ch} u - \cos v)(\operatorname{ch} u + \operatorname{ch} w)\,du^2 \right.$$
$$+ (\operatorname{ch} u - \cos v)(\cos v + \operatorname{ch} w)\,dv^2 \tag{105}$$
$$\left. + (\operatorname{ch} u + \operatorname{ch} w)(\cos v + \operatorname{ch} w)\,dw^2 \right\}.$$

$$\nabla^2\Phi \equiv \left(\frac{4}{a^2 - b^2}\right)^2 \frac{1}{(\operatorname{ch} u - \cos v)(\operatorname{ch} u + \operatorname{ch} w)(\cos v + \operatorname{ch} w)}$$
$$\times \left\{ (\cos v + \operatorname{ch} w)\frac{\partial^2\Phi}{\partial u^2} + (\operatorname{ch} u + \operatorname{ch} w)\frac{\partial^2\Phi}{\partial v^2} \right. \tag{106}$$
$$\left. + (\operatorname{ch} u - \cos v)\frac{\partial^2\Phi}{\partial w^2} \right\}.$$

12. Bi-spherical coordinates ξ, η, φ

Let

$$x^2 + y^2 = \rho^2, \quad x = \rho\cos\varphi, \quad y = \rho\sin\varphi \tag{107}$$

$$\rho^2 + (z - \lambda)^2 = \lambda^2 - a^2, \quad (\rho - \mu)^2 + z^2 = \mu^2 + a^2. \tag{108}$$

Take λ, μ, ν as bi-spherical coordinates; the ranges of variation are (a, ∞) and $(-\infty, -a)$ for λ, $(-\infty, \infty)$ for μ. For $\lambda > a$, we solve from (108),

$$\rho = \frac{a^2\sqrt{\lambda^2 - a^2}}{\lambda\sqrt{\mu^2 + a^2} - \mu\sqrt{\lambda^2 - a^2}},$$
$$z = \frac{a^2\sqrt{\mu^2 + a^2}}{\lambda\sqrt{\mu^2 + a^2} - \mu\sqrt{\lambda^2 - a^2}}. \tag{109}$$

For $\lambda < -a$, λ is replaced by $-\lambda$ and z by $-z$ in the above formulae; or, replace $\sqrt{\lambda^2 - a^2}$ with $-\sqrt{\lambda^2 - a^2}$.

$$ds^2 = H_\lambda^2 d\lambda^2 + H_\mu^2 d\mu^2 + \rho^2 d\varphi^2, \tag{110}$$

where

$$H_\lambda^2 = \frac{z^2}{\lambda^2 - a^2}, \quad H_\mu^2 = \frac{\rho^2}{\mu^2 + a^2}. \tag{111}$$

Apply the transformation

$$\lambda = \frac{a\xi}{\sqrt{\xi^2 - 1}}, \quad \mu = \frac{a\eta}{\sqrt{1 - \eta^2}}; \tag{112}$$

the ranges of variation are $(1, \infty)$ for ξ, $(-1, 1)$ for η. Substituting in (109), we obtain

$$\rho = \frac{a\sqrt{1 - \eta^2}}{\xi - \eta}, \quad z = \frac{a\sqrt{\xi^2 - 1}}{\xi - \eta}; \tag{113}$$

corresponding to $\lambda < 0$, $\sqrt{\xi^2 - 1}$ assumes negative value.

$$ds^2 = \frac{a^2}{(\xi^2 - 1)(\xi - \eta)^2}d\xi^2 + \frac{a^2}{(1 - \eta^2)(\xi - \eta)^2}d\eta^2 + \frac{a^2(1 - \eta^2)}{(\xi - \eta)^2}d\varphi^2. \tag{114}$$

$$\nabla^2\Phi \equiv \frac{(\xi - \eta)^3}{a^2}\left\{\sqrt{\xi^2 - 1}\frac{\partial}{\partial\xi}\left(\frac{\sqrt{\xi^2 - 1}}{\xi - \eta}\frac{\partial\Phi}{\partial\xi}\right) + \frac{\partial}{\partial\eta}\left(\frac{1 - \eta^2}{\xi - \eta}\frac{\partial\Phi}{\partial\eta}\right)\right.$$
$$\left. + \frac{1}{(1 - \eta^2)(\xi - \eta)}\frac{\partial^2\Phi}{\partial\varphi^2}\right\}. \tag{115}$$

Another kind of bi-spherical coordinates are u, v, φ; their relations to ξ, η are

$$\xi = \mathrm{ch}\, u, \quad \eta = \cos v, \tag{116}$$

the ranges of variation are $(-\infty, \infty)$ for u and $(0, \pi)$ for v. Their relations to λ, μ are

$$\lambda = \frac{a\, \mathrm{ch}\, u}{\mathrm{sh}\, u}, \quad \mu = a \cot v, \tag{117}$$

$$\rho = \frac{a \sin v}{\mathrm{ch}\, u - \cos v}, \quad z = \frac{a\, \mathrm{sh}\, u}{\mathrm{ch}\, u - \cos v}. \tag{118}$$

$$ds^2 = \frac{a^2}{(\mathrm{ch}\, u - \cos v)^2}\{du^2 + dv^2 + \sin^2 v\, d\varphi^2\}. \tag{119}$$

$$\nabla^2\Phi \equiv \frac{(\mathrm{ch}\, u - \cos v)^2}{a^2}$$
$$\times \left\{(\mathrm{ch}\, u - \cos v)\frac{\partial}{\partial u}\left(\frac{1}{\mathrm{ch}\, u - \cos v}\frac{\partial\Phi}{\partial u}\right)\right.$$
$$+ \frac{\mathrm{ch}\, u - \cos v}{\sin v}\frac{\partial}{\partial v}\left(\frac{\sin v}{\mathrm{ch}\, u - \cos v}\frac{\partial\Phi}{\partial v}\right)$$
$$\left. + \frac{1}{\sin^2 v}\frac{\partial^2\Phi}{\partial\varphi^2}\right\}. \tag{120}$$

We can also introduce the variables α, β:

$$\frac{\lambda}{a} = \frac{1+\alpha^2}{2\alpha}, \quad \frac{\mu}{a} = \frac{1-\beta^2}{2\beta} ; \tag{121}$$

the ranges of variation are $(-1, 1)$ for α, $(0, \infty)$ for β.

$$\alpha = \text{th}\, \frac{u}{2}, \quad \beta = \tan\frac{v}{2}, \quad \xi = \frac{1+\alpha^2}{1-\alpha^2}, \quad \eta = \frac{1-\beta^2}{1+\beta^2} . \tag{122}$$

$$\frac{\rho}{a} = \frac{\beta(1-\alpha^2)}{\alpha^2+\beta^2}, \quad \frac{z}{a} = \frac{\alpha(1+\beta^2)}{\alpha^2+\beta^2} . \tag{123}$$

$$ds^2 = \frac{a^2}{(\alpha^2+\beta^2)^2} \left\{ (1+\beta^2)^2 d\alpha^2 + (1-\alpha^2)^2 d\beta^2 + \beta^2(1-\alpha^2)^2 d\varphi^2 \right\} . \tag{124}$$

$$\begin{aligned}
\nabla^2 \Phi \equiv{} & \frac{(\alpha^2+\beta^2)^3}{a^2(1-\alpha^2)^2(1+\beta^2)^2} \frac{\partial}{\partial\alpha} \left[\frac{(1-\alpha^2)^2}{\alpha^2+\beta^2} \frac{\partial\Phi}{\partial\alpha} \right] \\
& + \frac{(\alpha^2+\beta^2)^3}{a^2\beta(1-\alpha^2)^2(1+\beta^2)} \frac{\partial}{\partial\beta} \left[\frac{\beta(1+\beta^2)}{\alpha^2+\beta^2} \frac{\partial\Phi}{\partial\beta} \right] \\
& + \frac{(\alpha^2+\beta^2)^2}{a^2\beta^2(1-\alpha^2)^2} \frac{\partial^2\Phi}{\partial\varphi^2} .
\end{aligned} \tag{125}$$

13. Toroidal coordinates ξ, η, φ

In the bi-spherical coordinates, interchanging ρ and z, we obtain

$$x^2 + y^2 = z^2, \quad x = z\cos\varphi, \quad y = z\sin\varphi . \tag{126}$$
$$(\rho-\lambda)^2 + z^2 = \lambda^2 - a^2, \quad \rho^2 + (z-\mu)^2 = \mu^2 + a^2 . \tag{127}$$

Take λ, μ, φ as toroidal coordinates with the ranges of variation: (a, ∞) for λ and $(-\infty, \infty)$ for μ. We solve from (127), for $z > 0$,

$$\rho = \frac{a^2\sqrt{\mu^2 + a^2}}{\lambda\sqrt{\mu^2 + a^2} - \mu\sqrt{\lambda^2 - a^2}} ,$$

$$z = \frac{a^2\sqrt{\lambda^2 - a^2}}{\lambda\sqrt{\mu^2 + a^2} - \mu\sqrt{\lambda^2 - a^2}} . \tag{128}$$

For $z < 0$, the factor $\sqrt{\lambda^2 - a^2}$ in the above equations is to be replaced by $-\sqrt{\lambda^2 - a^2}$. (128) are the same as (109) (ρ and z interchanged), but now, the range of variation of λ is different. λ always being positive and greater than a.

$$ds^2 = H_\lambda^2 d\lambda^2 + H_\mu^2 d\mu^2 + \rho^2 d\varphi^2 , \tag{129}$$

$$H_\lambda^2 = \frac{\rho^2}{\lambda^2 - a^2} , \qquad H_\mu^2 = \frac{z^2}{\mu^2 + a^2} . \tag{130}$$

Apply the transformation

$$\lambda = \frac{a\xi}{\sqrt{\xi^2 - 1}} , \qquad \mu = \frac{a\eta}{\sqrt{1 - \eta^2}} ; \tag{131}$$

the range of variation for ξ being $(1, \infty)$ and that for η being $(-1, 1)$. Substituting in (128) gives

$$\rho = \frac{a\sqrt{\xi^2 - 1}}{\xi - \eta} , \qquad z = \frac{a\sqrt{1 - \eta^2}}{\xi - \eta} . \tag{132}$$

For $z < 0, \eta$ and $\sqrt{1 - \eta^2}$ in the above formulae have to be replaced at the same time by $-\eta$ and $-\sqrt{1 - \eta^2}$ respectively. But, as the range of variation of η contains negative values, it is only necessary to change the sign of the square root $\sqrt{1 - \eta^2}$.

$$ds^2 = \frac{a^2}{(\xi^2 - 1)(\xi - \eta)^2} d\xi^2 + \frac{a^2}{(1 - \eta^2)(\xi - \eta)^2} d\eta^2 + \frac{a^2(\xi^2 - 1)}{(\xi - \eta)^2} d\varphi^2 . \tag{133}$$

$$\nabla^2 \Phi \equiv \frac{(\xi - \eta)^3}{a^2} \left\{ \frac{\partial}{\partial \xi} \left(\frac{\xi^2 - 1}{\xi - \eta} \right) \frac{\partial \Phi}{\partial \xi} + \sqrt{1 - \eta^2} \frac{\partial}{\partial \eta} \left(\frac{\sqrt{1 - \eta^2}}{\xi - \eta} \frac{\partial \Phi}{\partial \eta} \right) \right.$$

$$\left. + \frac{1}{(\xi^2 - 1)(\xi - \eta)} \frac{\partial^2 \Phi}{\partial \varphi^2} \right\} . \tag{134}$$

Another kind of toroidal coordinates are u, v, φ; their relations to ξ and η are

$$\xi = \operatorname{ch} u, \qquad \eta = \cos v ; \tag{135}$$

the range of variation for u is $(0, \infty)$ and that for v is $(-\pi, \pi)$ or $(0, 2\pi)$. Their relations to λ, μ are the same as (117).

$$\rho = \frac{a \operatorname{sh} u}{\operatorname{ch} u - \cos v} , \qquad z = \frac{a \sin v}{\operatorname{ch} u - \cos v} . \tag{136}$$

$$ds^2 = \frac{a^2}{(\operatorname{ch} u - \cos v)^2} \{ du^2 + dv^2 + \operatorname{sh}^2 u \, d\varphi^2 \} . \tag{137}$$

$$\nabla^2 \Phi \equiv \frac{(\text{ch } u - \cos v)^2}{a^2} \left\{ \frac{\text{ch } u - \cos v}{\text{sh } u} \frac{\partial}{\partial u} \left(\frac{\text{sh } u}{\text{ch } u - \cos v} \frac{\partial \Phi}{\partial u} \right) \right.$$

$$\left. + (\text{ch } u - \cos v) \frac{\partial}{\partial v} \left(\frac{1}{\text{ch } u - \cos v} \frac{\partial \Phi}{\partial v} \right) + \frac{1}{\text{sh}^2 u} \frac{\partial^2 \Phi}{\partial \varphi^2} \right\} . \tag{138}$$

We can further introduce the variables α and β given by (121) and (122). However, the ranges of variation are changed; $(0,1)$ for α and $(-\infty, \infty)$ for β.

$$\frac{\rho}{a} = \frac{\alpha(1+\beta^2)}{\alpha^2 + \beta^2}, \quad \frac{z}{a} = \frac{\beta(1-\alpha^2)}{\alpha^2 + \beta^2} . \tag{139}$$

$$ds^2 = \frac{\alpha^2}{(\alpha^2 + \beta^2)^2} \left\{ (1+\beta^2)^2 d\alpha^2 + (1-\alpha^2)^2 d\beta^2 + \alpha^2(1+\beta^2)^2 d\varphi^2 \right\} . \tag{140}$$

$$\nabla^2 \Phi \equiv \frac{(\alpha^2 + \beta^2)^3}{a^2 \alpha(1-\alpha^2)(1+\beta^2)^2} \frac{\partial}{\partial \alpha} \left[\frac{\alpha(1-\alpha^2)}{\alpha^2 + \beta^2} \frac{\partial \Phi}{\partial \alpha} \right]$$

$$+ \frac{(\alpha^2 + \beta^2)^3}{a^2 (1-\alpha^2)^2(1+\beta^2)^2} \frac{\partial}{\partial \beta} \left[\frac{(1+\beta^2)^2}{\alpha^2 + \beta^2} \frac{\partial \Phi}{\partial \beta} \right] \tag{141}$$

$$+ \frac{(\alpha^2 + \beta^2)^2}{a^2 \alpha^2(1+\beta^2)^2} \frac{\partial^2 \Phi}{\partial \varphi^2} .$$

BIBLIOGRAPHY

M. Abramowitz and I.A. Stegun: *Handbook of Mathematical Functions* (NBS, 1964).

P. Appell and J. Kampé de Fériet: *Fonctions Hypergéométriques et Hyperspheriques. Polynomes d'Hermite* (Gauthier-Villars, 1926).

W.N. Bailey: Generalized Hypergeometric Series (Cambridge, 1935).

L. Bieberbach: *Theorie der Gewöhnlichen Differentialgleichungen* (Berlin, 1953).

T.J.I. Bromwich: *An Introduction to the Theory of Infinite Series*, second edition (1925).

H. Buchholz: *Die Konfluente Hypergeometrische Funktion* (Berlin-Göttingen-Heidelberg, 1953).

P.F. Byrd and M.D. Friedman: *Handbook of Elliptic Integrals for Engineers and Physicists* (Springer, 1954).

A. Erdélyi: *Higher Transcendental Functions*, Vol. I, II, III (New York-Toronto-London, 1953).

E.W. Hobson: *Spherical and Ellipsoidal Harmonics* (Cambridge, 1931).

E.L. Ince: *Ordinary Differential Equations* (London, 1927).

N.W. Mclachlan: *Theory and Application of Mathieu Functions* (Oxford, 1947).

J. Meixner and F.W. Schäfke: *Mathieusche Funktionen und Sphäroidalfunktionen* (Berlin-Göttingen-Heidelberg, 1954).

L.J. Slater, *Confluent Hypergeometric Functions* (Cambridge, 1960).

G.N. Watson: *Theory of Bessel Functions* (Cambridge, 1944).

E.T. Whittaker and G.N. Watson: *Modern Analysis* (Cambridge, 1927).

GLOSSARY

k	modulus of Jacobian elliptical functions	511
k'	complementary modulus	511
K, K'	complete elliptic integrals (first kind)	511
$K_\nu(z)$	modified (imaginary argument) Bessel function (second kind)	374
$\mathrm{ker}_\nu(z)$ $\mathrm{kei}_\nu(z)$	Thomson (Kelvin) functions	376
$\mathrm{li}(z)$	logarithmic integral	334
$L_n(z)$	Laguerre polynomial	327
$L_n^\mu(z)$	generalized Laguerre polynomial (also, Sonine polynomial)	327
$M_{k,m}(z)$	Whittaker function	301
$[n/2]$	largest integer not greater than $n/2$	6
$\binom{n}{k}$	binomial expansion coefficient	6
$n_l(z)$	spherical Bessel function (second kind)	376
nc, nd, ns	Jacobian elliptic functions	532
$N_\nu(z)$	Neumann function (see $Y_\nu(z)$)	366
$O_n(t)$	Neumann polynomial	413
$\wp(z)$	Weierstrass elliptic function	466
$P_\lambda(x)$	periodic function	10
$P_n(x)$	Legendre polynomial	213
$P_\nu(z)$	Legendre function (first kind)	250
$P_l^m(x)$	associate Legendre function (first kind)	233
$P_\nu^\mu(z)$	general associate Legendre function (first kind)	248
$Q_n(x)$	Legendre function (second kind)	224
$Q_l^m(x)$	associate Legendre function (second kind)	234
$Q_\nu^\mu(z)$	general associate Legendre function (second kind)	251
$\mathrm{Re}(z)$	real part of z	35
sech z	hyperbolic secant	6
sc, sd	Jacobian elliptic functions	532
$\mathrm{se}_m(z, q)$	Mathieu function	624
sh z	hyperbolic sine	130
sn u	Jacobian elliptic function	530
$\mathrm{Se}_m(z, q)$	modified Mathieu function	651
$S_l(\theta, \varphi)$	surface (spherical) harmonics	242
$\mathrm{Si}(z)$	sine integral	344
$S_\mu^n(z)$	Sonine polynomial	327
th z	hyperbolic tangent	387
$T_n(x)$	Chebyshev polynomial of the first kind	173

INDEX